# Master Handbook of 1001 Practical Electronic Circuits

## Other TAB books by Kendall Webster Sessions, Jr.

| | |
|---|---|
| Book No. 590 | Practical Solid-State Principles & Projects |
| Book No. 613 | New IC FET Principles & Projects |
| Book No. 621 | The 2-Meter FM Repeater Circuits Handbook |
| Book No. 647 | Stereo/Quad Hi-Fi Principles & Projects |
| Book No. 667 | Miniature Projects for Electronic Hobbyists |
| Book No. 673 | How to Be A Ham—Including Latest FCC Rules |
| Book No. 722 | Amateur FM Conversion & Construction Projects |
| Book No. 756 | Four-Channel Stereo—From Source to Sound—2nd Edition |
| Book No. 827 | Amateur Radio Advanced Class License Study Guide |
| Book No. 851 | Amateur Radio General Class License Study Guide |
| Book No. 873 | Amateur Radio Novice Class License Study Guide |

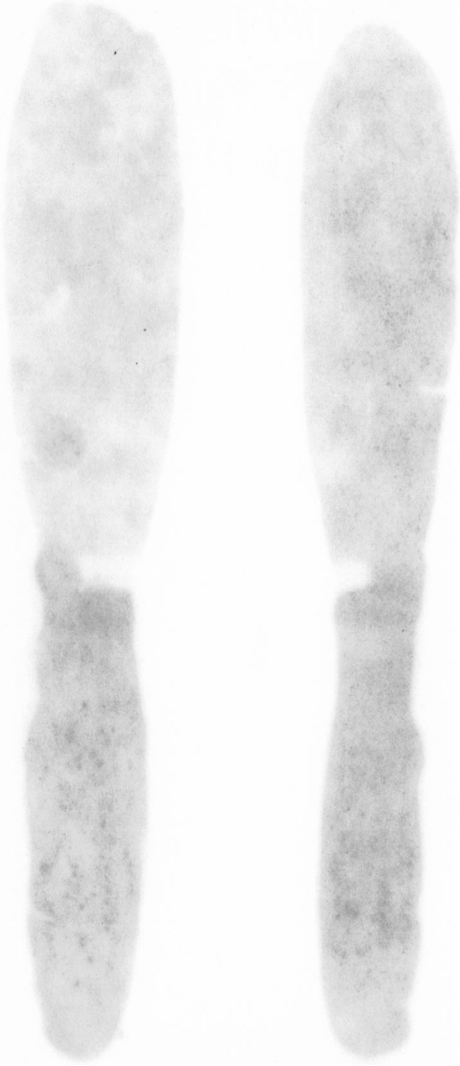

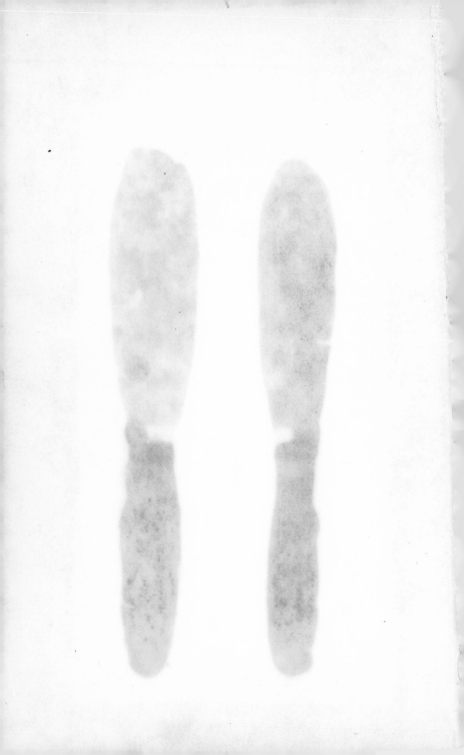

No. 800
$12.95

# Master Handbook of 1001 Practical Electronic Circuits

Edited by Kendall Webster Sessions

TAB BOOKS
Blue Ridge Summit, Pa. 17214

FIRST EDITION

FIRST PRINTING—NOVEMBER 1975
SECOND PRINTING—OCTOBER 1976



Printed in United States
of America



Hardbound Edition: International Standard Book No. 0-8306-5800-9

Paperbound Edition: International Standard Book No. 0-8306-4800-3

Library of Congress Card Number: 75-31458

# Contents

# Preface

Ever wanted a circuit diagram you couldn't find? You're not alone. It happens to all of us who work in electronics—and it happens all the time! The more active we are, the more we need schematics for circuits we want to build or adapt. Usually, we either end up buying a whole book just to get a single circuit diagram or we just forget the whole thing. Or we waste a lot of precious hours trying to build a working breadboard right from scratch. Usually, we find out too late that the circuit we want has already been designed, built, debugged, and perfected by someone else!

This is where *Master Handbook of 1001 Practical Electronic Circuits* comes in. But this book is more than merely a reproduction of one-thousand-and-one popular circuits. It is a collection of the circuits themselves as well as the peripheral information needed to put the circuits to work! Every circuit is accompanied by a "caption" that contains all the information you need to get it working in your own personal application. Where there are coils to be wound, you'll find full and complete coil-winding details.

Here are circuit diagrams for anything and everything, each with every component carefully labeled; I've made an effort to leave nothing to chance. The constant aim in preparation of this book has been to leave no questions unanswered—to make every single circuit complete in every detail, exactly as its designer intended!

The circuits in this handbook are completely up-to-date. They're transistor and integrated-circuit schematic diagrams that have been breadboarded, tested, and simplified by their designers. Most of these practical circuits have appeared in 73 Magazine, one of the most respected independent amateur radio journals. Which means *they've been built and used, checked out and corrected as necessary by thousands of active hobbyists prior to publication in this compendium*.

Whatever your forte, regardless of your electronic specialty, you'll find any circuit you're ever likely to need in these pages. Want to build a fire alarm or moisture detector? A 4-channel stereo decoder? A roomful of practical test instruments? A complete repeater? A digital computer? Audio or RF filters? Or how about a burglar alarm—you'll find a wide range of them here for home, shop, or car!

This is not a book of words. The only reading you'll have to do will be within the captions of only those circuits of direct and specific interest to YOU.

Appendix A includes base diagrams for most popular integrated circuits and lists a wide variety of IC substitutions; Appendix B is a pictorial listing of common electronic symbols.

One final comment: The more observant reader will note (and hopefully overlook) certain unavoidable inconsistencies of style in the diagrams which have been photographically reproduced from a wide variety of sources (73 Magazine, National Semiconductors, Calectro, Motorola, TAB books, GE, and others).

Kendall Webster Sessions

# Section A

# Alarms, Sensors, Triggering Circuits

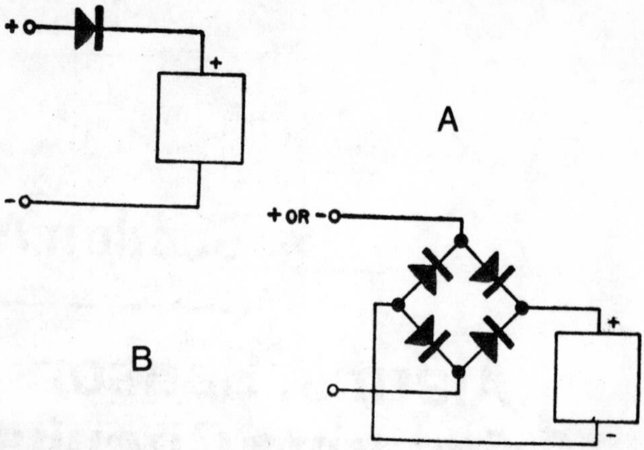


These two circuits protect equipment from incorrectly polarized voltage. The single diode keeps the equipment from working when the polarity is wrong, while the bridge automatically selects the proper polarity.

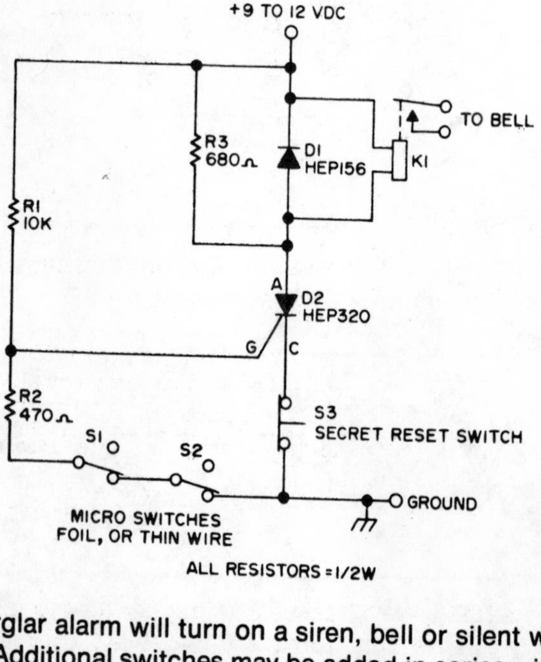


This burglar alarm will turn on a siren, bell or silent warning device. Additional switches may be added in series with SW1 and SW2; foil or thin wire may be used instead of SW1 and/or SW2. Courtesy Motorola Construction Projects HMA-37.

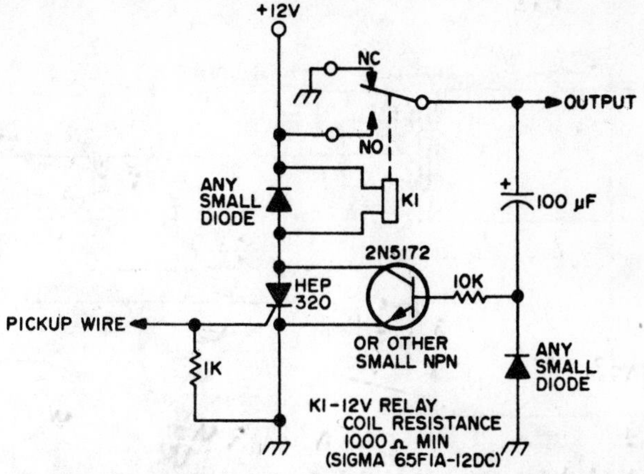


Lightning detector disconnects circuits until lightning storm passes; for use in remote radio systems where personal attendance is not possible.

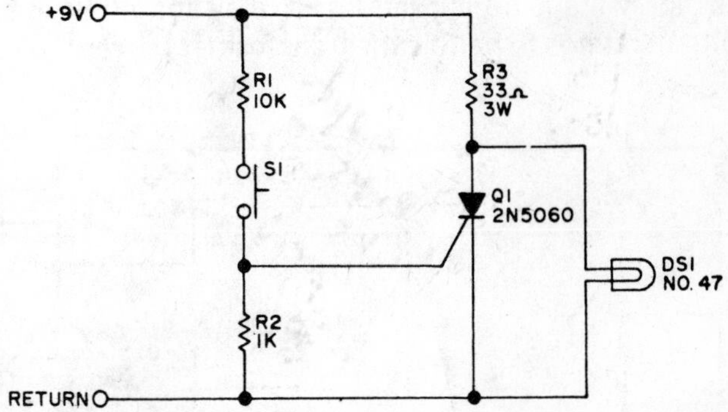


Interrupted-power indicator. Circuit courtesy Motorola Semiconductor Power Circuits Manual.

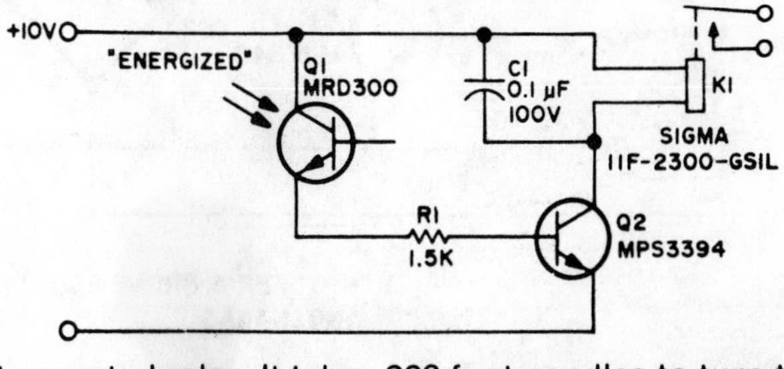


Light operated relay. It takes 220 foot candles to turn this on. Turning off the light turns off the relay. Circuit courtesy Motorola Semiconductor Power Circuits Handbook.

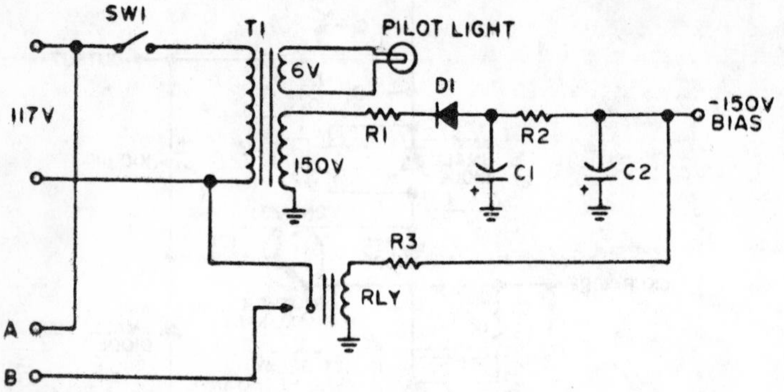


A protection circuit that will turn off the high voltage power supply in a linear amplifier when bias voltage fails. D1, 400 PIV ½A; R1, 10Ω 1W: R2, 1K 2W; R3 depends on relay voltage; C1 and C2, 4 μF 200V. Points A and B go to the 117V primary of the high voltage transformer. The relay should have adequate contact ratings to handle the transformer current.

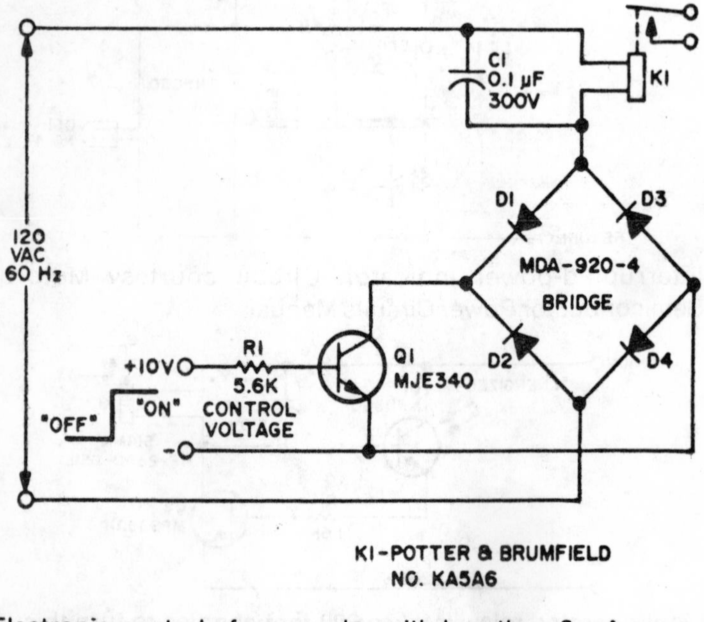


Electronic control of an ac relay with less than 2 mA current required to operate. Circuit courtesy Motorola Semiconductor Power Circuits Handbook.

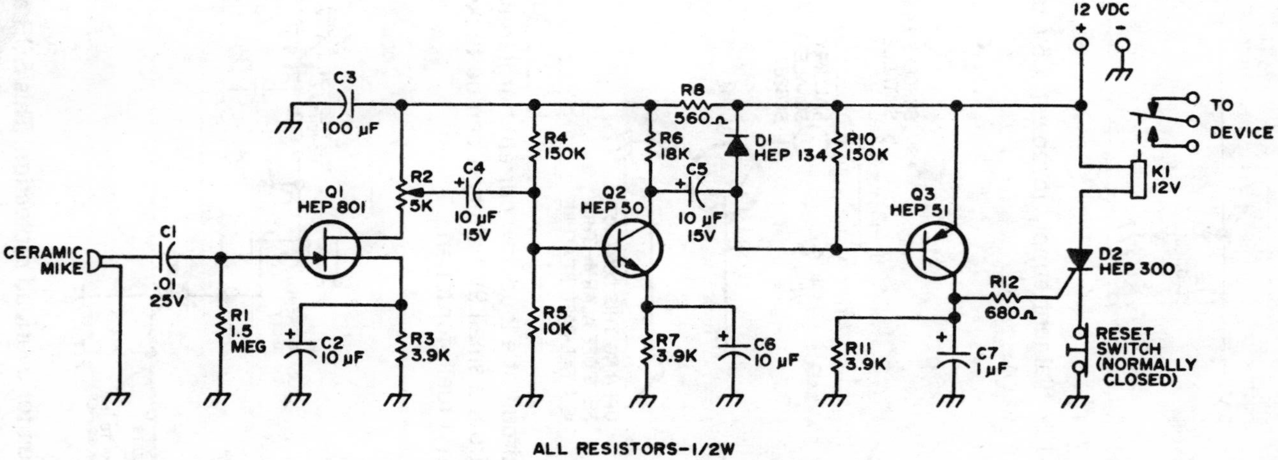


Sound-activated relay. Control any circuit with a clap of the hands or any sharp sound. The circuit remains activated until manually reset. It has adjustable selectivity. Circuit courtesy Motorola HMA-33 "Tips on Using FETs."

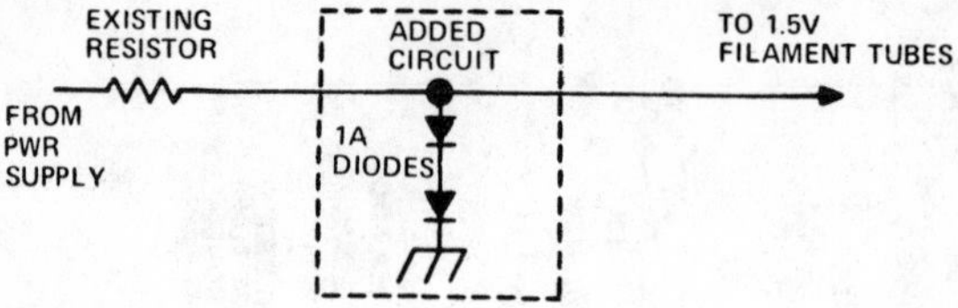


Use 6 or 12V filament supply to power 1.5V tubes. Diodes drop 700 mV each.

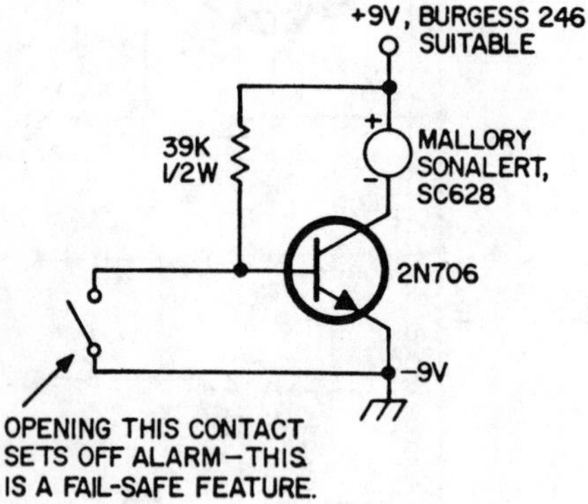


This is a circuit of a loud, low current burglar alarm. Since it operates from a small 9V battery it can be tucked away in a corner and virtually forgotten.

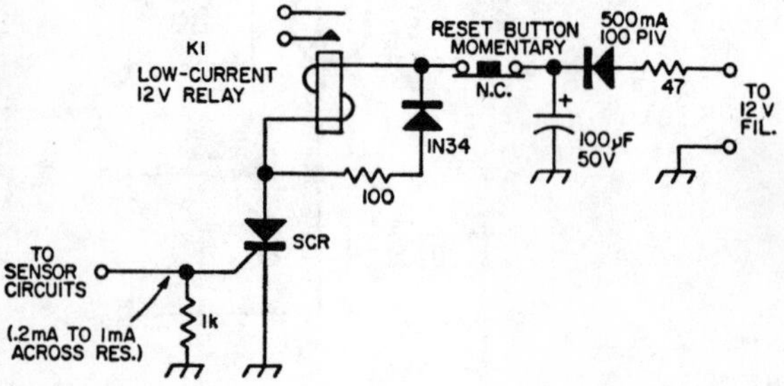


Basic circuit for overload protection. This SCR and relay will turn off any circuit (or ring a bell) when almost anything goes wrong with your station.

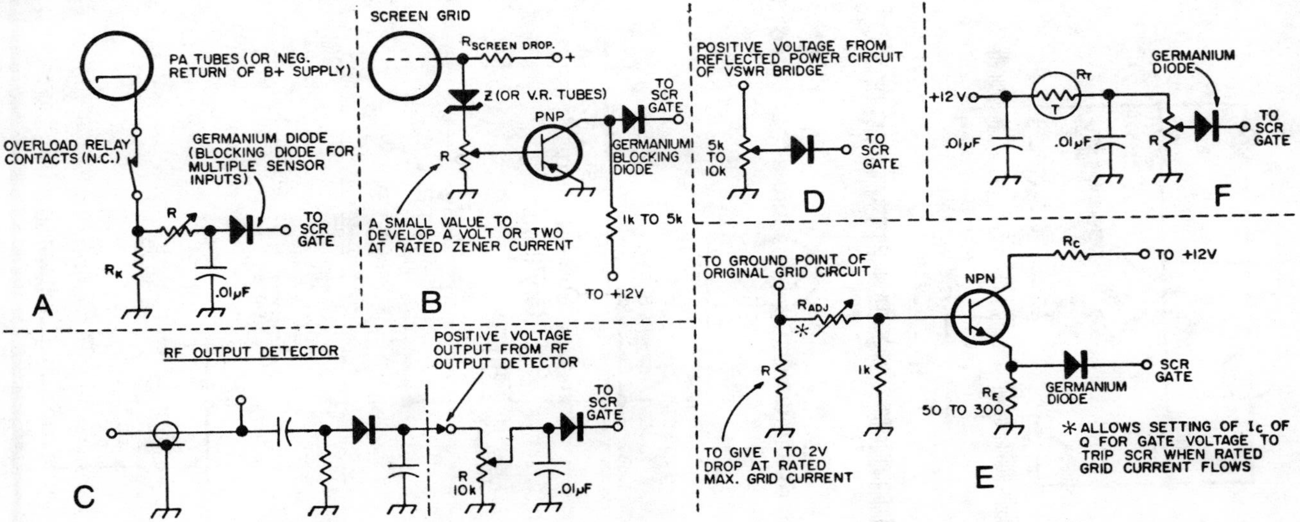


Sensors for transmitter overload circuit of P.14. A. Protection against too much plate current. B. Screen current protection. C. Excessive output protection. D. High VSWR protection. E. Protection against loss of grid drive. F. Excessive temperature protection.

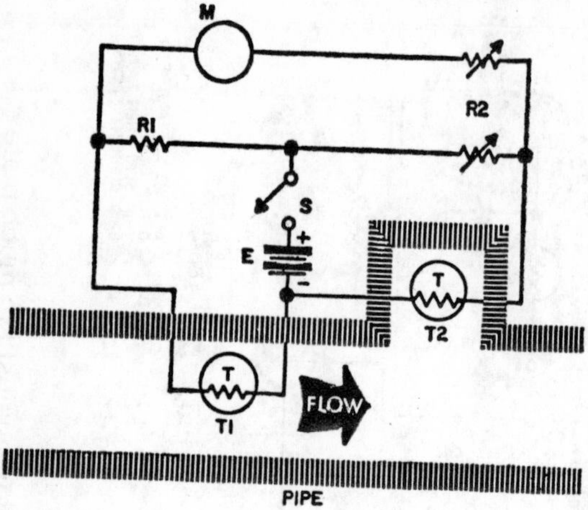


Typical circuit of a flow meter to measure either liquid or gas flow.

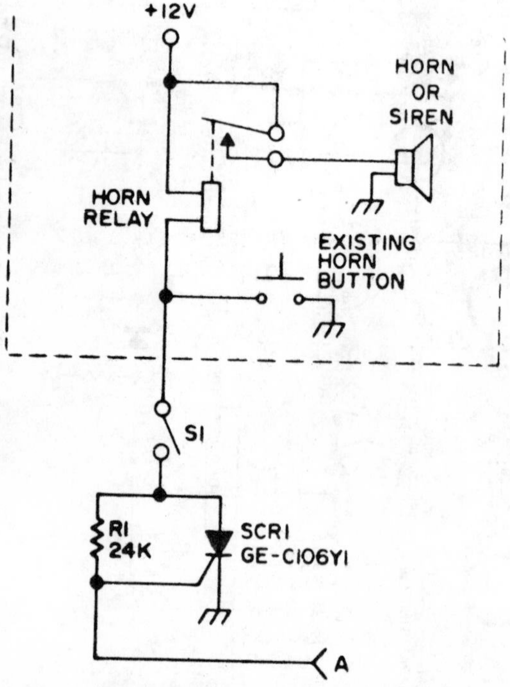


Simple SCR mobile theft alarm circuit.

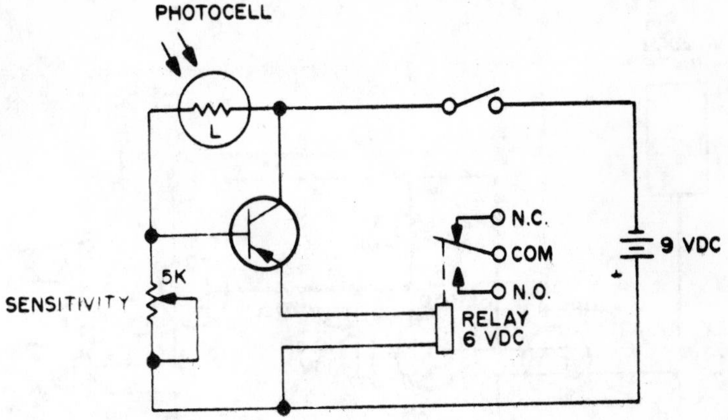


Turn anything on—or off—with light. Use it to trigger an alarm, to turn on lights at dusk, activate a counter as people pass. . .or whatever. Easy-to-build versatile circuit courtesy of Calectro Handbook. Transistor is a Calectro K4-505.

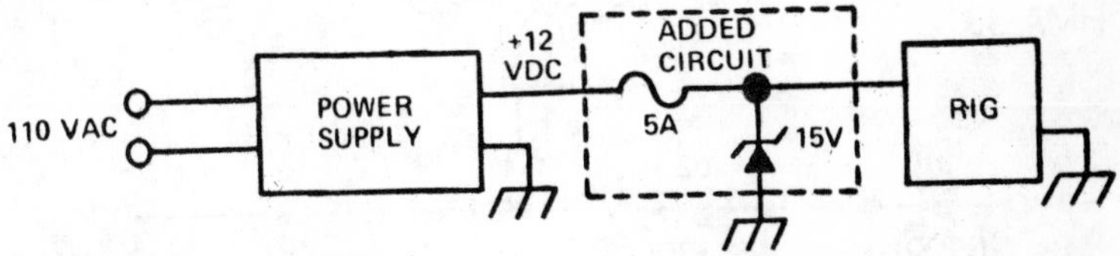


Protect transistor circuits from overvoltage with this zener clamp. (The zener should be at least 1W and preferably 10W.)

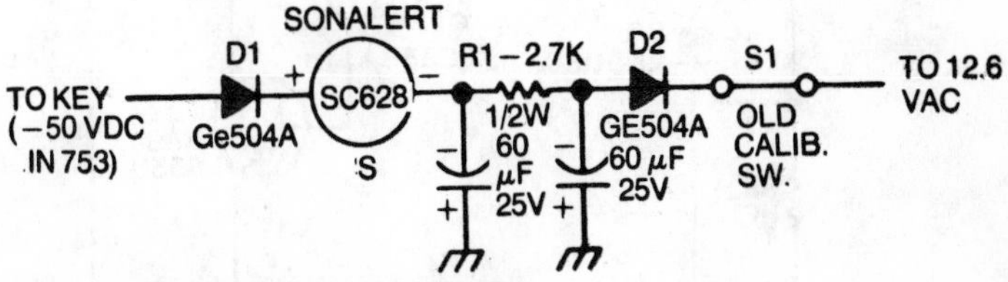


S – MALLORY SC628 SONALERT
(AUDIBABLE SIGNALING DEVICE).
S1 – S.P.S.T. SWITCH (CALIBRATOR SWITCH IN 753
IF CALIBRATOR IS UNUSED).

CW sidetone using Mallory SC628 Sonalert. For positive key voltage, reverse all polarities.

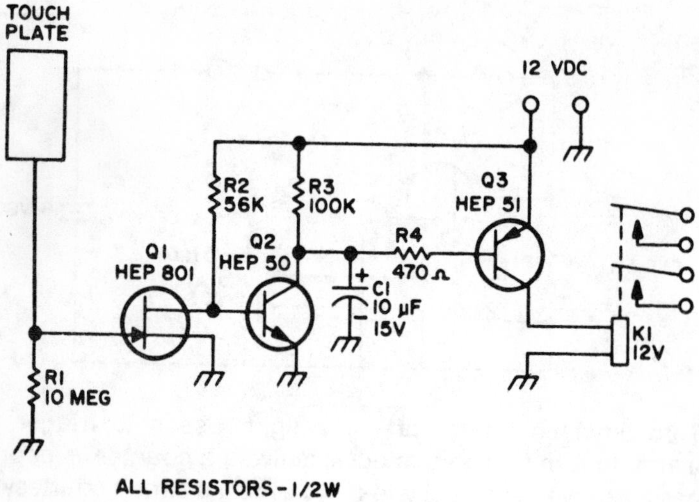


Touch switch. This switch is operated by body capacity. . .just touch the plate to operate the relay. Circuit courtesy Motorola HMA-33.

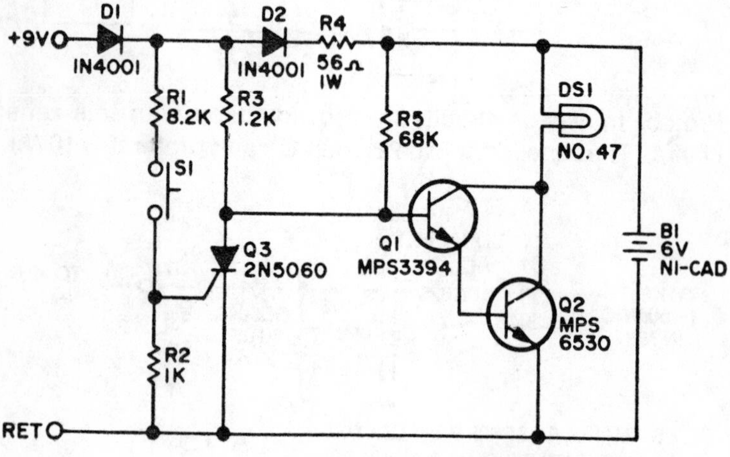


Power failure indicator. Indicates either momentary or continuous power failure. Circuit courtesy Motorola Semiconductor Power Circuits Handbook. When circuit is reset it will trickle-charge the battery, keeping it at full charge.

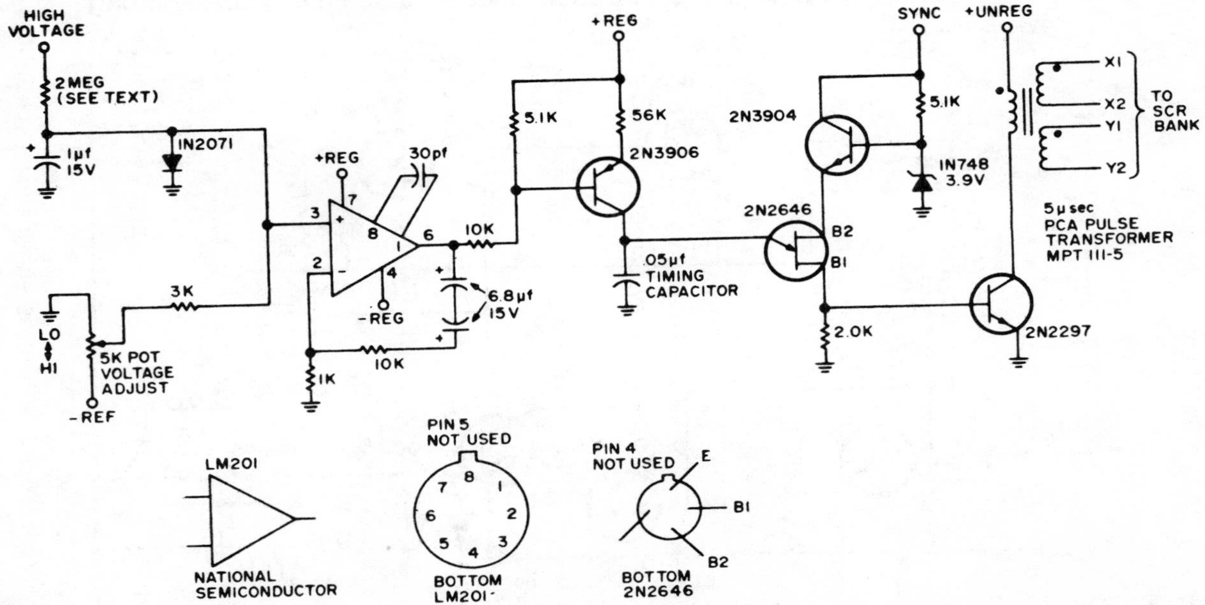


Error amplifier and trigger generator.

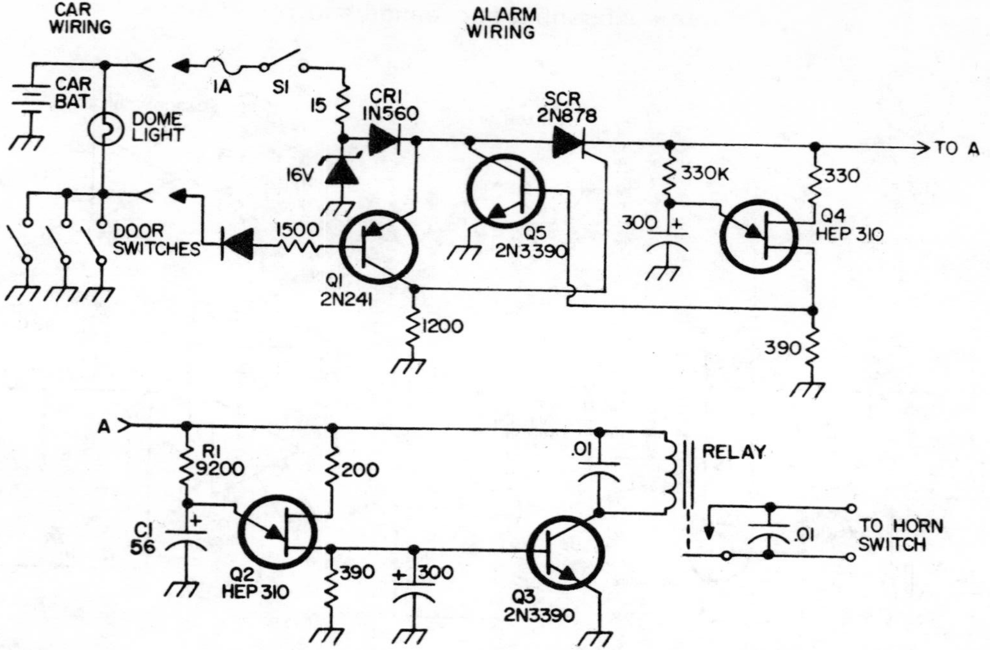


Burglar alarm for automotive applications resets itself after 1.5 minutes. Sounds horn when activated.

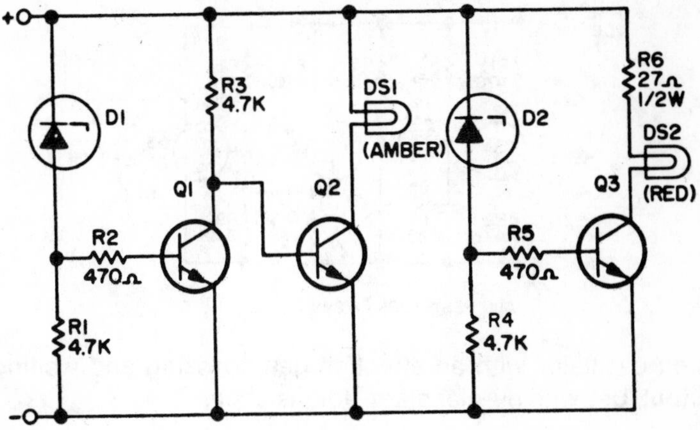


Voltage limit sensor continuously monitors automotive electrical system and indicates voltage variance outside a prescribed range. Amber indicator stays on when all is well; red comes on when trouble exists.

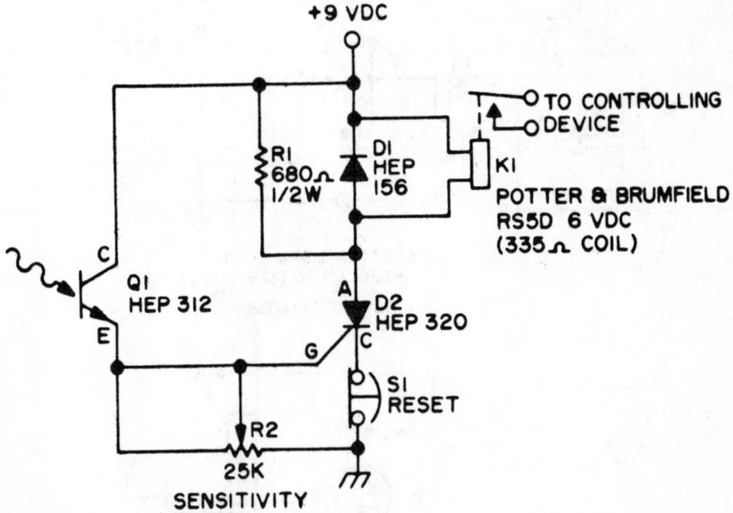


Headlight operated garage door or light switch. Relay contact will close when light hits photodiode. Sensitivity adjustment is provided to prevent false triggering. Circuit courtesy of Motorola construction projects.

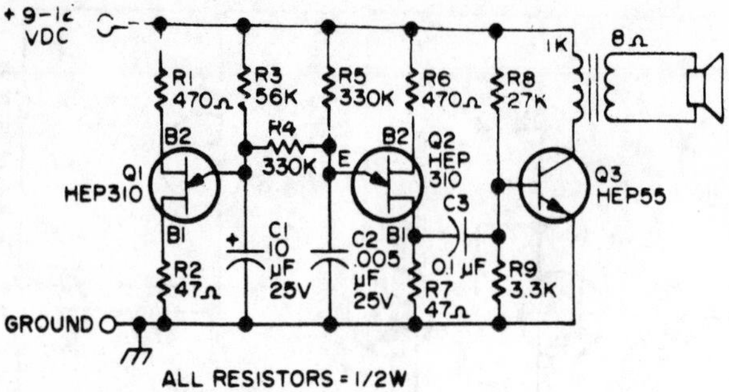


Siren oscillator with an attention-getting rising and wailing output. Use with burglar alarm, for instance.

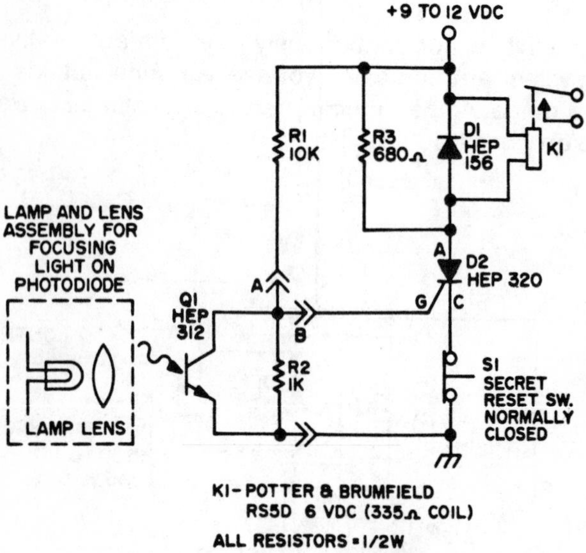


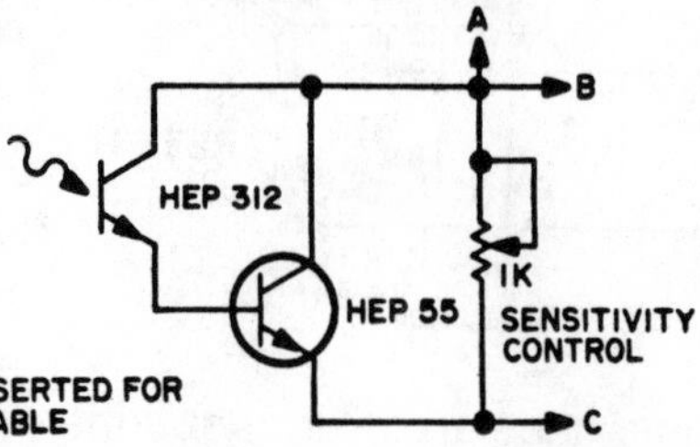


THIS CIRCUIT MAY BE INSERTED FOR INCREASED AND ADJUSTABLE SENSITIVITY.

Burglar alarm (photoelectric) will actuate alarm when light beam is broken. Courtesy Motorola construction projects.

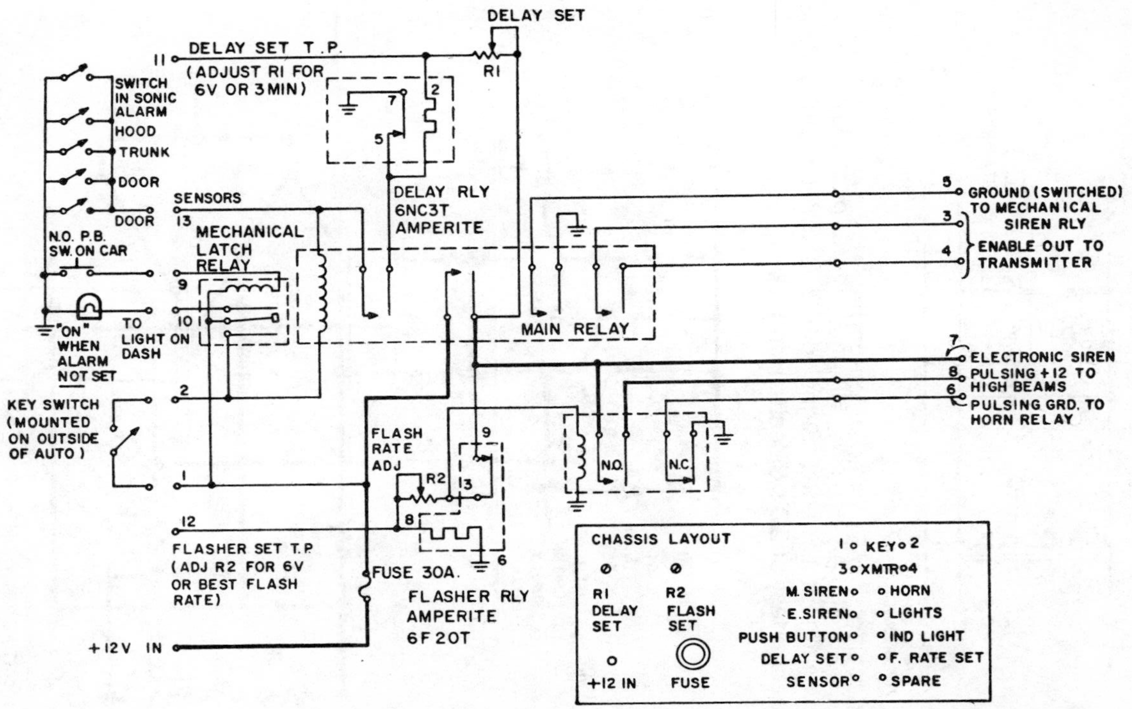


Control system for automobile burglar alarm; connects to external audible sounding device such as horn or siren. Heavy lines indicate No. 16 wire. R1 = 40Ω.

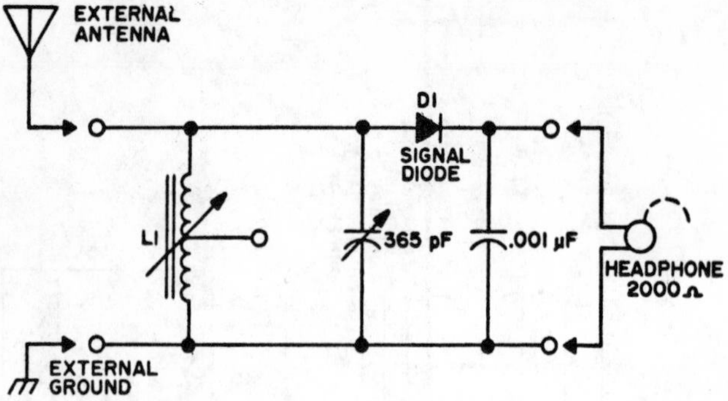


Crystal diode receiver, the simplest radio possible, runs forever without batteries. Signal diode is Calectro K4-550. Circuit courtesy of GC's Calectro Handbook.

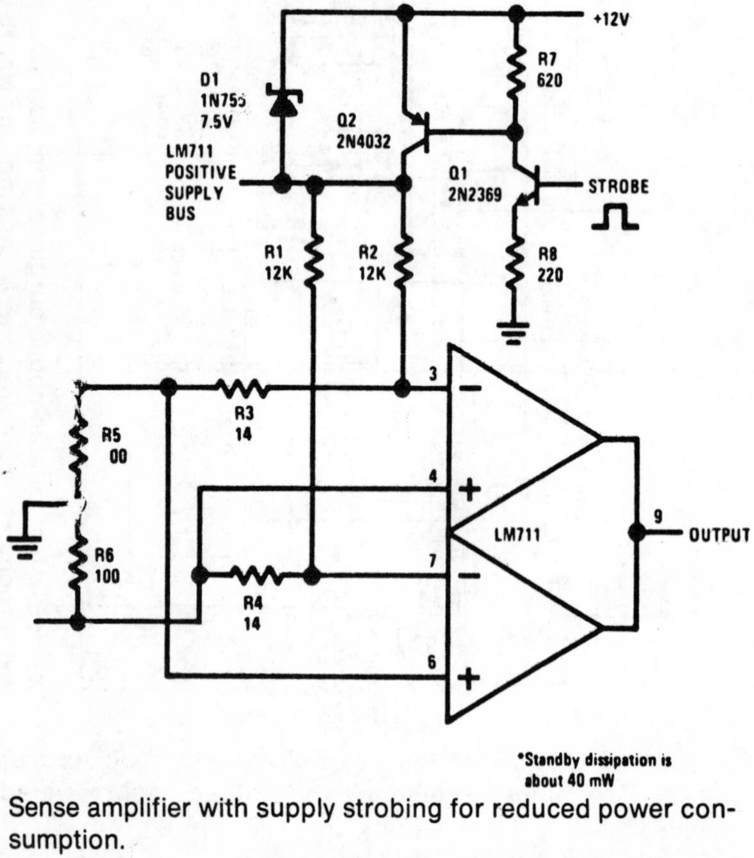


Sense amplifier with supply strobing for reduced power consumption.

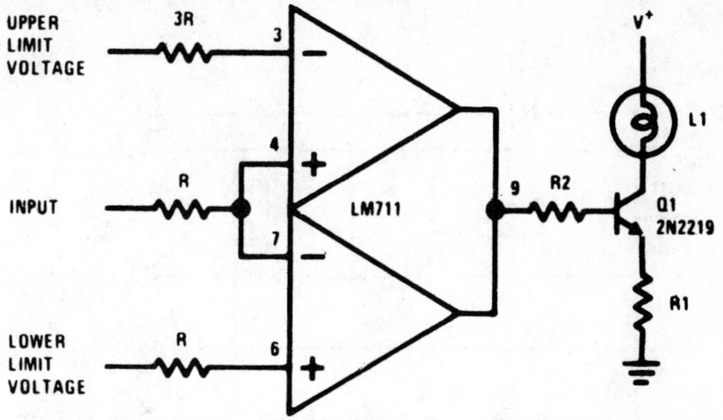


Double-ended limit detector with lamp driver.

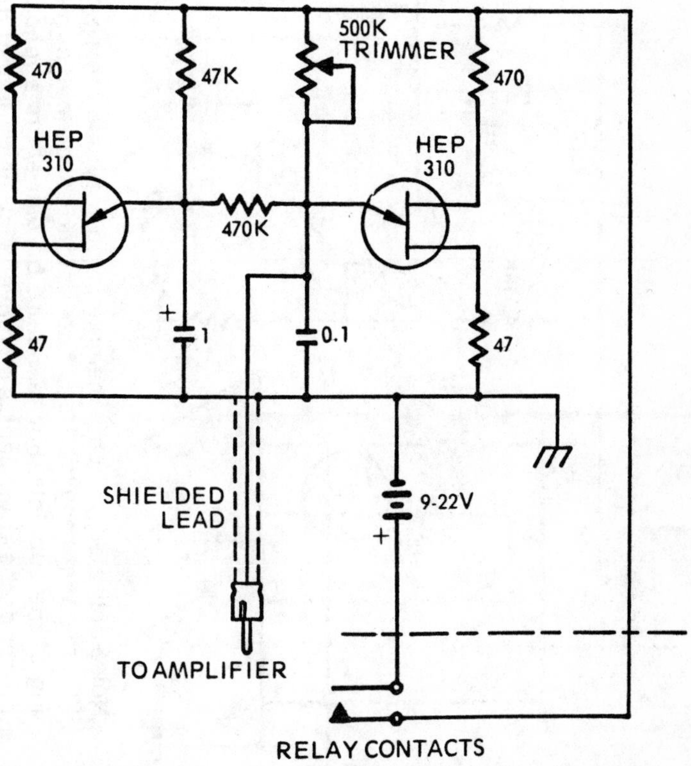


Schematic diagram of the basic dual-unijunction warbler alarm. The output should be fed into a high-powered amplifier.

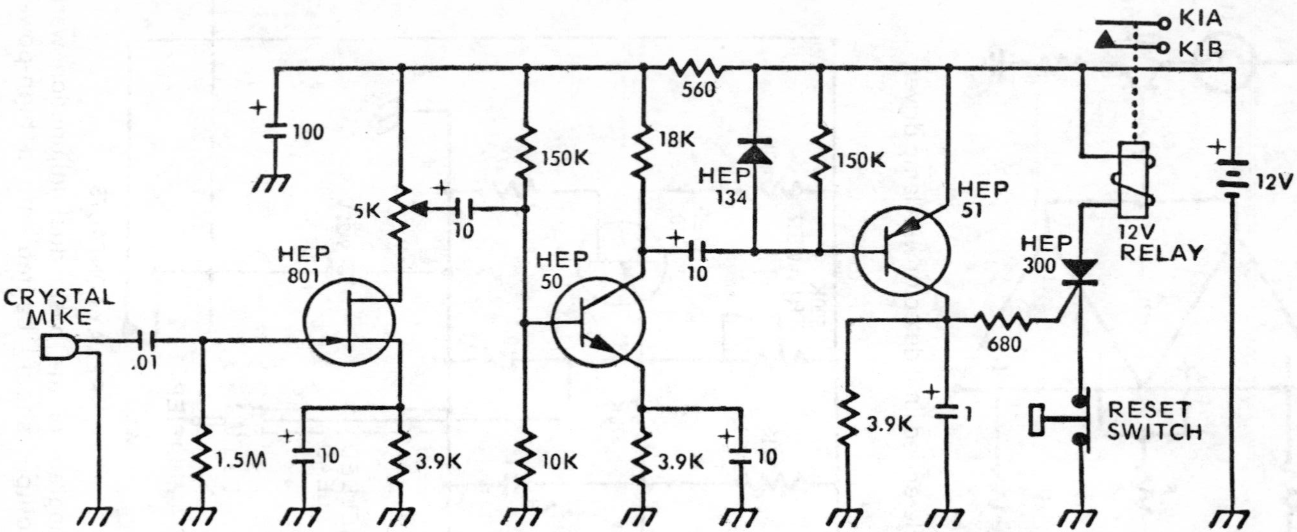


Noise-activated switch schematic. You may use the contacts of the relay (KIA and KIB) to control any load, ac or dc, as long as the load current does not exceed the rating of the contacts. For miniature burglar-alarm applications, where a low-current buzzer or the equivalent is employed, you can connect the buzzer in place of the relay coil.

# Section B

# Audio Conditioning Circuits

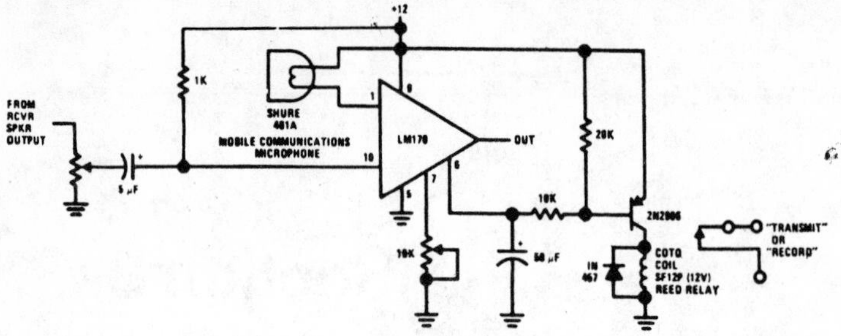


VOX/mike preamp with antitrip.

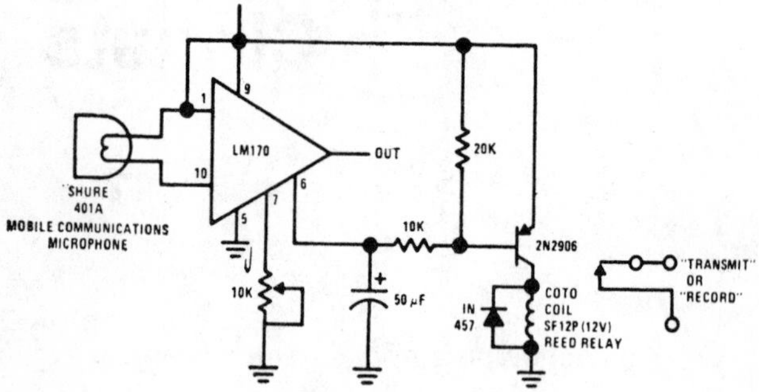


VOX/mike preamp.

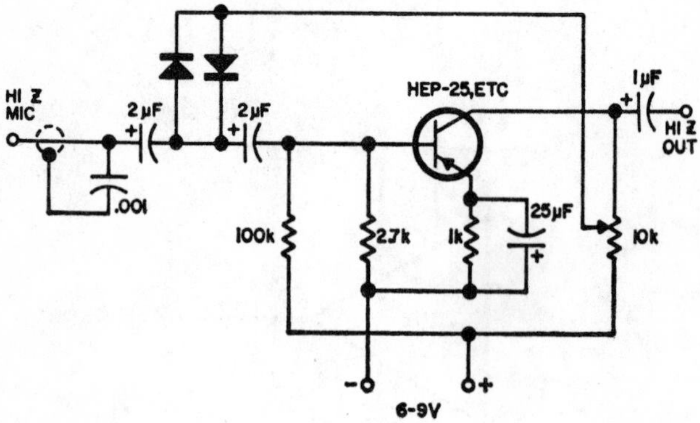


Circuit of preamplifier clipper circuit. Potentiometer adjusts clipping level and may be replaced by fixed resistors once desired level is found.

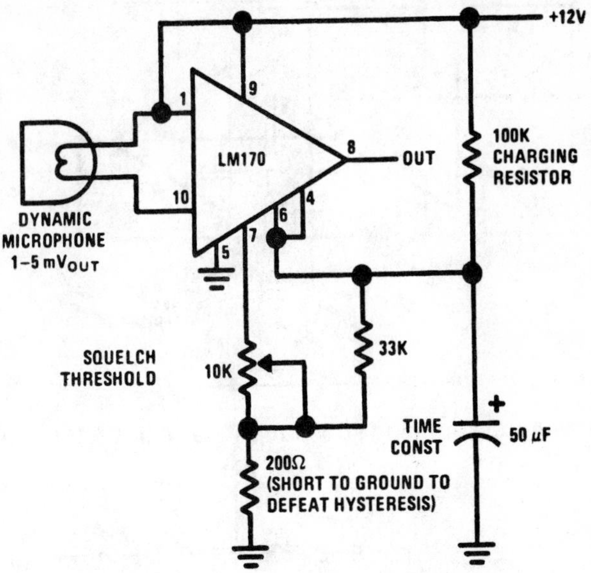


Squelched preamplifier with hysteresis.

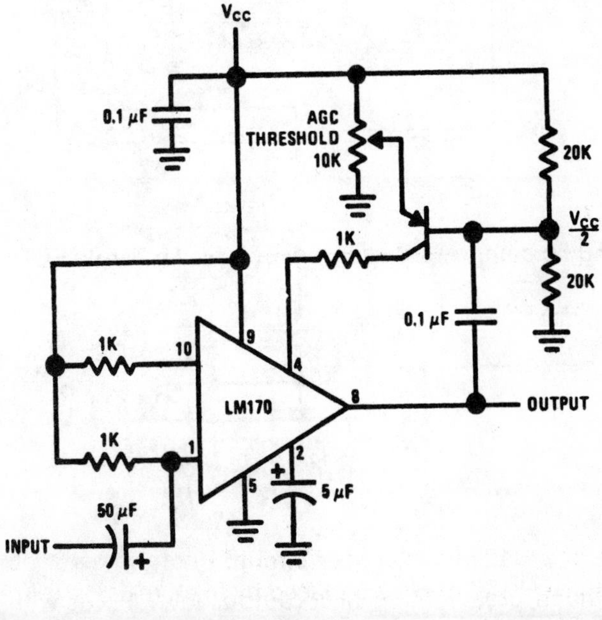


Speech compressor.

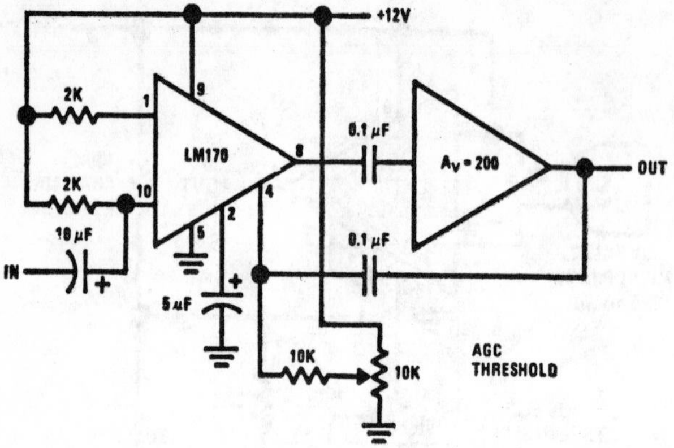


Speech compressor using subsequent gain for better control.

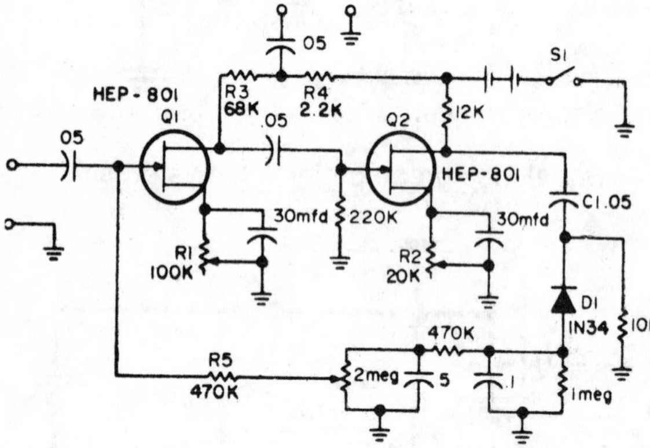


Audio compressor uses inexpensive Motorola FETs.

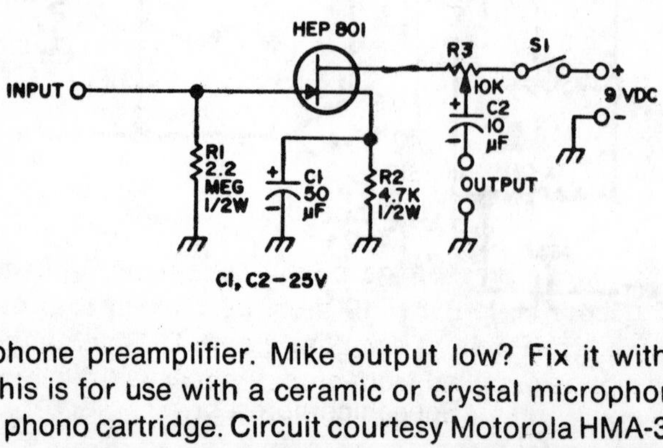


Microphone preamplifier. Mike output low? Fix it with this one. This is for use with a ceramic or crystal microphone or even a phono cartridge. Circuit courtesy Motorola HMA-33.

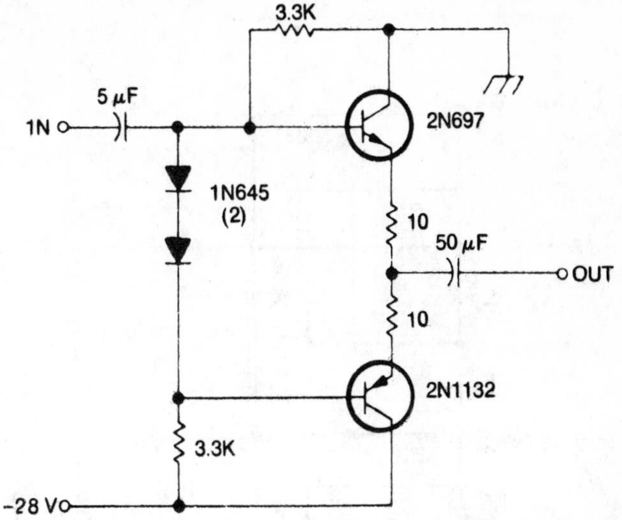


This schematic from the U.S. Navy's handbook of "preferred circuits," shows an emitter follower that provides 12 dB gain.

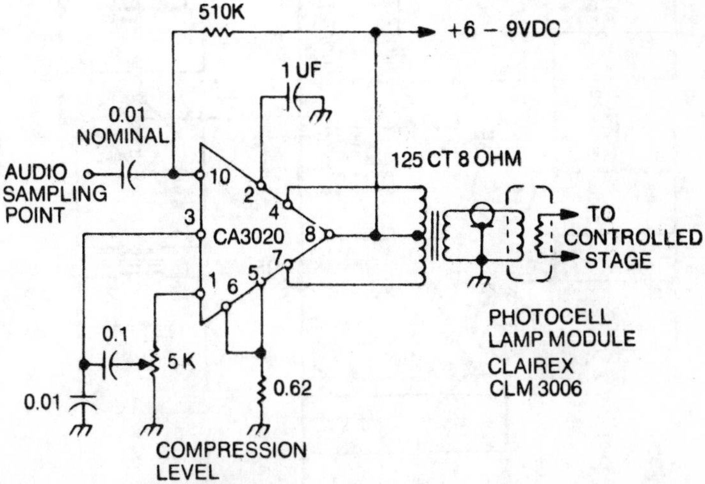


Photocell compressor/agc circuit schematic. Voltage rating of capacitor to terminal 10 must be chosen to protect unit from voltage found at sampling point. Dc operating voltage need not be supplied from an extremely well filtered source since audio quality of amplifier is not significant.

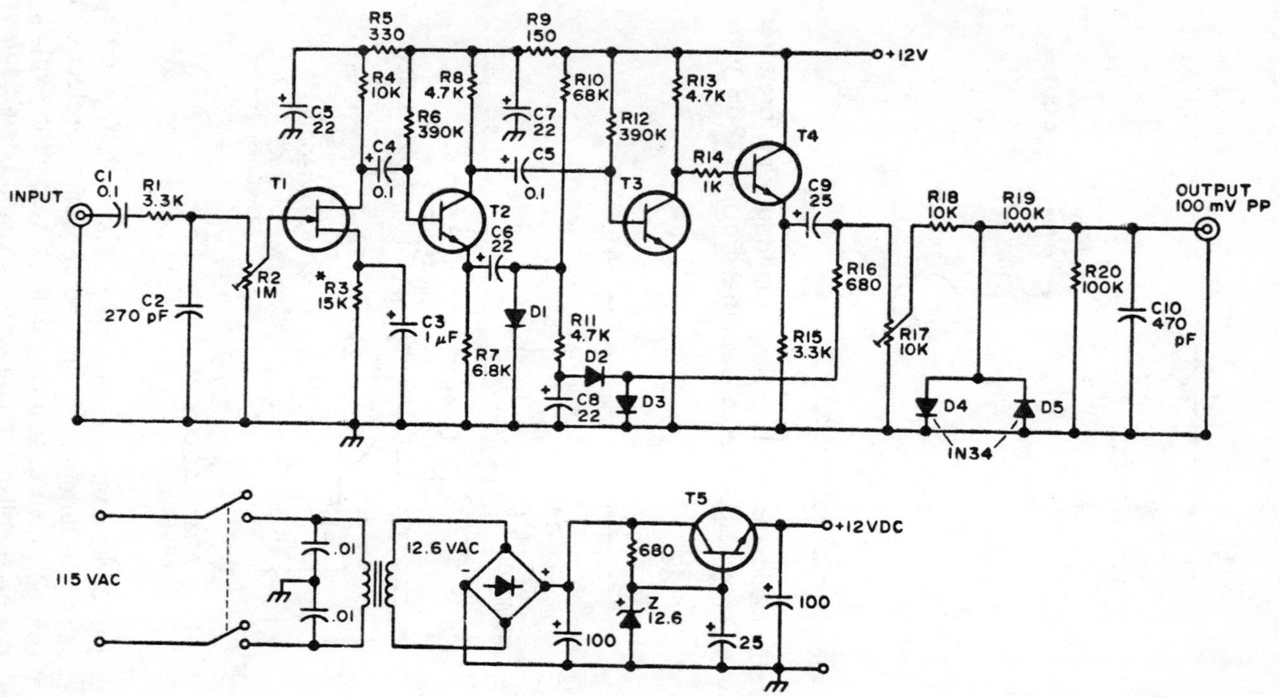


Compressor schematic.

Shaping circuitry to be added ahead of existing compressor.

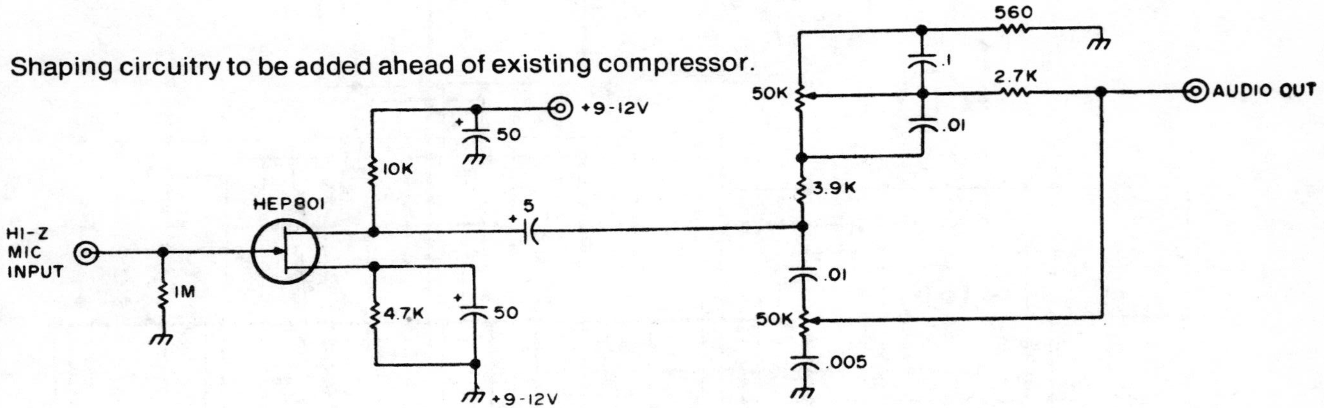


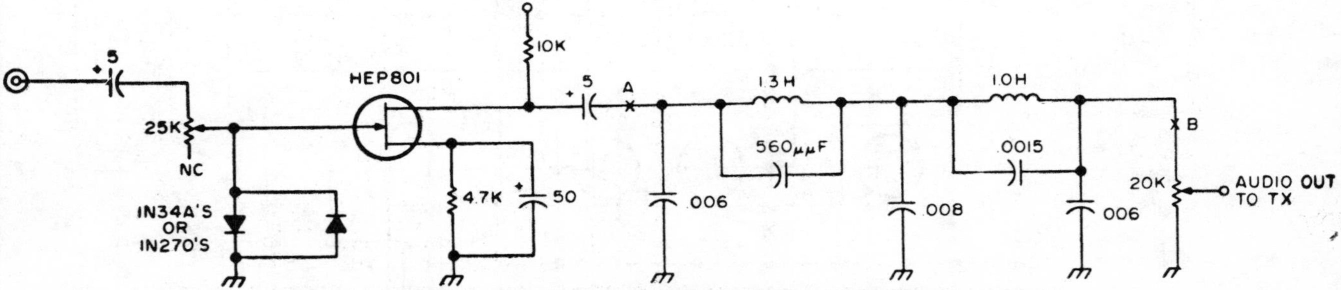


Clipper and filter for use at output of existing audio compressor.

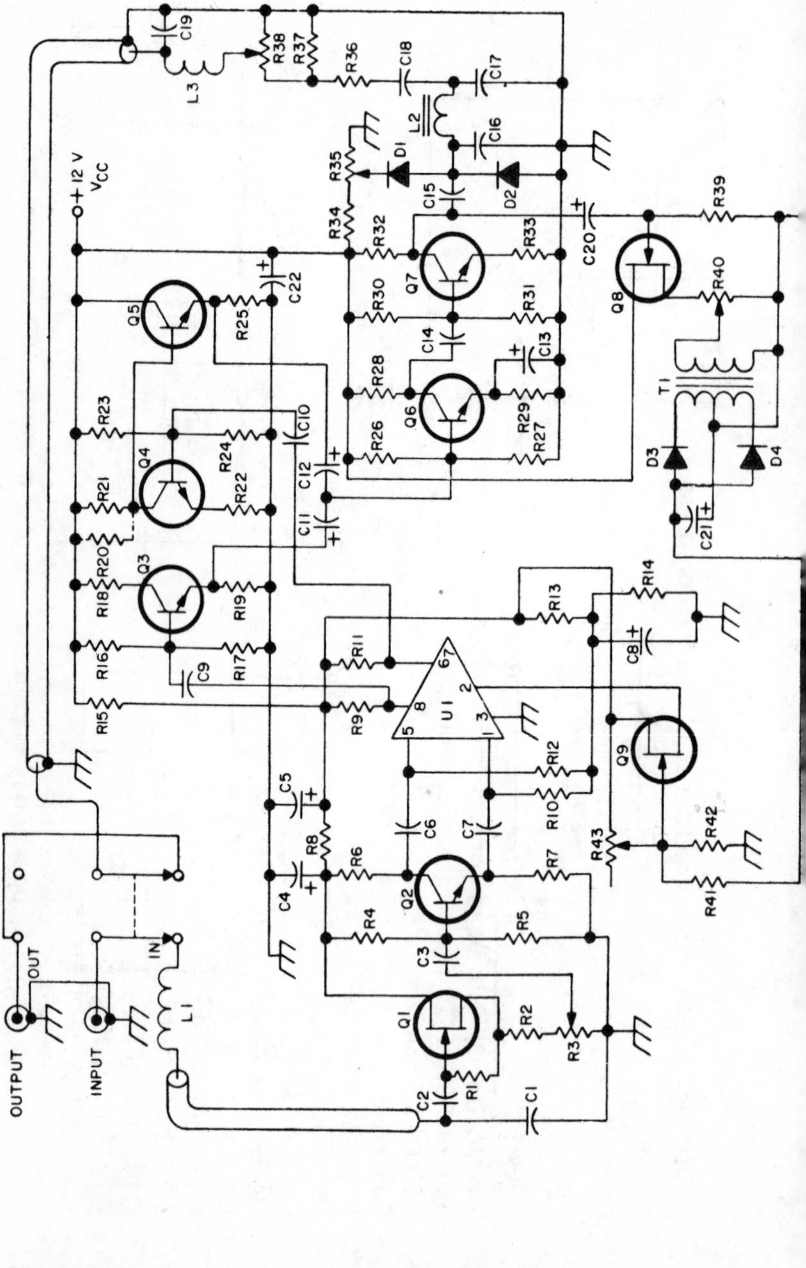
OUTPUT
INPUT
OUT
IN
+12 V
VCC
L1
L2
L3
Q1
Q2
Q3
Q4
Q5
Q6
Q7
Q8
Q9
U1
T1
D1
D2
D3
D4
C1
C2
C3
C4
C5
C6
C7
C8
C9
C10
C11
C12
C13
C14
C15
C16
C17
C18
C19
C20
C21
C22
R1
R2
R3
R4
R5
R6
R7
R8
R9
R10
R11
R12
R13
R14
R15
R16
R17
R18
R19
R20
R21
R22
R23
R24
R25
R26
R27
R28
R29
R30
R31
R32
R33
R34
R35
R36
R37
R38
R39
R40
R41
R42
R43

## PARTS LIST

**Components**

Q1,Q8,Q9 FET-1, HEP 802, 2N3819, etc.
Q2-Q7 2N697, HEP 54
U1 CA3028A D1-D4 1N270
L1-L3 2.5 mH rf choke
L2 3.5 H miniature audio choke, UTC DOT-8
T1 GC Co. D1-728 transformer. For primary, use half of 500Ω CT secondary. For secondary, use 1000Ω CT primary.

**Capacitors**

C1, C19 0.001 μF
C2 0.005 μF
C3 0.05 μF
C4, C5, C8, C22, 100μF 15V
C6, C7, C9, C10, C15, C18 0.1μF
C11, C12, 10 μF 15V
C13 30 μF 15V
C14, 0.02μF
C16, C17 0.0015μF
C20, C21 6 μF 1.5V

**Resistors (all except potentiometers 1/2 watt)**

R1 2.2M
R2 91-R3 10K audio-taper pot
R4, R16, R23, R26 180K
R5, R17, R24 56K
R6, R7, R18, R19, R21, R22, R25, R34 4.7K
R8, R14, R15, R29, R33, R41 1.0K
R9, R11 10K
R10, R12, R37 47K
R13 2K
R28, R32, 3.9K
R30 120K
R35 5K pot
R36 24K
R38 50K audio taper pot
R39 100K
R40 2K trimmer pot
R42 30K
R27, R31 33K

Speech processor increases effectiveness off SSB signal by compression, clipping, and filtering before modulation.

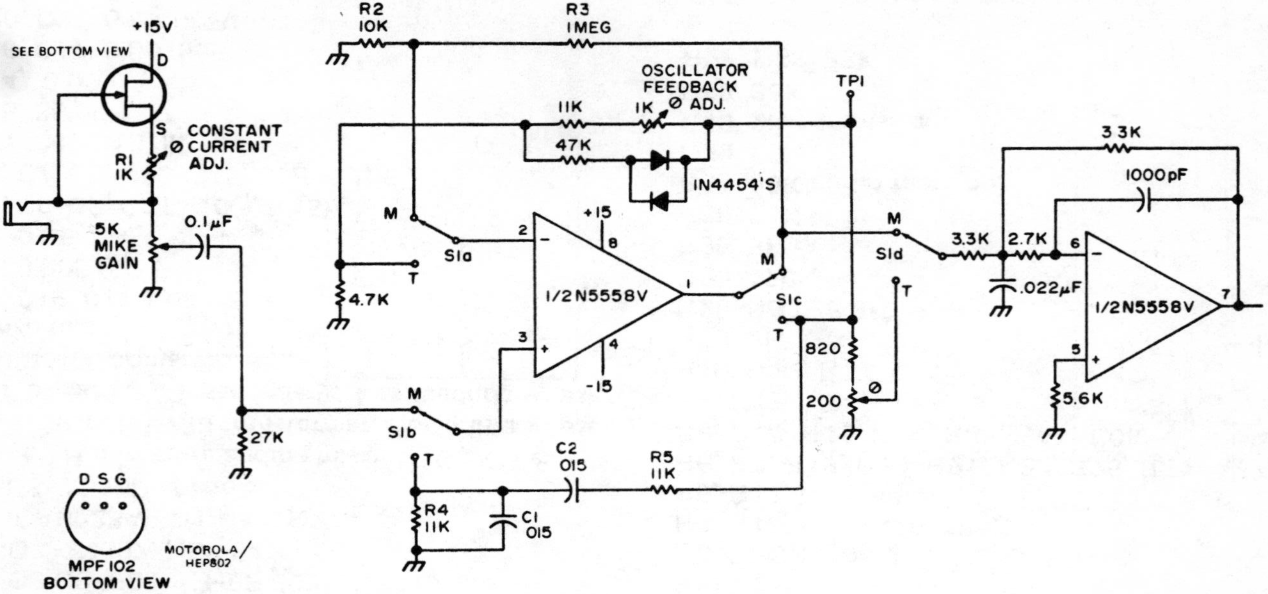


Combination preamp and tone generator.

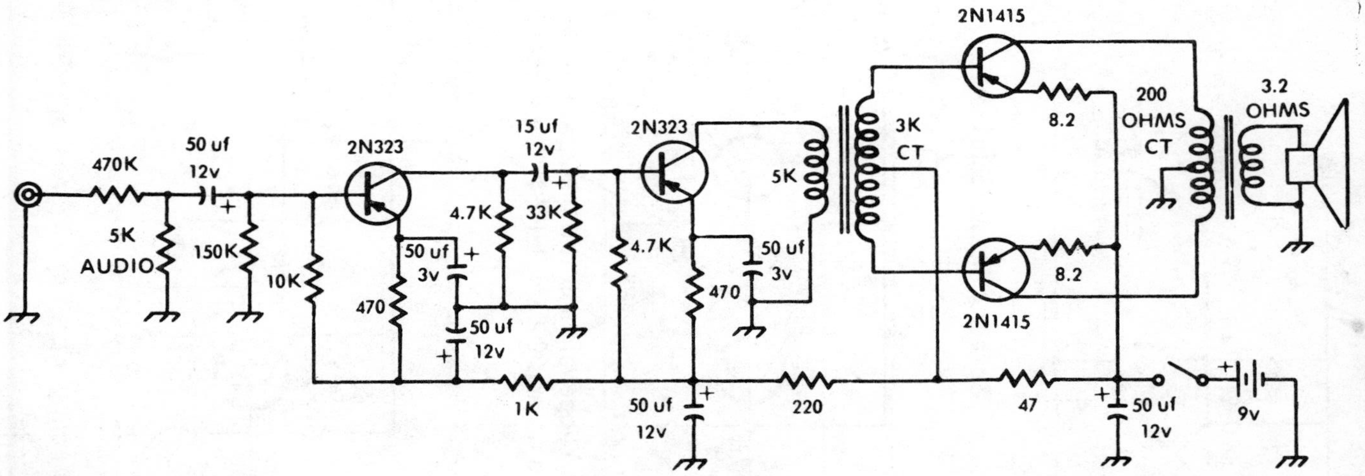


his low-power general-purpose amplifier employs a push-pull stage, which offers less distortion, better hum rejection, and higher overall efficiency than a single-ended stage of equivalent power capability.

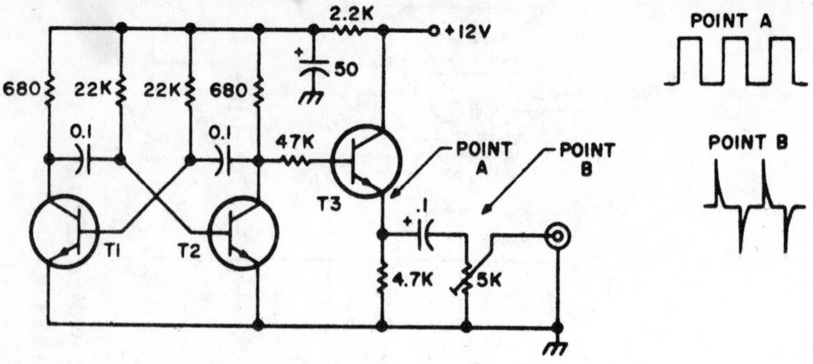


Speech simulator schematic.

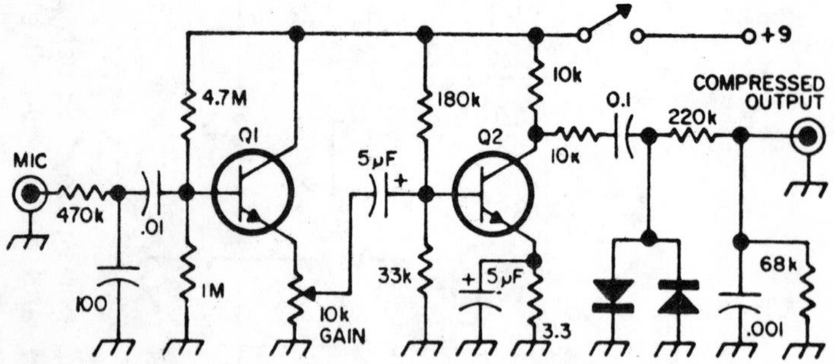


Two-stage clipper/preamp will increase the talk of any rig. Transistors Q1 and Q2 are HEP 54. The diodes are 1N456 or HEP 158.

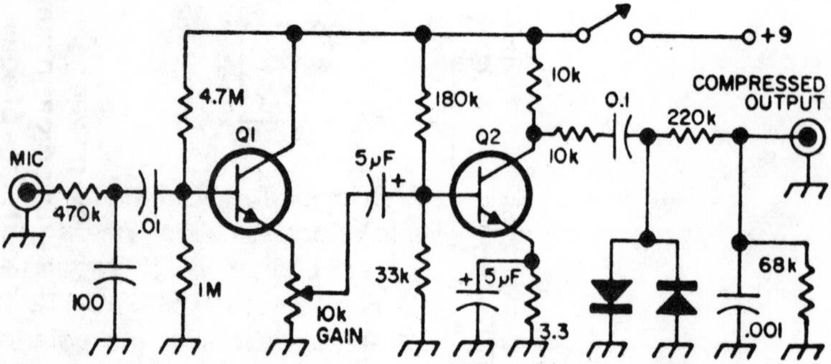


Two-stage clipper/preamp. Transistors Q1 and Q2 are 2N1304, 2N2926, 2N3391, SK3011, or HEP 54. The diodes are 1N456 or HEP 158.

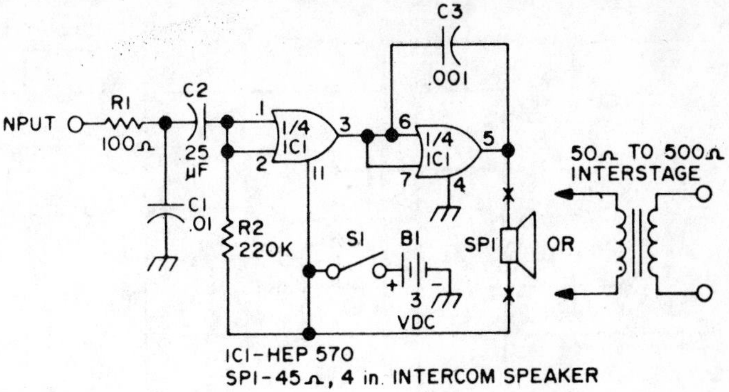


Microphone or audio amplifier, class A, high gain and compact. Circuit courtesy Motorola, from Radio Amateur's IC Pro jects, HMA-36.

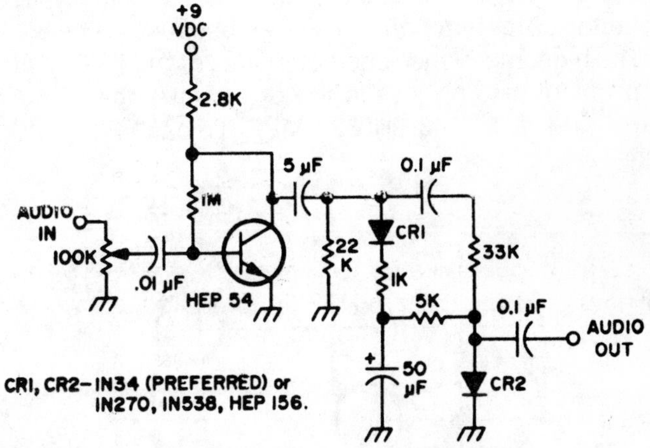


Audio preamp compressor.

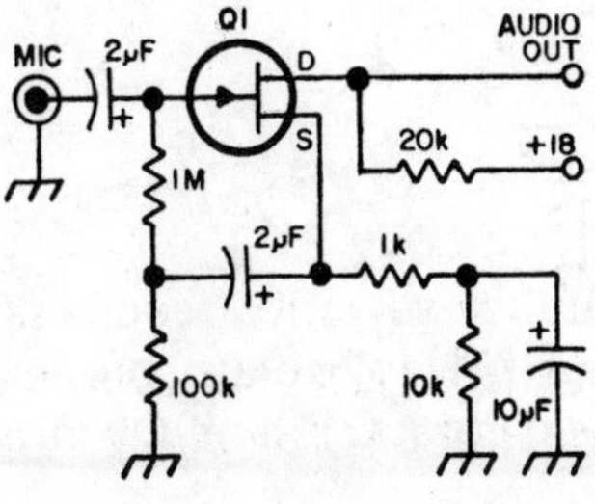


Microphone amplifier using a field-effect transistor has an input impedance of 5 megohms. Q1 is a 2N4360, TIM12, U-112 or U-110. By reversing the polarity of the supply voltage, a 2N3820, MPF 104 or HEP 801 may be used.

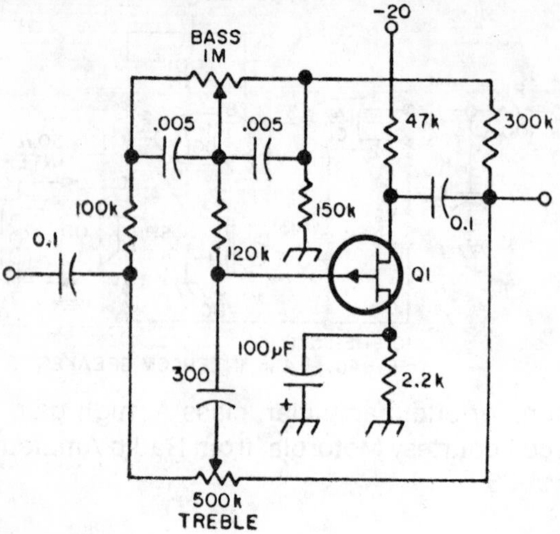


Resistance—capacitance tone controls are usually not too satisfactory with junction transistors because of heavy loading. The high impedance characteristics of the FET eliminates this problem with no loss in the dynamic range of the tone control. Q1 is a 2N2943, 2N3820, MPF 105 or HEP 801.

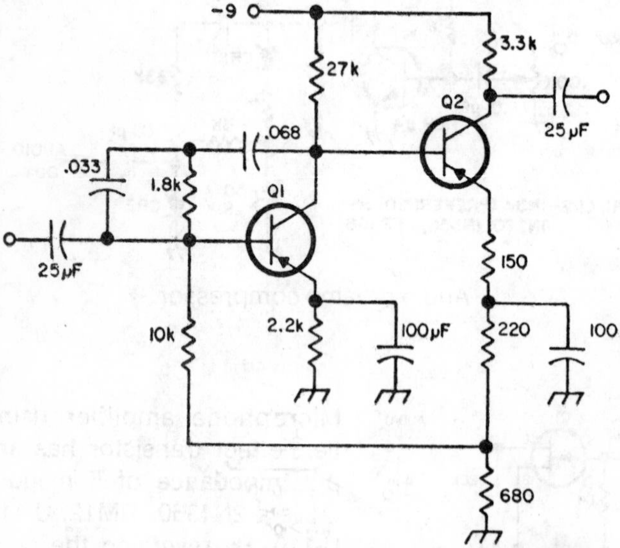


This phono preamplifier uses frequency-selective feedback between the collector and base of Q1 to obtain proper equalization during playback. Transistors Q1 and Q2 are 2N584, SK3003, GE-2 or HEP 254.

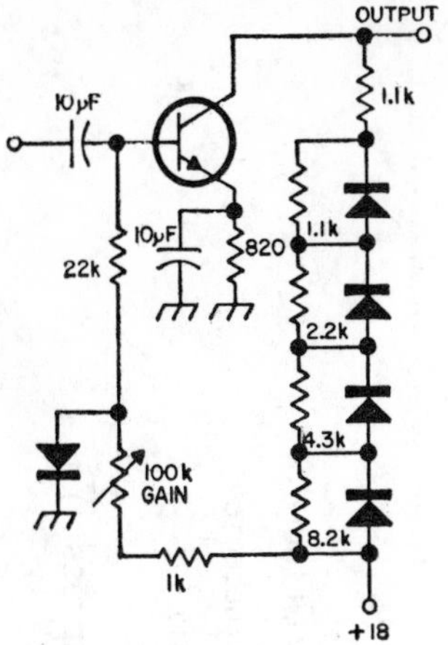


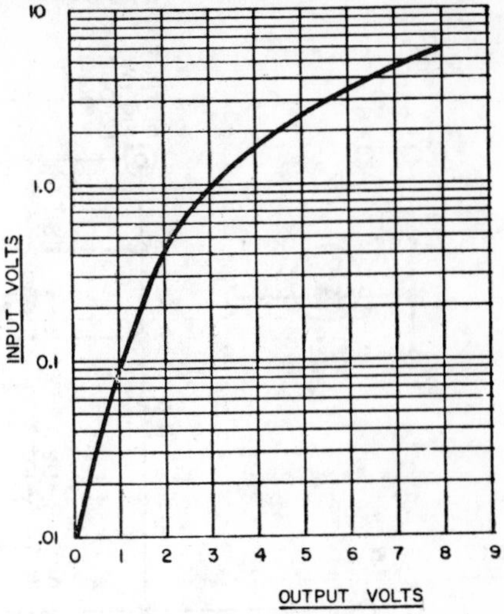


This simple dynamic range compressor provides 50 dB range; it exhibits gain with a 20 millivolt signal but will saturate with input voltages up to 6 to 7 volts. All the diodes are 1N914; transistor Q1 should be a 2N2926, 2N3391, SK3010, GE-8 or HEP 54.

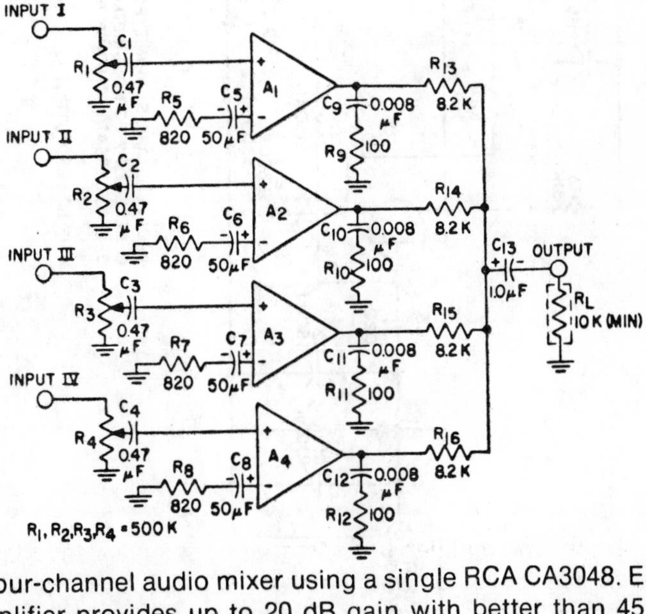


A four-channel audio mixer using a single RCA CA3048. Each amplifier provides up to 20 dB gain with better than 45 dB isolation between channels.

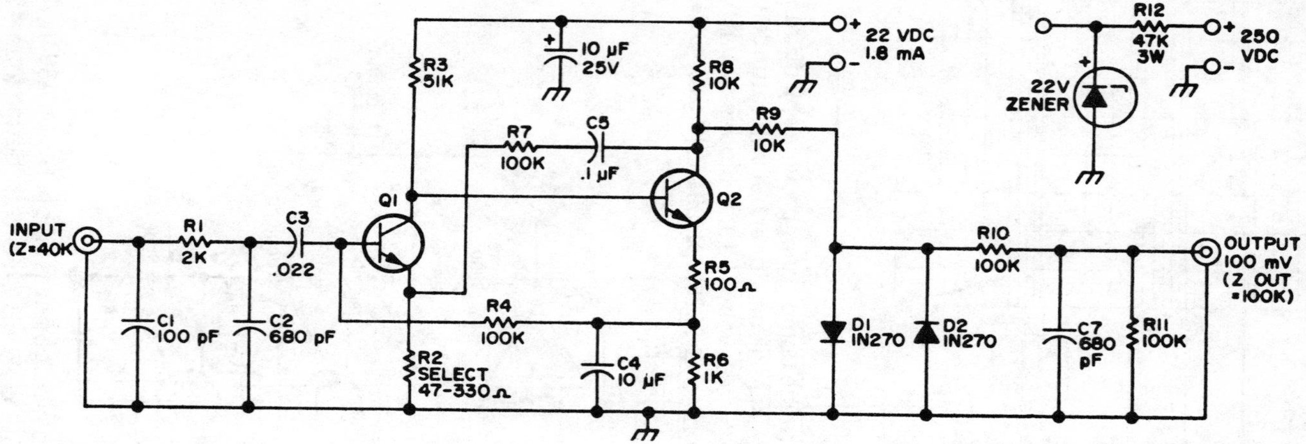


1- Q1 & Q2 ARE Si NPN TYPES, hfe 30-120, VCBo 45V OR EQUIV.
2-SELECT R2 FOR DESIRED GAIN.
3-IF NECESSARY, CHANGE R4 TO YIELD 10-12 VDC AT COLLECTOR OF Q2.
4-IF LOWER OUTPUT IS DESIRED, CHANGE R10-R11 RATIO.

Audio conditioning unit (preamplifier/compressor).

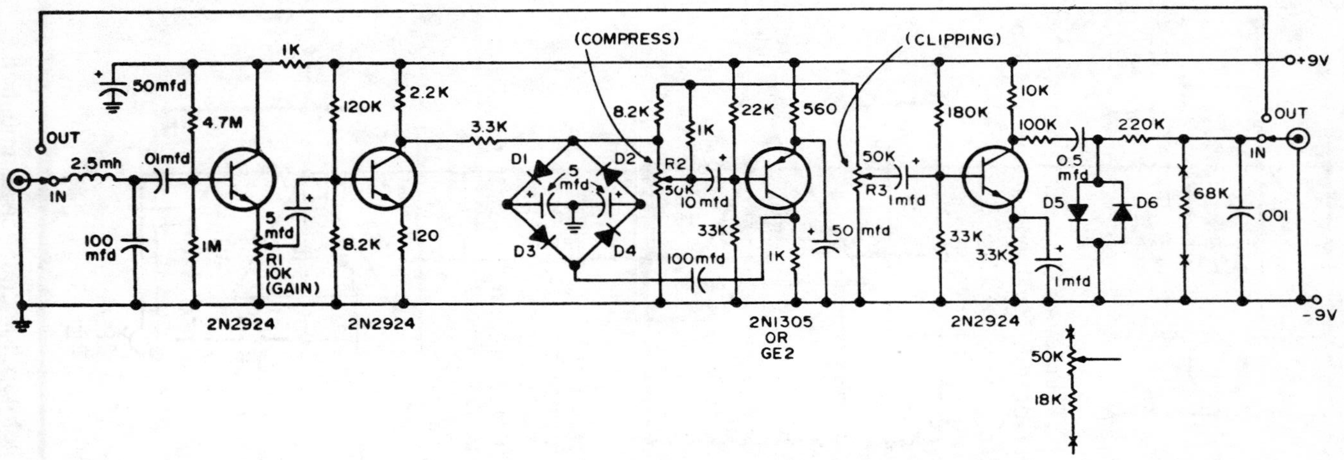


Versatile premodulation speech processor.

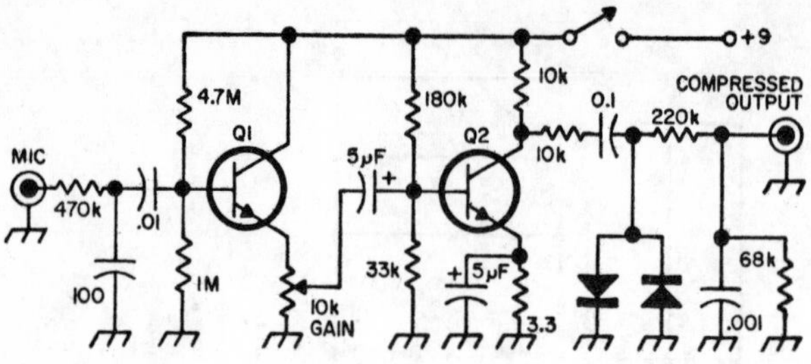


Clipper/preamp. Transistors Q1 and Q2 are 2N1304, 2N2926, 2N3391, SK3011, or HEP 54. The diodes are IN456 or HEP 158.

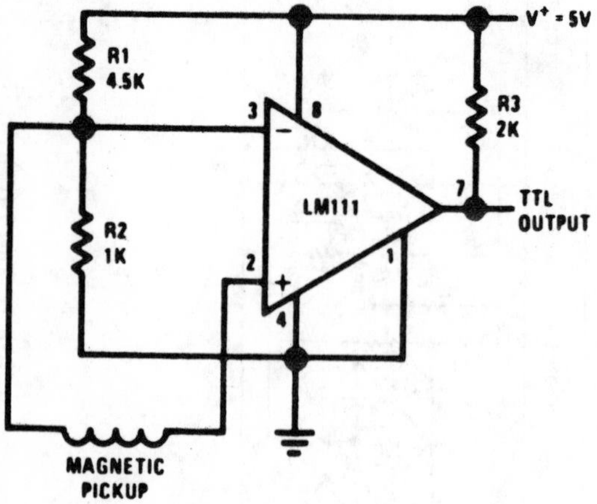


Detector for magnetic transducer.

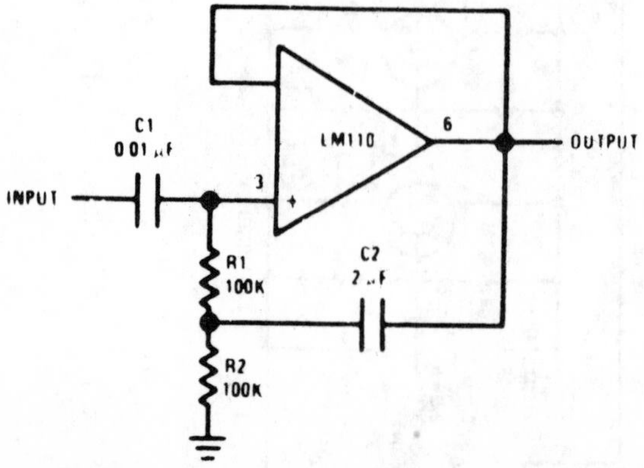


High-input-impedance ac amplifier.

# Section C

# Audio Amplifiers

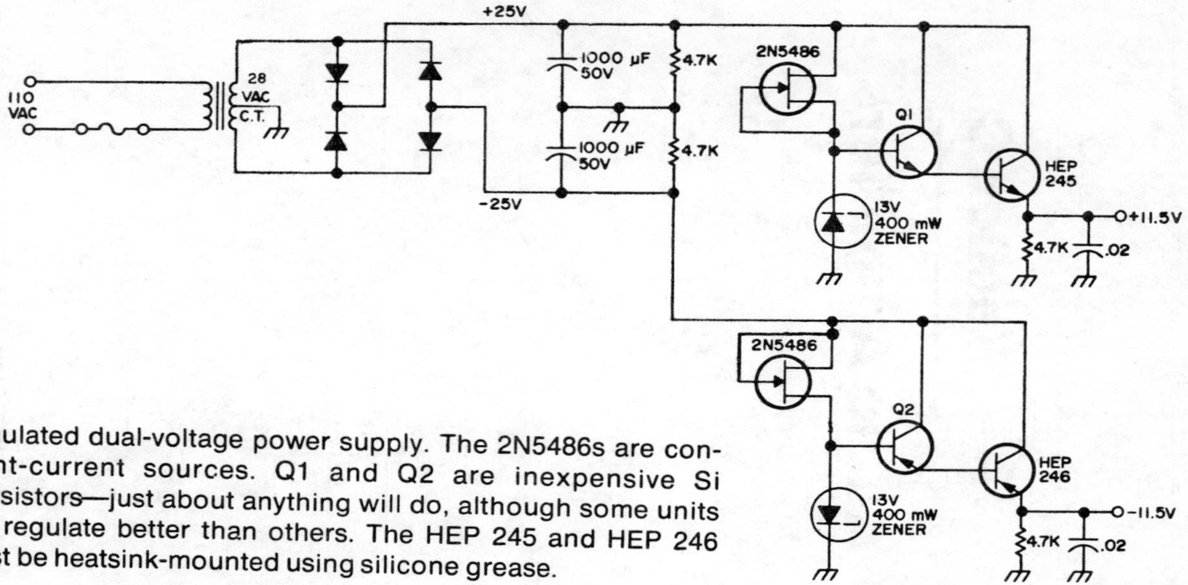


Regulated dual-voltage power supply. The 2N5486s are constant-current sources. Q1 and Q2 are inexpensive Si transistors—just about anything will do, although some units will regulate better than others. The HEP 245 and HEP 246 must be heatsink-mounted using silicone grease.

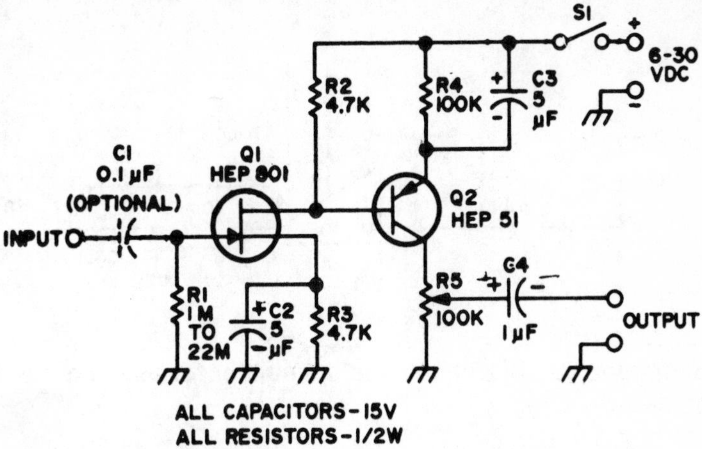


Audio amplifier. This circuit has a high-impedance input, low current drain (0.2 mA), low-impedance output, wide range (10–30,000 Hz) and a gain of 200 to 400. Circuit courtesy Motorola HMA-33.

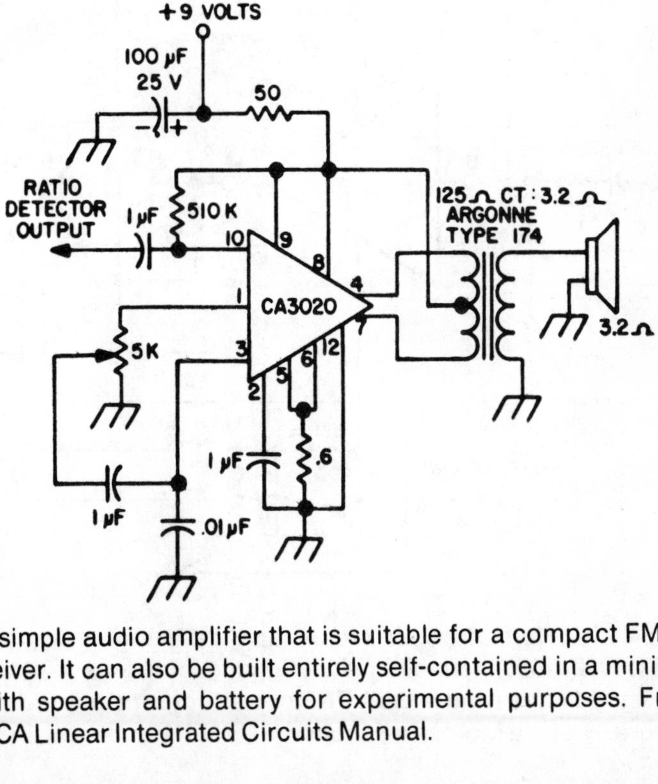


A simple audio amplifier that is suitable for a compact FM receiver. It can also be built entirely self-contained in a minibox with speaker and battery for experimental purposes. From RCA Linear Integrated Circuits Manual.

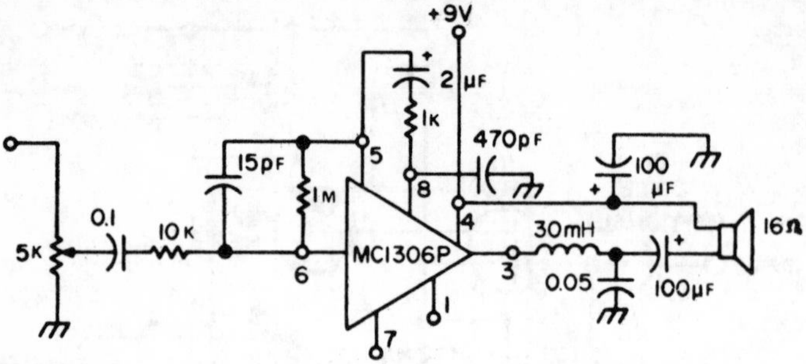


High-gain (30 dB) 250 mW audio amplifier for use in portable receiver.

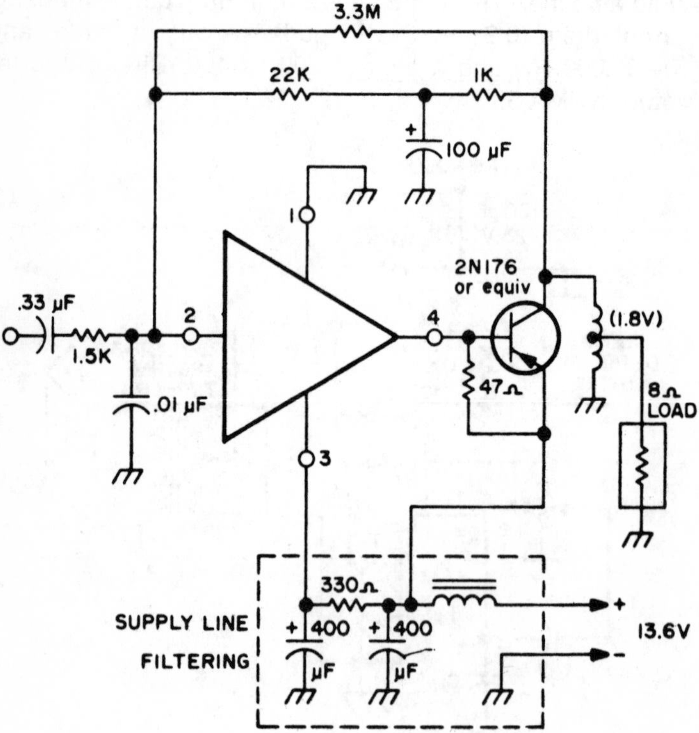


Typical 4 watt amplifier circuit application of a Motorola MFC4050 silicon monolithic functional circuit, an audio driver designed for driving class A PNP power output stages. Circuit courtesy of the Motorola Functional Circuit Handbook.

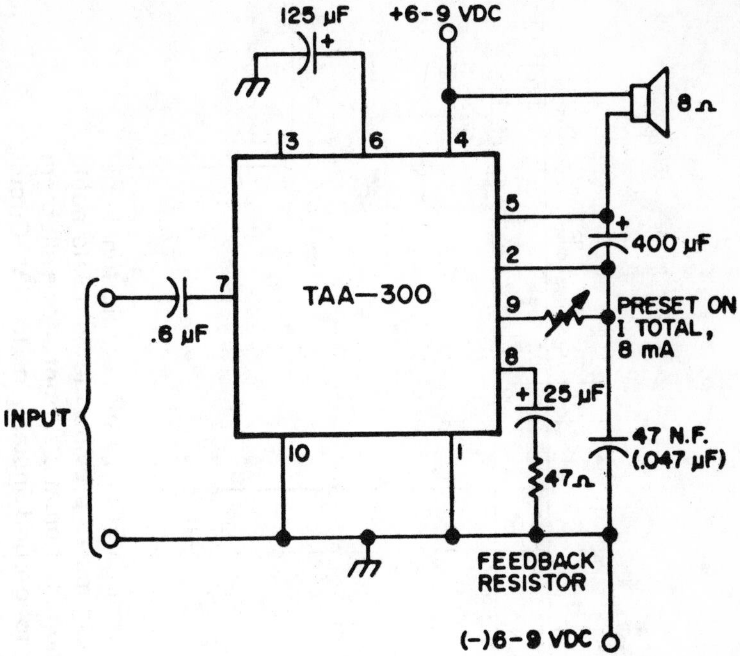


External schematic of a complete audio amplifier using the TAA-300 IC.

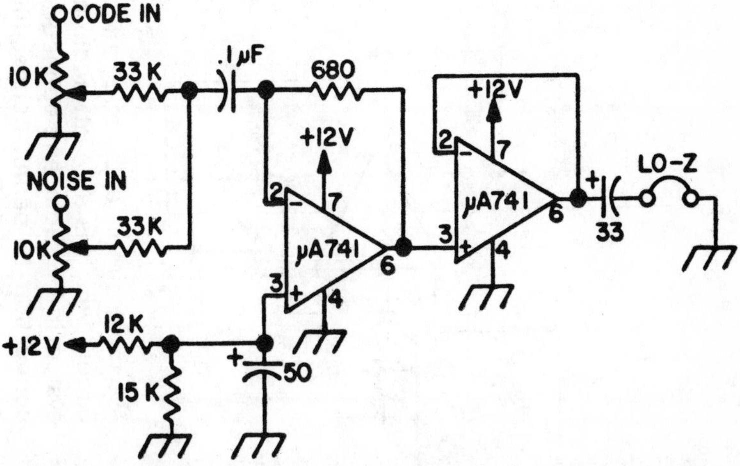


A code practice audio mixer. Feed code into one input and random CW noise from the Novice bands into the other while adjusting the controls for a realistic sound.

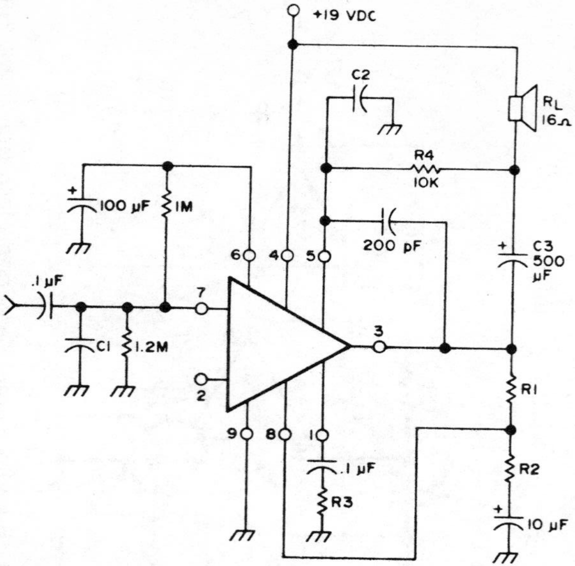


| SENSITIVITY FOR $P_o$ mV | C1 pF | C2 pF | R1 k OHMS | R2 OHMS | R3 OHMS | $R_L$ OHMS | $P_o$ W |
|---|---|---|---|---|---|---|---|
| 600 | 0 | 0 | 10 | 1.0K | 82 | 16 | 2.0 |
| 15 | 100 | 100 | 51 | 100 | 2.2K | 16 | 2.0 |

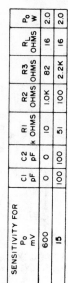

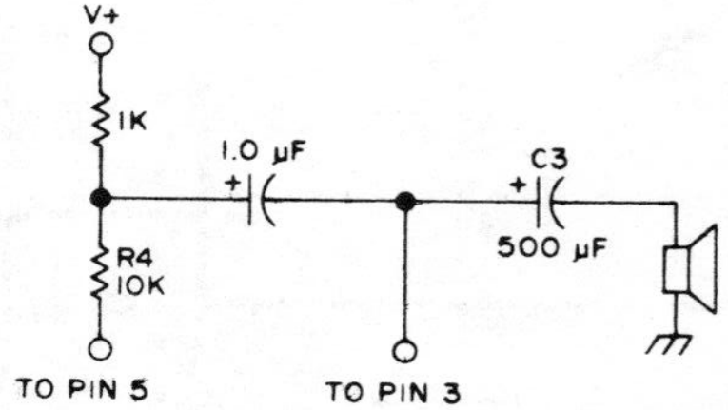


A typical circuit application utilizing a Motorola MFC9010 monolithic functional circuit, which is a 2-watt audio amplifier designed to provide the complete audio system in TV, radio, and phonograph equipment. Schematic at right shows alternate connection to permit connecting speaker to ground instead of to +V. Circuit courtesy Motorola Functional Circuits handbook.

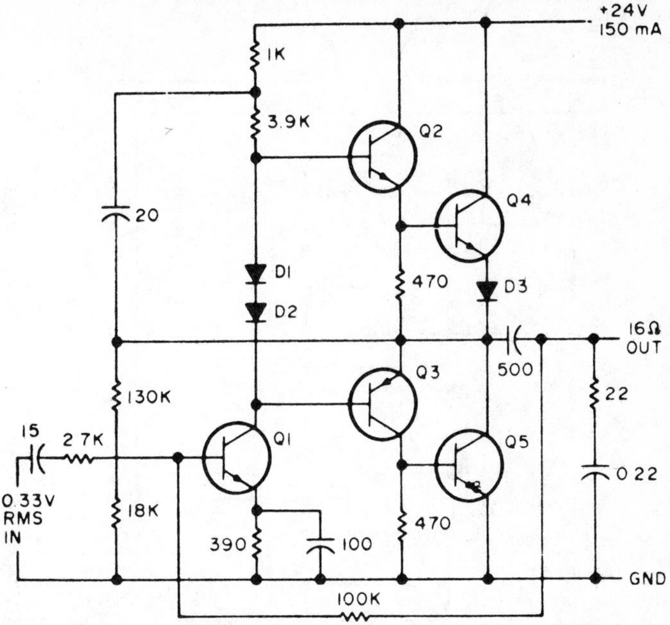


This circuit for a 2.5W audio power amplifier illustrates complementary-symmetry phase inverters (Q2 and Q3) as well as totem-pole output circuit (Q4 and Q5). Transistor Q1 is simply a preamp to provide drive to the phase inverter stage. A 100K resistor at bottom provides negative feedback. Diodes compensate for base–emitter voltage drops of transistors, making dc coupling safe to use. Transistors Q4 and Q5 must handle 800 mW dissipation; 2N3402 transistors are simply 2N3414s with built-in heatsinks to raise power rating and are recommended for Q4 and Q5.

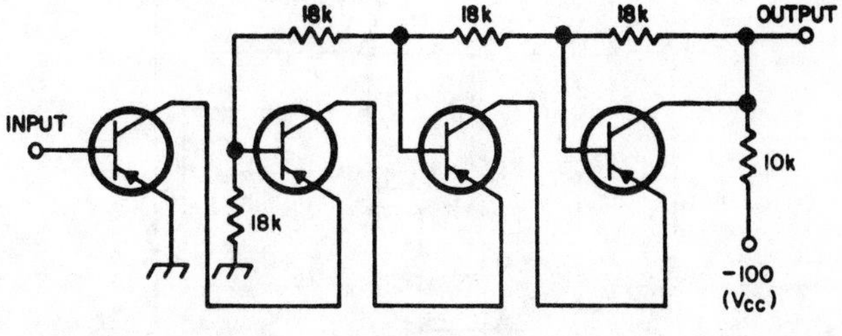


High-voltage direct coupled amplifier. The gain of this amplifier is equivalent to both sections of a 12AU7. All transistors are SK3008, GE-9 or HEP 51.

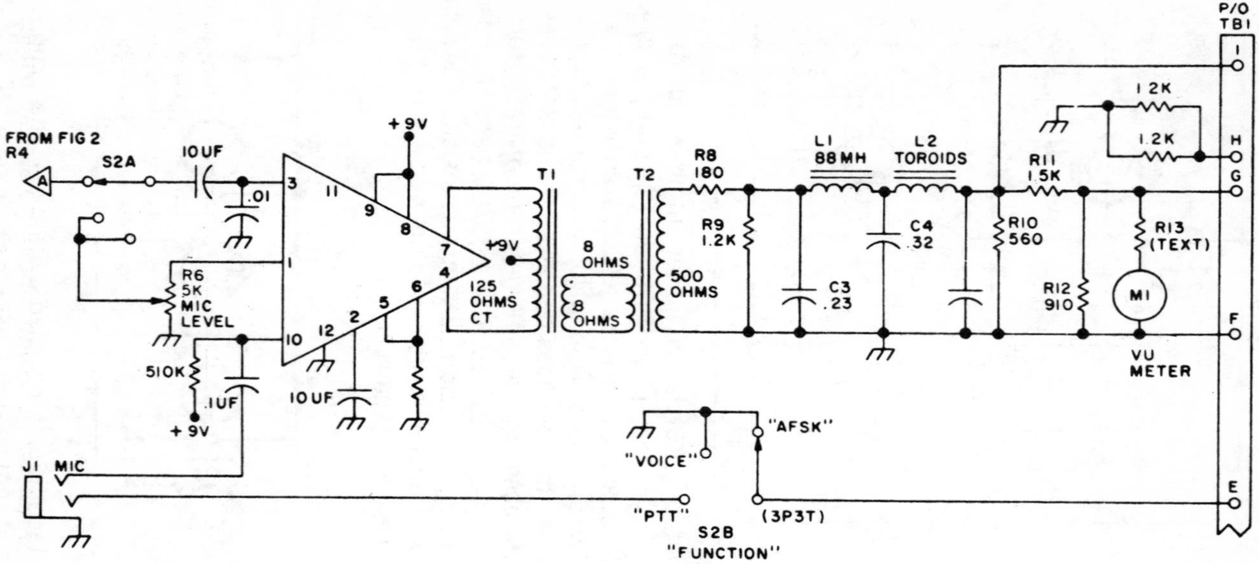


Audio amplifier and output network schematic.

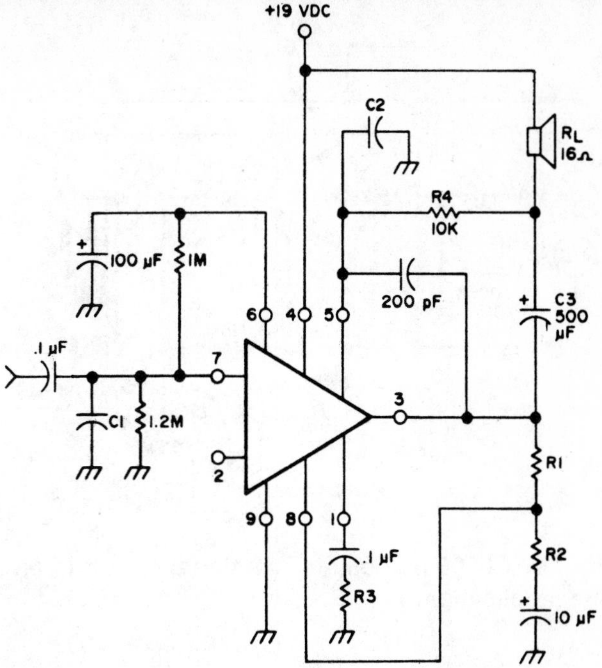


| SENSITIVITY FOR 1 WATT mV | C1 pF | C2 pF | R1 k OHMS | R2 OHMS | R3 OHMS |
|---|---|---|---|---|---|
| 400 | 0 | 0 | 10 | 1.0K | 82 |
| 10 | 100 | 100 | 51 | 100 | 2.2K |

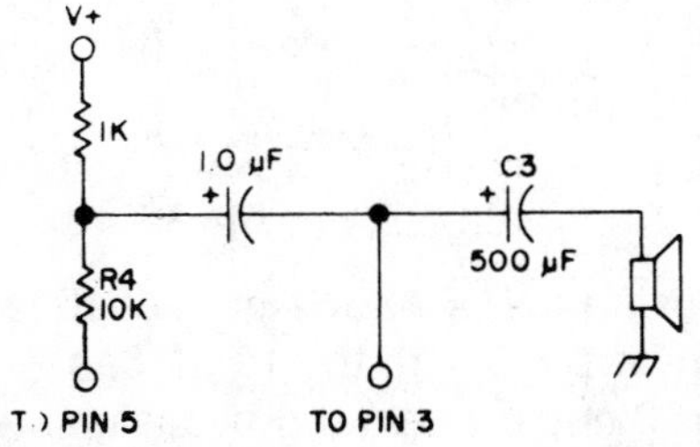


Alternate connection to permit connecting speaker to ground instead of to V+.

Typical circuit for a 1-watt audio power amplifier, using Motorola's Functional Circuit MFC8010.

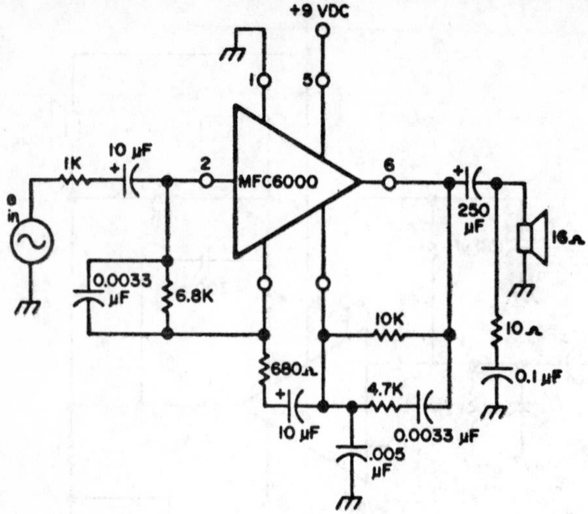


Motorola MFC6000 monolithic functional circuit in an audio amplifier application.

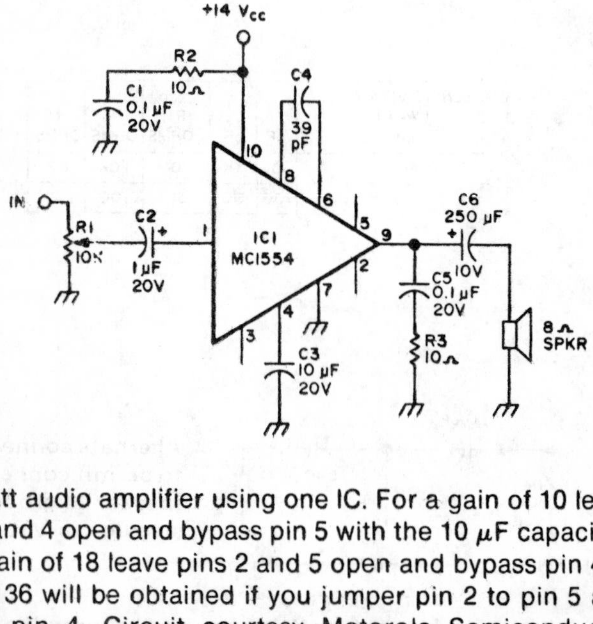


One-watt audio amplifier using one IC. For a gain of 10 leave pins 2 and 4 open and bypass pin 5 with the 10 $\mu$F capacitor. For a gain of 18 leave pins 2 and 5 open and bypass pin 4. A gain of 36 will be obtained if you jumper pin 2 to pin 5 and bypass pin 4. Circuit courtesy Motorola Semiconductor Power Circuits Handbook.

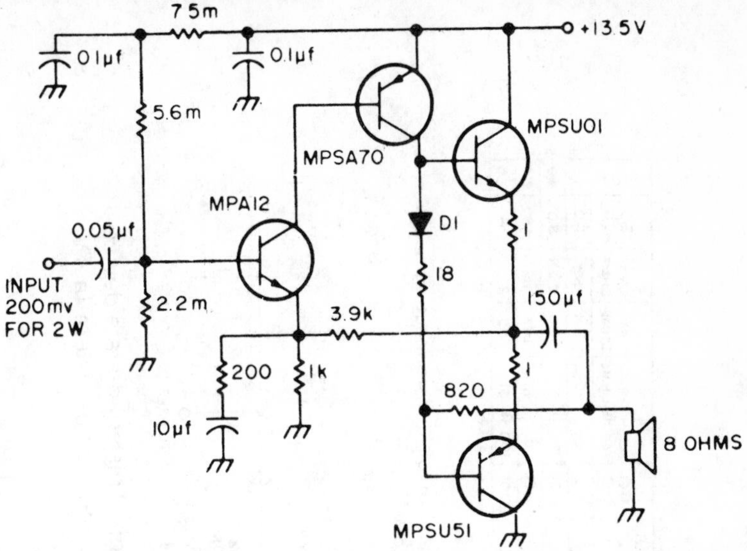


Audio amplifier. Motorola Application Note AN-426A.

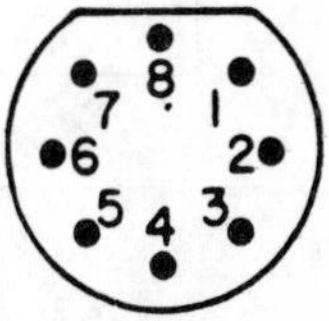


UL914
BOTTOM VIEW

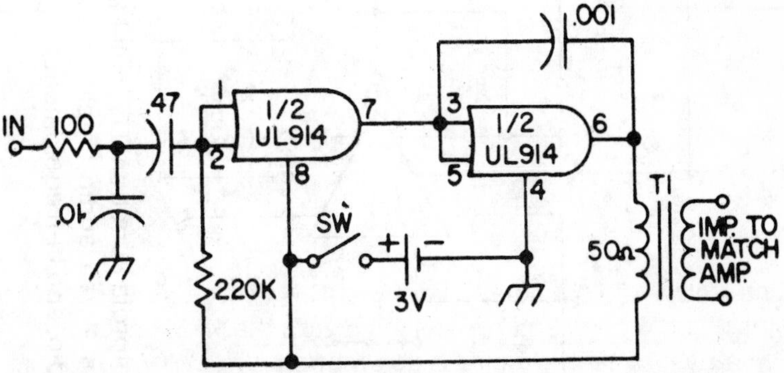


A mike or phono amp using the 914. Since the power requirements are minimal this is a good circuit to mount right in the microphone case along with two tiny batteries.

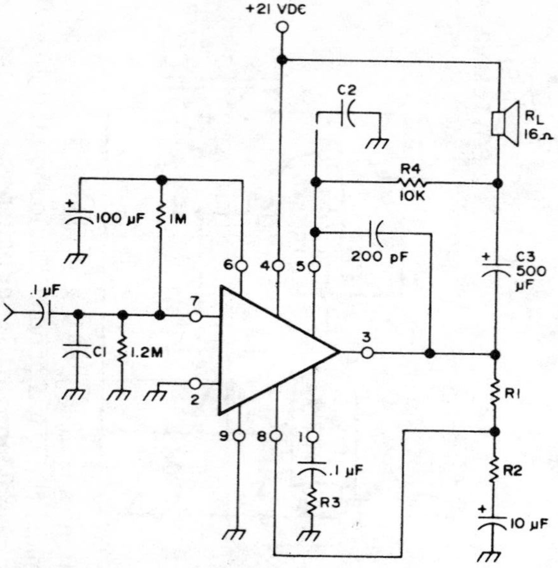


| SENSITIVITY FOR $P_o$ mV | C1 pF | C2 pF | R1 k OHMS | R2 OHMS | R3 OHMS | $R_L$ OHMS | $P_o$ W |
|---|---|---|---|---|---|---|---|
| 560 | 0 | 0 | 10 | 1.0K | 82 | 8.0 | 4.0 |
| 15 | 100 | 100 | 51 | 100 | 2.2K | 8.0 | 4.0 |
| 630 | 0 | 0 | 10 | 1.0K | 82 | 16 | 2.5 |
| 17 | 100 | 100 | 51 | 100 | 2.2K | 16 | 2.5 |

Alternate connection to permit connecting speaker to ground instead of to $V^+$:

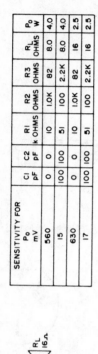

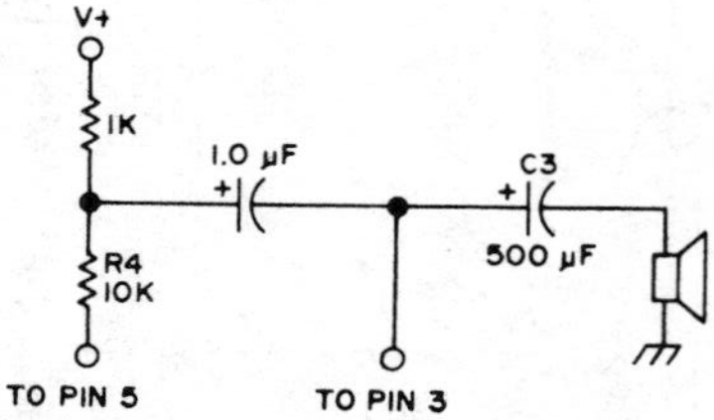


Motorola Functional Circuit MFC9000 in a typical application. This solid-state device is a 4-watt audio amplifier designed to provide the complete audio system in TV, radio, and hi-fi equipment.

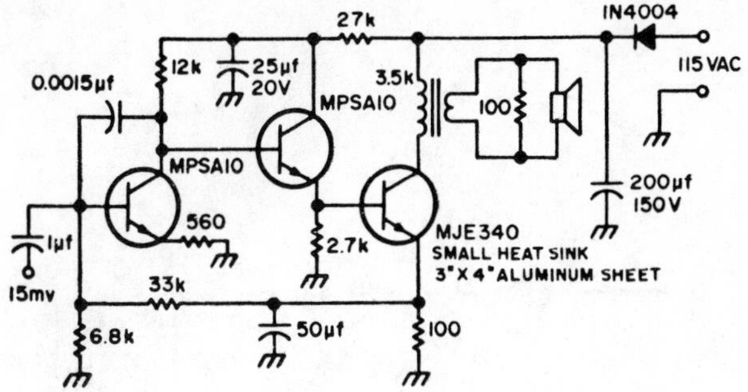


2½W line-operated amplifier. Motorola Application Note AN-426A.

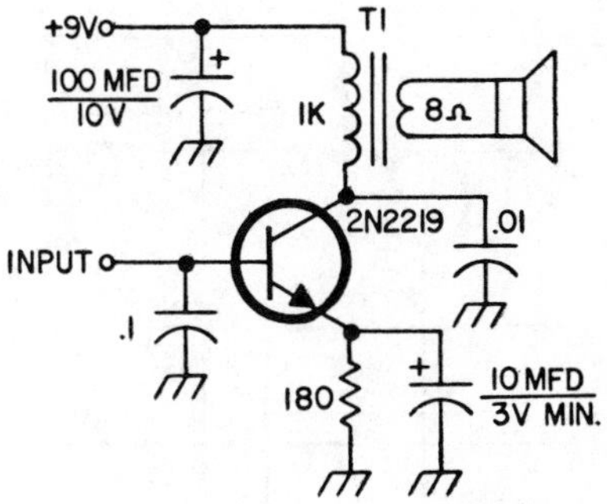


T1 = 1K:8Ω AUDIO TRANSFORMER

Simple audio amplifier provides comfortable gain when driven with modest varying signal input.

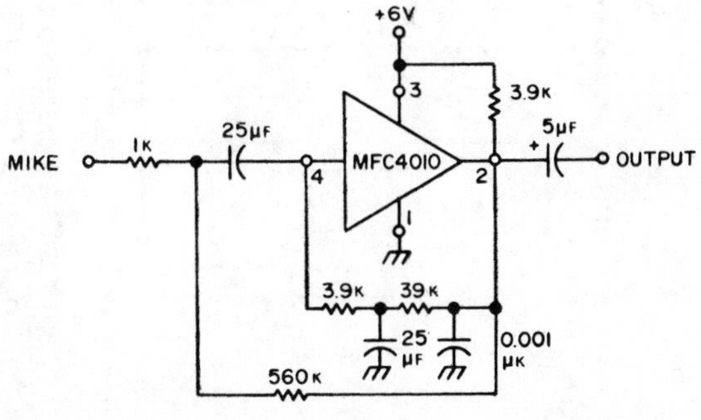


High-gain audio amplifier using Motorola MFC4010 wideband amplifier.

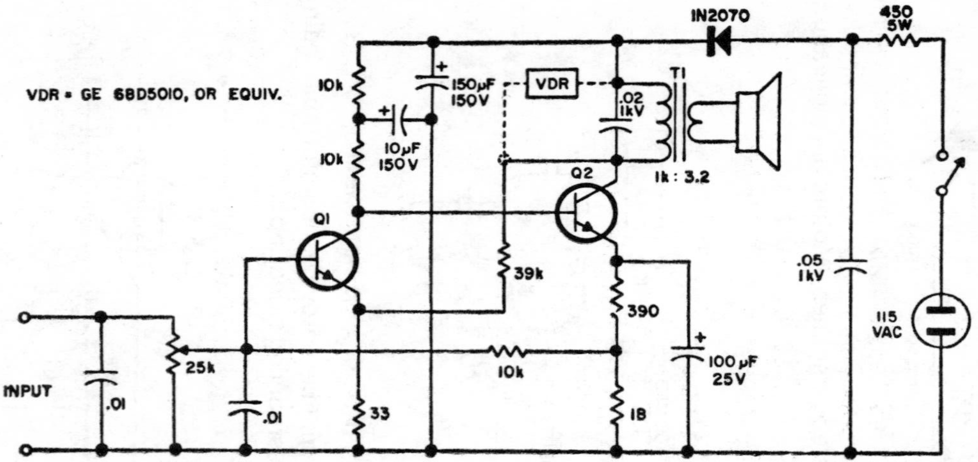


This line-operated audio amplifier provides about 500 mW output with an 80 mV input signal. Q1 is a 2N3565, SE4002, SK3020, or HEP 54; Q2 is a 2N3916 or SE7005.

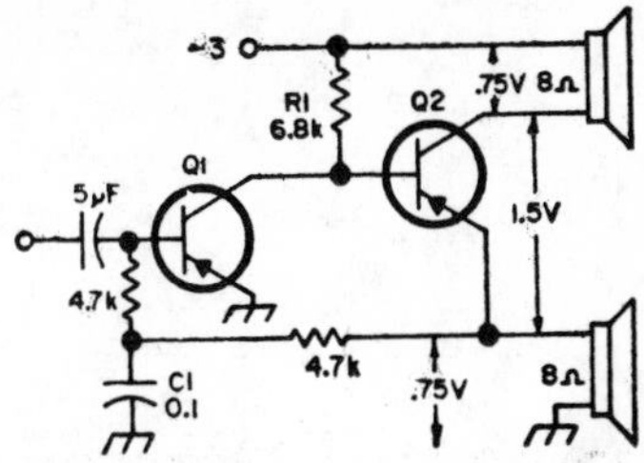


An audio power amplifier with push-pull output using a single transistor in the final stage may be obtained with this simple circuit. Only about 50 mW is available from this amplifier, but the gain is flat up to 30 kHz. Both Q1 and Q2 should be germanium audio transistors such as the 2N404, SK3004, or HEP 253.

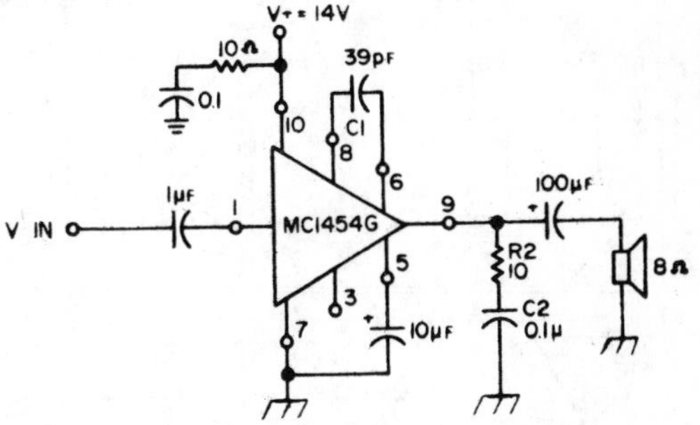


Audio power amplifier has 1W gain of 10; input impedance of 10K. An external heatsink is required. The HEP 593 can also be used in this circuit with similar results.

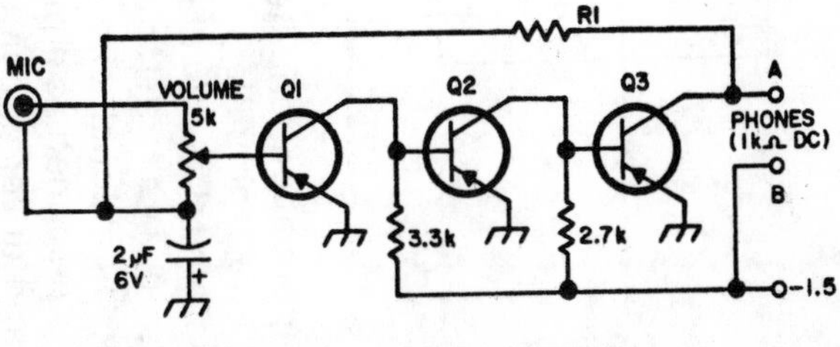


Simple direct-coupled amplifier. Transistors Q1, Q2, and Q3 should be the 2N207, 2N584, 2N1098, SK3003, GE-2, or HEP 254.

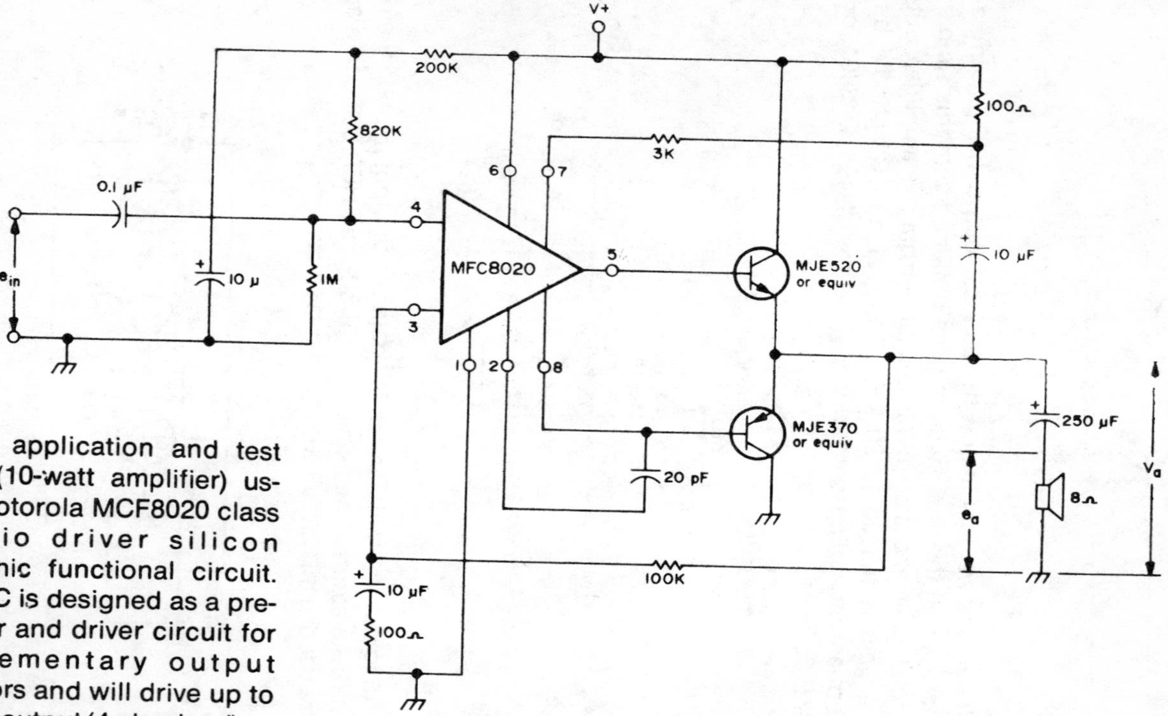


Typical application and test circuit (10-watt amplifier) using a Motorola MCF8020 class B audio driver silicon monolithic functional circuit. The MFC is designed as a pre-amplifier and driver circuit for complementary output transistors and will drive up to 15 watts output (4 ohm load).

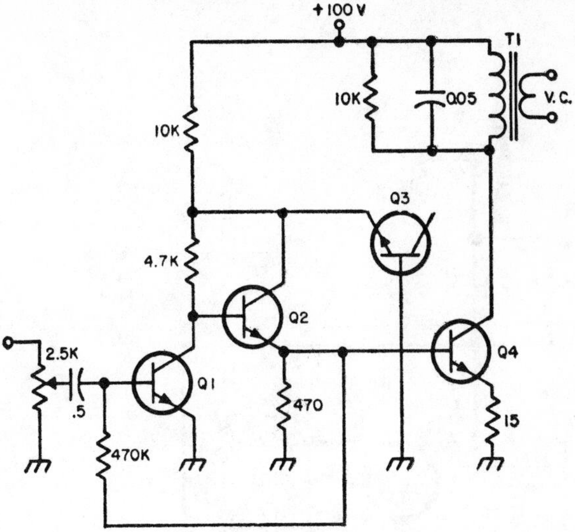


Q1 - 2N2925, 2N3391, 2N3392
Q2 - 2N2924, 2N2712, 2N3393
Q3 - 2N2923, 2N2926
Q4 - 2N4054
T1 - 2500 ohms TO VOICE COIL

High performance with low cost is obtained with this line-operated audio power amplifier because expensive electrolytic capacitors are eliminated by direct coupling between stages. This circuit delivers 1 watt to the speaker with 3 mW input.

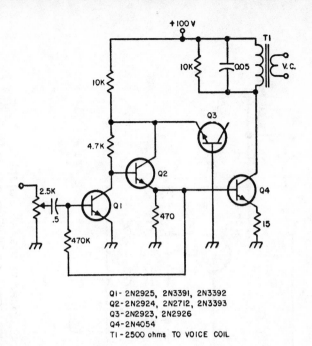

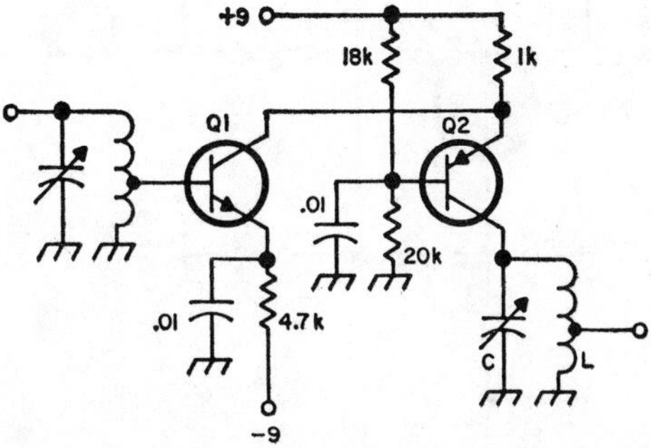


This cascode amplifier is extremely useful because it provides high gain without the need for neutralization. Q1 is a 2N918, 2N3464, 2N3478, MPS918, 40235 or HEP 56; Q2 is a 2N1742, 2N2398, 2N2894, 2N3399, TIM10 or HEP 2.

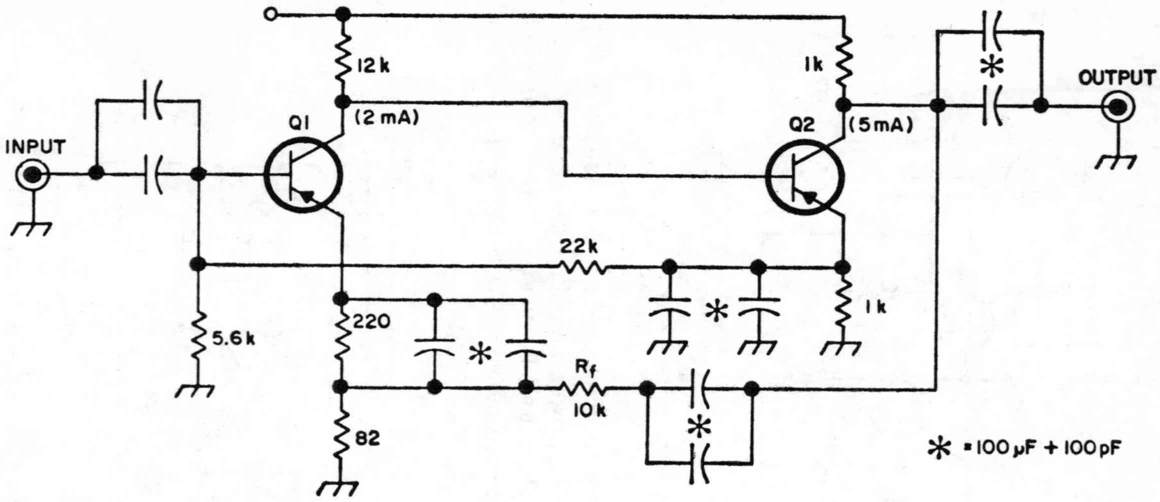


The gain of this wideband amplifier may be controlled by the value of the feedback resistor $R_f$. The 10K resistor shown here provides more than 30 dB gain from 10 Hz to 17 MHz. Q1 and Q2 are 2N2188, SK3006, GE-9, or HEP 2.

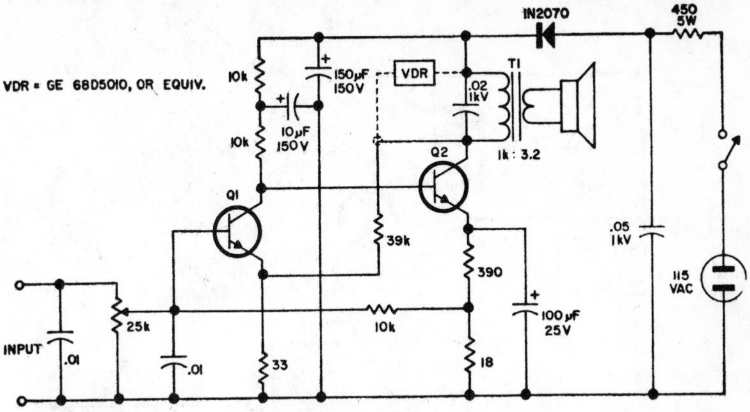


This line-operated audio power amplifier provides about 500 mW output with an 80 mV input signal. Q1 is a 2N3565, SE4002, SK3020, or HEP 54; Q2 is a 2N3916 or SE7005.

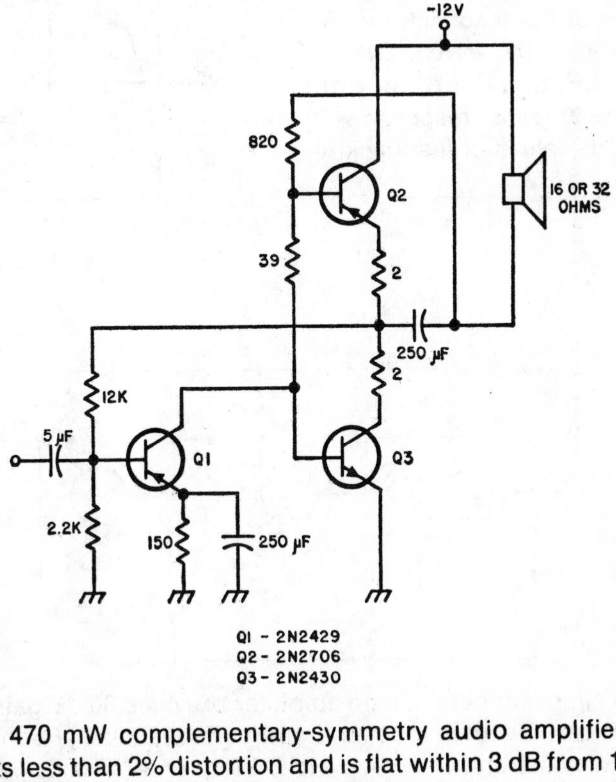


This 470 mW complementary-symmetry audio amplifier exhibits less than 2% distortion and is flat within 3 dB from 15 Hz to 130 kHz.

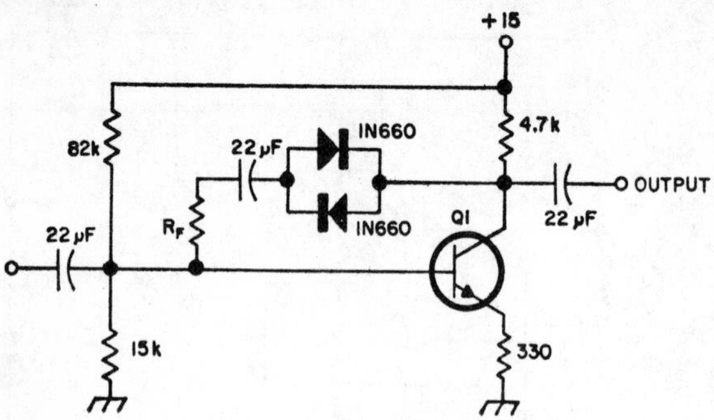


The gain of this amplifier is controlled by the nonlinear feedback provided by two back-to-back diodes and the value of the feedback resistor $R_f$ is a 2N706, 2N708, 2N3394, or HEP 50. Chart at right shows amplifier response with various values of feedback resistance ($R_F$).

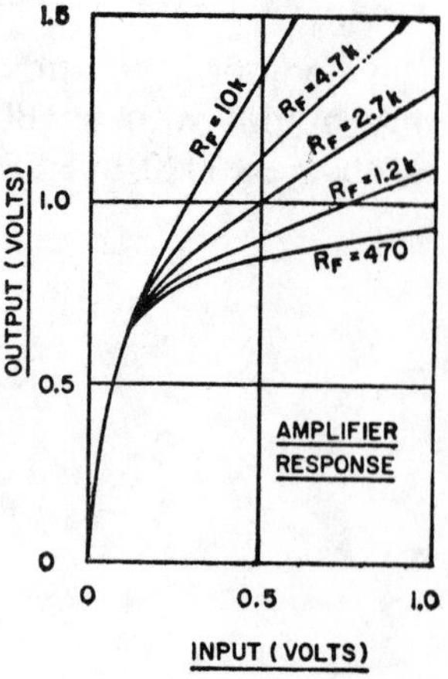


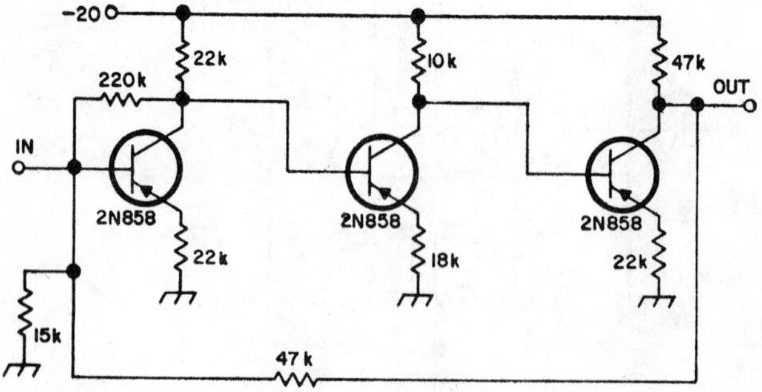


This simple direct-coupled amplifier provides 30 dB gain and identical 1500-ohm input and output impedances. For higher gain, similar units may be cascaded up until 10 volts peak to peak is obtained at the output.

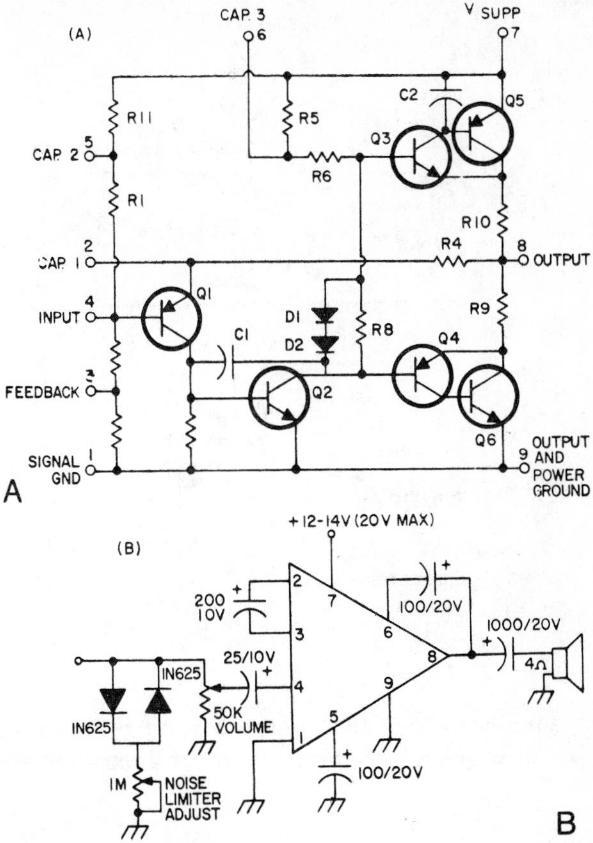


Miniature audio amplifier for mobile PA applications and where receiver must be heard at some distance from parked car. Unit delivers 6W into speaker. At top, circuit shows internal schematic of the hybrid chip used in this device; lower schematic shows connection of chip in actual service.

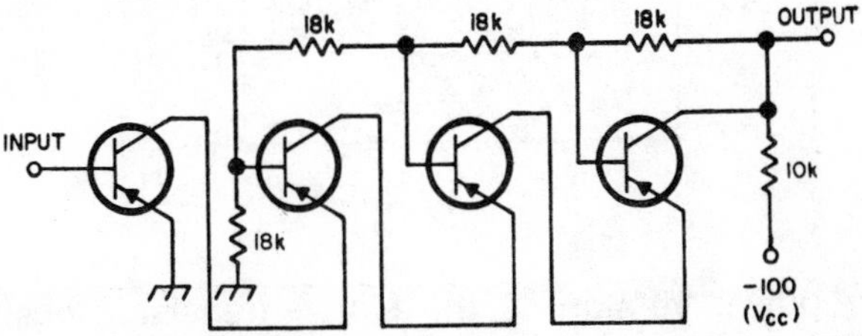


High-voltage direct-coupled amplifier. The gain of this amplifier is equivalent to both sections of a 12AU7. All transistors are 2N384, SK3008, GE-9, or HEP 51.

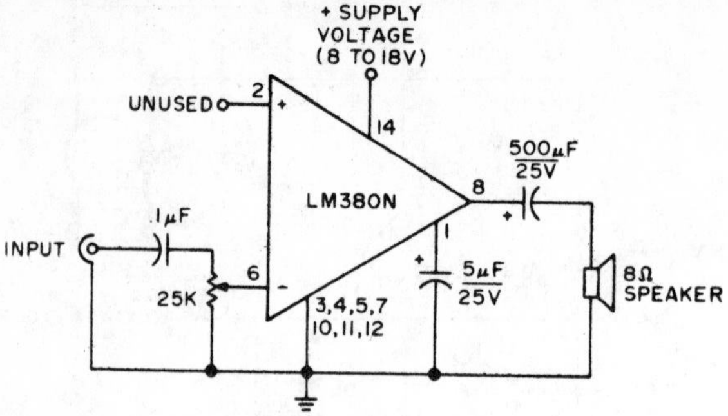


A

NTE-740A

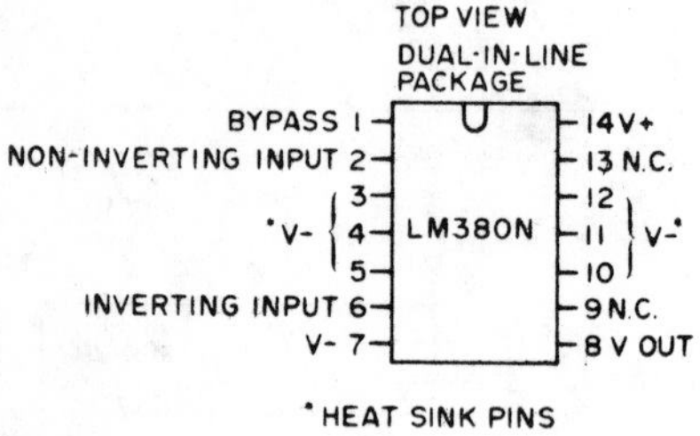


B

Audio amplifier provides 2 watts to speaker when supplied with 18V dc input. IC is National Semiconductor LM380N.

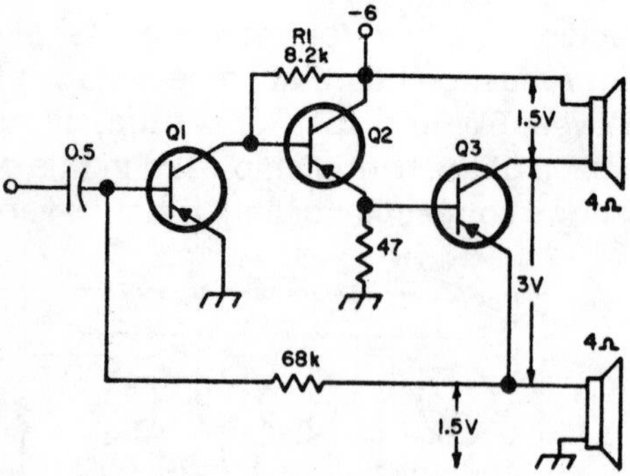


A larger power version of the single transistor push-pull circuit. The operating characteristics are similar to the 50 mW circuit except that approximately 1 watt may be obtained. Transistors Q1 and Q2 are 2N215, 2N404, 2N2953, SK3004, or HEP 253; Q3 is a 2N554, 2N1032, 2N1666, SK3009, or HEP 232.

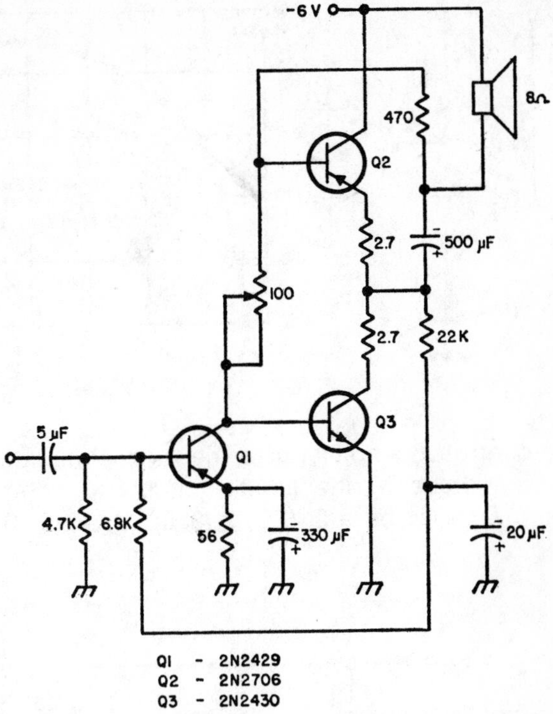


This complementary-symmetry amplifier provides up to 220 mW output with a frequency response from 90 Hz to 12.5 kHz. Although matched transistors are not required for Q2 and Q3, they are available as the 2N2707.

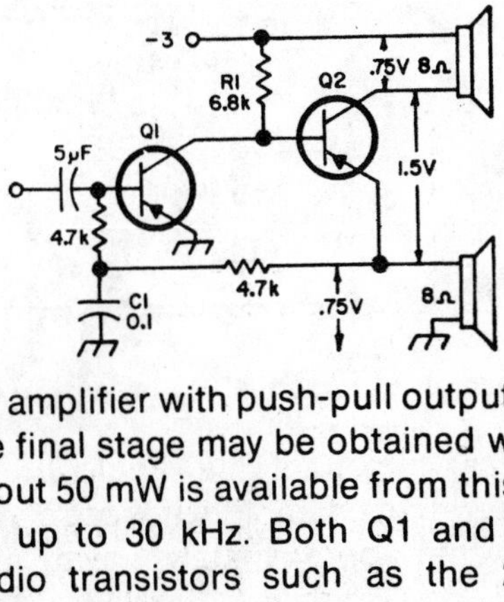


An audio power amplifier with push-pull output using a single transistor in the final stage may be obtained with this simple circuit. Only about 50 mW is available from this amplifier, but the gain is flat up to 30 kHz. Both Q1 and Q2 should be germanium audio transistors such as the 2N215, 2N404, 2N2953, SK3004, or HEP 253.

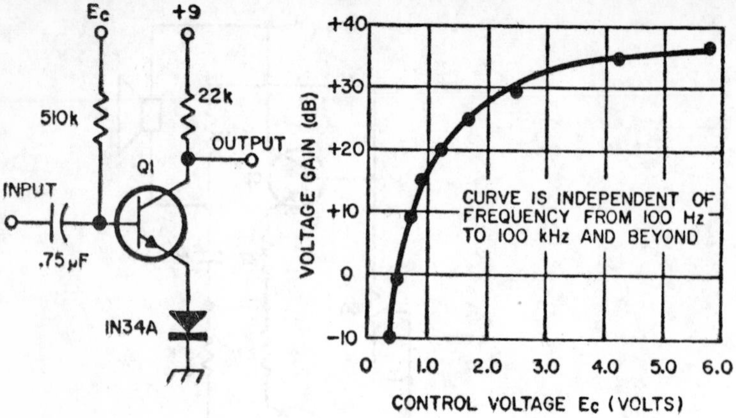


Voltage-controlled amplifier uses the varying impedance of a germanium diode in the emitter circuit to control gain. Transistor Q1 may be a 2N696, 2N3564, SK3019, GE-10, or HEP 54.

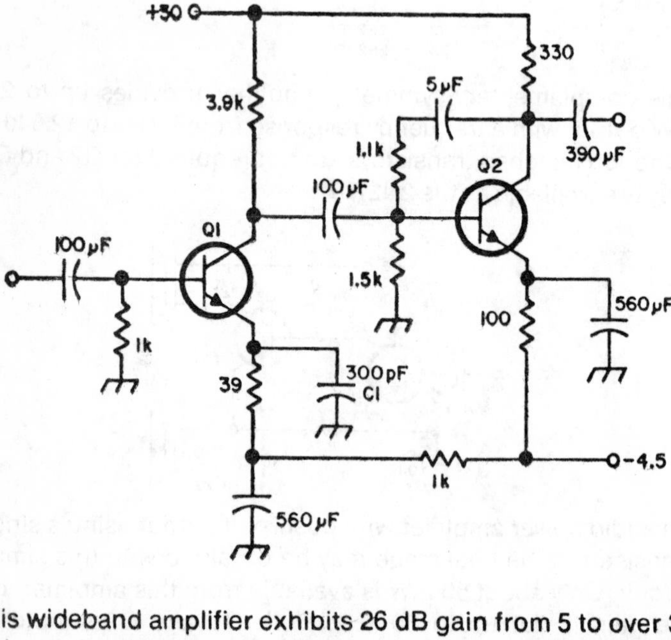


This wideband amplifier exhibits 26 dB gain from 5 to over 30 MHz and will deliver a 7V signal into a 100-ohm load. Transistors Q1 and Q2 are 2N2218.

# Section D

# Automotive Circuits

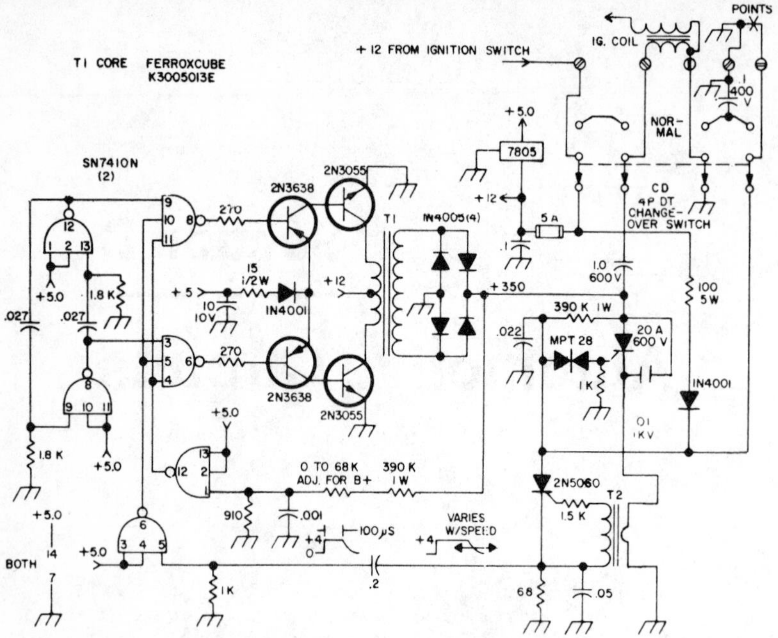


Capacitive-discharge ignition system. Transformers T1 and T2 must be fabricated, but construction is easy. T1 secondary has 15.24 meters of 26-gage Formvar insulated wire put on in 6 bank-winding sections (insulate with plastic tape); add 20 turns of 14-gage insulated wire evenly around core and centertap. T2 uses CTC coil form 5 mm dia by 10 mm long; it has 30-turn secondary of 36-gage hookup wire. T2 can also be made from an unshielded iron-core rf choke of 30 – 100 μH: just wind a couple of turns over it.

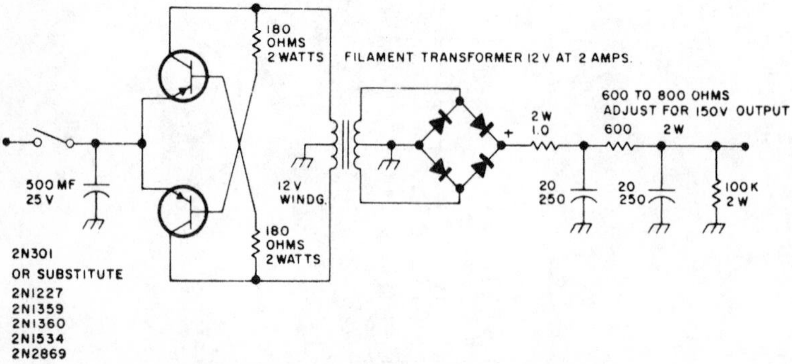


Mobile power supply for CW receiver.

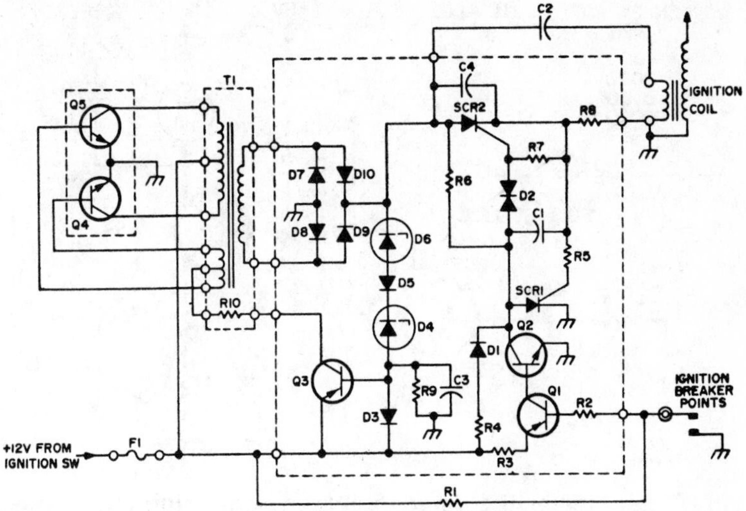


**Parts List**

C1 – .1,100V
C2 – 1.0 μF, 600V oil paper
C3 – .01
C4 – .01, 1 kV
R1 – 68, 5.0W
R2 – 4.7K
R3 – 2.2K
R4 – 220
R5 – 680
R6 – 150K, 2.W
R7 – 680
R8 – .5 (3 ft #32 E. wire)
R9 – 4.7K
R10 – 180/270 2.W See text
D1 – 1N914
D2 – MPT28
D3 – IN914
D4 – 200.V, 10.W zener
D5 – .75A, 400.V
D6 – 150.V, 10.W zener
D7–D10 – .75A, 1 kV
Q1 – 2N1132
Q2 – 2N697
Q3 – 2N1132
Q4, Q5 – 2N3055
SCR1 – Small 100.V sen. gate like C106F2
SCR2 – 500.V, 20.A
F1 – 4.A
Q4, Q5 heatsink – Birtcher #4AL-6-0-0
Q3 heatsink – Fuse clip chassis 4x6x2 w/bottom cover
T1 – Core = Ferrox cube #K3-005-01-3E Wire = 55' #26 Formvar, 4' Ω14 Formvar Ins. tape Epoxy

Schematic diagram of improved low-cost CD ignition.

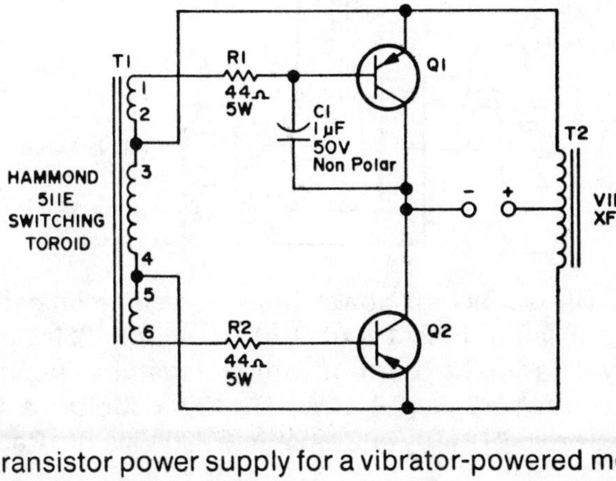


A transistor power supply for a vibrator-powered mobile rig.

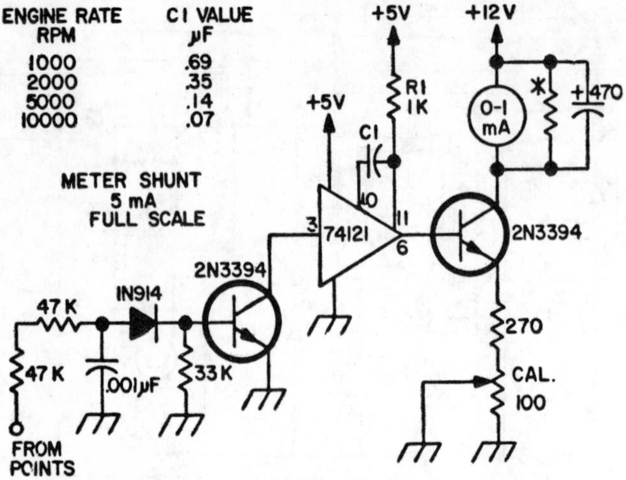


An IC tachometer that is adaptable, by changing C1, to different rpm rates. In the event there is no other tachometer handy with which to calibrate the unit, temporarily disconnect the three components at the cathode of the 1N914 and connect it to a signal generator through a 5.6K resistor.

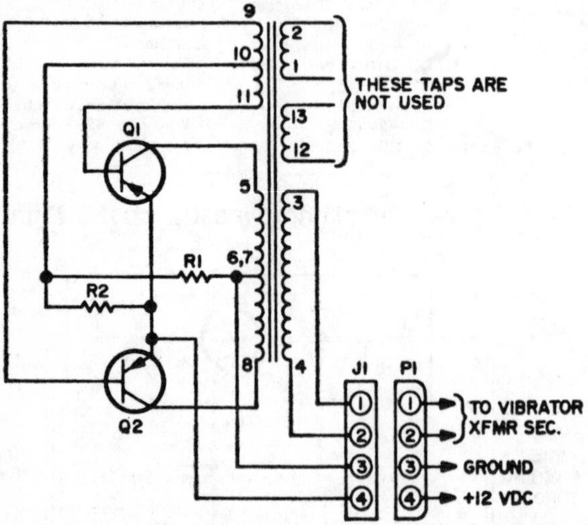


Transistor B+ supply for power tube-type communications equipment in auto. Pins 1 and 2 of P1 supply 250V high-frequency alternating current which must be rectified. R1 = 200Ω 10W; R2 = 10Ω 10W; Q1, Q2 = Motorola HEP 233; T1 = Toroid VT2 (Tower Commun. 1220 Villa St., Racine, WI 53403).

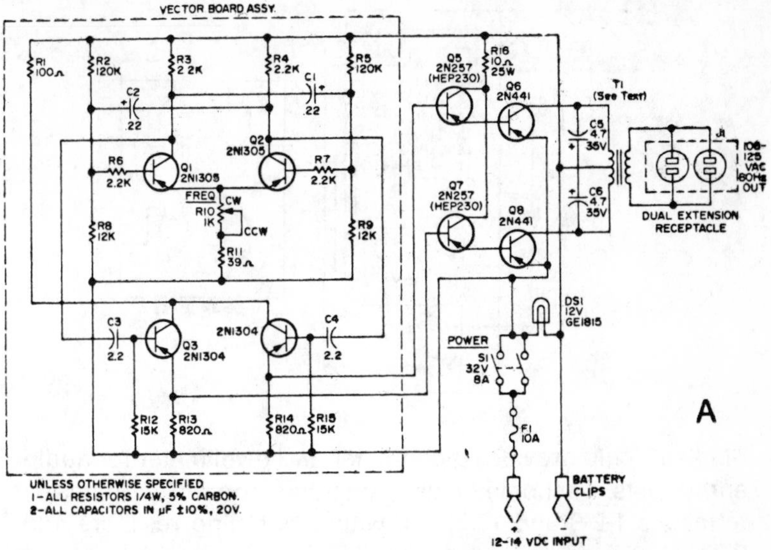


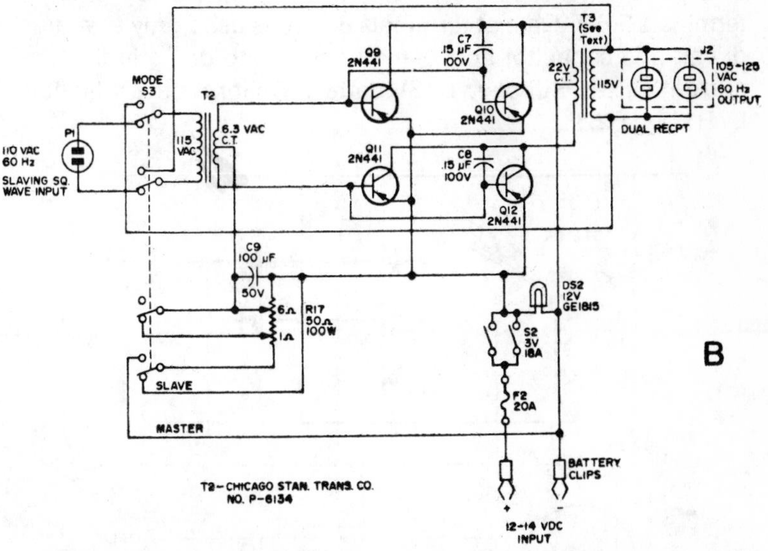


Two 115V, 60 Hz inverters for operation from 12V dc. Unit shown in lower schematic drives heavy loads and is slaved to upper unit for total output of several hundred watts. T1 is TV transformer with 5V windings as well as a centertapped 12.6V winding. One end of each 5V winding is connected to opposite ends of 12.6V winding. T3 is modified TV transformer: Remove end bells, unwrap filament windings, and splice additional turns for 23V.

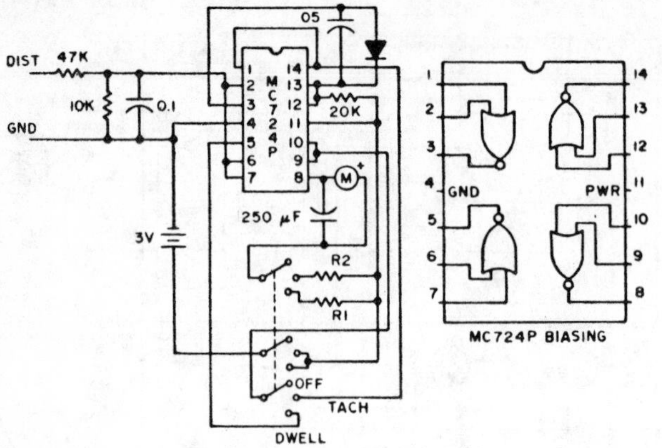


Single IC chip provides tachometer and dwellmeter for audio enthusiasts. Tachometer uses one-shot formed by gates at terminals 1-2-3 and 12-13-14, with 20K timing resistors and 0.05 μF timing capacitor. Dwellmeter is driven by gate at terminals 5-6-7. Gate at terminals 8-9-10 is used only as meter driver, and capacitor across meter serves to damp individual current pulses. Supplied by 3V battery. Calibration is adjusted by R1 and R2.

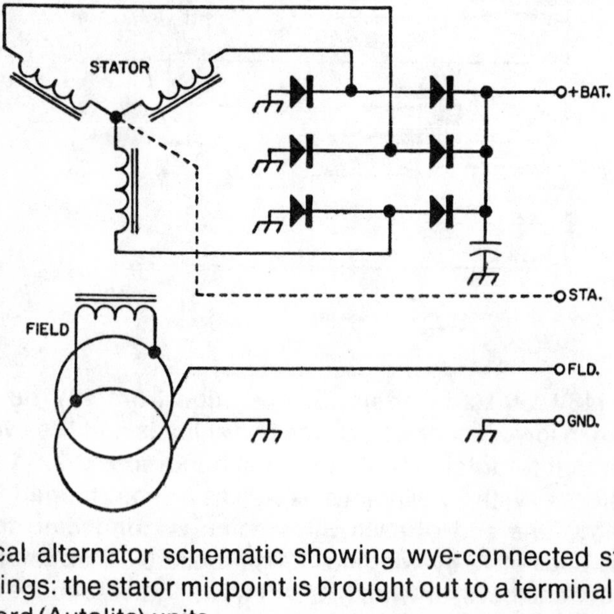


Typical alternator schematic showing wye-connected stator windings: the stator midpoint is brought out to a terminal only on Ford (Autolite) units.

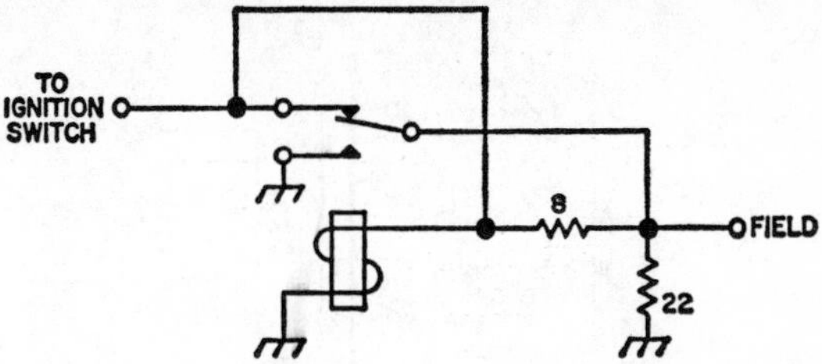


Typical automotive single-relay voltage regulator.

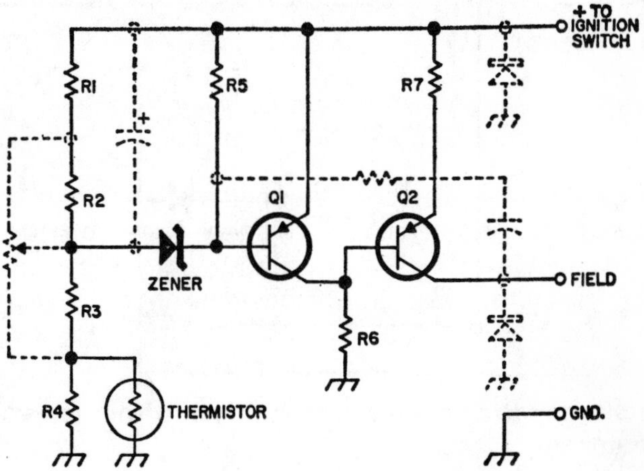


Schematic of a typical solid-state regulator. Components shown connected by dashed lines are used by some manufacturers, but not all of them.

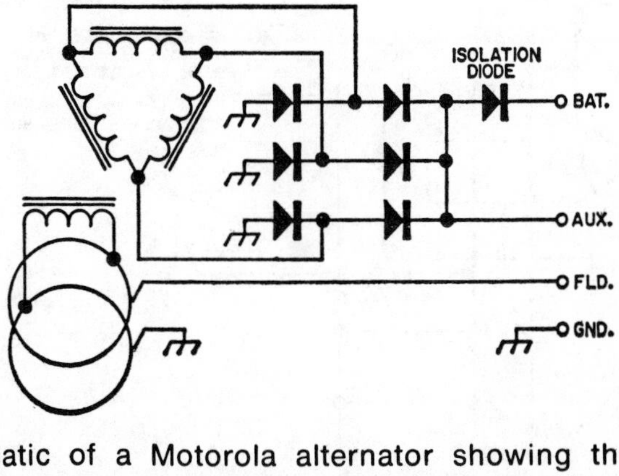


Schematic of a Motorola alternator showing the delta-connected stator used in the 55-ampere model. Also shown is the isolation diode and auxiliary terminal connection.

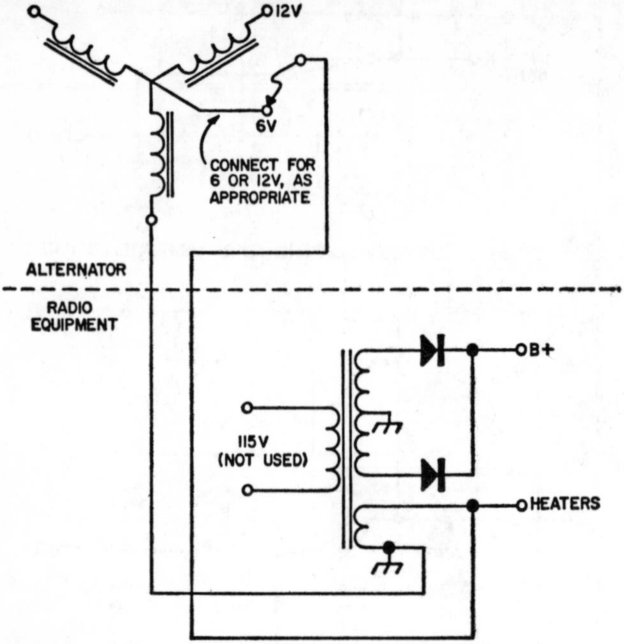


A simple way to operate radio equipment from low-voltage alternator ac.

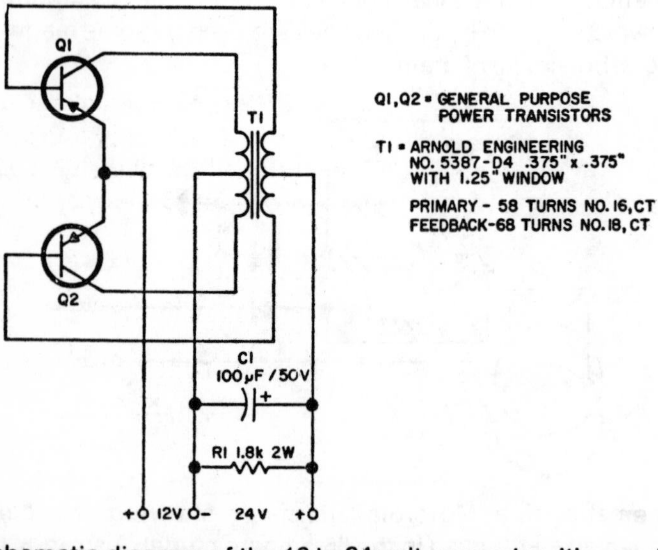


Schematic diagram of the 12 to 24 volt converter. It's good for about 1 ampere output at 24 volts.

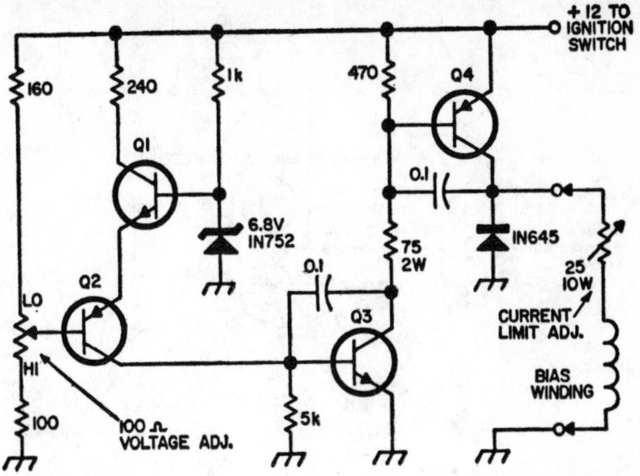


| Transistor | Type | Silicon | Germanium |
|---|---|---|---|
| $Q_1$ | NPN | 2N1711 | None recommended* |
| $Q_2$ | PNP | 2N1132 | None recommended* |
| $Q_3$ | NPN | 2N657<br>2N3738<br>2N3739 | None recommended* |
| $Q_4$ | PNP | 2N3790<br>2N3789 | 2N174, 2N1537,<br>2N3611, 2N3612 |

*No germanium transistors are suggested for $Q_1$, $Q_2$ and $Q_3$ because excellent silicon units should cost less than $6.00. If you use silicon all the way through, the cost should be less than $13.00 total.

Circuit diagram of the solid-state automotive voltage regulator. This circuit will provide excellent regulation to over 100 amperes. Although it was designed for use with Leece-Neville alternators, it will work with just about any alternator made.

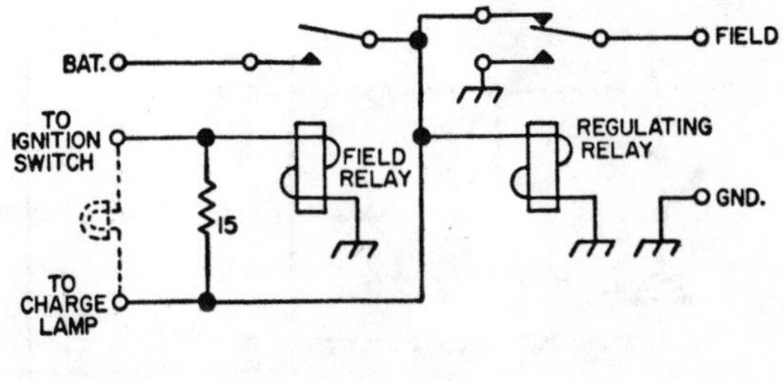


Schematic of the two-relay regulator with a terminal for operating a charge-indicator lamp.

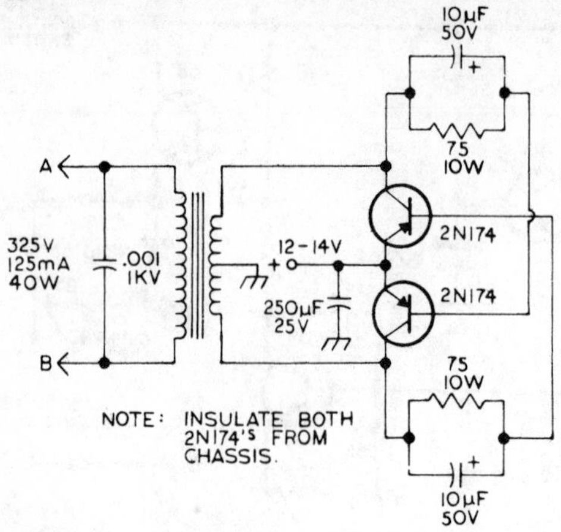


Mobile power supply schematic with an RC network establishing feedback. This unit is capable of 40 watts output, 525V at 125 mA.

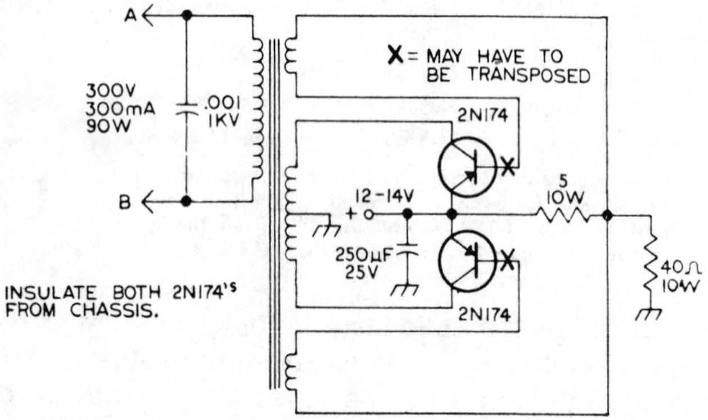


Schematic of second mobile power supply, delivering 90 watts output, 300V at 300 mA.

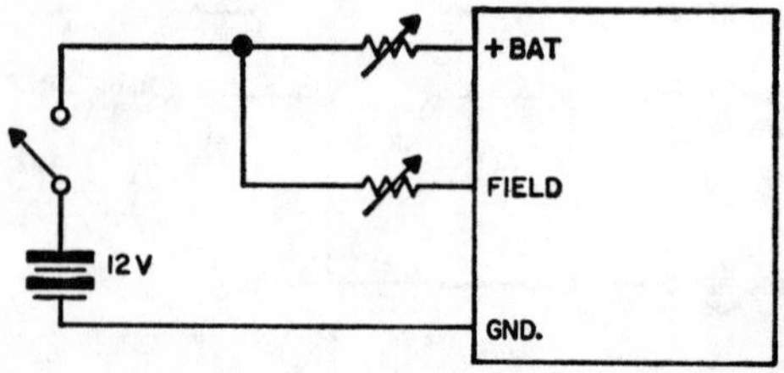


Arrangement for adjusting field current to a constant value when alternator is used to supply ac.

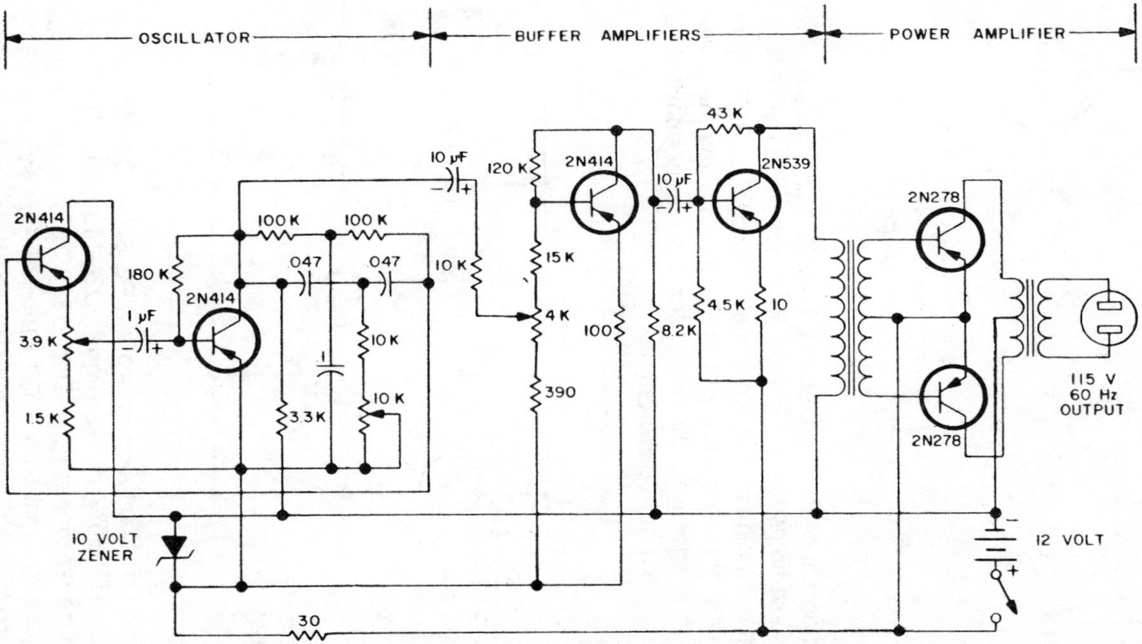


100-watt 115V inverter produces fair sine wave at 60 Hz when driven with 12V source. Excellent for operating home gear in car or van.

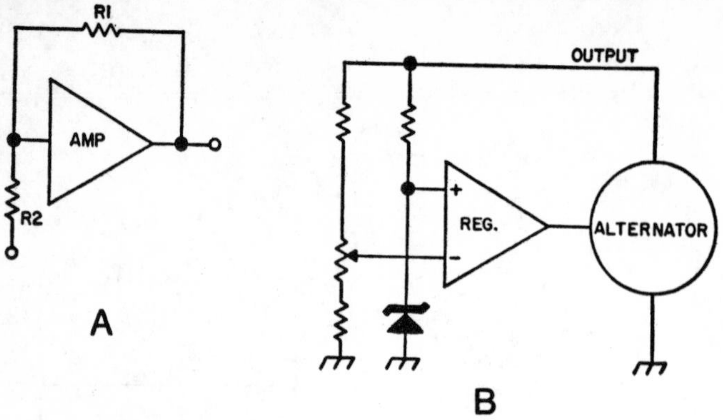


Block diagram of the solid-state alternator. The system used closely resembles the simple dc feedback amplifier in A. In this type of amplifier, the gain closely approaches $R_1/R_2$. If the gain is very high, the regulator in B will not allow the preset voltage to change until the alternator is no longer capable of delivering the required current.

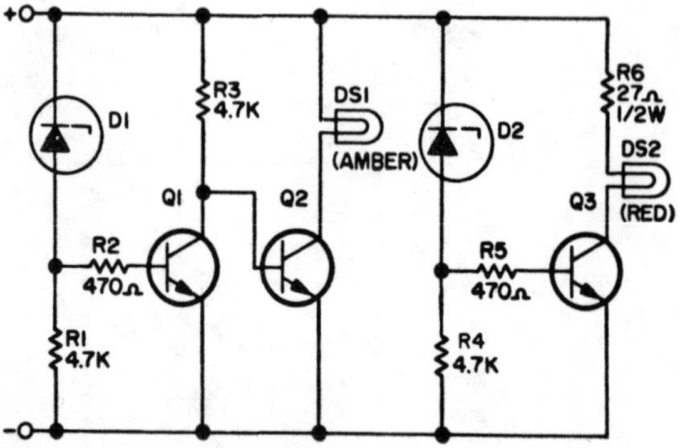


A sensor that continuously monitors your mobile's electrical system and indicates voltage variance outside a prescribed range.

Parts List: Q1, Q2, Q3—Motorola MPS 3704; D1—Motorola 1N5243B, 13V±5% zener diode; D2—Motorola 1N5245B, 15V±5% zener diode; PL1—Dialco MS 25256 pilot lamp assembly (amber lens) with V330 Bulb (T1 ¾, 14V at 80 mA); PL—Dialco Ms 25256 pilot lamp assembly (red lens) with V330 bulb (T1 ¾, 14 at 80 mA); R1, R3, R4—4.7K ¼W; R2, R5—470Ω, ¼W; R6—27Ω ½W; misc. hardware—CU 2101A minibox, terminal strips, rubber grommet, etc.

A

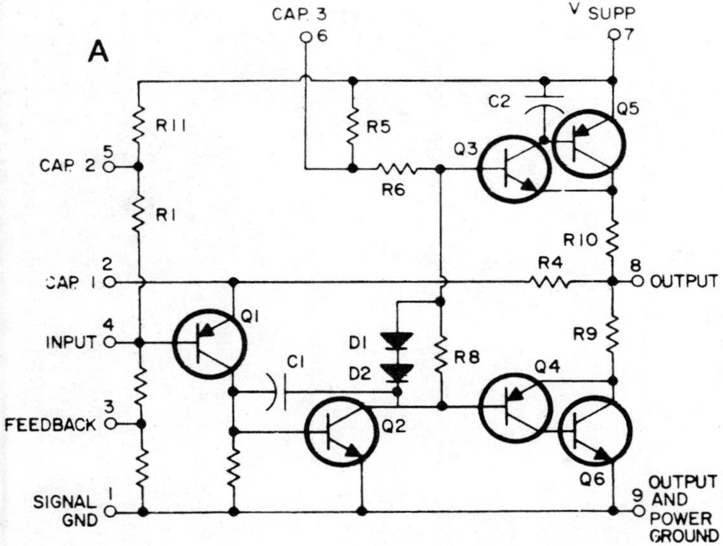


B

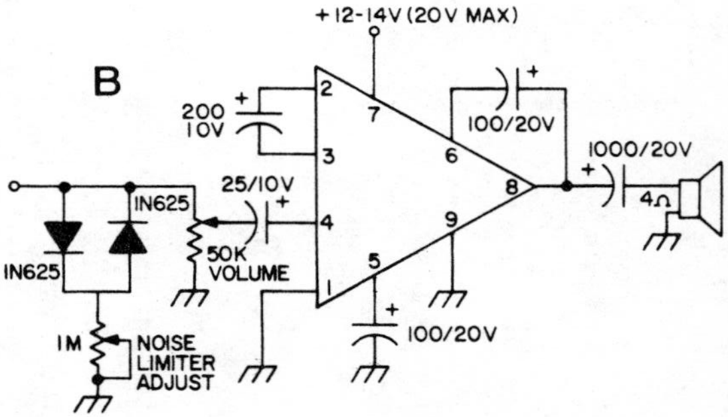


Miniature audio amplifier for mobile PA applications and where receiver audio must be heard at some distance from your parked car. Unit delivers 6W into speaker. At left circuit shows internal schematic of the hybrid chip used in this device; schematic at right shows connection of chip in actual service.

# Section E

# Filters

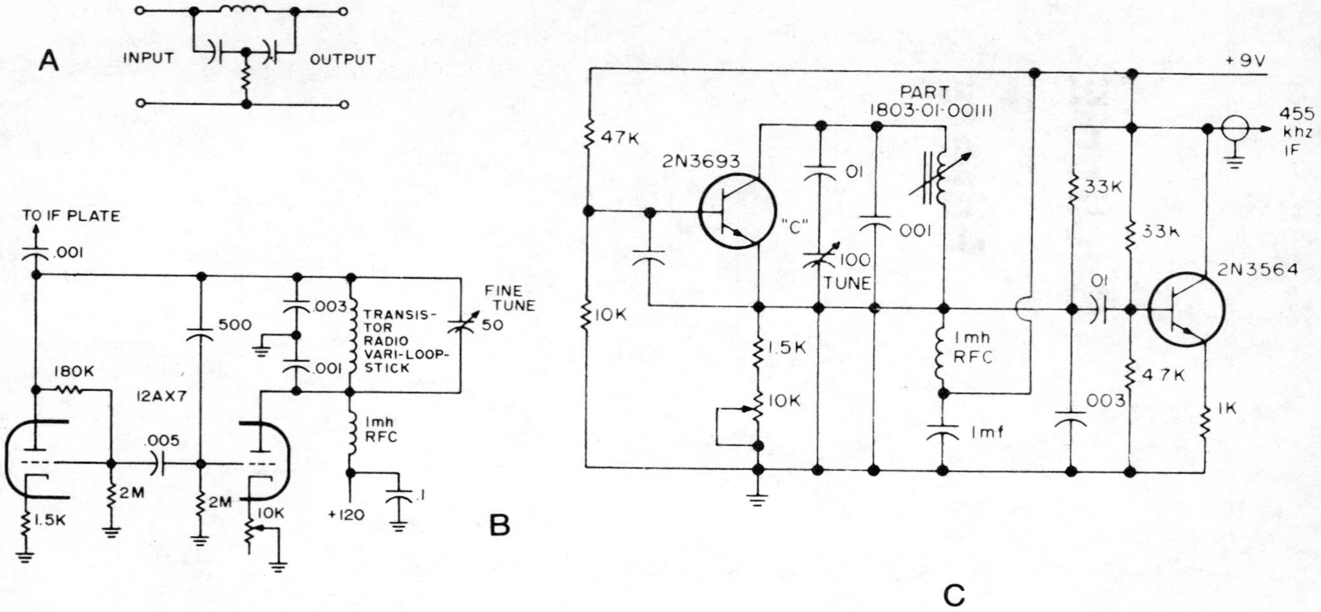


Bridged-T network (A) is basis of most notch filters. Q multiplier in notch function uses network in feedback circuit. (C) varies notch frequency approximately ±6 kHz from 455 kHz center frequency.

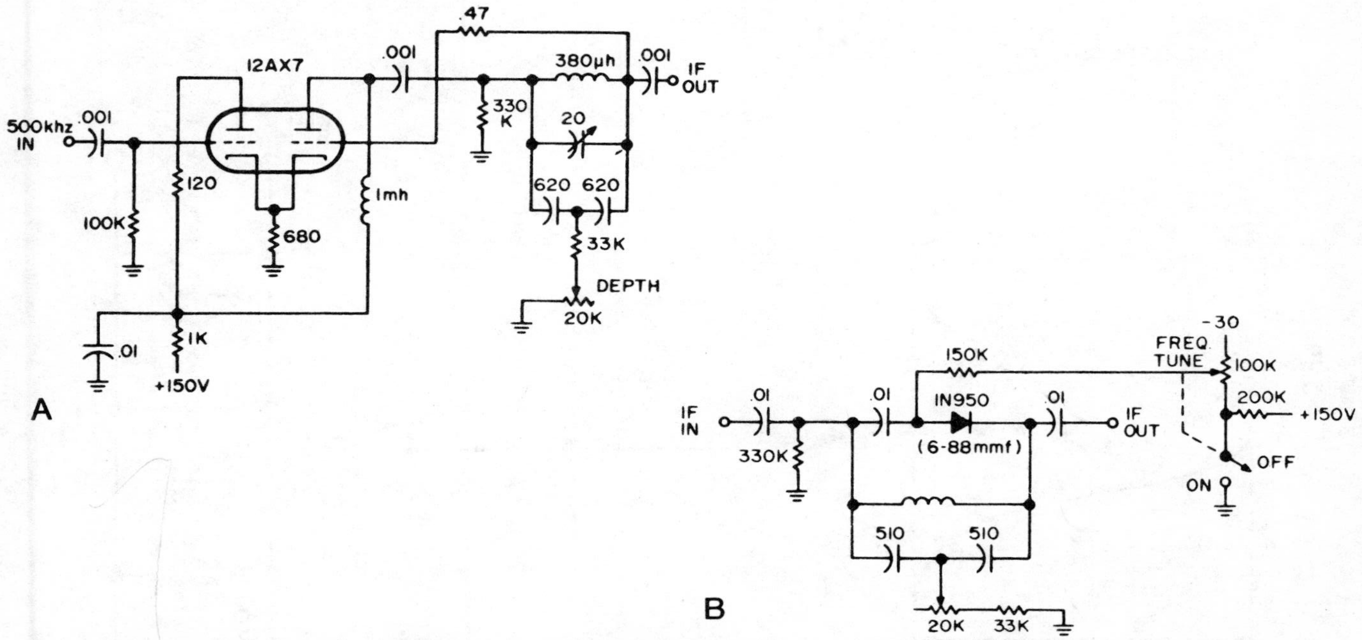


Notch filter using a feedback amplifier to increase Q of series bridged-T network. Instead of air variable capacitor, varactor diode can be used (B) controlled by potentiometer.

$$y = \frac{3.15}{0.93} = 3.39$$

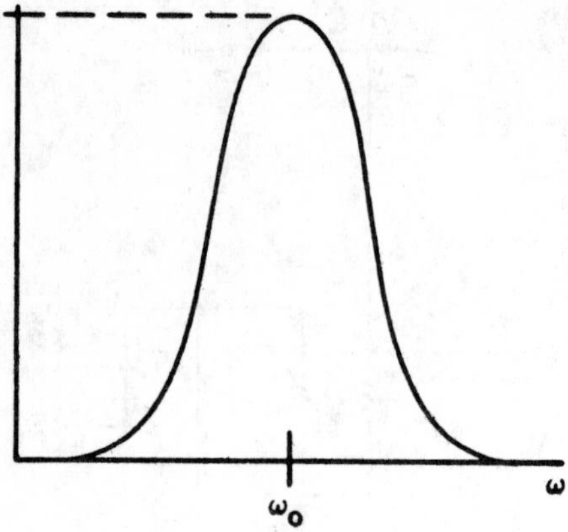


A

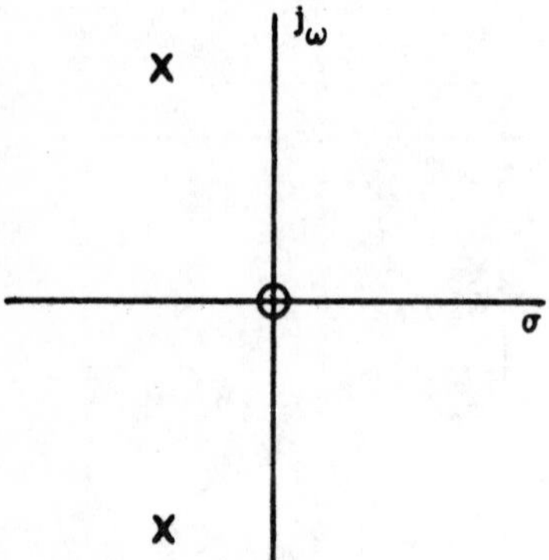


B

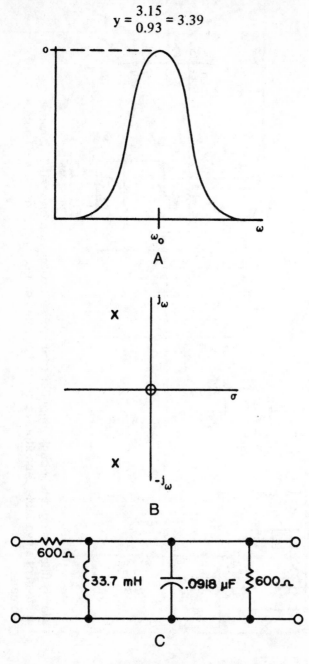

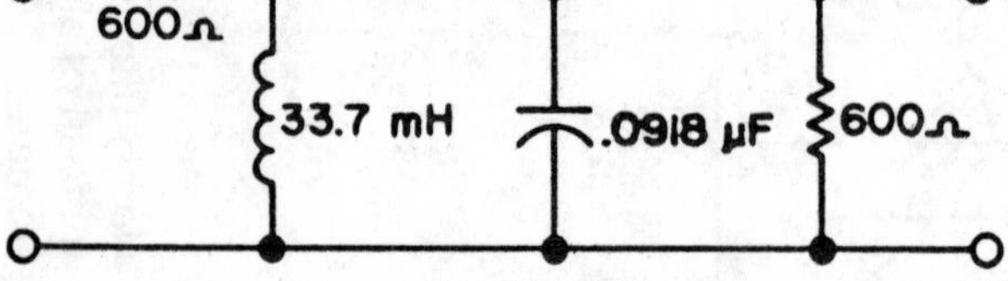


C

Bandpass filter.

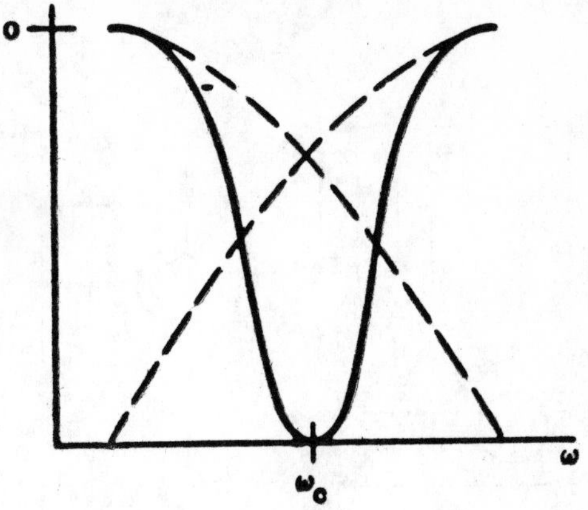


A

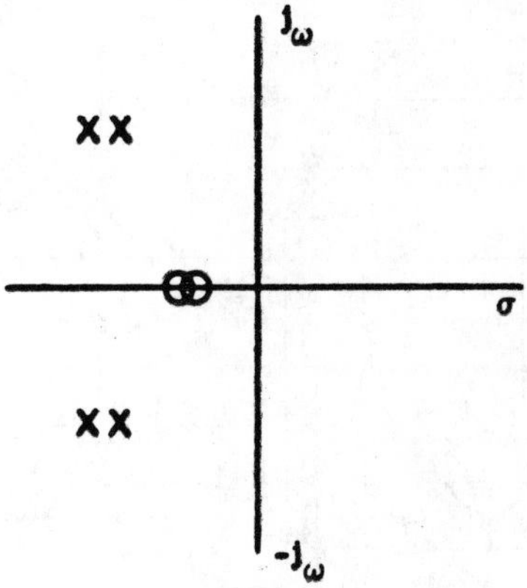


B

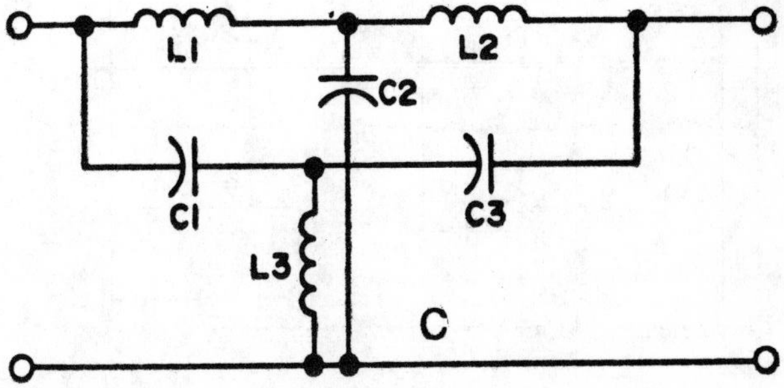


Bandstop filter.

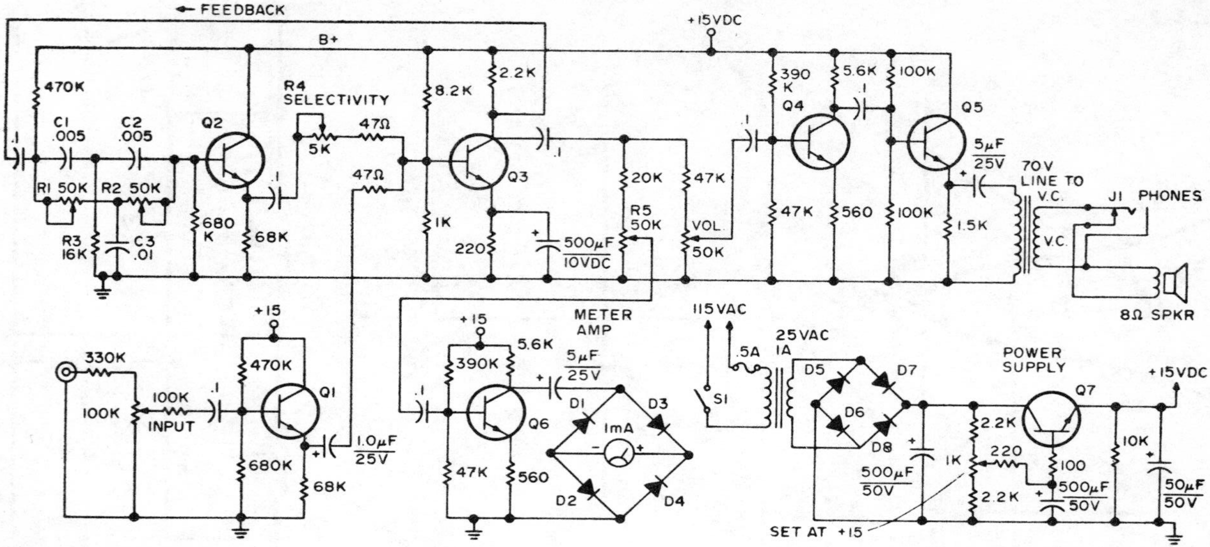


Schematic of the variable Q filter. Although it is designed around a center frequency of 1000 Hz, information is given in the text to modify the frequency to suit any need. Q1 through Q6 are GE-20 transistors, and Q7 may be a GE-28.

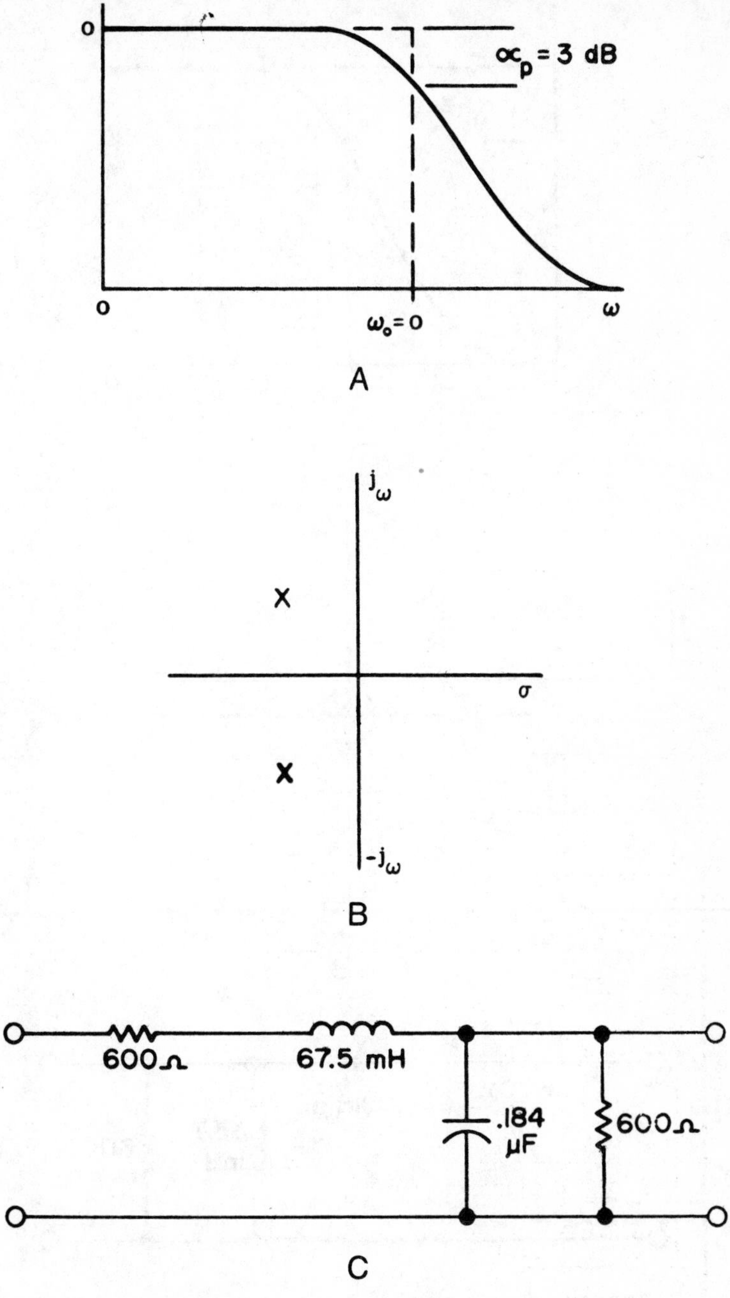


Low-pass Butterworth (ideal response) filter.

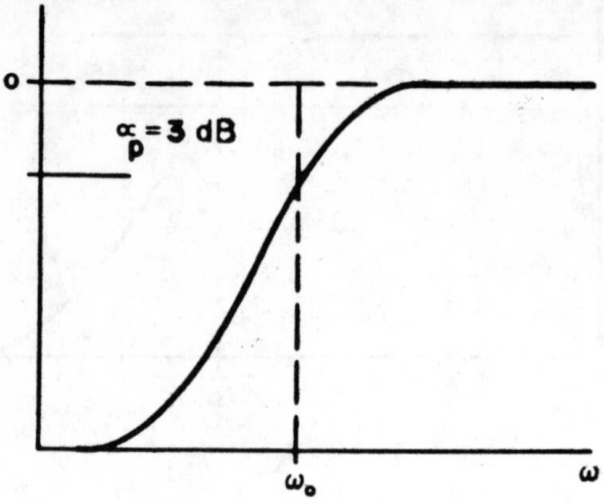


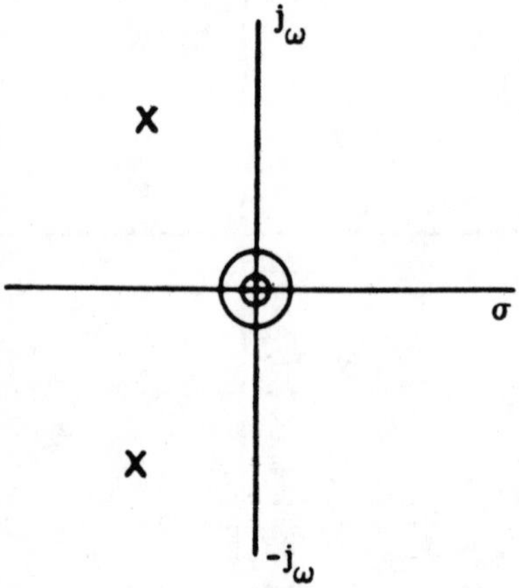


B

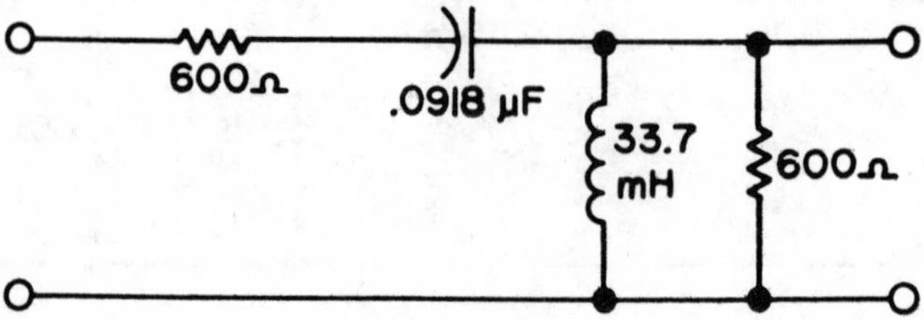


C

High-pass filter.

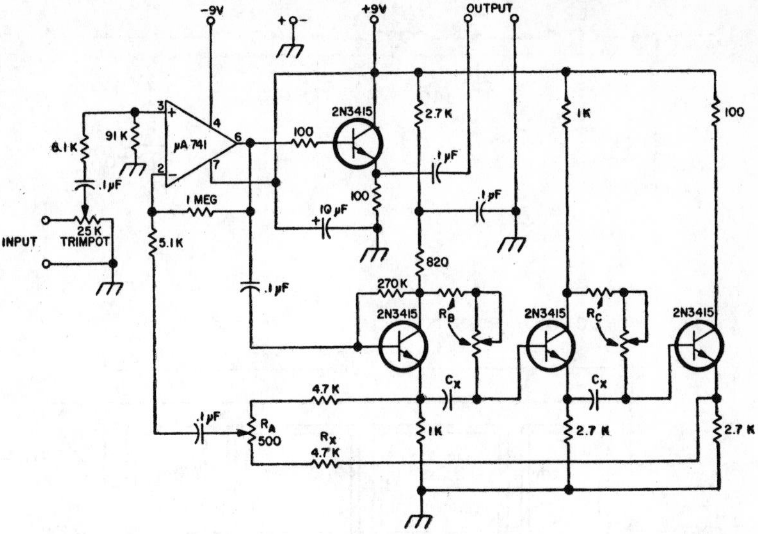


A

Phase-shift filter. Chart shows component values for various frequencies.

| f, kHz | Rb | Rc | Cx | Gain | 3 dB Bandwidth |
|---|---|---|---|---|---|
| 2.125 | 200K | 200K | 370 pF | 45 dB | 350 Hz |
| 2.975 | 140K | 140K | 370 pF | 44 dB | 380 Hz |
| 10. | 50K | 50K | 330 pF | 44 dB | 1.4 kHz |
| 50. | 10K | 10K | 300 pF | 44 dB | 2.4 kHz |
| 60. | 11K | 11K | 300 pF | 48 dB | 2.0 kHz |
| 85. | 15K | 15K | 100 pF | 44 dB | 2.0 kHz |
| 100 | 15K | 15K | 90 pF | 44 dB | 2.5 kHz |
| 1 MHz | 5K | 5K | 30 pF | 36 dB | 4.5 kHz |
| 5 MHz | 1K | 1K | 30 pF | 27 dB | 8.9 kHz UNSTABLE |

B

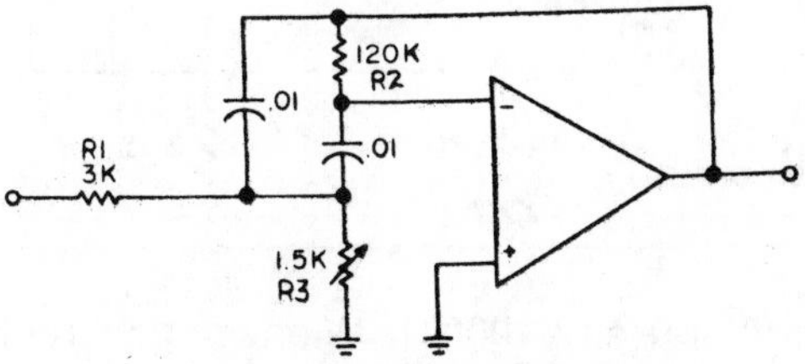


Variable bandpass filter.

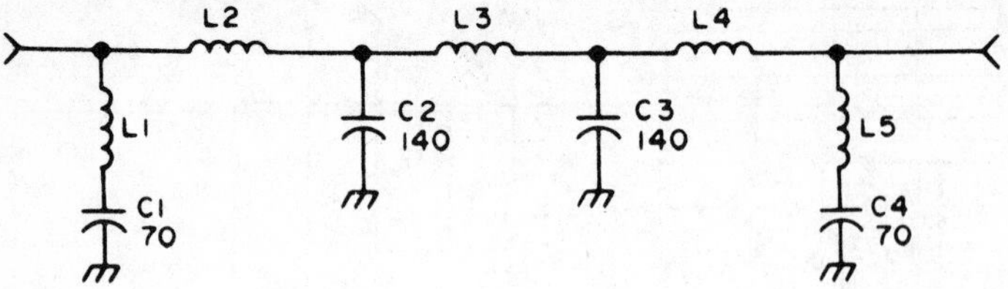


A

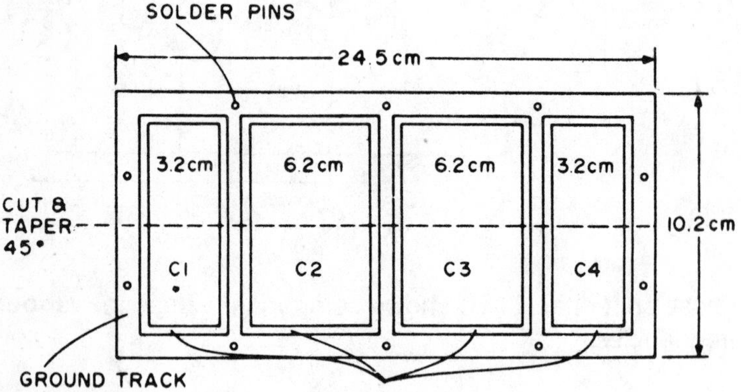


B

Isolating tracks 3 cm (cut or etch).

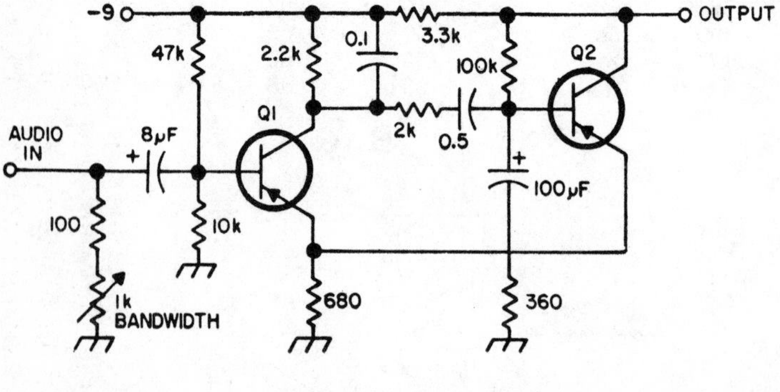


This audio filter uses a 1000 Hz. Wien bridge circuit to provide bandwidths from 70 to 600 Hz Q1 and Q2 are 2N408, 2N2613, SK3004, GE-2 or HEP 254.

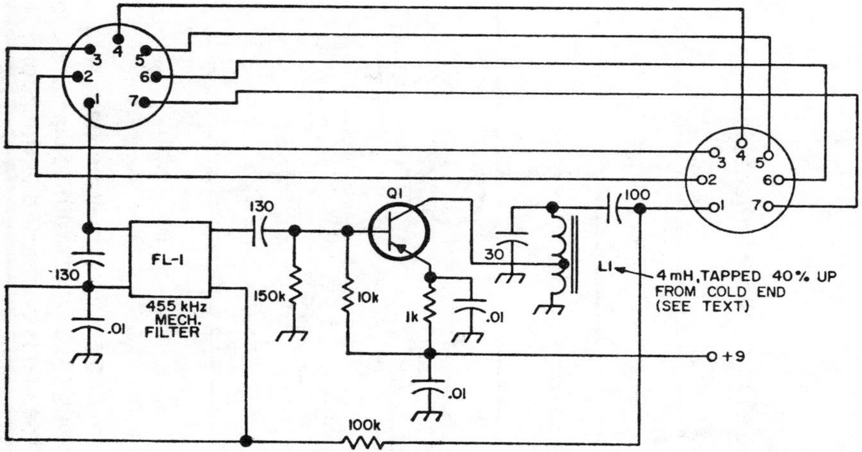


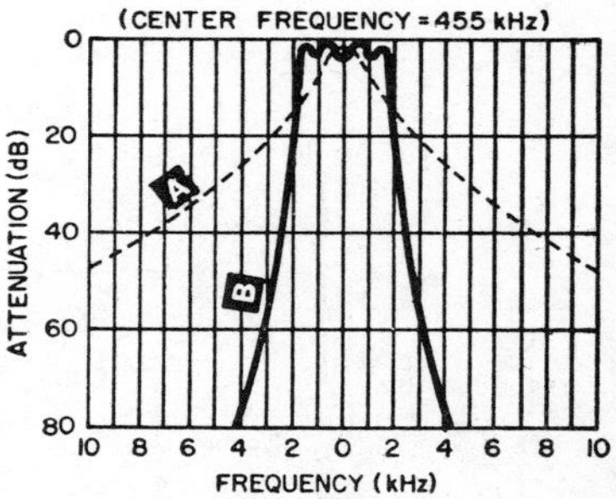


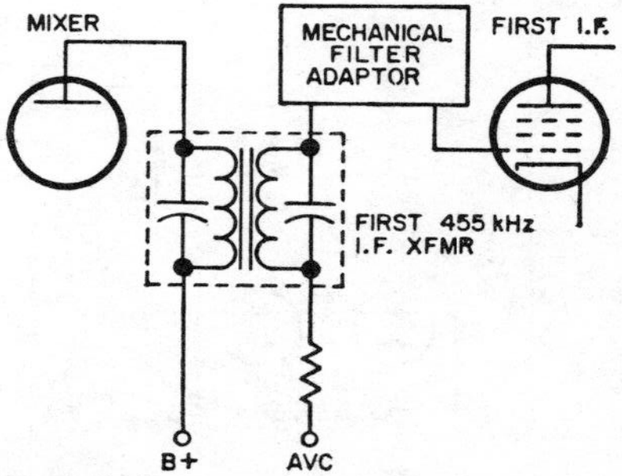


The selectivity of inexpensive communications receivers may be substantially increased by the addition of this mechanical filter adapter. The transistor is used to make up for the 10 dB loss through the filter. The typical passband of a receiver without the filter is shown by A in the frequency response curve; the mechanical filter adapter results in curve B. Q1 should be a 2N1638, 2N1727, SK3008, or HEP 3.

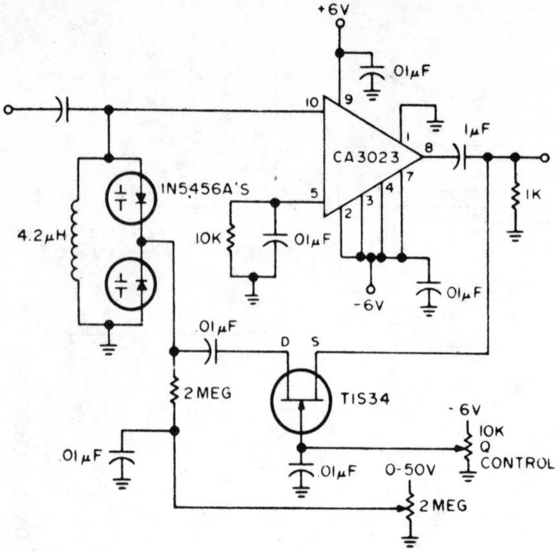


Tunable i-f amplifier with variable Q.

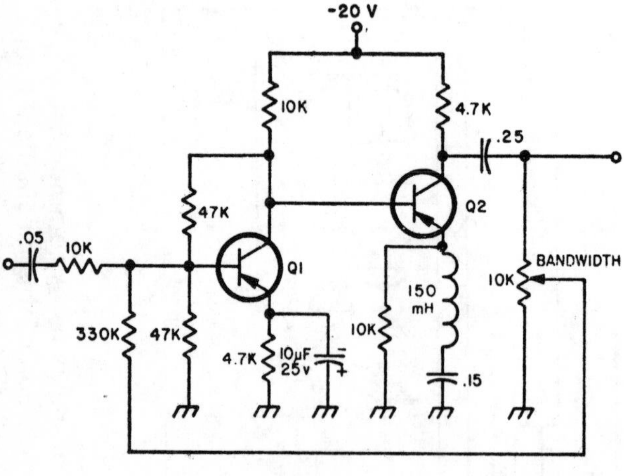


This simple audio bandpass filter may be narrowed to the limits of unintelligibility. At a bandwidth of 80 Hz, it provides about 20 dB gain. The unit is connected to the phone jack on receiver while headphones are connected across the output.

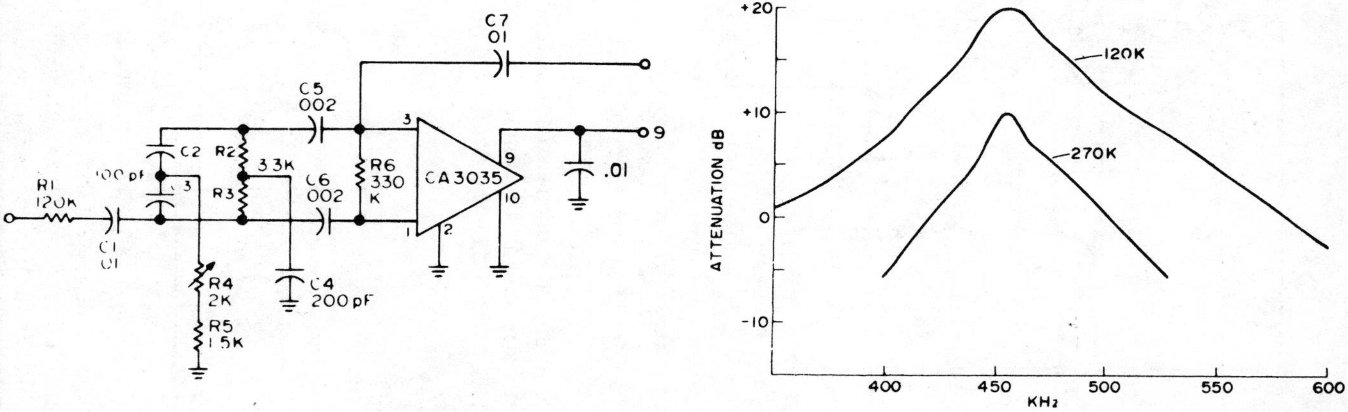


455 kHz buffer amplifier/filter, with response curves.

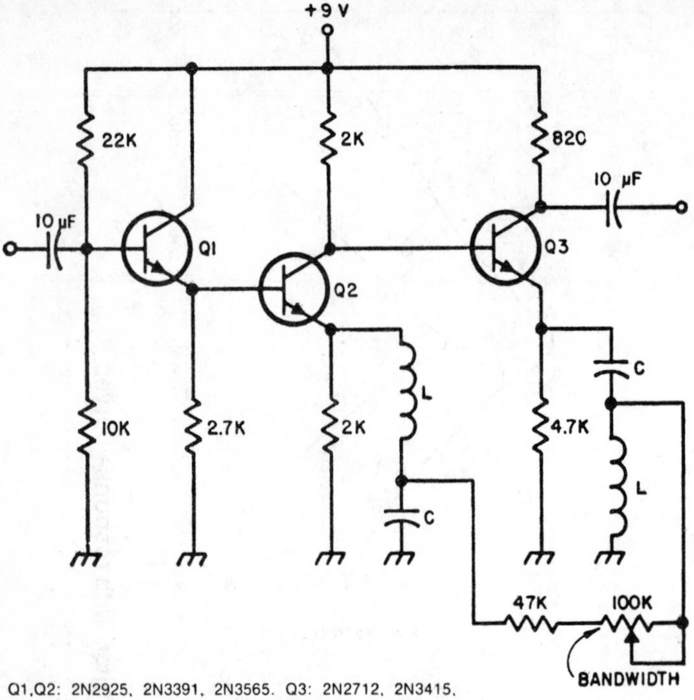


Q1,Q2: 2N2925, 2N3391, 2N3565. Q3: 2N2712, 2N3415, 2N3566. Choose L and C for resonance at the desired frequency. For center frequency of 1000 Hz, use 250 mH and 0.1 µF

This three-stage audio filter uses two series resonant circuits to provide a very narrow audio passband. The Q of the circuits and bandwidth is controlled by the amount of feedback.

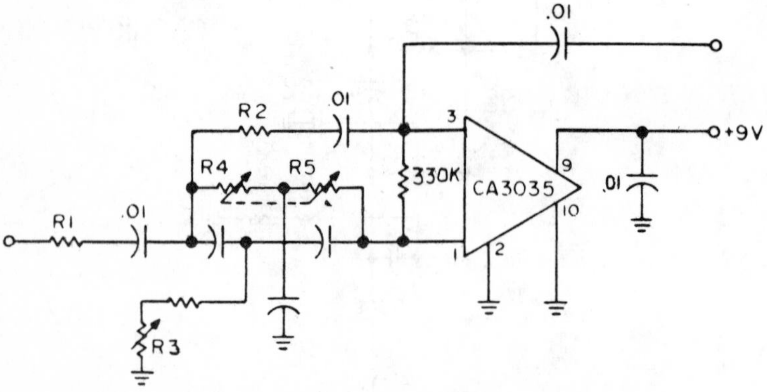


Variable notch filter.

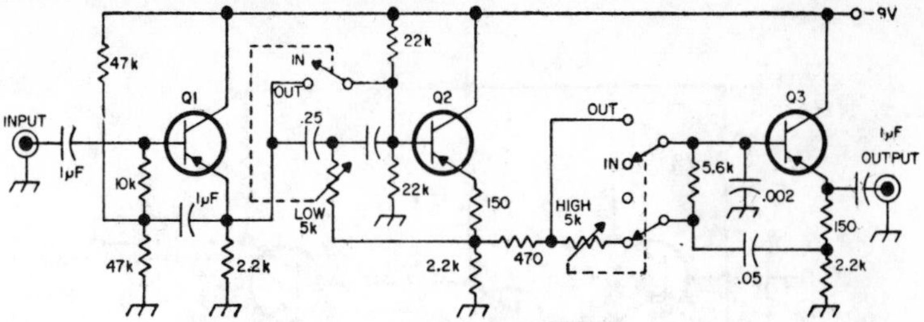


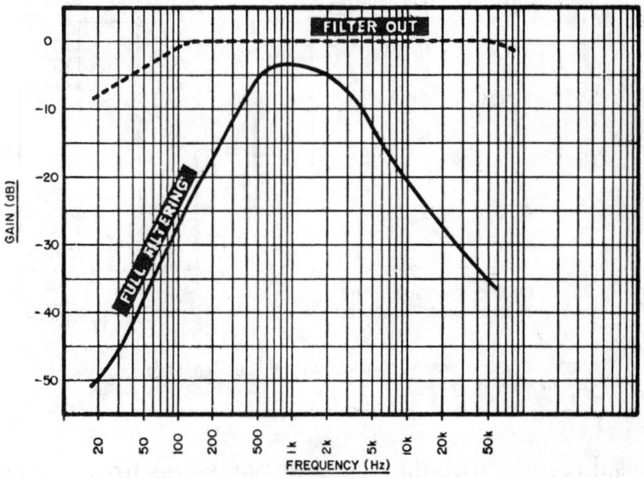


This highly versatile audio filter may be used to completely shape the audio spectrum of receiver or transmitter; it may be used with the filters out, or with the variable low- and high-pass filter networks connected. All the transistors are low-cost audio types such as the 2N1305, 2N1380, GE-2 or HEP 253.

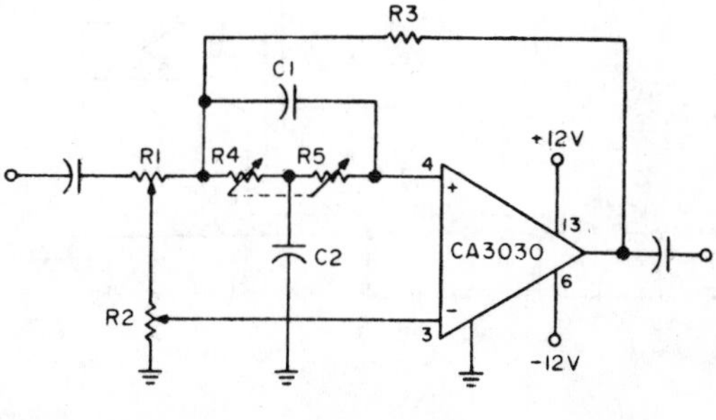


Deep notch filter with variable Q, frequency.

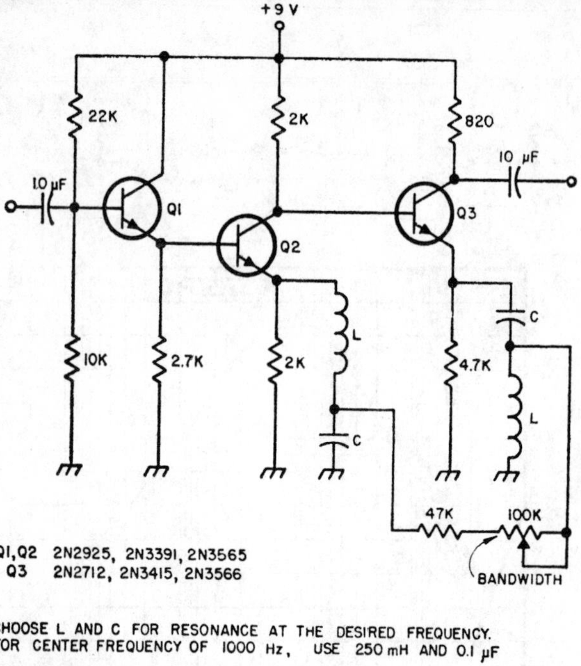


This three-stage audio filter uses two series resonant circuits to provide a very narrow audio passband. The Q of the circuits, and therefore the bandwidth, is controlled by the amount of feedback.

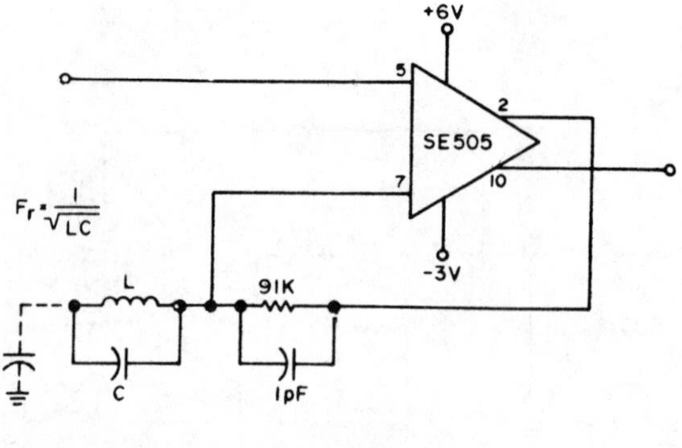


$F_r = \frac{1}{\sqrt{LC}}$

Band reject amplifier.

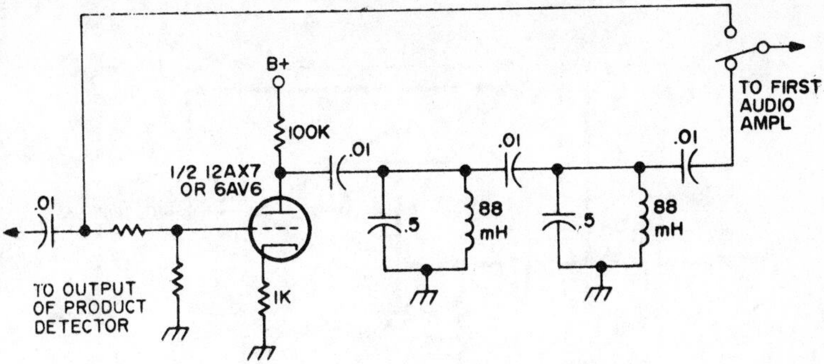


Ultrasimple selective audio filter.

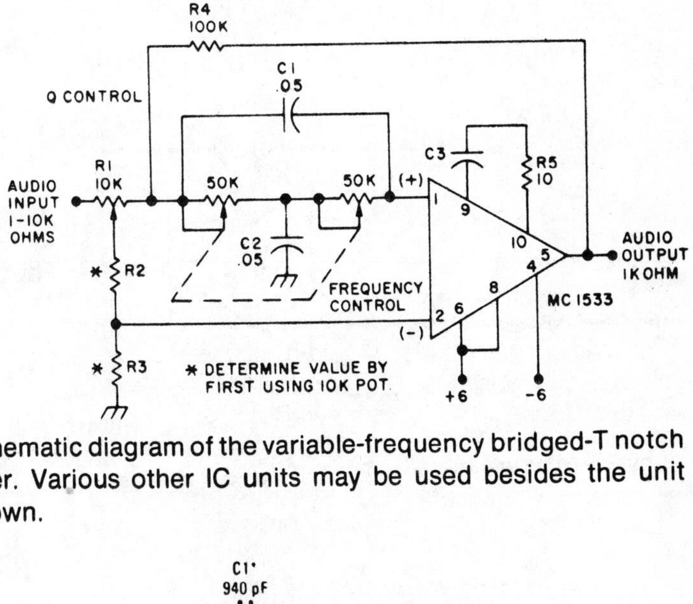


Schematic diagram of the variable-frequency bridged-T notch filter. Various other IC units may be used besides the unit shown.

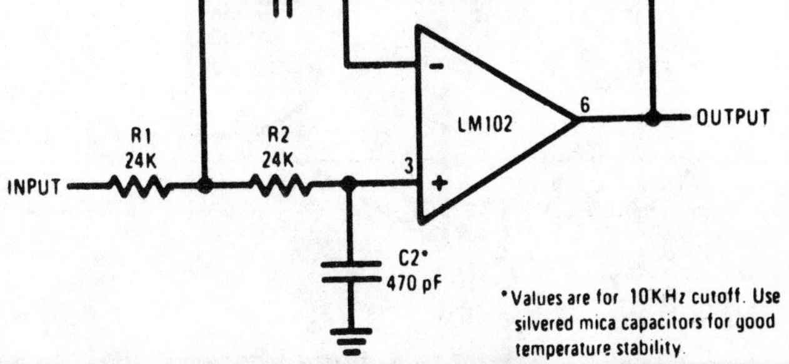


Low-pass active filter.

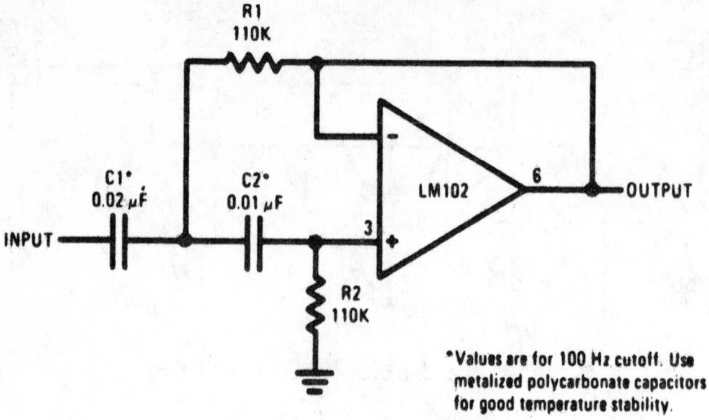


High-pass active filter.

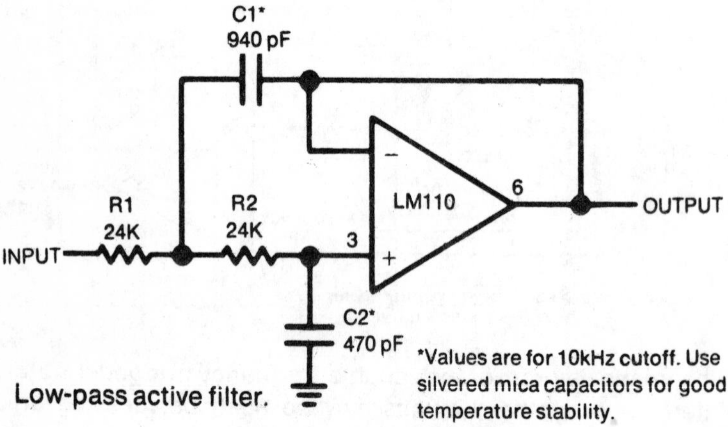


*Values are for 10kHz cutoff. Use silvered mica capacitors for good temperature stability.

Low-pass active filter.

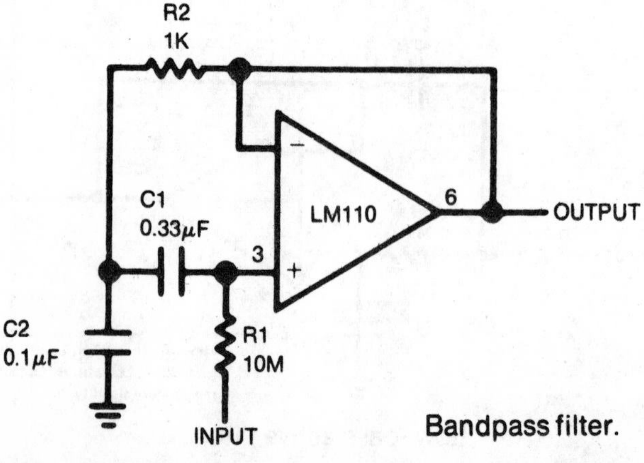


Bandpass filter.

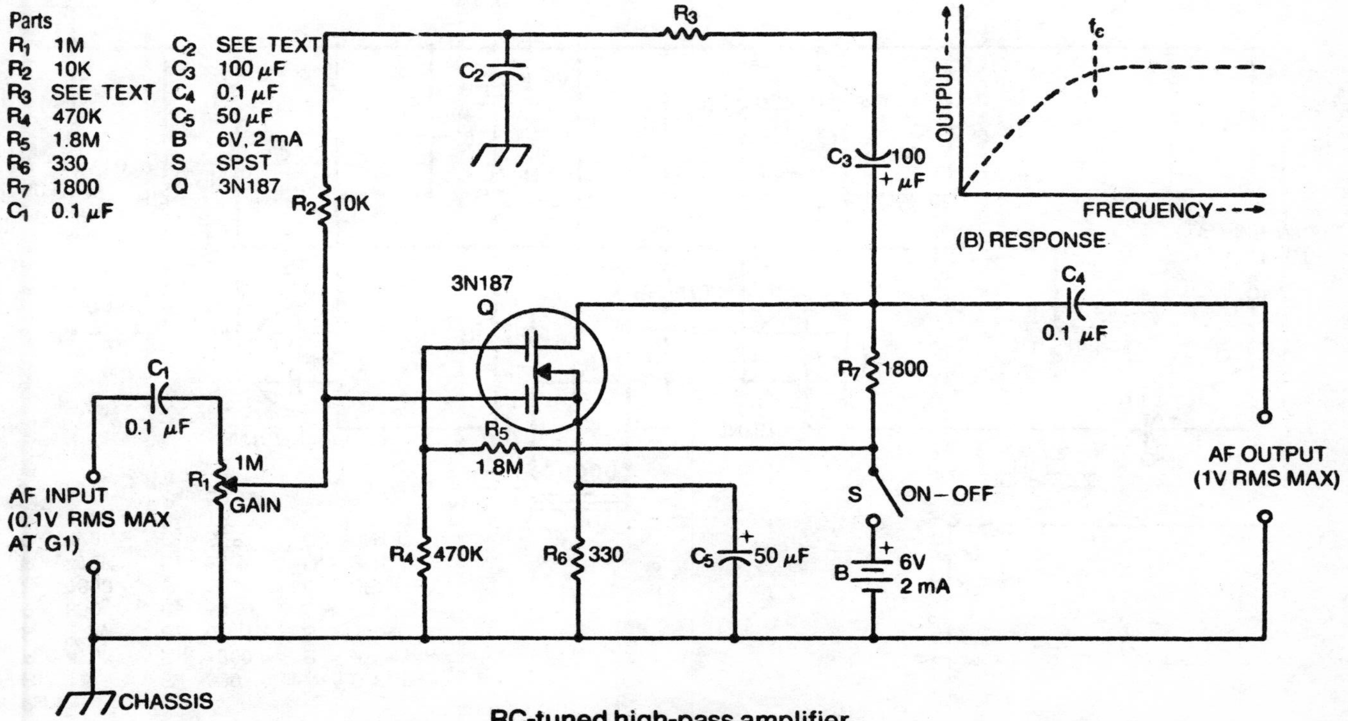


RC-tuned high-pass amplifier.

Parts

| | | | | | |
|---|---|---|---|---|---|
| $R_1$ | 1M | $R_9$ | 330 | $C_7$ | 0.1 μF |
| $R_2$ | 470K | $R_{10}$ | 1800 | B | 6V, 4 mA |
| $R_3$ | 1.8M | $C_1$ | 0.1 μF | S | SPST |
| $R_4$ | 330 | $C_2$ | 50 μF | S.B. | 0.1 μF |
| $R_5$ | 1800 | $C_3$ | 10 μF | L | SEE TEXT |
| $R_6$ | 470K | $C_4$ | 1 μF | $Q_1$ | 3N187 |
| $R_7$ | 470K | $C_5$ | SEE TEXT | $Q_2$ | 3N187 |
| $R_8$ | 1.8M | $C_6$ | 50 μF | | |

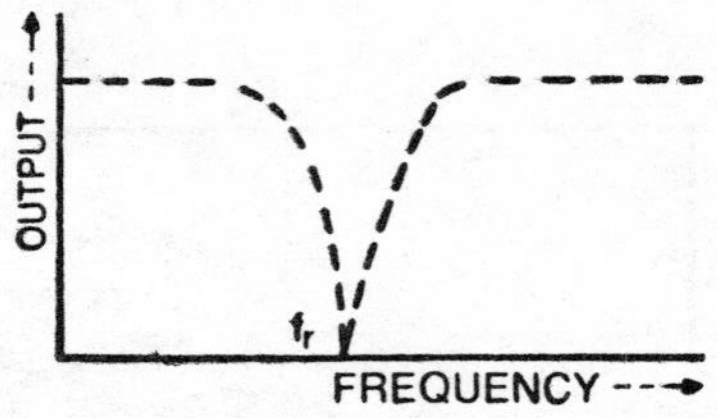

(B) RESPONSE

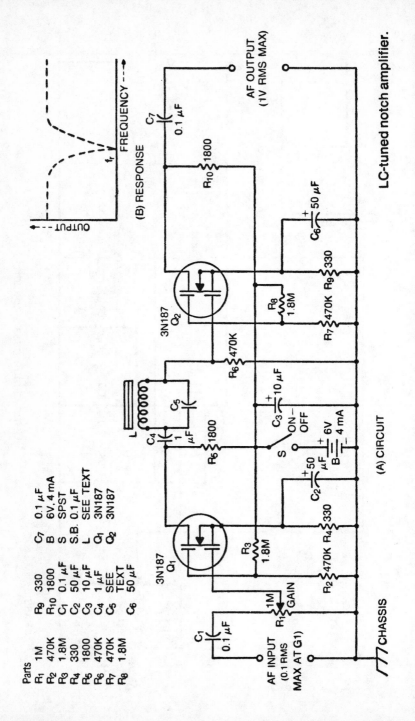

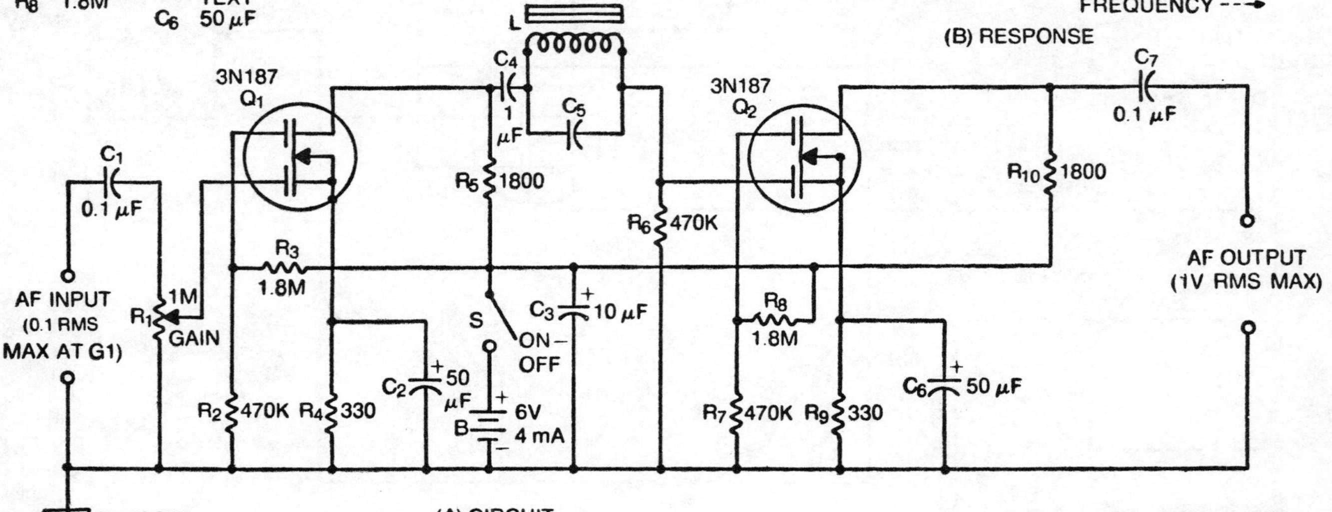

(A) CIRCUIT

LC-tuned notch amplifier.

# Section F

# Logic Circuits, Counters, Clocks

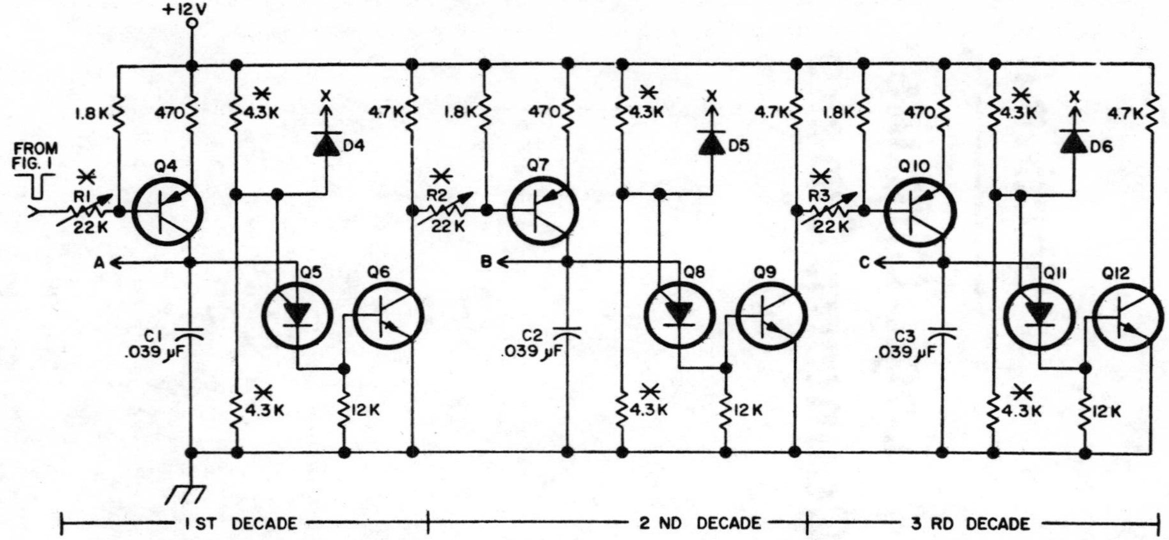


✱ SEE TEXT

Q4, Q7, Q10 2N4249
Q6, Q9, Q12 2N2924

Q5, Q8, Q11 PUT RADIO SHACK #276-119
D4, D5, D6 IN4148

A

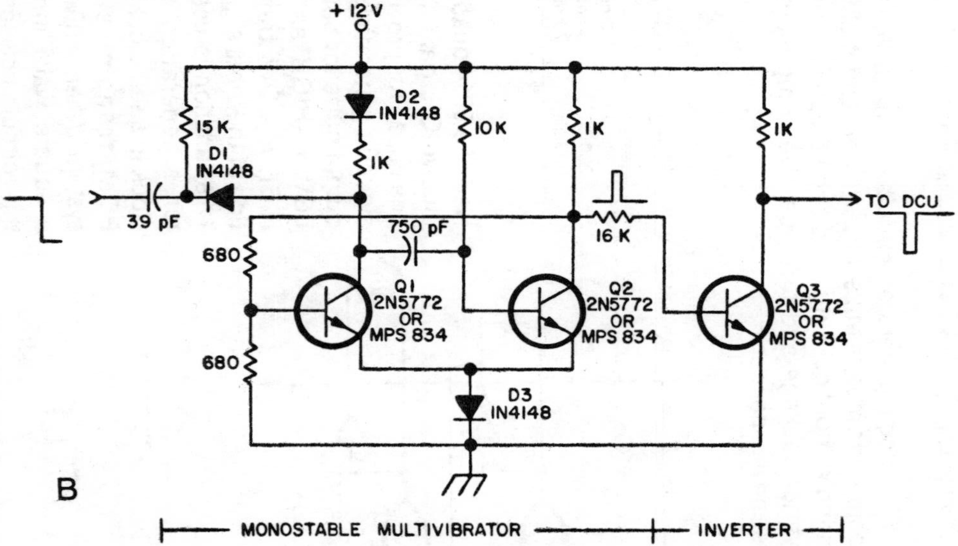


B

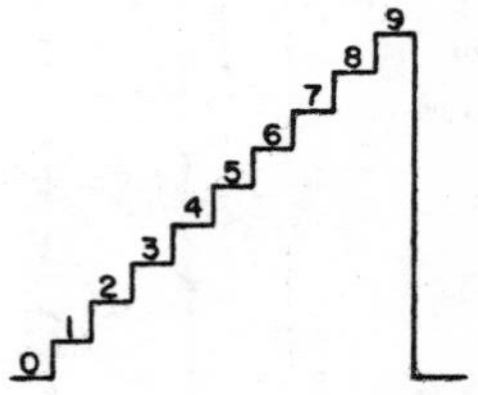


C

Decimal counting unit. The input stage (left) converts negative-going pulses into uniform signals. Counting decades are shown at right. The value of R1 is adjusted to produce staircase whose voltage value at the source of the output FET represents the count.

| C | f |
|---|---|
| 200 pF | 5 MHz |
| 1600 pF | 1 MHz |
| 0.018 μF | 100 KHz |
| 0.18 μF | 10 KHz |

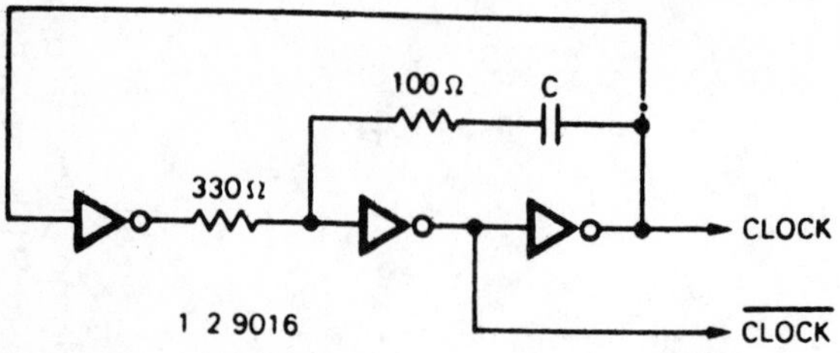


The simple TTL clock generator circuit shown provides a clock satisfactory for most simple TTL systems and it always starts oscillating without coaxing. This circuit requires only ½ of a hex inverter package and three passive components—two resistors and a capacitor.

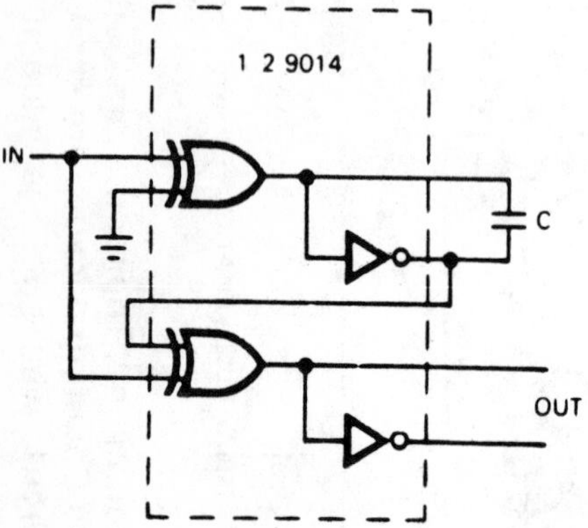


| C | $t_P$ |
|---|---|
| 0 | 10 ns |
| 200 pF | 30 ns |
| 1000 pF | 70 ns |

Half of a 9014 quad exclusive-OR gate with one capacitor provides a circuit generating an output pulse for both a LOW-to-HIGH and a HIGH-to-LOW transition of the input signal. This function is useful for regenerating the clock in a self-clocking PDM transmission system. When fed with a square wave input, this circuit acts as a frequency doubler.

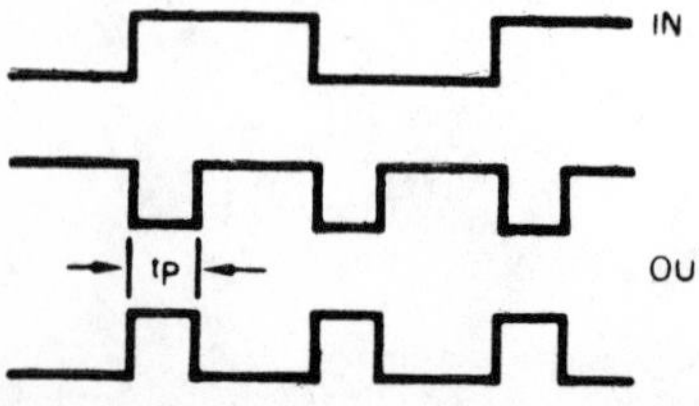

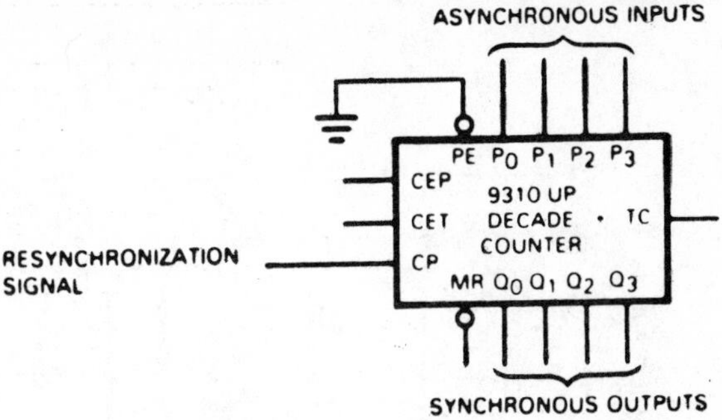


A resynchronizer using a 9310 (or 9316) as four D-input flip flops is shown. In this circuit the PE input is grounded, and the resynchronizing input is applied to the CP input. In most cases, the 9300 universal shift register is preferable for this function.

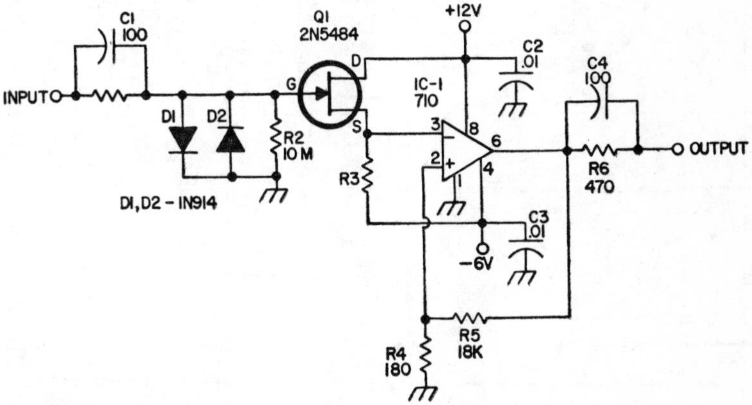


Buffering circuit isolates frequency counter from signal source, offers high-impedance input, relatively low-impedance output. The value of the series input resistor is 100K; R3 is 1800 ohms.

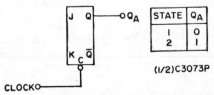

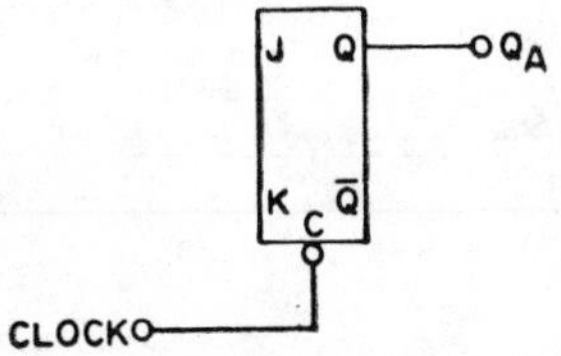


| STATE | $Q_A$ |
|---|---|
| 1 | 0 |
| 2 | 1 |

Synchronous divide-by-2 up counter.

See p. 150

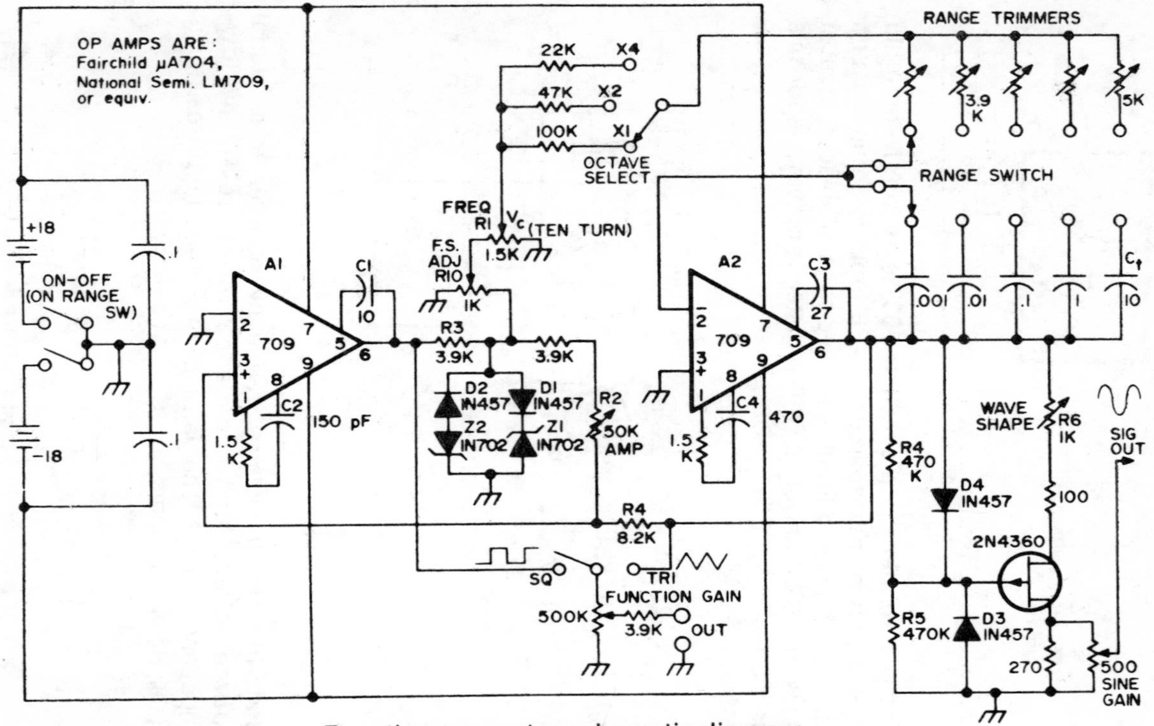


Function generator schematic diagram.

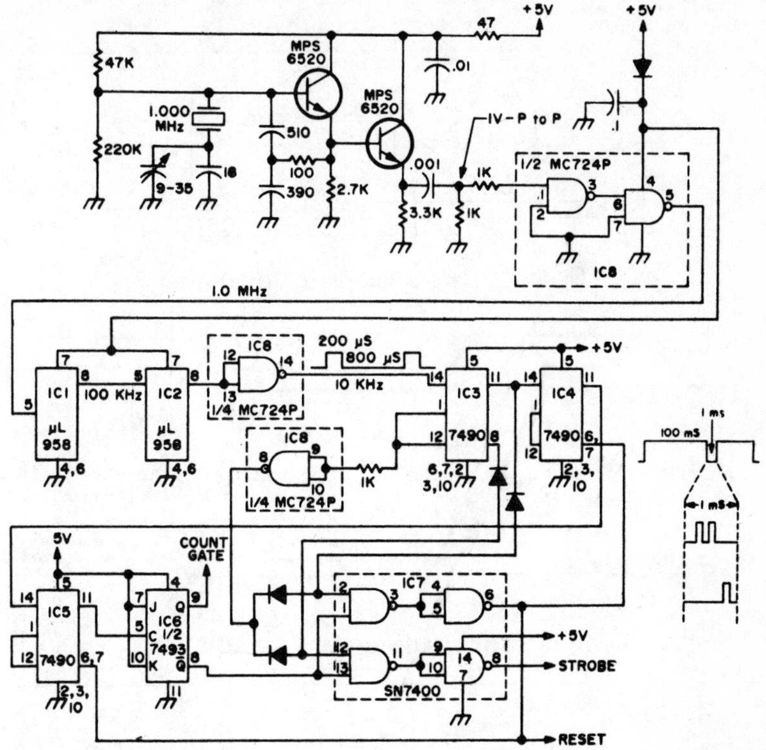


Frequency-standard oscillator and counter control circuitry with timing diagrams.

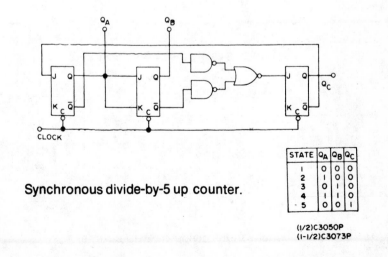

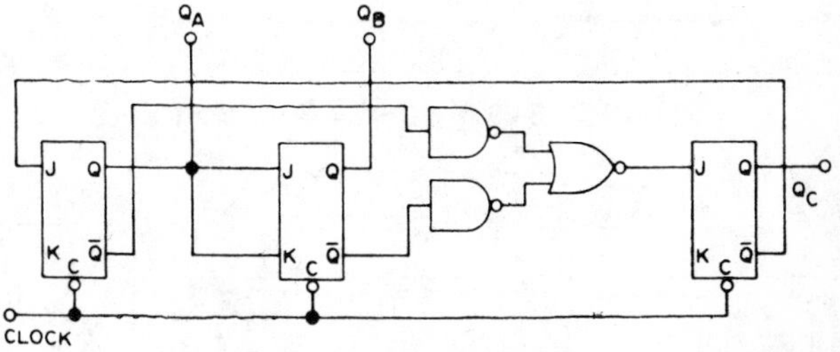


Synchronous divide-by-5 up counter.

| STATE | $Q_A$ | $Q_B$ | $Q_C$ |
|---|---|---|---|
| 1 | 0 | 0 | 0 |
| 2 | 1 | 0 | 0 |
| 3 | 0 | 1 | 0 |
| 4 | 1 | 1 | 0 |
| 5 | 0 | 0 | 1 |

(1/2)C3050P
(1-1/2)C3073P

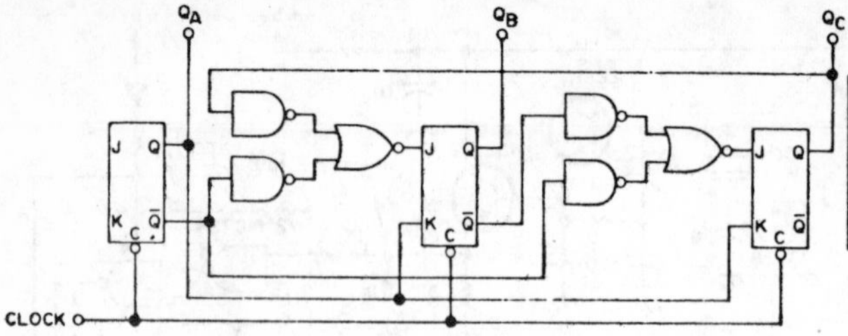


| STATE | QA | QB | QC |
|---|---|---|---|
| 1 | 0 | 0 | 0 |
| 2 | 1 | 0 | 0 |
| 3 | 0 | 1 | 0 |
| 4 | 1 | 1 | 0 |
| 5 | 0 | 0 | 1 |
| 6 | 1 | 0 | 1 |

(1)C3050P
(1-1/2)C3073P

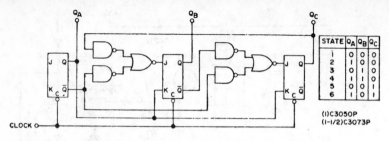

Synchronous divide-by-6 up counter.

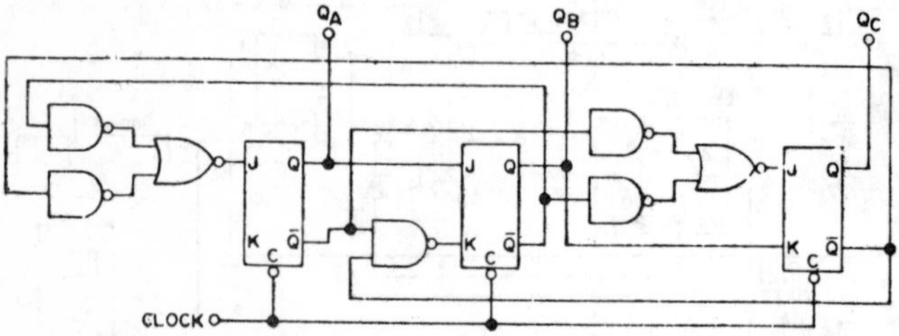


| STATE | QA | QB | QC |
|---|---|---|---|
| 1 | 0 | 0 | 0 |
| 2 | 1 | 0 | 0 |
| 3 | 0 | 1 | 0 |
| 4 | 1 | 1 | 0 |
| 5 | 0 | 0 | 1 |
| 6 | 1 | 0 | 1 |
| 7 | 0 | 1 | 1 |

(1/4)C3000P
(1)C3050P
(1-1/2)C3073

Synchronous divide-by-7 up counter.

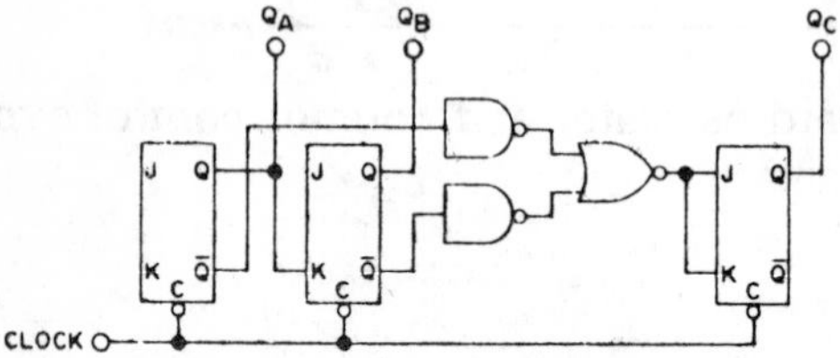


| STATE | QA | QB | QC |
|---|---|---|---|
| 1 | 0 | 0 | 0 |
| 2 | 1 | 0 | 0 |
| 3 | 0 | 1 | 0 |
| 4 | 1 | 1 | 0 |
| 5 | 0 | 0 | 1 |
| 6 | 1 | 0 | 1 |
| 7 | 0 | 1 | 1 |
| 8 | 1 | 1 | 1 |

(1/2)C3050P
(1-1/2)C3073P

Synchronous divide-by-8 up counter.

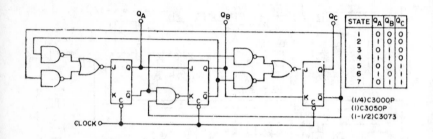

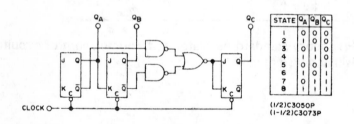

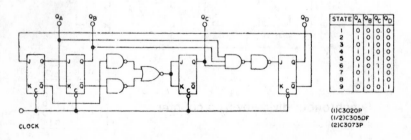

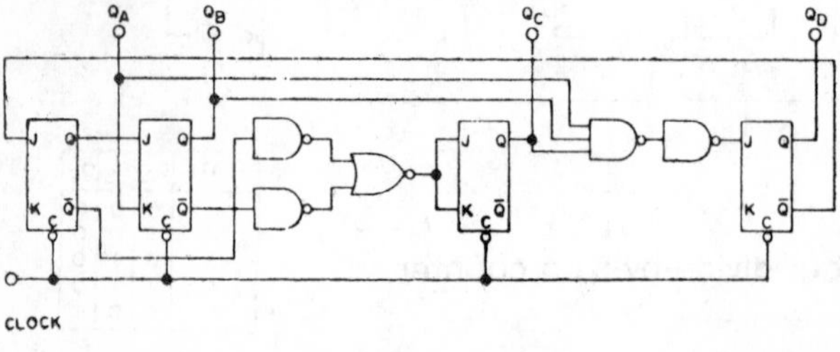


| STATE | QA | QB | QC | QD |
|---|---|---|---|---|
| 1 | 0 | 0 | 0 | 0 |
| 2 | 1 | 0 | 0 | 0 |
| 3 | 0 | 1 | 0 | 0 |
| 4 | 1 | 1 | 0 | 0 |
| 5 | 0 | 0 | 1 | 0 |
| 6 | 1 | 0 | 1 | 0 |
| 7 | 0 | 1 | 1 | 0 |
| 8 | 1 | 1 | 1 | 0 |
| 9 | 0 | 0 | 0 | 1 |

(1)C3020P
(1/2)C305DF
(2)C3073P

Synchronous divide-by-9 up counter.

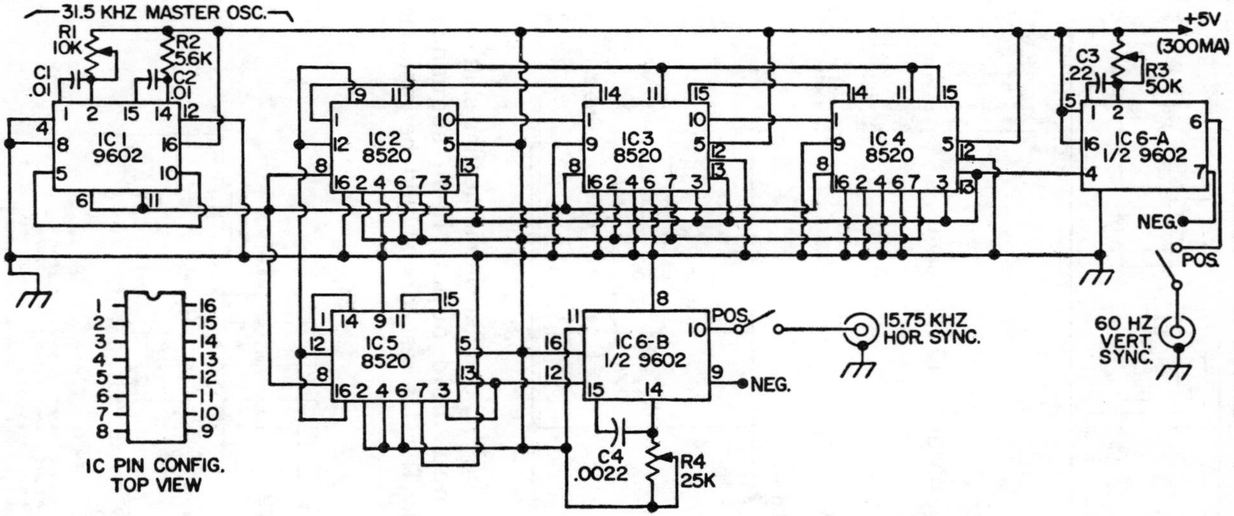


Schematic of the digital interlaced sync generator. R2 is 5.6K, ½W, 5%; potentiometers are miniature and capacitors are Mylar. IC-1 and 6, 9602PC, are available from Schweber Electronics, Syosset NY, $3.00 each plus postage. IC-2 through –5, DM8520, are available from JTM Associates, P.O. Box 843, Manchester MO 63011, $1.90 each plus postage or from Babylon Electronics, P.O. Box J, Carmichael CA 95608, $2.00 each plus postage.

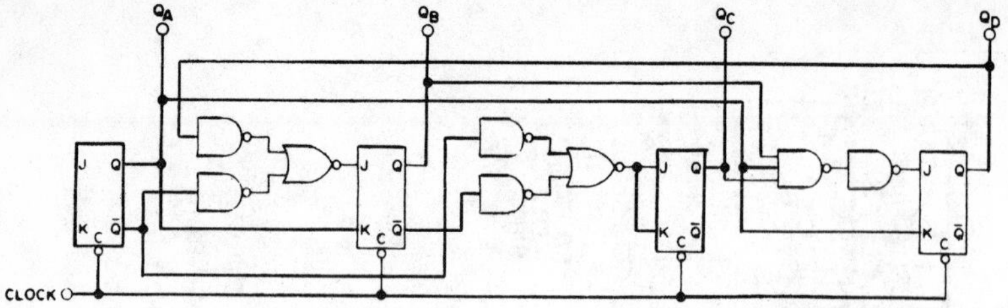


| STATE | $Q_A$ | $Q_B$ | $Q_C$ | $Q_D$ |
|---|---|---|---|---|
| 1 | 0 | 0 | 0 | 0 |
| 2 | 1 | 0 | 0 | 0 |
| 3 | 0 | 1 | 0 | 0 |
| 4 | 1 | 1 | 0 | 0 |
| 5 | 0 | 0 | 1 | 0 |
| 6 | 1 | 0 | 1 | 0 |
| 7 | 0 | 1 | 1 | 0 |
| 8 | 1 | 1 | 1 | 0 |
| 9 | 0 | 0 | 0 | 1 |
| 10 | 1 | 0 | 0 | 1 |

(1)C3020P
(1)C3050P
(2)C3073P

Synchronous divide-by-10 up counter.

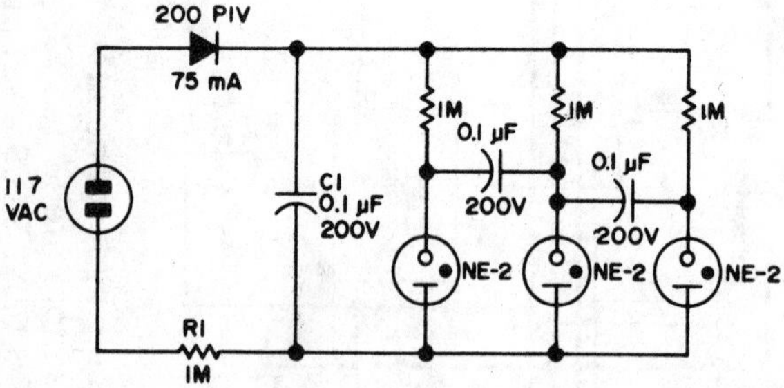


Three-stage oscillating ring counter with indicating shift register plus into household outlet; flash sequence: 2-12-12-32-12-12-3. (The 1M resistors should be matched fairly closely.)

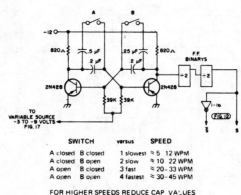

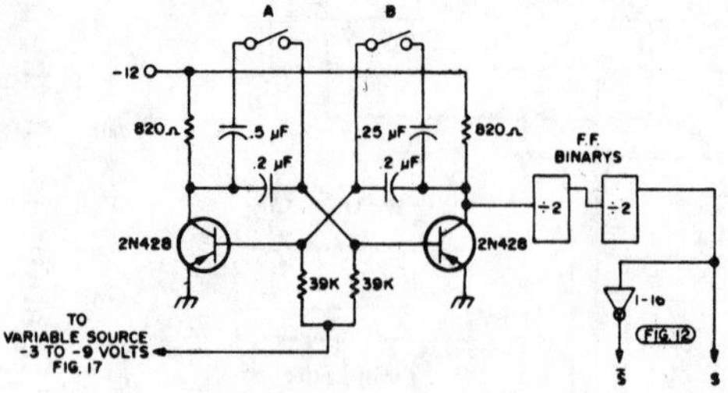


| SWITCH | | versus | SPEED |
|---|---|---|---|
| A closed | B closed | 1 slowest | ≈ 5  12 WPM |
| A closed | B open | 2 slow | ≈ 10  22 WPM |
| A open | B closed | 3 fast | ≈ 20 - 33 WPM |
| A open | B open | 4 fastest | ≈ 30 - 45 WPM |

FOR HIGHER SPEEDS REDUCE CAP VALUES

Variable clock.

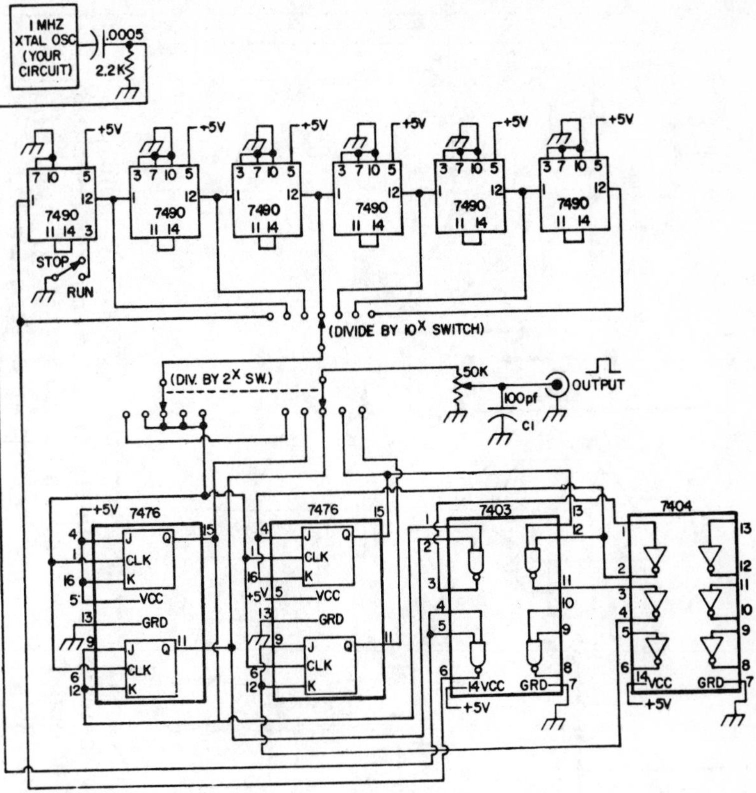


Frequency standard wiring diagram.

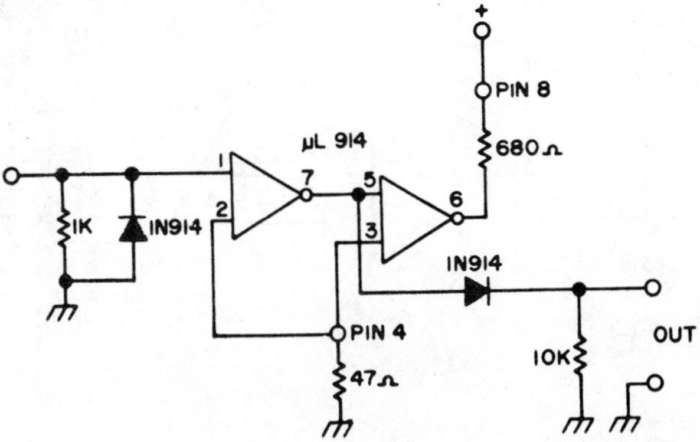


Schmitt trigger.

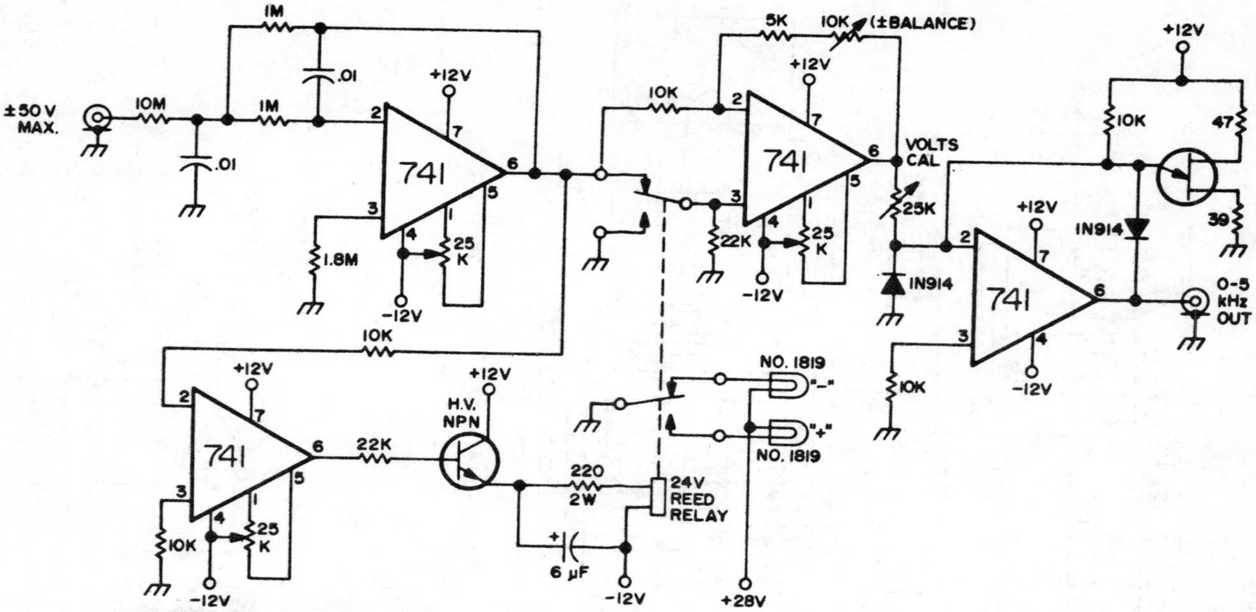


DVM adapter for a frequency counter. This circuits consists of 200K Ω/V input low-pass active filter stage, a polarity detector and automatic switcher/indicator, and a voltage-to-frequency converter. The output frequency is adjusted so that 50V input will give a frequency of 5000 out. The three 25K pots are used for offset balancing. Any counter capable of readout to Hz is fine.

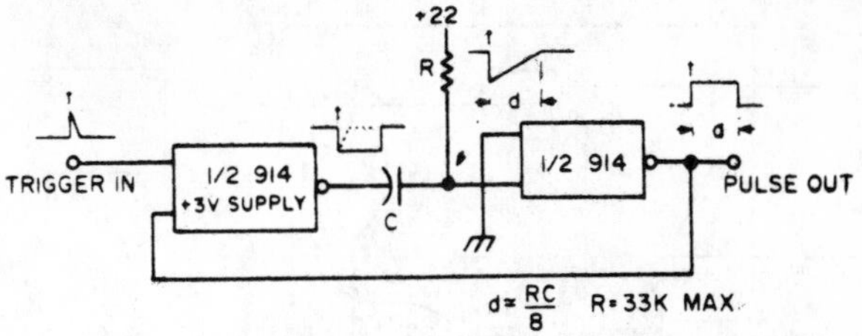


One-shot circuit is easily built from 914 IC by adding single capacitor and external resistor. If 3V supply is used for IC, and 22V supply to drive external timing resistor, duration of output pulse will be almost exactly ⅛ the RC time constant, with R in ohms and C in farads. Maximum resistance usable for R is 33K to permit turn-on current to flow. Capacitance C, however, may be any value desired to achieve required output pulsewidth.

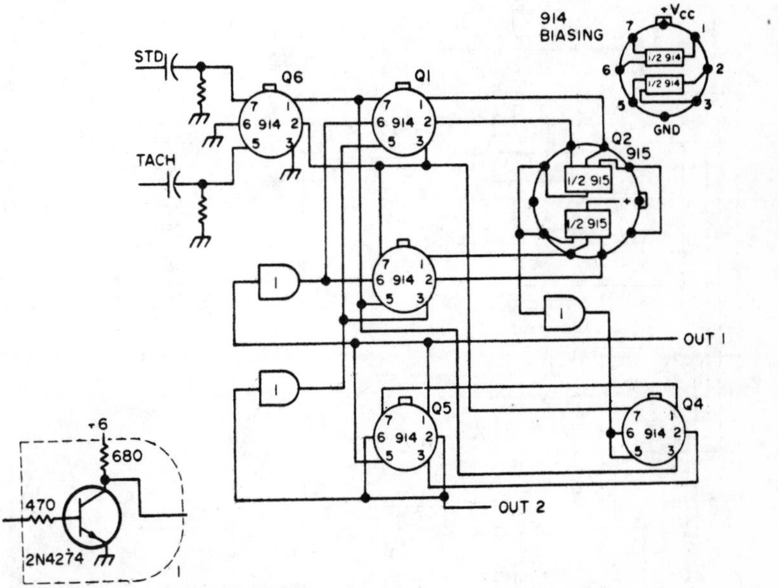


This rather complex arrangement of ICs is a frequency comparator and phase detector. The unknown frequency is fed into the TACH input and the standard to which it is to be compared goes to the STD input. Output level indicates whether TACH is higher or lower in frequency than STD. OUT 1 and OUT 2 are of opposite polarity. Type 915 IC (Q2) is same as 914 but has 3-input gates rather than 2-input elements.

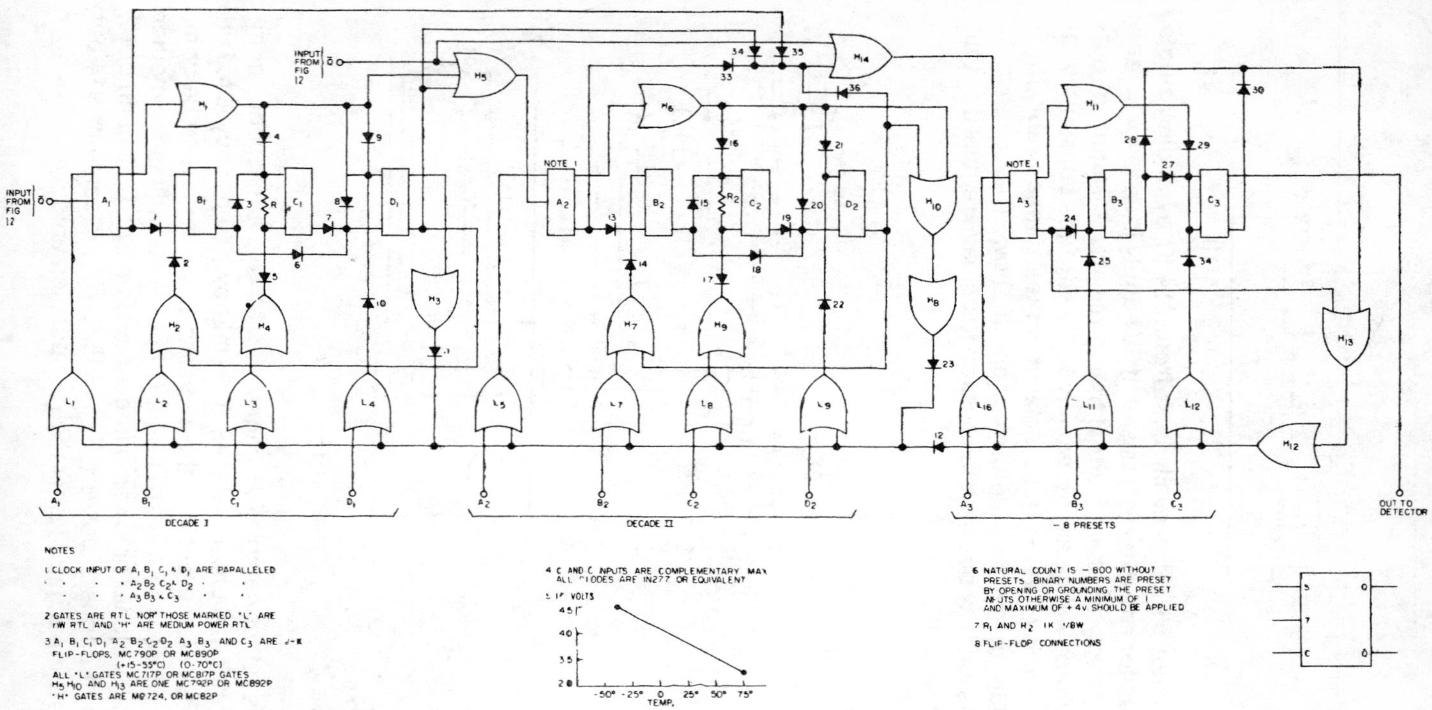
INPUT FROM FIG 12
OUT TO DETECTOR
DECADE I
DECADE II
– 8 PRESETS
NOTE 1
NOTES
1 CLOCK INPUT OF A1 B1 C1 & D1 ARE PARALLELED
A2 B2 C2 & D2
A3 B3 & C3
2 GATES ARE RTL 'NOR' THOSE MARKED 'L' ARE mW RTL AND 'H' ARE MEDIUM POWER RTL
3 A1 B1 C1 D1 A2 B2 C2 D2 A3 B3 AND C3 ARE J-K FLIP-FLOPS, MC790P OR MC890P
(+15-55°C) (0-70°C)
ALL 'L' GATES MC717P OR MC817P GATES
H5 H10 AND H13 ARE ONE MC792P OR MC892P
'H' GATES ARE MC724, OR MC82P
ALL DIODES ARE IN277 OR EQUIVALENT
VOLTS
TEMP.
-50° -25° 0 25° 50° 75°
6 NATURAL COUNT IS – 800 WITHOUT PRESETS. BINARY NUMBERS ARE PRESET BY OPENING OR GROUNDING THE PRESET INPUTS OTHERWISE A MINIMUM OF 1 AND MAXIMUM OF + 4v SHOULD BE APPLIED
8 FLIP-FLOP CONNECTIONS

Divide-by-N circuit.

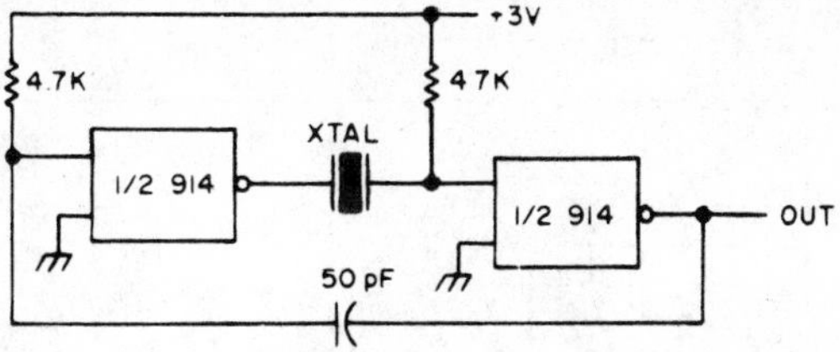


Minor modifications to one-shot circuit, including substitution of a crystal for the timing capacitor and insertion of a capacitor in the dc feedback loop, turn it into a crystal-controlled oscillator which may be used for a frequency standard. Output is rich in harmonics, and this circuit is not recommended for transmitter use for that reason.

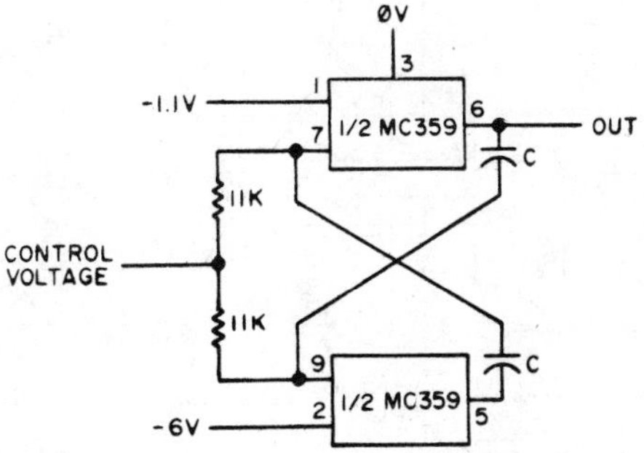


Voltage-controlled oscillator can be varied over nearly 10-to-1 frequency range simply by varying control voltage. This circuit may be used as part of phase-locked detector.

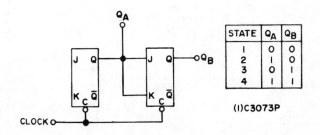

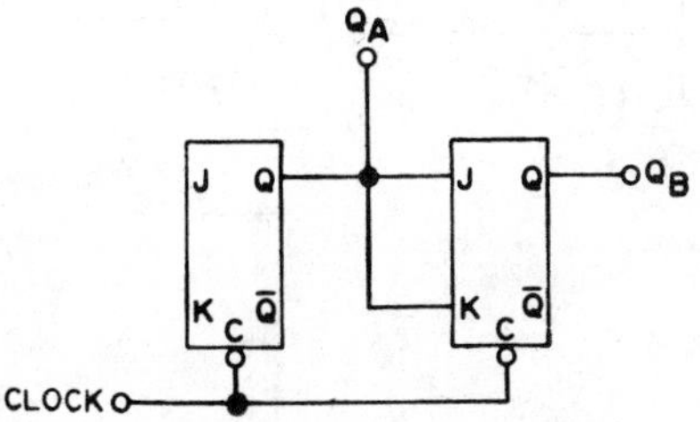


| STATE | $Q_A$ | $Q_B$ |
|---|---|---|
| 1 | 0 | 0 |
| 2 | 1 | 0 |
| 3 | 0 | 1 |
| 4 | 1 | 1 |

(1)C3073P

Synchronous divide-by-4 up counter.

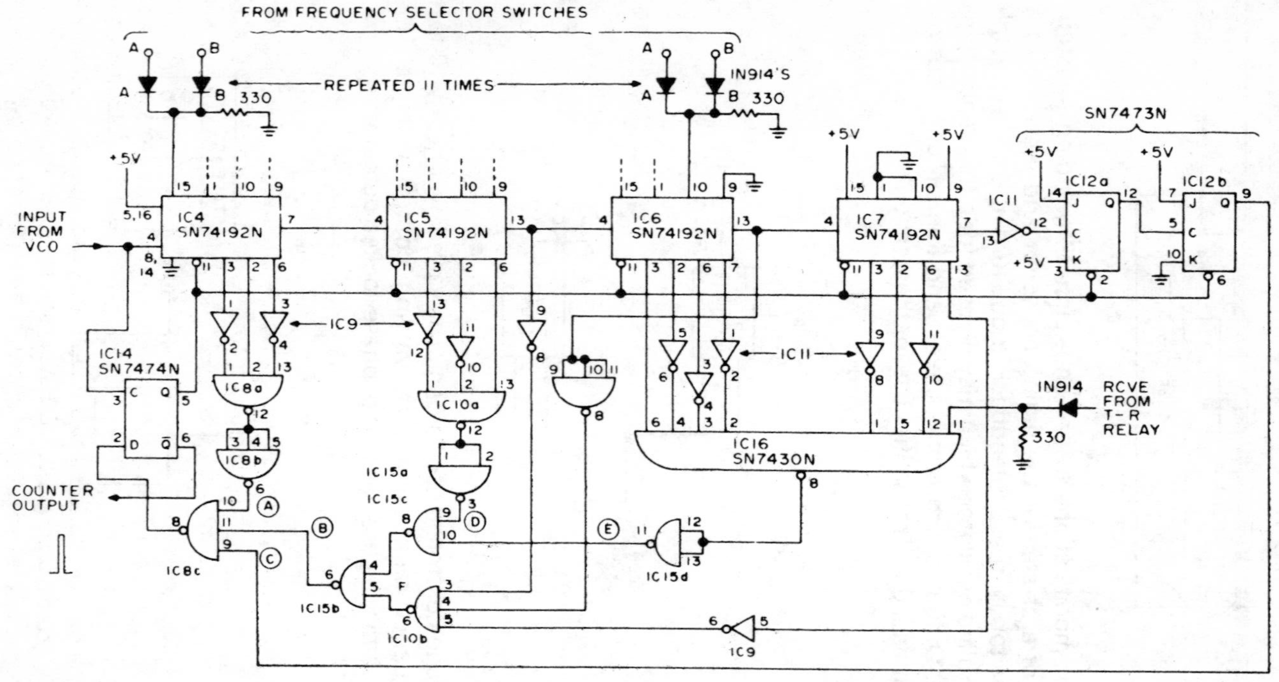


Programmable divider (÷ N).

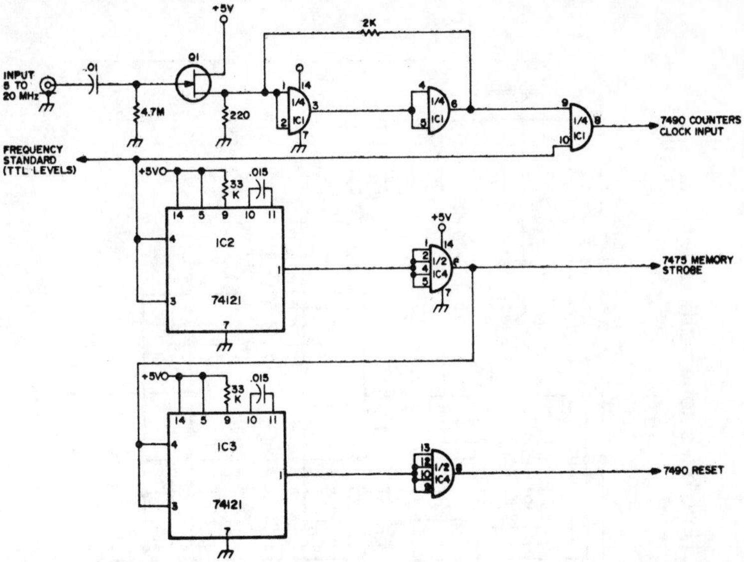


Frequency counter input: gating, strobing, and resetting. The sensitivity is set by the ratio of the 220 to 2K resistors.

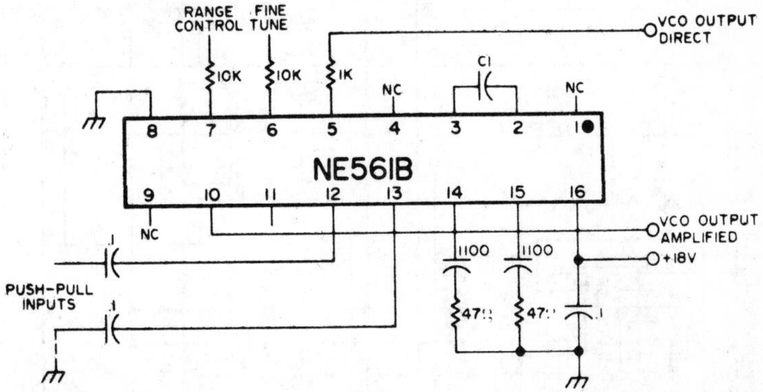


PLL makes fine frequency multiplier or divider. For this application, audio output connections are ignored and the VCO output is used instead. If input is single-ended, one of the two push-pull input leads should be bypassed to ground as shown by dotted lines. Circuit will multiply up to 10 times, and divide input frequency by 3, 5, 7, or 9. C1 and fine-tuning adjustment must be set for operation near desired output frequency. When input is applied, VCO will lock to exact multiple or odd submultiple of input if it is within locking range and of adequate strength.

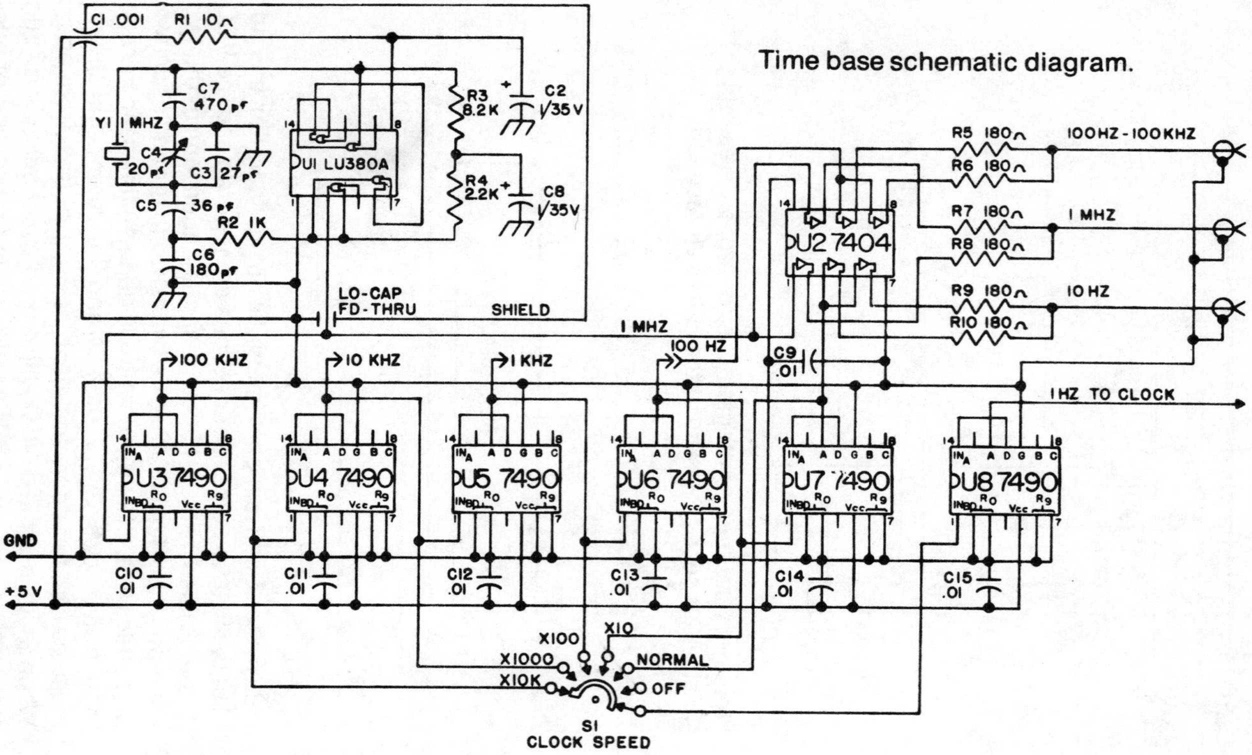

Time base schematic diagram.

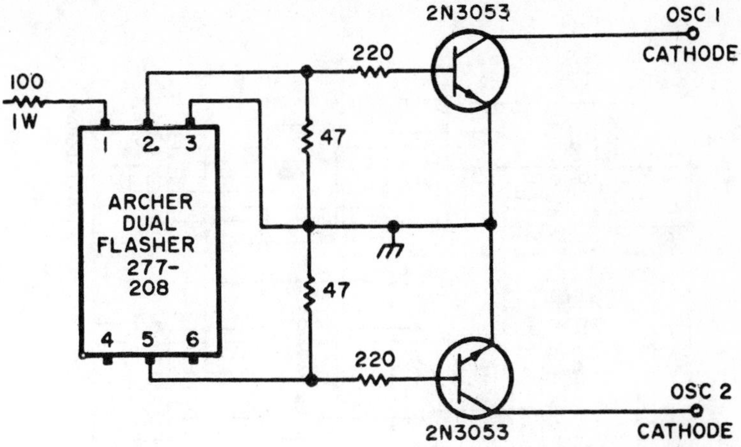


A nonlocking scanner can be made easily by using an Archer dual flasher in conjunction with an external switching circuit.

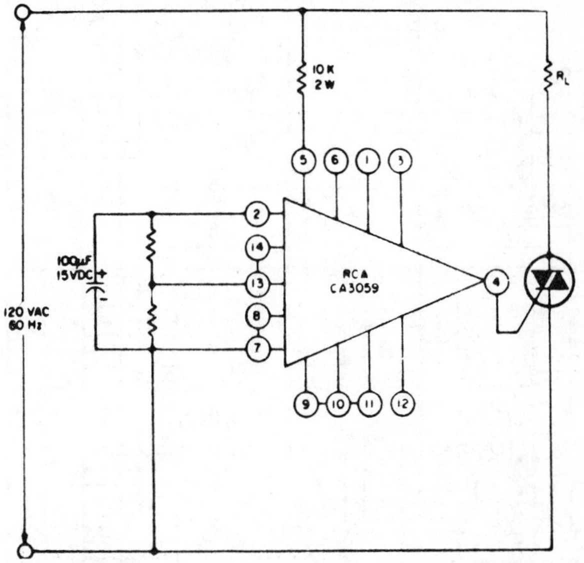


Controller for resistive loads. The RCA 2N5444 triac can be used for load currents up to 40A. The RCA 40668 triac will switch intermediate loads and the 40526 will handle lighter loads and those which are somewhat inductive.

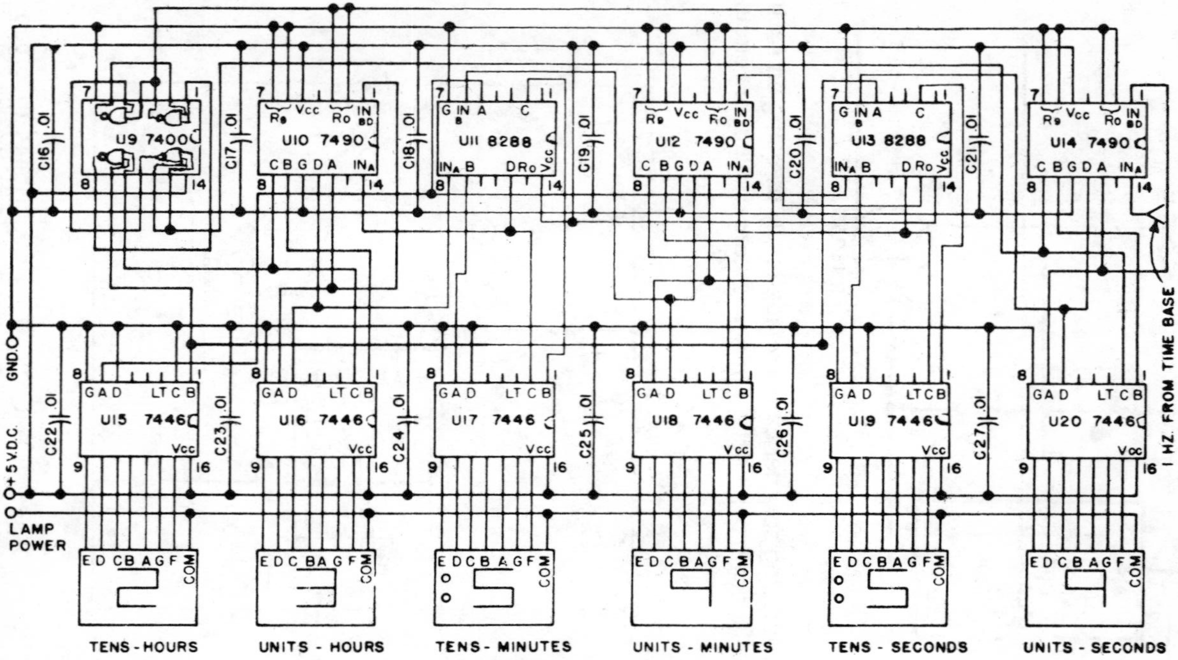


Clock schematic diagram.

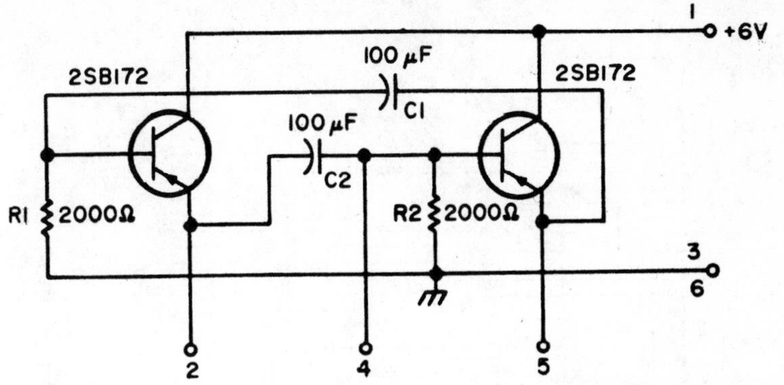


$$f = \frac{1}{R.C. + R2C2}$$

2SB172 → RCA SK3004
AF OUTPUT
PNP GERMANIUM

Circuit diagram of the flasher module. Changing the RC values here (test) will permit variation of the switching frequency.

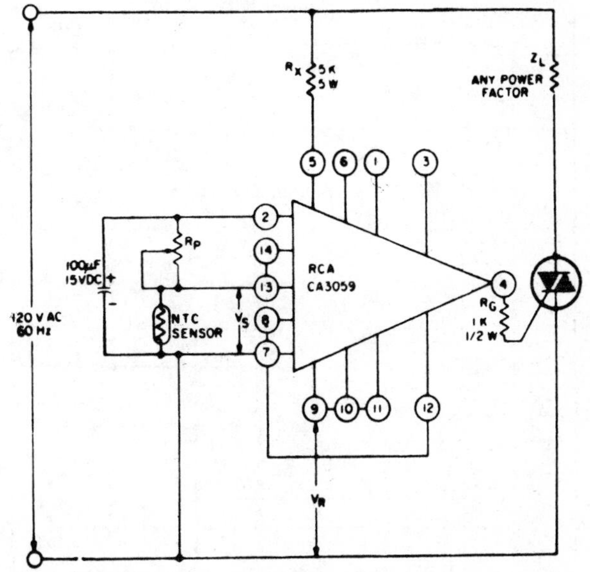


Differential comparator circuit. The load is switched on when the voltage difference between $V_S$ and $V_R$ becomes less than 50 $\mu$V. Note the jumper between terminals 7 and 12, which deactivates the anti-RFI feature.

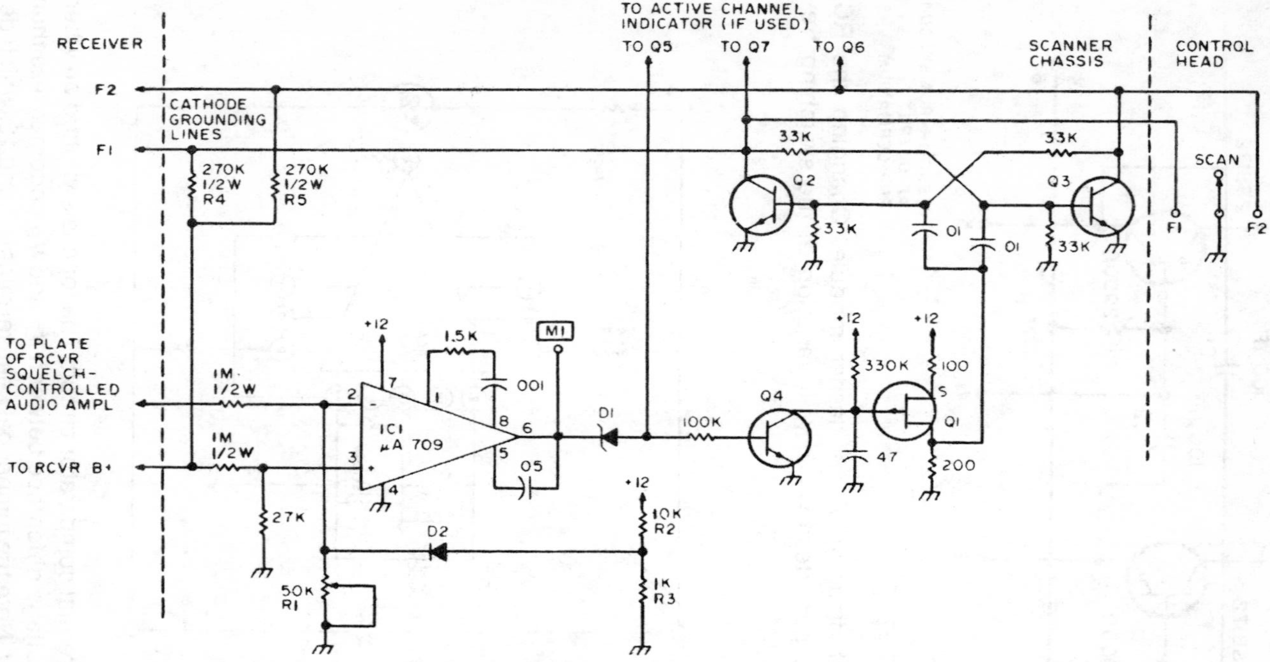


Two-channel search-lock for FM receivers. The switch is mounted on the control head; everything else may be mounted inside the radio cabinet.

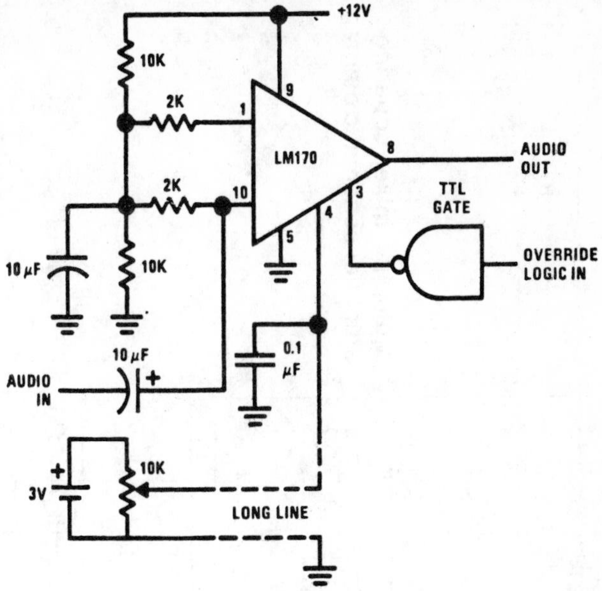


Remote or digital control amplifier.

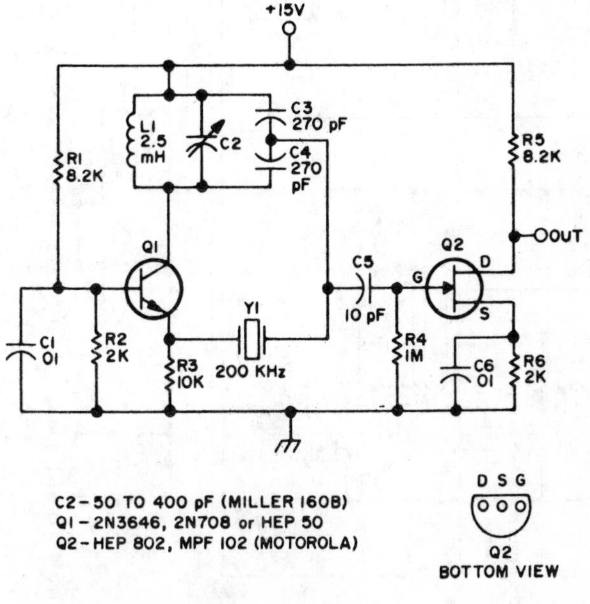


200 kHz crystal standard for counter time base.

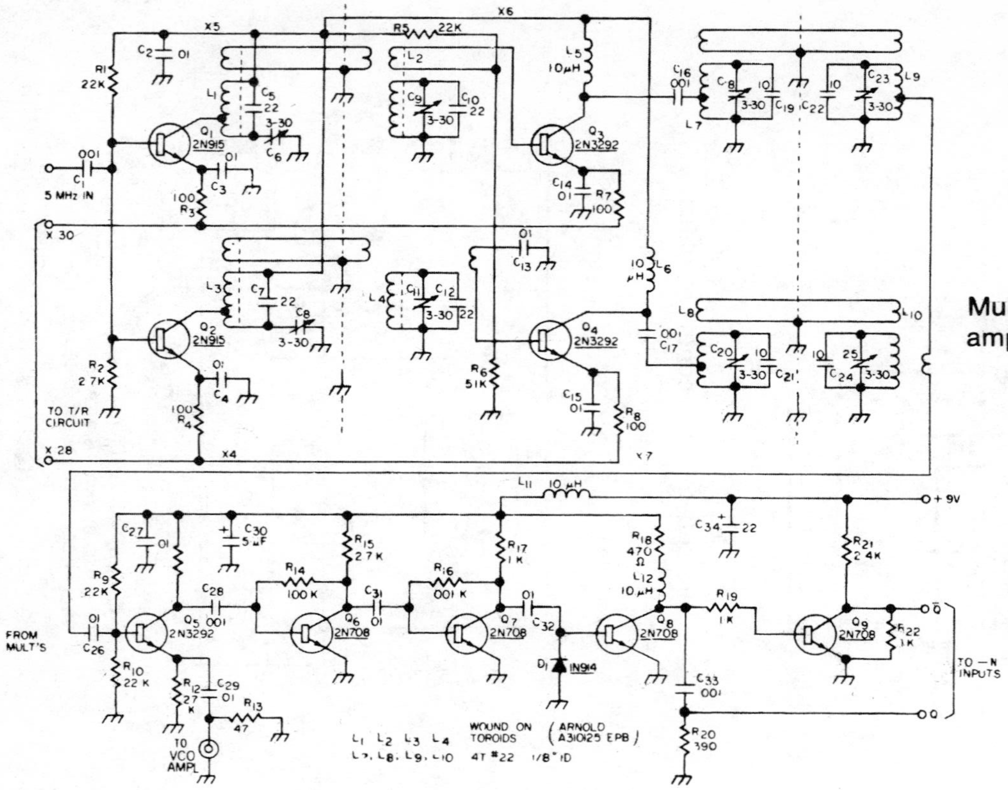


Multiplier, mixer, squaring amplifier schematic diagram.

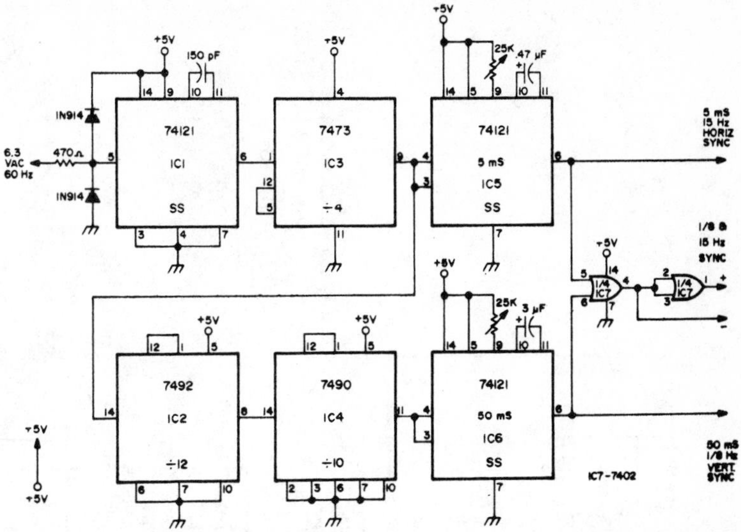


SSTV frequency standard. This simple circuit provides the need pulses for slow-scan TV cameras, flying-spot scanners, and pattern generators.

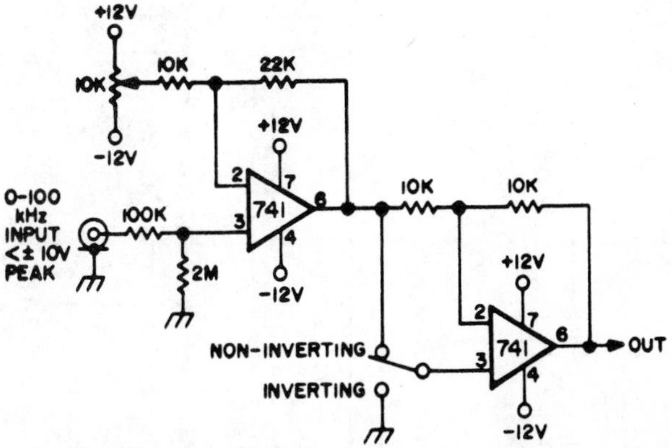


Low-speed counter, scope or audio dc input level and polarity switcher. This circuit will change the level of the input signal by means of the pot. The polarity (rise and fall) may be inverted by means of the switch.

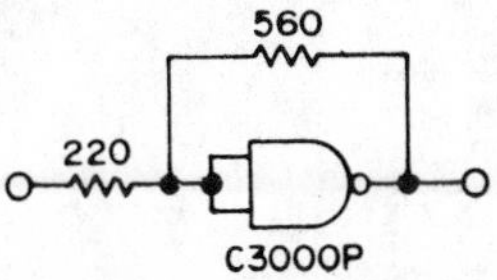


How to linearize TTL gates with a feedback bias resistor.

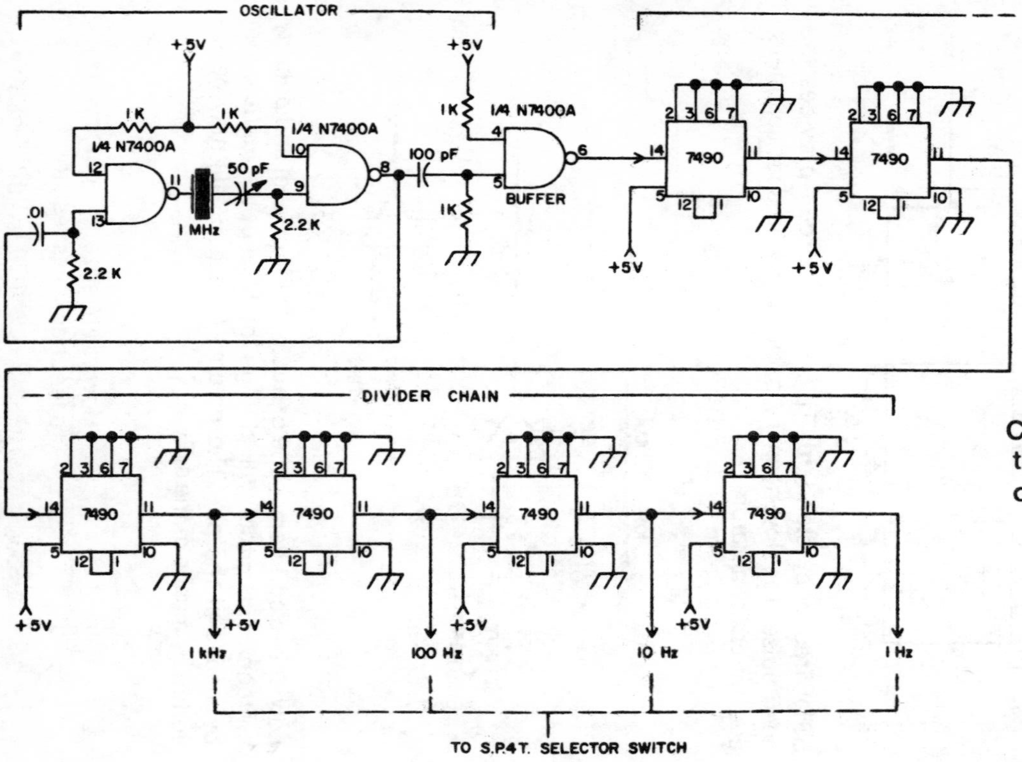


Clock generates an accurate timing signal of 1, 0.1, 0.01, or 0.001 second.

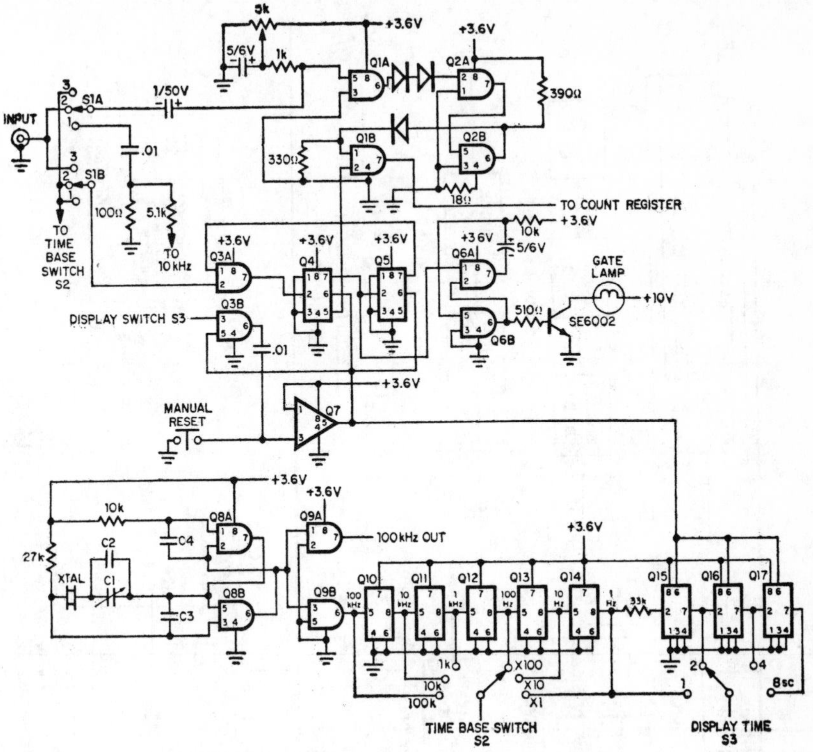


| | |
|---|---|
| Q1,2,3,6,8,9 | = μL914 |
| Q4,5,15,16,17 | = μL923 |
| Q7 | = μL900 |
| Q10–Q15 | = μL958 |
| S1, pos. 1 | = PERIOD |
| S1, pos. 2 | = FREQ. |
| S1, pos. 3 | = CHECK |
| S1 | = 2 pole, 3 position |

| | |
|---|---|
| S2 | = 1 pole, 6 position |
| S3 | = 1 pole, 4 position |
| Xtal | = 100kHz |
| C1 | = 2–18 pF |
| C2 | = 24 pF |
| C3 | = 180 pF |
| C4 | = .002 μF |

Counter operates up to 10 MHz with 0.001% tolerance of error. Output is binary, for bank of panel lamps.

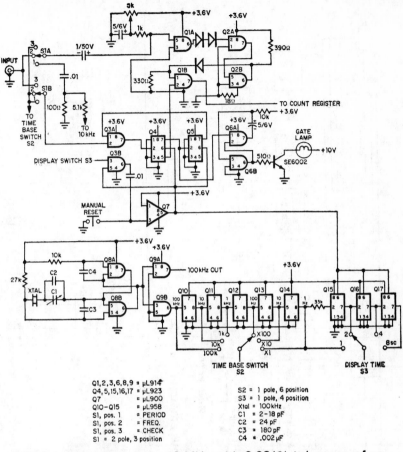

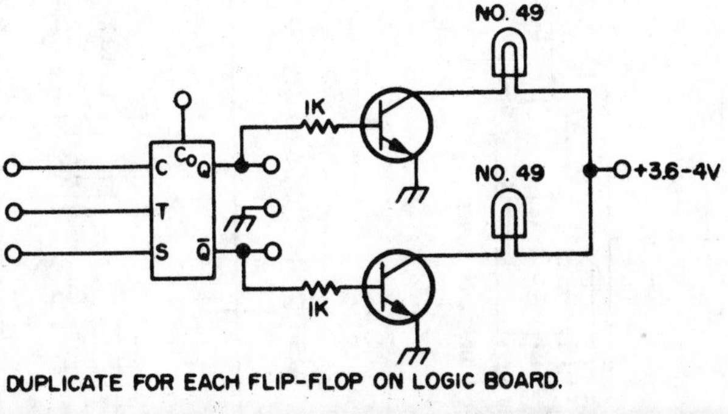


DUPLICATE FOR EACH FLIP-FLOP ON LOGIC BOARD.

Sample log c board flip-flop with pilot light display.

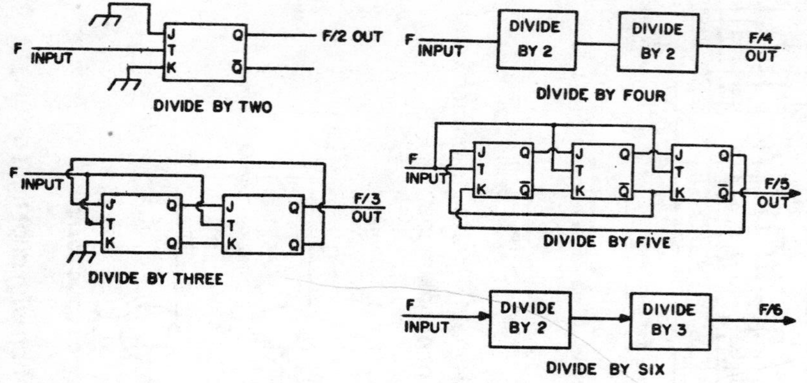


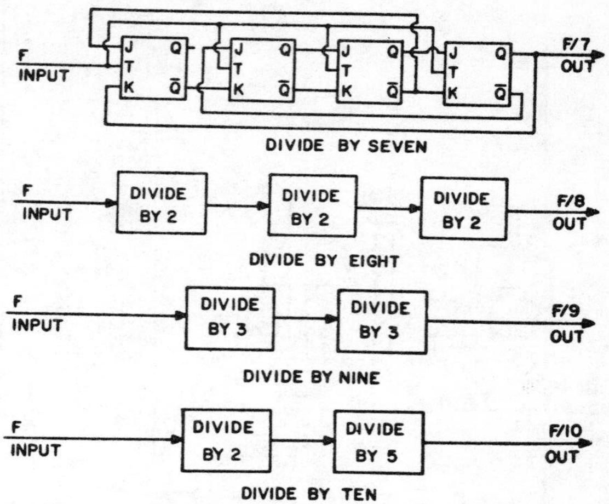


Using flip-flops to divide by various amounts.

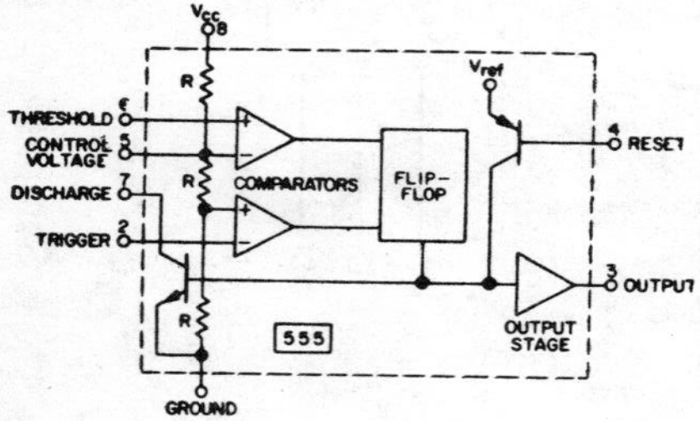


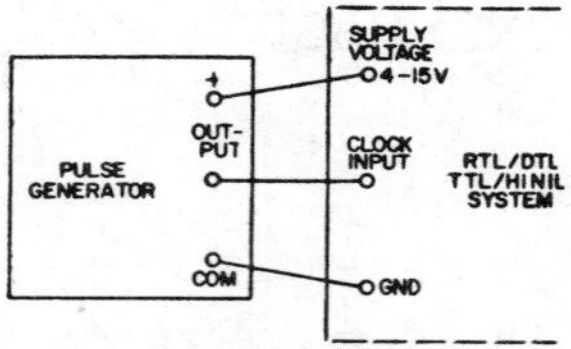


CONNECTING THE PULSE GENERATOR TO A DIGITAL SYSTEM

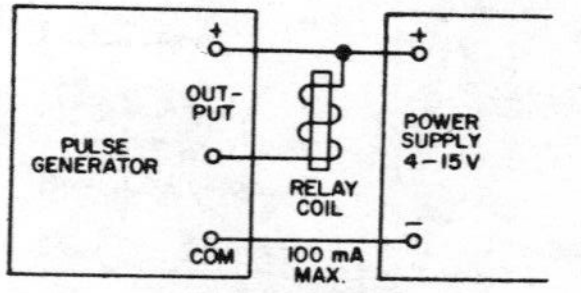


DRIVING A RELAY

This pulse generator is a variable clock testing digital IC systems, especially at low clocking rates. Due to the wide range supply voltage allowed for the SE/NE555 it can be used with RTL, DTL, TTL, and HiNil. It uses anything between +4 and +15V. This circuit comprises three 555 timers. The first 555 is connected as an astable clock. The leading dege of the negative output pulse is used to trigger the second 555. This delivers a positive-going pulse used as the positive output. The third 555 is connected as an inverter to generate the negative-going output pulse. This pulse appears about 4 $\mu$sec after the positive pulse has started. The rise and fall times are about 100 nsec. With a load of 1K between output and $+V_{CC}$ the current drawn by the complete circuit is 17 mA at 5V, and 52 mA at 15V. Table shows the capacitor values and the corresponding PRI and PL ranges achieved. Sketches above show 3 applications.

| C: | PRI | PL |
|---|---|---|
| .0022 µF | 70 µS – 1 6 mS | 15 µS – 580 µS |
| .0047 µF | 130 µS – 3,7 mS | 34 µS – 1,3 mS |
| .01 µF | 280 µS – 8,1 mS | 71 µS – 2,8 mS |
| .022 µF | 600 µS – 17 mS | 165 µS – 6,6 mS |
| .047 µF | 1,4 mS – 40 mS | 350 µS – 14 mS |
| .1 µF | 2,8 mS – 78 mS | 770 µS – 30 mS |
| .22 µF | 6,1 mS – 172 mS | 1,7 mS – 67 mS |
| .47 µF | 12 mS – 360 mS | 3,1 mS – 122 mS |
| 1 µF | 30 mS – 810 mS | 8 mS – 330 mS |
| 2,2 µF | 62 mS – 1,76 S | 17 mS – 700 mS |

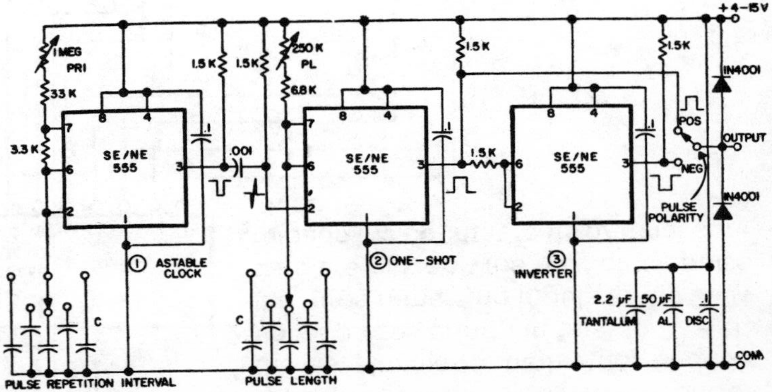

Programmable counter counts in modulo $2^n$, where n is the programmable input. Input n drives the selected output low so that when a parallel load occurs, all highs are written into the register except at the stage represented by the address n. At condition 0000111111111, the terminal count of the last stage goes high. After 14 additional pulses bring the total to $2^n - 1$, the three remaining inputs to the 9004 gate are hih, and the next clock pulse reloads the counter to its original condition.

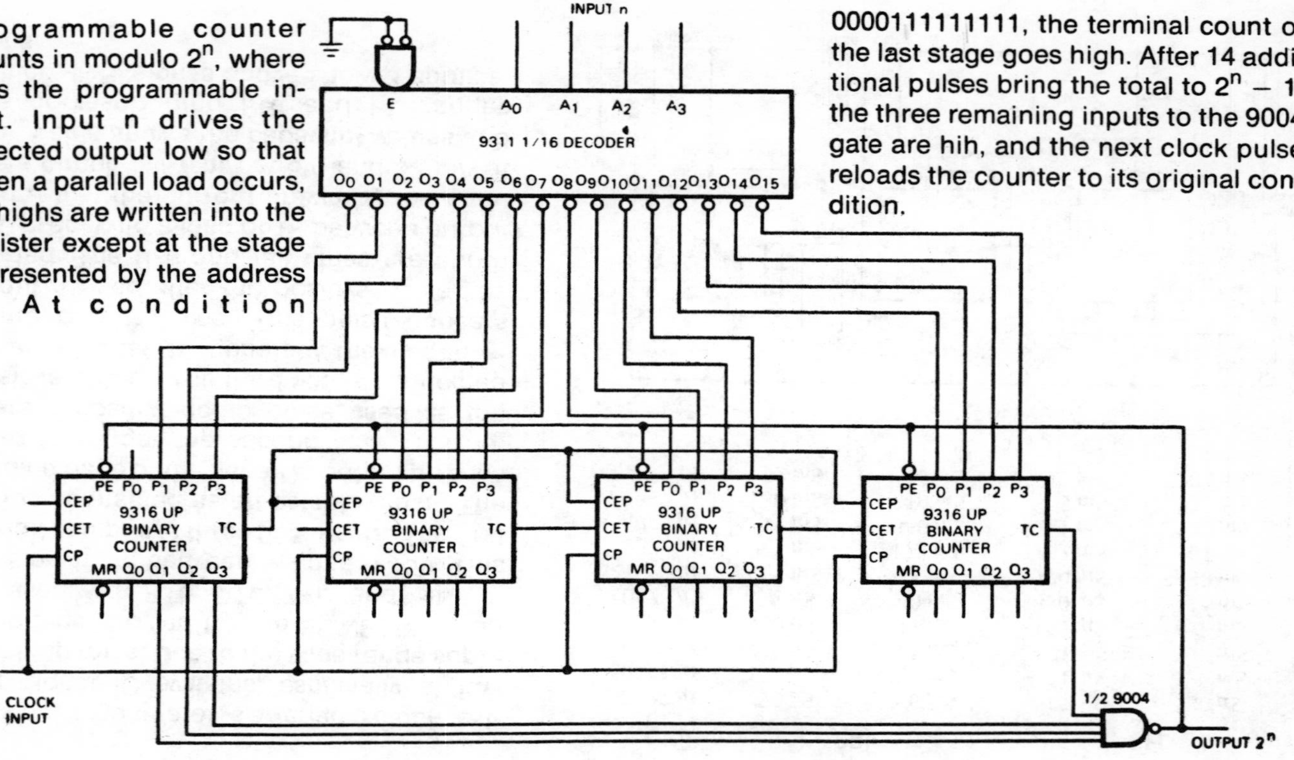

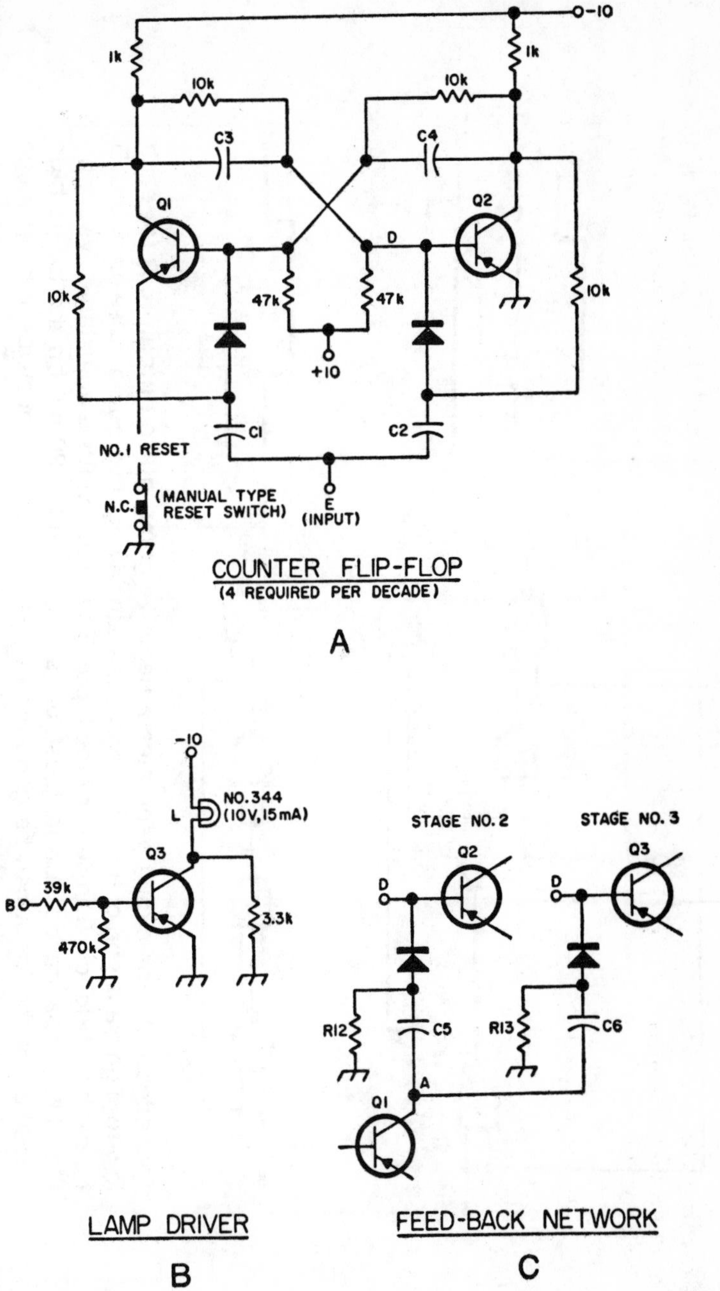


Counter flip-flop, lamp driver, and feedback network.

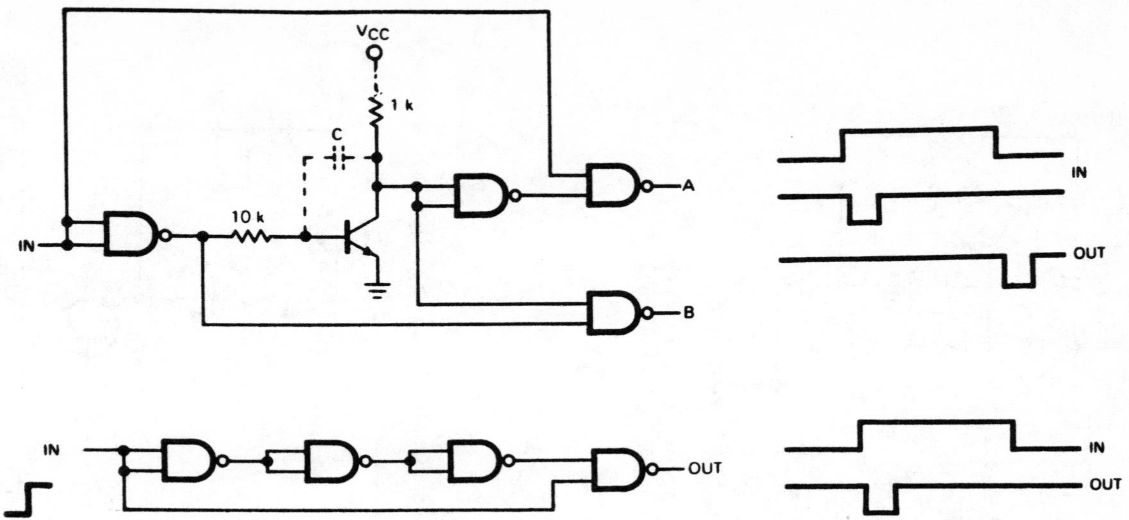


This edge detector circuit generates a negative-going pulse on output A for each low-to-high transition of the input, and generates a negative-going pulse on output B for each high-to-low transition of the input. The pulse width is adjustable by varying the Miller capacitance. A nonadjustable short pulse ($\approx$ 20 ns) on the low-to-high transition of the input can be generated by replacing the transistor inverter stage with the unused fourth NAND gate.

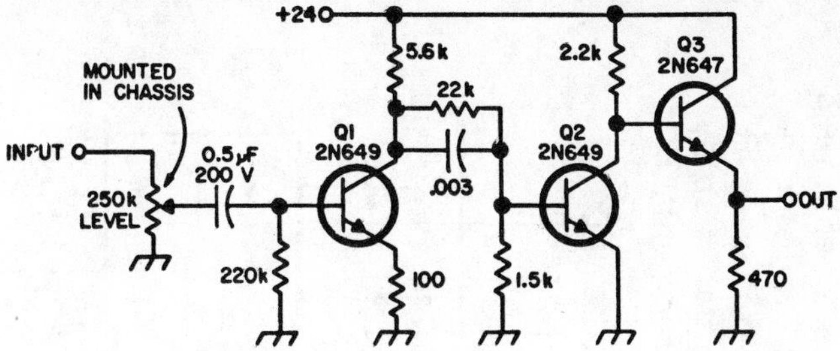


Pulse shaper.

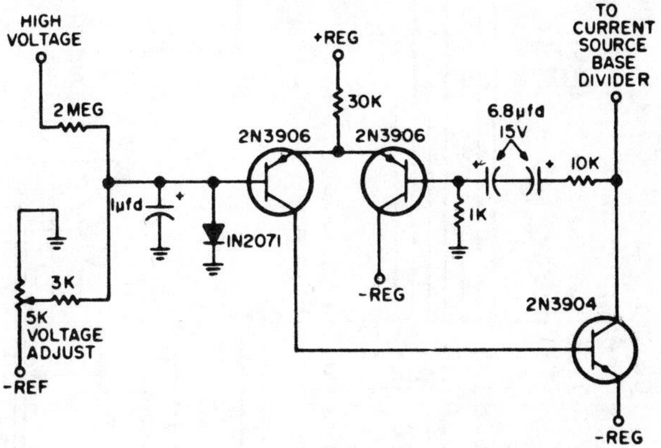


Cheaper differential amplifier.

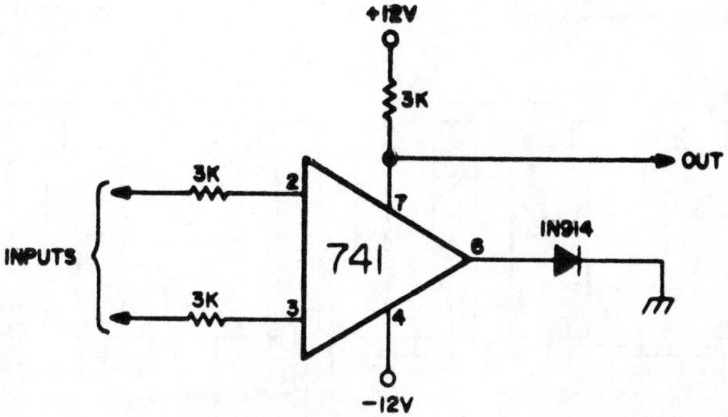


741 used as high-speed comparator. The output of this circuit varies from +12 to +2 volts and back with extremely fast rise and fall times.

This circuit stores 9 digits bit-serially in a 36-bit shift register comprised of two 9328 dual 8-bit shift registers and a 9300 4-bit universal shift register incrementing or decrementing with an exclusive-OR. This counter offers very economical display multiplexing and is shown driving 7-segment LED

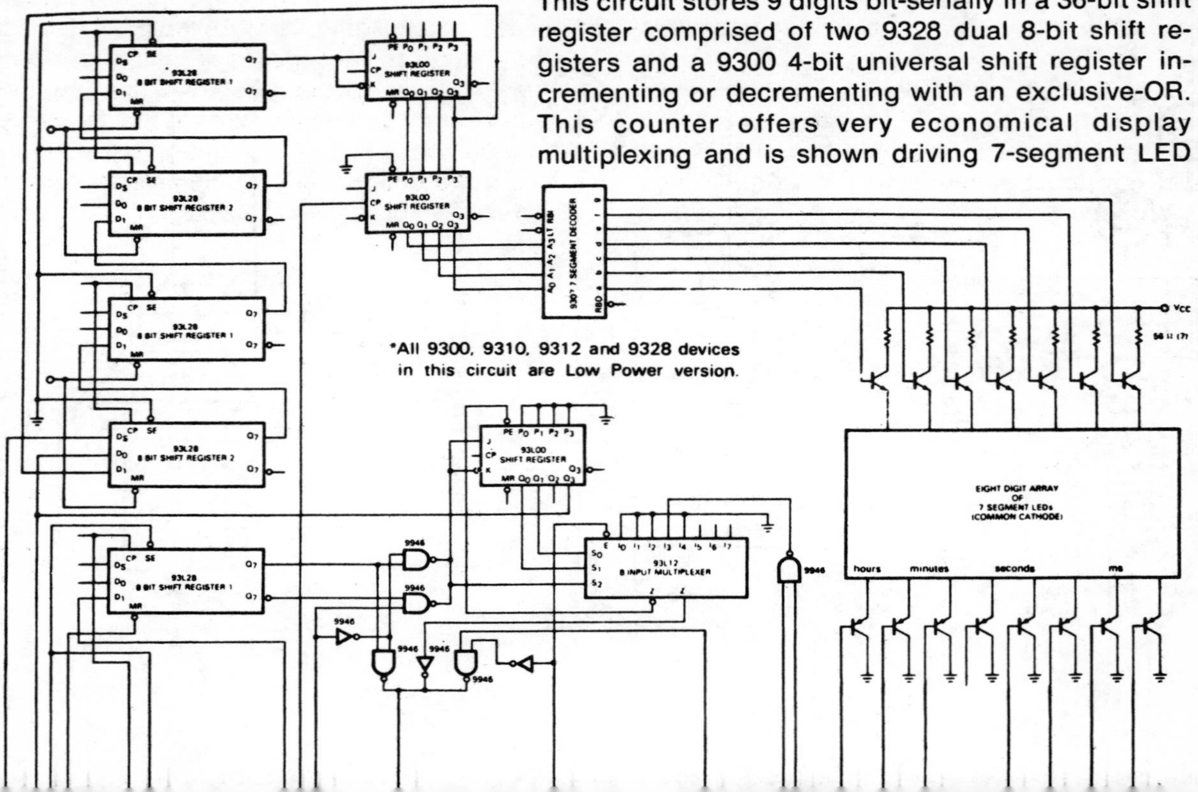


*All 9300, 9310, 9312 and 9328 devices in this circuit are Low Power version.

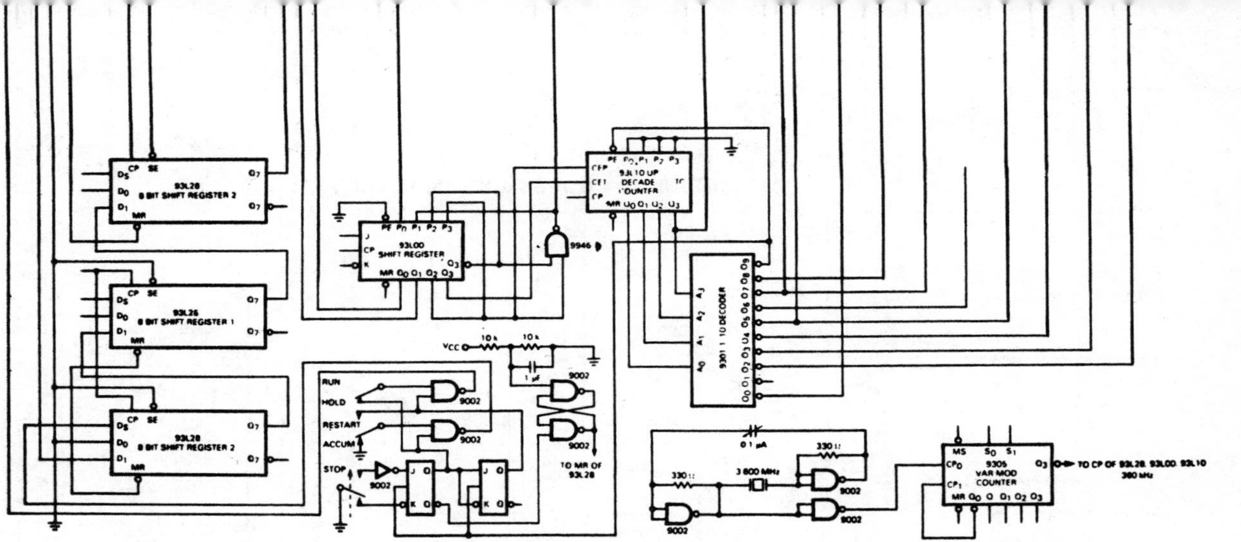


displays. This circuit operates as a crystal-controlled stopwatch, displaying milliseconds to hours. The time counter is a 36-bit (9-digit) bit/serial incrementer controlled by a 3.6 MHz crystal oscillator and time base so that the 10-second and 10-minute digits are counted modulo 6. A second set of shift registers stores display data independently when the **stop** contact is activated. The contents of the storage register are strobed every four clock pulses into a 93L00 feeding 9307 7-segment decoder. This decoder, through current-limited buffers, drives the anodes of the 8-digit LED display matrix. Since this counter requires 36 clock pulses to increment the least significant digit, the shift frequency is 360 kHz, derived from a 3.6 MHz oscillator through a 9305 decade counter. In this case, the low count rate inherent to serial increments is advantageous, resulting in a shorter divide chain for the time base.

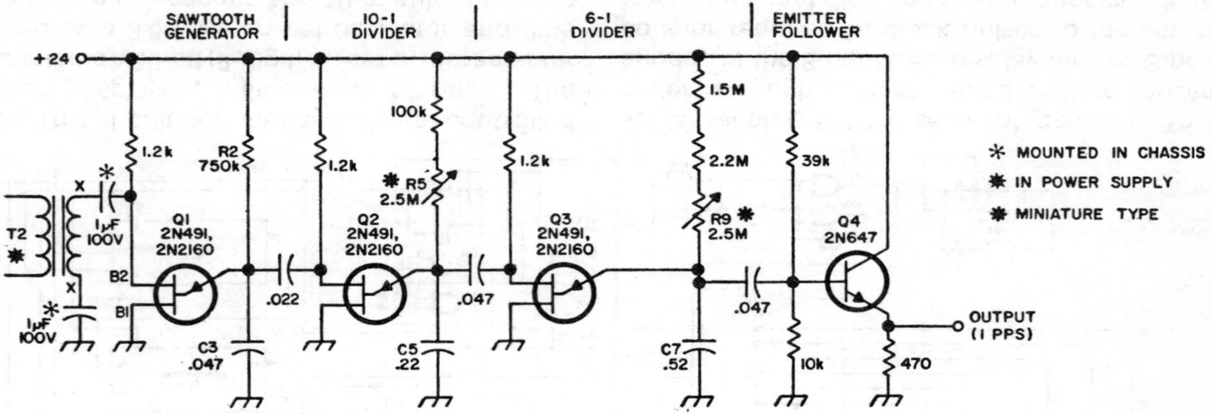


One-pulse-per-second generator.

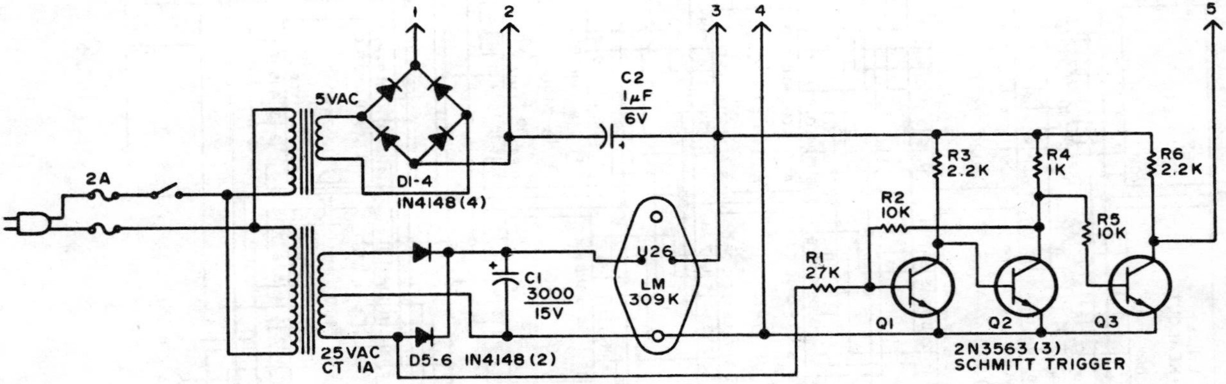


Counter power supply. Connections are made via the points numbered 1—5.

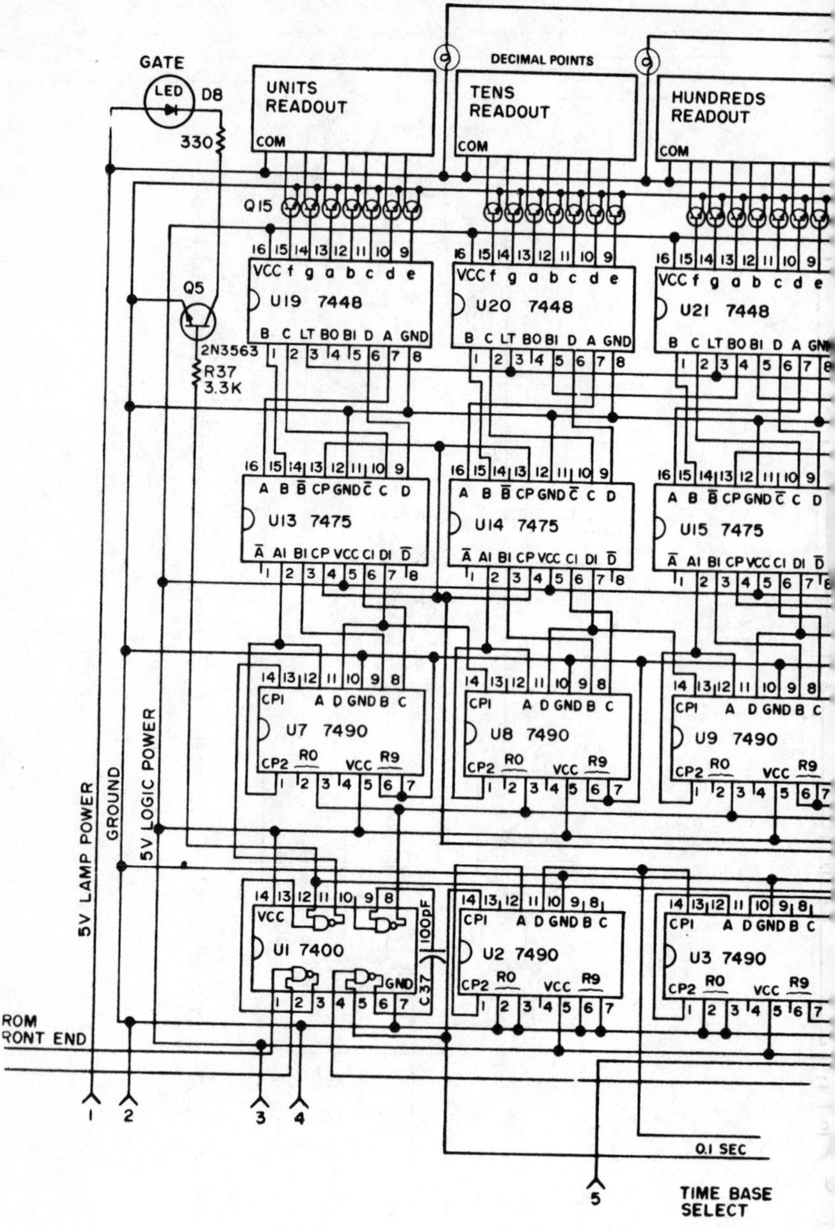
GATE
LED
D8
330
UNITS READOUT
TENS READOUT
HUNDREDS READOUT
DECIMAL POINTS
COM
Q15
Q5
2N3563
R37
3.3K
U19 7448
U20 7448
U21 7448
U13 7475
U14 7475
U15 7475
U7 7490
U8 7490
U9 7490
U1 7400
U2 7490
U3 7490
C37
100pF
5V LAMP POWER
GROUND
5V LOGIC POWER
ROM
RONT END
0.1 SEC
TIME BASE SELECT

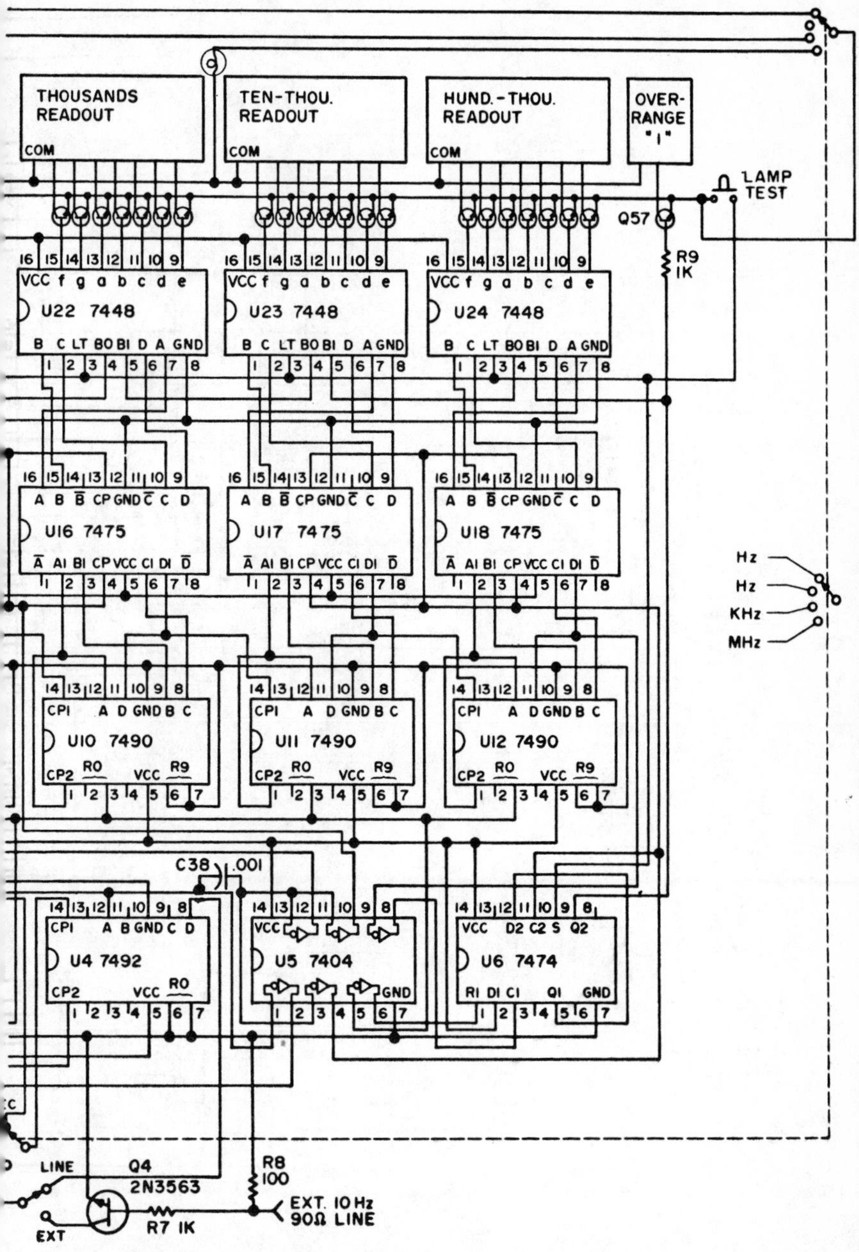


Schematic of frequency counter. Note: C3—C26 are not shown. These are 0.01 disc ceramics connected directly from $V_{CC}$ to ground at each IN.

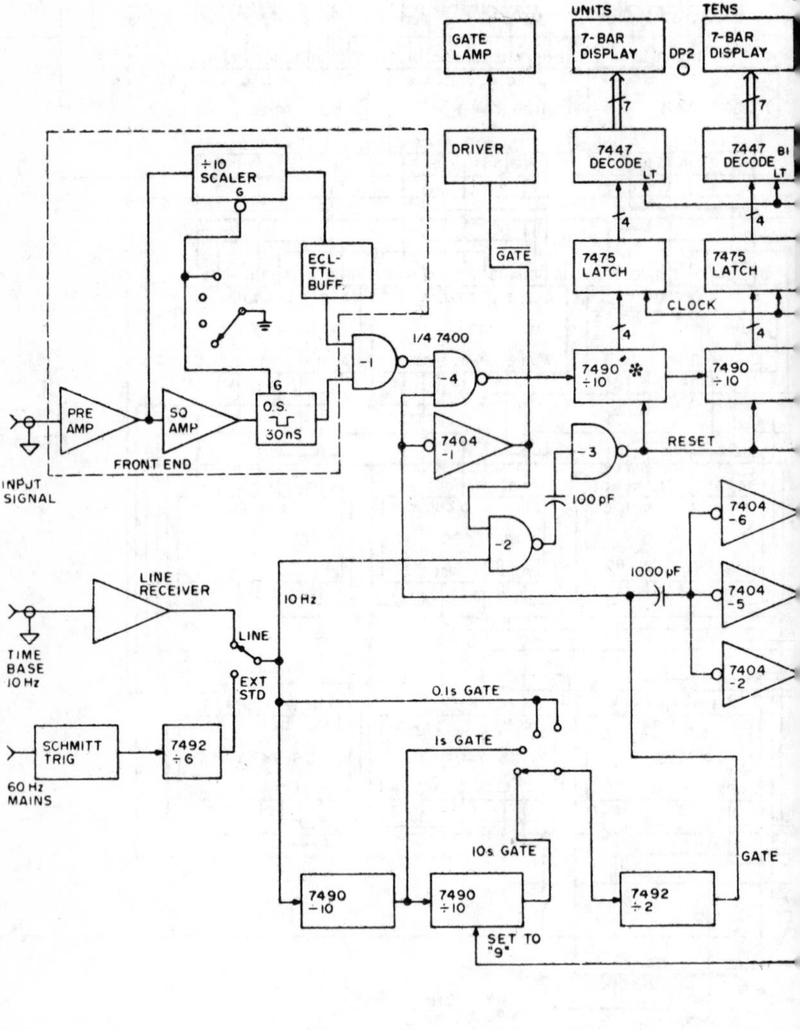

GATE LAMP
DRIVER
GATE
UNITS
7-BAR DISPLAY
DP2
7
7447 DECODE
LT
4
7475 LATCH
CLOCK
7490 ÷10 *
TENS
7-BAR DISPLAY
7447 DECODE
BI
LT
7475 LATCH
7490 ÷10
÷10 SCALER
G
ECL-TTL BUFF.
G
O.S. 30nS
PRE AMP
SQ AMP
FRONT END
INPUT SIGNAL
-1
1/4 7400
-4
7404 -1
-3
RESET
100 pF
-2
7404 -6
7404 -5
7404 -2
1000 pF
LINE RECEIVER
TIME BASE 10 Hz
LINE
EXT STD
10 Hz
SCHMITT TRIG
7492 ÷6
60 Hz MAINS
0.1s GATE
1s GATE
10s GATE
GATE
7490 ÷10
7490 ÷10
7492 ÷2
SET TO "9"

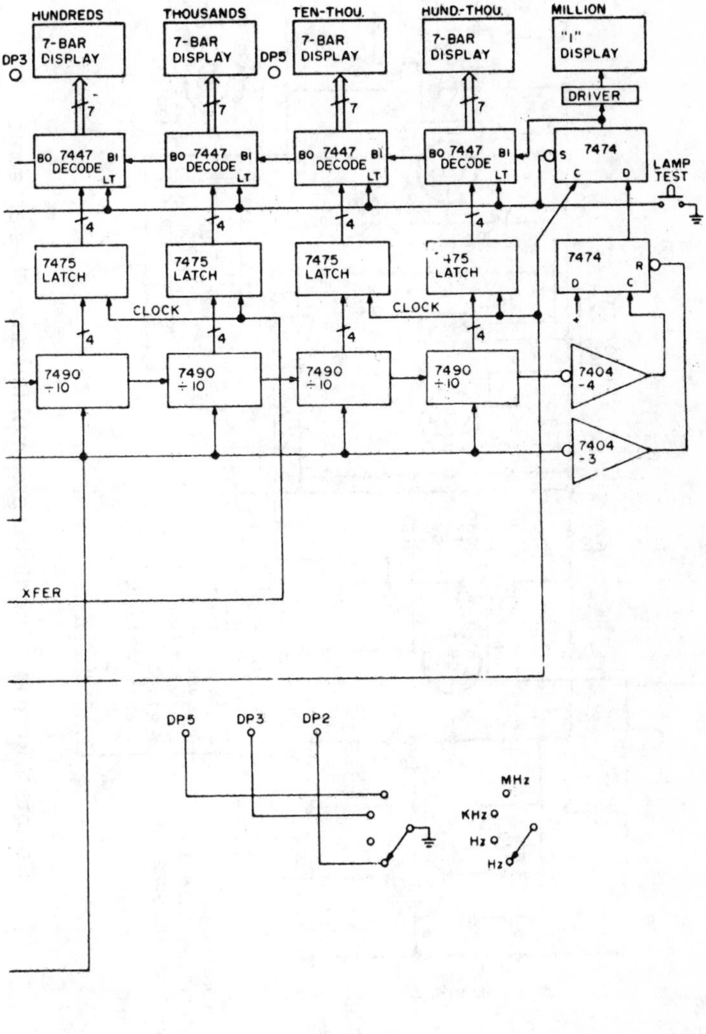


Detailed block diagram of frequency counter. The 7490 identified by * should be selected for >20 MHz switching speed.

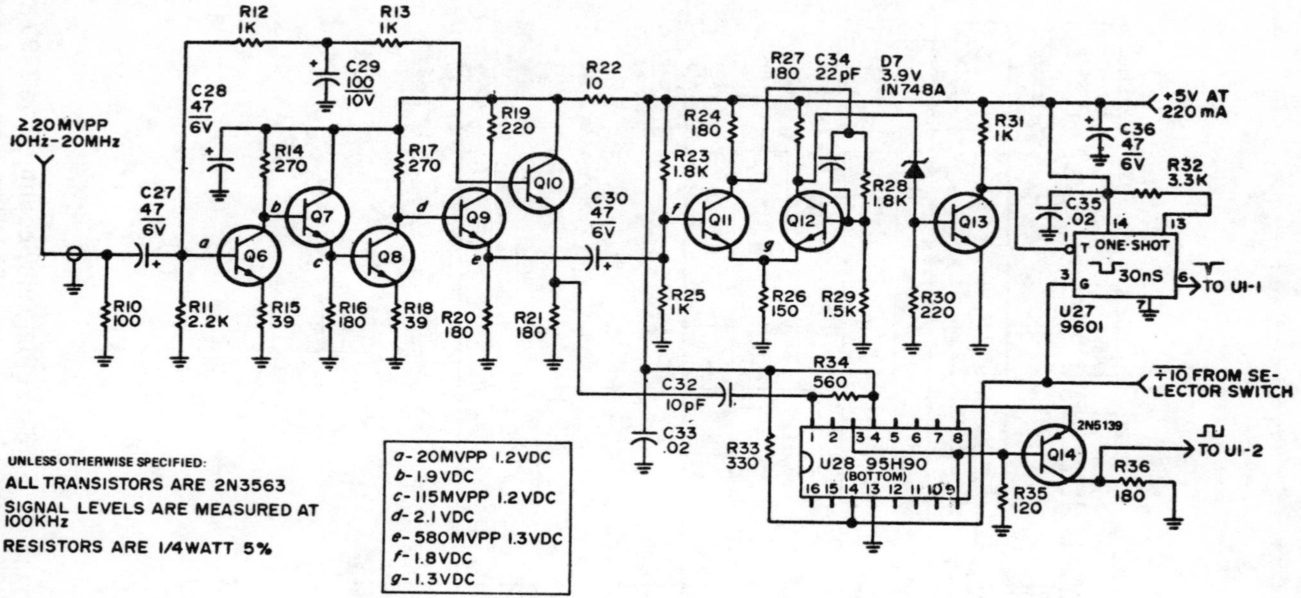


Counter "front end" contains preamplifier as well as prescaler.

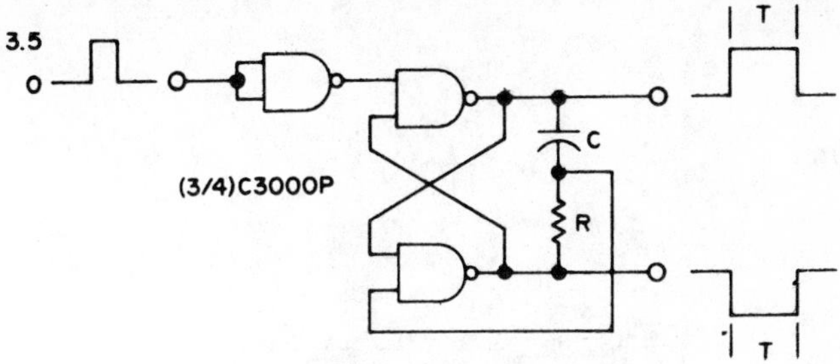


A

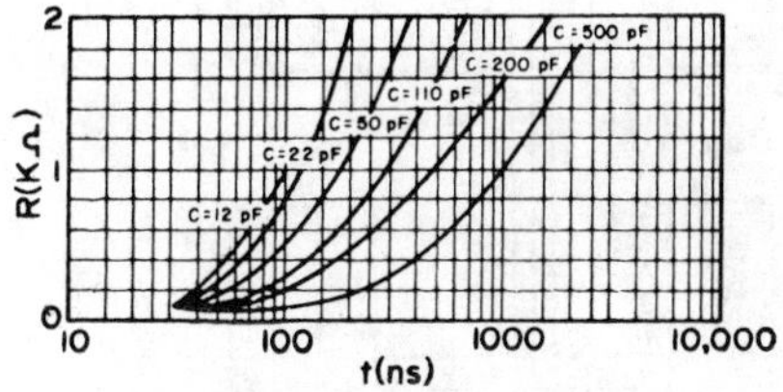


B

TTL one-shot used to develop wider pulses. Chart shows values for C and R for various output pulse widths.

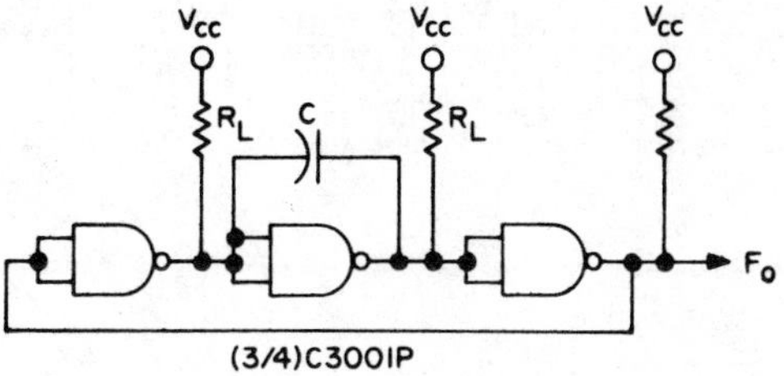


A

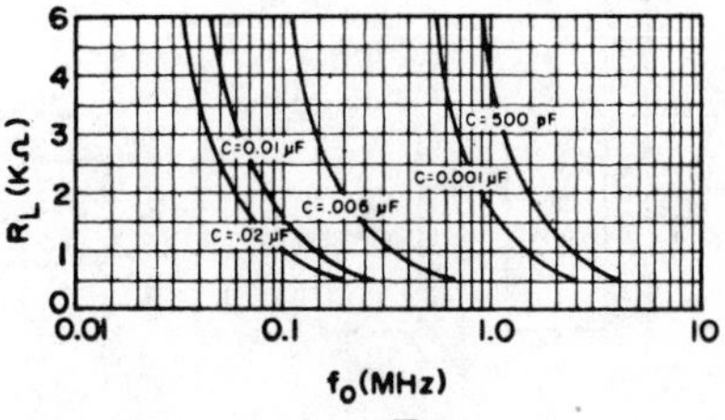


B

Square wave generator. Frequency is determined by $R_L$ and C as shown in the chart.

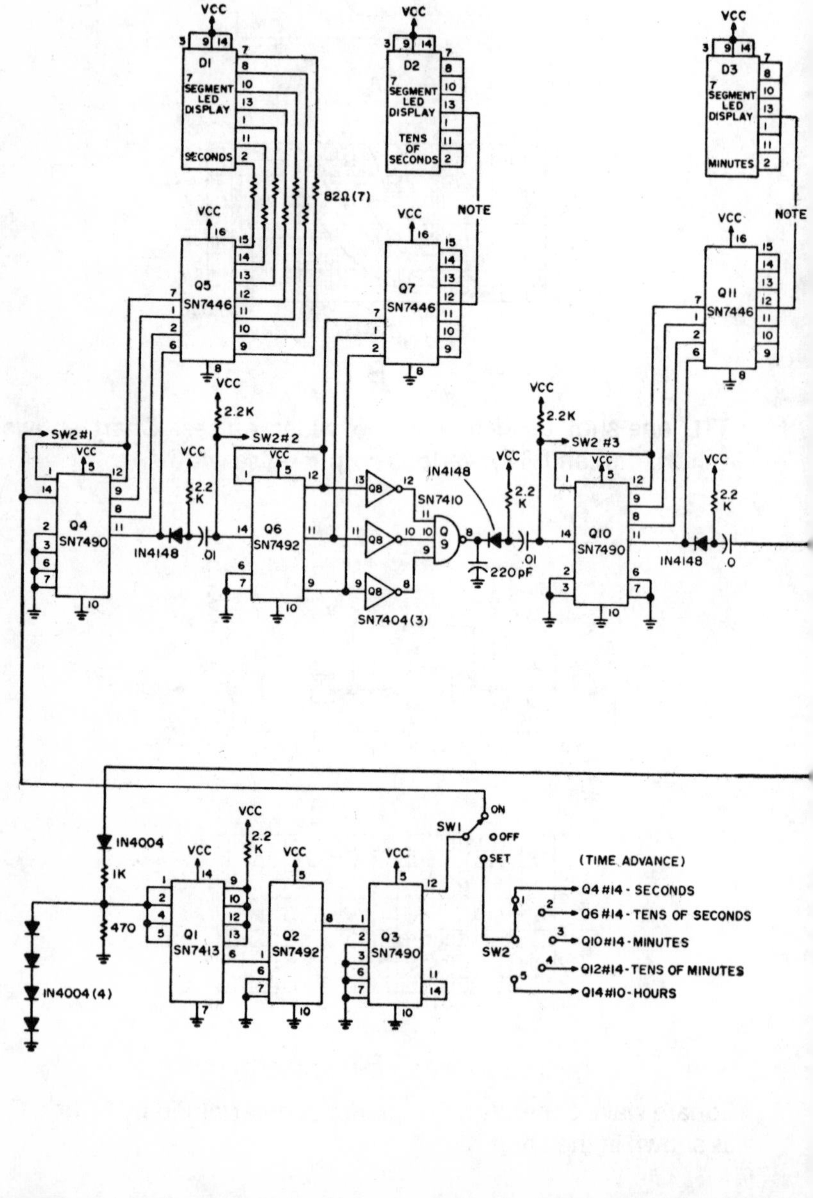
VCC
D1
7 SEGMENT LED DISPLAY
SECONDS
82Ω (7)
Q5
SN7446
D2
7 SEGMENT LED DISPLAY
TENS OF SECONDS
NOTE
Q7
SN7446
D3
7 SEGMENT LED DISPLAY
MINUTES
Q11
SN7446
2.2K
SW2#1
SW2#2
SW2 #3
Q4
SN7490
IN4148
.01
Q6
SN7492
Q8
SN7410
Q9
SN7404(3)
220 pF
Q10
SN7490
2.2 K
IN4004
1K
470
IN4004(4)
Q1
SN7413
Q2
SN7492
Q3
SN7490
SW1
ON
OFF
SET
(TIME ADVANCE)
Q4 #14 - SECONDS
Q6 #14 - TENS OF SECONDS
Q10 #14 - MINUTES
Q12#14 - TENS OF MINUTES
Q14 #10 - HOURS
SW2

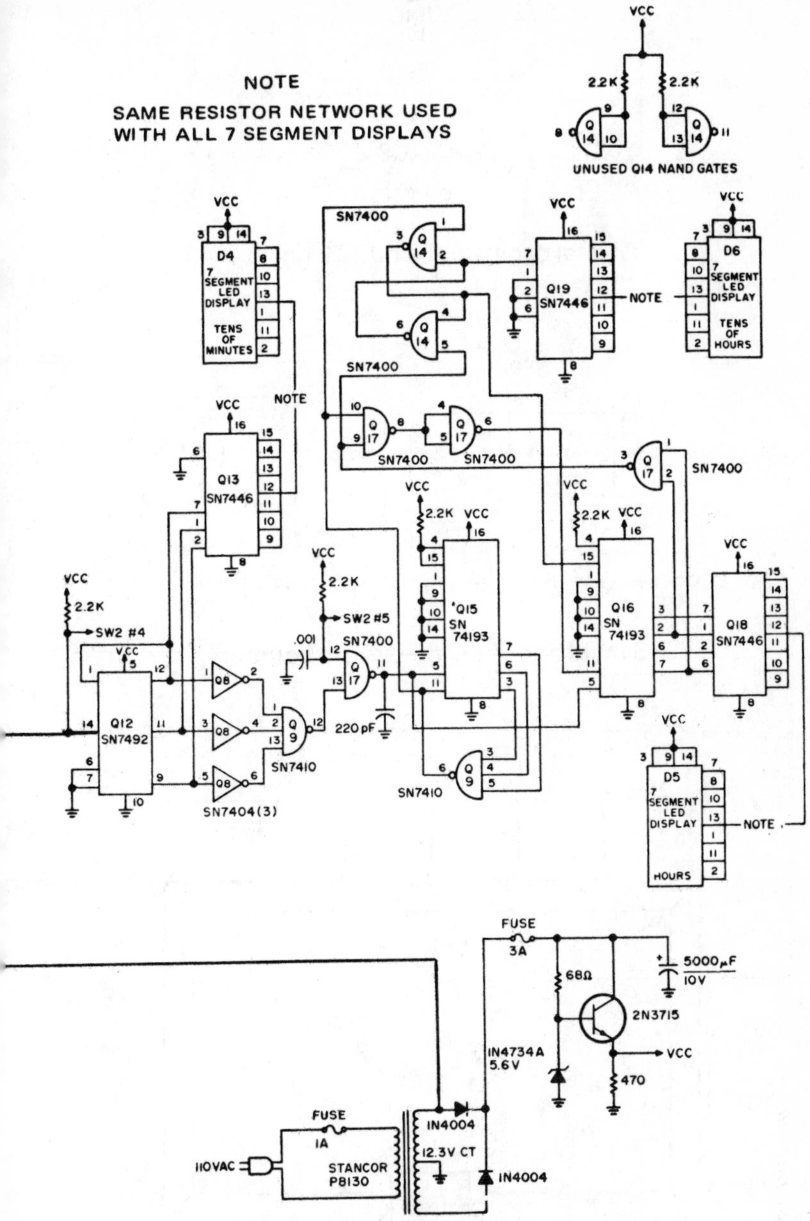


Digital clock uses 19 inexpensive ICs.

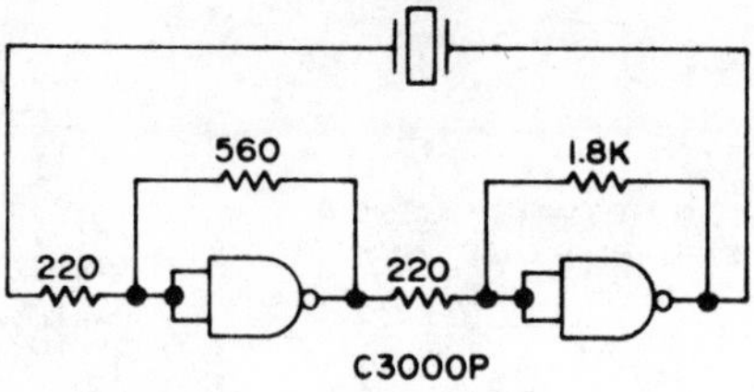


Crystal oscillator using TTL gates.

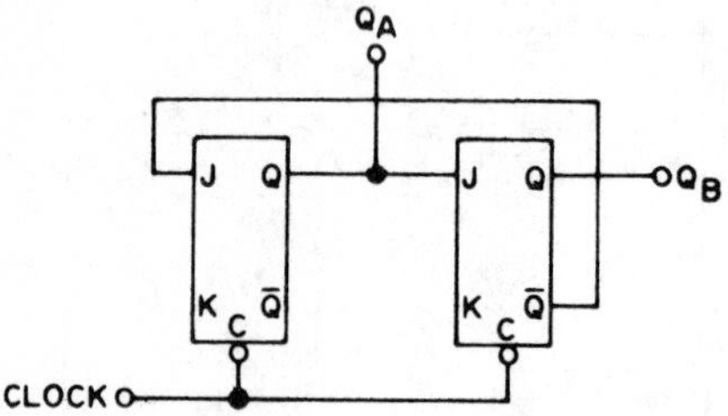


| STATE | $Q_A$ | $Q_B$ |
|---|---|---|
| 1 | 0 | 0 |
| 2 | 1 | 0 |
| 3 | 0 | 1 |

(1)C3073P

Synchronous divide-by-3 up counter.

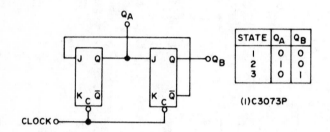

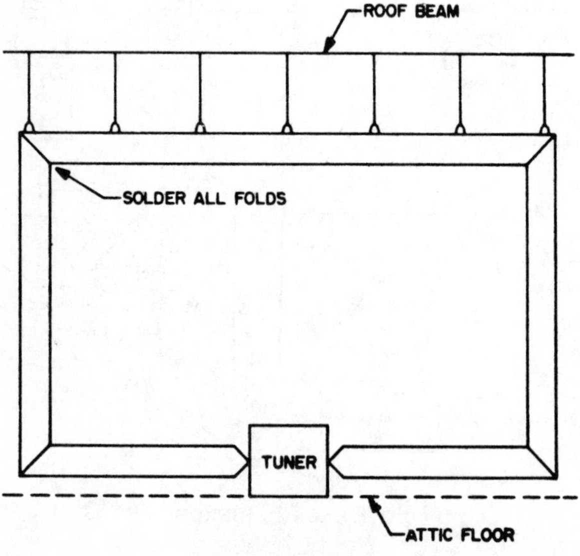


A multiband loop for suspension in an attic.

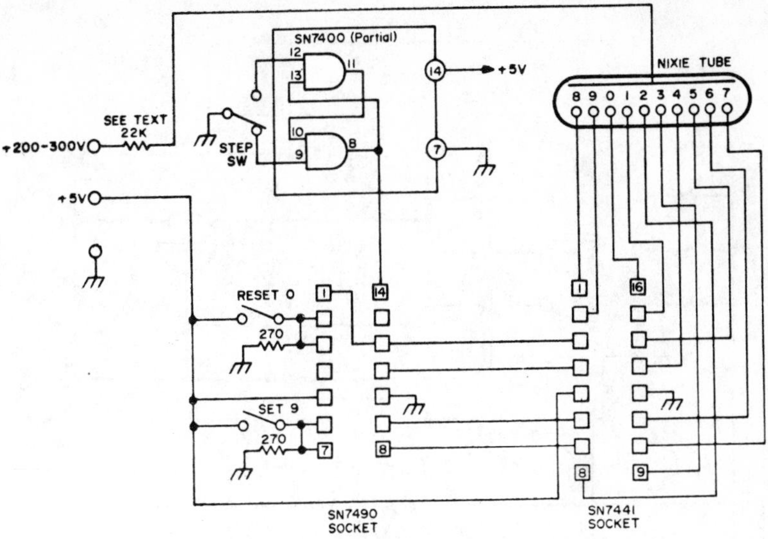


A

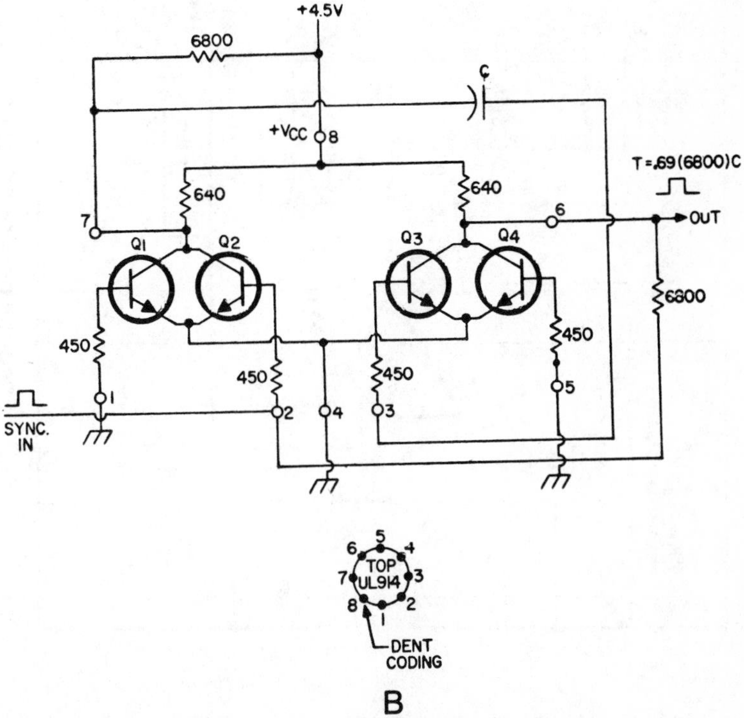


B

The 914 as a monostable multivibrator.

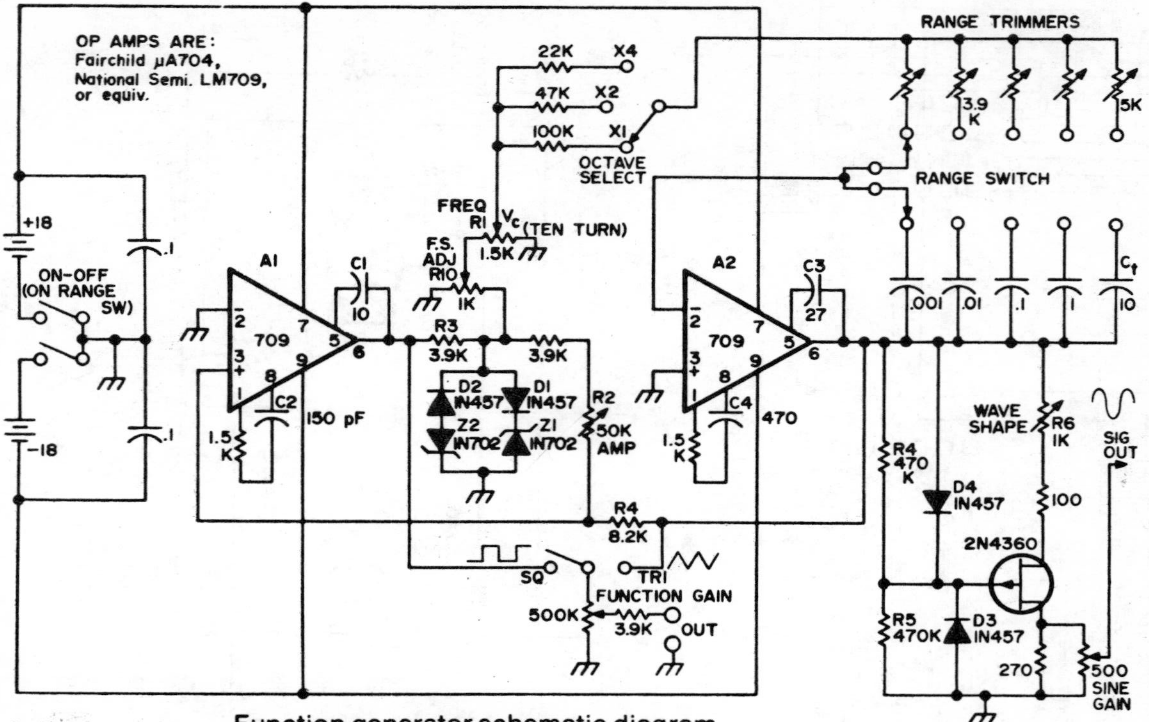

Function generator schematic diagram.

see p 108

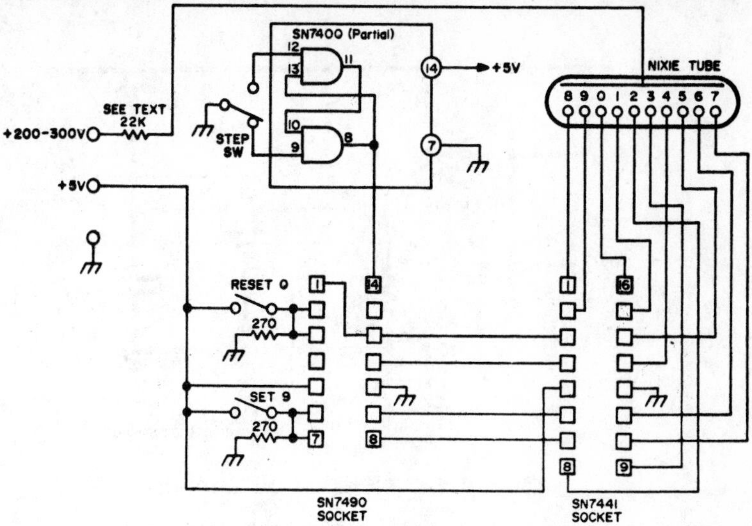


Decade counter and Nixie driver tester.

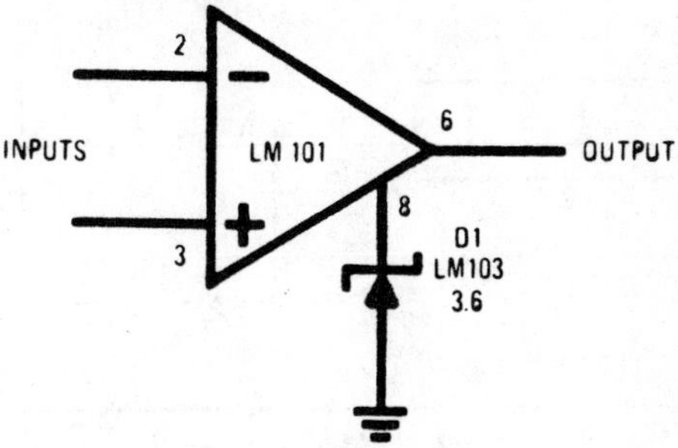


Voltage comparator for driving DTL or TTL integrated circuits.

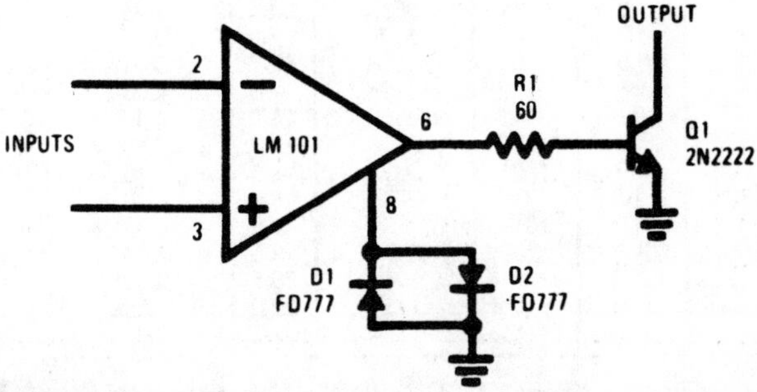


Voltage comparator for driving RTL logic or high-current driver.

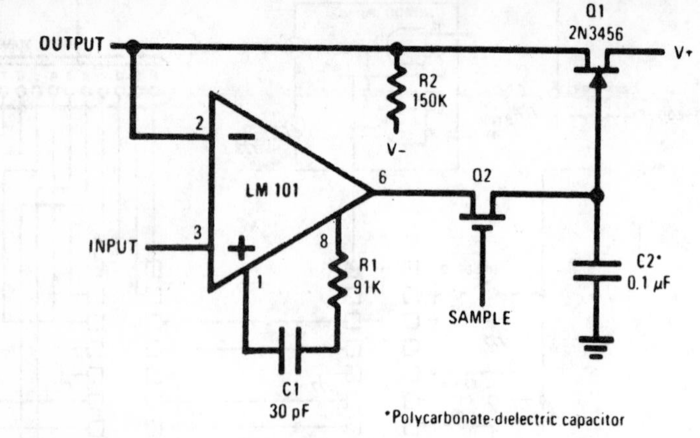


Low-drift sample-and-hold circuit.

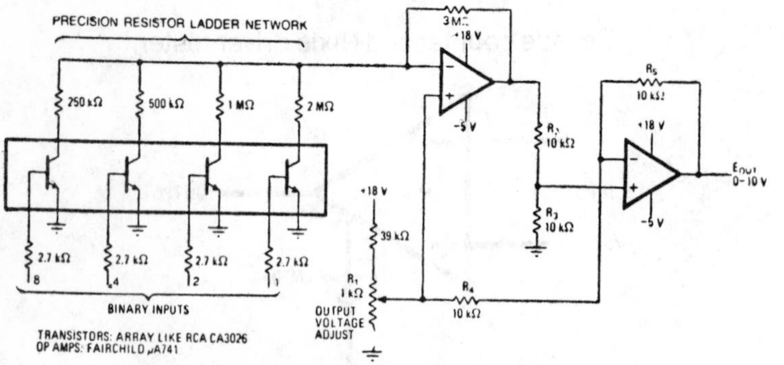


| Input Word Decimal | Binary | Ladder Resistance | Gain | Output ($V_{ref}$ = 0.425V) |
|---|---|---|---|---|
| 0 | 0000 | 0 | 1 | 0.425 |
| 1 | 0001 | 2M | 2.5 | 1.06 |
| 2 | 0010 | 1M | 4.0 | 1.70 |
| 3 | 0011 | 667K | 5.5 | 2.34 |
| 4 | 0100 | 500K | 7.0 | 2.97 |
| 5 | 0101 | 400K | 8.5 | 3.61 |
| 6 | 0110 | 333K | 10.0 | 4.25 |
| 7 | 0111 | 286K | 11.5 | 4.80 |
| 8 | 1000 | 250K | 13.0 | 5.53 |
| 9 | 1001 | 222K | 14.5 | 6.16 |
| 10 | 1010 | 200K | 16.0 | 6.80 |
| 11 | 1011 | 182K | 17.5 | 7.44 |
| 12 | 1100 | 167K | 19.0 | 8.08 |
| 13 | 1101 | 154K | 20.5 | 8.72 |
| 14 | 1110 | 143K | 22.0 | 9.35 |
| 15 | 1111 | 133K | 23.5 | 9.99 |

In this digital-to-analog converter, the 4-bit parallel binary word is programed via the transistor switches and the analog output appears as a voltage. The table explains operation.

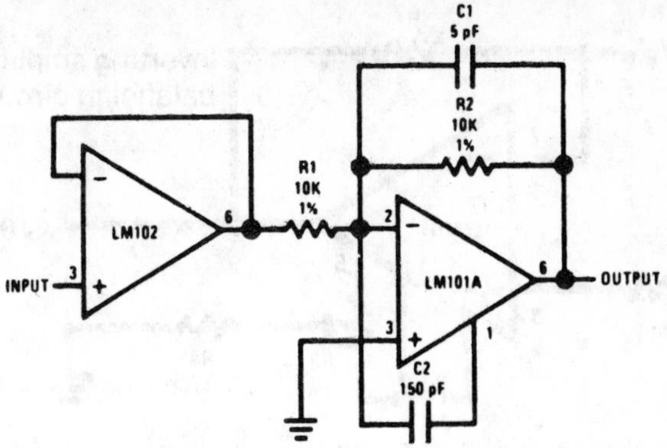


Fast inverting amplifier with high input impedance.

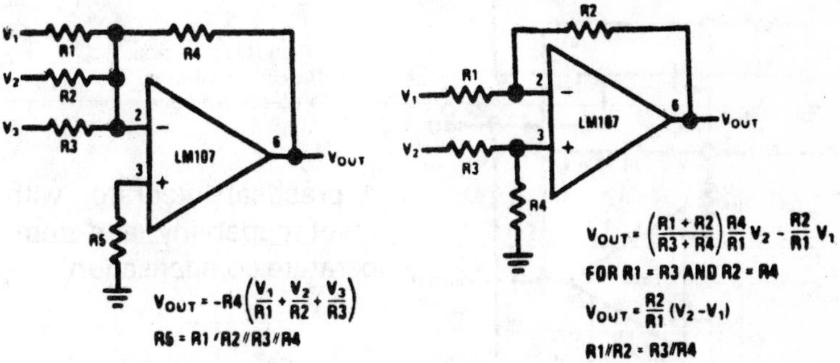


a. summer

b. difference amplifier

Two practical linear IC summer circuits.

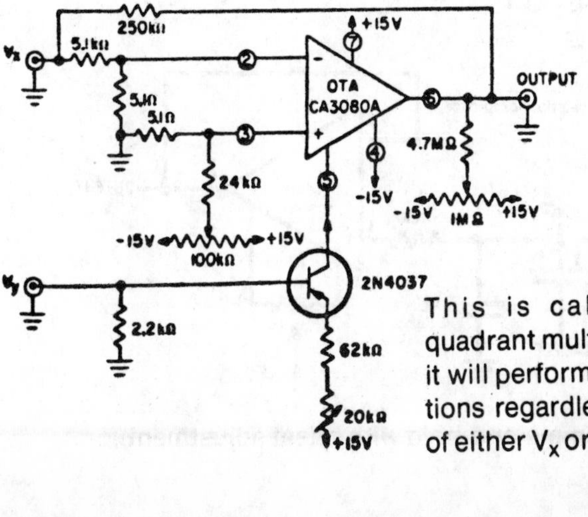


This is called a four-quadrant multiplier because it will perform its multiplications regardless of the sign of either $V_x$ or $V_y$.

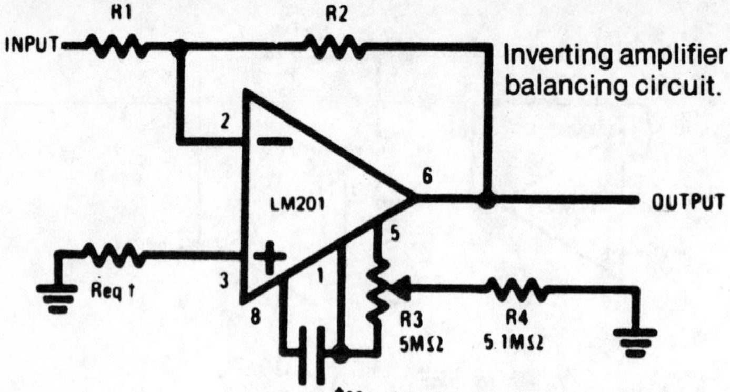


Inverting amplifier with balancing circuit.

†May be zero or equal to parallel combination of R1 and R2 for minimum offset.

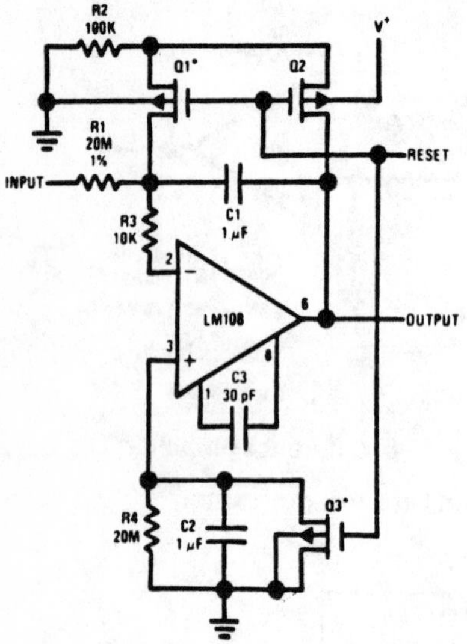


* Q1 and Q3 must not have internal gate protection diodes

A practical integrator with reset capability and temperature compensation.

*Polycarbonate-dielectric capacitor.

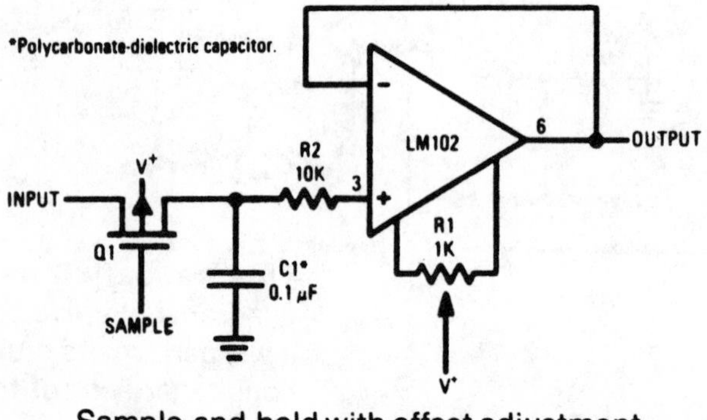


Sample-and-hold with offset adjustment.

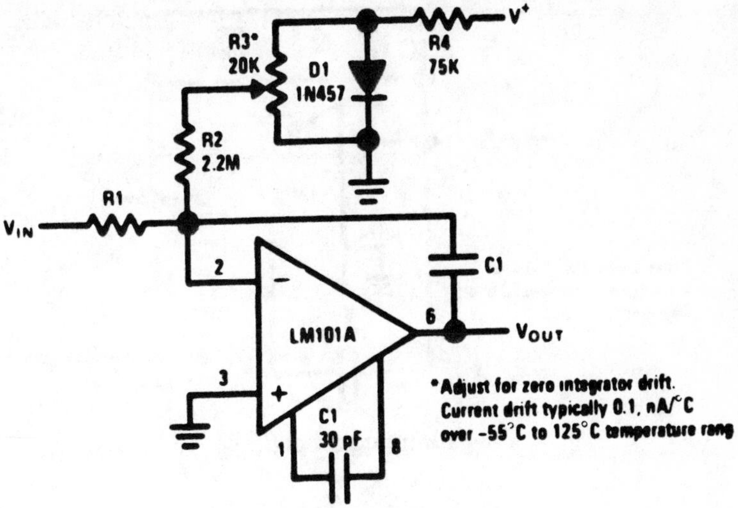


Fast voltage follower.

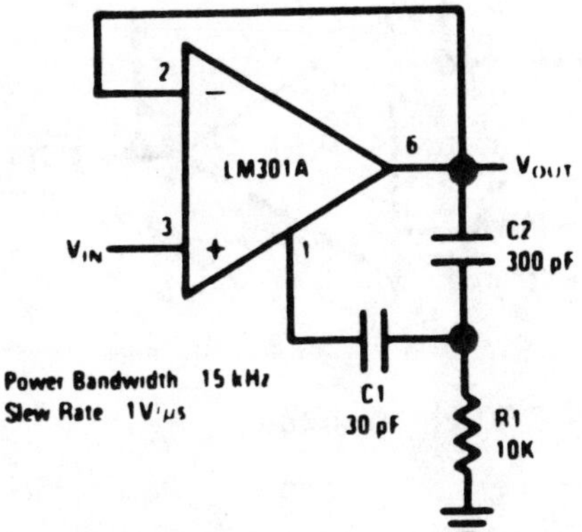


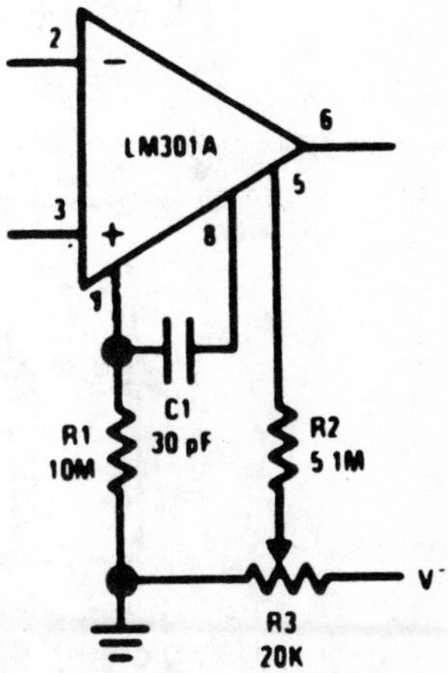


Standard compensation and offset balancing circuit.

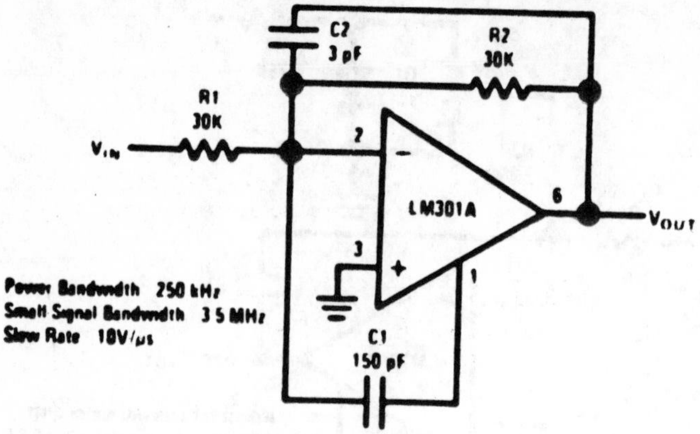


Fast summing amplifier.

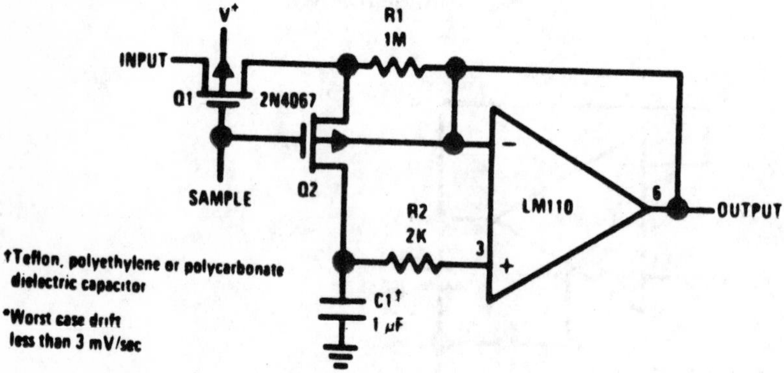


Low-drift sample-and-hold circuit.

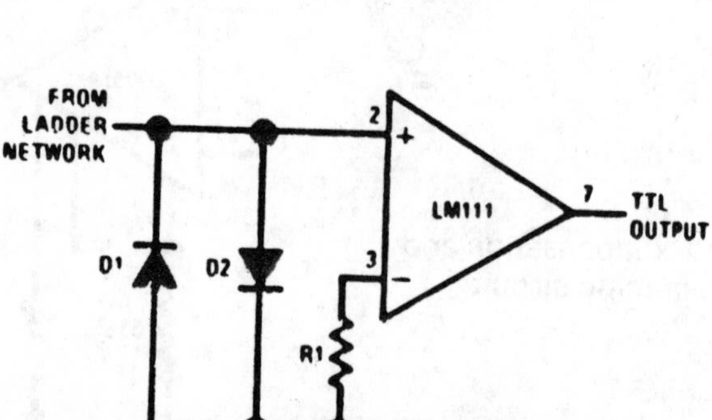


Using clamp diodes to improve response.

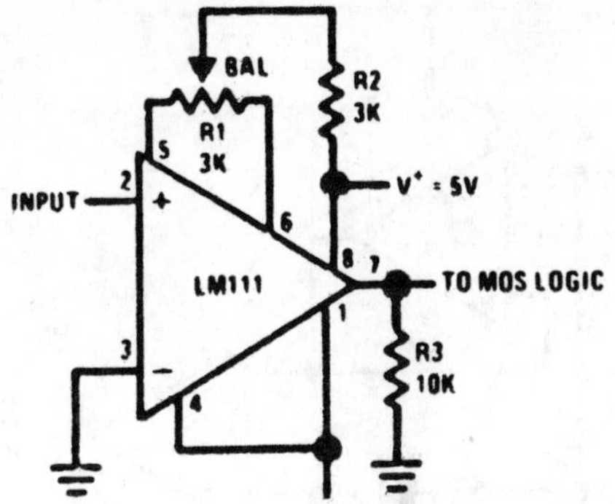


Zero-crossing detector driving MOS logic.

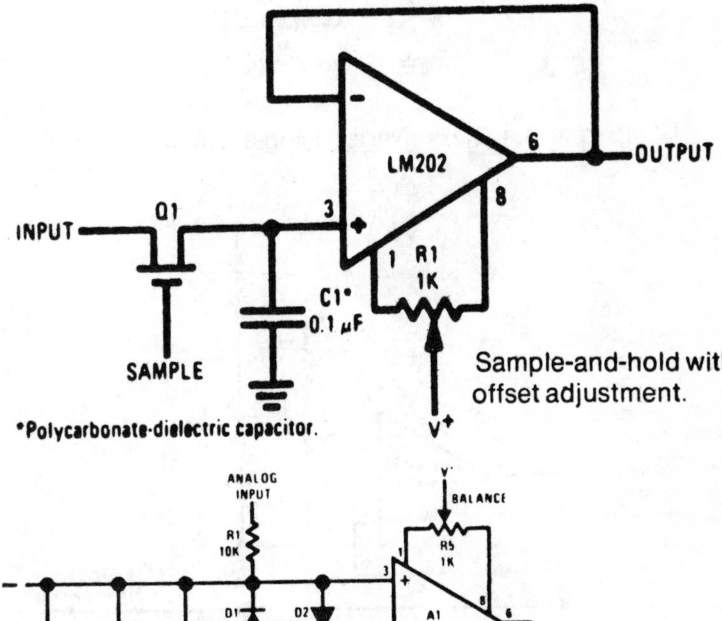


Sample-and-hold with offset adjustment.

*Polycarbonate-dielectric capacitor.

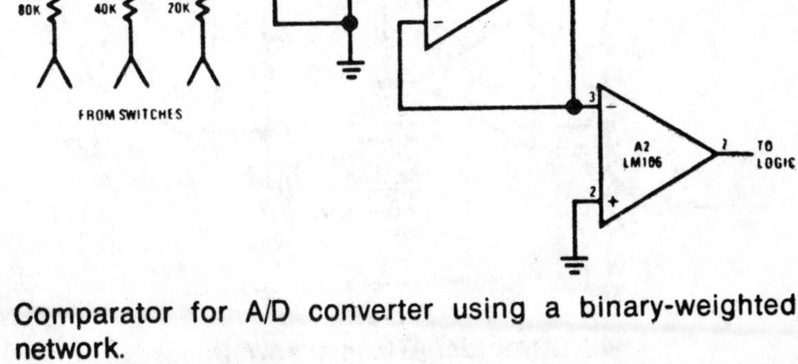


Comparator for A/D converter using a binary-weighted network.

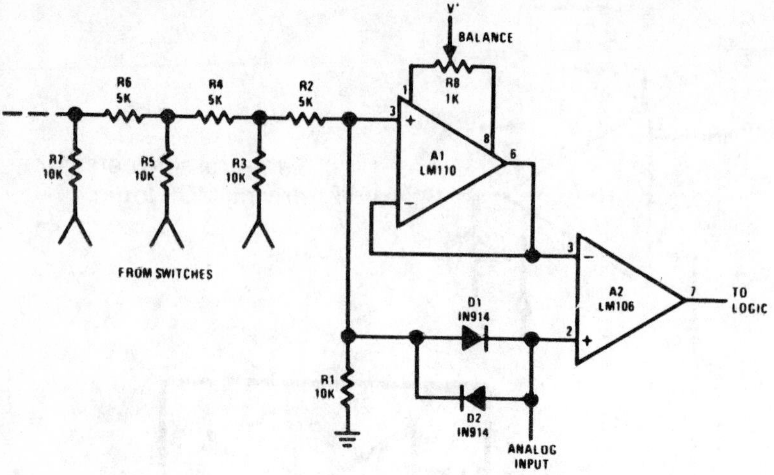


Comparator for A/D converter using a ladder network.

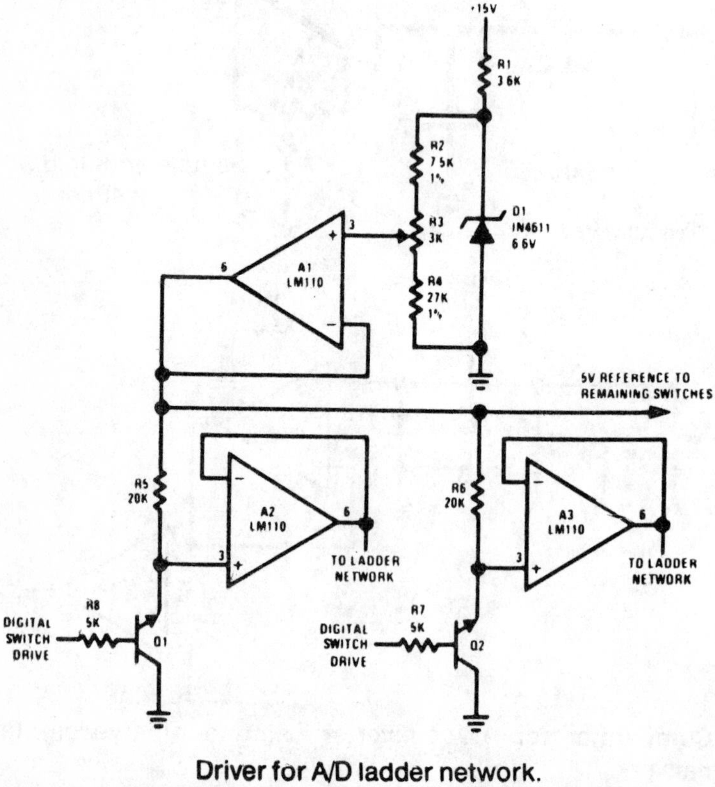


Driver for A/D ladder network.

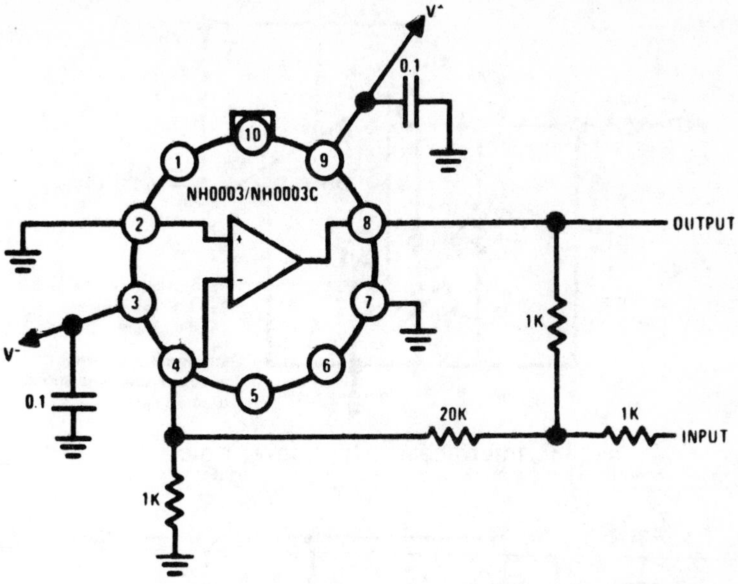


High-slew-rate unity-gain inverting amplifier.

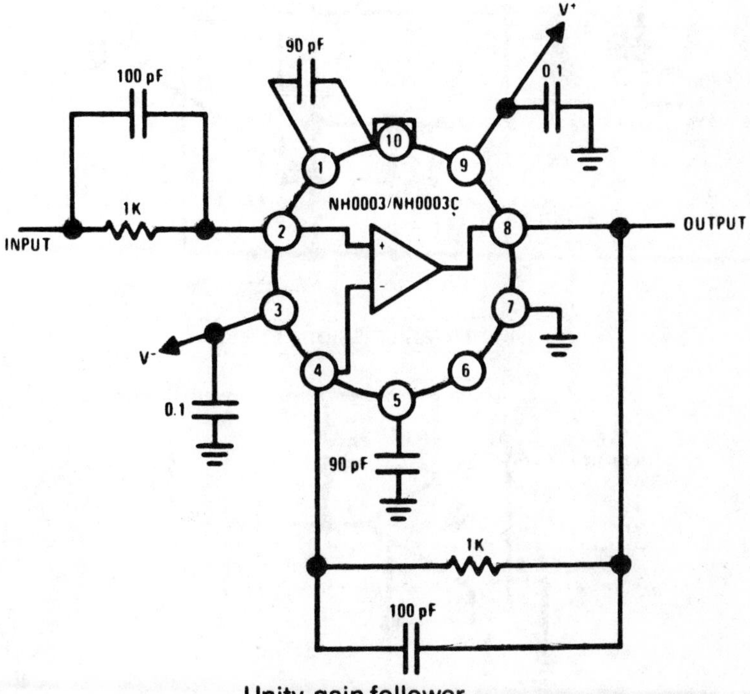


Unity-gain follower.

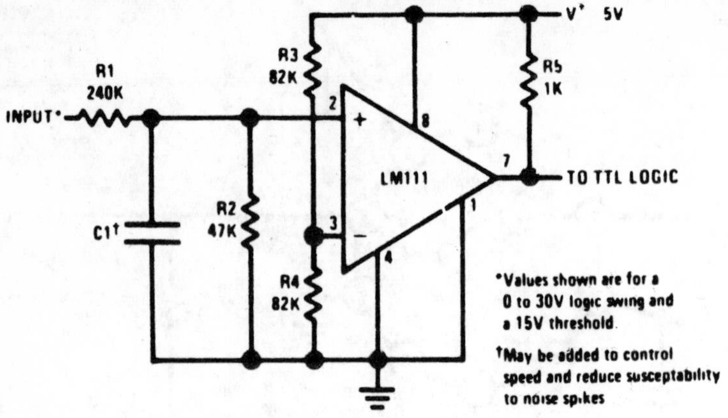


TTL interface with high-level logic.

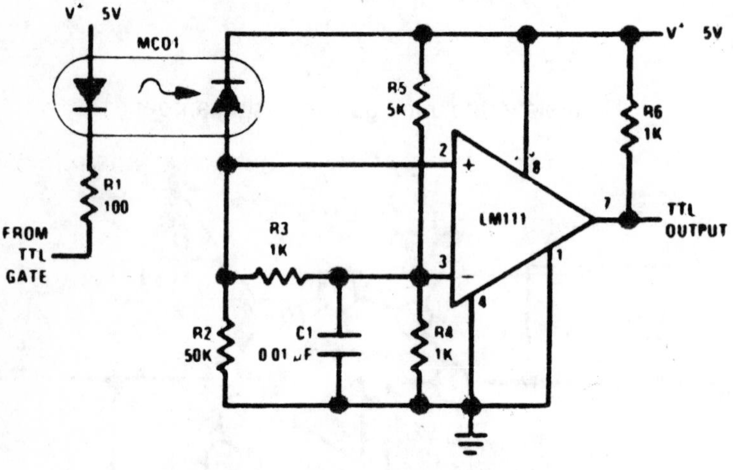


Digital transmission isolator.

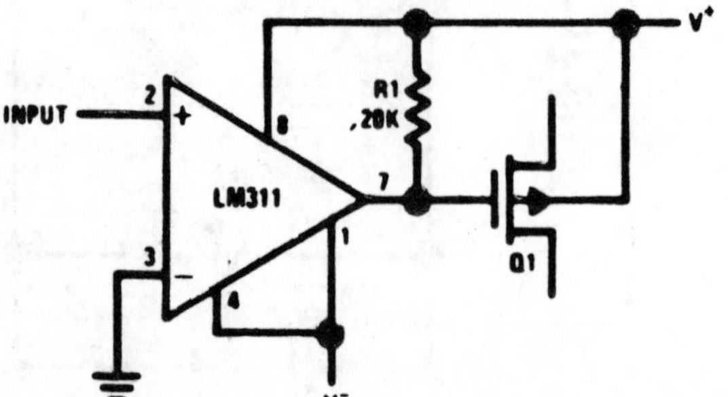


Zero-crossing detector driving MOS switch.

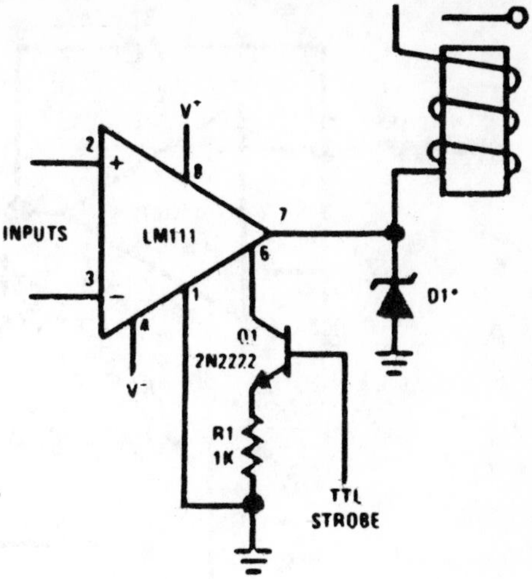


Relay driver with strobe.

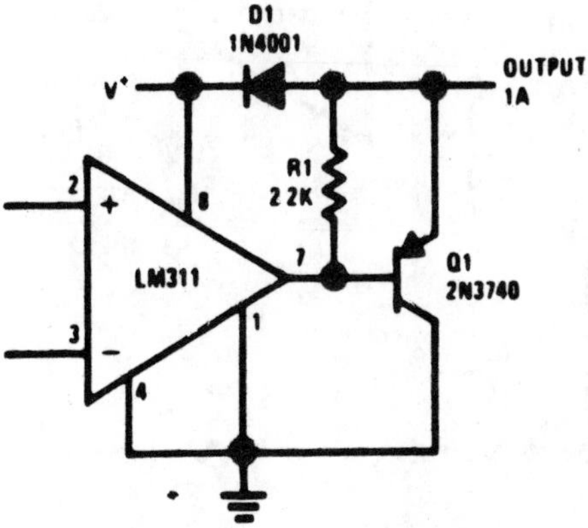


Comparator and solenoid driver.

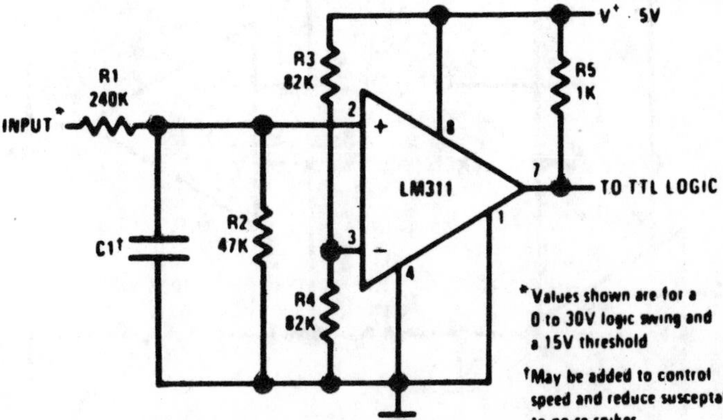


*Values shown are for a 0 to 30V logic swing and a 15V threshold

†May be added to control speed and reduce susceptability to noise spikes

TTL interface with high-level logic.

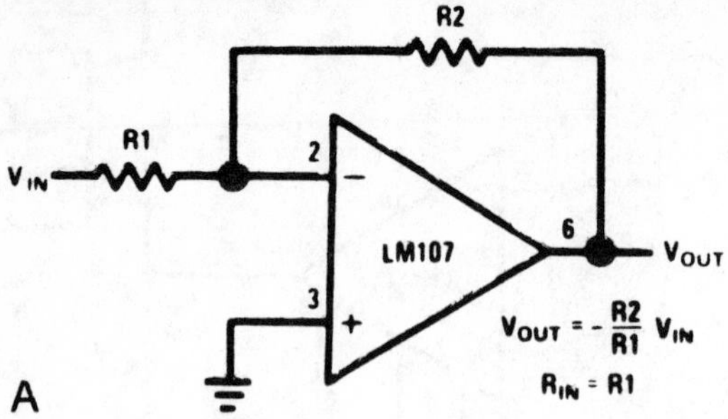

**A**

$$V_{OUT} = -\frac{R2}{R1} V_{IN}$$

$$R_{IN} = R1$$

**B**

$$V_{OUT} = \frac{R1 + R2}{R1} V_{IN}$$

$$R_{IN} = R3$$

$$R3 = R1 \quad R2$$

**C**

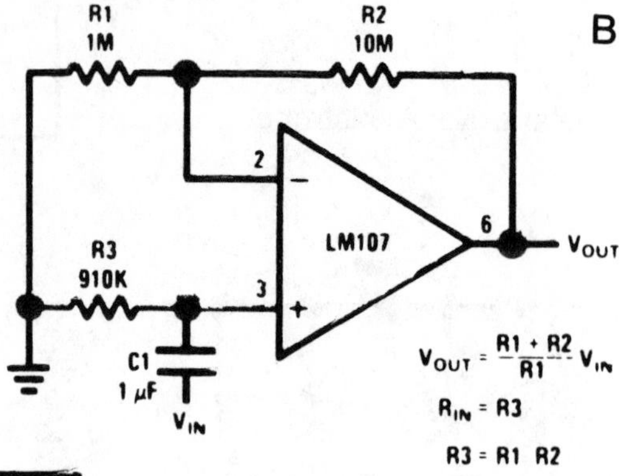

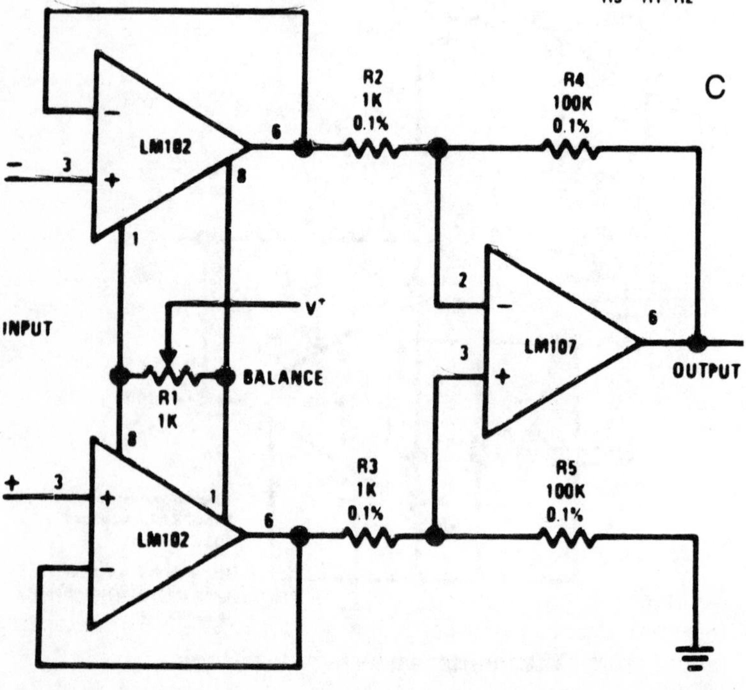

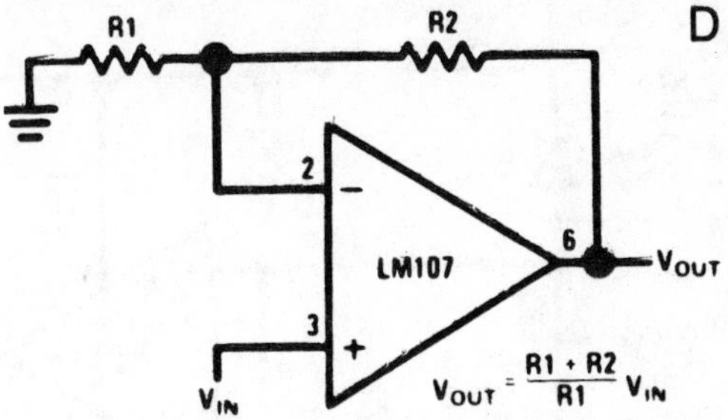


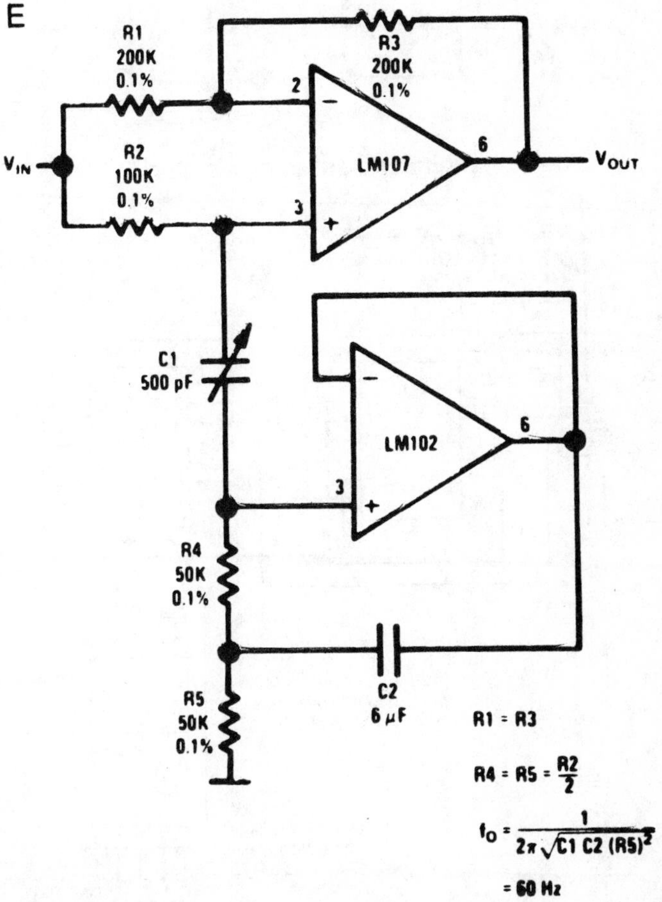


(A) Inverting amplifier, (B) noninverting SC amplifier, (C) differential input instrumentation amplifier, (D) noninverting amplifier, (E) tunable notch filter.

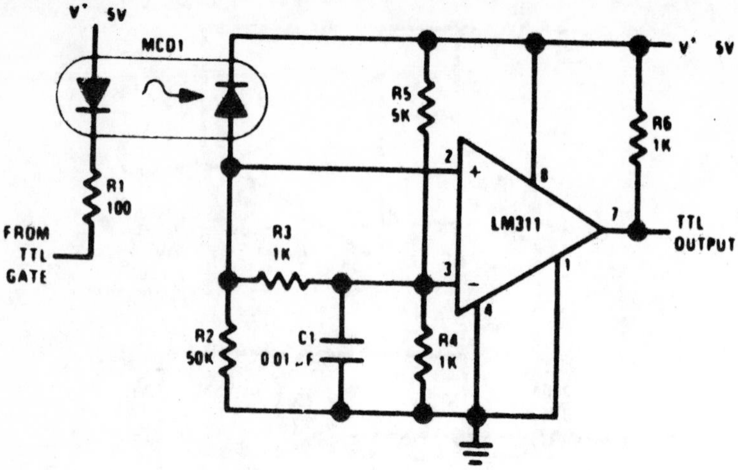


Digital transmission isolator.

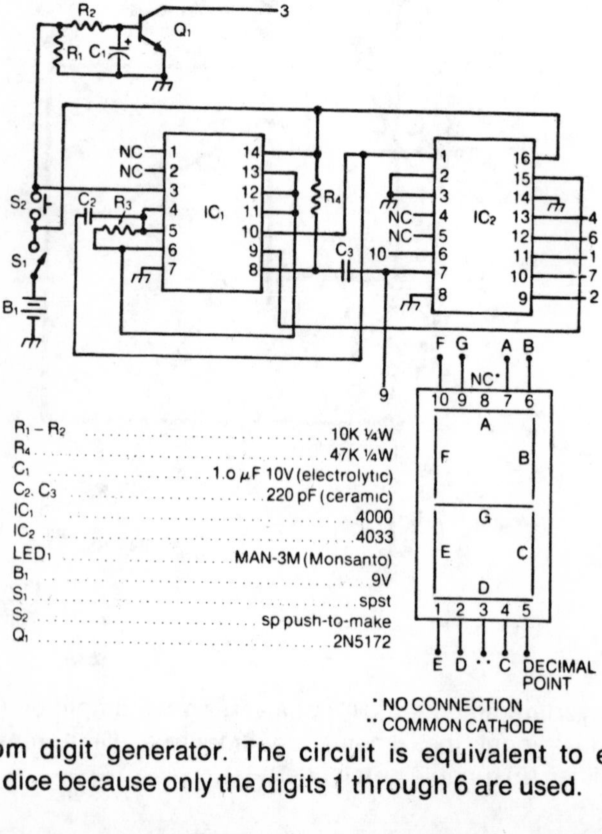


| | |
|---|---|
| $R_1 - R_2$ | 10K ¼W |
| $R_4$ | 47K ¼W |
| $C_1$ | 1.0 μF 10V (electrolytic) |
| $C_2$, $C_3$ | 220 pF (ceramic) |
| $IC_1$ | 4000 |
| $IC_2$ | 4033 |
| $LED_1$ | MAN-3M (Monsanto) |
| $B_1$ | 9V |
| $S_1$ | spst |
| $S_2$ | sp push-to-make |
| $Q_1$ | 2N5172 |

Random digit generator. The circuit is equivalent to electronic dice because only the digits 1 through 6 are used.

# Section G

# Power Supplies, Voltage Multipliers

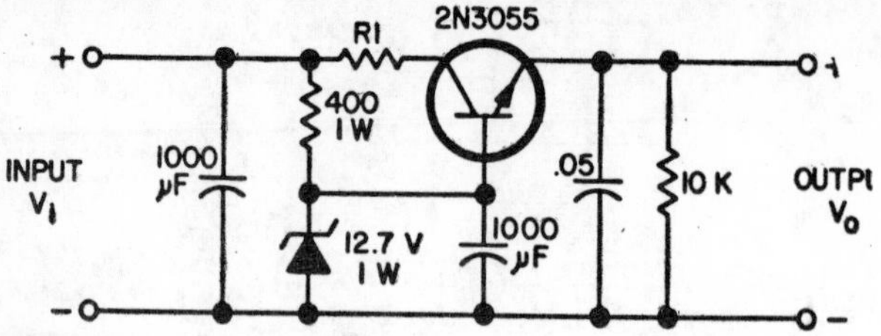


A power supply for a transistorized transceiver. Since rigs tend to draw much more current when transmitting, the transistor in the power supply must dissipate more current. The use of R1 lowers the amount of power dissipated and enables a smaller inexpensive transistor to be used. Choose R1 so the total voltage drop is about 2 volts.

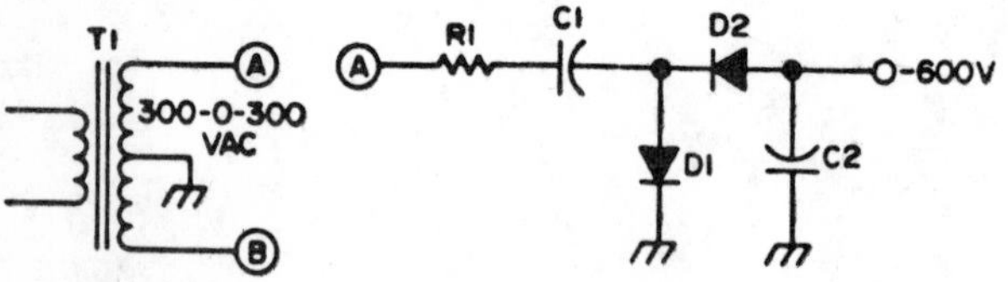


R1 – 1.5K
C1 – .25 µF, 1 kV
C2 – .5 µF, 1.5 kV
C3 – .25 µF, 1 kV

Standard voltage doubler circuit.

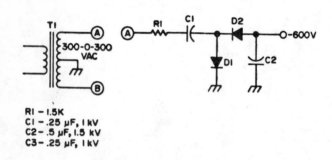

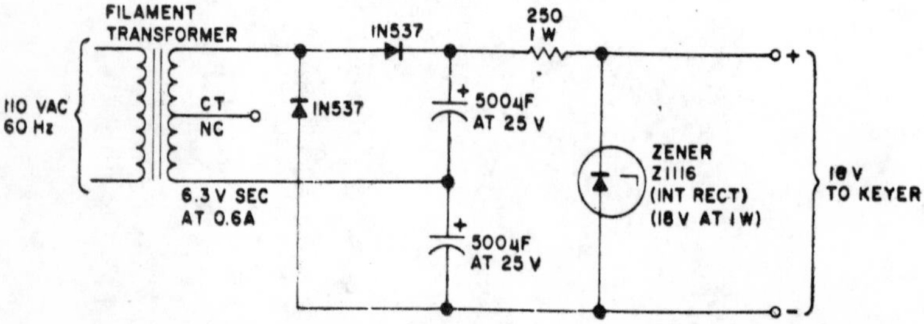


18V voltage-doubling power supply uses zener to achieve regulation.

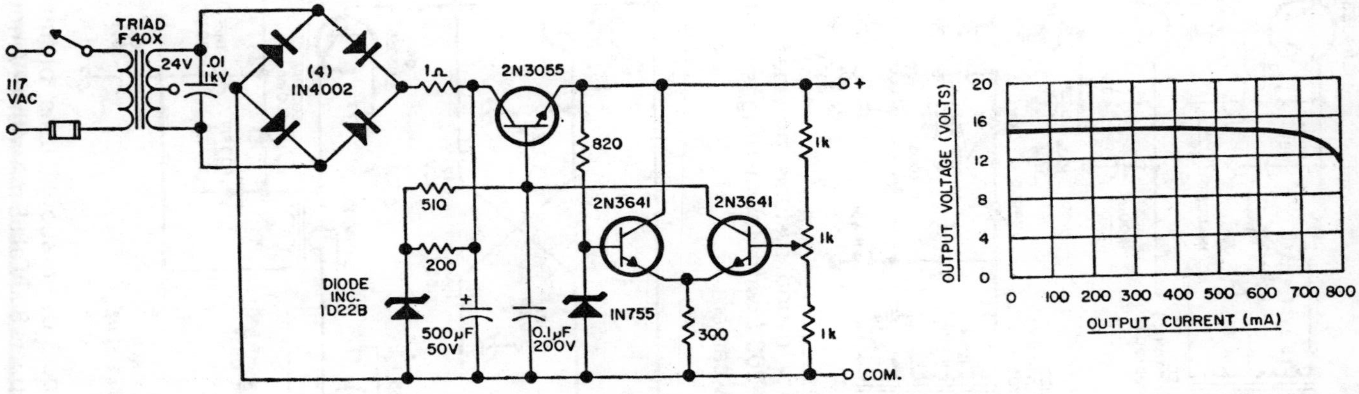


Practical supply using a differential amplifier pair.

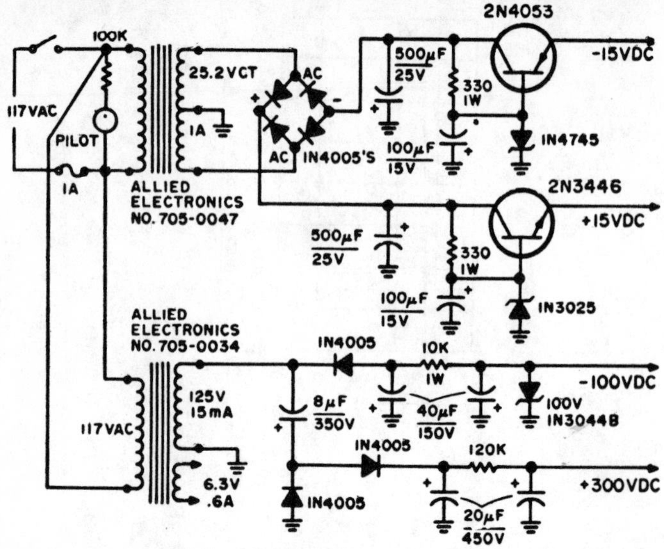


Power supply for hybrid systems provides +15V, -15V, −100V. and +300V with low voltages well regulated. Developed for TV camera.

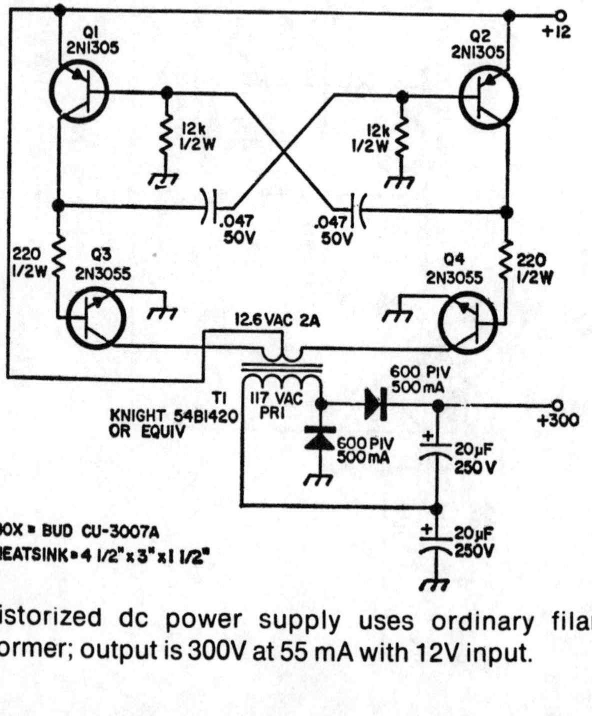


Transistorized dc power supply uses ordinary filament transformer; output is 300V at 55 mA with 12V input.

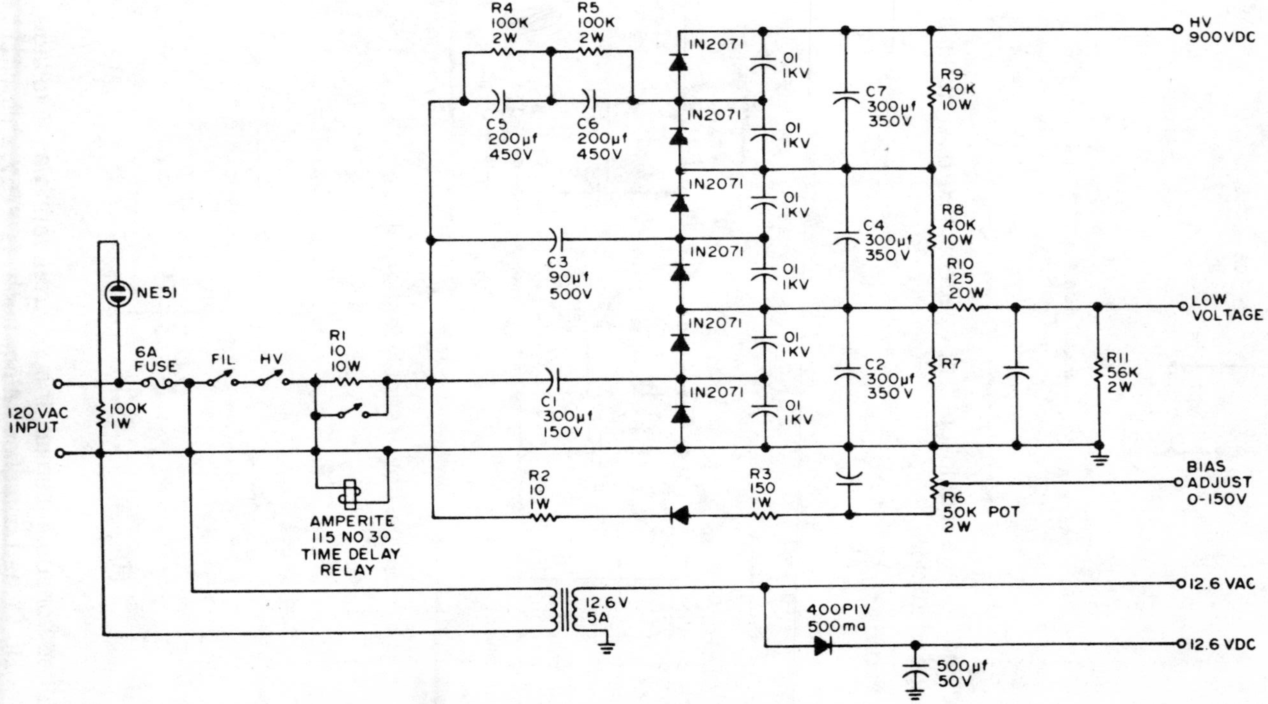


Complete schematic of sextupler power supply.

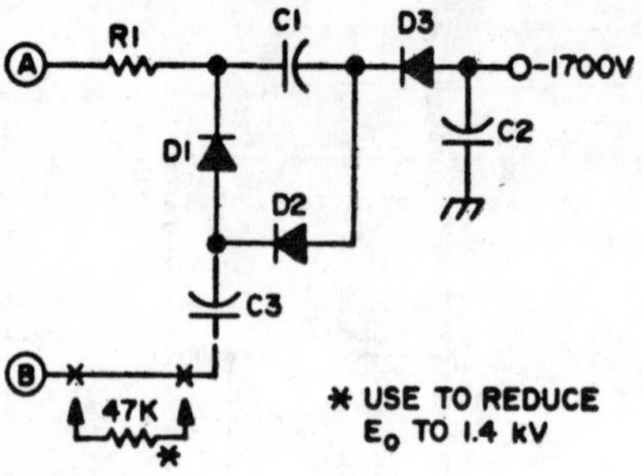


Modified voltage multiplier. It now appears to operate as a voltage sextupler.

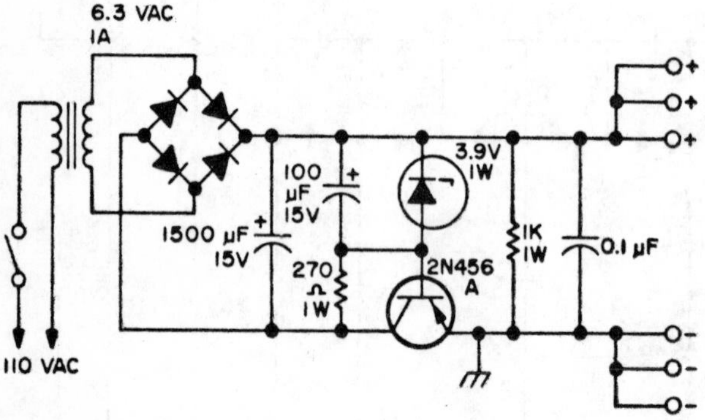


Simple power supply for IC projects.

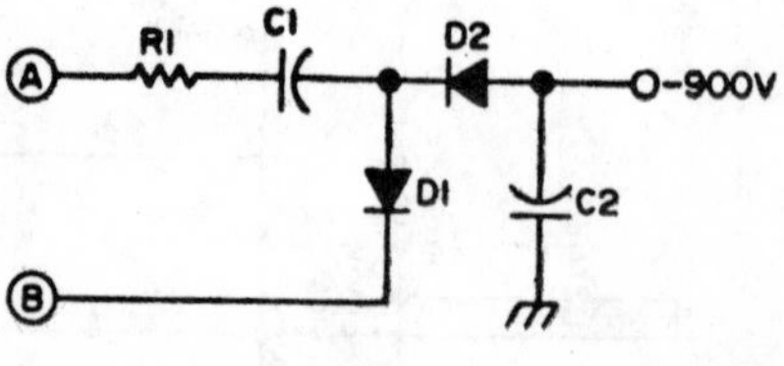


Modified voltage multiplier. It is now operating as a voltage tripler. If T1 had a higher rating, such as 375V, this circiit could well be adequate for most scopes.

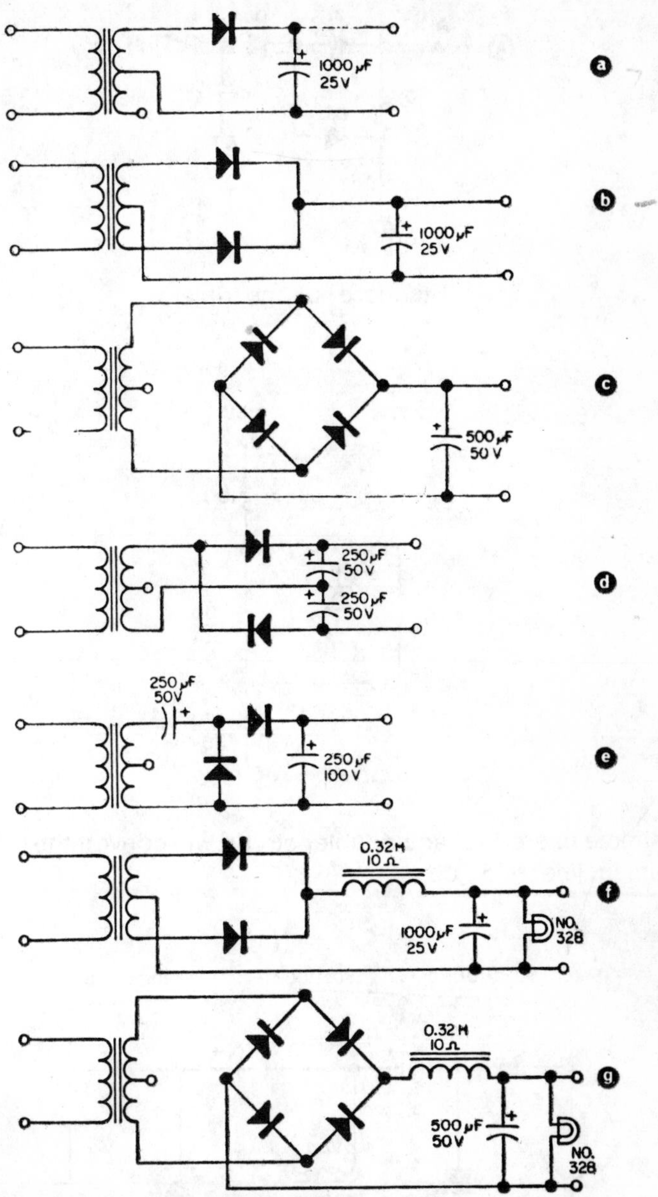


The six basic types of rectifiers. **a** is a simple half-wave rectifier; **b**, full wave; **c**, bridge; **d**, conventional (full-wave) doubler; **e**, cascade (half-wave) doubler. The preceding are all capacitor-input circuits. **f** is a full-wave rectifier with choke input, and **g** is a bridge rectifier with choke input.

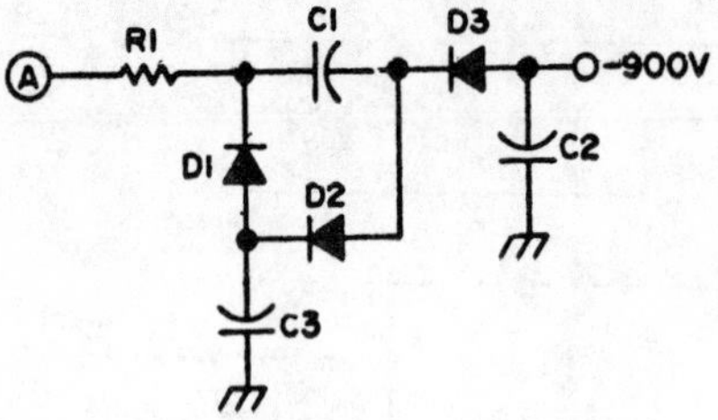


Standard voltage tripler.

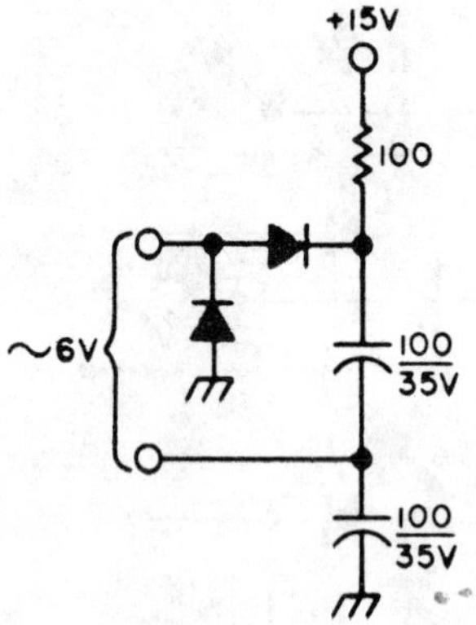


A simple filtered voltage doubler circuit will convert the 6V ac filament line to 15V dc.

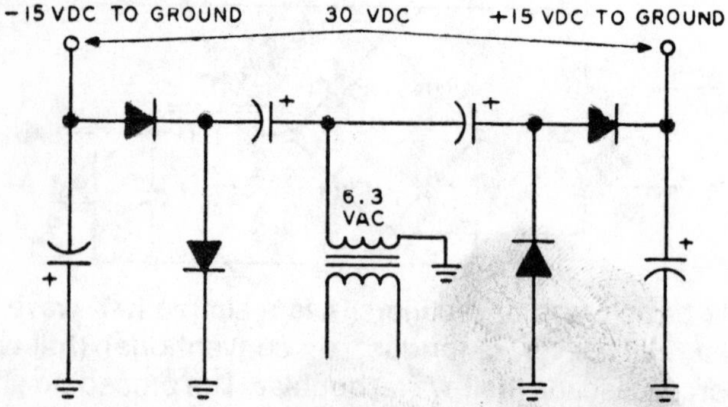


Dual-voltage power supply from single-winding secondary.

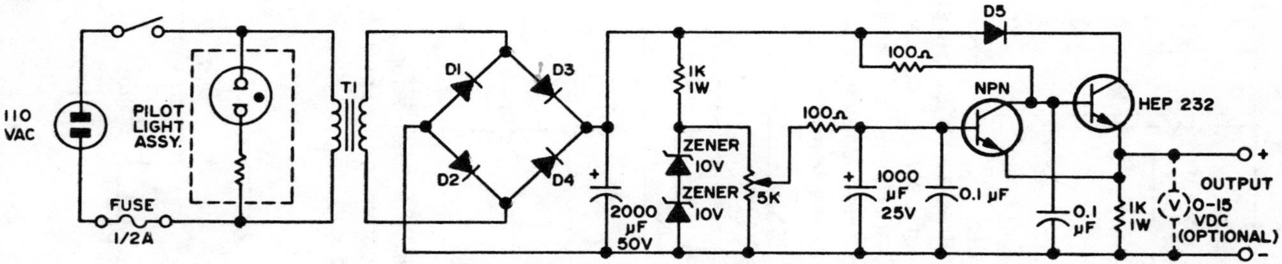


T1 – FILAMENT TRANSFORMER
24 VAC, 2 AMP

D1 THRU D5 – 50V PIV, 1 AMP

Power supply for 0–20 volts up to 1A. Run that tape recorder, transistor radio, preamplifier, or charge a battery. Circuit courtesy Calectro Handbook.

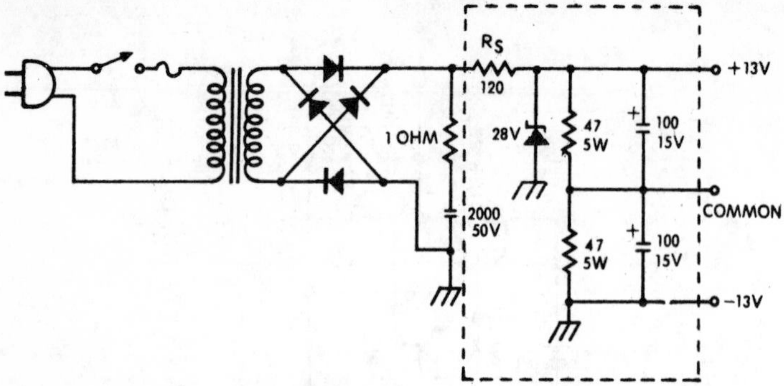


A split supply can be synthesized with any stable dc power source if a voltage-divider network with a single zener is added. The value of the series resistor in the dc line will depend on the voltage of the line. Values of all resistors in the broken-line box are based on a 36-volt dc supply.

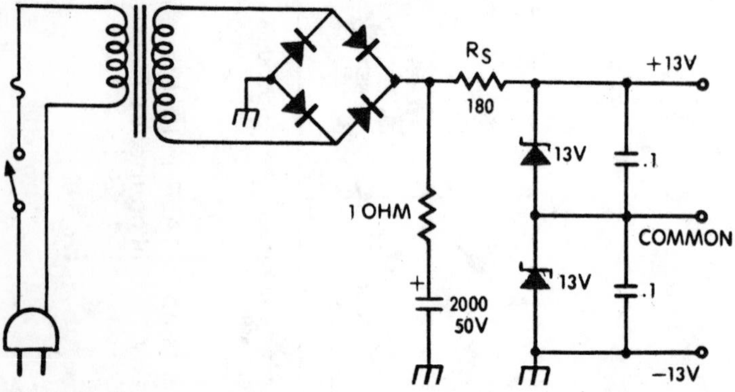


Two 13-volt zeners drop the total supply voltage to two matched voltages. The "common" line must be isolated from the power supply chassis.

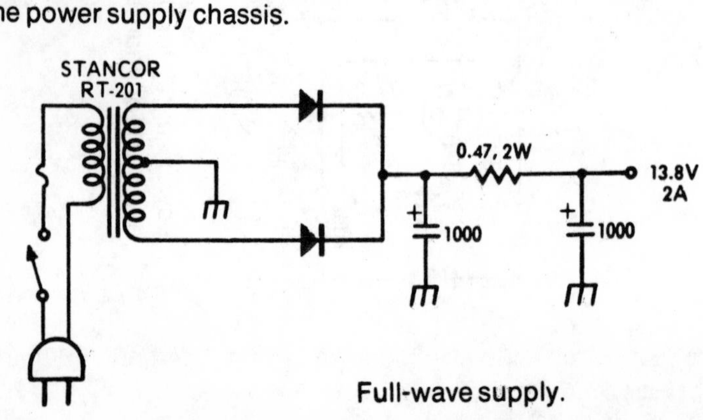


Full-wave supply.

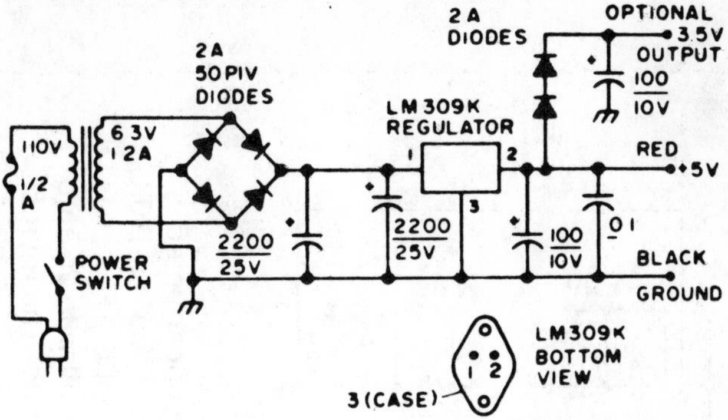


Power supply for integrated circuits uses LM309K regulator and features a 3V output that uses the forward voltage drop of series diodes to establish voltage level.

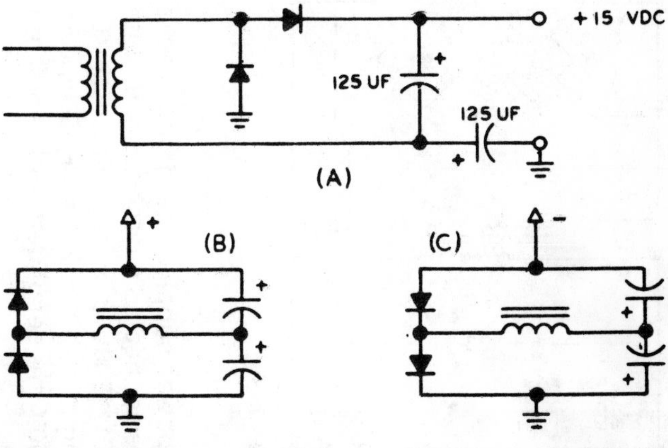


Voltage doubler circuit. 2B has negative ground. 2C has positive ground.

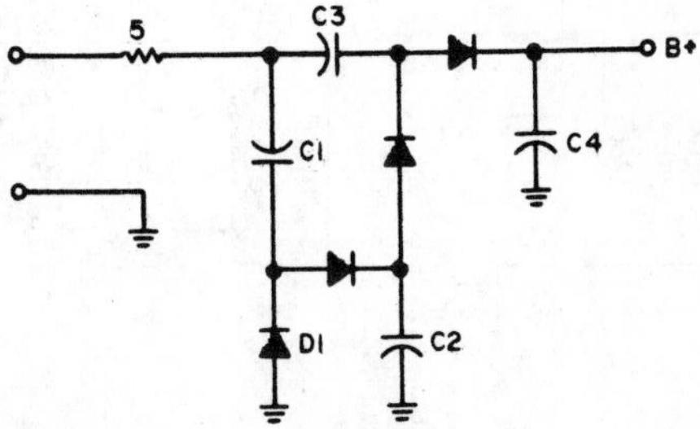


Conventional half-wave quadrupler. Full output voltage is across C4.

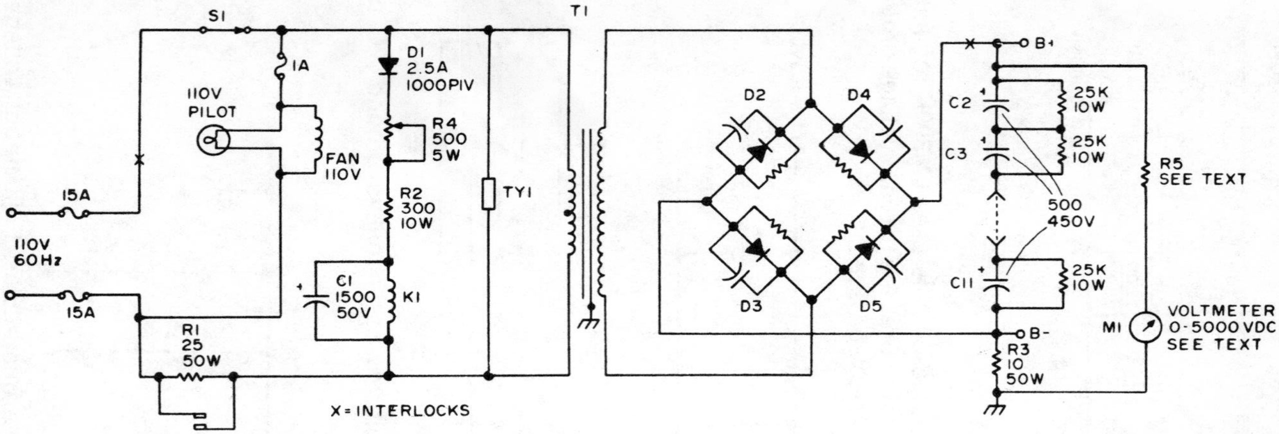


Schematic of 3000V power supply. The diode stacks D2–D5 are constructed of 8–2.5A 1000 PIV series connected diodes each. Shunted across each diode is a 470K 1W resistor and a 0.01 1000V disc capacitor. C2–C11 should be 500 $\mu$F with a minimum voltage rating of 450V dc. K1 is a P&B type PR3DY 24V dc coil with 25 amp contacts. T1 has a 2200V rms secondary with a 500 mA minimum rating. The thyrector is a GE 6R20SP4B4.

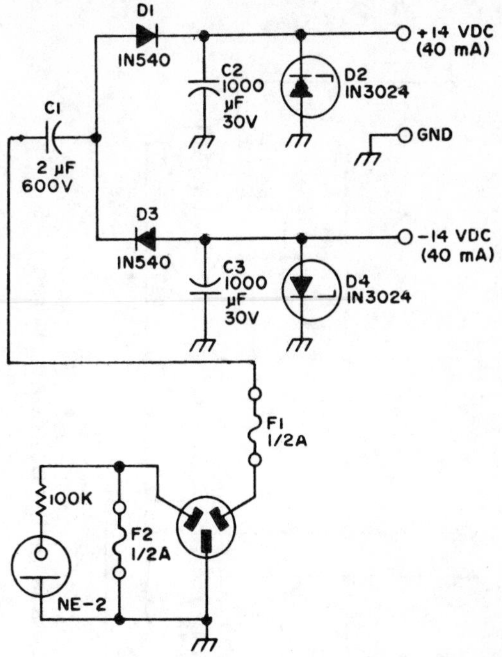


Low-current power supply.

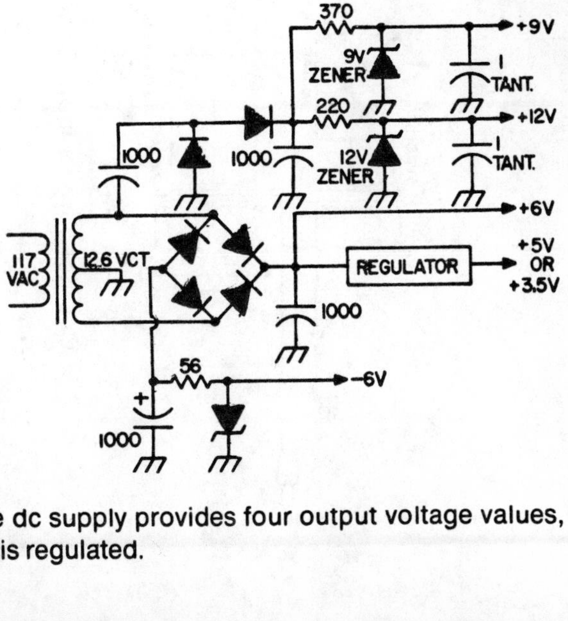


Simple dc supply provides four output voltage values, one of which is regulated.

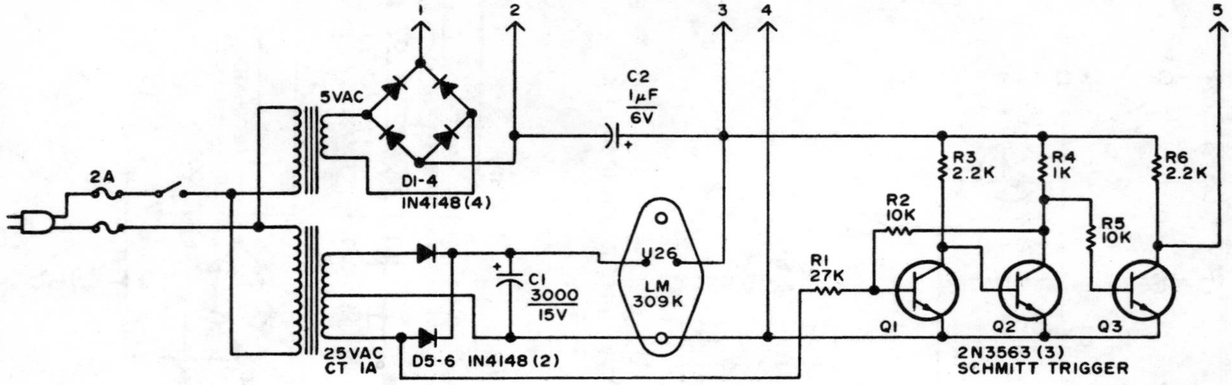


Counter power supply. Connections are made via the points numbered 1 — 5.

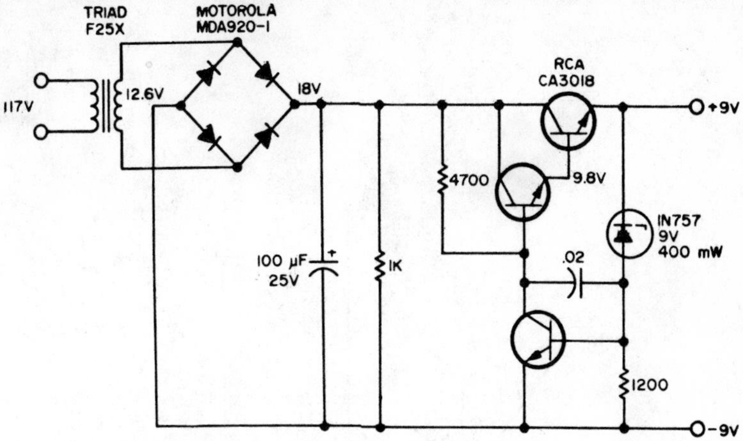


Regulated 9V 30 mA power supply using CA3018.

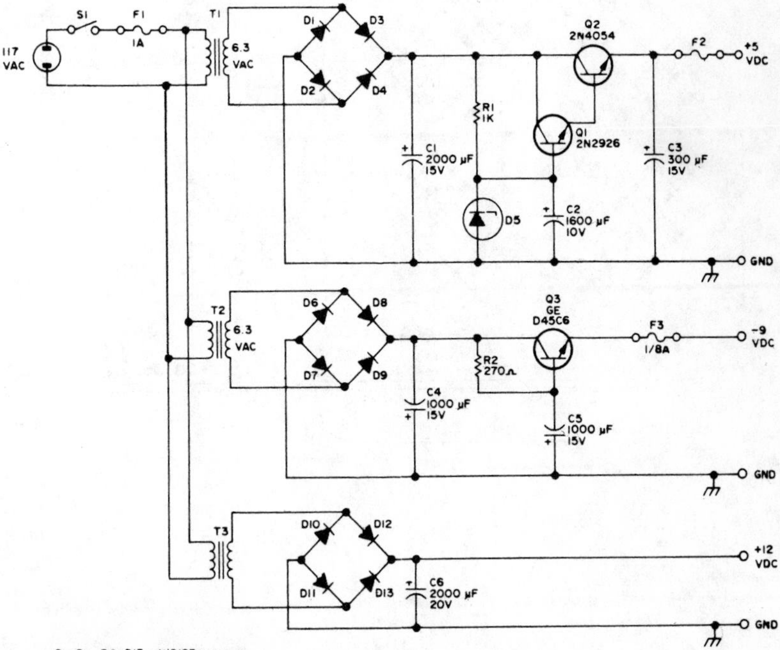


D1–D4, D6–D13 IN3193 or equiv.
D5 - IN5234, 6.2 ZENER

Three-in-one power supply for solid-state circuits.

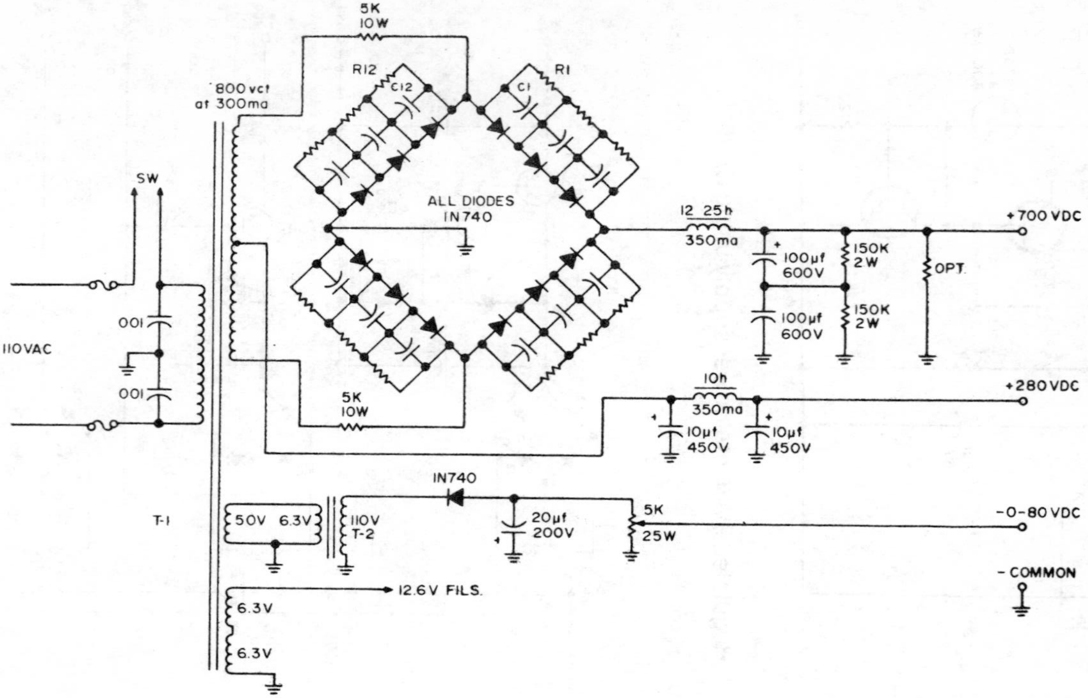


Cheap and easy power supply.

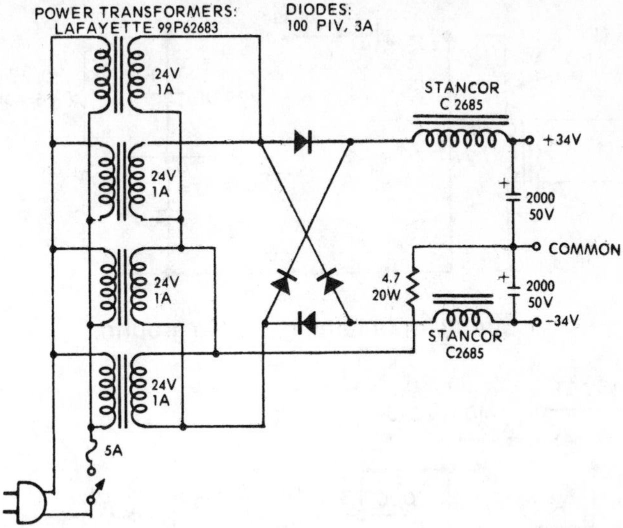


Multiwinding transformers make excellent high-current supplies. If all windings are identical, they may be paralleled. Here, top two windings and bottom two windings are paralleled, then the two paralleled sets are placed in series to give twice the voltage and twice the current of single winding.

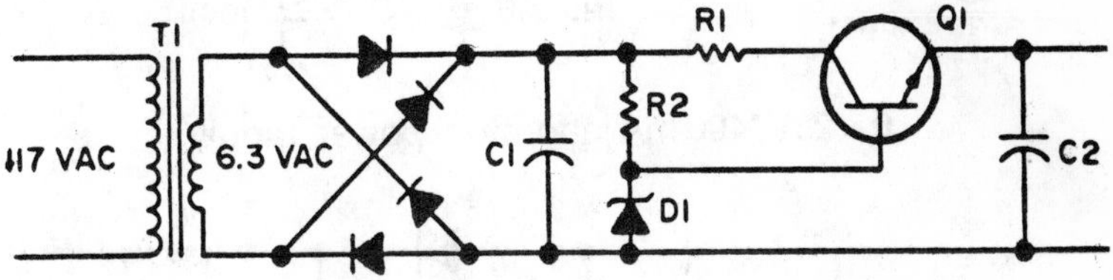


| | |
|---|---|
| $T_1$ | Stancor P6465 117 VAC to 6.3 VAC, 600 Ma. |
| $D_1$ | 3.9 Volt Zener Diode, 1N748 Motorola |
| D | 1 Amp, 12 Volt Diodes or HEP Bridge Rectifier |
| $C_1$ | 1000 ufd. at 12 Volts or greater |
| $C_2$ | 200 ufd. at 12 Volts or greater |
| $R_1$ | 10 ohm at 6.3 VAC input<br>5 ohm at 3.2 VAC input |
| $R_2$ | 220 ohms at 6.3 VAC input<br>110 ohms at 3.2 VAC input* |
| $Q_1$ | 2N4921, 2N3055 or HEP 245 |

*To be used with center tap 6.3 VAC and full wave rectifier.

5V regulated supply for powering ICs.

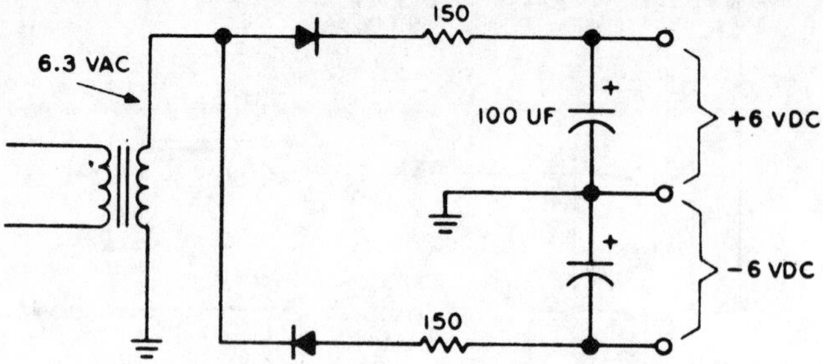


Circuit for obtaining ±6V to ground.

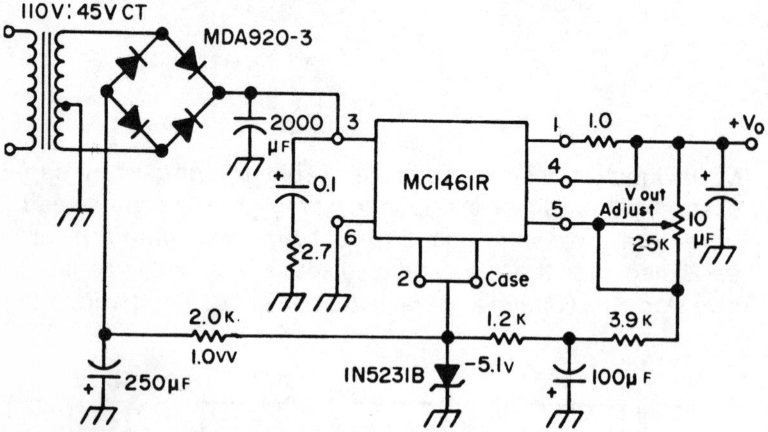


0–25V, 400 mA laboratory power supply.

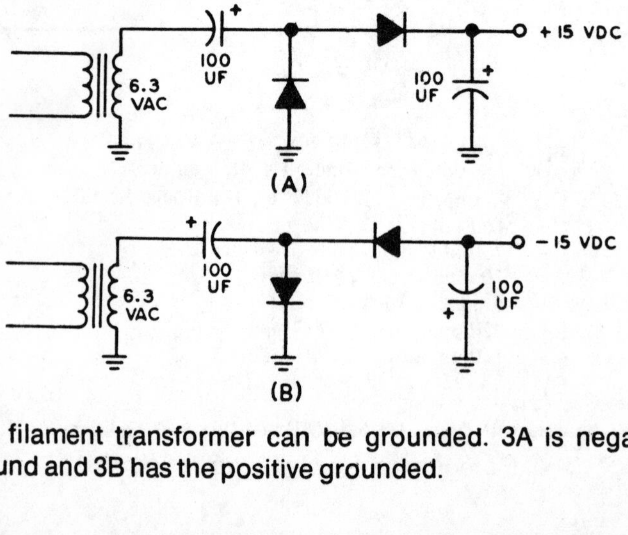


The filament transformer can be grounded. 3A is negative ground and 3B has the positive grounded.

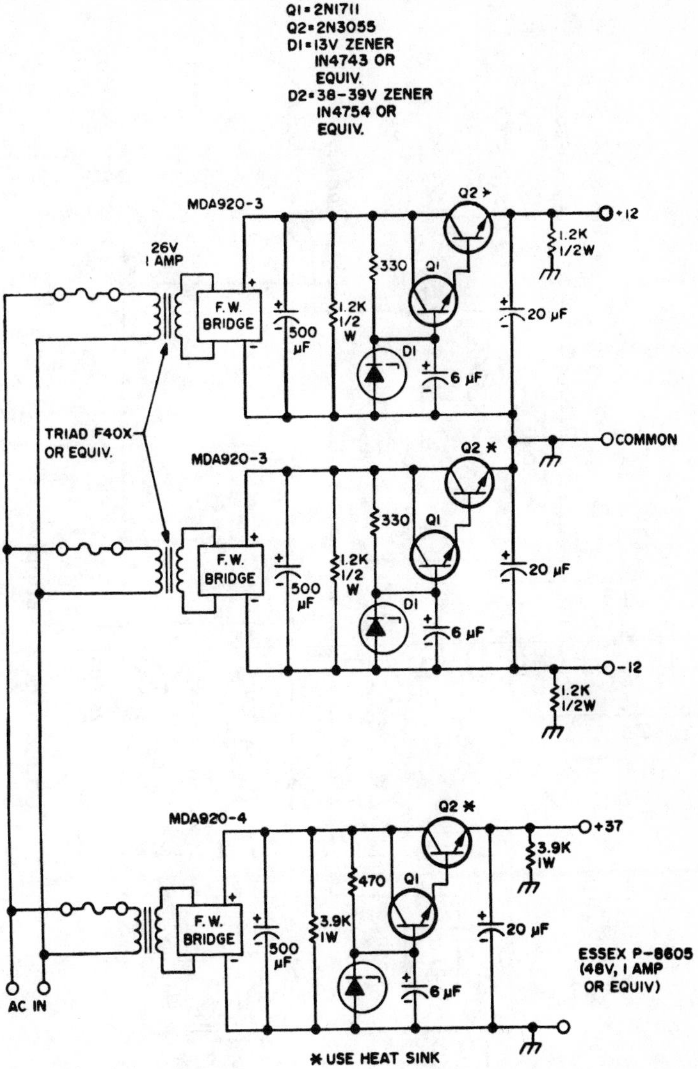


High-power regulated power supply shown here was designed for UHF transmitter. The 2N1711 and 2N 3055 (Q1 and Q2, respectively) are not critical, and can be replaced with comparable NPN units. The full-wave bridge may be a self-contained "black box" package or a group of individual silicon diodes of 50 PIV each. A heatsink must be used for each Q2 transistor. Transformers shown are Triad F40X units.

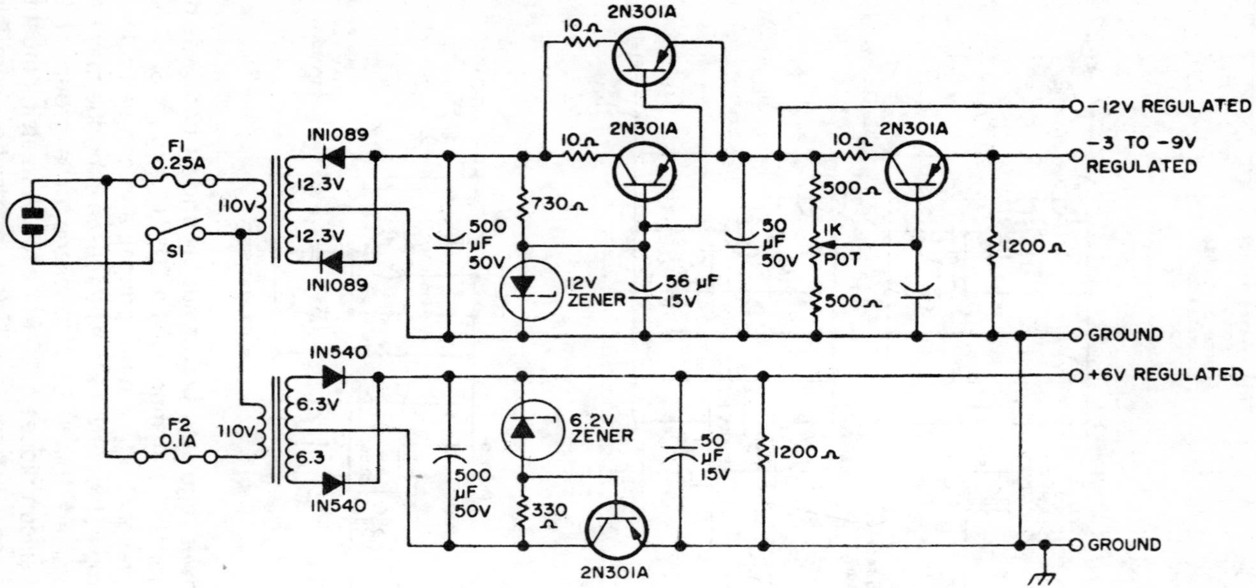


Regulated power supply.

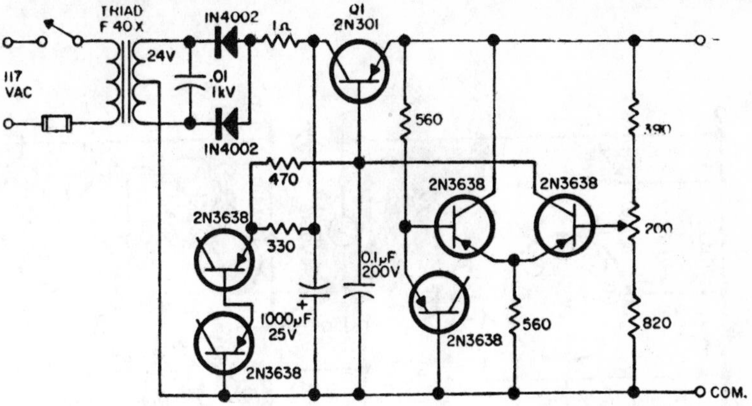


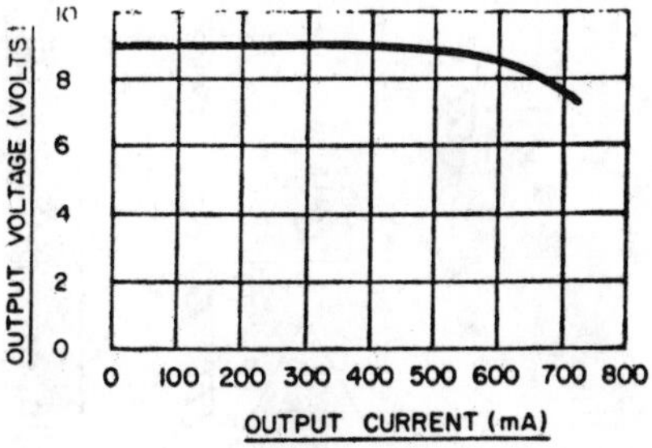


Another differential pair regulated supply. Here the base–emitter junctions of inexpensive transistors are used as zener diodes.

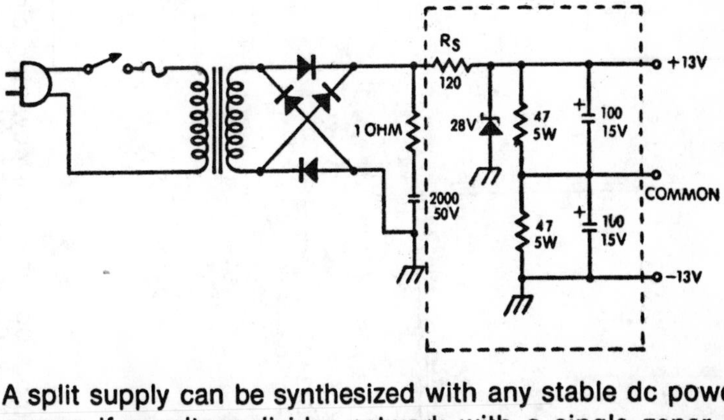


A split supply can be synthesized with any stable dc power source if a voltage-divider network with a single zener is added. The value of the series resistor in the dc line will depend on the voltage of the line. Values of all resistors in the broken-line box based on a 36-volt dc supply.

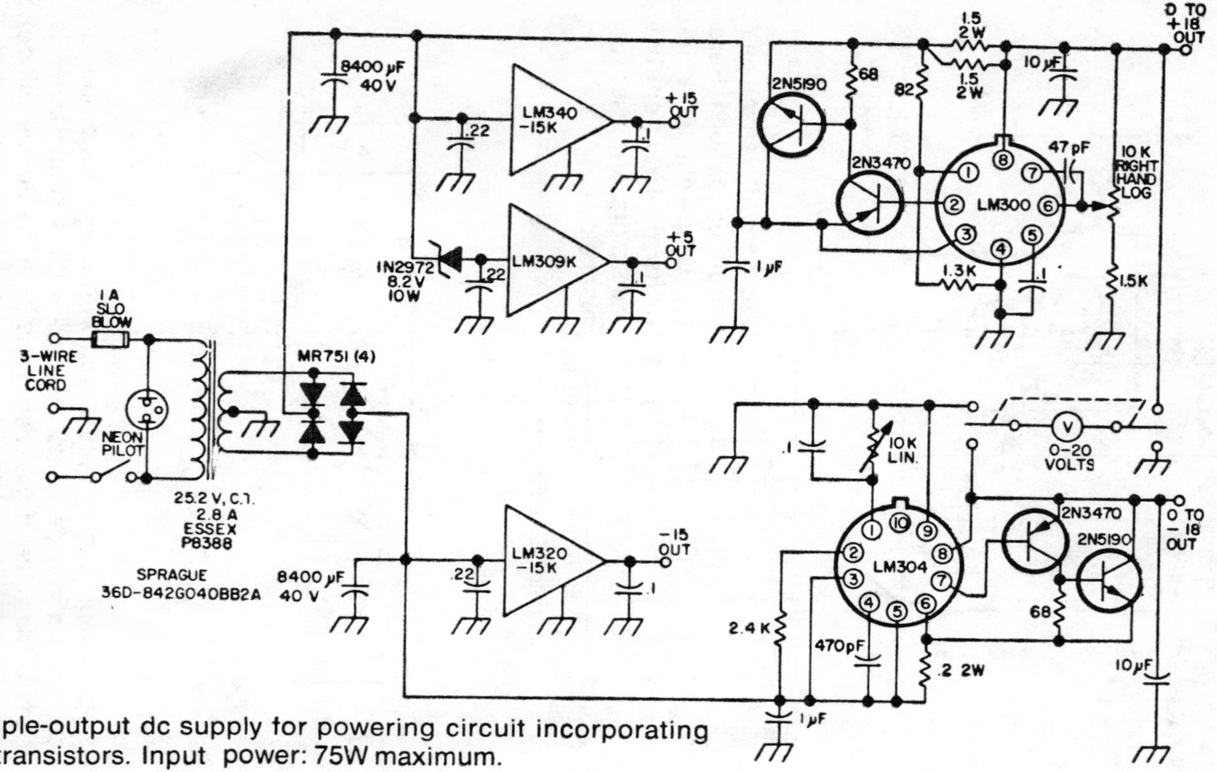


Multiple-output dc supply for powering circuit incorporating ICs, transistors. Input power: 75W maximum.

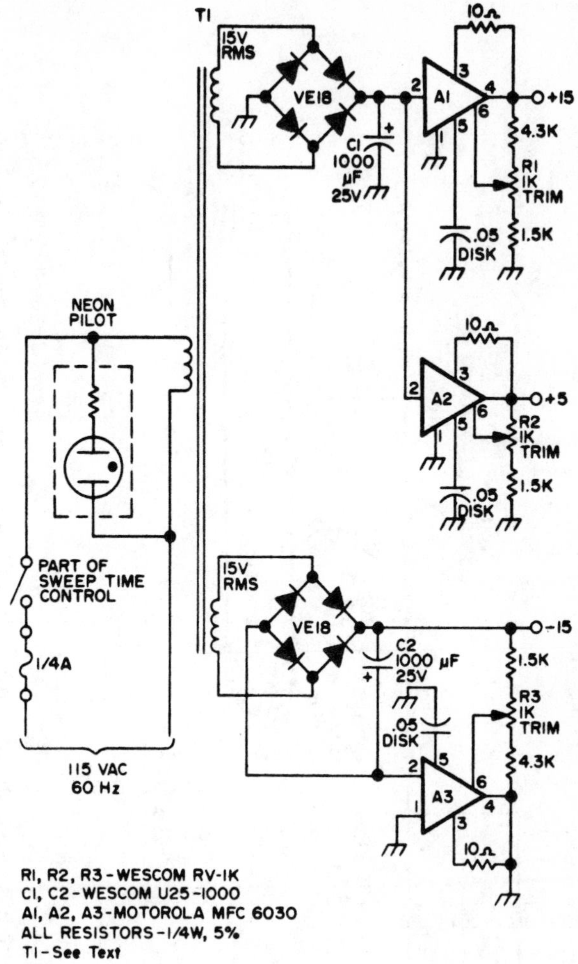


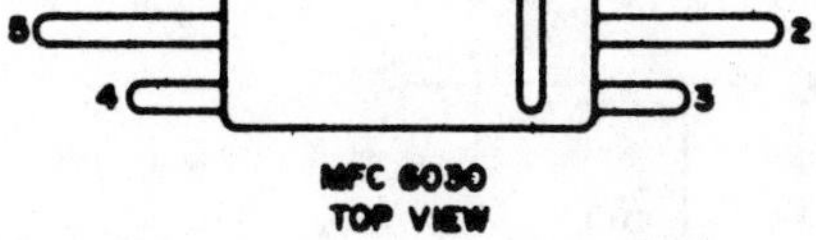


Regulated 15V power supply. Transformer T1 can be a small pair of transformers or one unit with 2 secondaries.

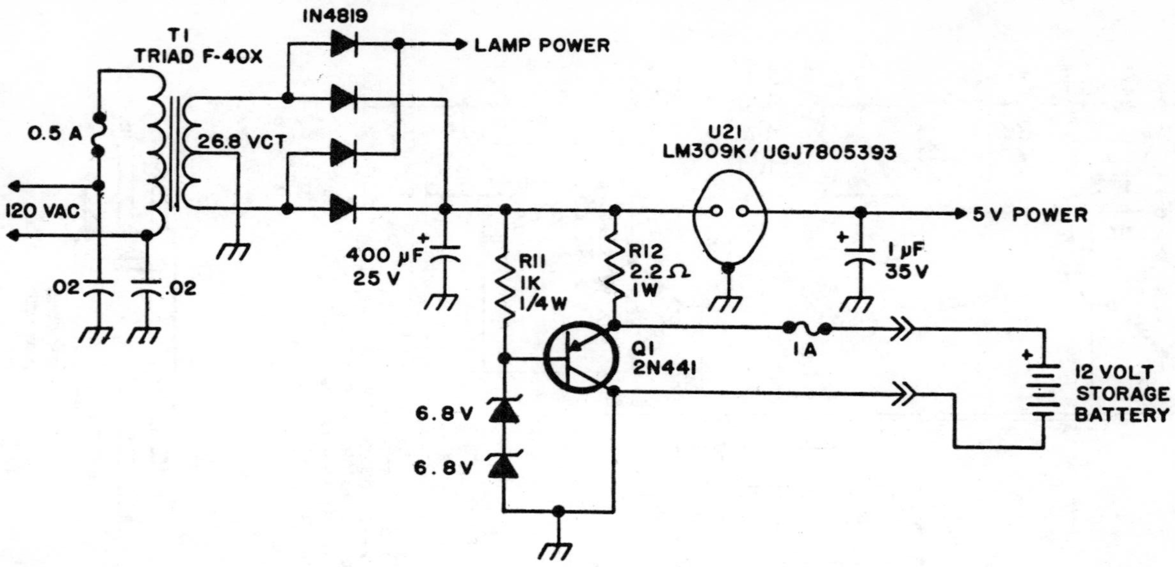


Dc power supply for ICs uses shunt regulator and is not adjustable; but simplicity justifies inconvenience.

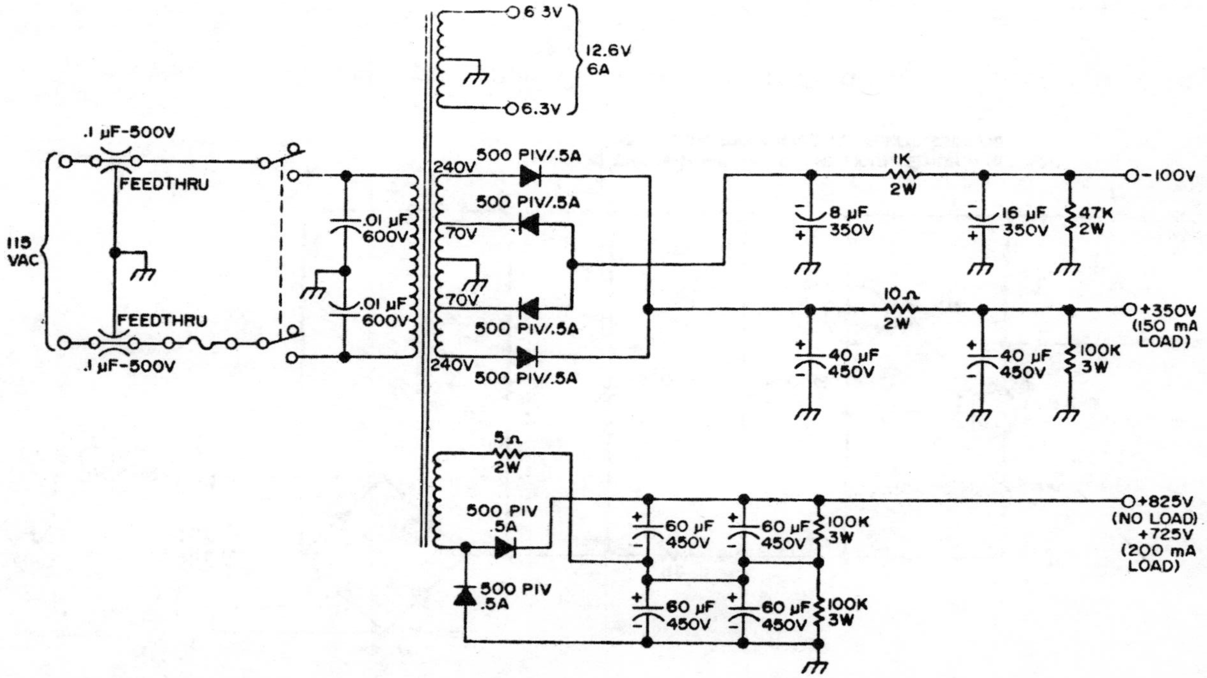


High-voltage power supply for heavy-current tube-type applications. Outputs are −100V, +350V, and 750V (0.2A).

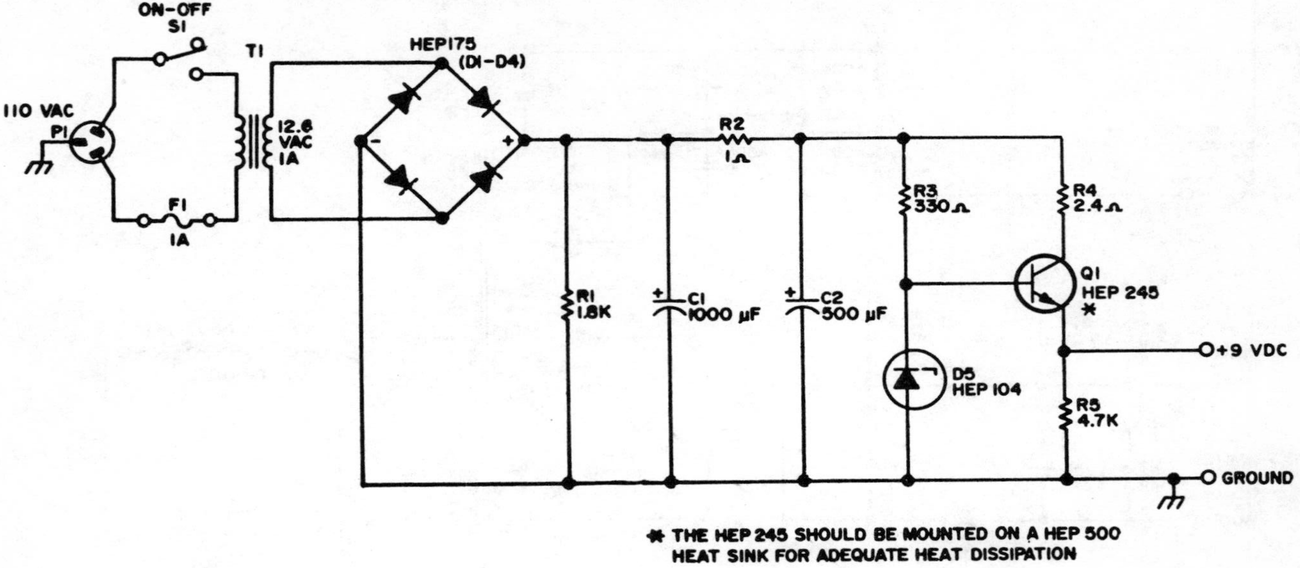


Simple power supply. Circuit courtesy of Motorola.

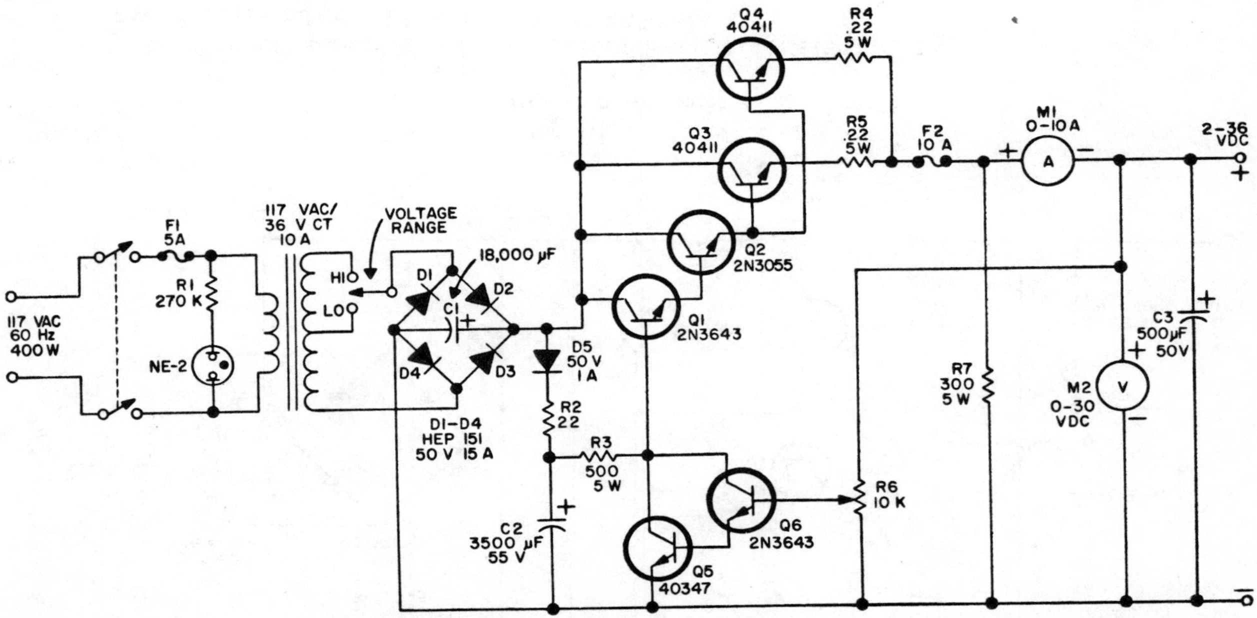


Dc power supply provides from 1 to 10 amperes at 2 to 36V.

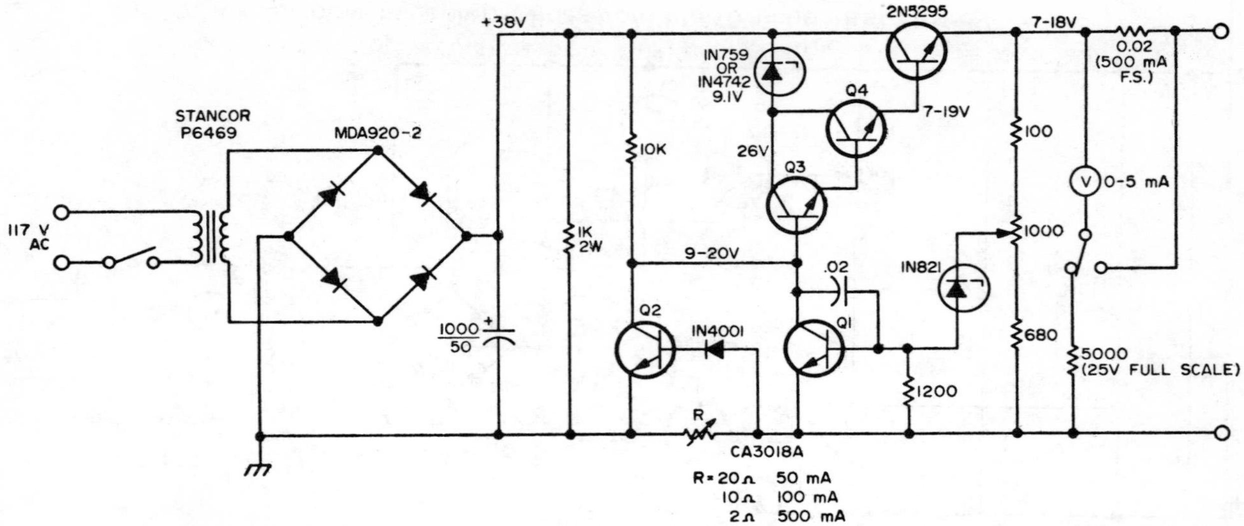


500 mA 7 TO 18V

UTILITY POWER SUPPLY

"Utility" power supply with 7 – 18V output. Resistor R determined by current requirements: 20Ω – 50 mA; 10Ω – 100 mA; 2Ω – 500 mA.

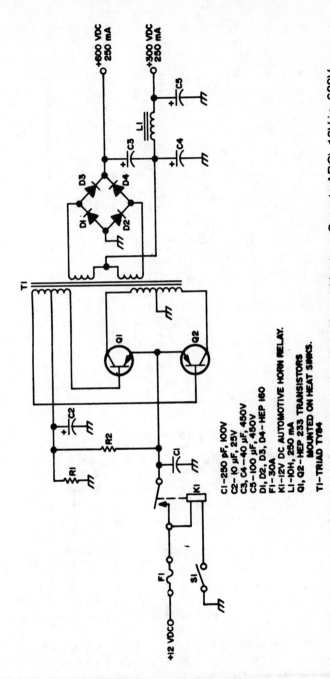

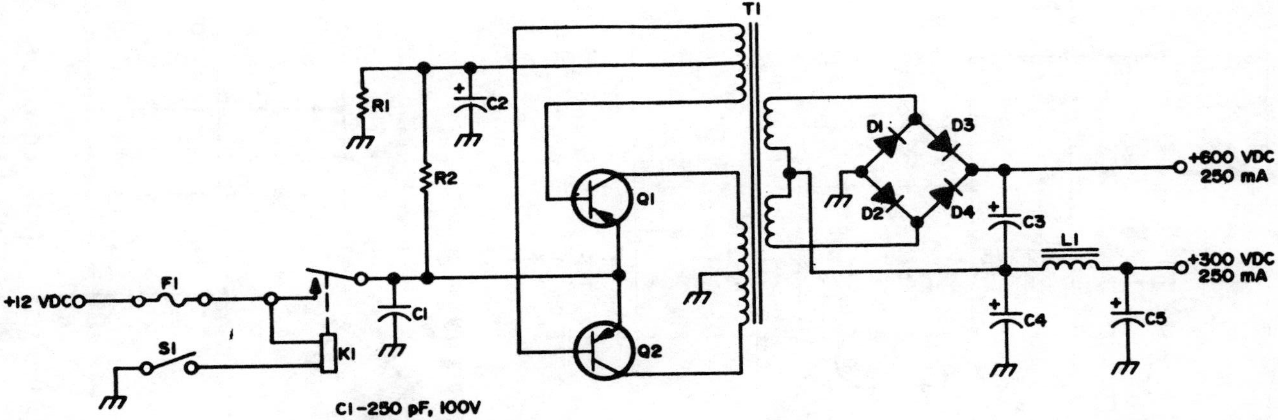


C1–250 pF, 100V
C2–10 µF, 25V
C3, C4–40 µF, 450V
C5–100 µF, 450V
D1, D2, D3, D4–HEP 160
F1–30A
K1–12V DC AUTOMOTIVE HORN RELAY.
L1–10H, 250 mA
Q1, Q2–HEP 233 TRANSISTORS
MOUNTED ON HEAT SINKS.
T1–TRIAD TY84

Mobile transceiver power supply (from The Keyer, Ventura County ARC). 12V in, 600V at 250 mA and 300V at 250 mA out. Mount transistors on heatsinks.

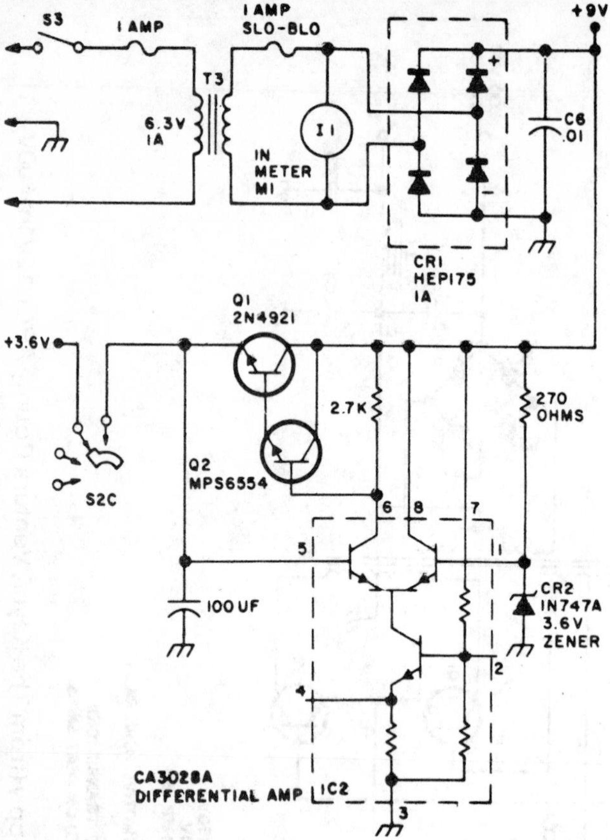


Power supply schematic.

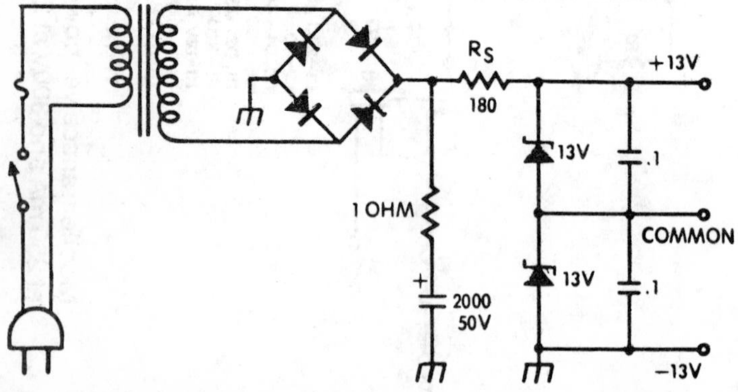


Two 13-volt zeners drop the total supply voltage to two matched voltages. The "common" line must be isolated from the power supply chassis.

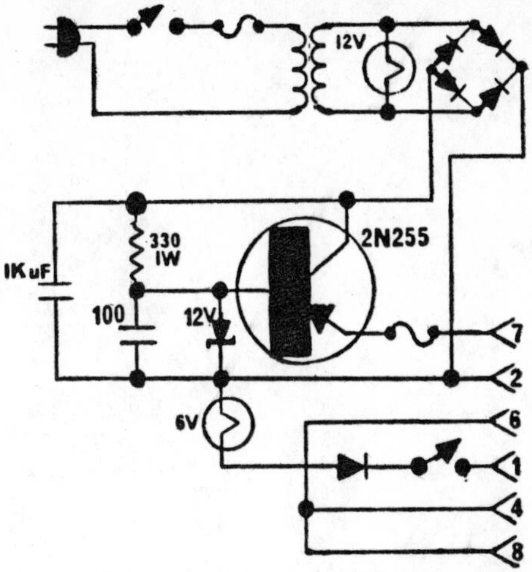


Power supply designed to operate Motorola tube-type H23 transceiver from household 115V outlet. Dc output of supply is about +12V.

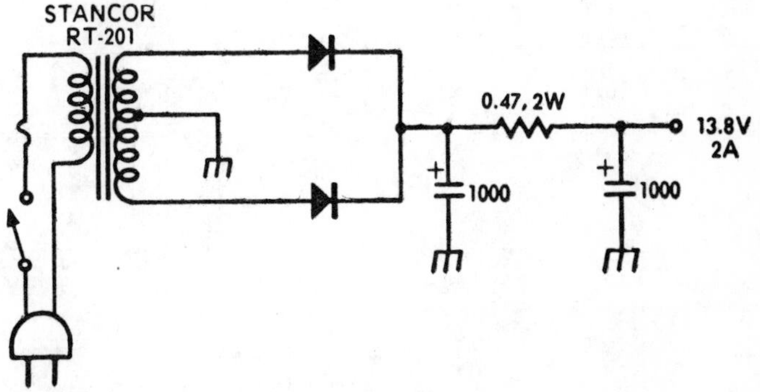


Low-voltage adjustable reference supply. Note: For other power supply ideas, see **REGULATORS** section, page 257.

# Section H

# Receivers, RF Preamps, Converters, Etc.

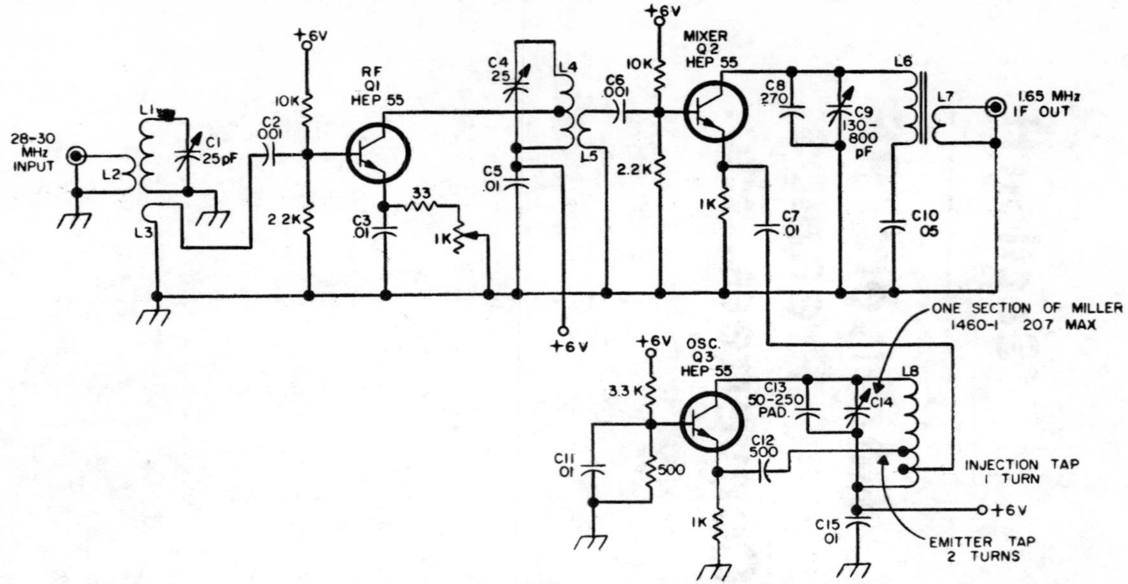


Overall converter schematic. L1, 17 turns, 15 mm OD, 25 mm long; L2, 3 turns on cold end of L1; L3, 1 turn on cold end of L1; L4, 11 turns, 8 mm OD, 35 mm long; L5, 2 turns on cold end of L4; L6, 21 turns No. 30 inside cup core from Millen 10C; L7, 4 turns No. 30 wound over L6; L8, 5 turns, 15 mm OD, 8mm long, mixer tap at 1 turn, and emitter tap at 2 turns from ground.

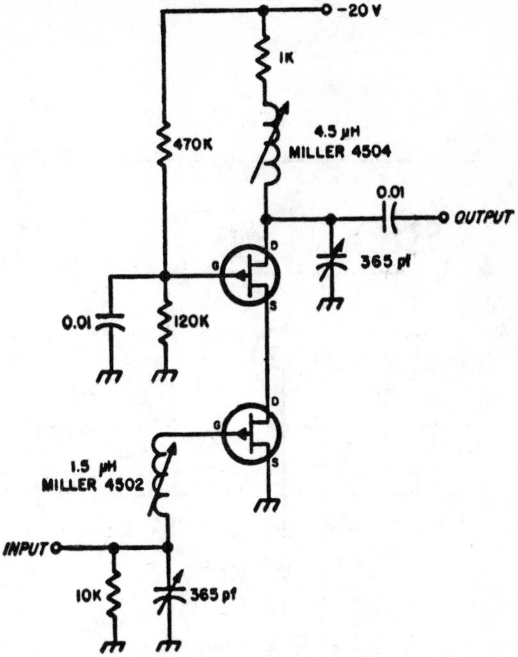


This 30 MHz i-f stage uses two FETs connected in the cascode arrangement to provide 20 dB gain without neutralization; the bandwidth is 4 MHz. Both FETs in this circuit are 2N3819, MPF105 or TIS34. With a negative supply voltage, the 2N4360 would be suitable.

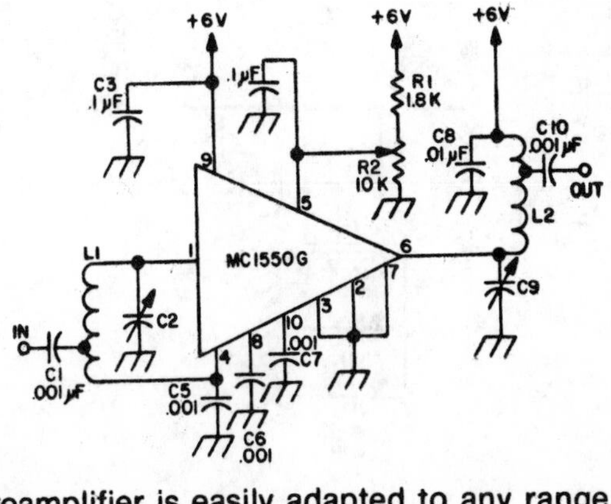


This rf preamplifier is easily adapted to any range under 30 MHz simply by choosing appropriate input and output tuned circuits.

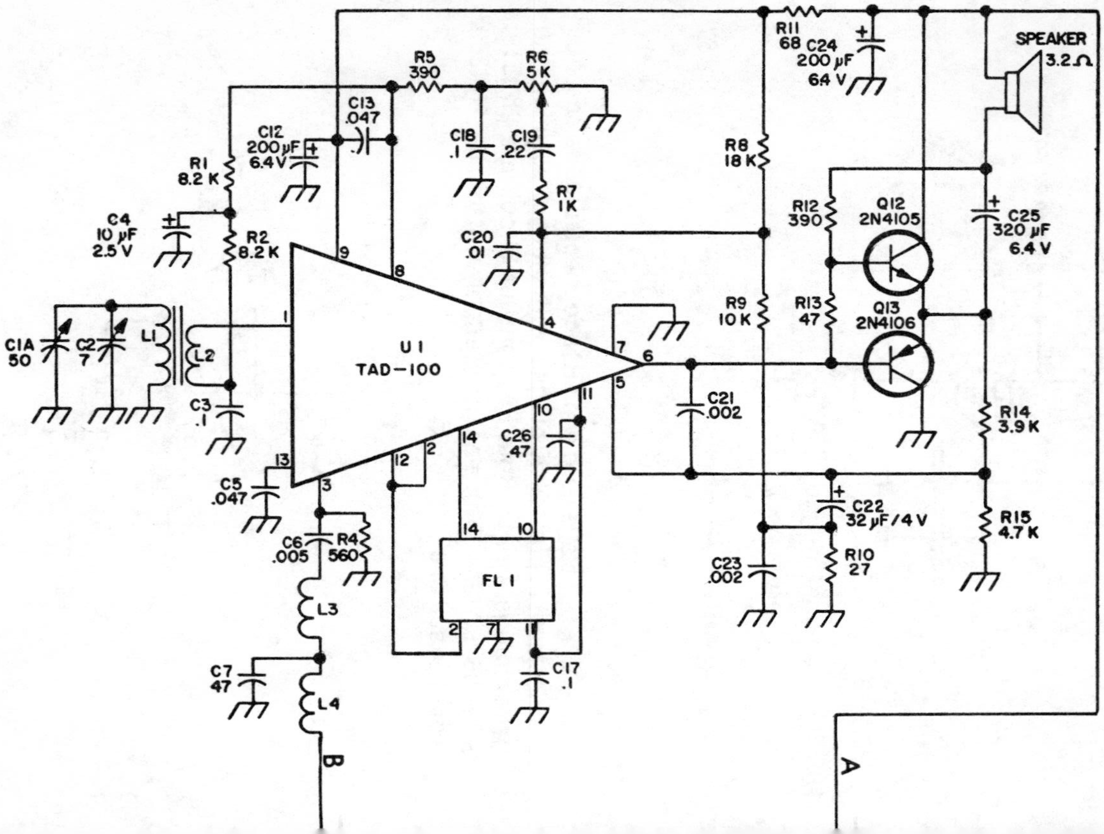
SPEAKER
3.2 Ω
R11
68
C24
200 µF
64 V
R5
390
R6
5 K
C13
.047
C12
200 µF
6.4 V
C18
.1
C19
.22
R1
8.2 K
R7
1 K
R8
18 K
R12
390
Q12
2N4105
C25
320 µF
6.4 V
C4
10 µF
2.5 V
R2
8.2 K
C20
.01
U1
TAD-100
C1A
50
C2
7
L1
L2
C3
.1
R9
10 K
R13
47
Q13
2N4106
C21
.002
R14
3.9 K
C26
.47
C5
.047
C22
32 µF / 4 V
R15
4.7 K
C6
.005
R4
560
FL 1
C23
.002
R10
27
L3
C7
47
L4
C17
.1
B
A

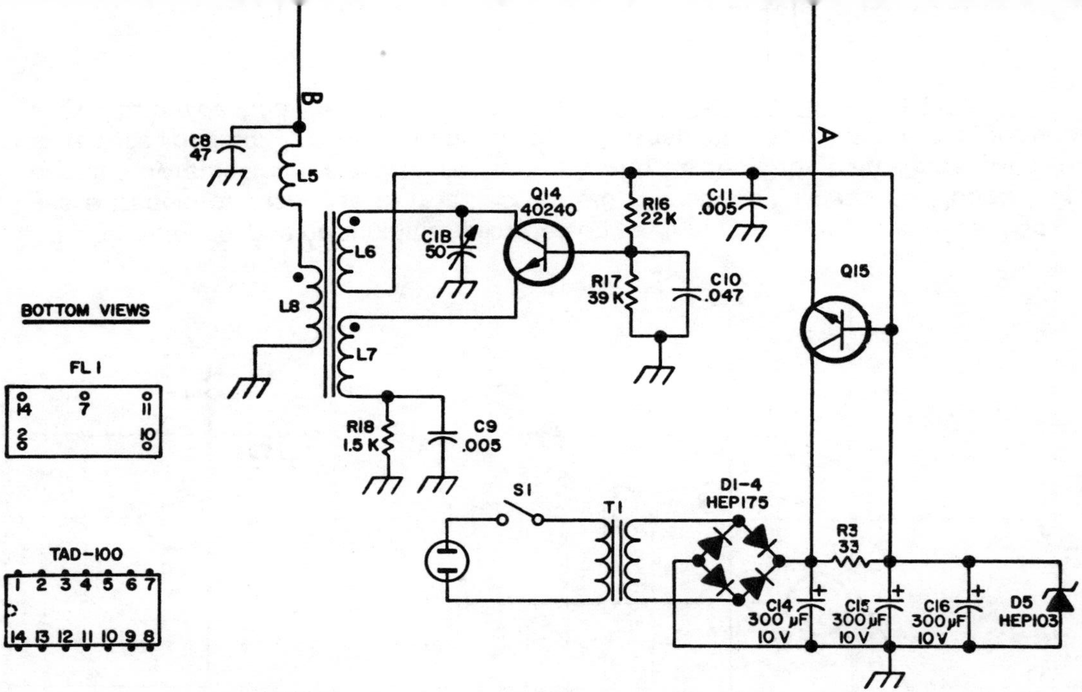


Circuit diagram of the SWL receiver incorporating the TAD-100 IC chip

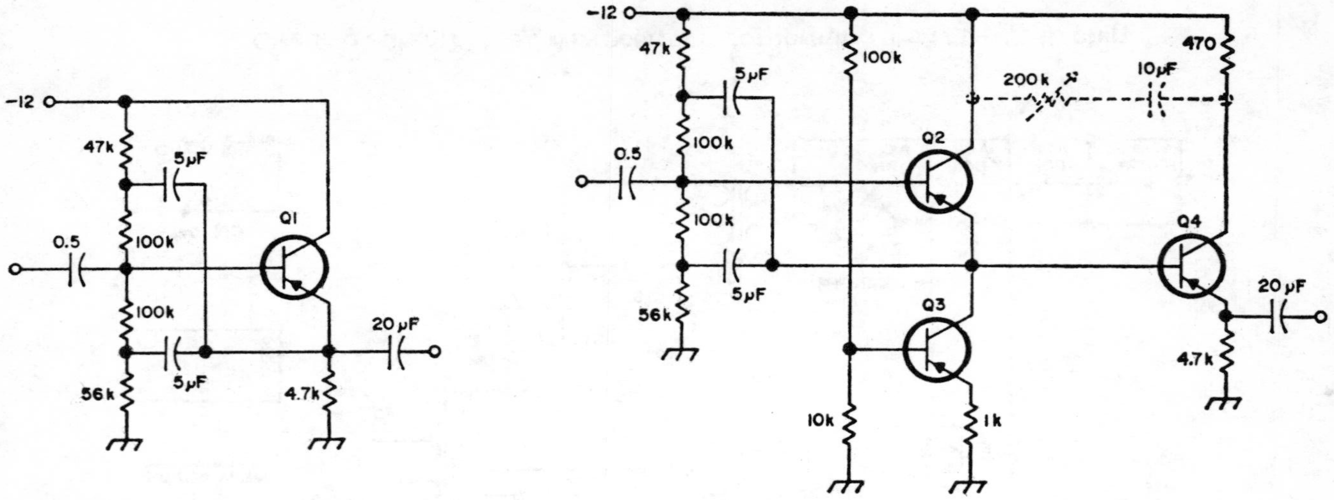


This high-impedance preamplifier provides up to 20 megohms input impedance and has a frequency response from 10 Hz to 220 kHz. Circuit B was developed from circuit A by replacing the emitter resistor in A with Q3 and adding an emitter follower to reduce loading. The input impedance is further increased by the components shown by the dashed line. All transistors are 2N2188, SK3005, GE-9, or HEP 2.

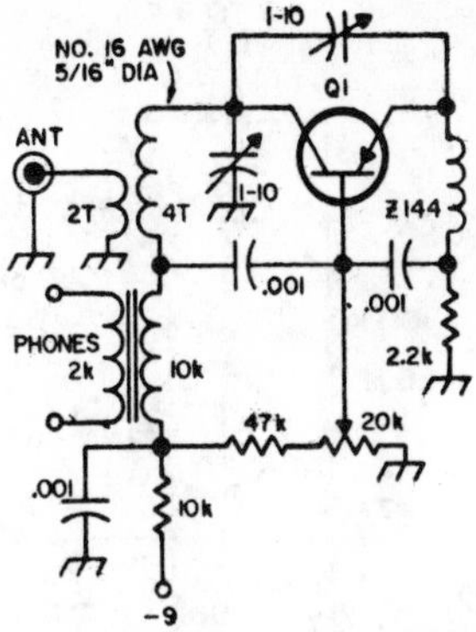


This simple one-transistor superregenerative receiver for two meters may be used for copying many local signals. With the components shown this receiver will tune from about 90 to 150 MHz. It may be used on other frequencies by changing the inductor and capacitor Q1 is a GE-9 or HEP 2.

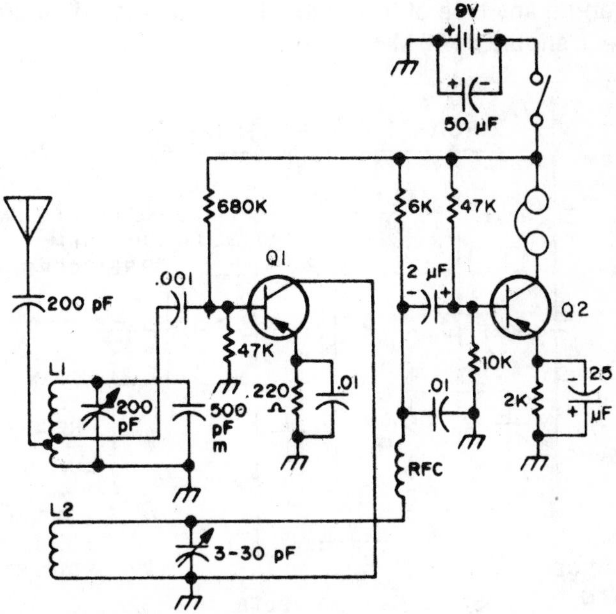


Regenerative receiver for 3.5 MHz optimizes LC ratio for better tuning and controllability. Q1 = 2N370, Q2 = OC70 or 71.

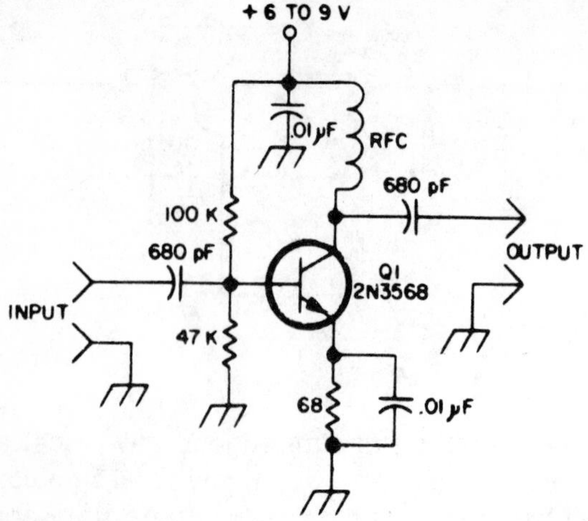


This unit offers 12–15 dB gain on 10m, and about 20 dB or more from 15m and down. All leads should be as short as possible. Building it on a PC board should give good results. Q1 can be any type of transistor. The beta should be around 15o+. $f_t$ should be 60 MHz or better.

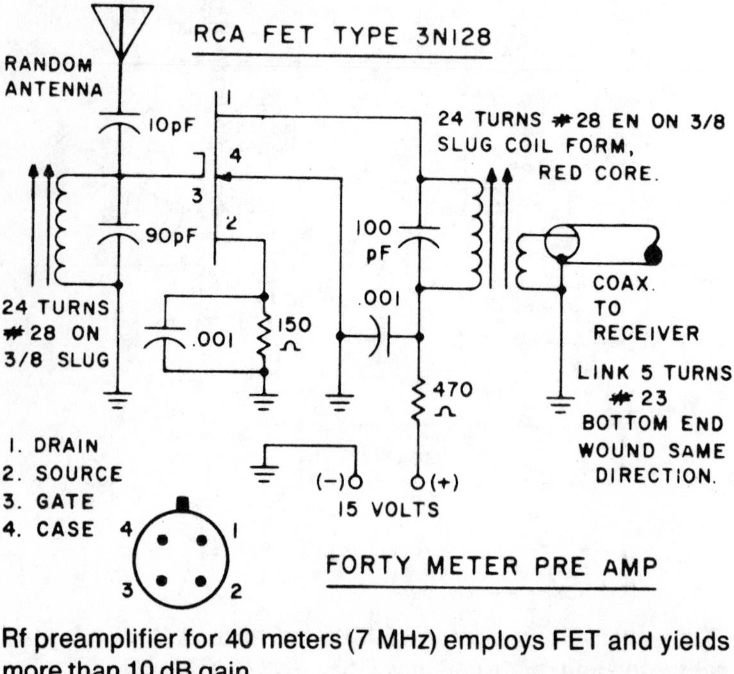


Rf preamplifier for 40 meters (7 MHz) employs FET and yields more than 10 dB gain.

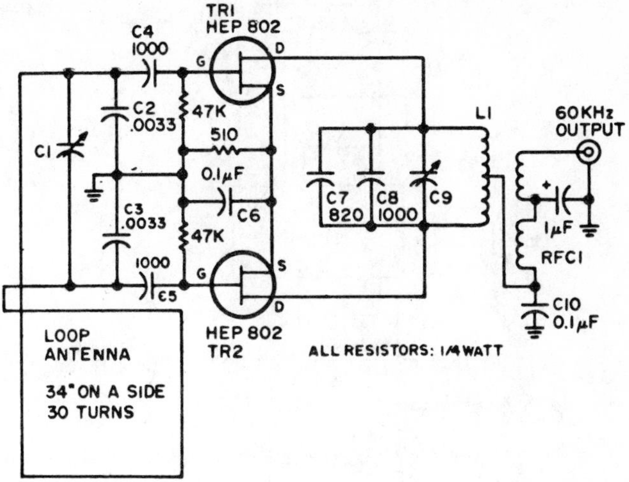


Preamp. C, C9—275—970 pF Elmenco padder 306; C2, C3—0.0033 µF Mylar (2200 pF for 6 ft loop); C4, C5, C8—1000 pF dipplied silver/mica; C6, C10—0.1 µF Centralab CK-104; C7—820 pF dipped silver/mica; L1—3 mH CT. Wound with No. 26 magnet wire approximately 85 turns on an Indiana General Corp. cup core TC7-04-400. Link is 3 turns of insulated wire; RFC1—10 mH Ferrite core rf choke; TR1, TR2—Motorola HEP 802 transistors or RCA 3N128.

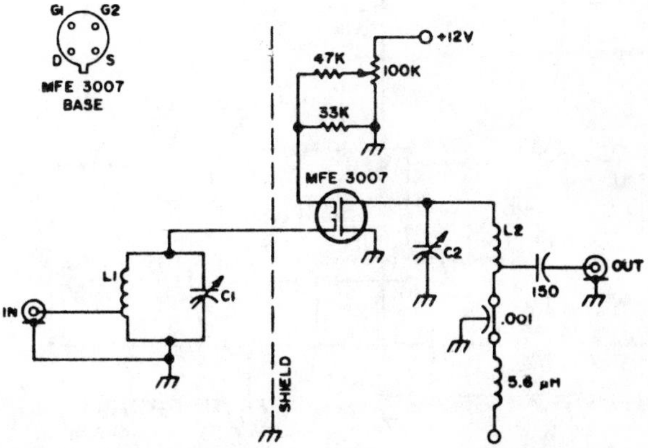


Dual-gate FET preamplifier for 150 MHz range ups received signal strength by 20 dB.

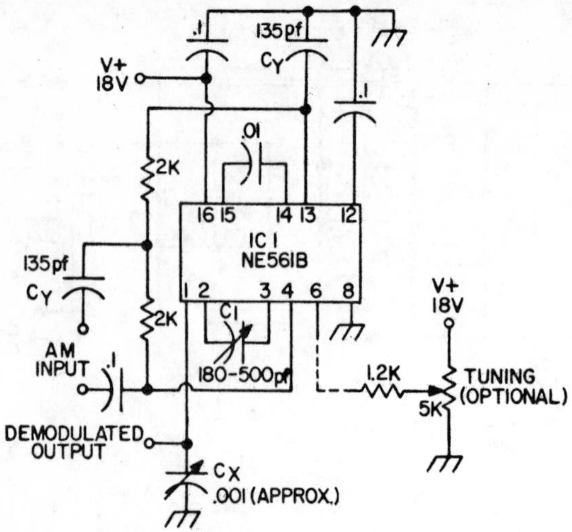


Phase-locked loop allows this AM broadcast-band receiver to be tuned with a 5K pot. Can be adapted to any frequency from 1 to 15 MHz by changing values of $C_Y$ and C1; $C_Y = (f_{hi} - f_{10})/(f_{hi} \times f_{lo})$; C1 = 300 pF/f (MHz).

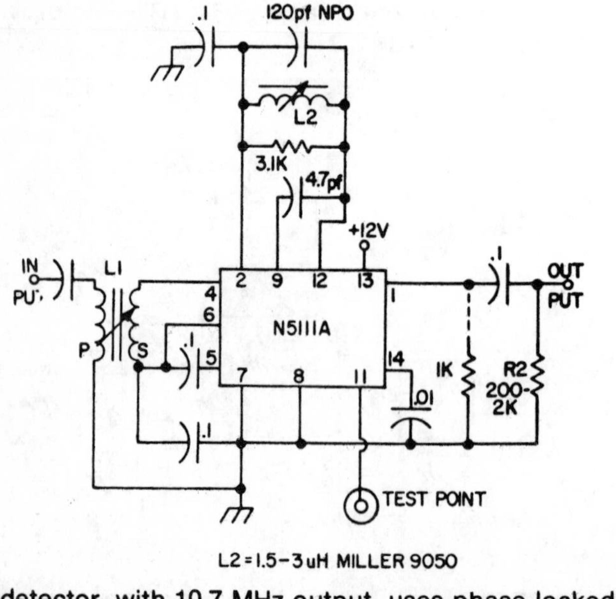


L2 = 1.5–3 uH MILLER 9050

FM detector, with 10.7 MHz output, uses phase-locked loop. Part values are shown.

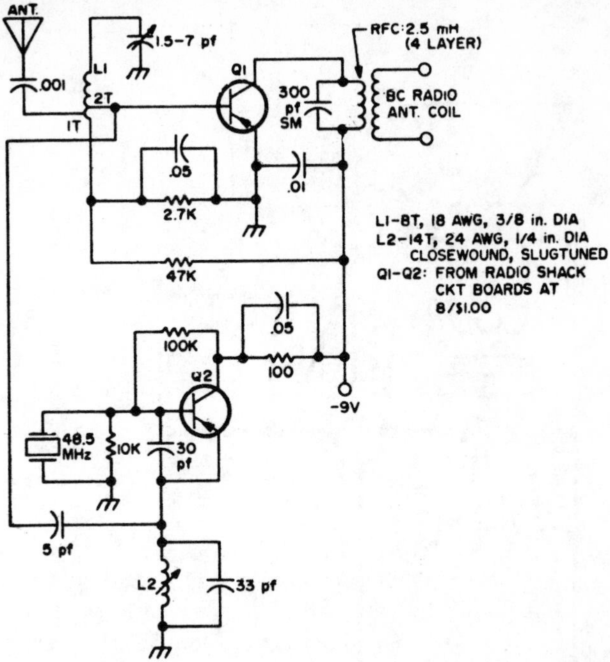


Here is a circuit of a simple 2m converter that works in a pocket AM radio. Since it is crystal controlled, the receiver must tune to a frequency that equals the desired frequency minus 3 × 48.5. Substitute a different frequency crystal if a strong BC station happens to heterodyne with the desired 2m signal.

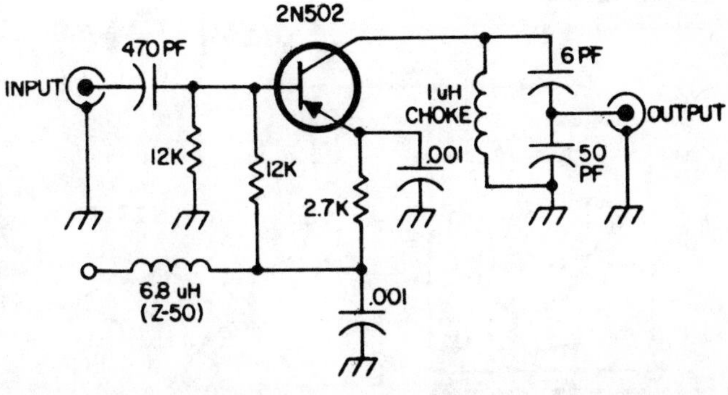


Simple rf preamplifier for 6 meters, TV channel 2 or 3, or any frequency between 50 and 60 MHz. No tuning is required because of broad bandwidth (10 MHz); output impedance is about 50 ohms.

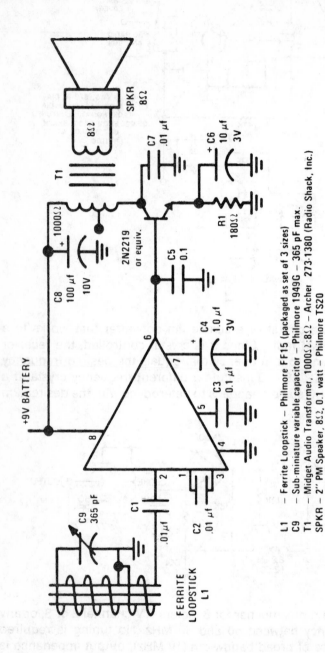

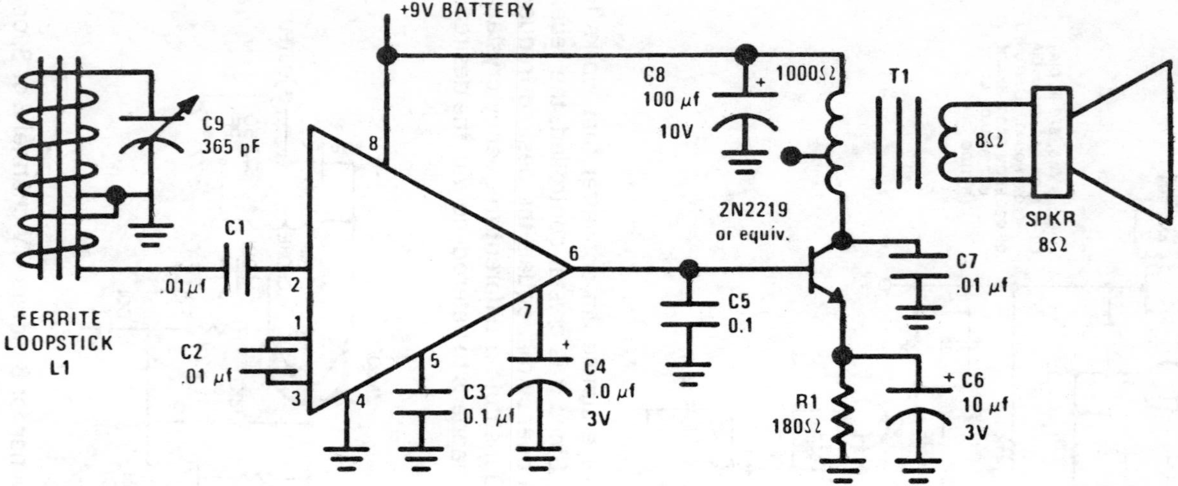


L1 – Ferrite Loopstick – Philmore FF15 (packaged as set of 3 sizes)
C9 – Sub-miniature variable capacitor – Philmore 1949G – 365 pF max.
T1 – Midget Audio Transformer, 1000Ω:8Ω – Archer 273-1380 (Radio Shack, Inc.)
SPKR – 2" PM Speaker, 8Ω, 0.1 watt – Philmore TS20

TRF receiver for standard broadcast band uses LM372 integrated circuit.

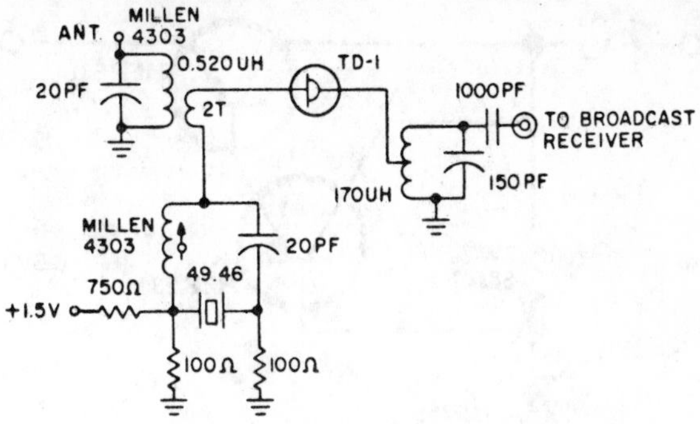


Tunnel-diode converter changes AM broadcast radio to receiver capable of detecting 50 MHz signals.

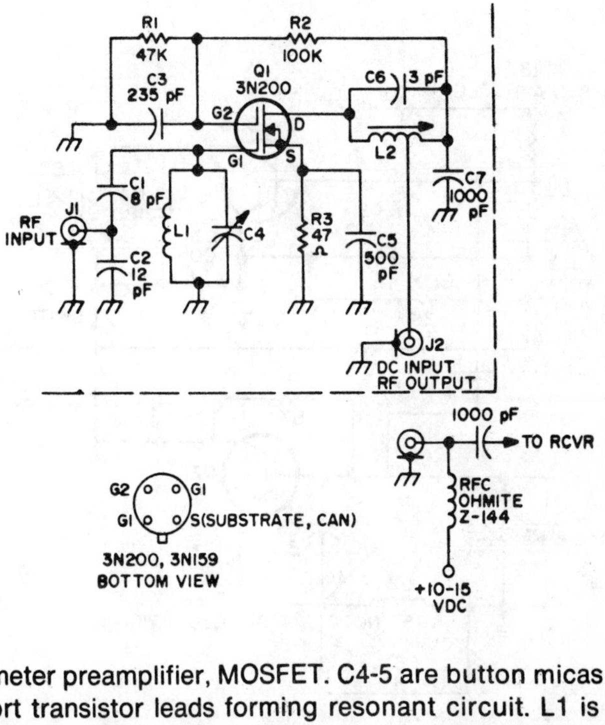


Two-meter preamplifier, MOSFET. C4-5 are button micas and support transistor leads forming resonant circuit. L1 is JFD LC374 tank circuit which contains C1. L2 is 6T 22-gage enamel on 5 mm slug-tuned form, tap at 1 turn. 3N159 will also work in circuit. Circuit courtesy of RTTY Journal, P.O. Box 837, Royal Oak MI 48068.

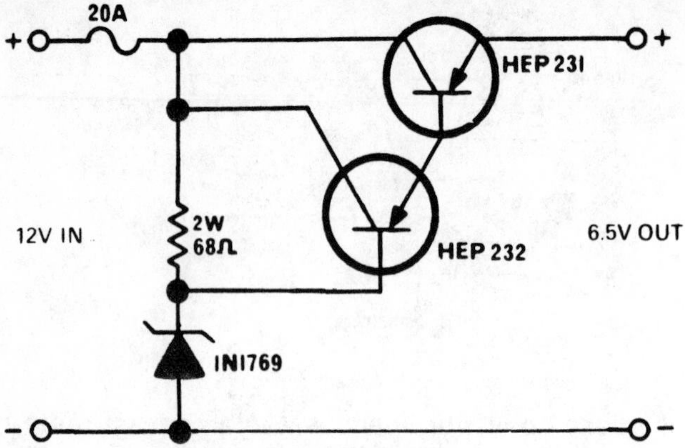


You say you got a real bargain on an old motorcycle FM rig, only to find out it is 6 volt? Fret no more.

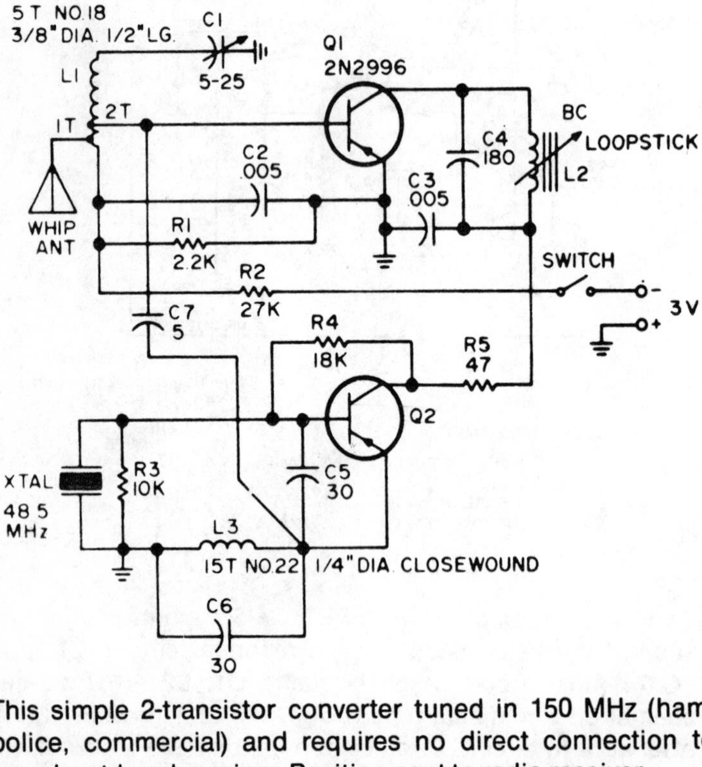


This simple 2-transistor converter tuned in 150 MHz (ham, police, commercial) and requires no direct connection to broadcast-band receiver. Position next to radio receiver.

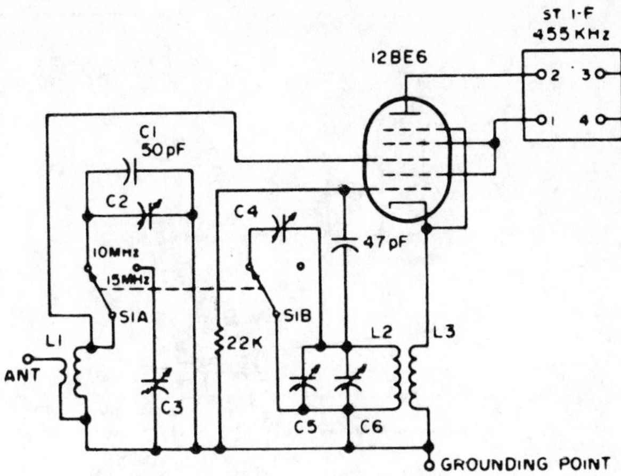


Tunable shortwave converter, designed for receiving WWV on ordinary table radio, allows table radio to receive any signal between 10 and 15 MHz. 15 MHz position must be calibrated with C6 to tune oscillator and C3 to peak rf amplifier; then switch S1 to 10 MHz and tune C4 and C2.

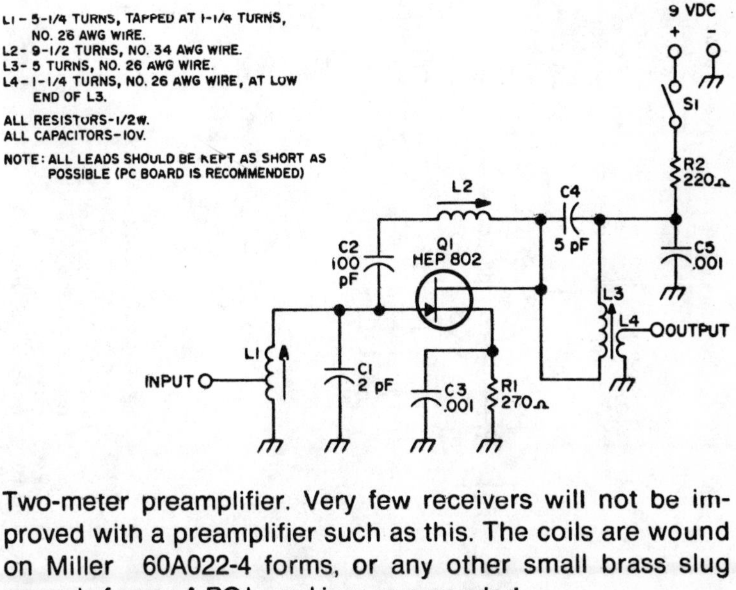


Two-meter preamplifier. Very few receivers will not be improved with a preamplifier such as this. The coils are wound on Miller 60A022-4 forms, or any other small brass slug ceramic forms. A PC board is recommended.

Coil Table, Fig. 2. (All taps counted from ground [cold] end.)

L1 7 turns, airwound, 8/per in. 5/8 O.D., antenna tap at 3 turns, 590 tap at 4 turns.
L2 6 turns, airwound, 4 per in., mixer tap at 2 turns.
L3 5 turns, airwound 8 per in., tap at 1 turn.
L4 35 turns (No. 30), in Miller cup core (from No. 10c 1.F.T.).

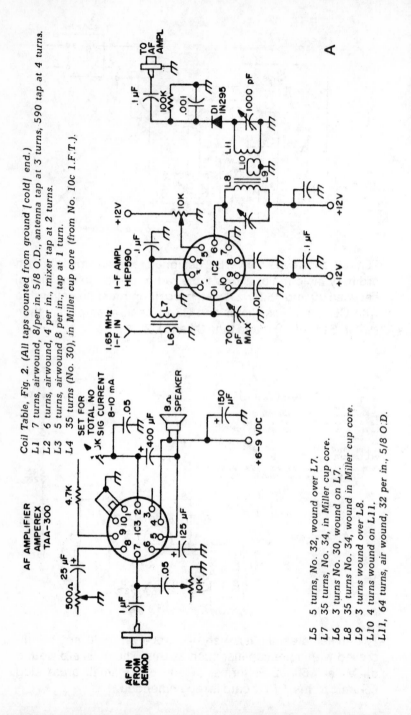

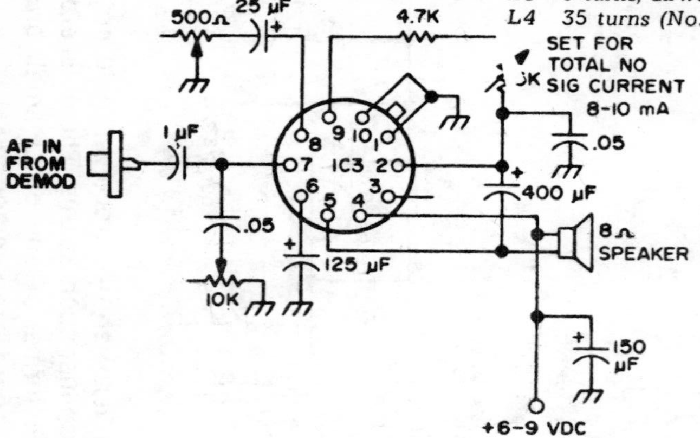


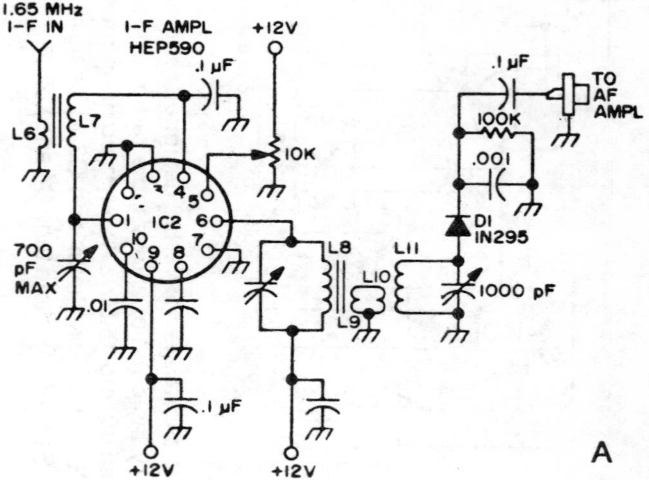


A

L5 5 turns, No. 32, wound over L7.
L7 35 turns, No. 34, in Miller cup core.
L6 3 turns No. 30, wound on L7.
L8 35 turns No. 34, wound in Miller cup core.
L9 3 turns wound over L8.
L10 4 turns wound on L11.
L11, 64 turns, air wound, 32 per in., 5/8 O.D.

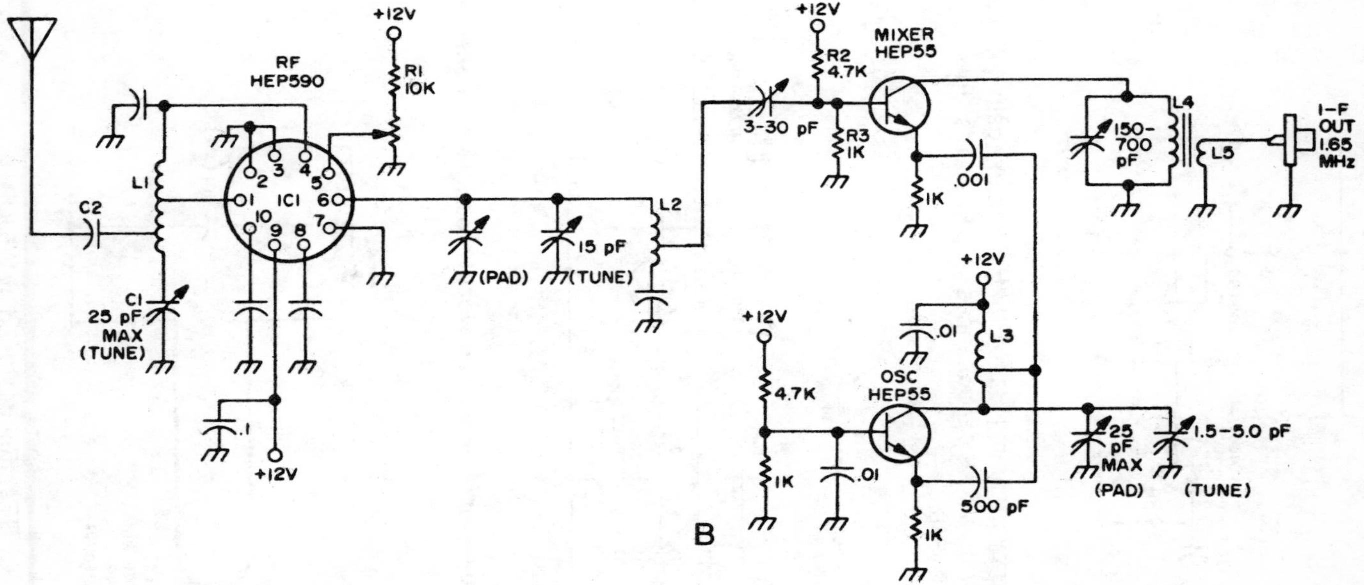


Single-conversion superhet receiver for 50 MHz uses three HEP 590 Motorola integrated circuits. Unit trades selectivity for extreme sensitivity.

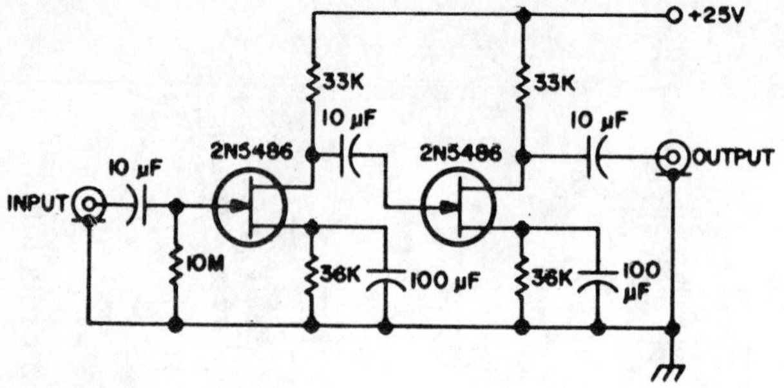


Low-frequency preamp 1 Hz to 50 kHz, voltage gain 400, extremely low noise, all capacitances in $\mu$F, all resistors ½W, transistors 2N5486.

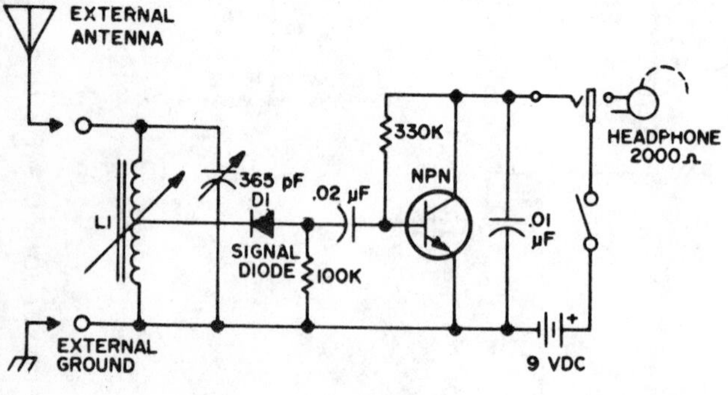


One-transistor radio picks up local broadcast stations.

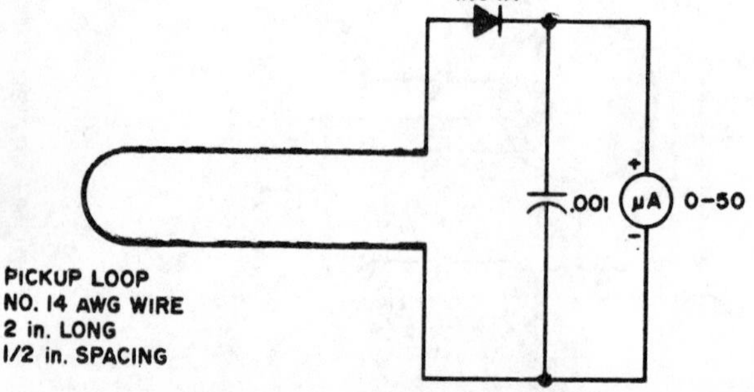


RF sniffer designed by W5JCB.

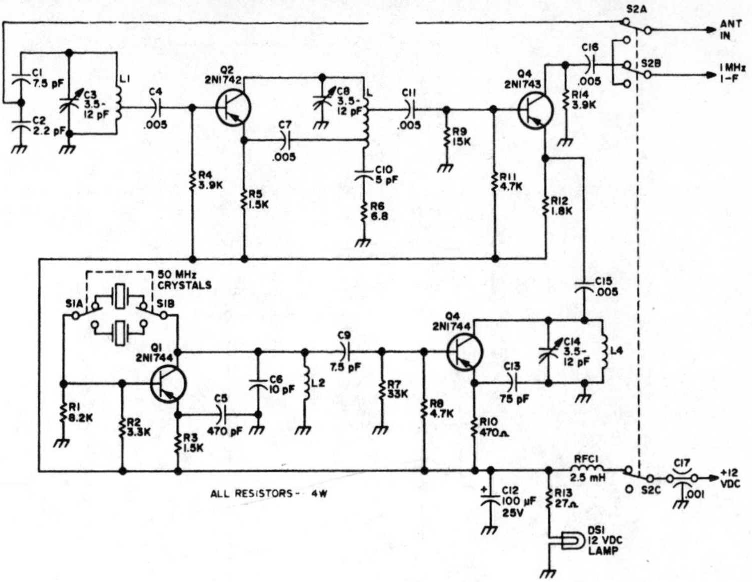


Car radio converter.

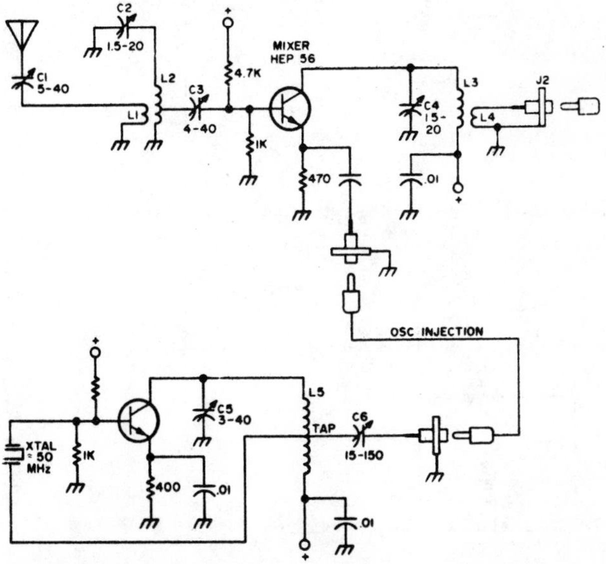


Simple converter allows 170 MHz receiver to be used for reception of 220 MHz signals.

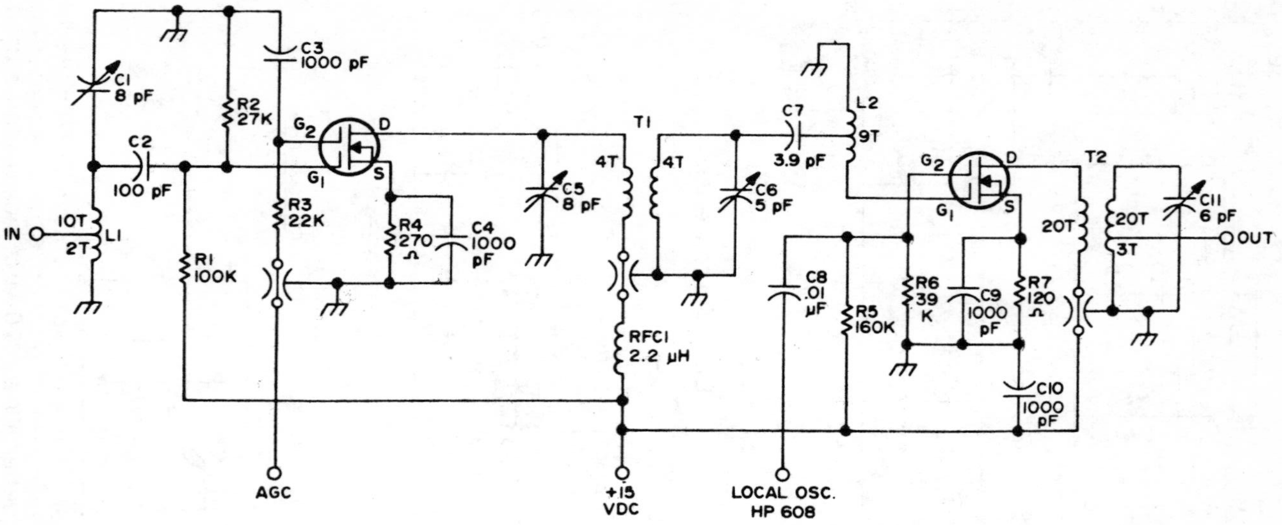


VHF TV tuner using the FT0601 dual-gate MOSFET in RF amplifier and mixer stages.

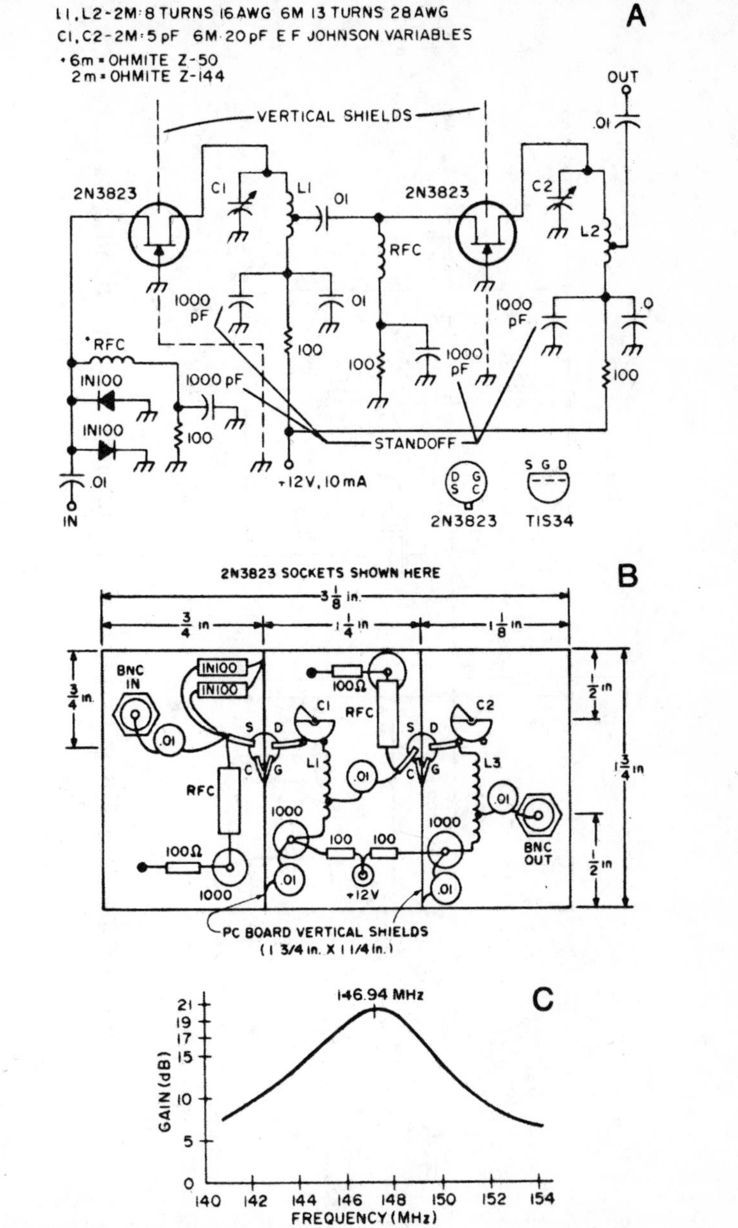


Cascade preamp circuit, using two grounded-gate FETS, provides plenty of rf gain on 6 or 2 meters. Frequency-sensitive values are listed at the upper left portion of the diagram.

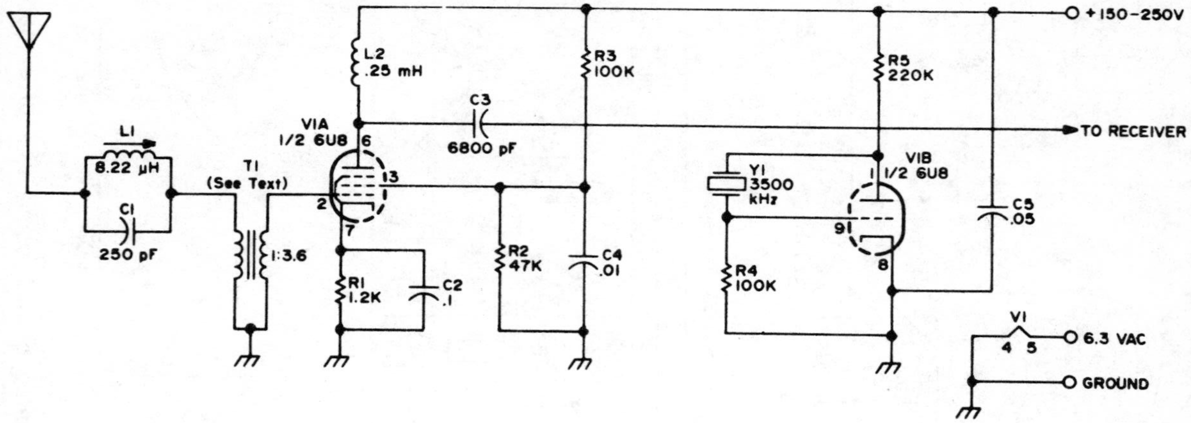


VLF converter, tuning 10 to 30 kHz. I-f is at the low end of the 80 meter band.

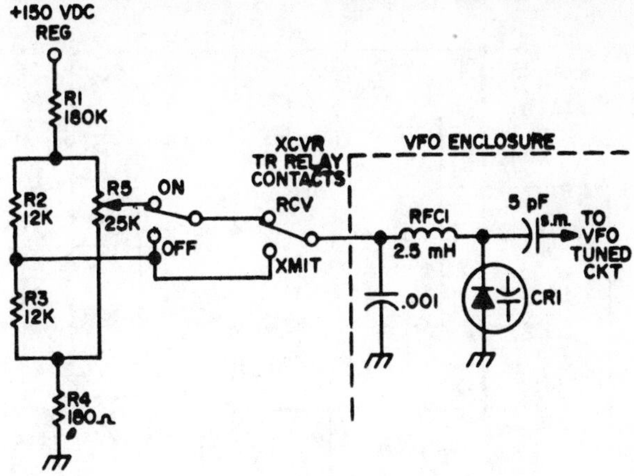


A

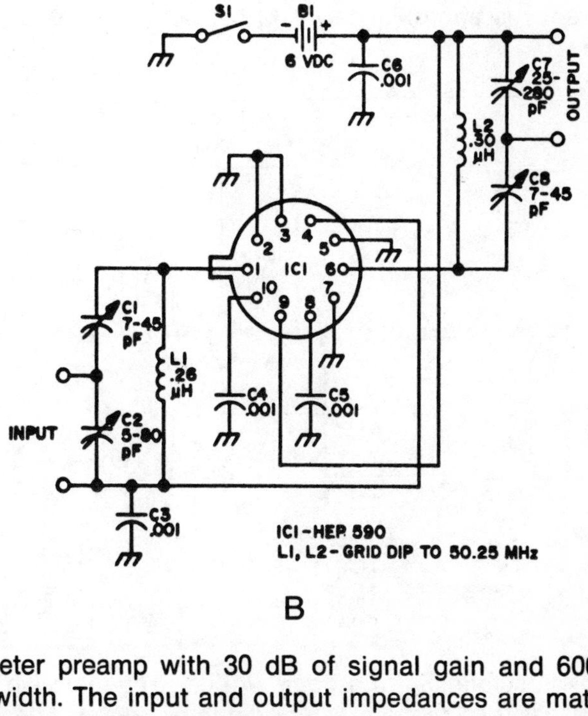


B

Six-meter preamp with 30 dB of signal gain and 600 kHz bandwidth. The input and output impedances are matched. AGC may be added to pin 5. For FM use, dip the coils to 52.5 MHz.

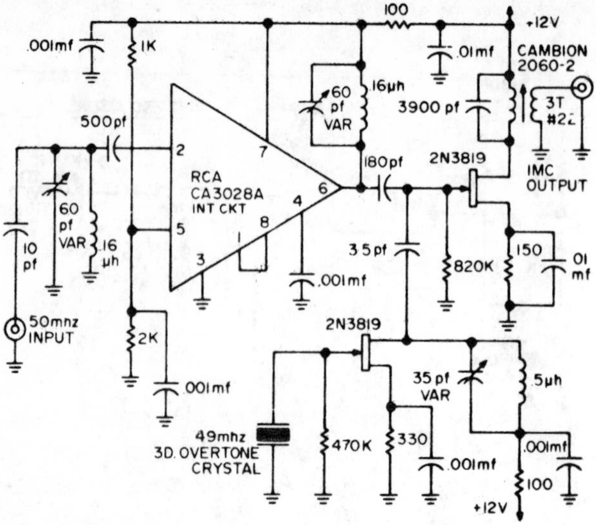


Receiver converter for 50 MHz uses integrated circuit. Output frequency is approximately 1 MHz (center of AM broadcast band).

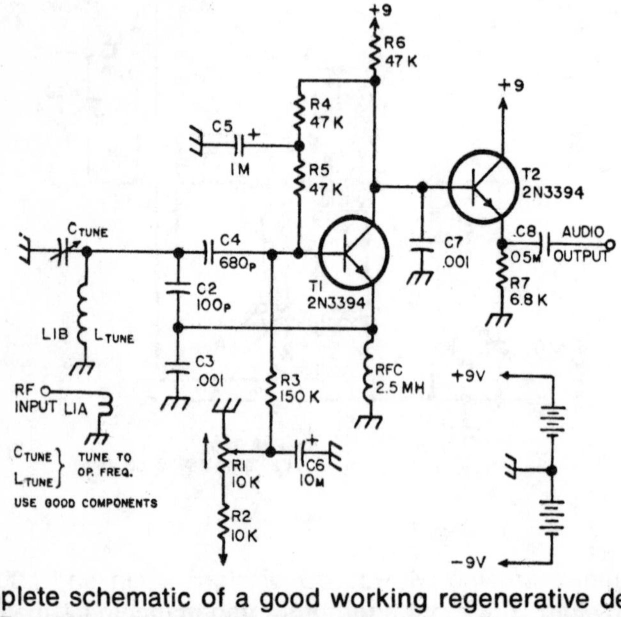


Complete schematic of a good working regenerative detector. Transistor T1 is the actual detector, which operates at very low power levels. T2 is an emitter follower, which copies out the signal with minimum loss.

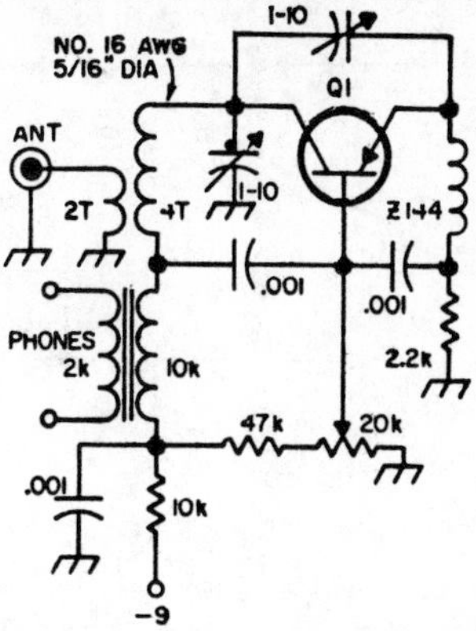


This simple one-transistor superregenerative receiver for 2 meters may be used for copying mnay local signals. With the components shown, this receiver will tune from about 90 to 150 MHz. It may be used on other frequencies by changing the inductor and capacitor. Q1 is 2N1742, 2N2398, 2N3399, GE-9, or HEP 2.

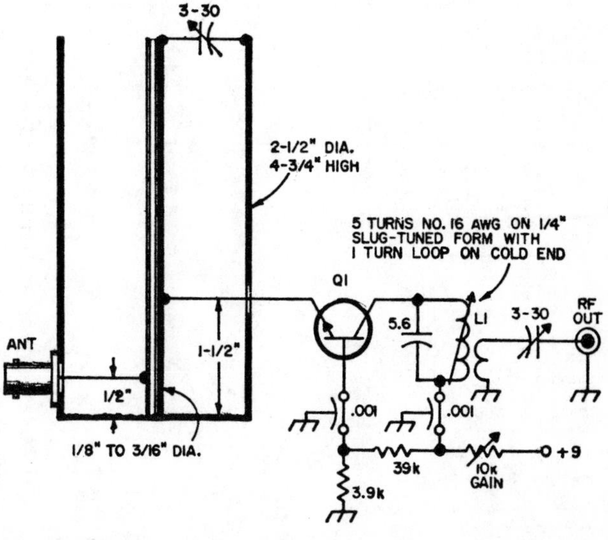


Low-noise 2-meter preamplifier uses a beer can cavity to provide excellent discrimination against nearby kilowatts. Q1 is a 2N3478, 2N3563, 2N3564, 40235, or SK3019.

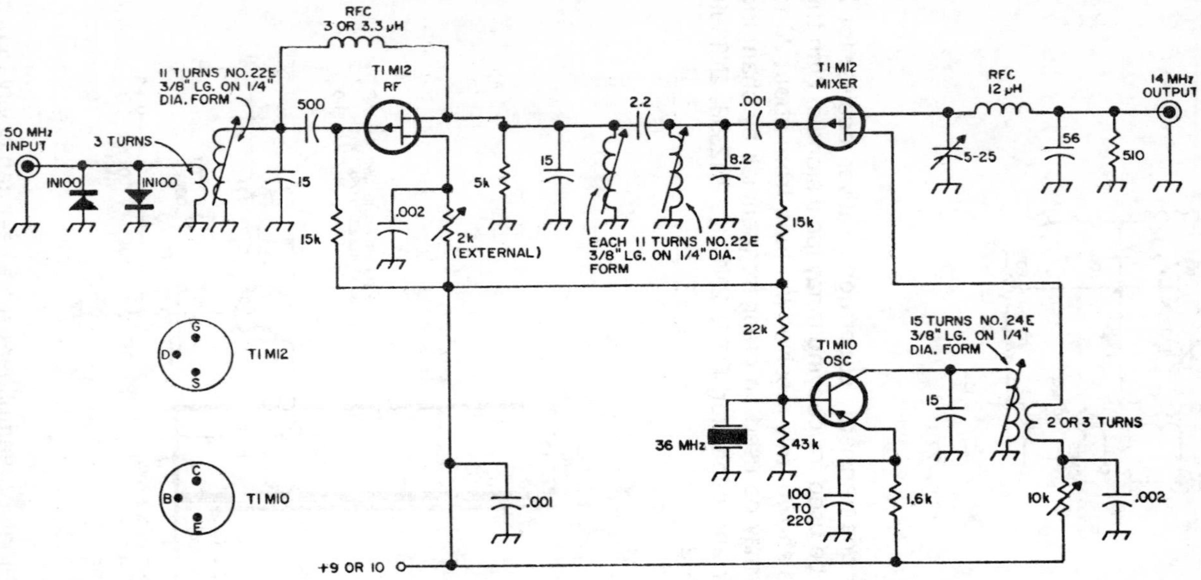


50 MHz converter using field effect transistor rf amplifier and mixer. The FETs cost about $1 each. This converter has excellent noise figure and great resistance to cross modulation.

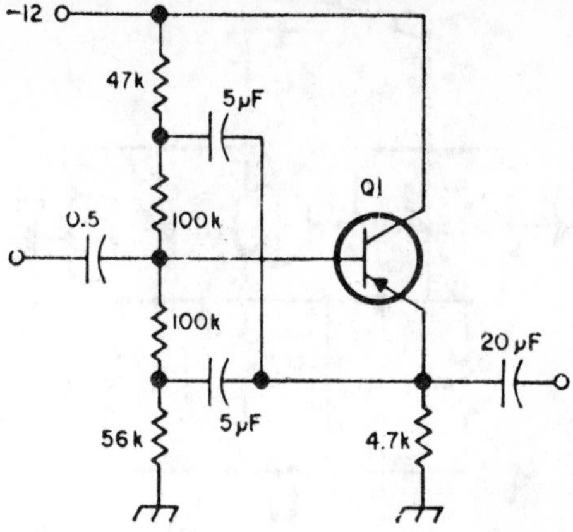


A

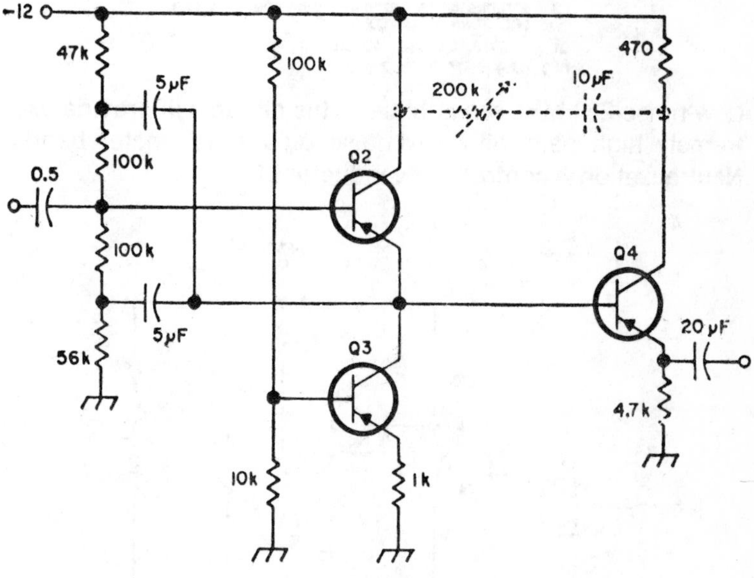


B

This high-impedance preamplifier provides up to 20 megohms input impedance and has a frequency response from 10 Hz to 200 kHz. Circuit B was developed from circuit A by replacing the emitter resistor in A with Q3 and adding an emitter follower to reduce loading. The input impedance is further increased by the components shown by the dashed line. All transistors are 2N2188, SK3005, GE-9, or HEP 2.

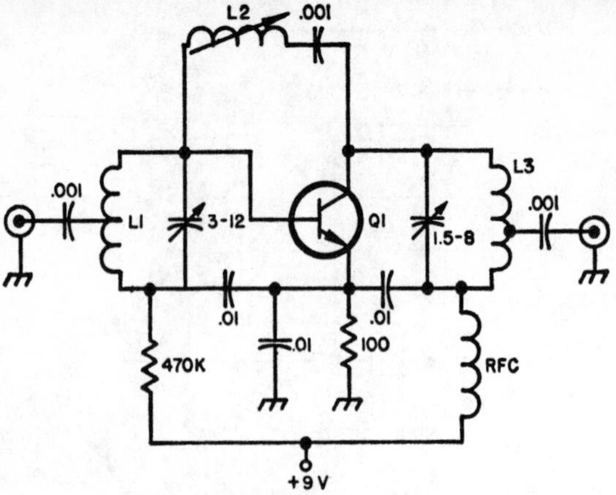


L1 3-1/2 TURNS NO. 16, 1/4" DIAM, 1/2" LONG. TAPPED AT CENTER.
L2 8-1/4 TURNS NO. 24 ON 1/4" SLUG-TUNED FORM.
L3 8 TURNS NO. 16, 1/8" DIAM, 7/8" LONG. TAPPED ONE TURN FROM COLD END.
Q1 2N3478, 2N3564, 40235
RFC 0.84 μH (OHMITE Z-220)

Low-noise 220 MHz preamplifier. This circuit will provide extremely high gain with low noise on the 1¼ meter band. Neutralization is controlled by inductor L2.

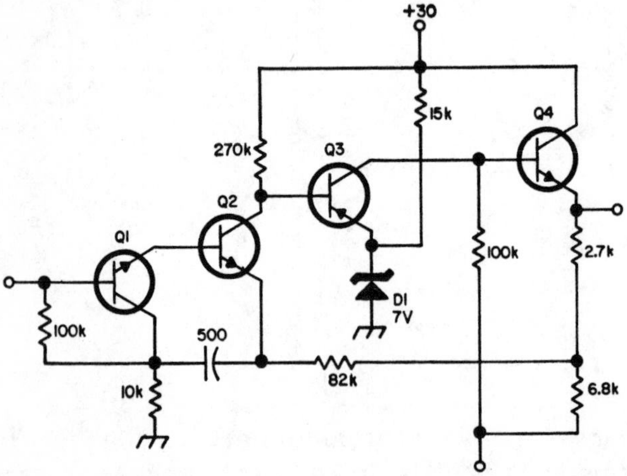


This preamplifier provides 11 dB gain from 0.5 Hz to 2 MHz and has an input impedance of 32 megohms. Transistors Q1, Q2 and Q4 are 2N338, SK3020, or HEP 53; Q3 is a 2N328, GE-2, or HEP 52.

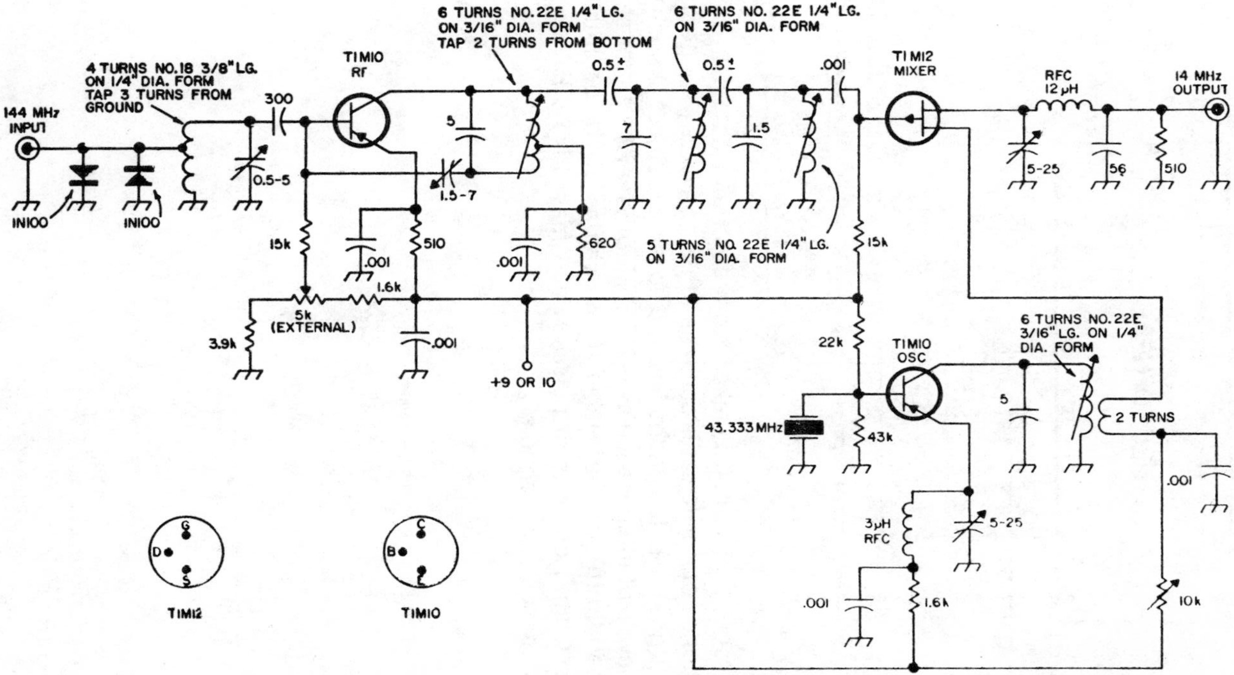


Schematic of low-cost, low-noise, low-cross-modulation 2-meter converter. Note that only the mixer uses an FET; the mixer is responsible for most cross modulation.

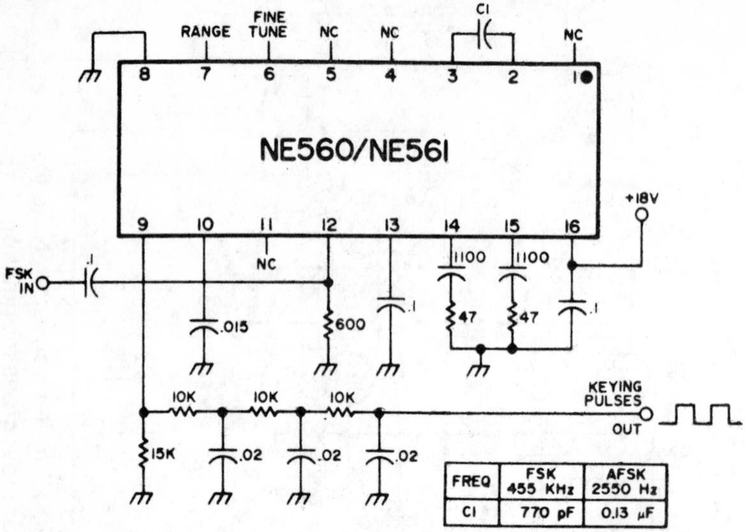


| FREQ | FSK 455 KHz | AFSK 2550 Hz |
|---|---|---|
| C1 | 770 pF | 0.13 μF |

RTTY converter circuit is taken from computer data set applications note; data set is same as AFSK converter, but gets input signal from telephone line and so is not subject to such high levels of interference as is RTTY. Input may be either at i-f or audio frequencies; table shows values of C1 for both cases. Output consists of pulses which may drive a keying circuit for selector magnets.

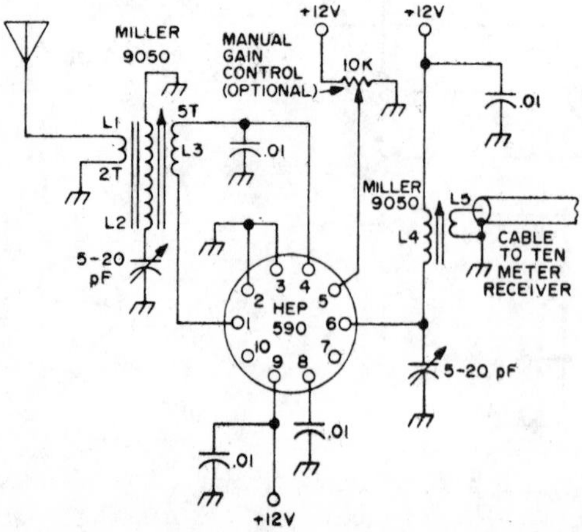


28 MHz rf preamplifier uses HEP 590 integrated circuit (Motorola); interconnection to receiver is via coaxial cable.

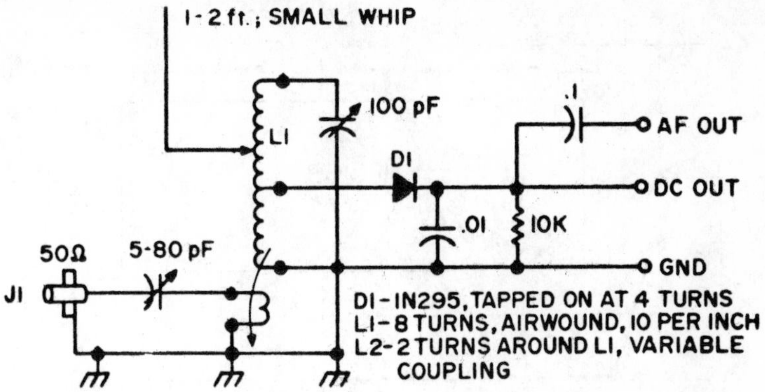


A diode detector makes a convenient and useful receiver for checking the performance of your postage-stamp transmitter while it's on the bench.

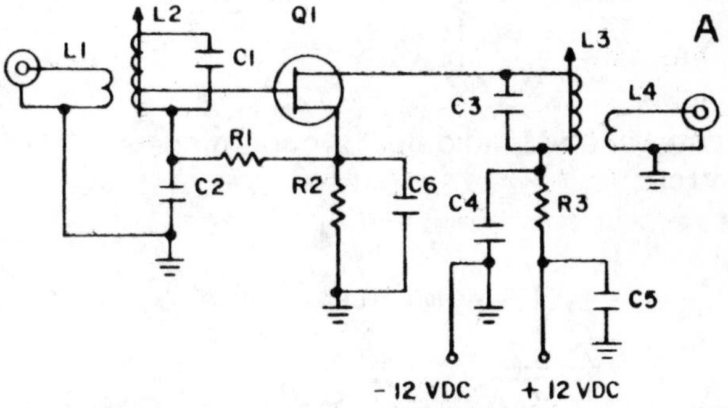


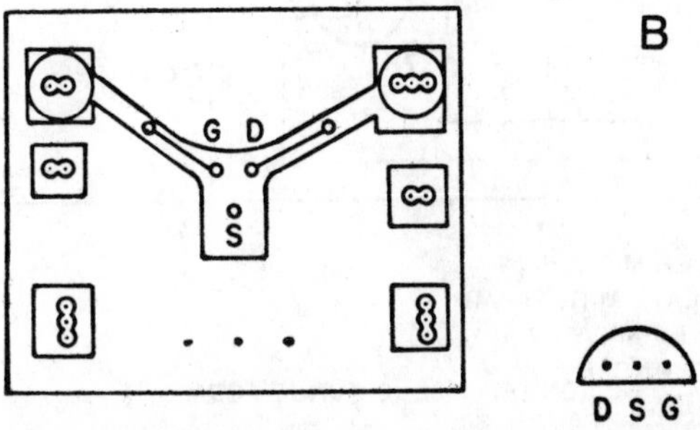


2m FM preamp. L1 & L4, 2T No. 22 hookup on cold end of L3; L2 & L3, 3½T No. 16 spaced the dia. L2 & L3 must be wound opposite directions: C1, 10 pF, C2, 470 pF; C3, 10 pF; C4 R1, 220 Ω ½W; R2, 370Ω ½W; R3, 22Ω ½W; Q1, Motorola MPF-107 or HEP 802.

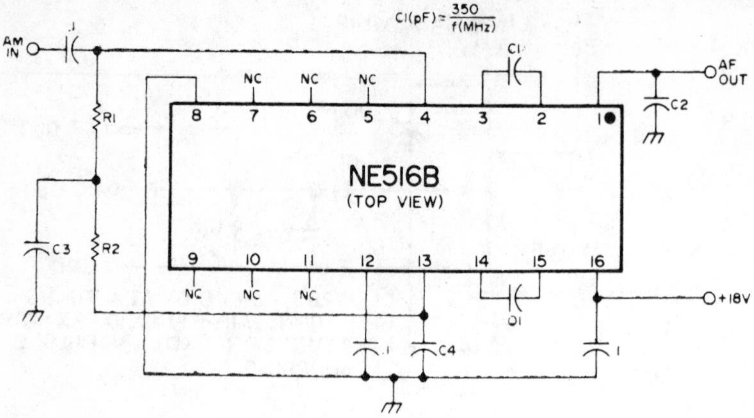


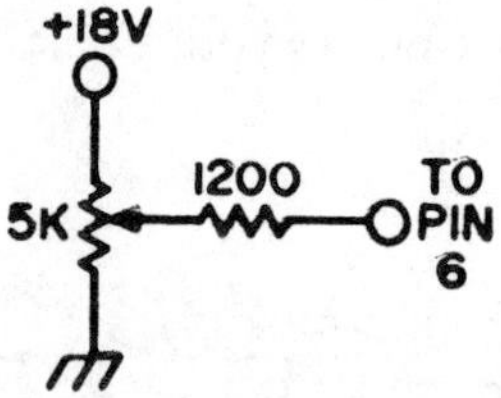


Broadcast-band AM radio built around Amperex TAD 100 integrated circuit.

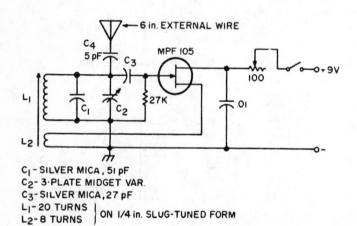

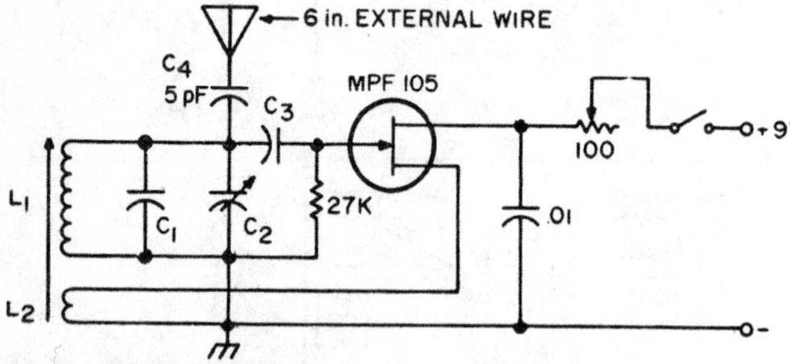


$C_1$ - SILVER MICA, 51 pF
$C_2$ - 3-PLATE MIDGET VAR.
$C_3$ - SILVER MICA, 27 pF
$L_1$ - 20 TURNS } ON 1/4 in. SLUG-TUNED FORM
$L_2$ - 8 TURNS }

The tiny oscillator circuit operates from a standard 9V battery. A small hunk of wire provides an antenna sufficient to insure healthy output for several feet. Placed close to a conventional all-wave receiver, the unit provides sufficient carrier injection for copying single sideband and CW.

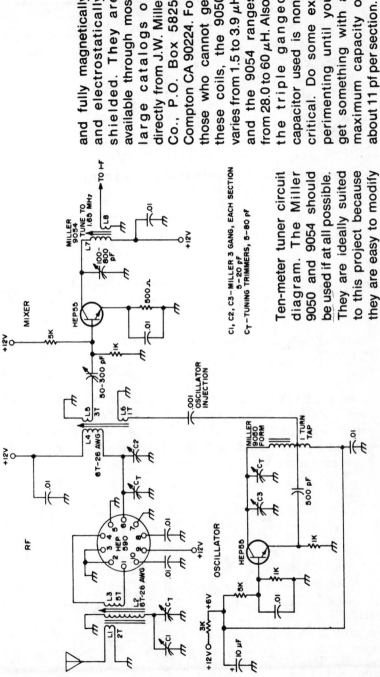

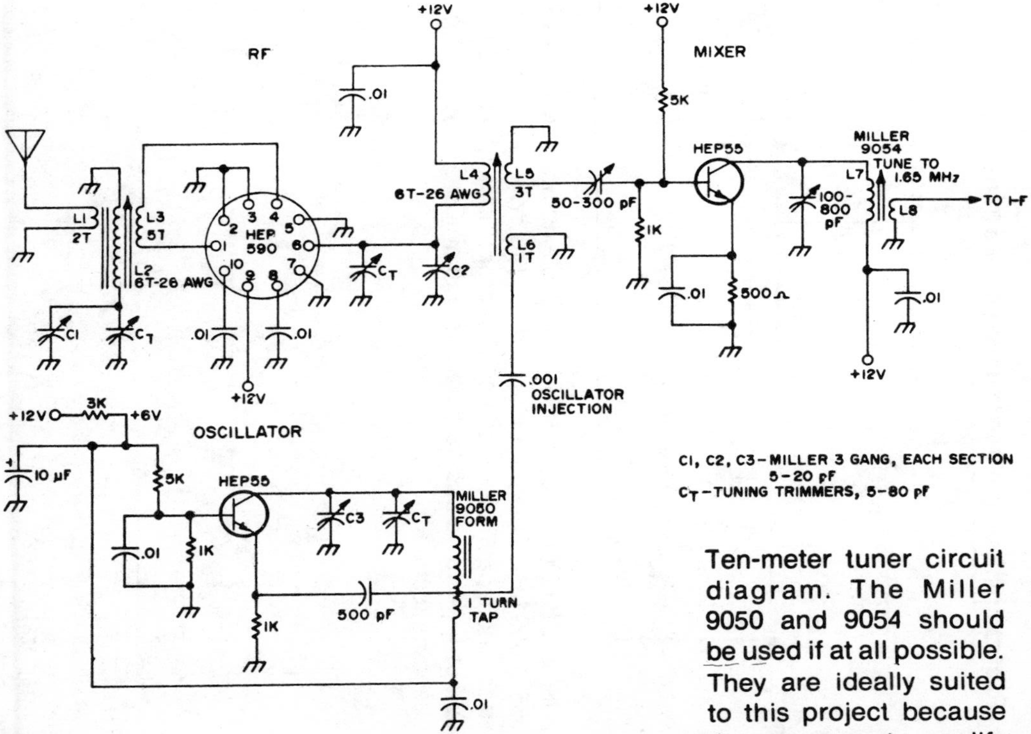


Ten-meter tuner circuit diagram. The Miller 9050 and 9054 should be used if at all possible. They are ideally suited to this project because they are easy to modify

and fully magnetically and electrostatically shielded. They are available through most large catalogs or directly from J.W. Miller Co., P.O. Box 5825, Compton CA 90224. For those who cannot get these coils, the 9050 varies from 1.5 to 3.9 $\mu$H and the 9054 ranges from 28.0 to 60 $\mu$H. Also, the triple ganged capacitor used is non-critical. Do some experimenting until you get something with a maximum capacity of about 11 pf per section.

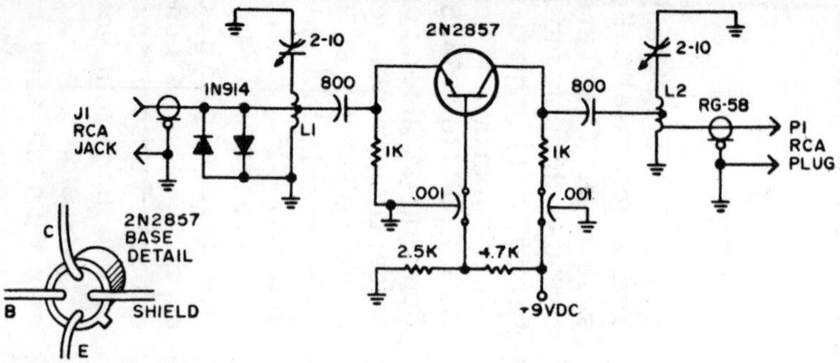


Rf preamplifier for 450 MHz. Insert shows transistor basing.

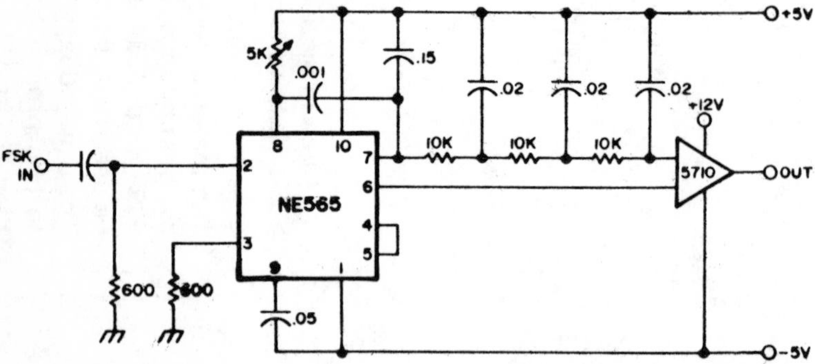


Alternate RTTY circuit uses NE565 IC. Maximum frequency of 565 is 500 kHz. This circuit is designed to drive digital IC devices, and type 5710 voltage comparator is included to adjust output level to values suitable for digital ICs. Pot is for frequency adjustment.

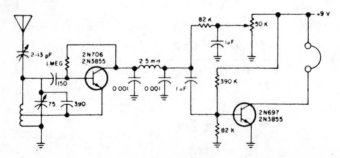

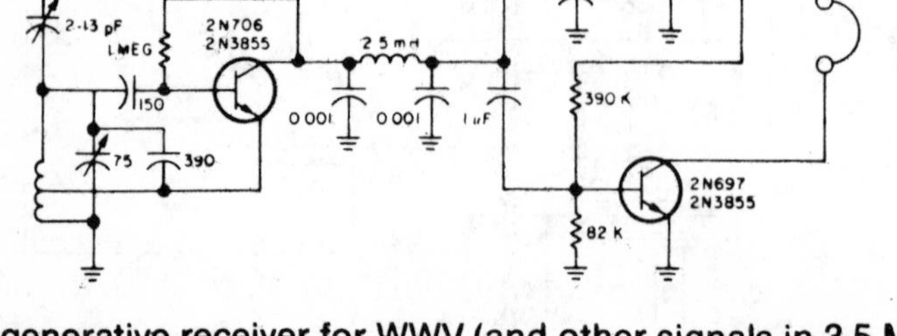


Regenerative receiver for WWV (and other signals in 3.5 MHz region).

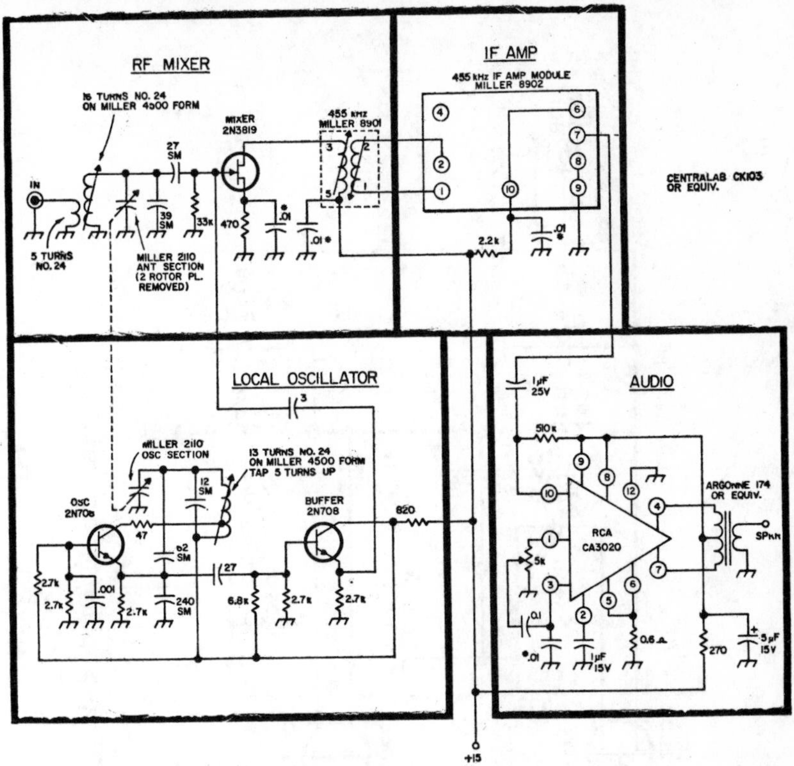


Building-block receiver is made from modules you put together separately.

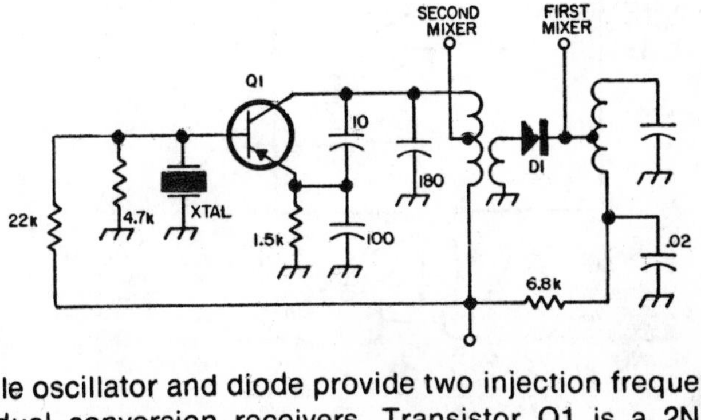


Single oscillator and diode provide two injection frequencies for dual conversion receivers. Transistor Q1 is a 2N1745, 2N2188, T1M10, GE-9 or HEP-2; the diode should be a 1N82A or similar.

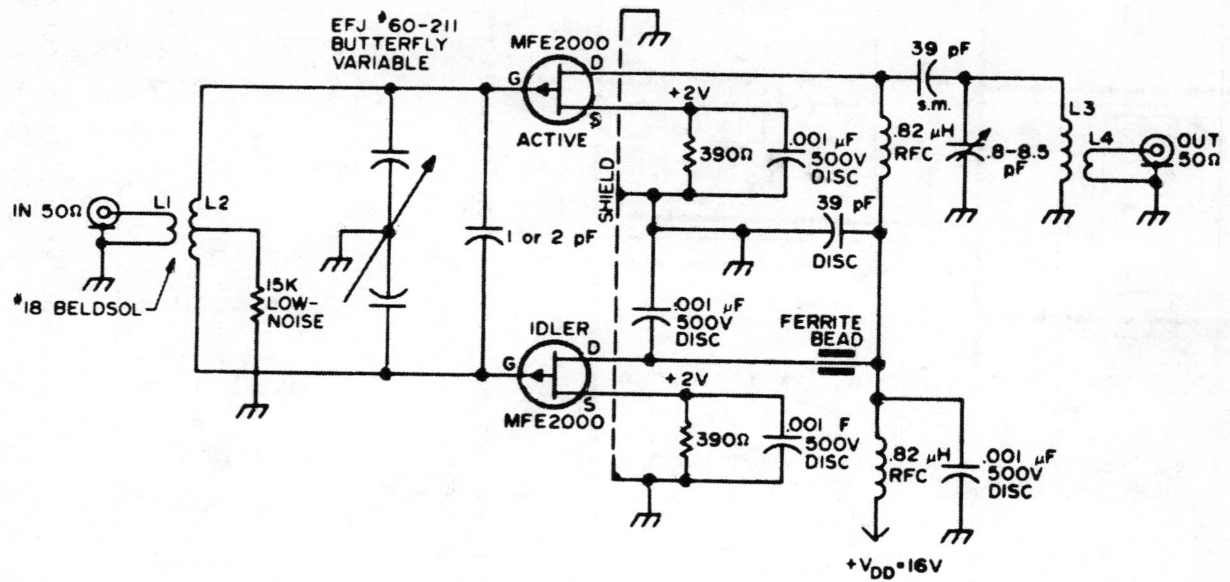


Low noise JFET preamplifier for 2 meters.

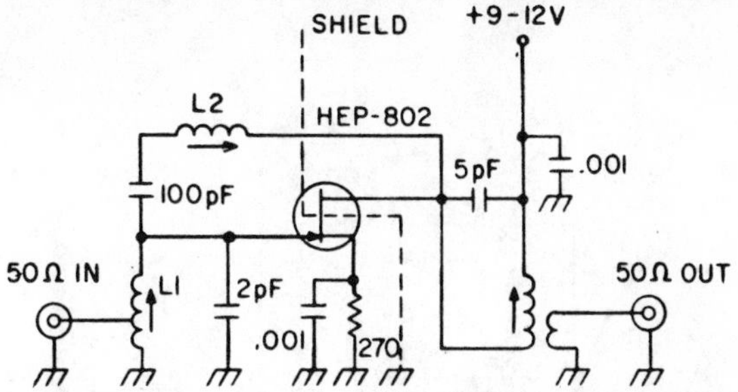


150 MHz rf preamplifier uses Motorola FET for true 14 dB gain (after factoring out noise). Coils should be wound on ¼ in. ceramic forms with brass slugs. L1 is 5.5 turns 26-gage tapped 1.25 turns from cold end; L2 is 9.5 turns 34-gage; L3 is 5 turns 26-gage, L4 is 1.5 turns 26-gage wrapped around lower end of L3. Shield well.

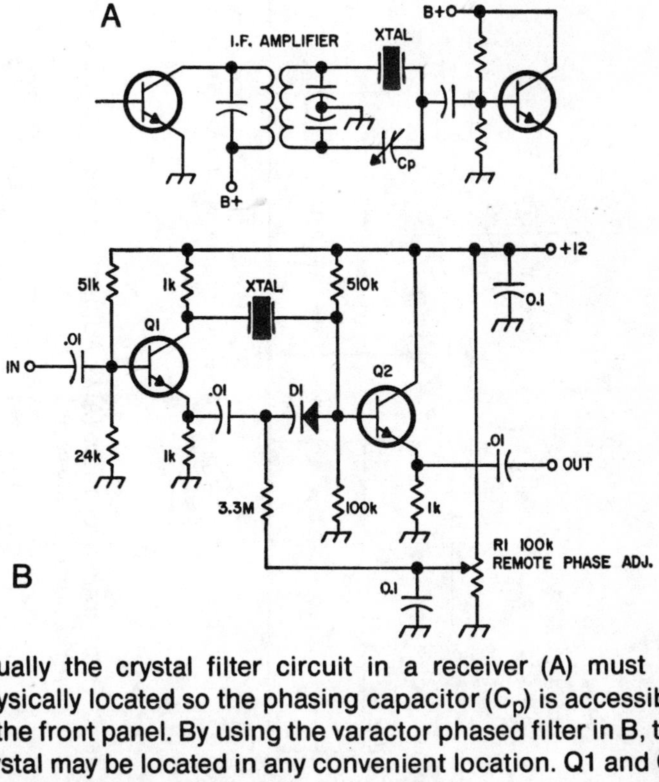


Usually the crystal filter circuit in a receiver (A) must be physically located so the phasing capacitor ($C_p$) is accessible to the front panel. By using the varactor phased filter in B, the crystal may be located in any convenient location. Q1 and Q2 are 2N3478, 2N3564, 2N3707, 40236 or HEP 50; D1 is a 20 pF varactor such as the IN954.

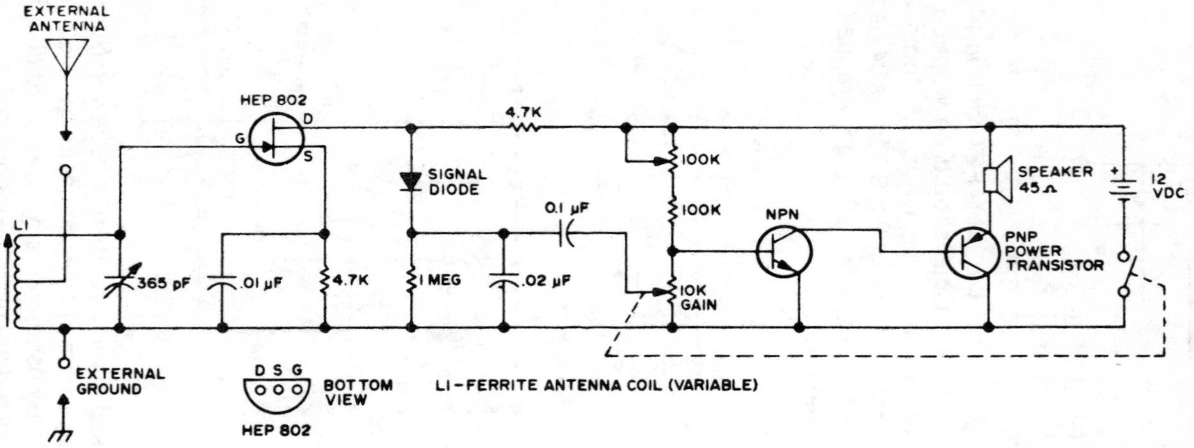


Three transistor radio (AM). Adjust R4 so voltage across speaker is ½ supply voltage. Works surprisingly well.

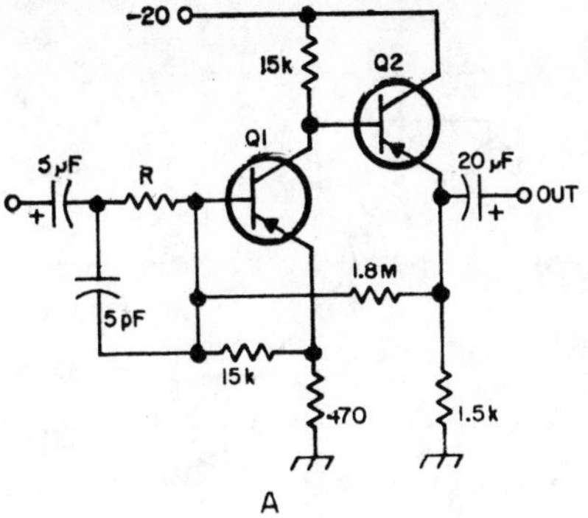


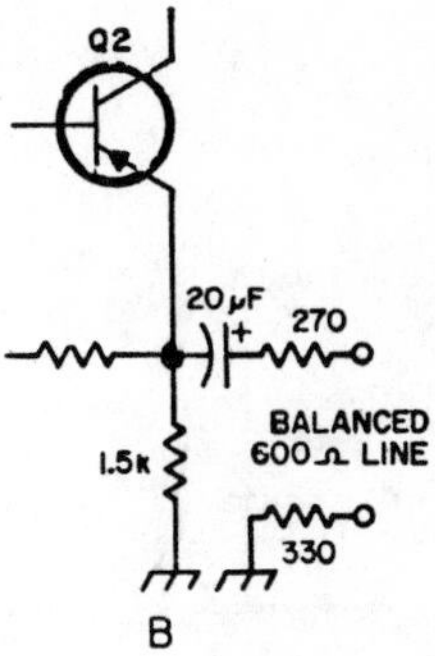


High-impedance preamplifier provides up to 1.2 megohms input impedance; the exact value depends upon resistor R. Both Q1 and Q2 should be a 2N2613, 2N2614, 2N2953, SK3004, GE-2, or HEP 254. A balanced output for reduced hum and noise may be obtained by using the padded output in B.

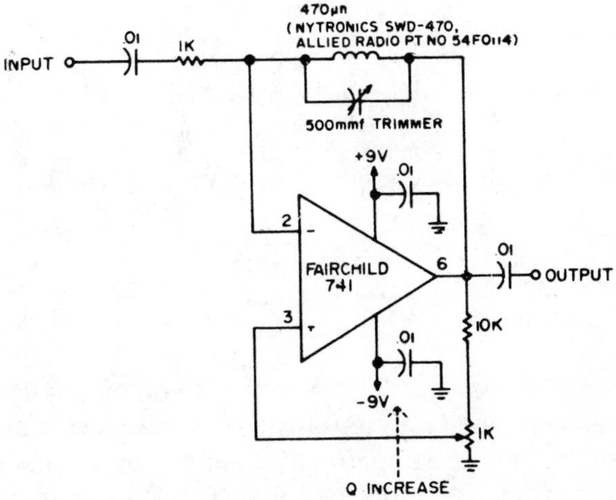


Circuit of the Q-multiplier as constructed for a 455 kHz i-f.

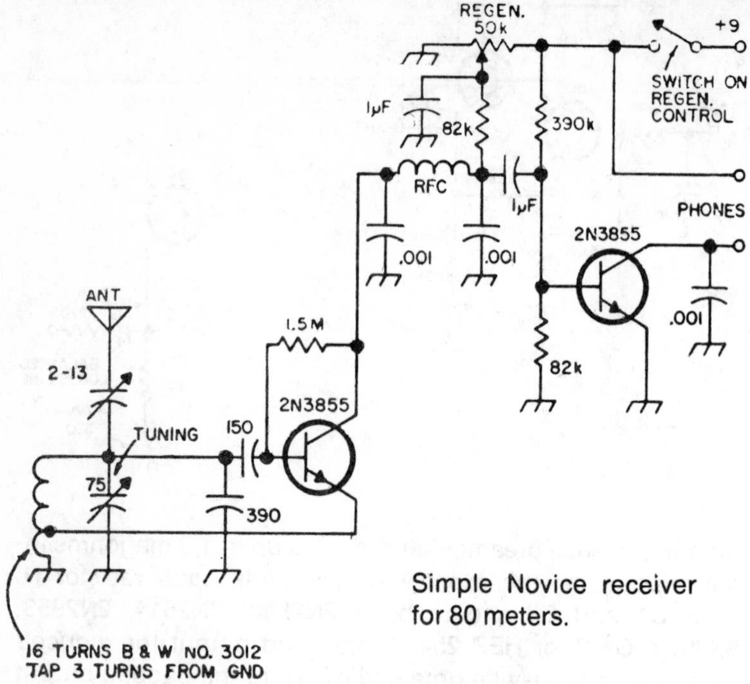


Simple Novice receiver for 80 meters.

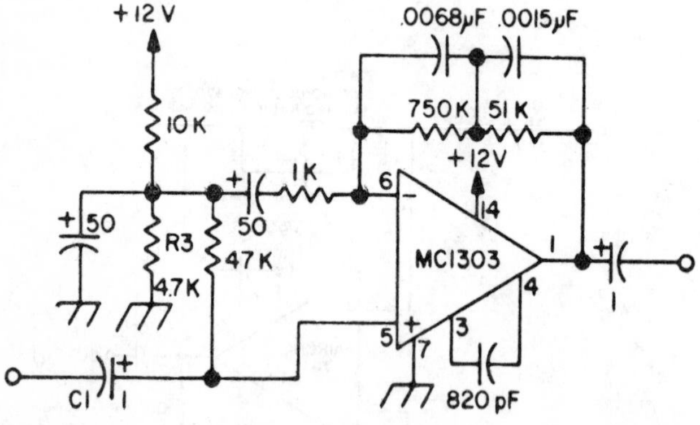


A preamplifier designed for use with a stereo magnetic phono cartridge using the MC1303. The IC is a dual device and only one-half of the preamp (one channel) is shown in the schematic. Build the other half exactly the same as the first but change the following pin numbers to those in parentheses: 6(5), 5(8), 3(11), 4(10) and 1(13).

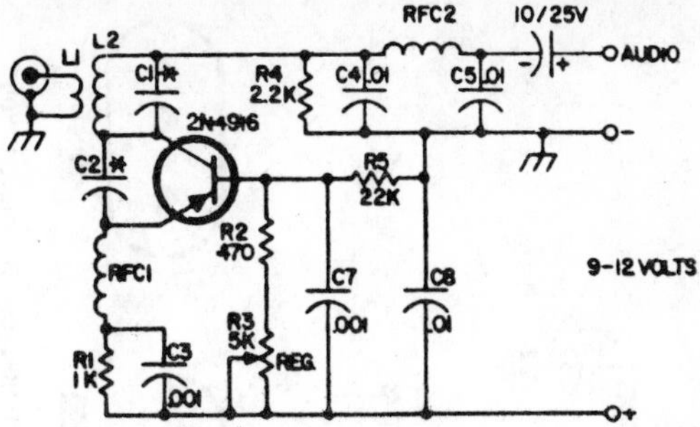


Schematic diagram of superregenerative receiver for the ¾ meter bana.

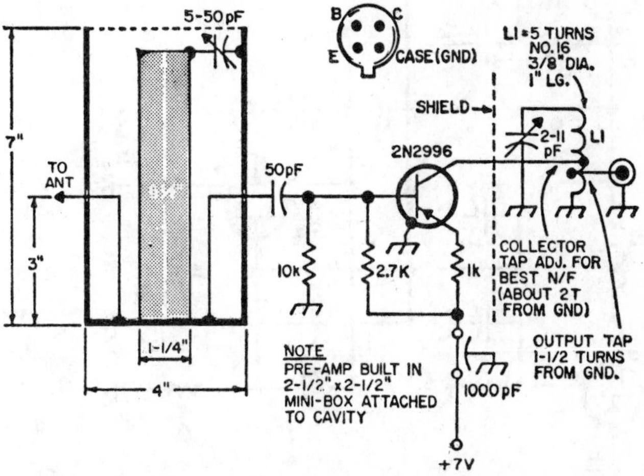


Simple coaxial cavity and transistor preamp for 150 MHz. Emitter should be bypassed with 1000 pF disc.

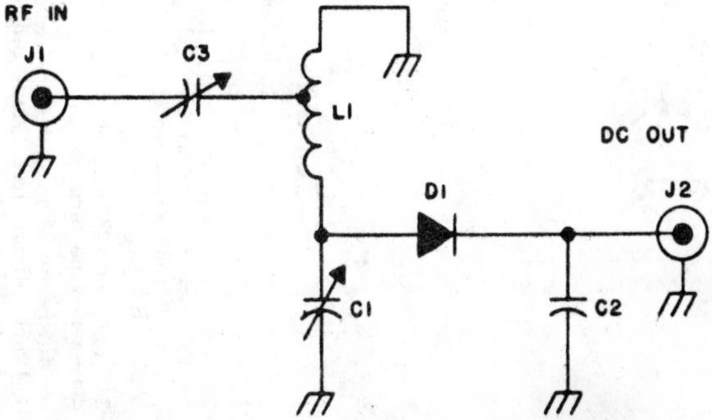


Schematic of diode receiver for 432 MHz.

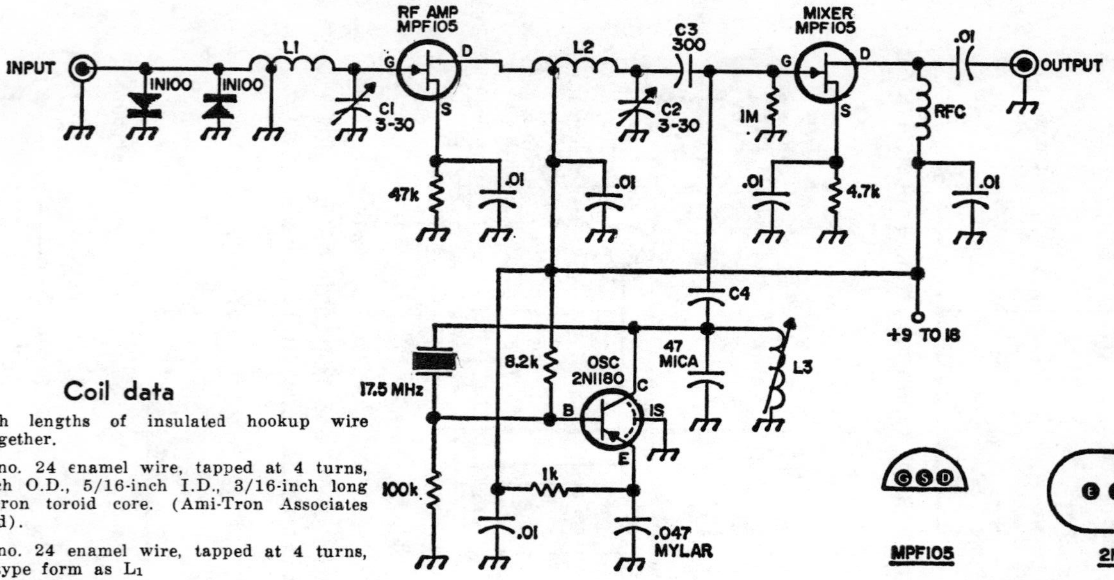


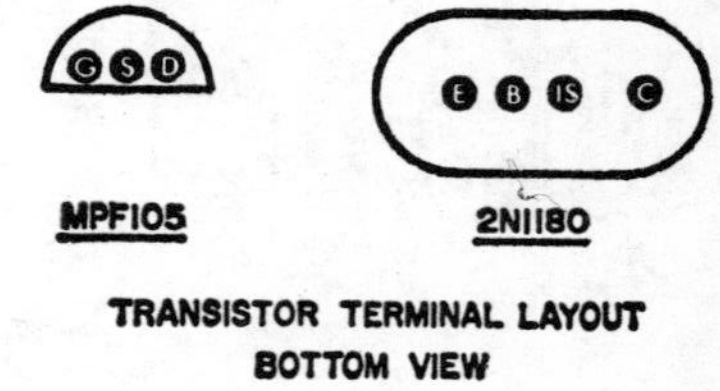


NOTE
ALL RESISTORS 1/4W
ALL CAPACITORS DISC CERAMIC EXCEPT UNLESS OTHERWISE SPECIFIED

## Coil data

$C_4$—two 2-inch lengths of insulated hookup wire twisted together.

$L_1$—23 turns no. 24 enamel wire, tapped at 4 turns, on 1/2-inch O.D., 5/16-inch I.D., 3/16-inch long powdered-iron toroid core. (Ami-Tron Associates T-50-2 Red).

$L_2$—21 turns no. 24 enamel wire, tapped at 4 turns, on same type form as $L_1$

$L_3$—25 turns no. 30 enamel wire, close-wound on 1/4-inch diam. iron slug tuned form (Miller 20A000RBI usable).

Receiver converter accepts 21 MHz signals and amplified them before converting them to about 3.75 MHz.

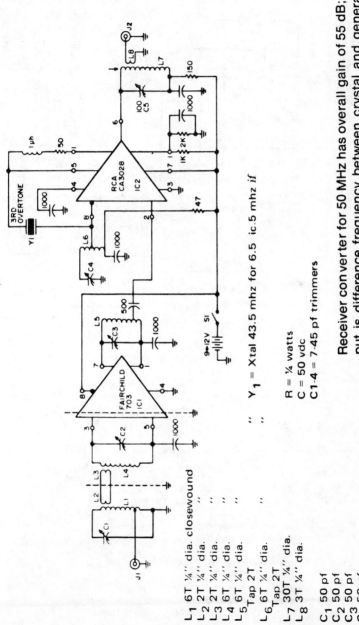

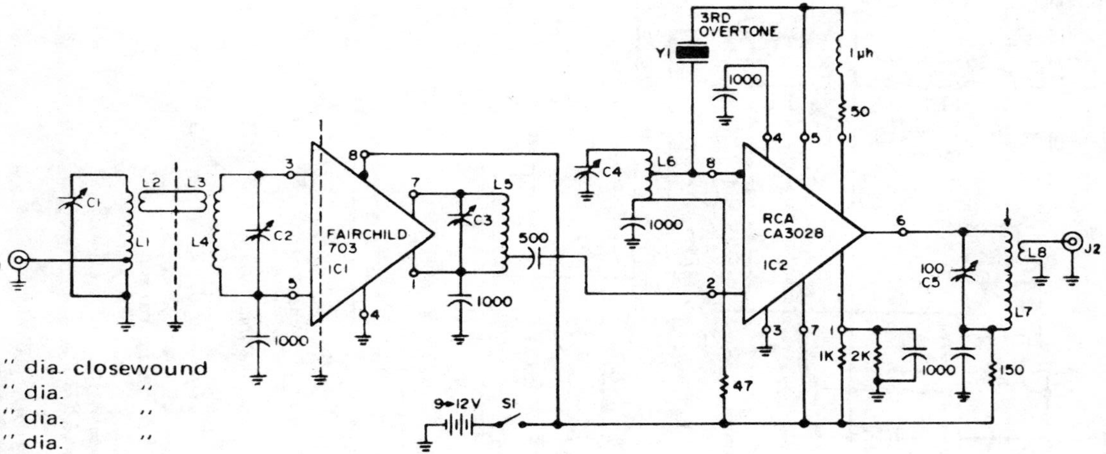


$L_1$ 6T ¼″ dia. closewound
$L_2$ 2T ¼″ dia. ″
$L_3$ 2T ¼″ dia. ″
$L_4$ 6T ¼″ dia. ″
$L_5$ 6T ¼″ dia. ″
Tap 2T ″
$L_6$ 6T ¼″ dia. ″ ″
Tap 2T
$L_7$ 30T ¼″ dia.
$L_8$ 3T ¼″ dia.

$Y_1$ = Xtal 43.5 mhz for 6.5 ic.5 mhz *if*

R = ¼ watts
C = 50 vdc
C1-4 = 7-45 pf trimmers

$C_1$ 50 pf
$C_2$ 50 pf
$C_3$ 50 pf
$C_4$ 50 pf
$C_5$ 100 pf

Receiver converter for 50 MHz has overall gain of 55 dB; output is difference frequency between crystal and generating frequency.

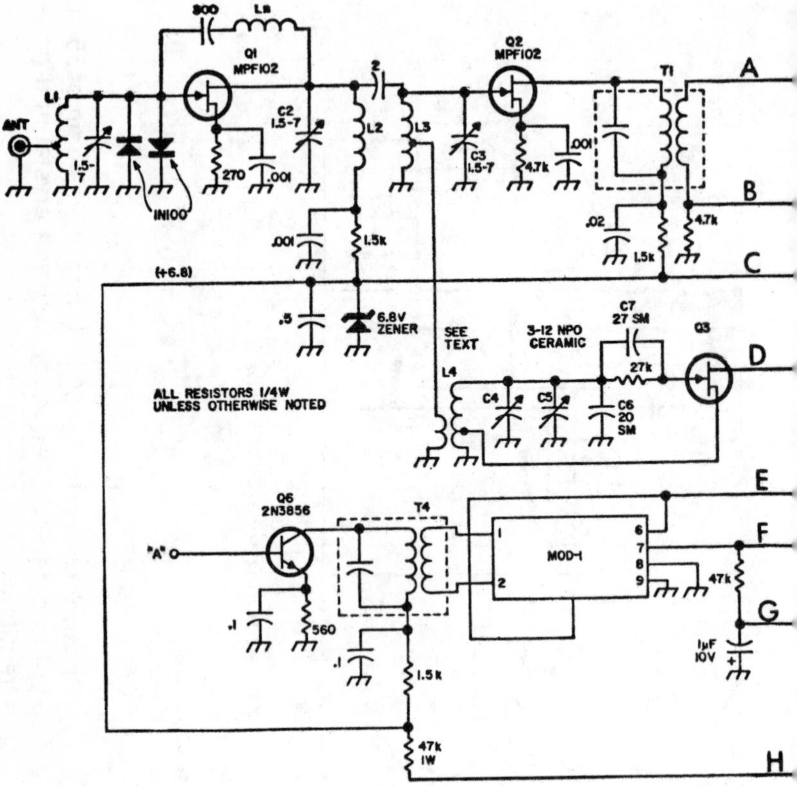


| | | |
|---|---|---|
| *T*1 | 10.7 *mc. FM input IF transformer, J. W. Miller* 2070 | *T*6 |
| *T*2 | 10.7 *mc. FM interstage transformer, J. W. Miller* 2071 | *T*7 |
| *T*3 | 455*khz input transformer, J. W. Miller* 2031 | *CH*1 |
| *T*4 | 455*khz transformer, supplied with IF module* | *L*1<br>*L*2 |
| *T*5 | *driver transformer,* 10*K to* 2*K c.t. Midland* 250633, *Calrad CR*75 | *L*3 |

*output transformer,* 500 *ohms c.t. to* 3.2 *ohms, Midland* 25-631

6.3*V*, .6 *amp or smaller filament transformer, Triad F*-13*X*

2 *hy,* 15 *ma. low resistance choke, Stancor C*2707

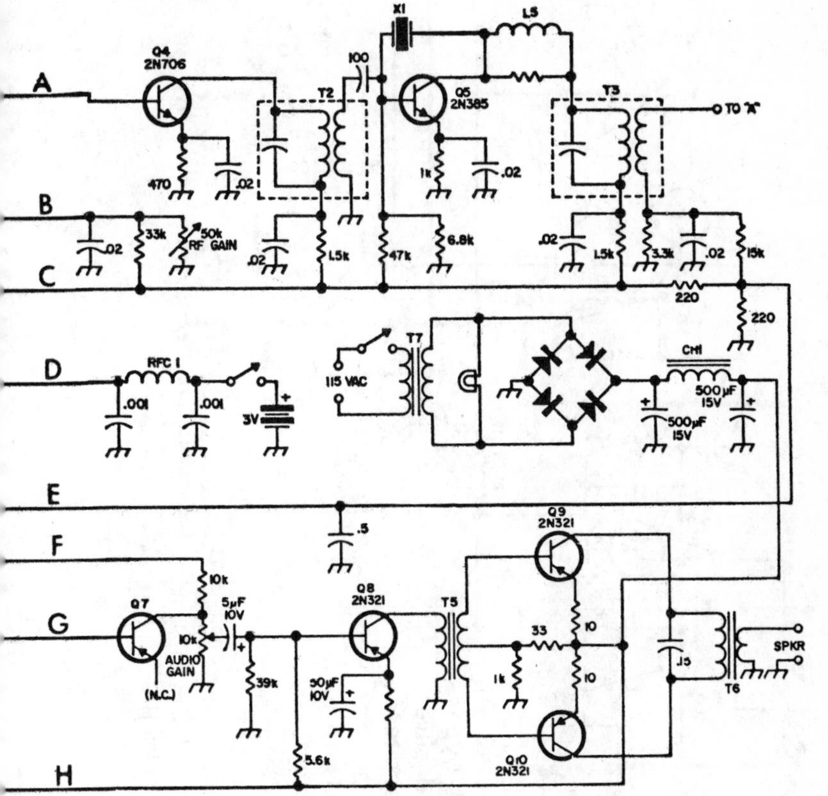


|  |  |
|---|---|
|  | *4t. #20 bare copper 5/16" ID 1/2" long* |
|  | *5t. #20 bare copper 5/16" ID 7/8" long* |
|  | *3t #BJ bare copper 5/16" ID 1/2" long tap 1t. from ground end* |
| *L4* | *13/4t. #14 bare copper 5/16" ID tap 3/4 turn from ground end link 1t. #20 bare copper 1/4" away from ground end* |
| *L5* | *200 microhenry RF choke, J. W. Miller 9210-90* |
| *Ln* | *8t. #26 enamel closewound at one end of 1/4" dia slug form* |
| *RFC-1* | *Ohmite Z-144 or 18 turns #24 wound on 1 meg 1 watt resistor* |
| *X1* | *crystal, 11.155 mc.* |

High-performance receiver for 150 MHz.

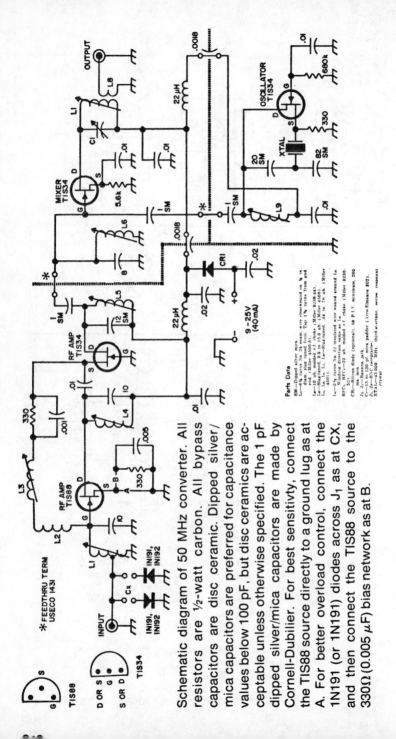

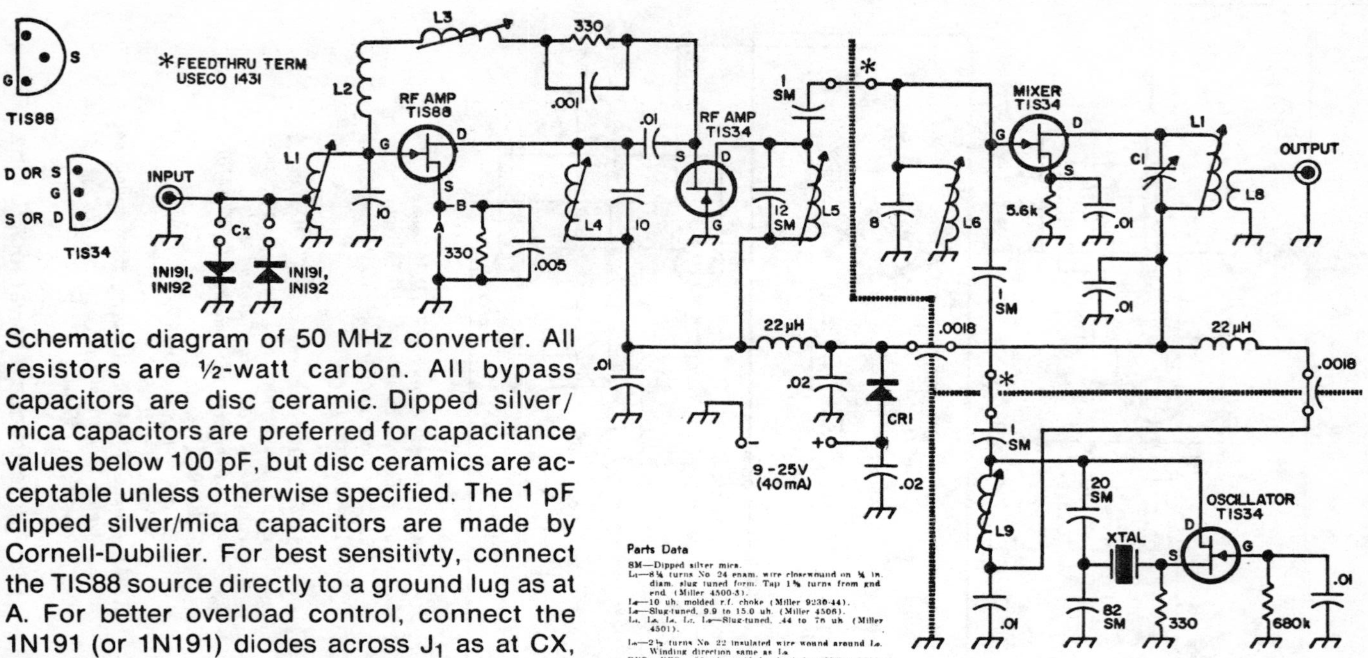


Schematic diagram of 50 MHz converter. All resistors are ½-watt carbon. All bypass capacitors are disc ceramic. Dipped silver/mica capacitors are preferred for capacitance values below 100 pF, but disc ceramics are acceptable unless otherwise specified. The 1 pF dipped silver/mica capacitors are made by Cornell-Dubilier. For best sensitivty, connect the TIS88 source directly to a ground lug as at A. For better overload control, connect the 1N191 (or 1N191) diodes across $J_1$ as at CX, and then connect the TIS88 source to the 330Ω (0.005 μF) bias network as at B.

**Parts Data**

SM—Dipped silver mica.
$L_1$—8¾ turns No. 24 enam. wire closewound on ¼ in. diam. slug tuned form. Tap 1⅝ turns from gnd end. (Miller 4500-3).
$L_2$—10 uh. molded r.f. choke (Miller 9230-44).
$L_3$—Slug-tuned, 9.9 to 15.0 uh. (Miller 4506).
$L_4$, $L_5$, $L_6$, $L_7$, $L_9$—Slug-tuned, .44 to 76 uh (Miller 4501).
$L_8$—2½ turns No. 22 insulated wire wound around $L_1$. Winding direction same as $L_1$.
$RFC_1$, $RFC_2$—22 uh. molded r.f. choke (Miller 9230-52).
$CR_1$—Silicon diode (optional), 50 P.I.V. minimum, 200 ma. min.
$J_3$, $J_4$—Banana jack.
$C_1$—15 to 130 pf. mica padder (Arco-Elmenco 302).
$J_1$, $J_2$—BNC receptacles.
XTAL—43.000 MHz, third overtone, series resonant crystal)

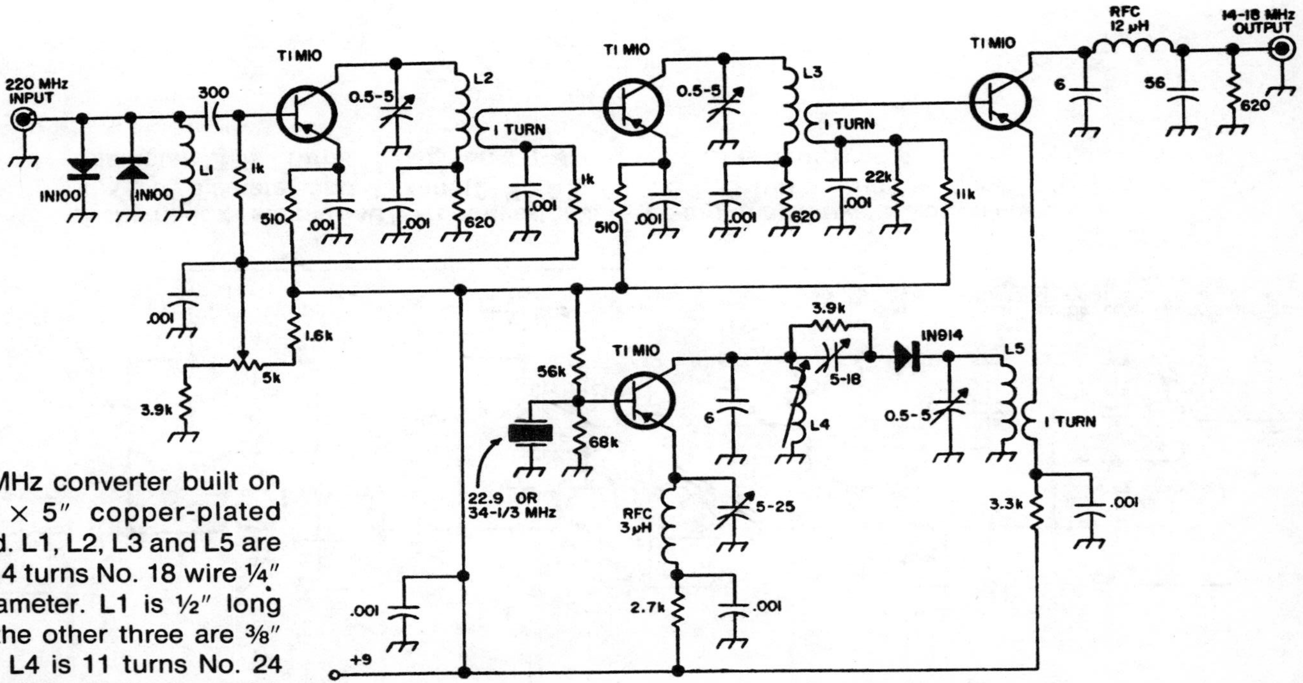


220 MHz converter built on a 2″ × 5″ copper-plated board. L1, L2, L3 and L5 are each 4 turns No. 18 wire ¼″ in diameter. L1 is ½″ long and the other three are ⅜″ long. L4 is 11 turns No. 24 enameled on a ¼″ form with a brass slug. The winding is ¼″ long.

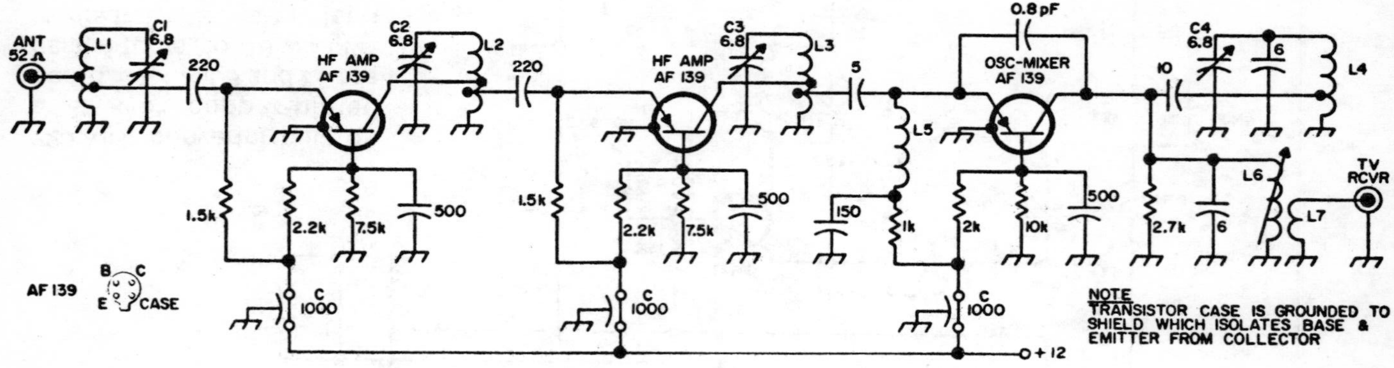


Schematic of simple 440 MHz converter. L1 – L4 are quarter-wave trough lines, 5 mm (1/4") in diameter and 68 mm (2 11/16") long. L5 is 3 turns 18-gage 7 mm (5/16") in diameter. L6 is 7 turns 18-gage on a 5/16" form and L7 is 3 turns on it.

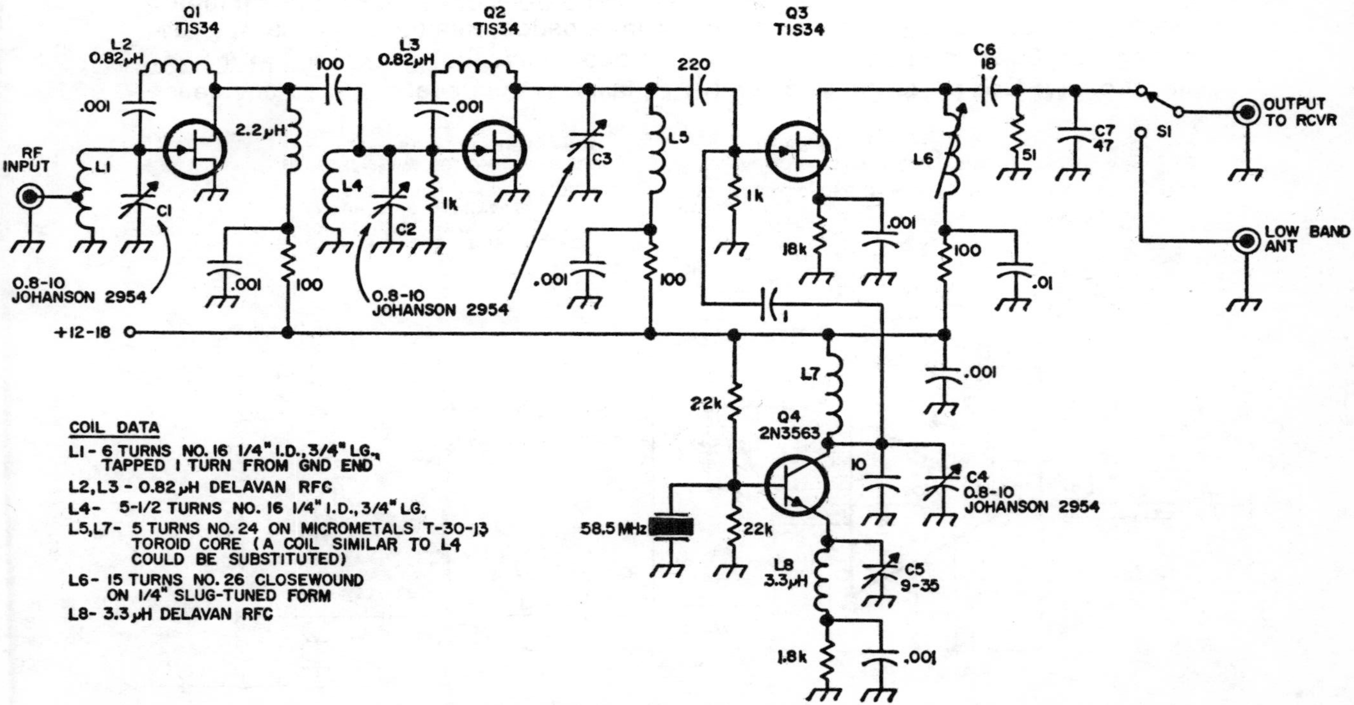


150 MHz converter uses cheap FETs. Noise figure of under 2.5 dB, gain of 27 dB. Excellent performer.

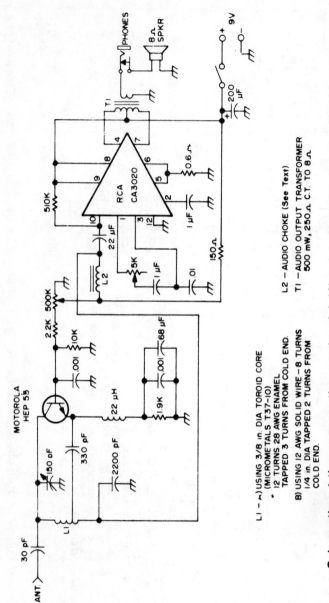

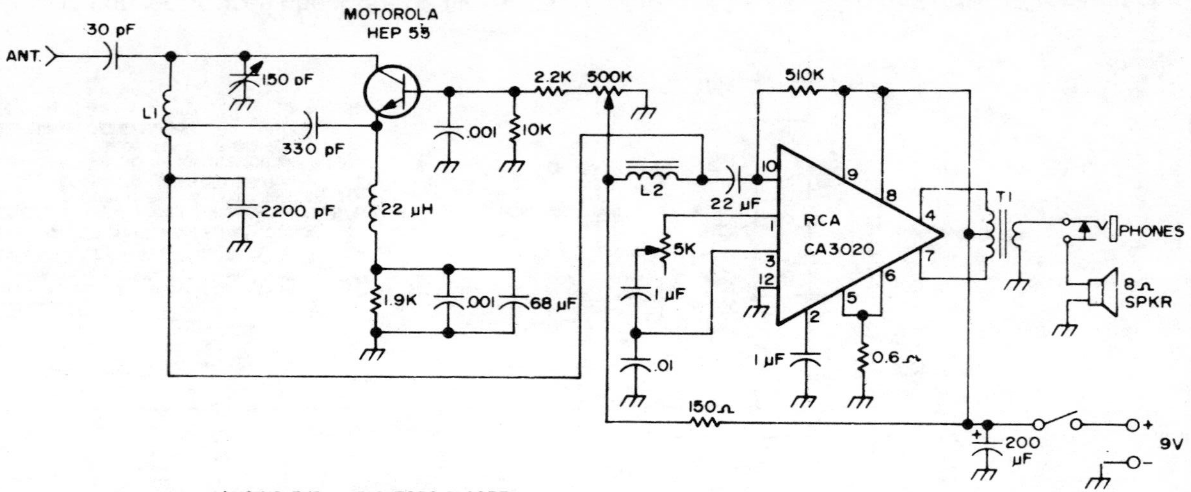


L1 – A) USING 3/8 in. DIA TOROID CORE
(MICROMETALS T37-10)
12 TURNS 28 AWG ENAMEL
TAPPED 3 TURNS FROM COLD END.

B) USING 12 AWG SOLID WIRE – 8 TURNS
1/4 in. DIA TAPPED 2 TURNS FROM
COLD END.

L2 – AUDIO CHOKE (See Text)

T1 – AUDIO OUTPUT TRANSFORMER
500 mW, 250Ω C.T. TO 8Ω

Schematic of the regenerative receiver L1. (A) using ⅜ in. toroid core (Micrometals T37-10) 12 turns No. 28 enamel tapped 3 turns from cold end; (B) using No. 12 solid wire 8 turns ¼ in. diameter tapped 2 turns from cold end; L2 audio choke T1 audio output transformer 500 mW 250Ω CT to 8 Ω.

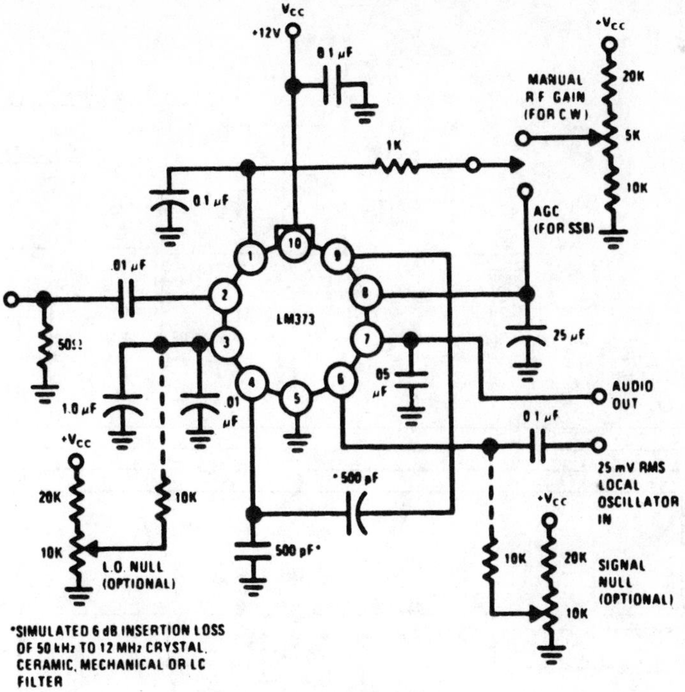


SSB/CW i-f strip

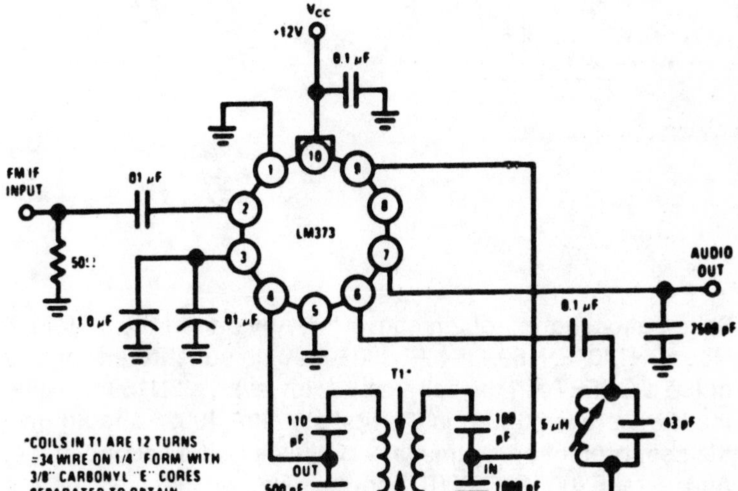

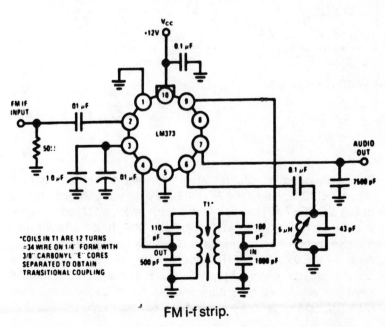

FM i-f strip.

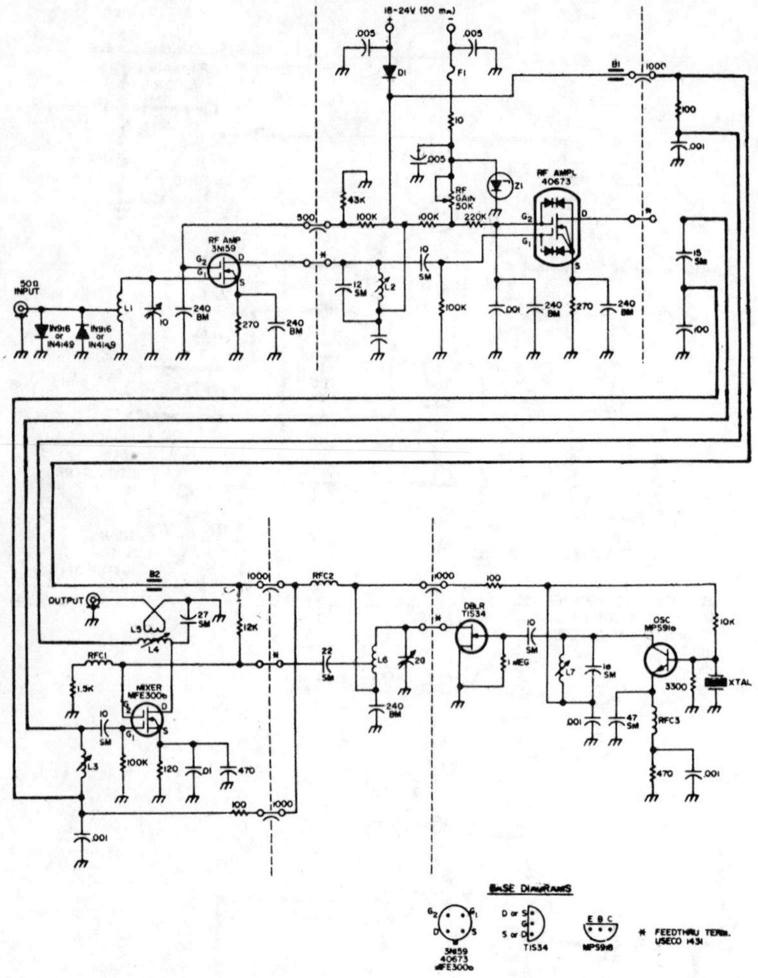

Schematic diagram of 2m converter. Any one of the MOSFET types 3N140, 3N159, or MFE3007 may be substituted for any of the MOSFETs in the schematic. However, a 3N159 will give the lowest noise figure in the first stage. A 40673 should give the best protection against any rf spikes in the second stage. And a MFE3006/MFE3007 should give the best protection against steady, high-voltage rf signals in the mixer stage. All resistors are ½-watt carbon, 5%. All fixed capacitors other than SM, BM, or feedthrough types are disc ceramic.

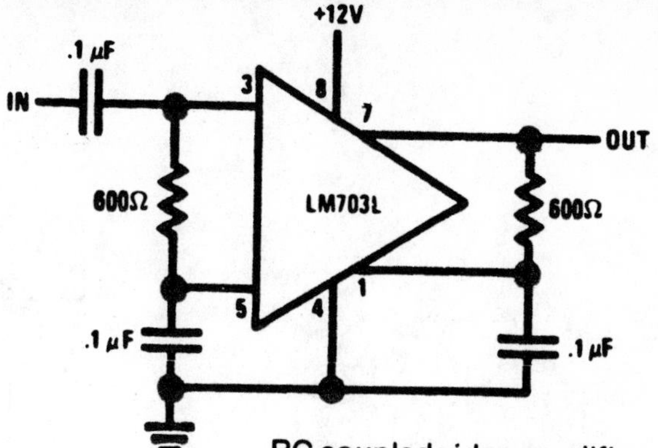


RC coupled video amplifier.

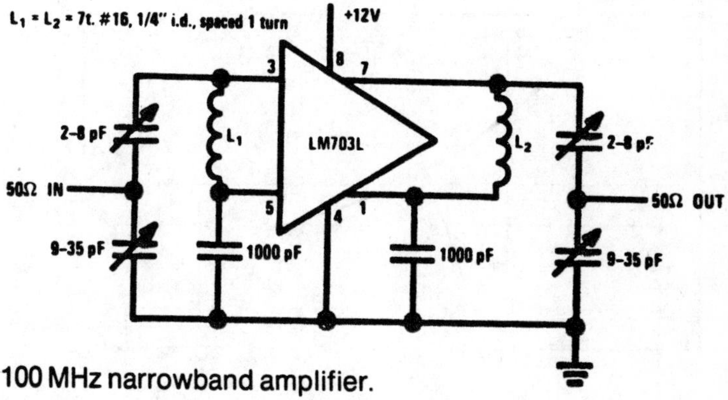


100 MHz narrowband amplifier.

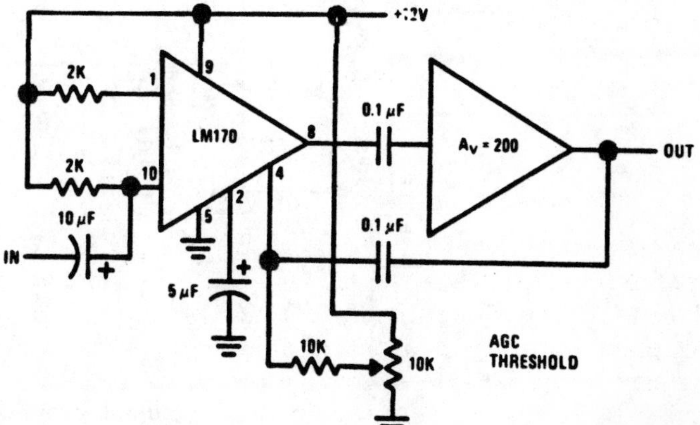


Agc using built-in detection, driven by additional system gain.

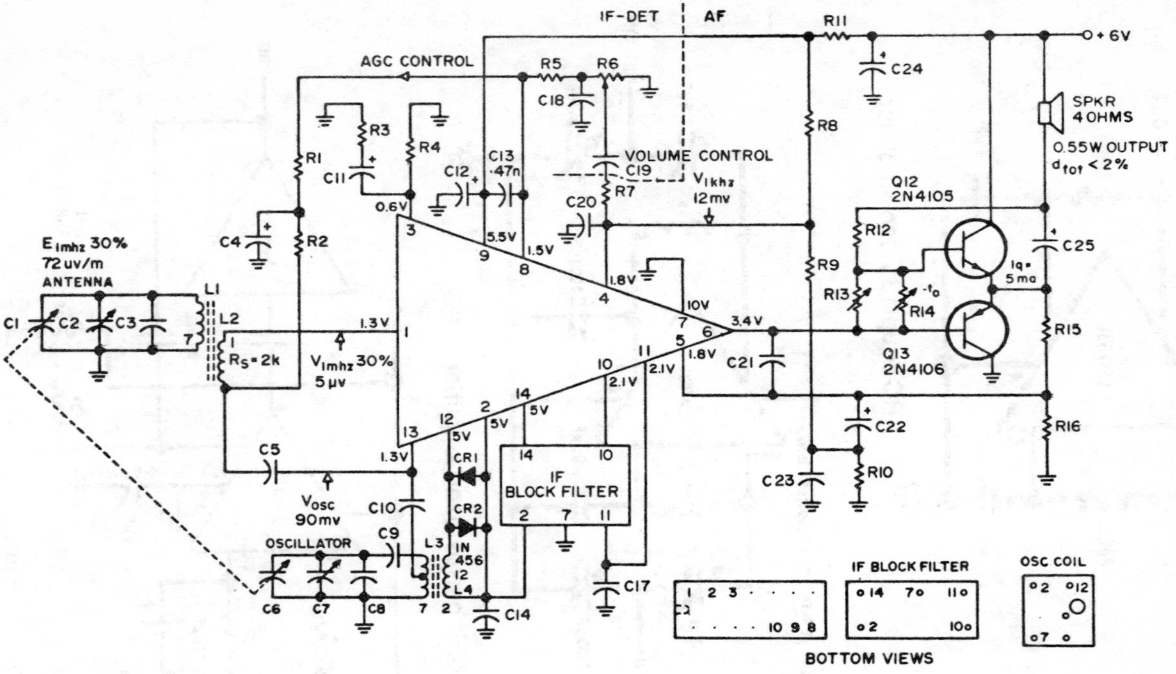


Broadcast-band AM radio built around Amperex TAD100 integrated circuit.

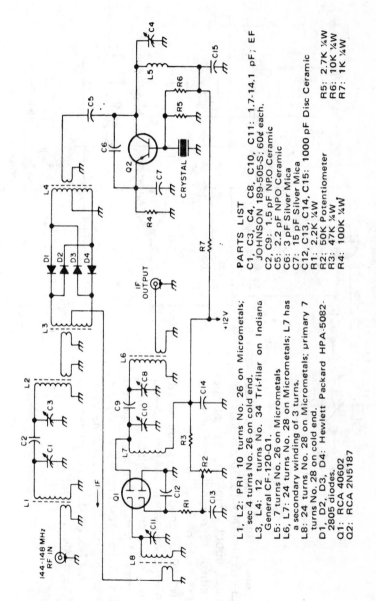

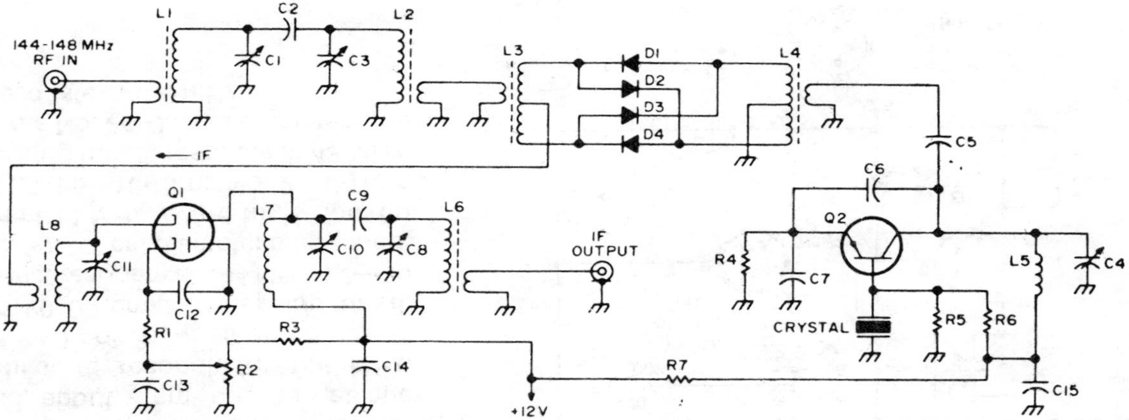


L1, L2: PRI 10 turns No. 26 on Micrometals; sec 4 turns No. 26 on cold end.
L3, L4: 12 turns No. 34 Tri-filar on Indiana General CF-120-Q1.
L5: 7 turns No. 26 on Micrometals
L6, L7: 24 turns No. 28 on Micrometals; L7 has a secondary winding of 3 turns.
L8: 24 turns No. 28 on Micrometals; primary 7 turns No. 28 on cold end.
D1, D2, D3, D4: Hewlett Packard HPA-5082-2805 diodes.
Q1: RCA 40602
Q2: RCA 2N5187

PARTS LIST
C1, C3, C4, C8, C10, C11: 1.7-14.1 pF; EF JOHNSON 189-505-S; 60¢ each.
C2, C9: 1.5 pF NPO Ceramic
C5: 2.2 pF NPO Ceramic
C6: 3 pF Silver Mica
C7: 15 pF Silver Mica
C12, C13, C14, C15: 1000 pF Disc Ceramic
R1: 2.2K ¼W
R2: 50K Potentiometer
R3: 47K ¼W
R4: 100K ¼W
R5: 2.7K ¼W
R6: 10K ¼W
R7: 1K ¼W

Two-meter hot-carrier-diode converter.

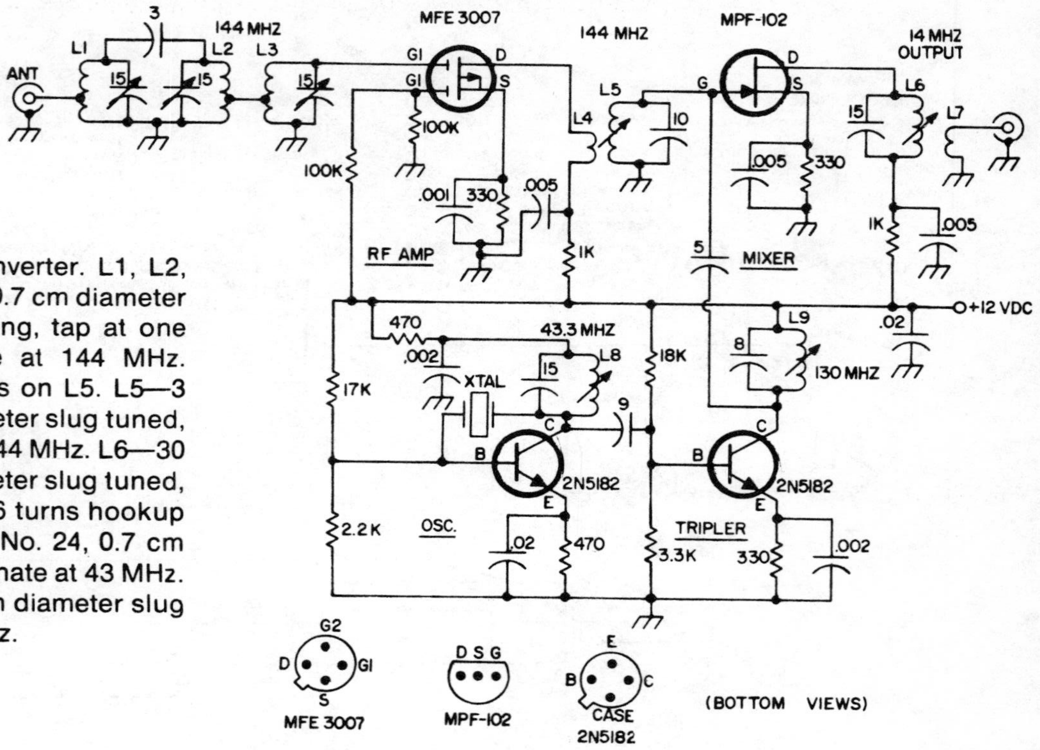


Schematic of the 2m converter. L1, L2, and L3—L5 turns No. 18, 0.7 cm diameter ariwound, about 2 cm long, tap at one turn, adjust to resonate at 144 MHz. L4—3 turns hookup wires on L5. L5—3 turns No. 20, 0.7 cm diameter slug tuned, 1.3 cm long, resonate at 144 MHz. L6—30 turns No. 30, 0.7 cm diameter slug tuned, resonate at 14 MHz. L7—6 turns hookup wire on L6. L8—10 turns No. 24, 0.7 cm diameter slug tuned, resonate at 43 MHz. L9—4 turns No. 24, 0.7 cm diameter slug tuned, resonate at 130 MHz.

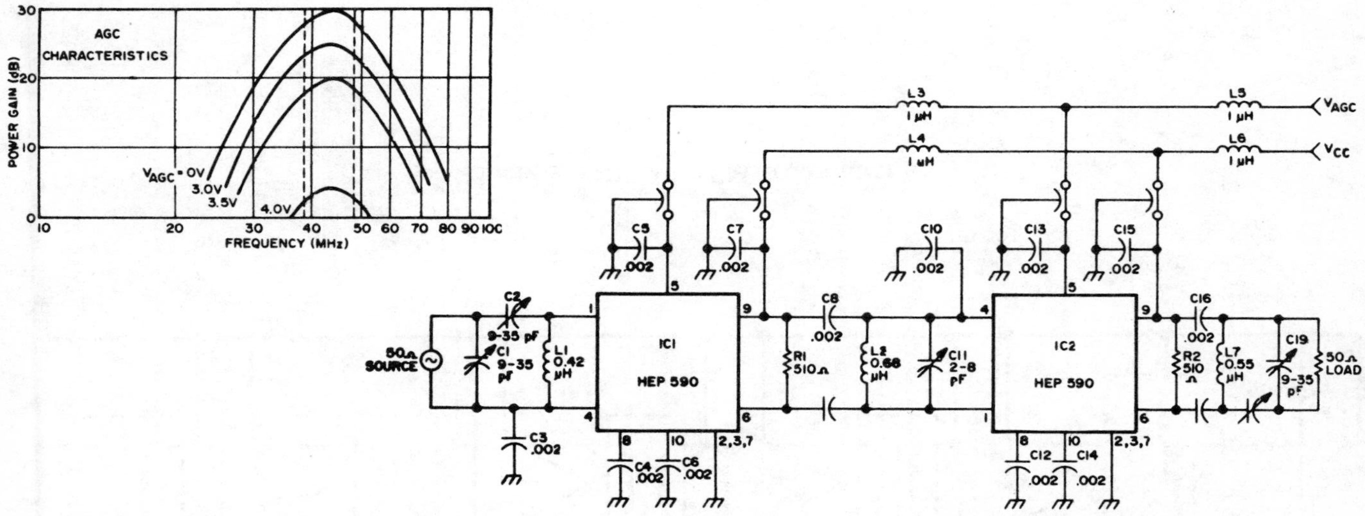


Wideband 45 MHz amplifier, with chart showing agc characteristics.

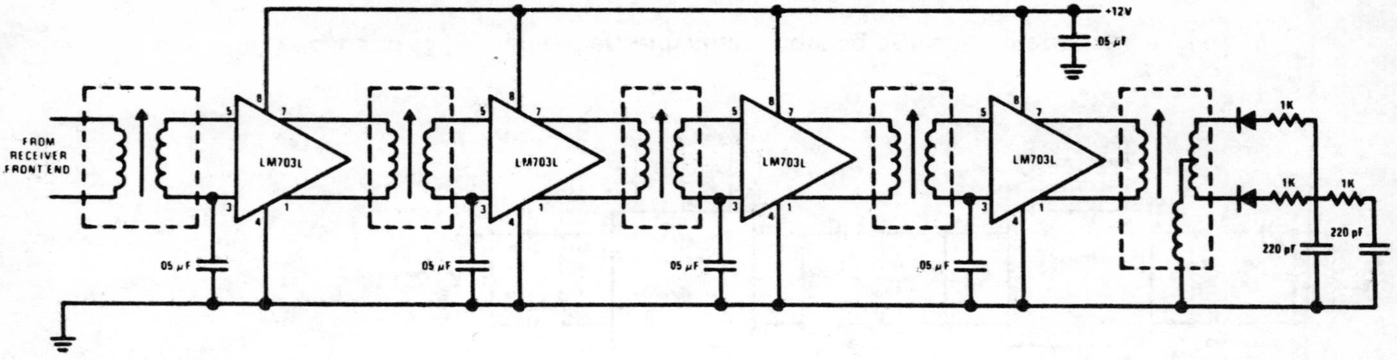


Four stage 10.7 MHz FM i-f amplifier.

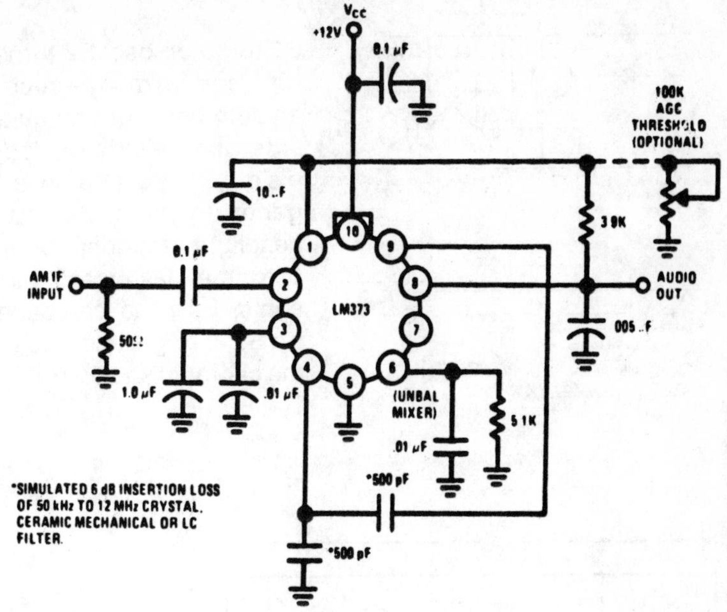


AM i-f strip.

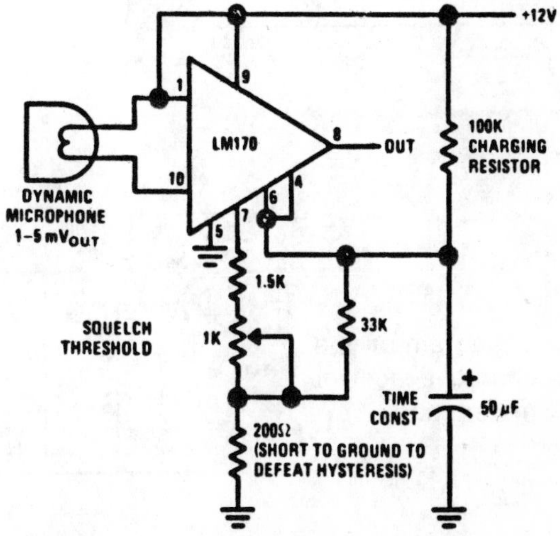


Squelched preamplifier with hysteresis.

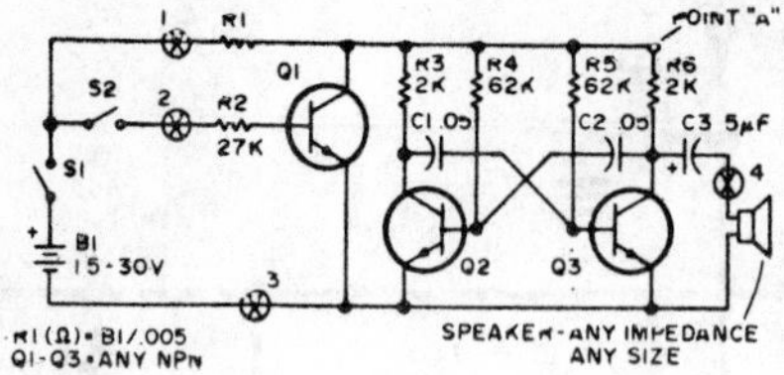


S1 turns on osc. S2 turns it off. Uses are many—such as an auto headlight reminder, sidetone oscillator, code osc., square wave generator, etc. For auto headlight reminder connect 1 to the dashboard panel lights, 2 to the car battery via S2, 3 to ground and 4 to the car radio speaker.

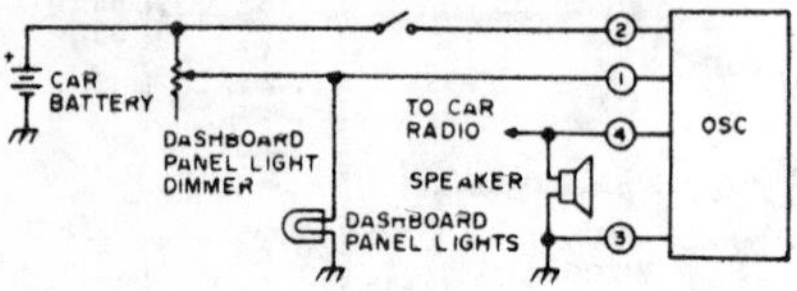


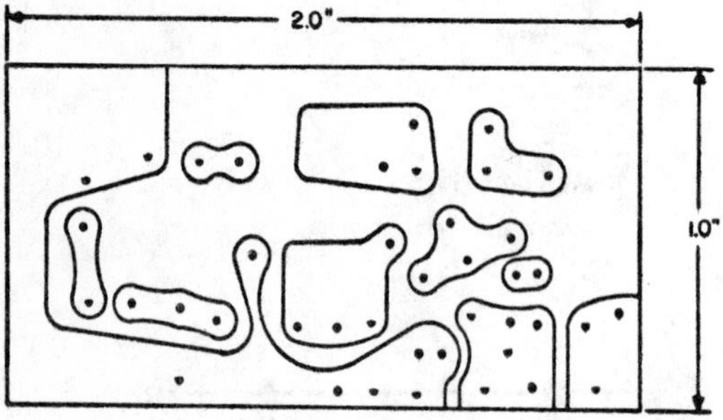


PC board, foil side.

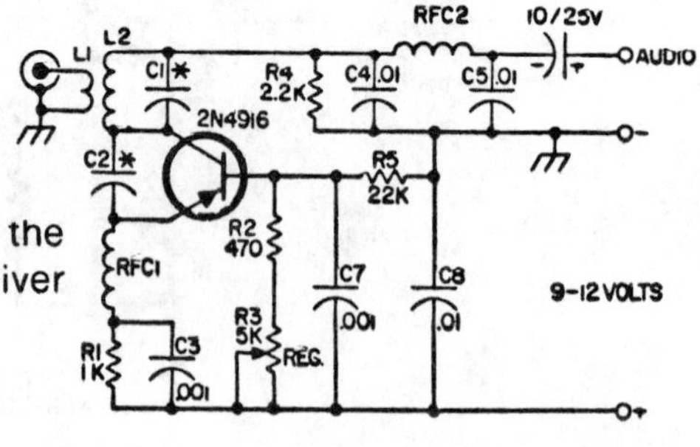


Schematic diagram of the superregenerative receiver for the ¾ meter band.

# Section I

# Regulators

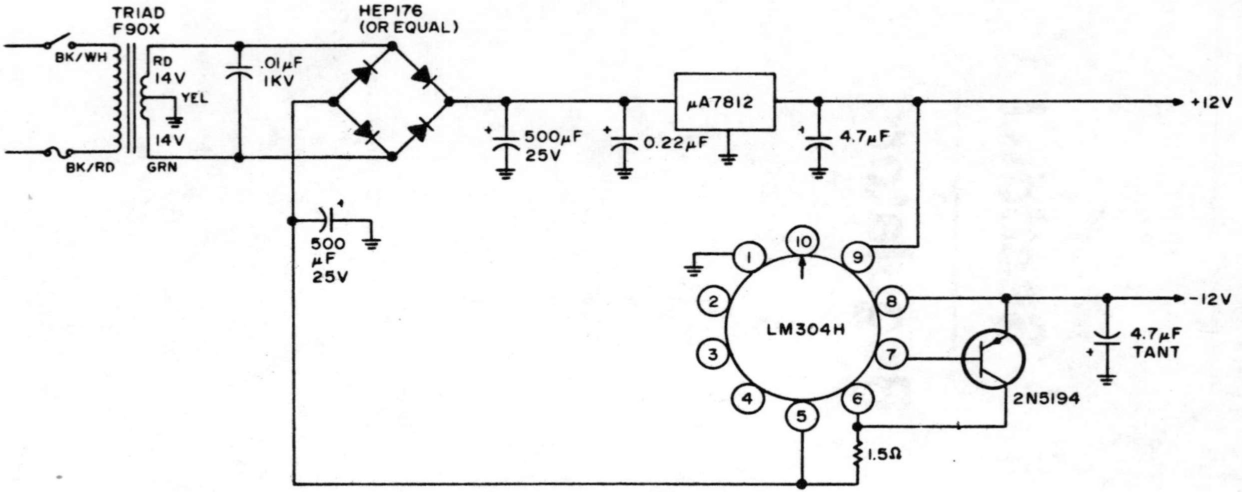


Regulated dc power supply uses 7812 IC and LM304H, and delivers 5% outputs of 12V both negative and positive.

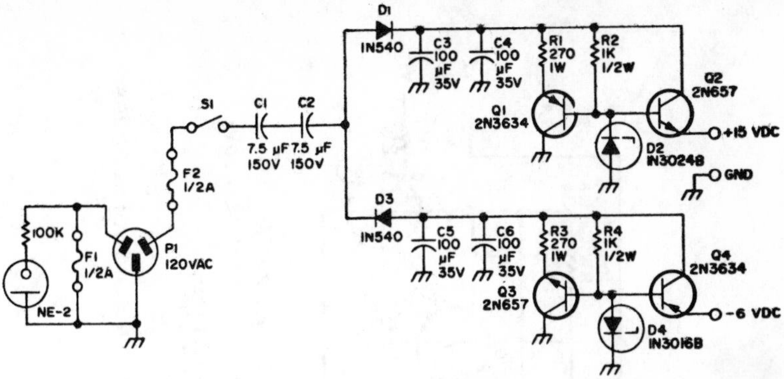


Low-voltage transformerless power supply.

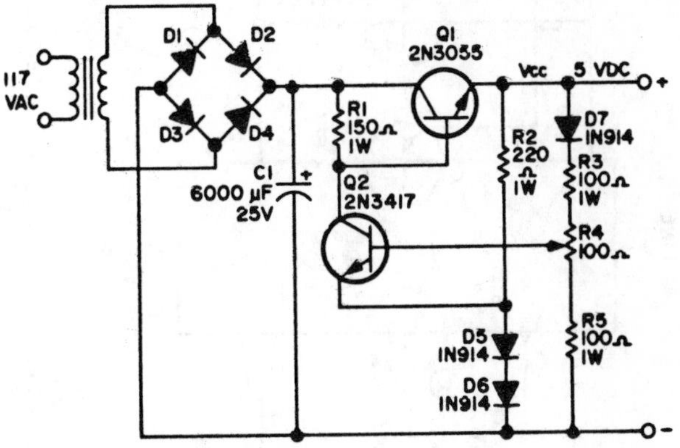


D1-D4 200V, 1A
T1-6.3 VAC, 1A

Regulated 5V power supply for ICs. Reprinted from "Zero Beat," published by the Victoria Short Wave Club, Victoria, British Columbia.

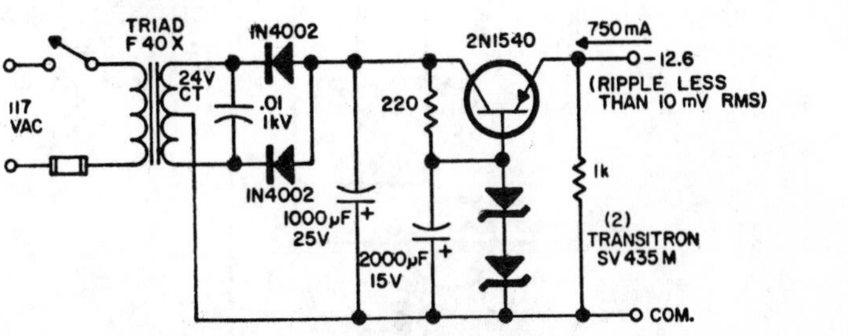


Simple emitter-follower regulated power supply for the heaters of five 12AX7s.

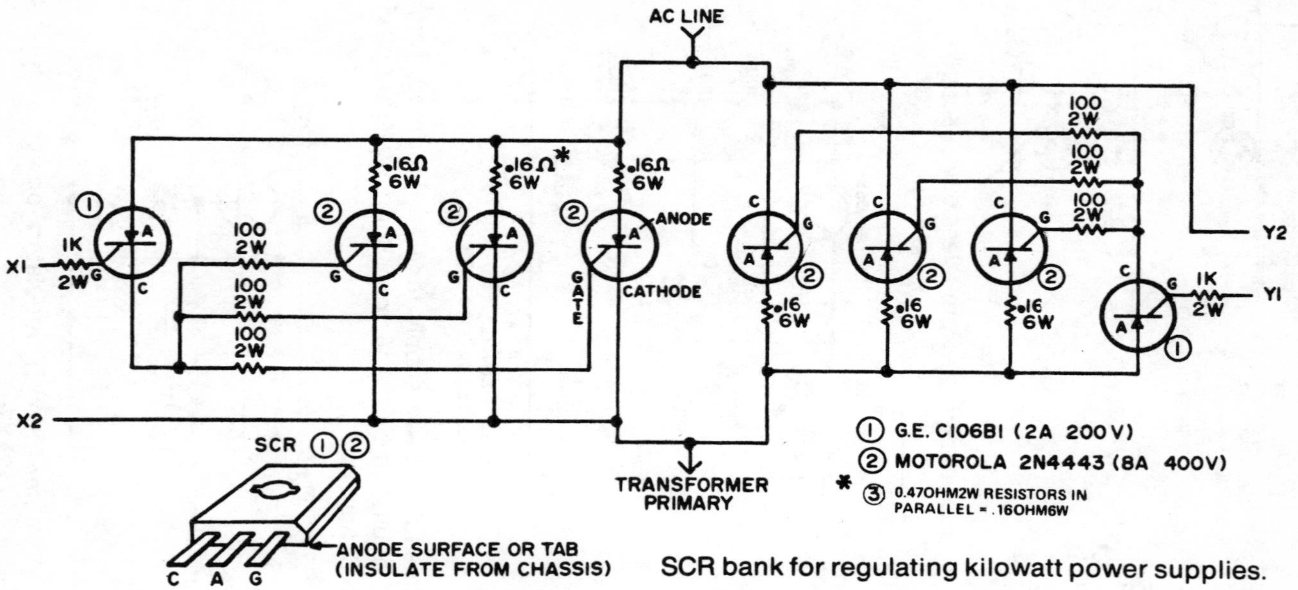


SCR bank for regulating kilowatt power supplies.

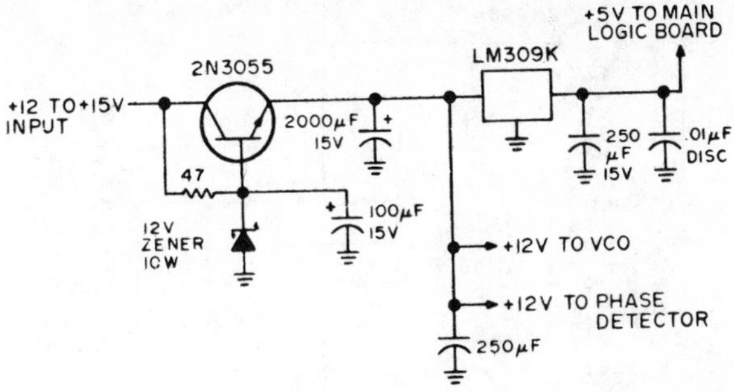


Voltage regulator for connection to existing power supply.

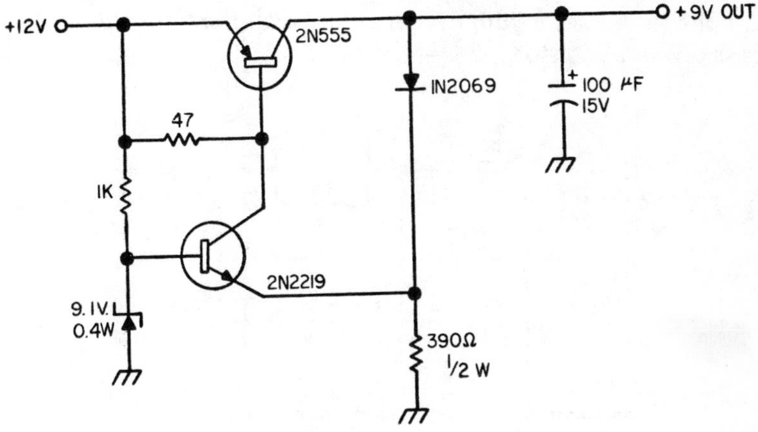


Voltage regulator, 9V dc.

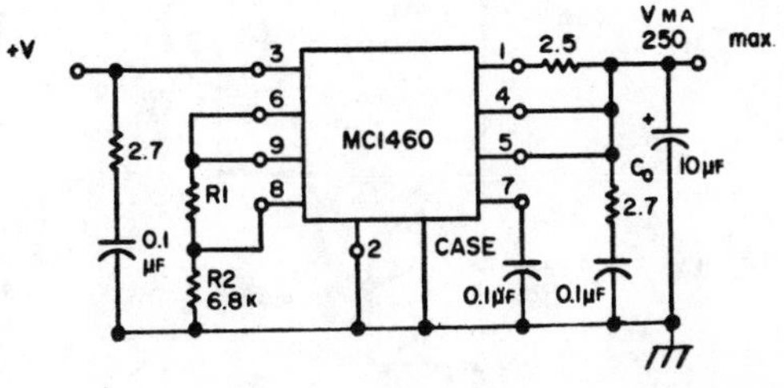


High-performance IC regulator using the MC1460. The input voltage $V_{IN}$ must be at least 3V greater than the output. Maximum input for the MC1460 is 20V. The MC1460 is good for 35V input. An etched circuit board is recommended for con struction. Use VHF techniques.

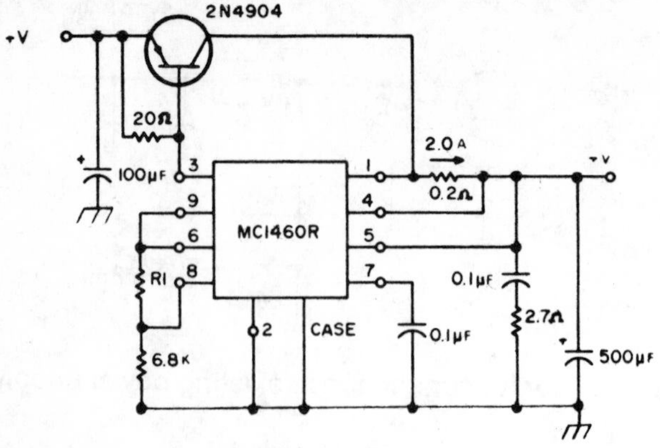


IC voltage regulator with 2A capability using external series-pass power transistor.

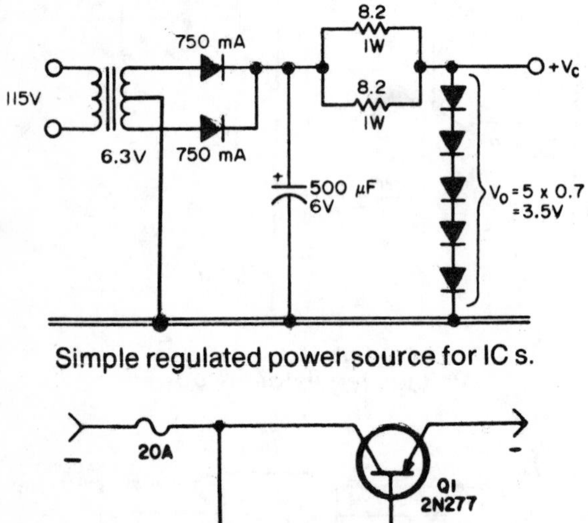


Simple regulated power source for IC s.

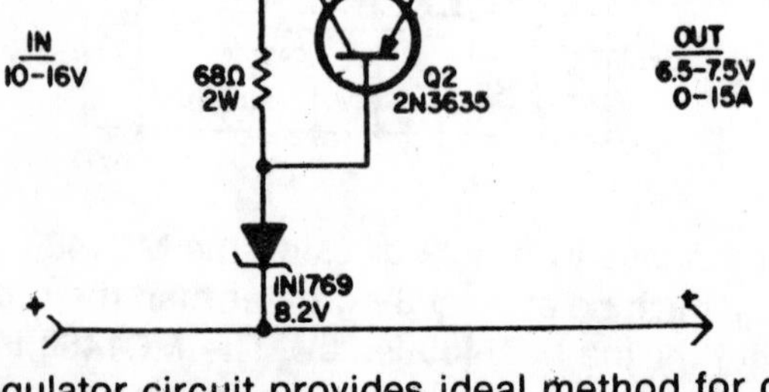


Simple regulator circuit provides ideal method for dropping 12V to 6V with current capability of up to 15 amperes.

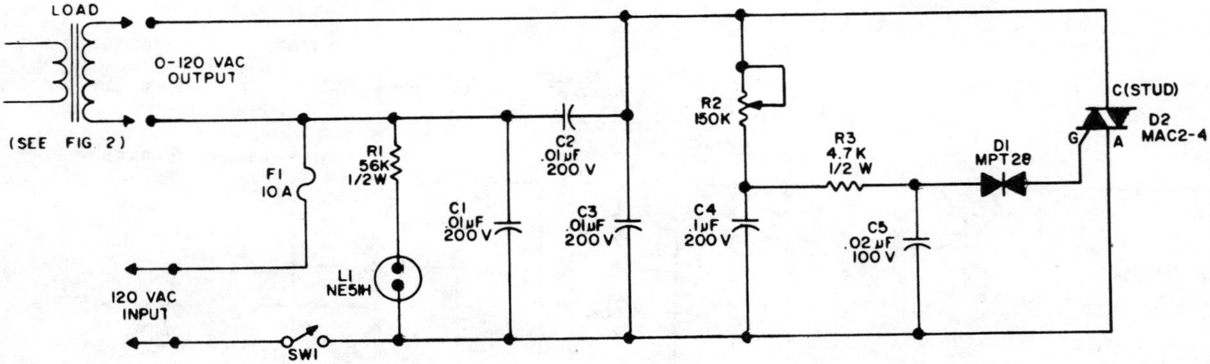


"Electronic Variac" schematic. C1 to 3—0.01$\mu$F 200V capacitor; C4—0.1$\mu$F 200V capacitor; C5—0.02$\mu$F 100V capacitor; D1—MPT28, 3-layer diode (Motorola); D2—MAC-4, 200 V triac (Motorola); F1—10 amp fuse; L1—NE51H neon lamp and socket; R1—56K, ½ watt ±10% resistor; R¼—150K pot, lin. taper, ½ watt; R3—4.7K, ½ watt ±10% SW1—SPST switch, 10A; misc.—line cord, terminal strip, and chassis.

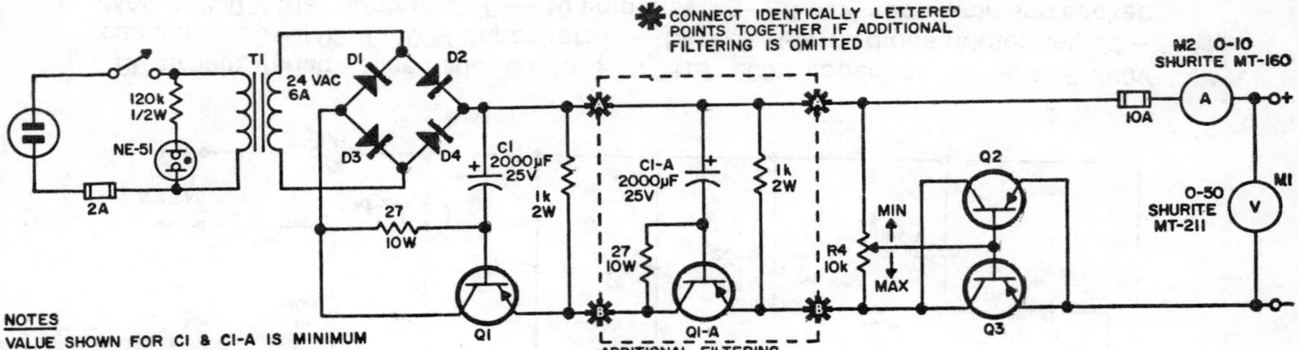


**NOTES**

- VALUE SHOWN FOR C1 & C1-A IS MINIMUM
- ALL TRANSISTORS ARE DELCO 2N1970
- ALL DIODES ARE MOTOROLA HEP-151
- FOLLOWING PARTS AVAILABLE AT INDICATED PRICES FROM BAY ELECTRONICS, 2315 ARTESIA BLVD, REDONDO BEACH, CALIFORNIA, 90278 (ADD POSTAGE)

| | |
|---|---|
| C1, C1-A (25,000 uF 25V) | 2.88 EA |
| D1,2,3,4 | .79 EA |
| Q1, 1-A, 2, 3 | .99 EA |
| T1 | 6.88 EA |

Versatile regulated variable power supply delivers several amps at 24V.

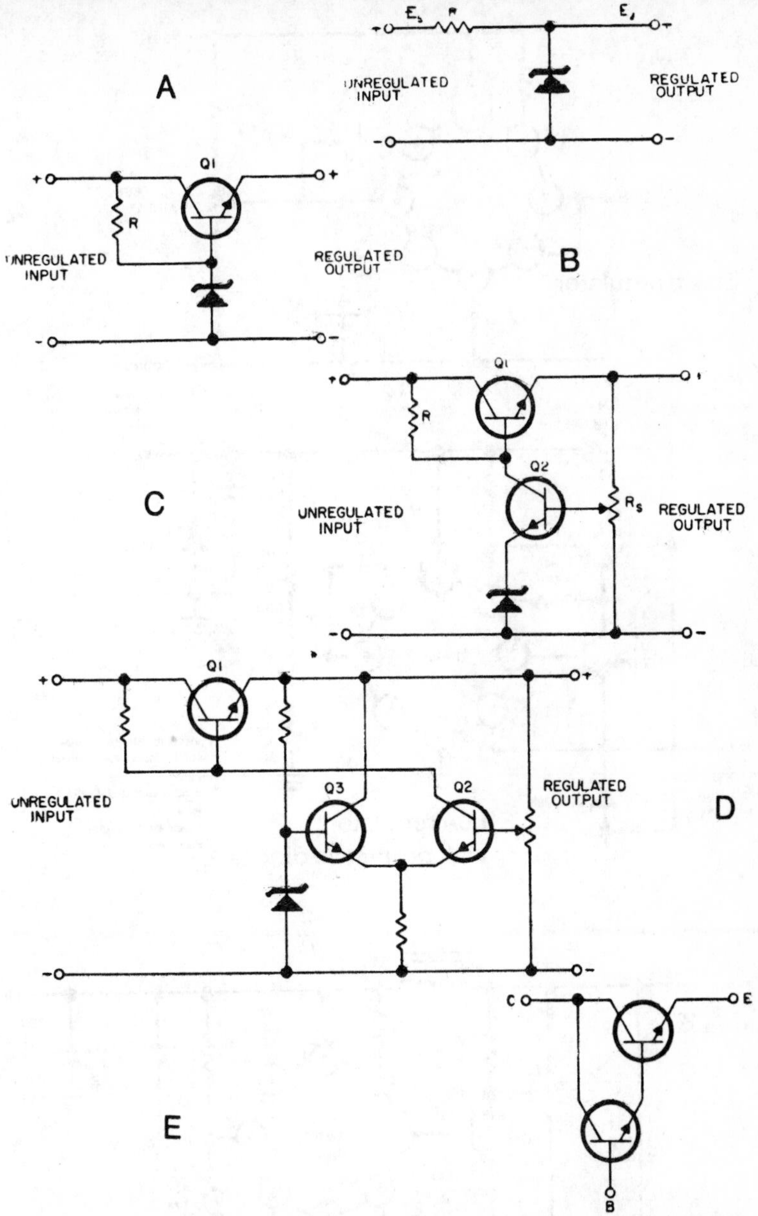


Regulator circuits. **a** shows a zener regulator; **b**, an emitter-follower regulator; **c**, an emitter follower with amplifier; **d**, emitter follower with full differential amplifier. The circuit in **e** is a Darlington pair which acts as a single transistor with higher gain.

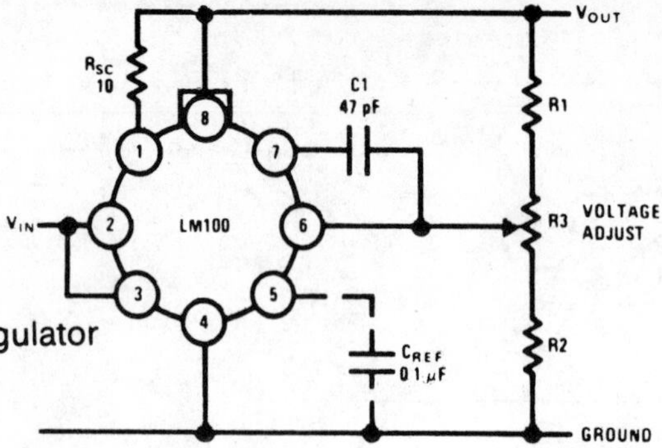


Basic regulator

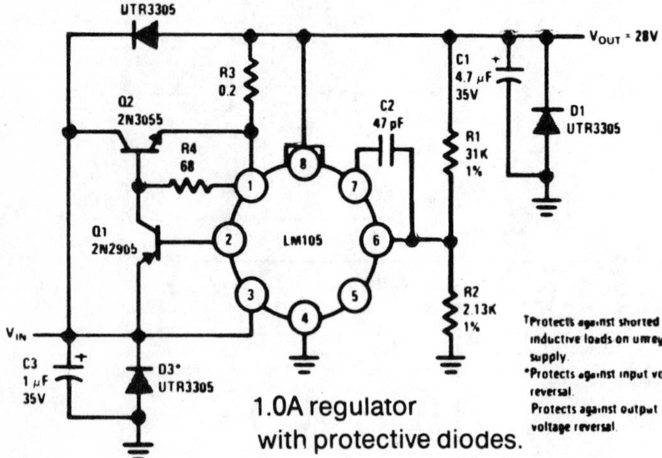


†Protects against shorted input or inductive loads on unregulated supply.
*Protects against input voltage reversal.
Protects against output voltage reversal.

1.0A regulator with protective diodes.

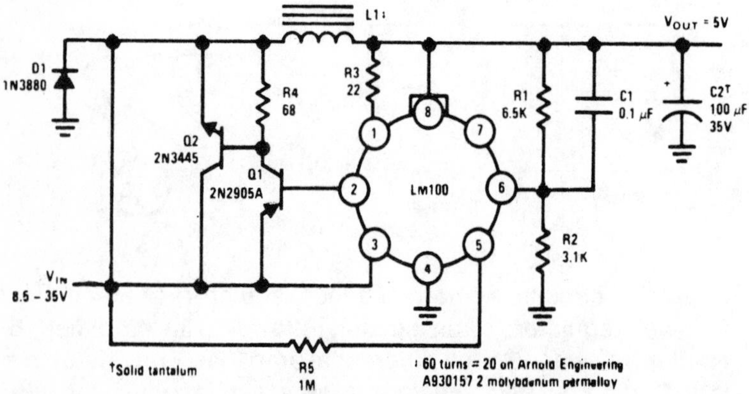


†Solid tantalum

‡60 turns = 20 on Arnold Engineering A930157 2 molybdenum permalloy core.

4A switching regulator.

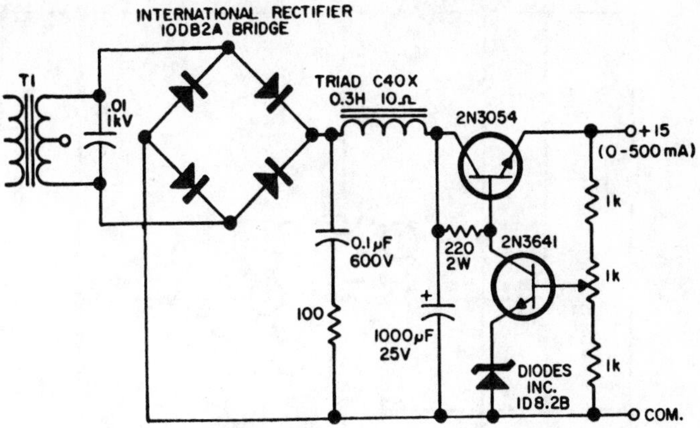


T1 is a surplus vibrator transformer marked 25.2V input and 135V dc, 118 mA output; vibrator frequency, 115 Hz. Half of the high voltage winding is used in this circuit.

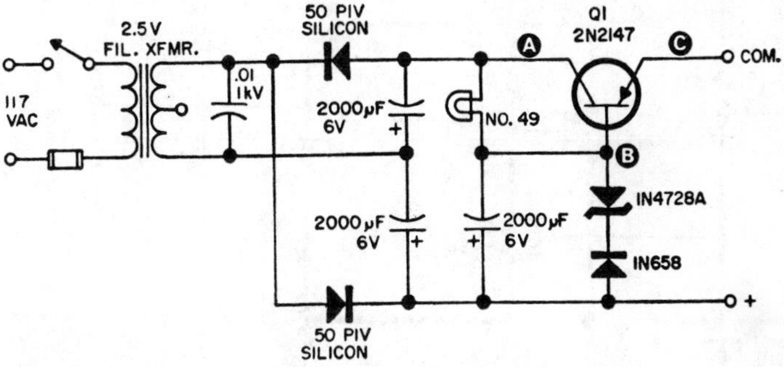


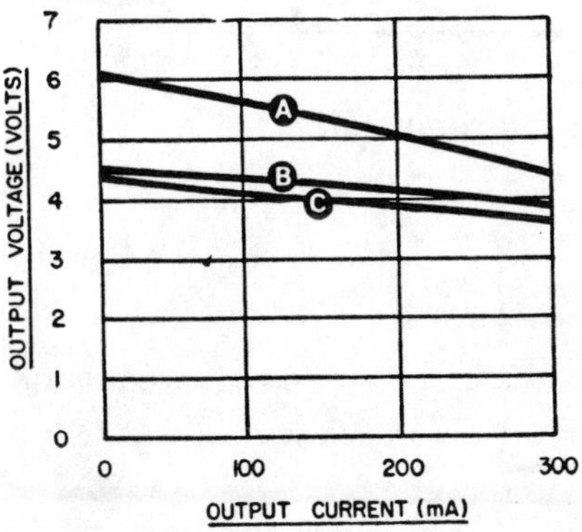


Regulated power supply designed for use with inexpensive Fairchild integrated circuits. The letters on the graph refer to the voltages at the points shown on the schematic.

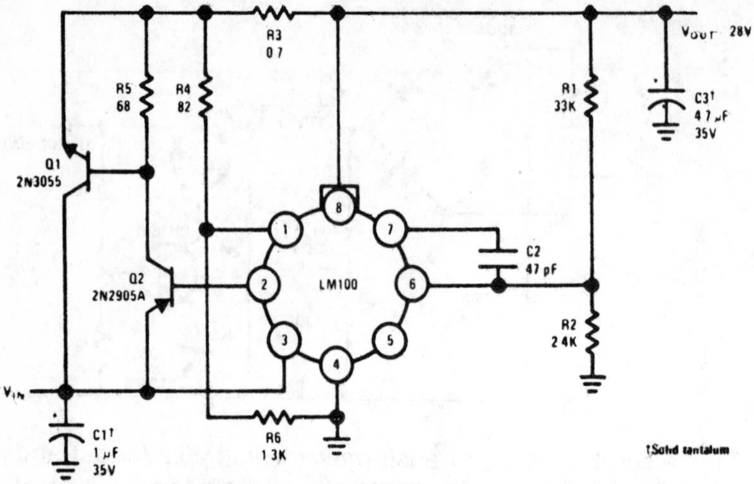


2A regulator with foldback current limiting.

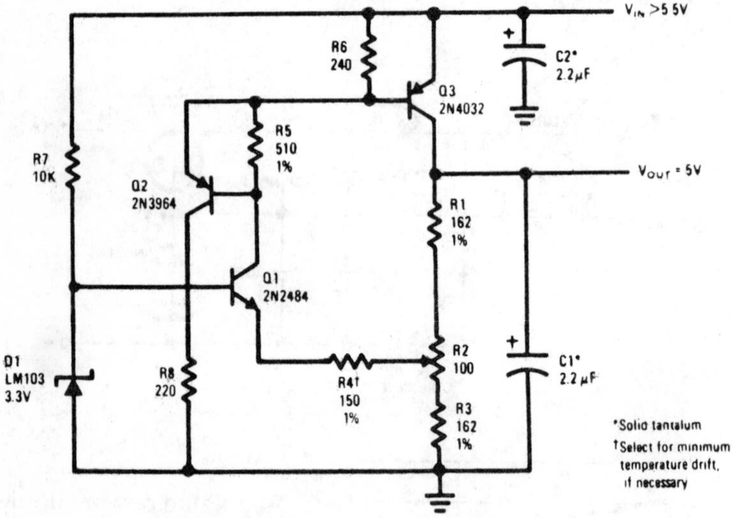


200 mA positive regulator.

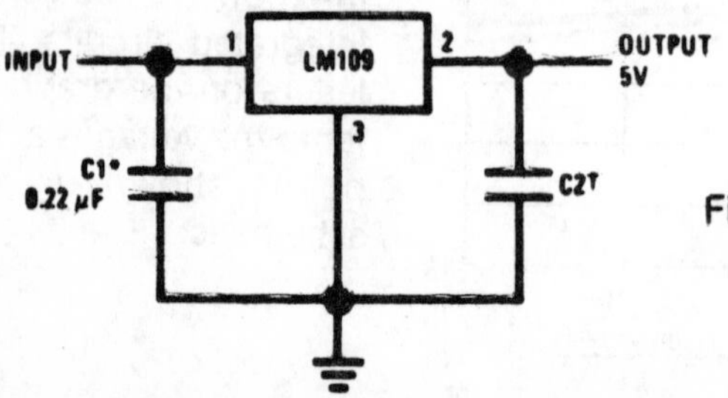


Fixed 5V regulator.

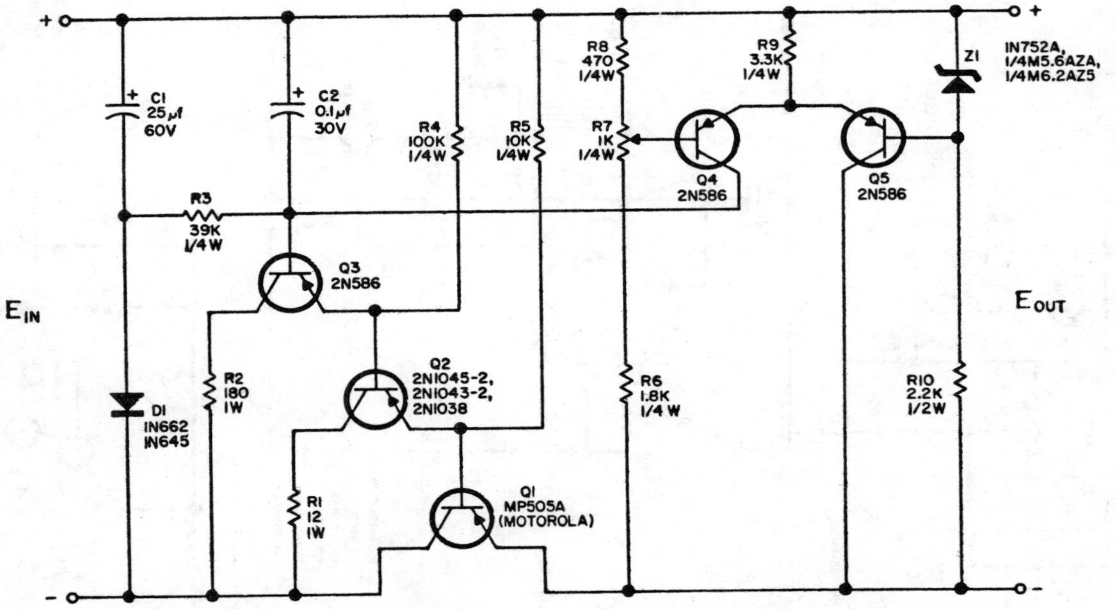


Transistor regulator for 9 to 28V output at up to 10A. Omit Q1 for a maximum current of 3A.

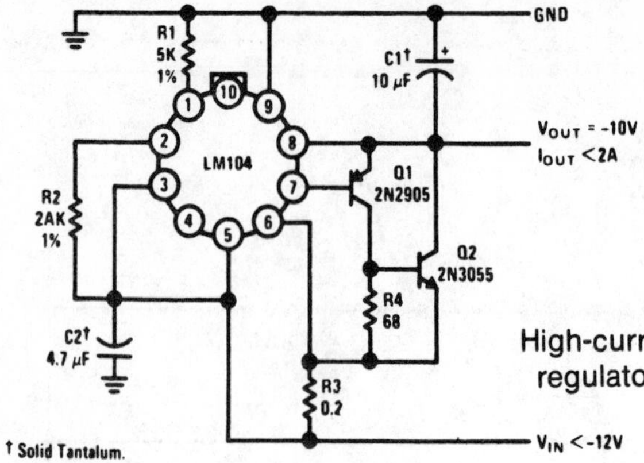


High-current regulator.

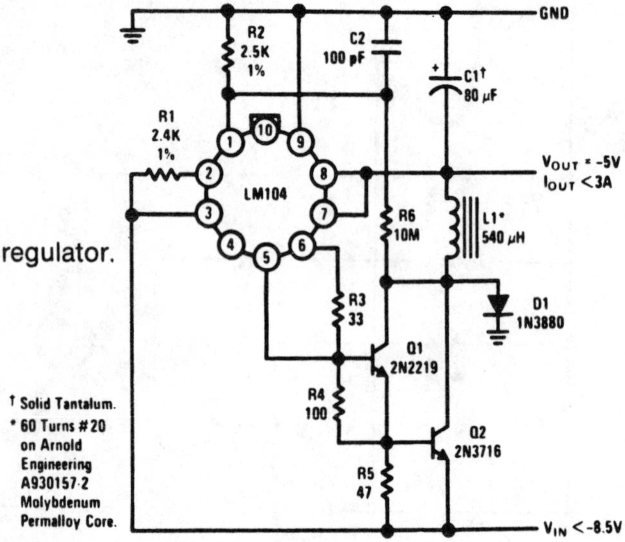


Switching regulator.

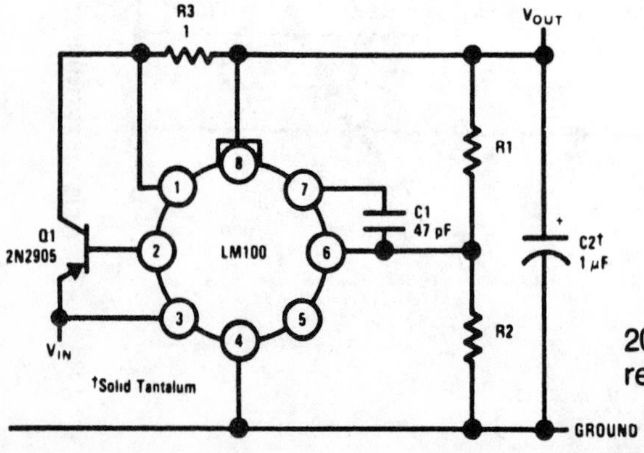


200 mA regulator.

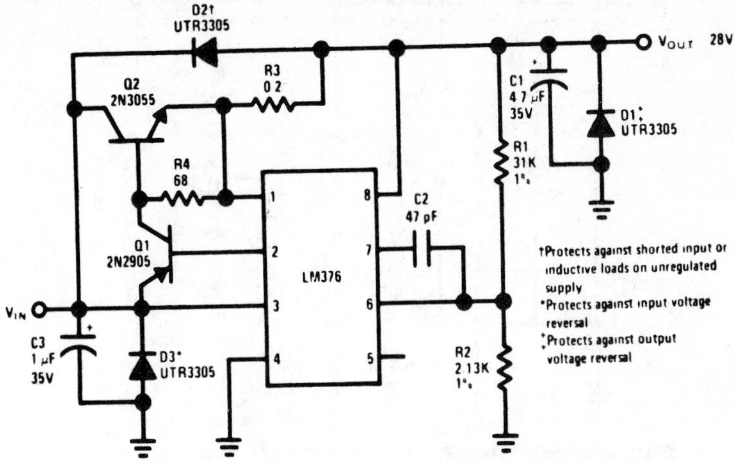


1.0A regulator with protective diodes.

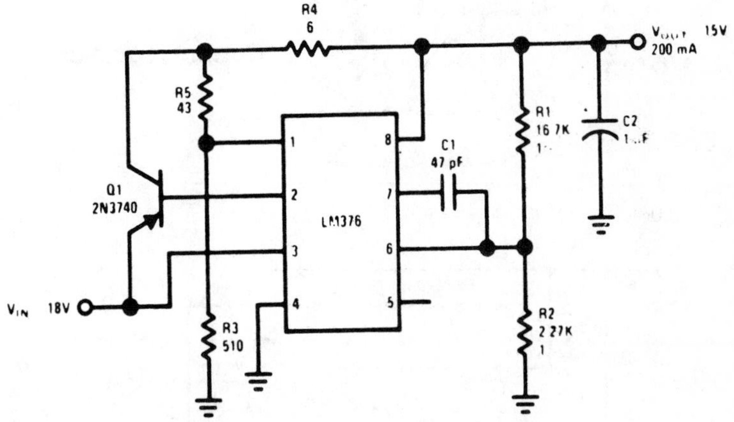


Linear regulator with foldback current limiting.

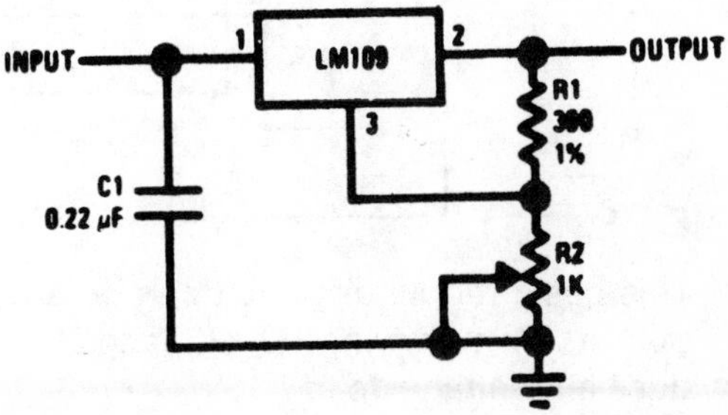


Adjustable-output regulator.

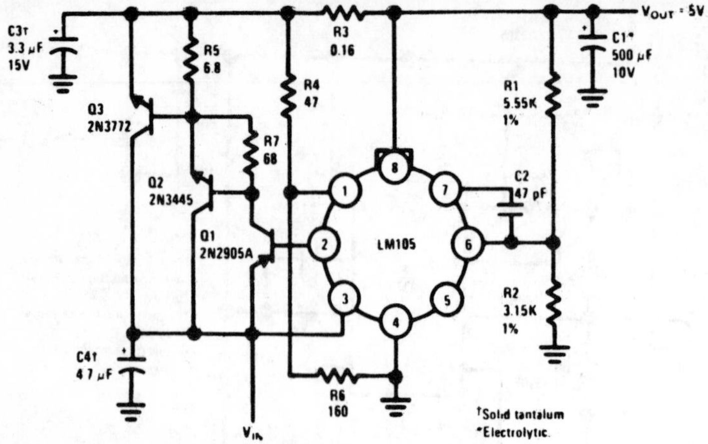


10A regulator with foldback current limiting.

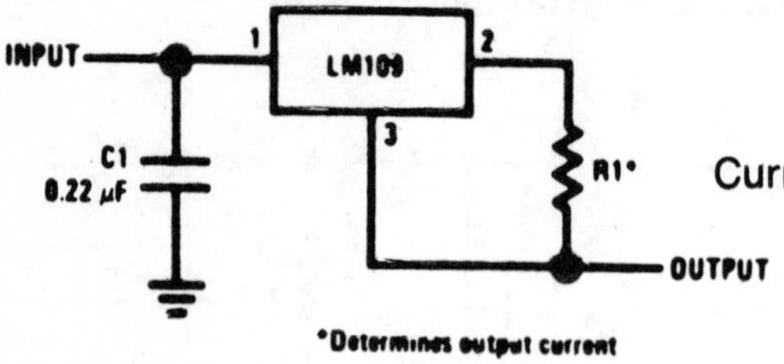


Current regulator.

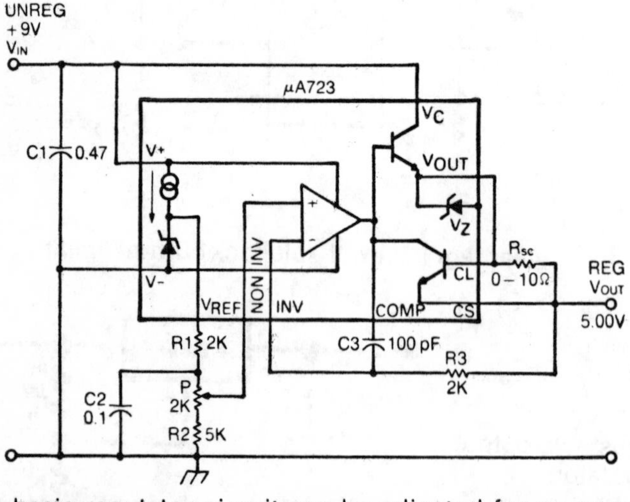


This basic regulator circuit can be adjusted for an exact 5V output. By modifying R1, R2, and P, it can be made to produce any voltage up to 7.15V.

# Section J

# RF Generators & Waveshapers

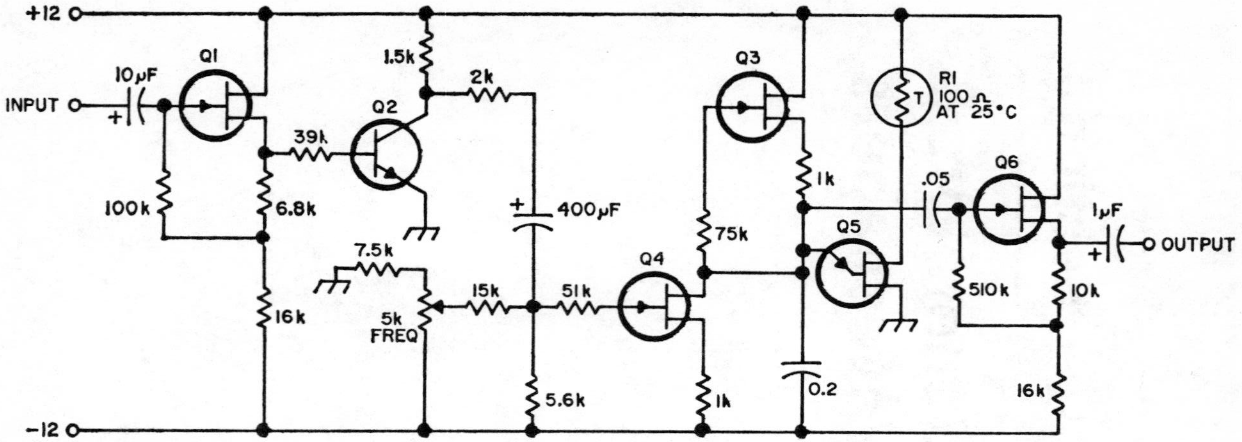


This simple sawtooth generator is linear within 2% and may be adjusted from 1 kHz to 3 kHz with the center frequency control. Q1, Q3, Q4, and Q6 are FETs such as the 2N3819, 2N3820, T1S34, MPF105 or HEP 801; Q2 is a 2N388, 2N2926, 2N3391, SK3011 or HEP 54; Q5 is a 2N1671, 2N2160, 2N2646, 2N3480, or HEP 310.

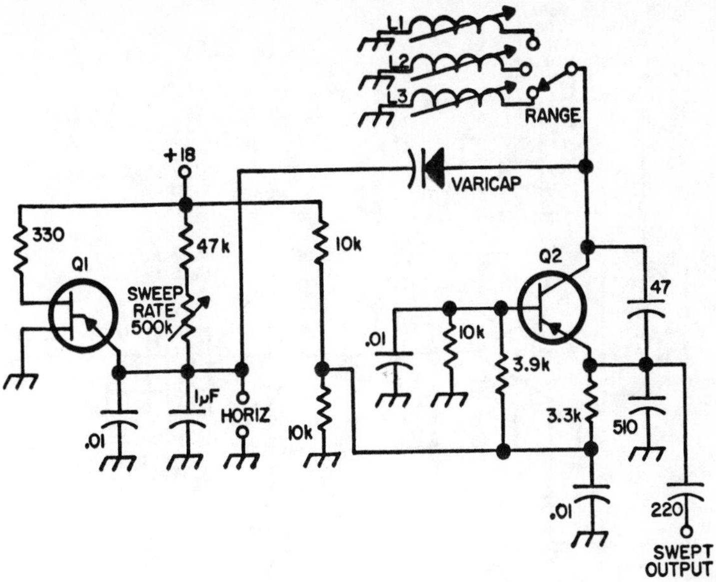


**COILS**

| Frequency | Miller No. |
|---|---|
| 65 kHz —140 kHz | 9007 |
| 95 kHz —190 kHz | 9006 |
| 150 kHz —300 kHz | 9005 |
| 190 kHz —550 kHz | 9004 |
| 380 kHz —1000 kHz | 9003 |
| 700 kHz — 1.8 MHz | 9002 |
| 1.4 MHz— 3.7 MHz | 9001 |
| 3.7 MHz— 4.7 MHz | 4508 |
| 4.7 MHz— 5.9 MHz | 4507 |
| 5.9 MHz— 7.5 MHz | 4506 |
| 7.5 MHz— 10 MHz | 4505 |
| 10 MHz— 14 MHz | 4504 |
| 14 MHz— 18 MHz | 4503 |
| 18 MHz— 23 MHz | 4502 |
| 23 MHz— 29 MHz | 4304 |
| 29 MHz— 36 MHz | 4303 |
| 36 MHz— 45 MHz | 4302 |
| 45 MHz— 60 MHz | 4301 |

A sweep frequency generator is a very handy gadget, but many times the commercial units are more complicated than required. This simple sweeper may be used at any spot frequency between 100 kHz and 60 MHz. By using a three-position range switch, the three most popular frequencies may be used, such as 455 kHz, 1600 kHz, and 10.7 MHz. Q1 is a 2N1671, 2N2160, 2N2646, 2N3480, or HEP 310; Q2 is a 2N741, 2N1747, 2N2188, GE-9 or HEP 2. The varactor is a 56 pF capacitance diode such as the 1N955 or TRW V56.

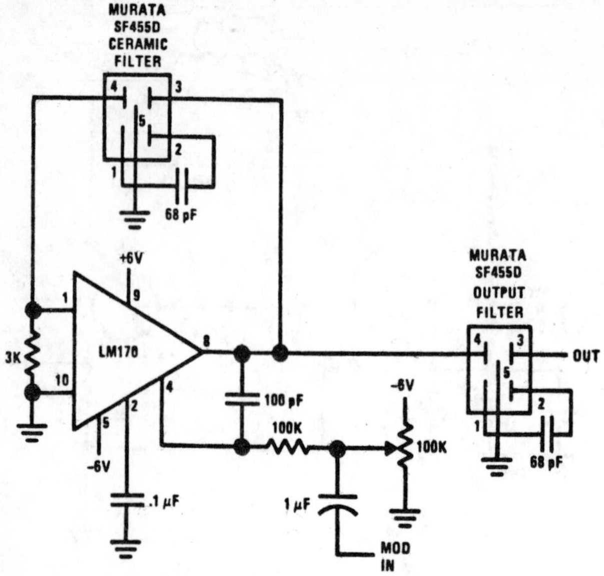


455 kHz modulated, regulated output signal generator.

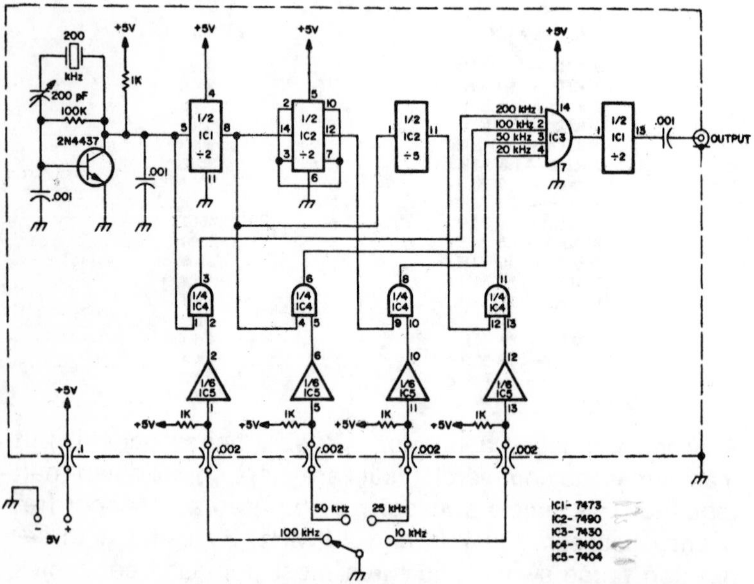


Crystal calibrator. This circuit gives symmetrical square waves out on 100, 50, 25, and 10 kHz. The frequency switch may be any distance from the totally shielded calibrator, as the lines have only dc levels.

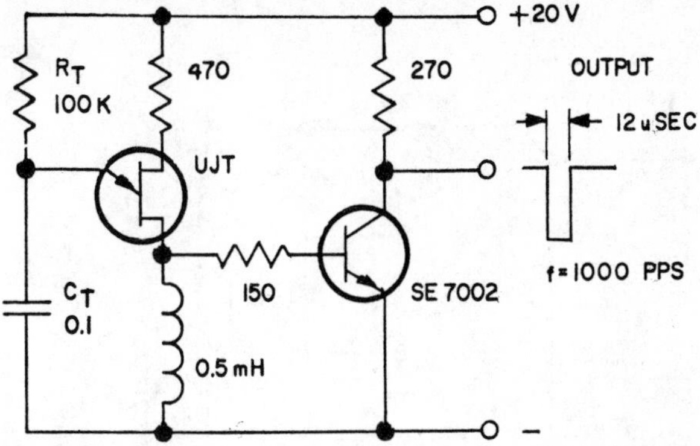


Pulse generator uses unijunction. Pulse width is determined by base 2 inductance. Rise and fall times will be 2–5% of pulse width.

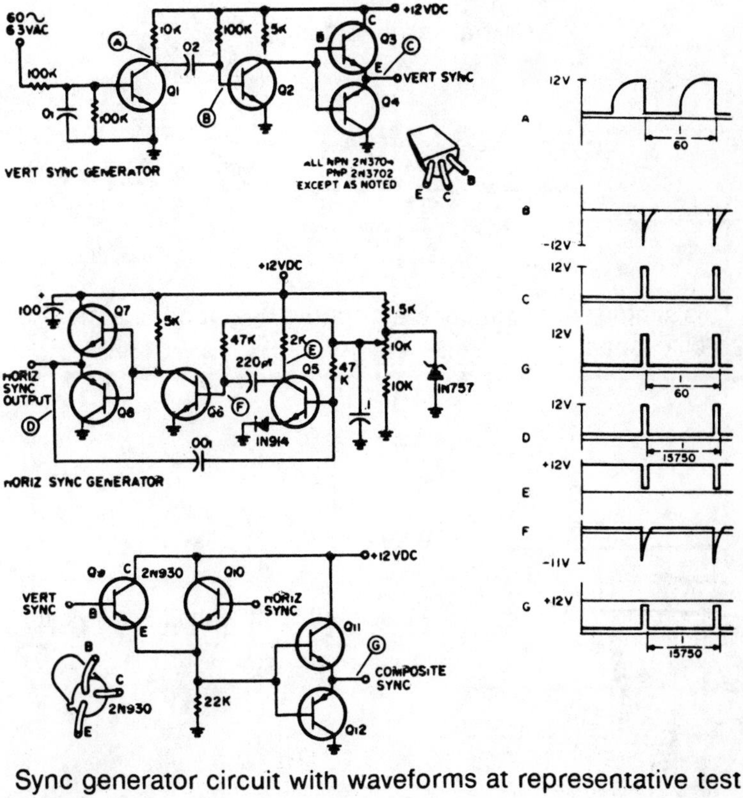


Sync generator circuit with waveforms at representative test points.

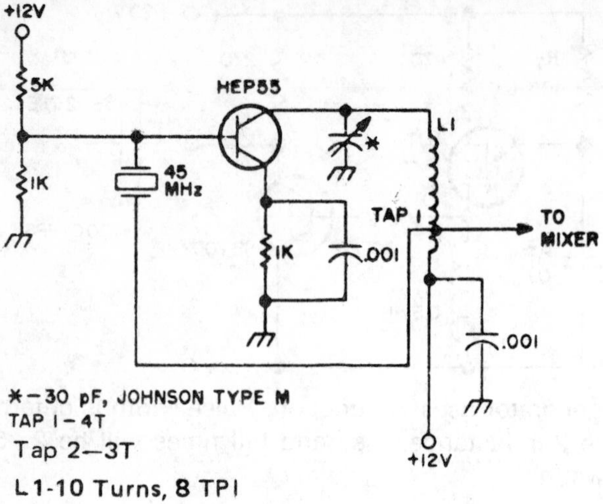


Crystal oscillator for 45 MHz.

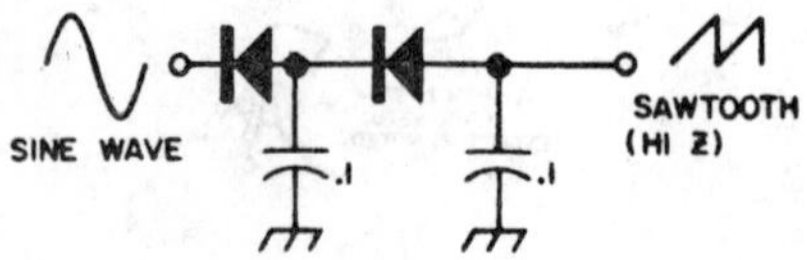


This simple sawtooth generator could be added to a monitor oscilloscope.

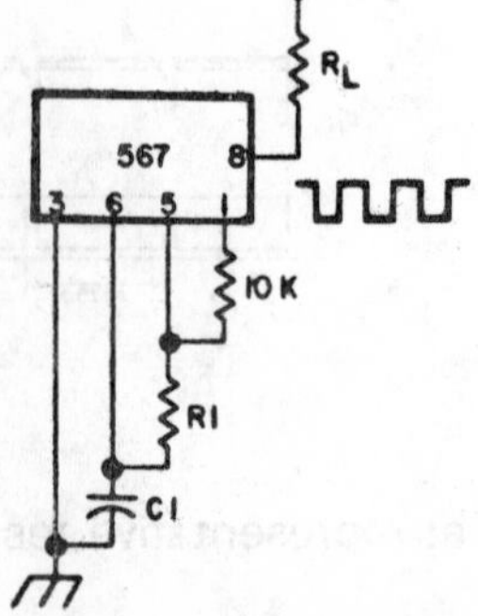


Pulse generator with 25% duty cycle.

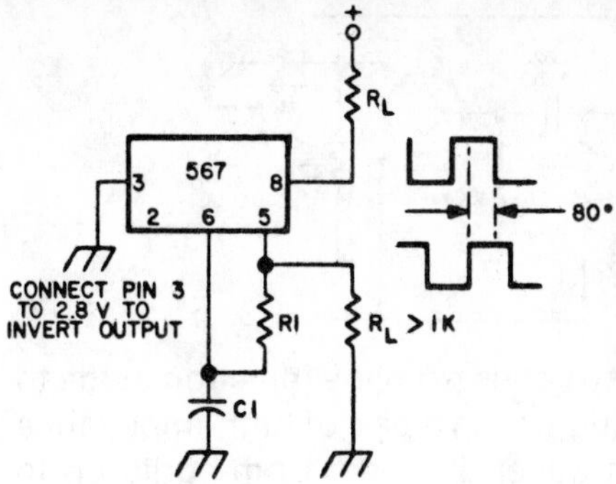


567 IC makes simple dual square-wave source. Note 80° phase shift between waves.

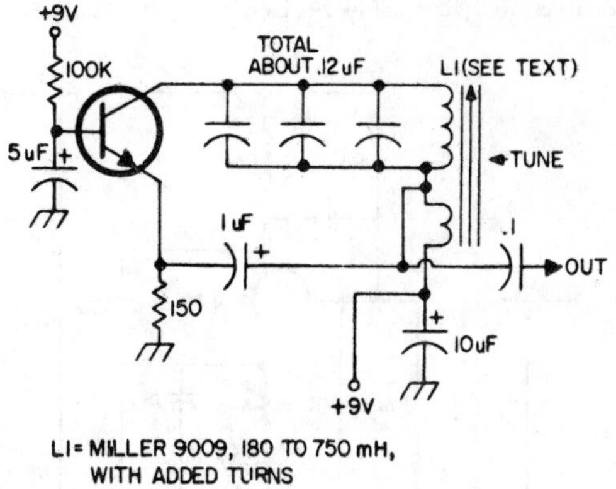


2000 Hz test oscillator. L1 is commercial 180 mH coil with about 80 added turns.

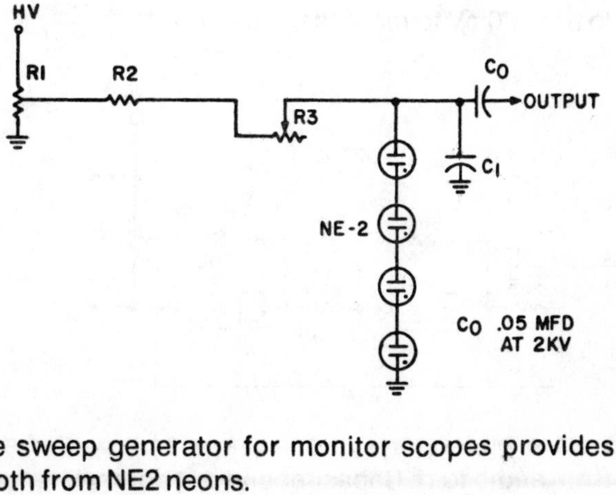


Simple sweep generator for monitor scopes provides 30 Hz sawtooth from NE2 neons.

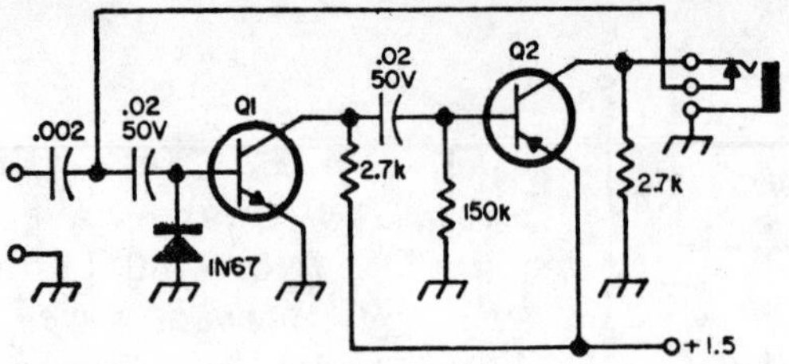


This signal injector/tracer switches from the injection mode to a signal tracer by simply plugging in a pair of high-impedance magnetic earphones. As a tracer, it works from audio up to 432 MHz. Transistor Q1 is a 2N170, 2N388A, 2N1605, SK3011, or GE-7; Q2 is a 2N188A, 2N404, 2N2953, SK3004 or HEP 253.

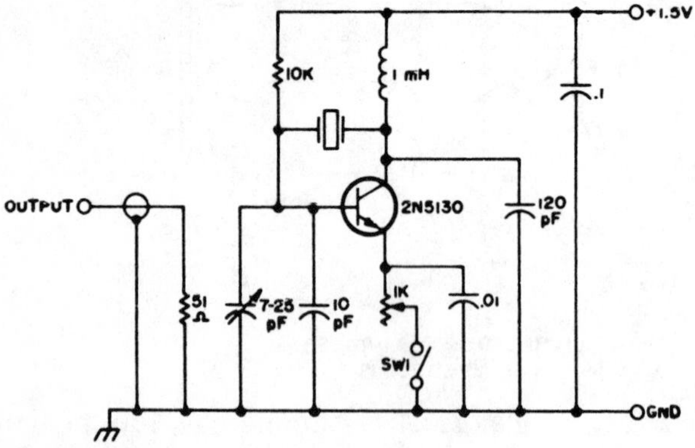


Extremely stable, signal generator provides output in range from 1.8 to 450 MHz at impedance of 50 ohms; output is adjustable from 80 nV to more than 50 mV rf.

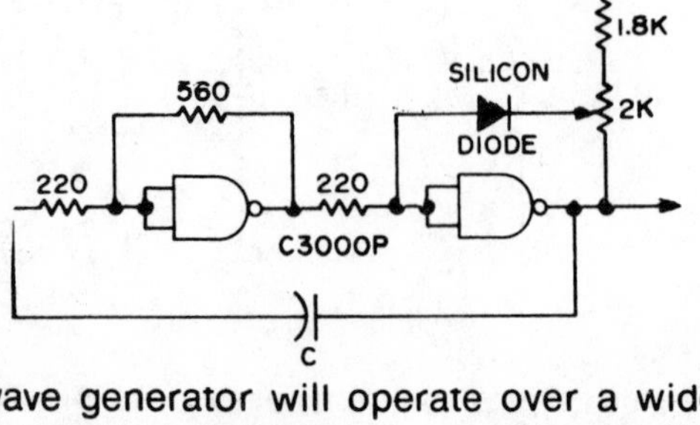


Square-wave generator will operate over a wide frequency range from audio to rf. Capacitor and pot control frequency.

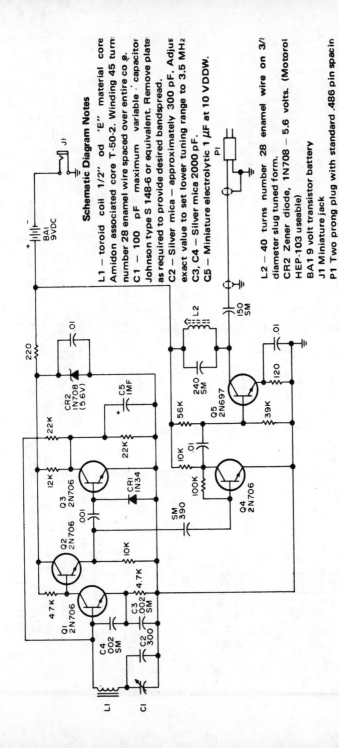

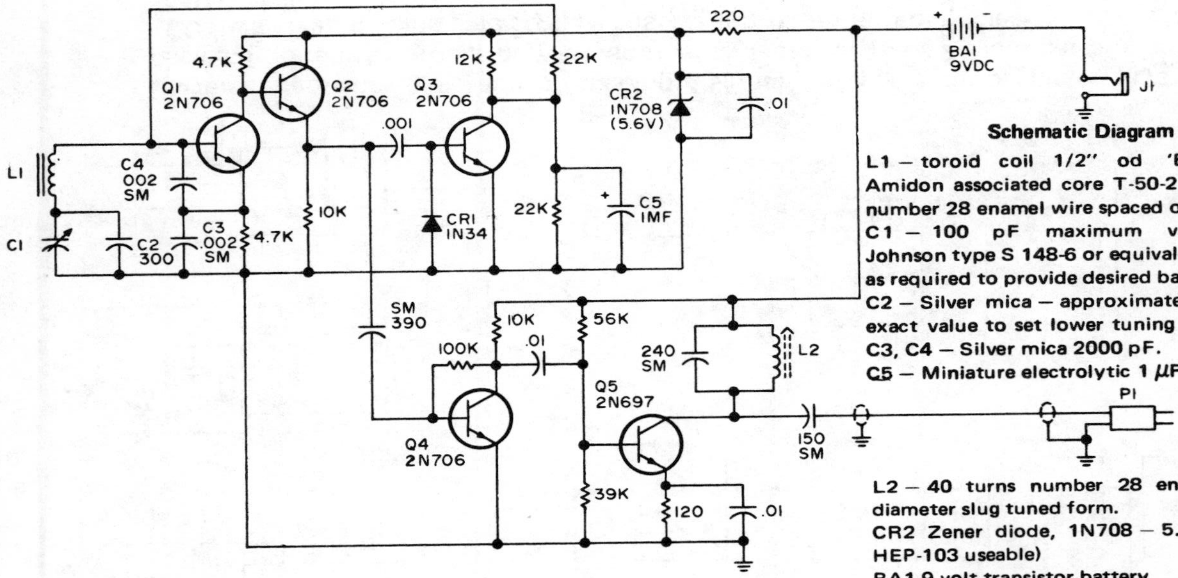


Variable-frequency oscillator operates over 75–80 meter range (about 3.5 MHz).

**Schematic Diagram Notes**

L1 – toroid coil 1/2" od 'E" material core Amidon associated core T-50-2. Winding 45 turn number 28 enamel wire spaced over entire co e.

C1 – 100 pF maximum variable capacitor Johnson type S 148-6 or equivalent. Remove plate as required to provide desired bandspread.

C2 – Silver mica – approximately 300 pF. Adjus exact value to set lower tuning range to 3.5 MHz

C3, C4 – Silver mica 2000 pF.

C5 – Miniature electrolytic 1 $\mu$F at 10 VDDW.

L2 – 40 turns number 28 enamel wire on 3/ diameter slug tuned form.

CR2 Zener diode, 1N708 – 5.6 volts. (Motorol HEP-103 useable)

BA1 9 volt transistor battery

J1 Miniature jack

P1 Two prong plug with standard .486 pin spacin for crystal sockets or other to match companio transmitter.

Q1, Q2, Q3, Q4 2N706

Q5 2N697

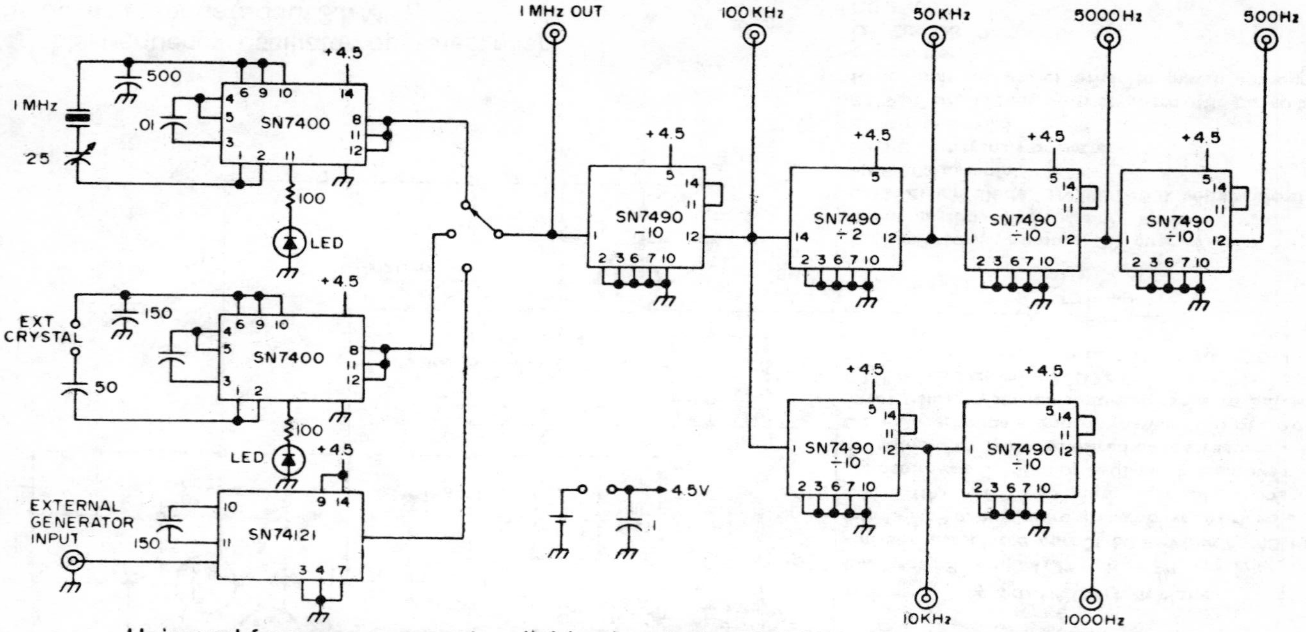


Universal frequency generator divides input signal (1 MHz master oscillator, crystal oscillator, or any external sine-wave source) by factor of 10 or 2. Note gates driving LEDs, which light to indicate crystal is oscillating. (Output frequencies shown are for 1 MHz oscillator.)

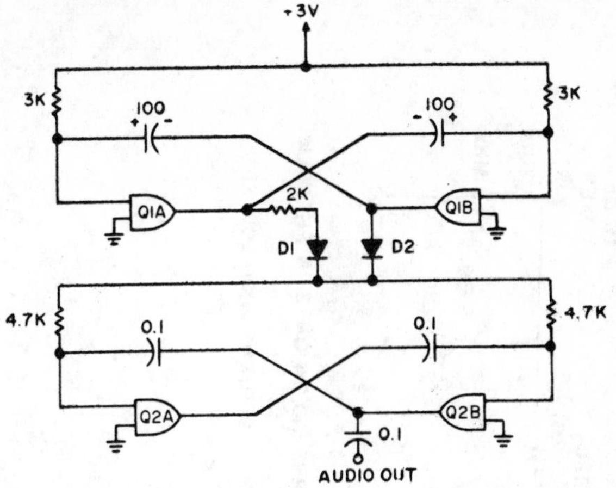


D1,D2 = 1N34 or other germanium.
Q1,Q2 = μL914 or equivalent.

Circuit shown above provides a square-wave output from Q2 whose audio frequency is changed alternately by action of multivibrator Q1. The resulting warbling note provides an excellent burglar alarm.

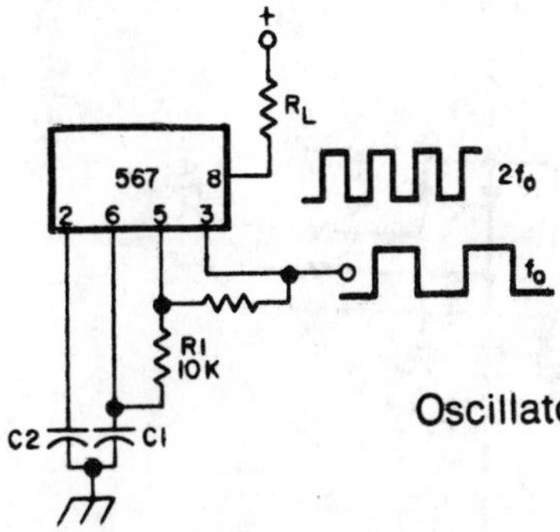


Oscillator with double frequency output.

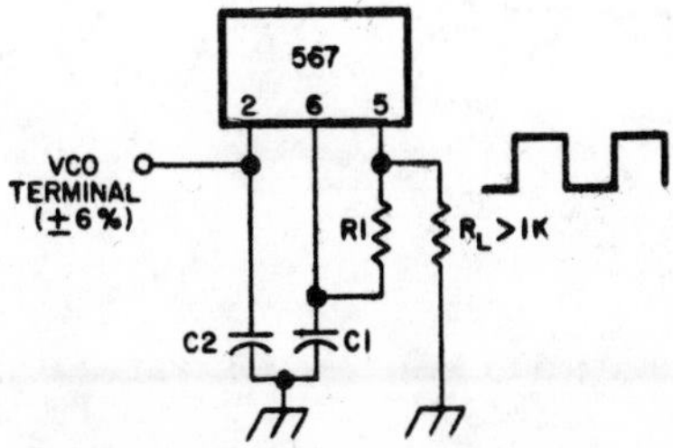


Precision oscillator with 20 nsec switching.

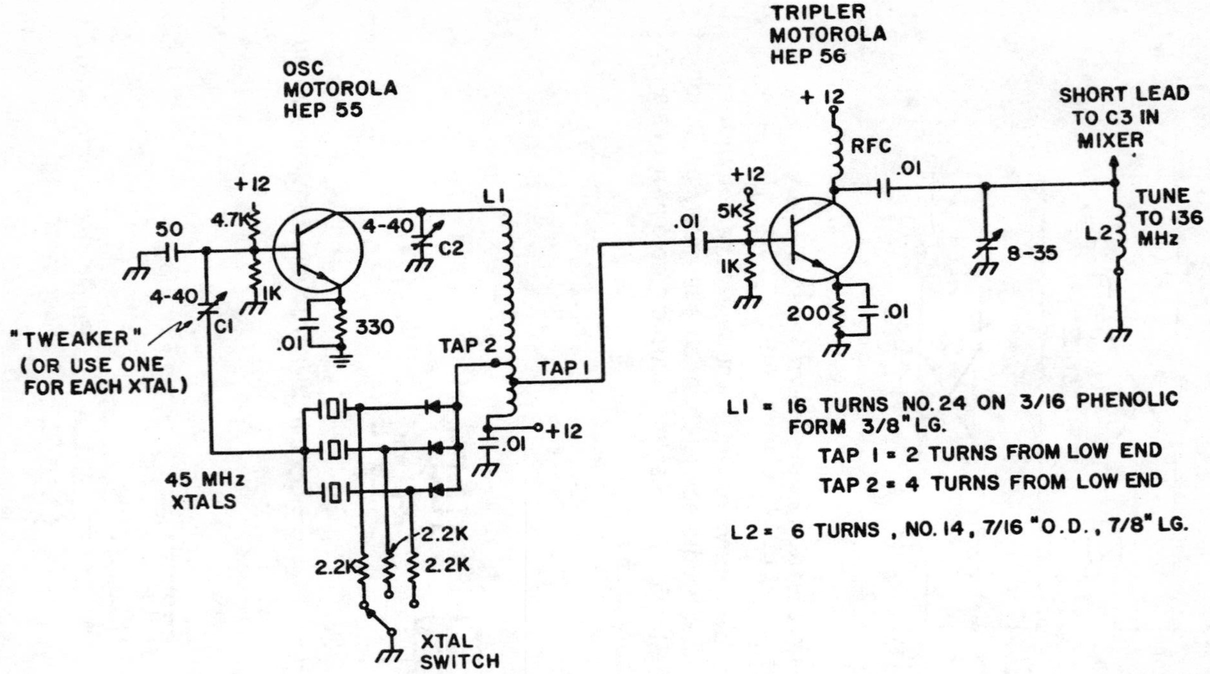


45 MHz oscillator and tripler section.

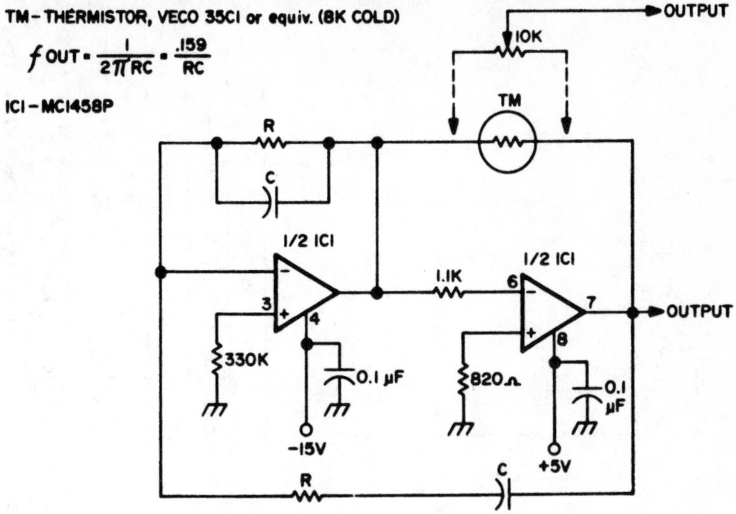


A simple sine-wave generator.

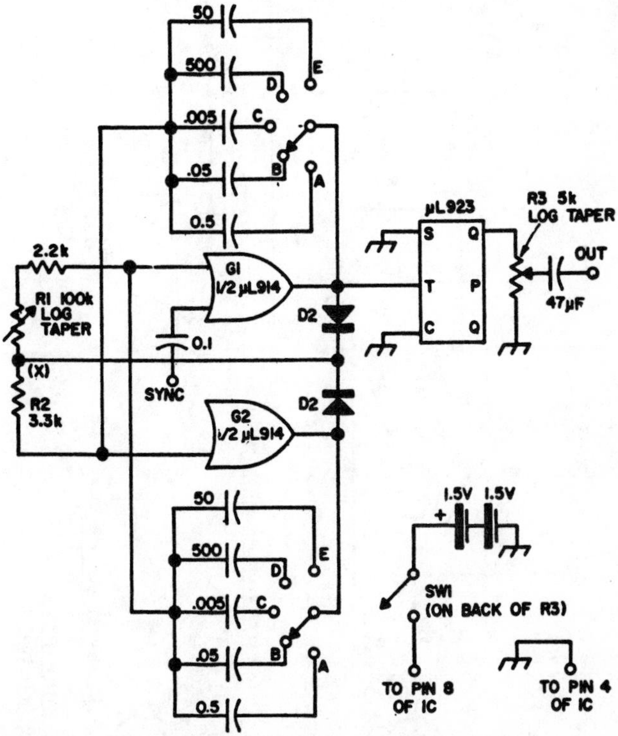


Complete square-wave generator. Bandswitching capacitors are 10% or better tolerance. Resistors are ¼ watt.

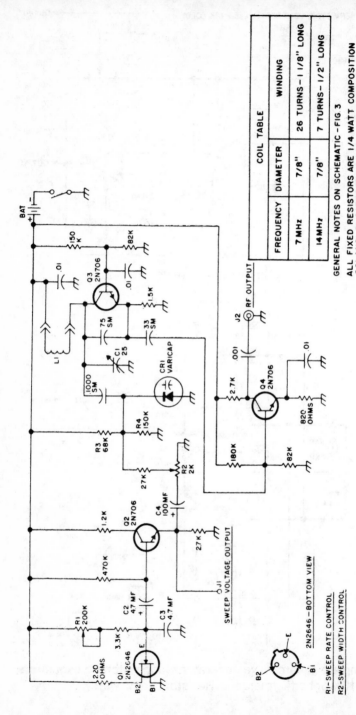

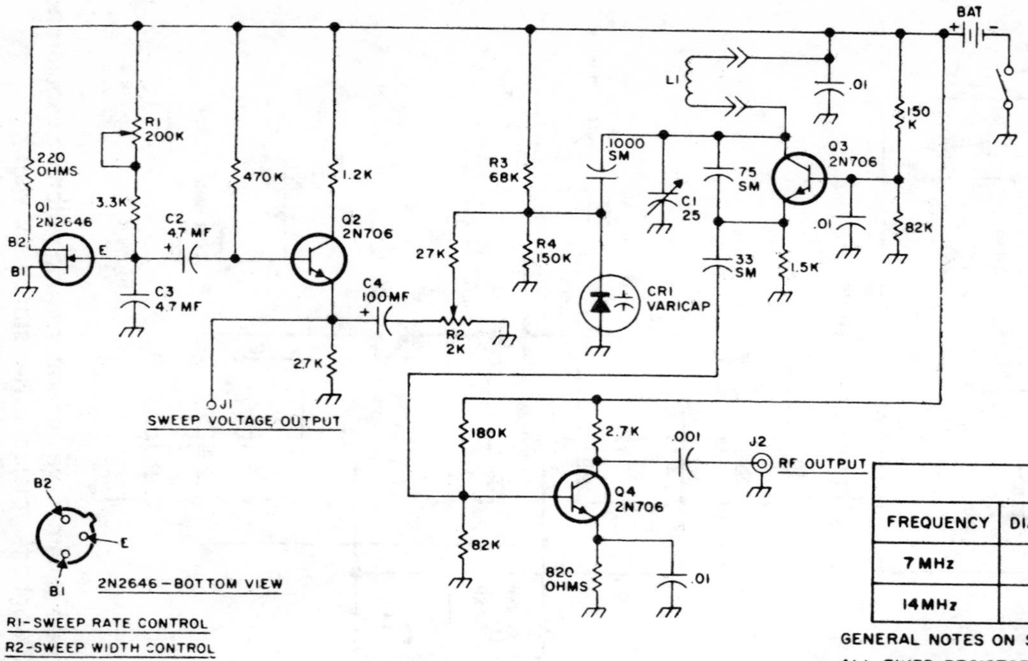


| COIL TABLE | | |
|---|---|---|
| FREQUENCY | DIAMETER | WINDING |
| 7 MHz | 7/8" | 26 TURNS – 1 1/8" LONG |
| 14 MHz | 7/8" | 7 TURNS – 1/2" LONG |

GENERAL NOTES ON SCHEMATIC – FIG. 3

ALL FIXED RESISTORS ARE 1/4 WATT COMPOSITION
.001 AND .01 CAPACITORS ARE DISC CERAMIC
CAPACITORS MARKED "SM" ARE SILVER MICA R1
AND R2 STANDARD CARBON ELEMENT POTENTIOMETERS
C2, C3, AND C4 ARE ELECTROLYTICS – 25WVDC OR MORE

Sweep oscillator circiit.

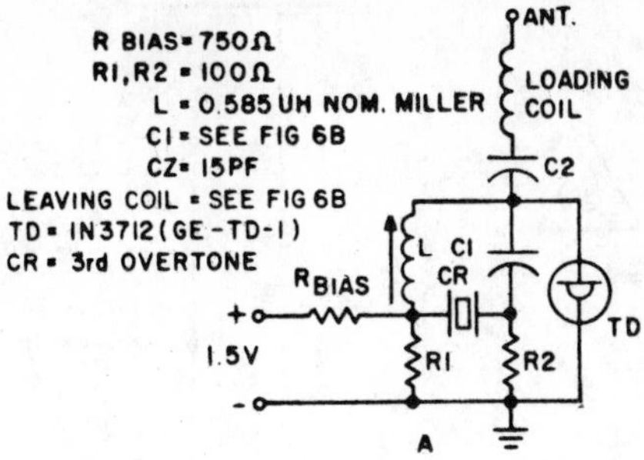


| C₁ = | | LOADING COIL UH |
|---|---|---|
| 144MHZ | 20PF | 0.1 |
| 50MHZ | 2.2 | 0.65 |
| 28MHZ | 58 | 2 |
| 14 MHZ | 240 | 9 |
| 7 MHZ | 900 | 32 |

B

Crystal-controlled oscillator uses tunnel diode.

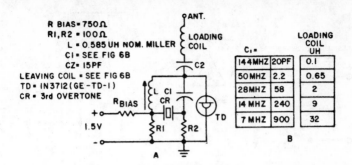

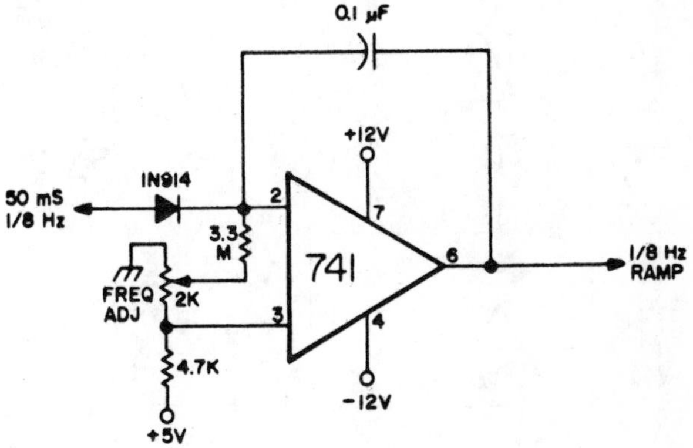


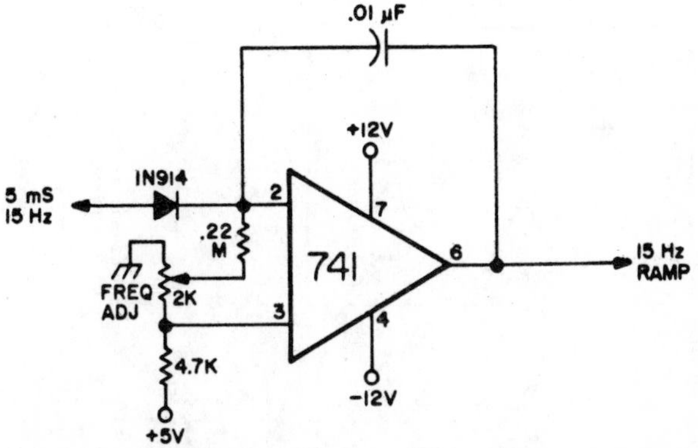


SSTV ramp generators. This circuit will give an extremely linear ramp for SSTV monitors, cameras, and flying spot scanners. The voltage varies from ≈ −10 to +10V. A positive going pulse of +2 to +5V amplitude resets the ramp for the next sweep.

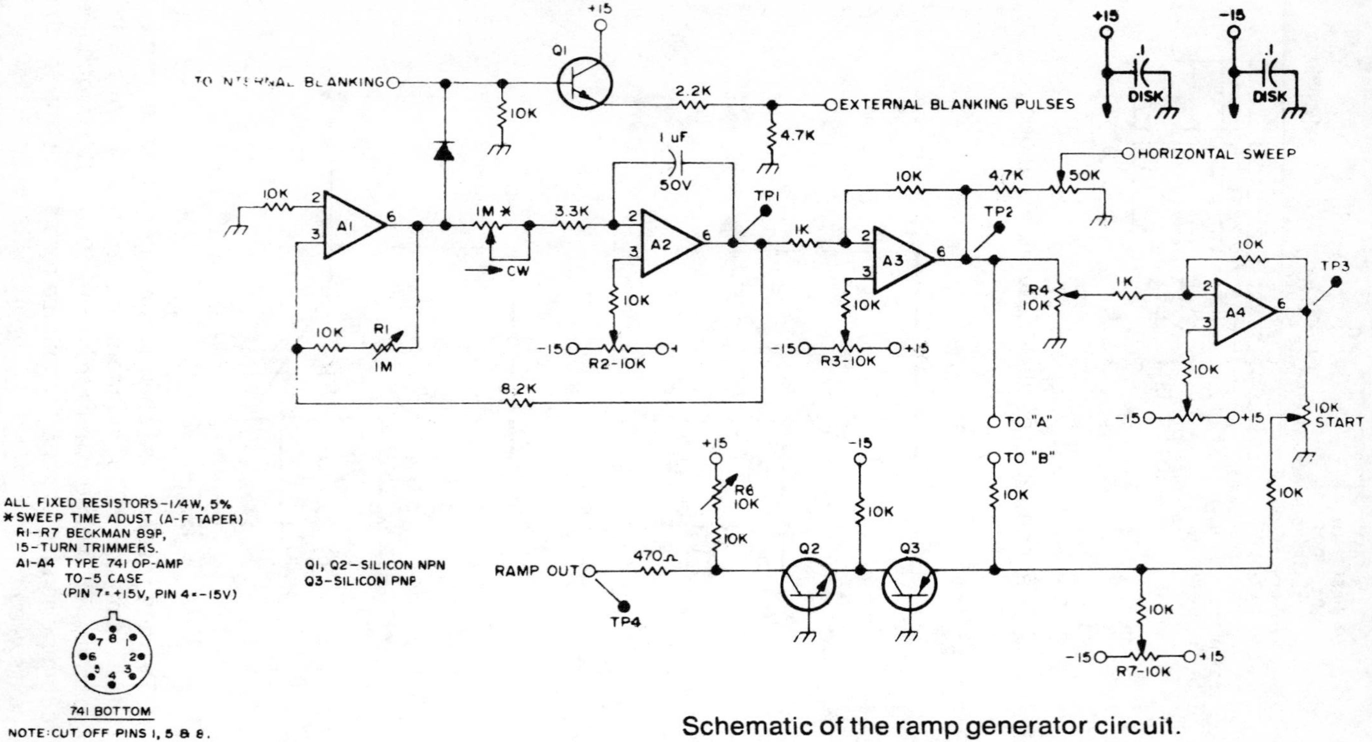


Schematic of the ramp generator circuit.

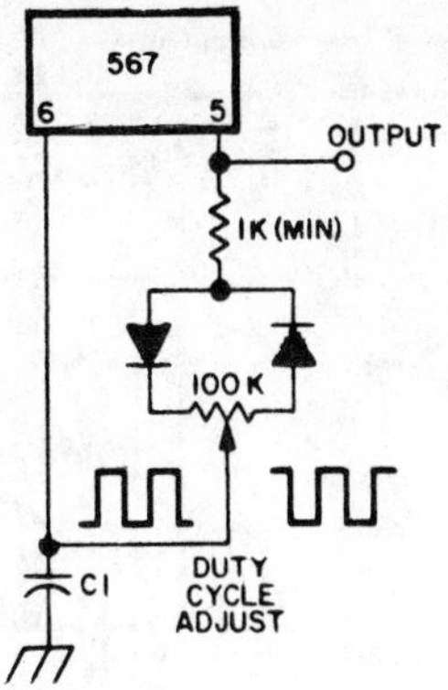


Pulse generator. Courtesy of Signetics catalog.

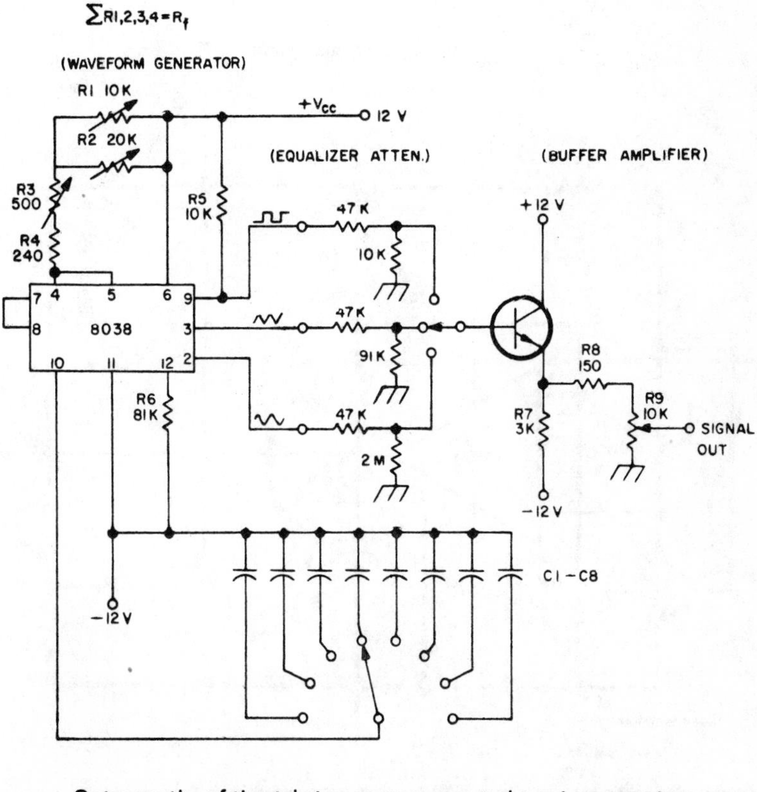


Schematic of the triple-wave output signal generator.

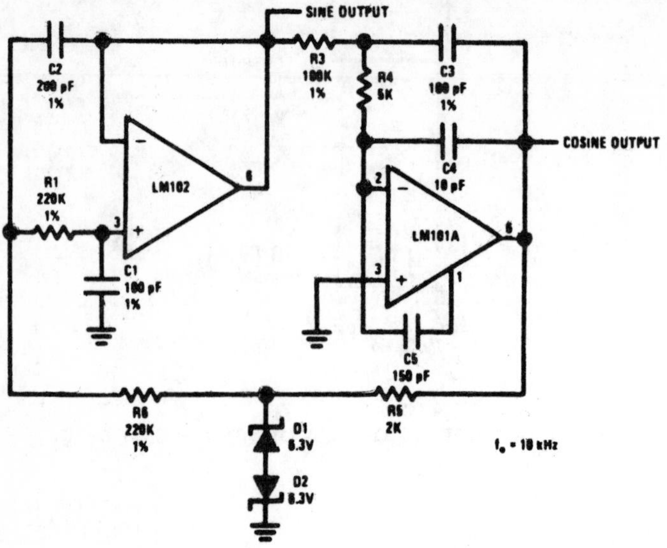


Sine-wave oscillator.

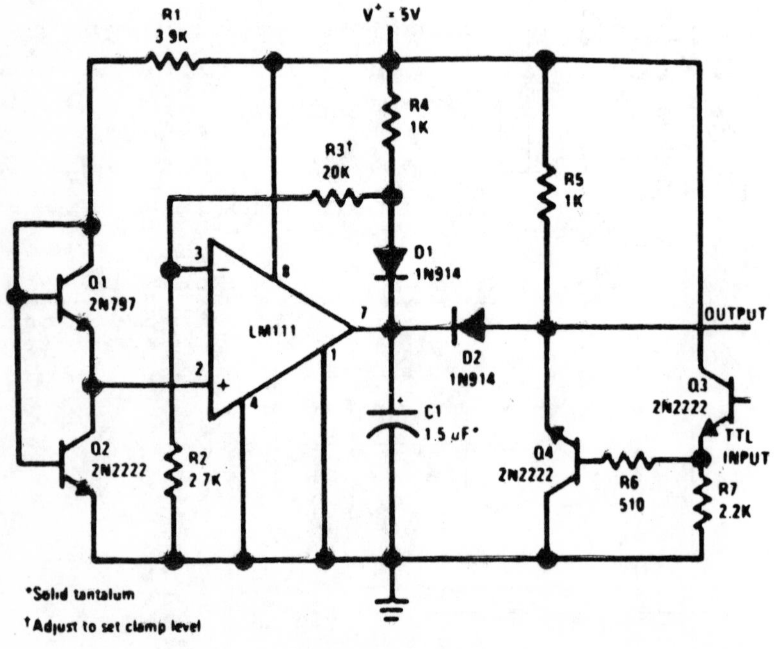


Precision squarer.

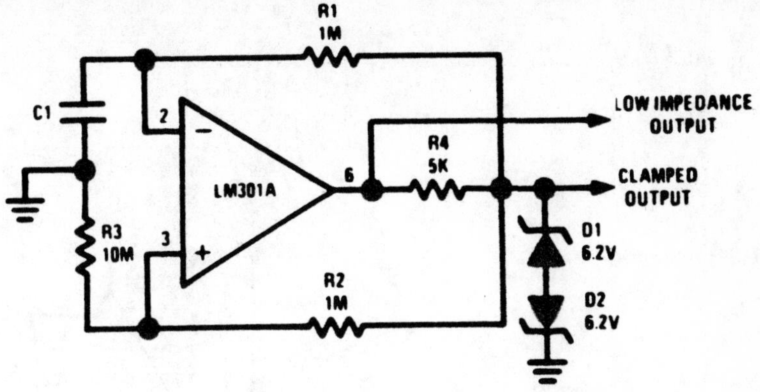


Low-frequency square-wave generator.

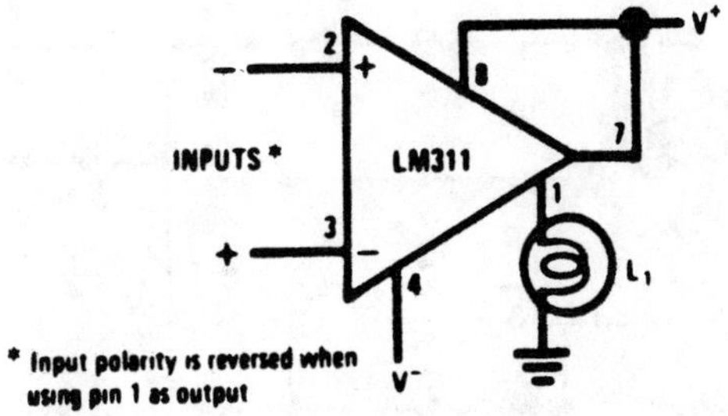


Driving ground-referred load.

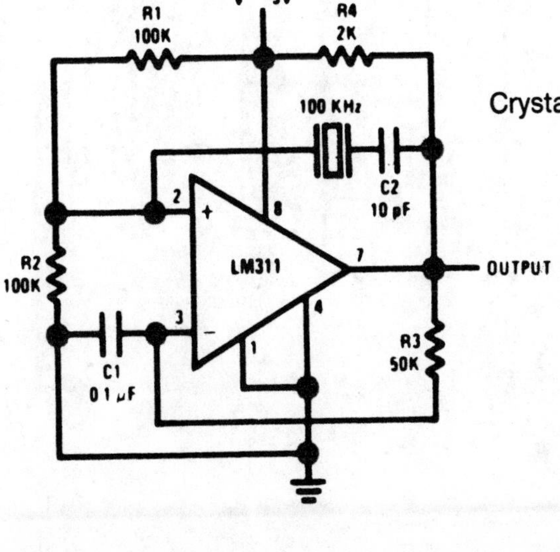


Crystal oscillator.

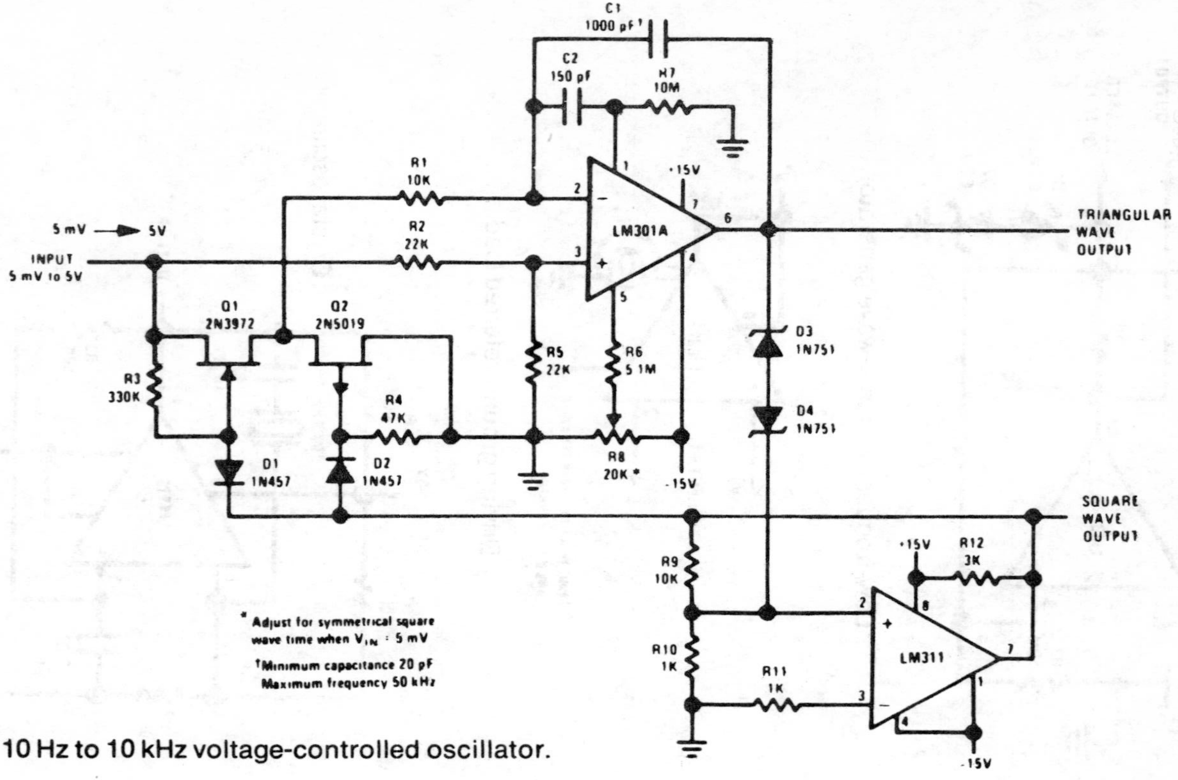


* Adjust for symmetrical square wave time when $V_{IN}$ = 5 mV

† Minimum capacitance 20 pF
Maximum frequency 50 kHz

10 Hz to 10 kHz voltage-controlled oscillator.

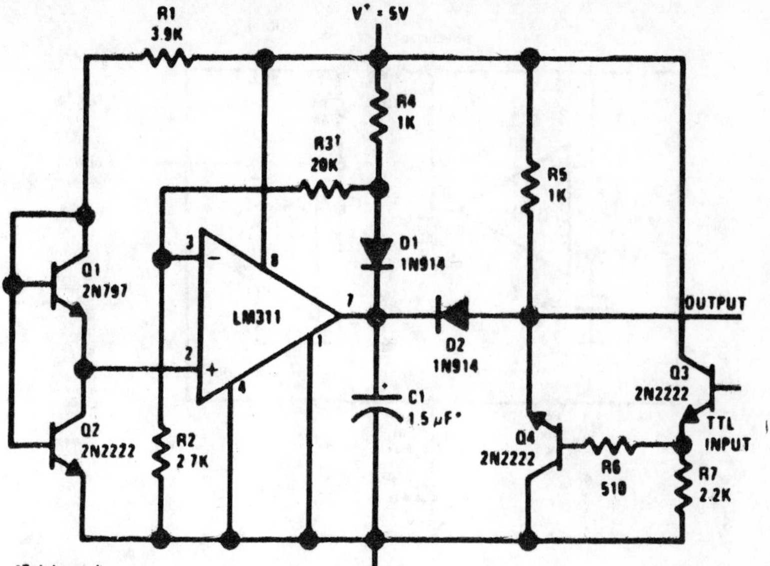


*Solid tantalum

†Adjust to set clamp level

Precision squarer.

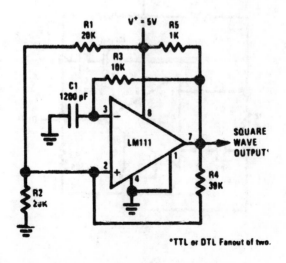

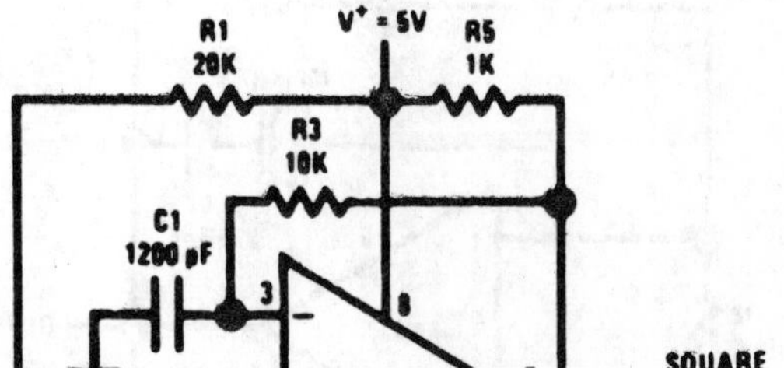


*TTL or DTL Fanout of two.

100 kHz free-running multivibrator.

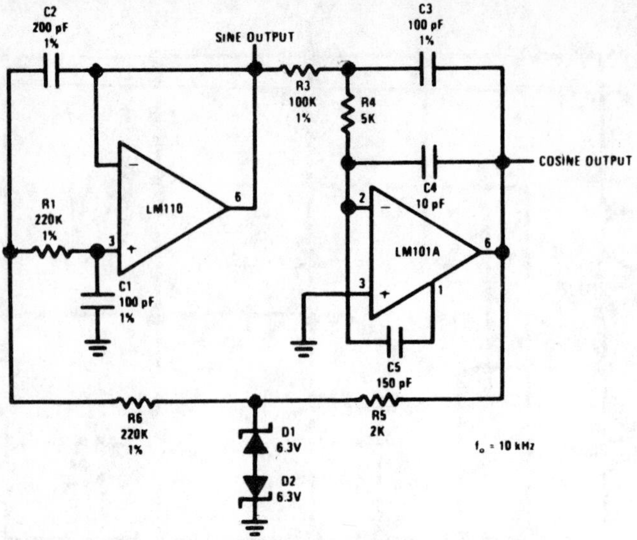


Sine-wave oscillator.

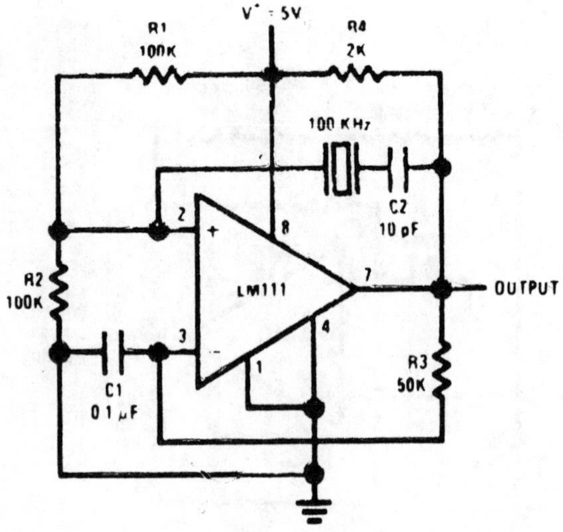


Crystal oscillator.

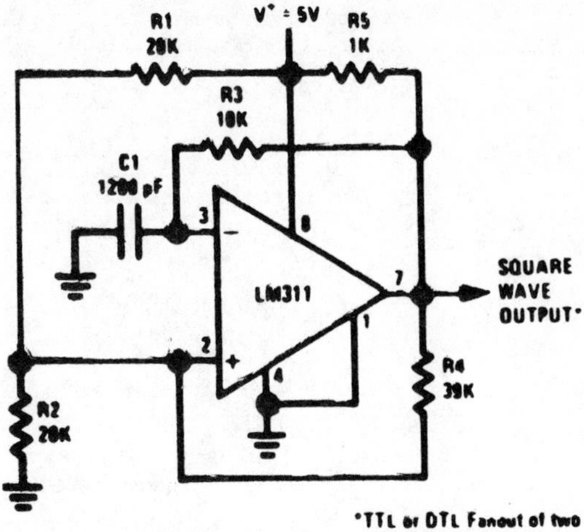


100 kHz free-running multivibrator.

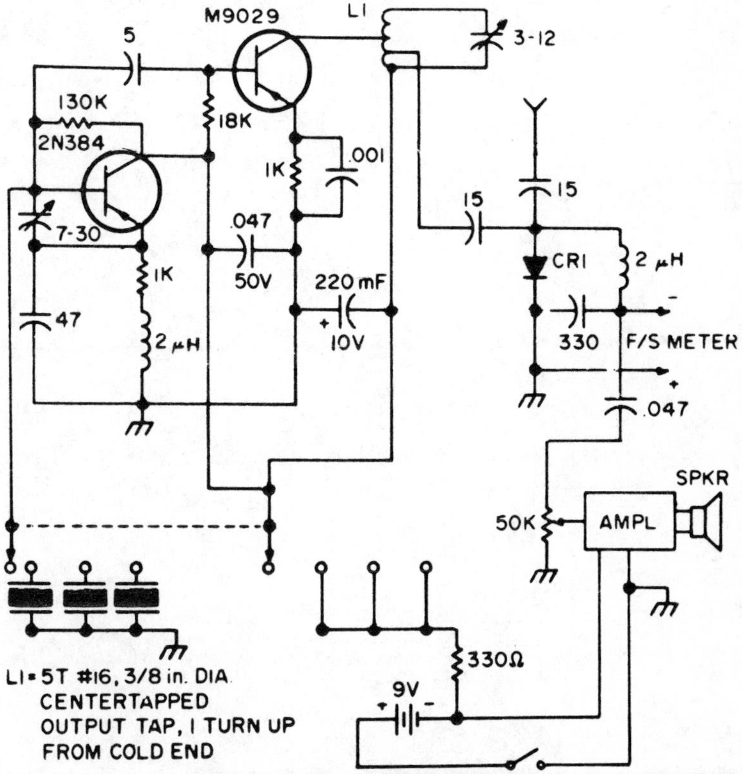


Signal generator/amplifier.

# Section K

# RF Power Amplifiers

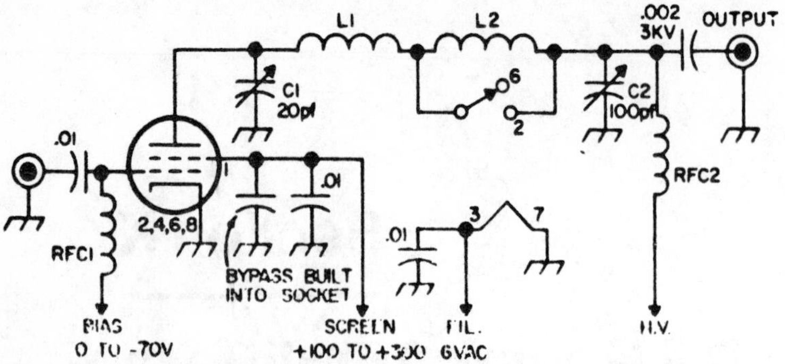


4CX250 amplifier for six and two meters, using single-pole switching. Simple modifications by K4ETZ [forget the 6 meter coil and use smaller HA tuning capacitors, class C bias and a grid tank] allow 10W drive, 200–250W out. Amplifier should have a screen clamp tube with G2 supply bled down from B+ for best results.

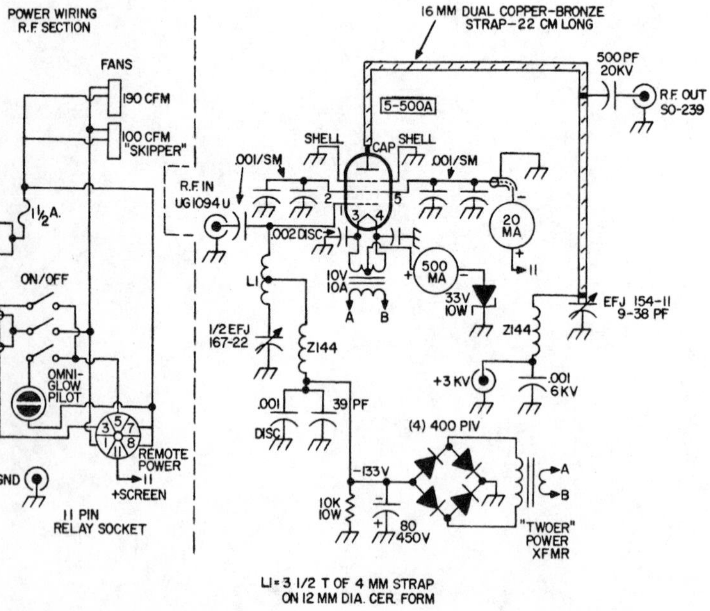


Kilowatt linear amplifier (rf) for 50 MHz. Requires 2500 to 3000V dc for 5–500A plate.

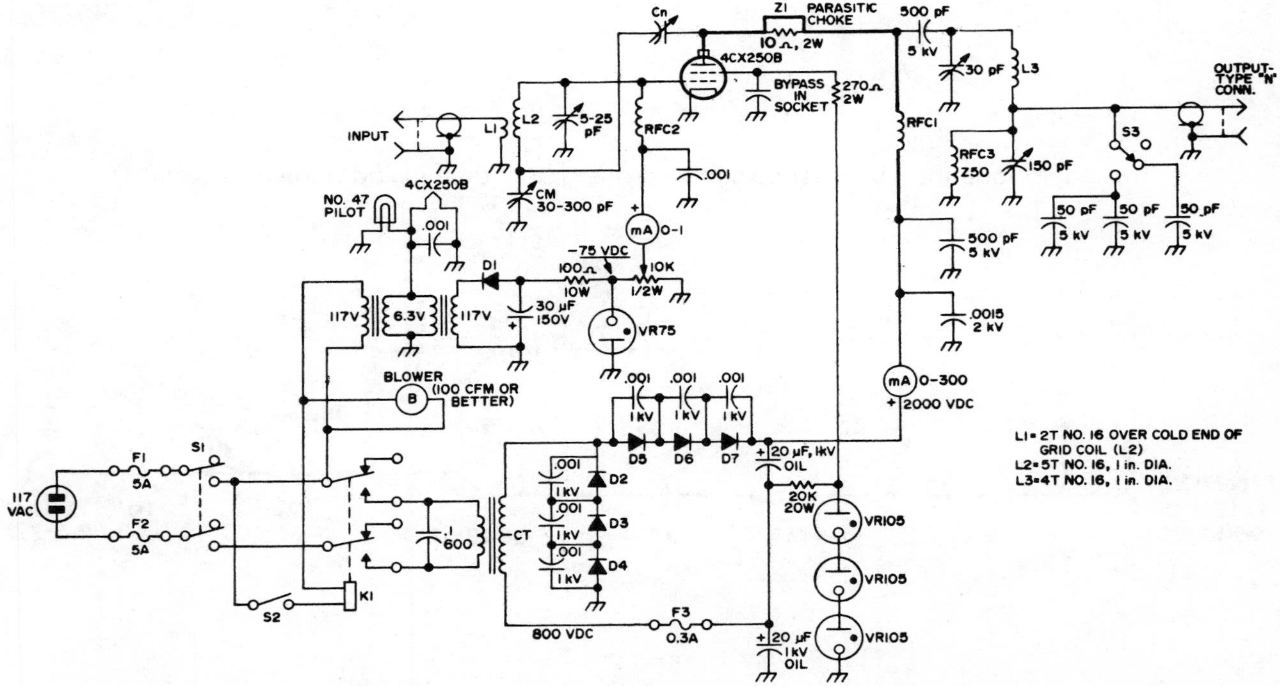


150 MHz linear amplifier. Capacitors should have 1 kV rating.

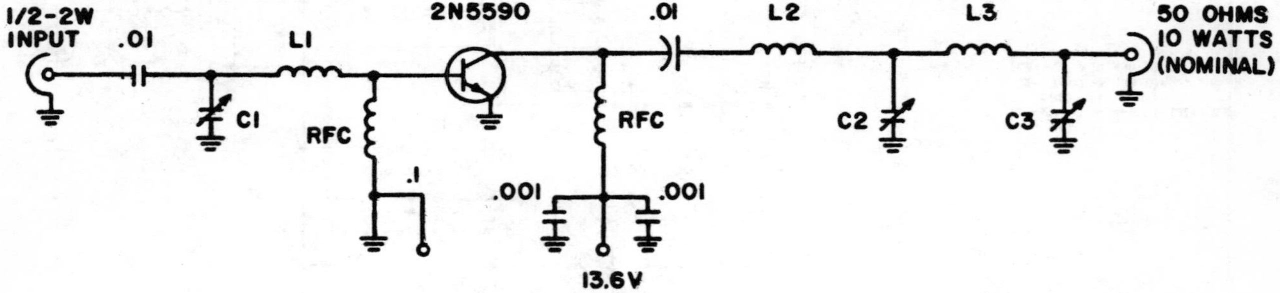


Typical transistor rf power amplifier; L and C values depend on frequency of interest.

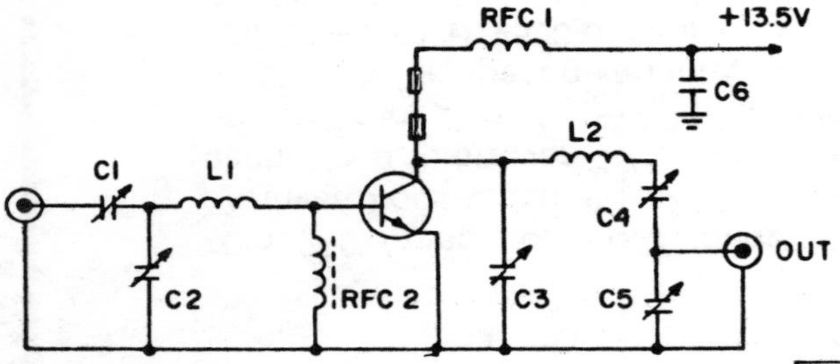


**PARTS LIST**

| | |
|---|---|
| $Q_1$ | 2N5591 or S3007 |
| $C_1$, $C_2$ | 7-60 pF (ARCO 462) |
| $C_3$, $C_4$ | 4-40 pF (ARCO 403) 20-250 pF |
| $C_5$ | (ARCO 424) |
| $C_6$ | .02 μF ceramic |
| $RFC_1$ | 10 T #18, ¼" I.D., tightly wound, 2 ferrite beads |
| $L_1$, $L_2$ | 4 T #14, ¼" I.D. x 5/8" long |
| *$R_1$ | 20-100Ω – use if unit is instable – start with 100Ω and reduce when stability is achieved; goes from base to ground. |
| $RFC_2$ | 2 turns in ferroxcube bead |

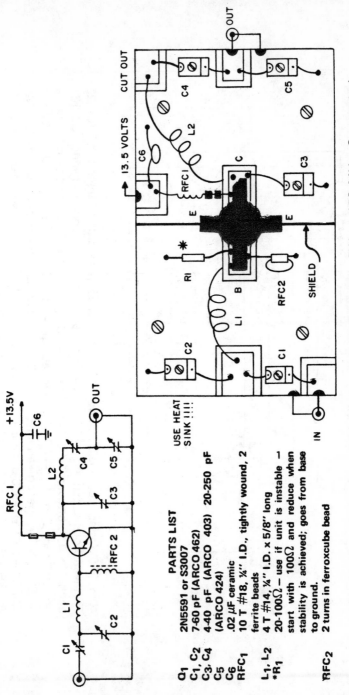

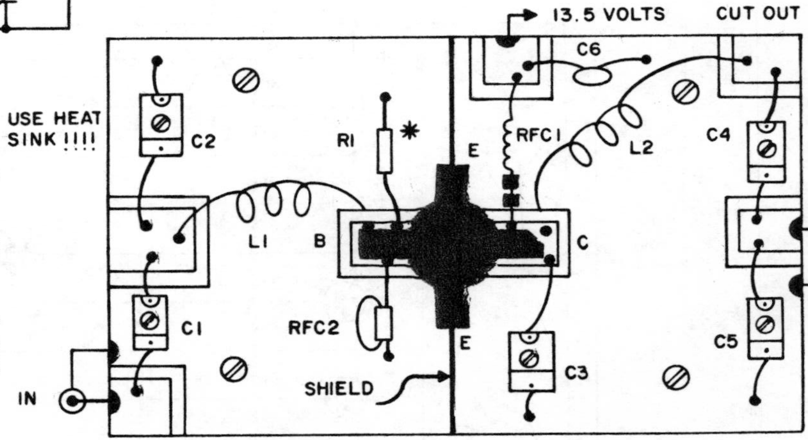


Rf amplifier produces 18W output when driven with 5W at 150 MHz. Suggested layout employing shielding is shown.

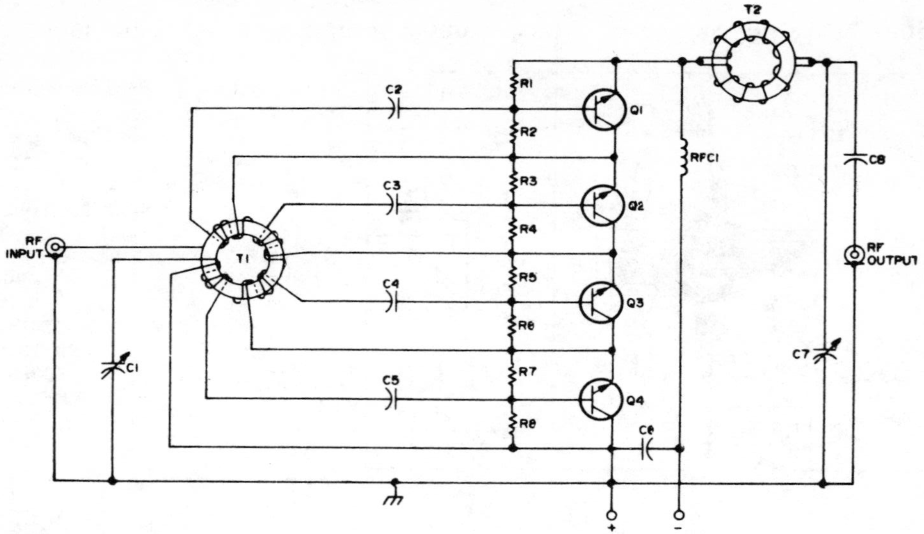


40m, 200W amplifier. T1: 2 Amidon 15/16 in. diameter toroids. Primary; 25T No. 22 Formvar. Four secondaries; 5T No. 22 Formvar each. T2: 2 airwound windings, 18T each, No. 14 Formvar, $1^{3}/_{16}$ diam. These windings are paralleled and shaped into a donut 3½ in. diam. C1—450 pF broadcast variable. C2 through 6—0.1 $\mu$F, 100V. C7 −450 pF wide spaced variable. C8—0.01 $\mu$F, 500V (vary for best coupling). Q1, Q2, Q3, Q4—TRW PT4526 or similar. R1, R3, R5 – 7 to 10$\Omega$, 2 to 5W wirewound. R2, R4, R6, R8—450$\Omega$, 1W. RFC ⅜" dia. ferrite rod, 50T No. 22 Formvar.

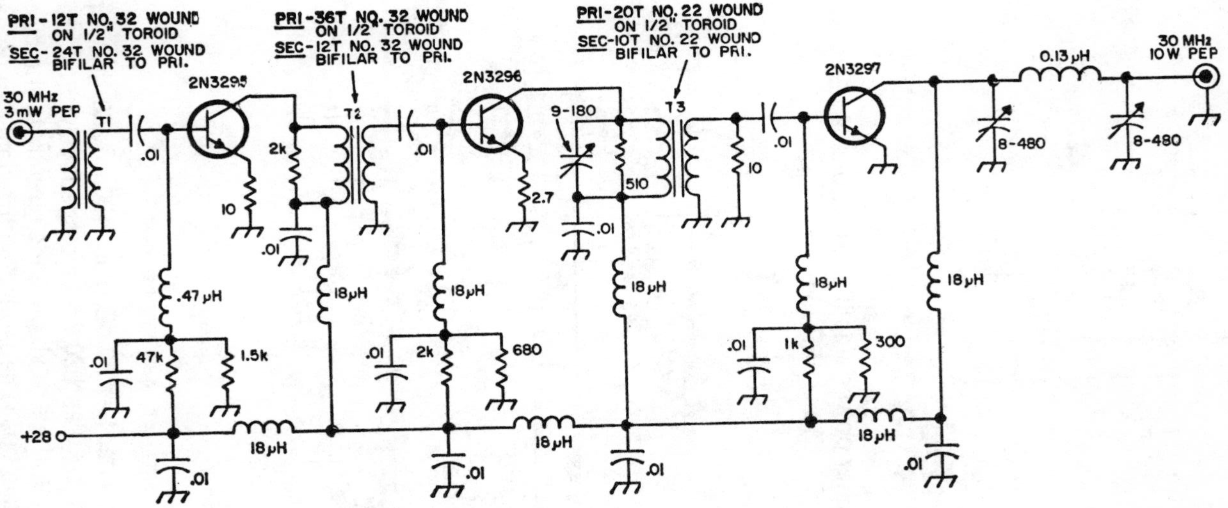


This 10-meter linear amplifier for SSB service uses transistors which were designed specifically for single-sideband linear operation. Many junction transistors cannot be used satisfactorily for this application, because linear amplification at low power levels is a serious problem.

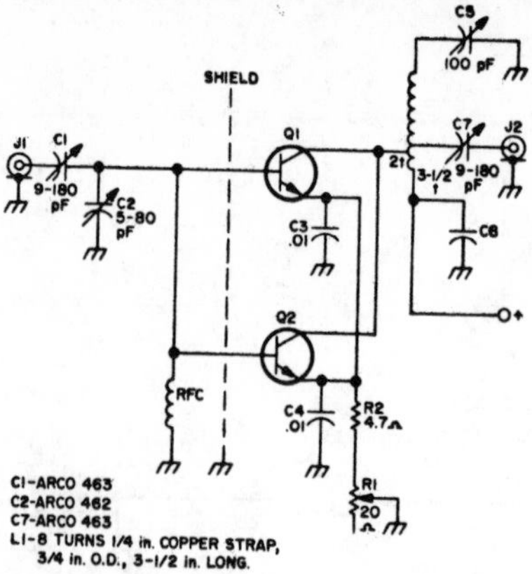


50 MHz transmitter power amplifier. Rf choke is 7 – 10 μH C6 is 1000 pF disc. Q1, Q) are HEP 75s.

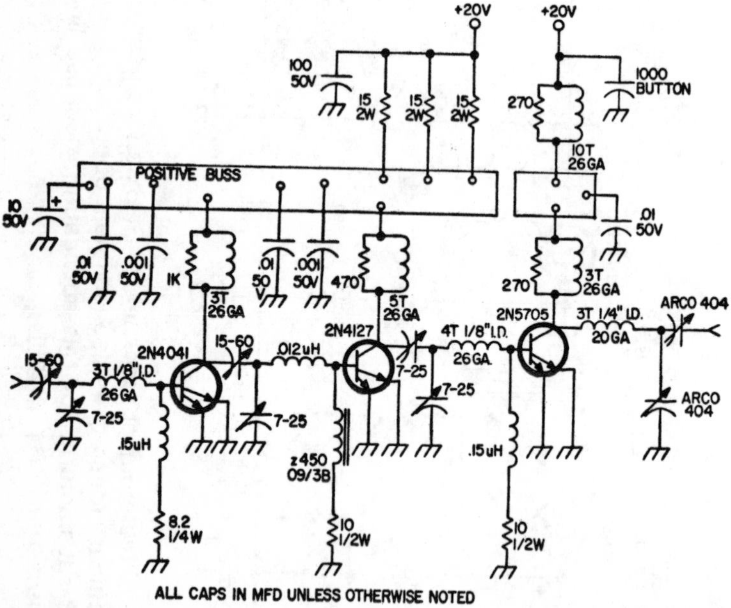


2-meter amplifier that has 20 dB gain. 100 or 200 mW input will be amplified to 10 and 20 watts, respectively. Be sure to keep leads short and make all grounds directly to the circuit board. The positive bus is separated so the final can be used for AM. For FM, just connect the two terminats together. Don't forget to use heatsinks!

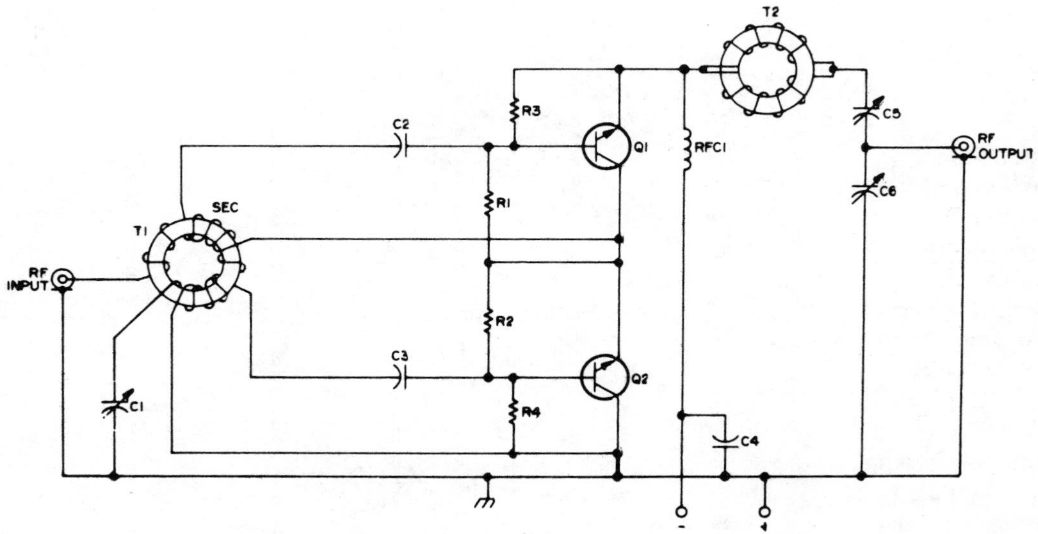


40m, 75W amplifier. T1: Amidon 15/16 in. toroid. Primary; 50T No. 24. 2 secondaries; 10T, No.24, solid insulated wire each. T2: Same type core except 2 used: 2 windings No. 16 Formvar 5T each connented in parallel. RFC: ⅜ dia. ferrite rod, 60T No. 22 Formvar. C1, C5: mica adjustable 200–800 pF. C2, C3: 1 μF, 100V disc. N4: 0.1 μF disc, 100V max. 25V min. C6: mica adjustable 400–1200 pF. R2, R3: 10Ω, 2–5W. R1, R4: 450Ω. Q1, Q2: TRW PT4526 or similar.

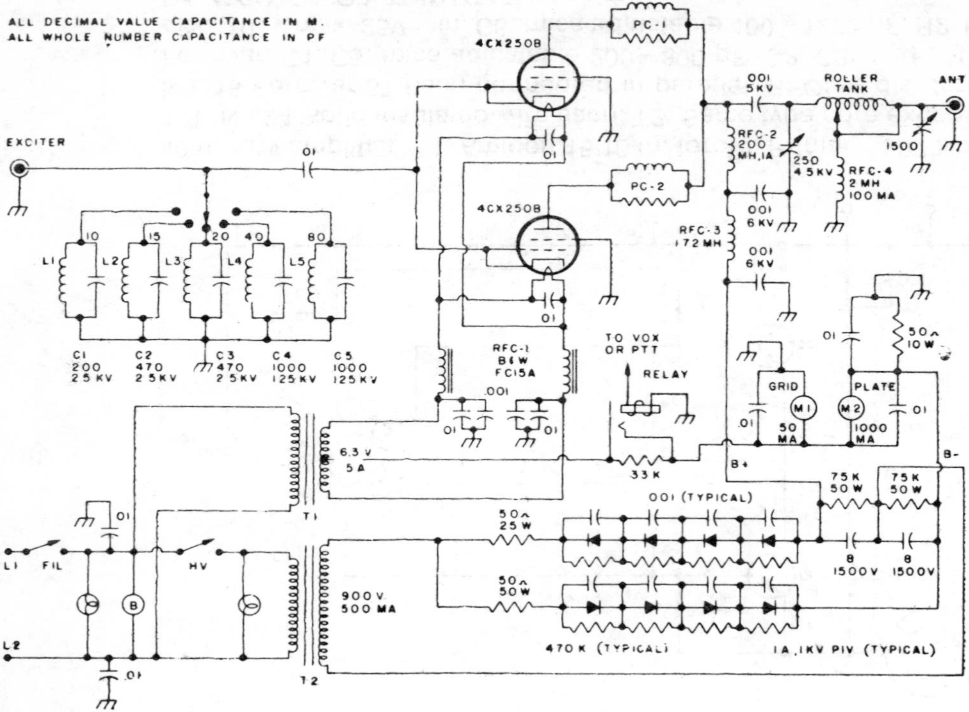


L1, L2 (0.15 $\mu$H): 4 turns No. 16 enamel on National XR50 ½ in. diameter. Slug removed from both coils. L3, L4 (0.31 $\mu$H) 6 turns No. 14 enamel on National XR50 form. Slug removed from L3, L5 (1.3 $\mu$H) 13 turns No. 18 enamel on National XR-50 form. PC1, PC2: 3 turns No. 16 enamel wound on 50Ω 2W carbon resistor.

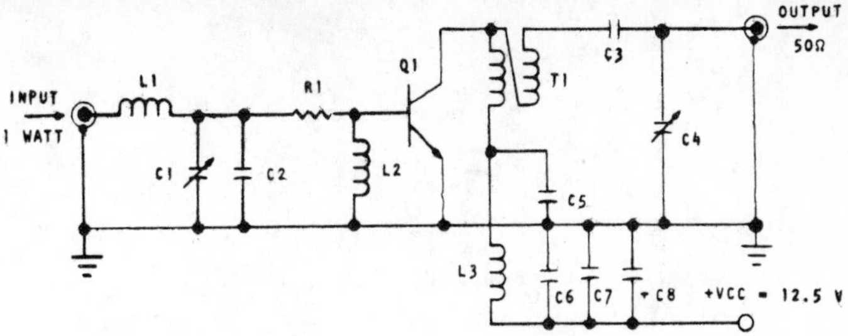


PARTS LIST
Q1 · TRW PT5757
C1 – 51 pF 5%, DIPMICA
C2, C4 – 8–60 pF
C3 – 100 pF DISC
C5 – .001 μF DISC
C6 – .01 μF DISC
C7 – .1 μF DISC
C8 – 10 μF 20V dc
L1 – 4 No. 20 0.3 cm I.D.
L2 – 12T no. 28 0.3 cm I.D.
L3 – 10 no. 20 0.3 cm I.D.
T1 – 4:1 transmission line transformer, made up of a 3" length of twisted pair, no. 20 enameled wire.
R1 – 1.0Ω 5% ½W

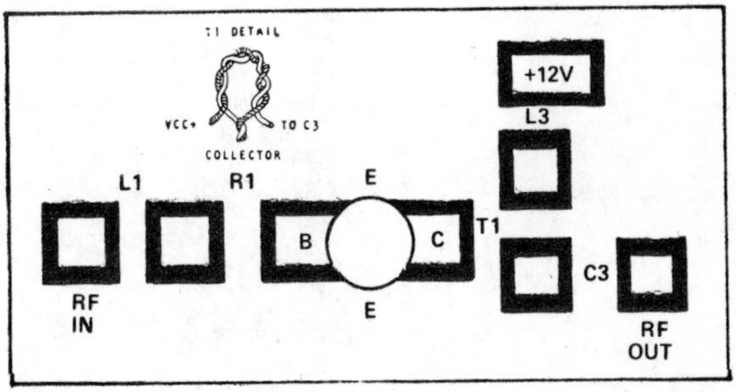


150 MHz 10-watt rf power amplifier schematic, parts list, and board layout.

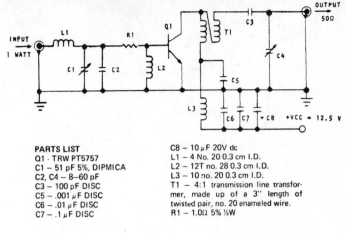

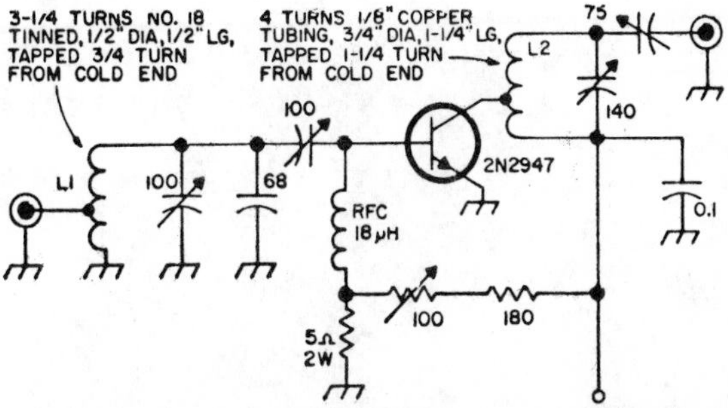


This 10 meter single-side band linear power amplifier will provide up to 8 watts PEP. The power gain of the 2N)947 is 13 dB at this frequency, and the odd order distortion products are at least 30 dB below the desired output.

# Section L

# Test Equipment

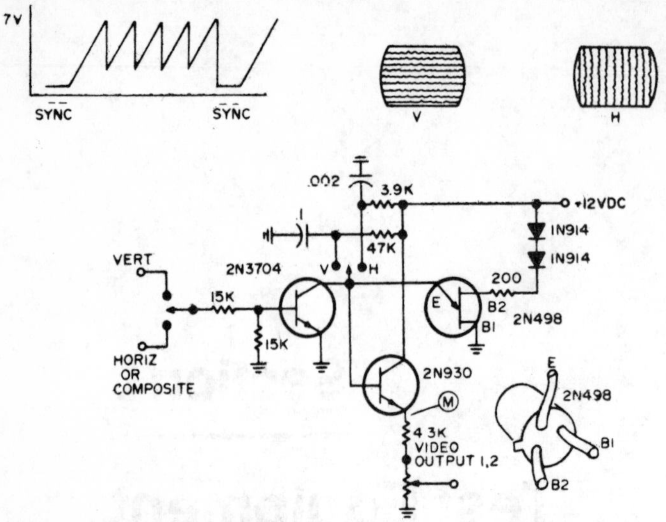


Vertical and horizontal bars can be generated with this circuit for checking TV sweep linearity.

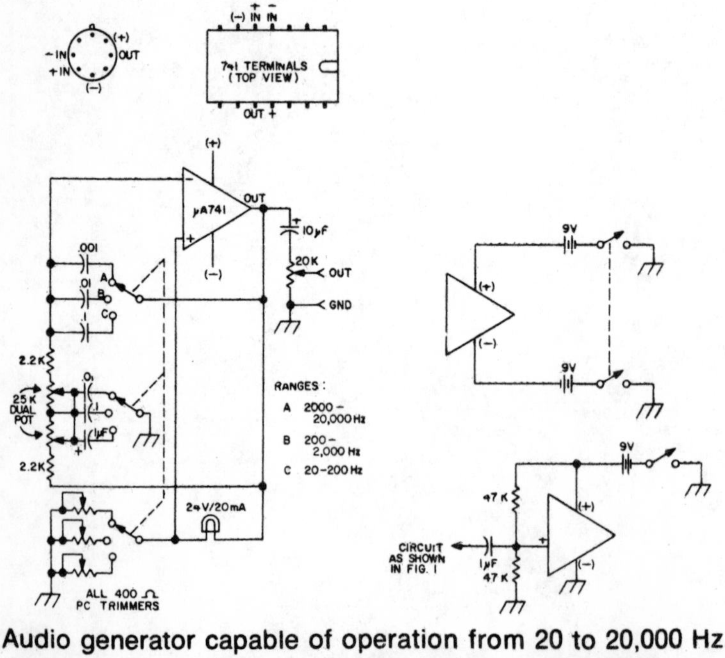


Audio generator capable of operation from 20 to 20,000 Hz. Optional connections shown for dual or single 9V dc supplies. The 25K dual potentiometer should have a linear (not audio) taper.

Sensitive rf voltmeter has full-scale ranges from 30 mV to 10V with a flat frequency response to 200 MHz. All FETs, transistors, and diodes used in pairs should be carefully matched.

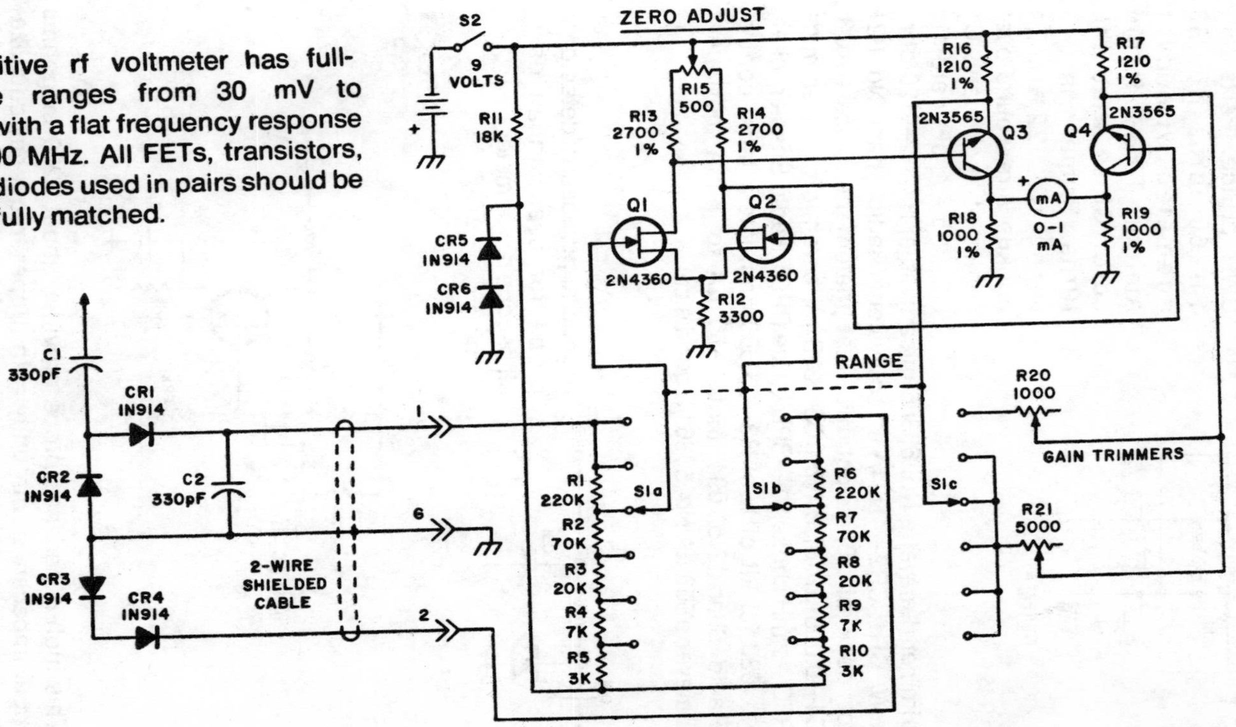

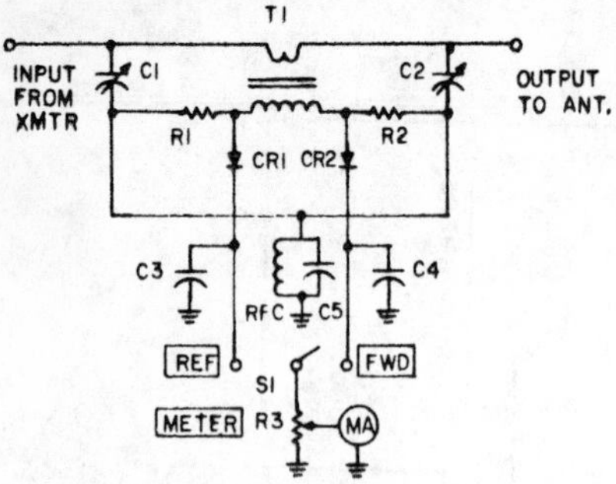


SWR bridge. Parts list: T1—60 turns No. 30 enameled wire over Amidon T-68-2 toroid core. Close wind the turns. Primary is two turns No. 22 or 24 hookup wire wound over center of secondary. C1 C2—1.5 – 7 pF ceramic trimmer (Lafayette No. 68386 mica usable) R1, R2—120 ohm ½W CR1, CR2—1N34A or equivalent (Radio Shack No. 821 for pack of 10; select two that match the closest). C3, C4—0.005 μF disc type C5—330 pF ceramic or silver mica RFC – 1 mH choke S1—SPDT (Use DPDT Radio Shack rocker; No. 030 for kit of two) R3—25,000 ohm linear taper control (Radio Shack No. 094) Meter—50 μA to 1 mA movement (Lafayette 500 μA No. 50361 a good size).

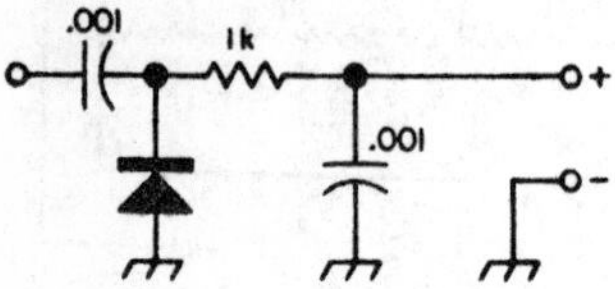


A general-purpose rf detector probe for use with an oscilloscope or voltmeter.

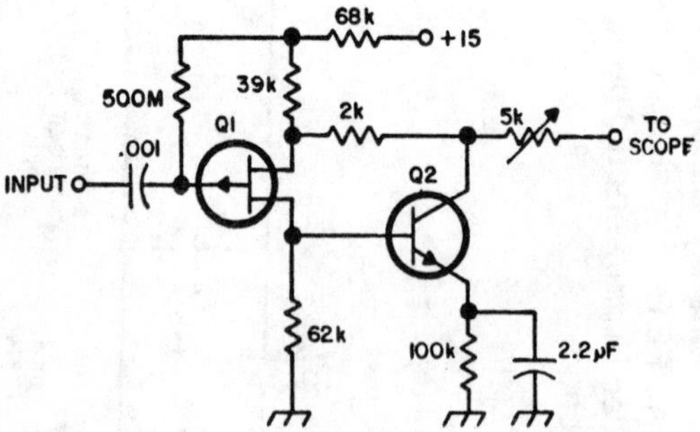


This high-impedance probe provides about 1200 megohms input impedance with unity gain. Upper frequency equalization is provided by the 5K pot. Q1 is a U112, 2N2607, 2N4360; Q2 is a 2N706, 2N708, 2N2926, 2N3394, or HEP 50.

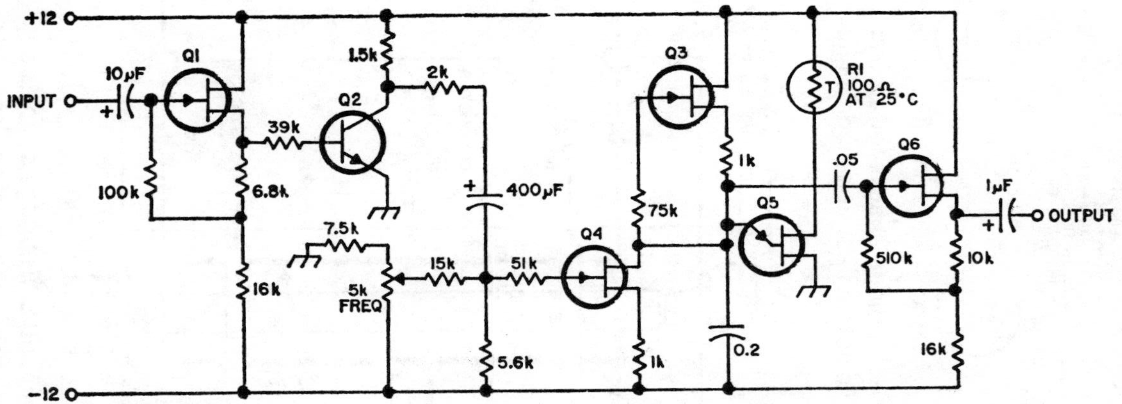


Although this capacitance meter will not measure electrolytic capacitors, it will measure any other type from zero to 0.1 μF with reasonable accuracy. On the lower end 4 pF can be read accurately and 2pF easily estimated. Transistors Q1 and Q2 are 2N168, 2N1605, 2N2926, SK3011, or HEP 54; the meter is a 0−50 microampere unit and the range switch a Centralab PA1021.

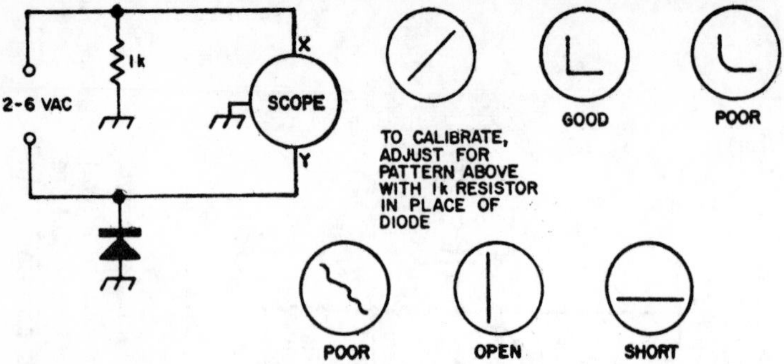


One of the easiest types of diode checks for a person with a scope is this, but it tells nothing about a diode's high voltage performance.

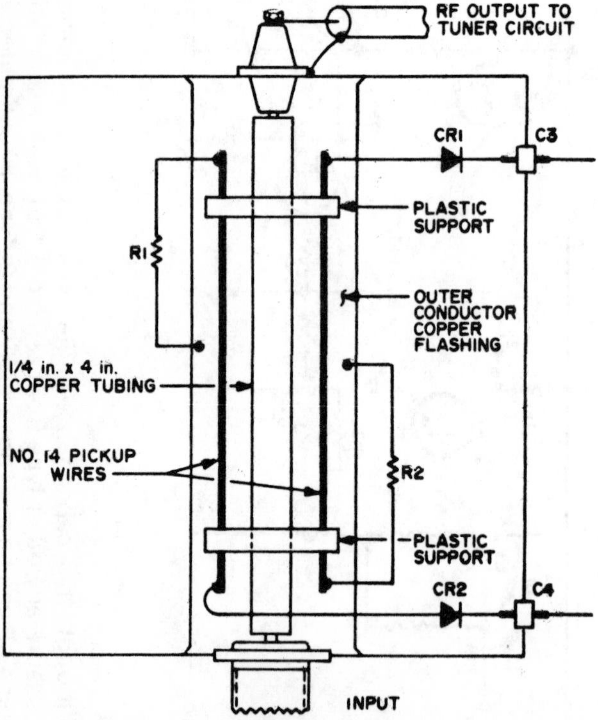


Z-match antenna tuner and SWR bridge couples any antenna to any transmitter or receiver, low frequencies to VHF. Allows smooth transition from series to parallel tuning without bandswitching. Construction details shown in accompanying figure.

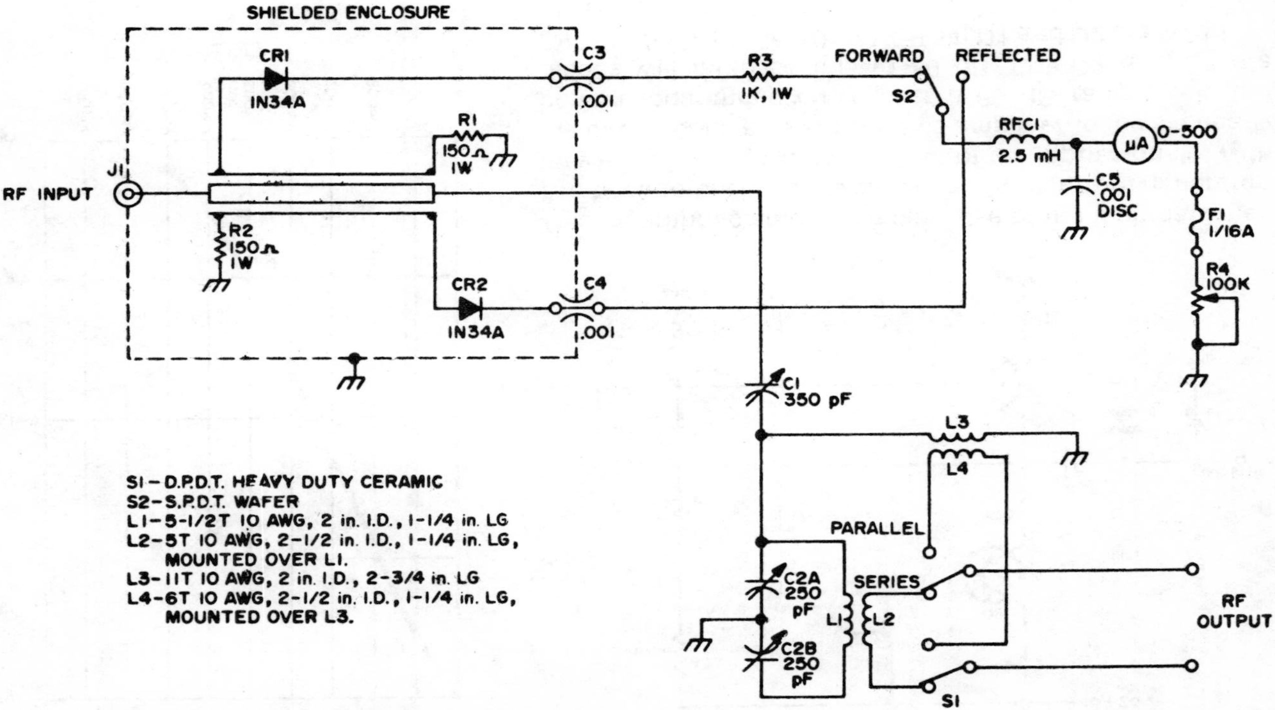


Overall schematic of the tuner.

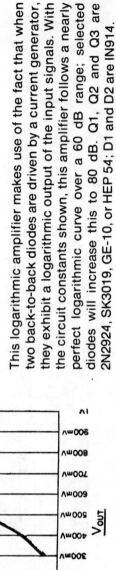

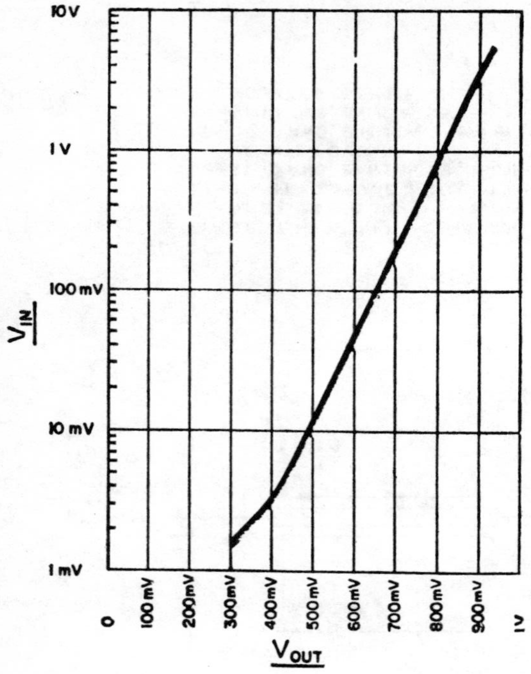


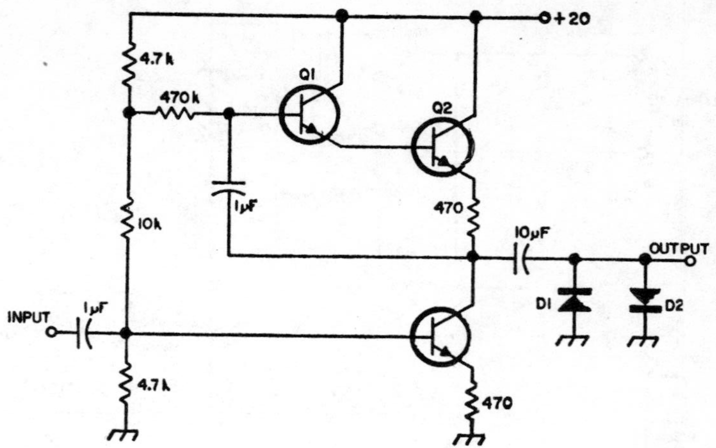


This logarithmic amplifier makes use of the fact that when two back-to-back diodes are driven by a current generator, they exhibit a logarithmic output of the input signals. With the circuit constants shown, this amplifier follows a nearly perfect logarithmic curve over a 60 dB range; selected diodes will increase this to 80 dB. Q1, Q2 and Q3 are 2N2924, SK3019, GE-10, or HEP 54; D1 and D2 are IN914.

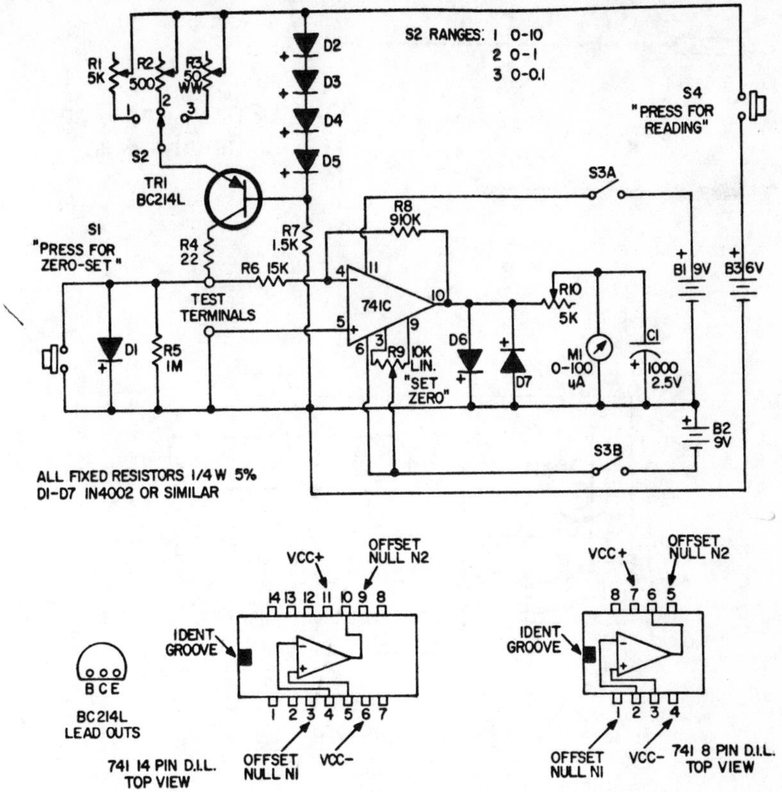


Full circuit diagram for ohmmeter capable of measuring resistance down to less than 0.001Ω. The two circuit points marked 'A' are wired directly together.

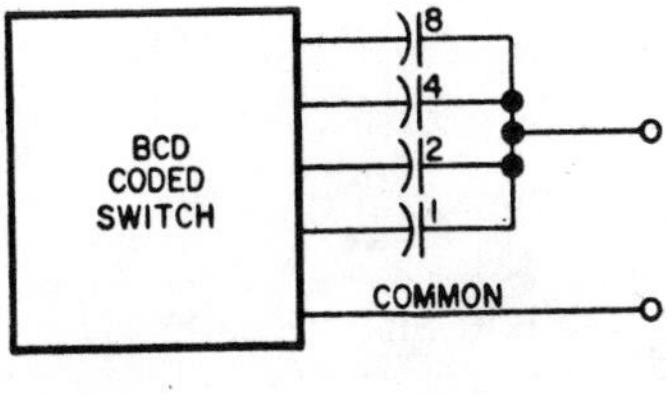


Simple decade capacitance. Using a BCD switch, capacitor in parallel can be switched in or out. By using 4 capacitors, any value from 1 to 16 can be obtained; however, since most BCD switches are from 0 to 10, a decade is obtained. Capacitors used are in the relation 1, 2, 4, 8 with the proper multiplier 1 μF, 2 μF, 4 μF, 8 μF. Keep all leads as short as possible.

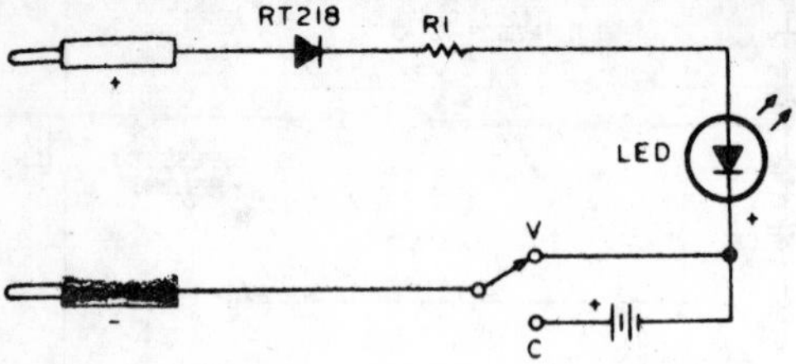


LED voltage and continuity tester.

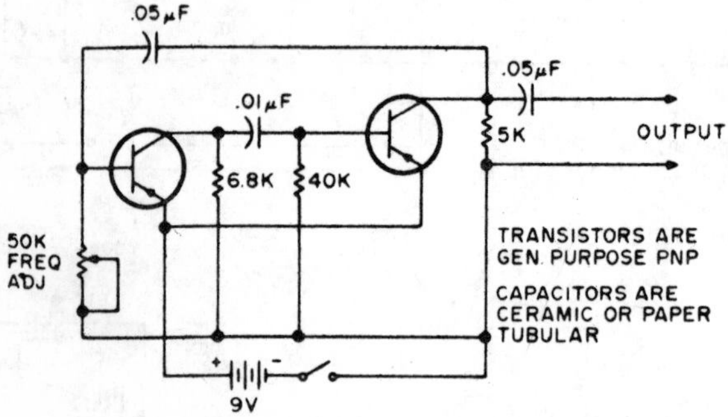


Signal injector.

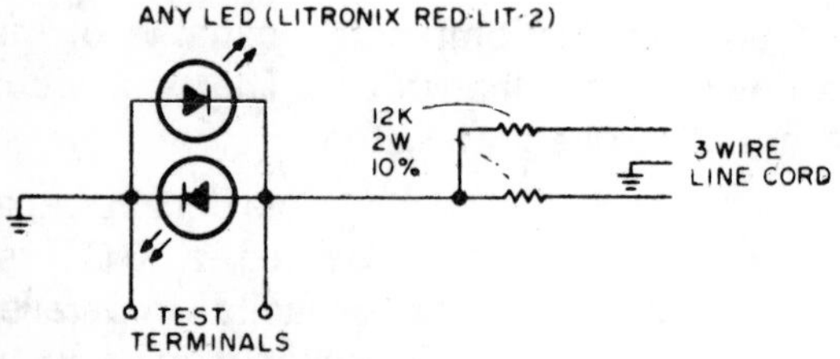


Super-simple diode checker.

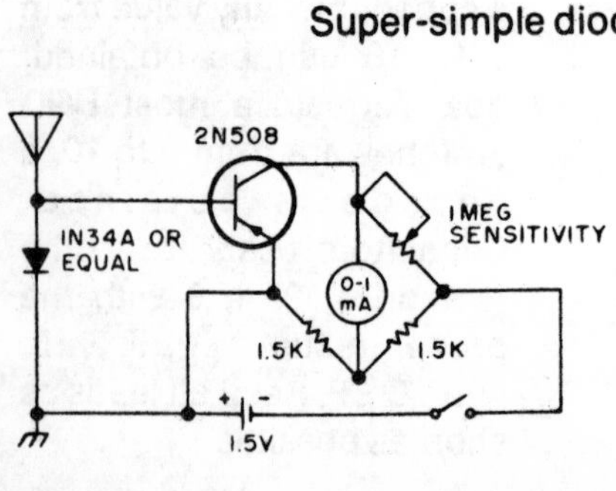


Field-strength meter.

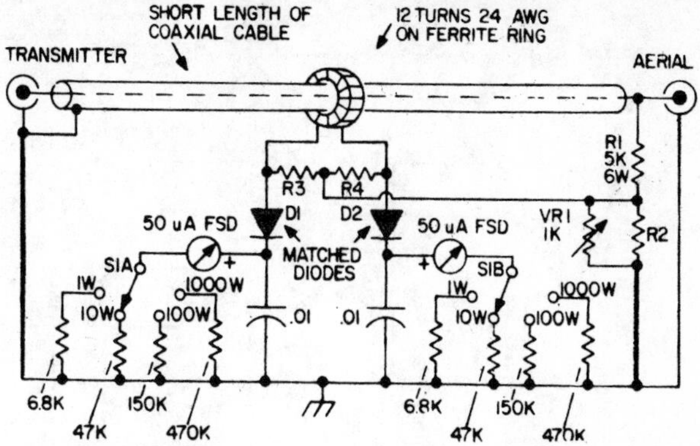


Wattmeter uses two 50 $\mu$A meters and indicates forward and reflected power simultaneously from 100 kHz to 70 MHz.

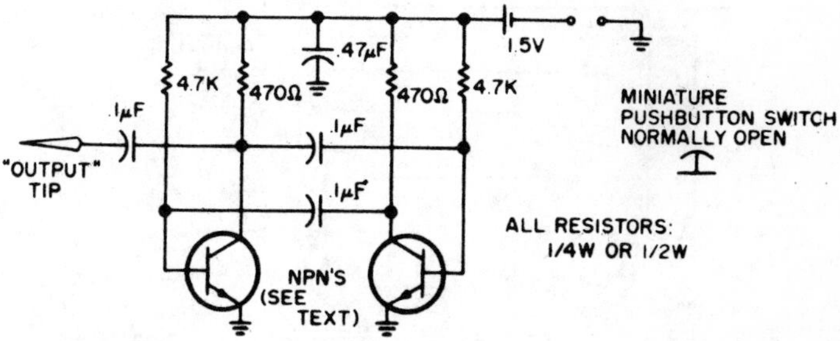


Astable multivibrator produces audio signal rich in harmonics for signal-tracing applications; may be fitted into plastic cigar container. Transistors are 2N3841; any general-purpose types will suffice.

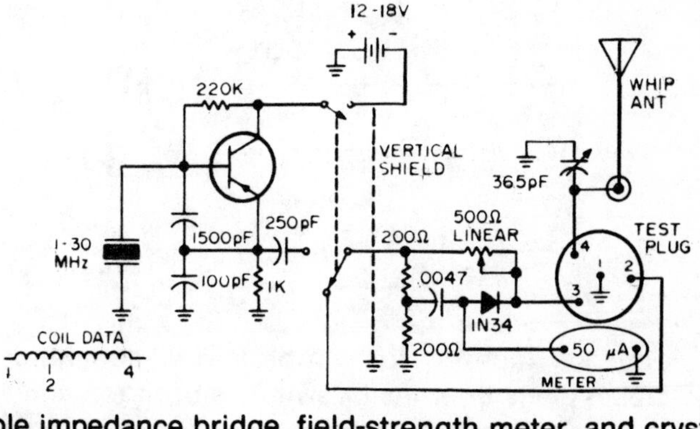


Portable impedance bridge, field-strength meter, and crystal calibrator.

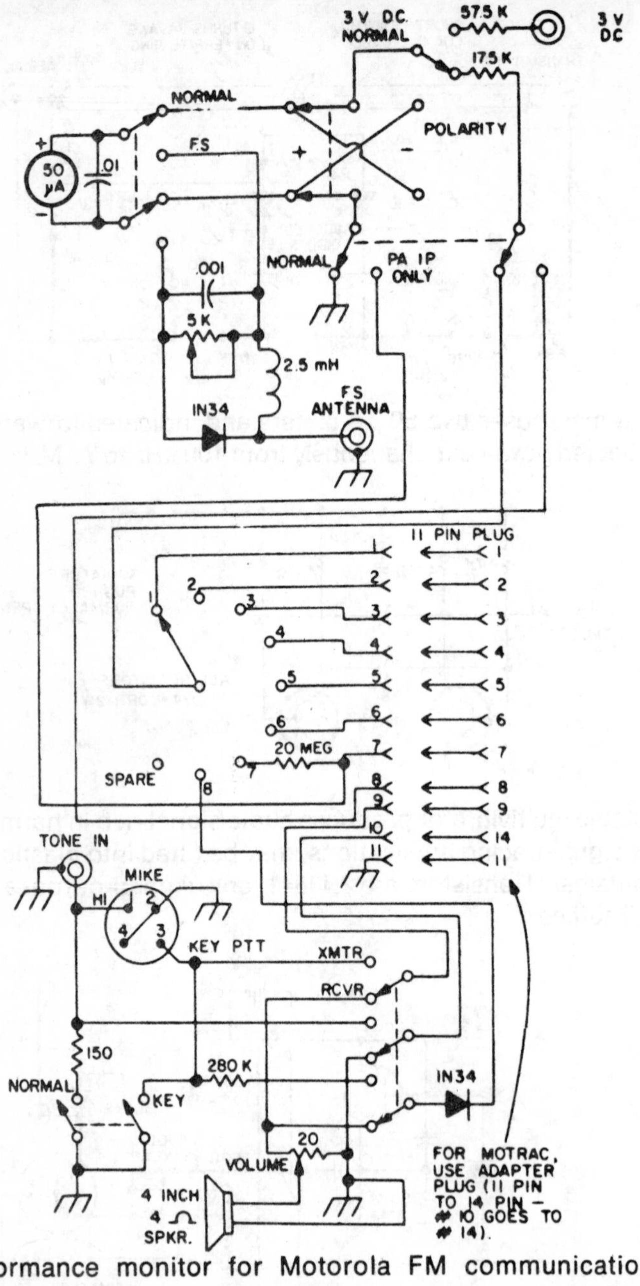


Performance monitor for Motorola FM communications equipment plugs in; selector switch is used to choose function to be monitored. Adapts to other units by changing connector type.

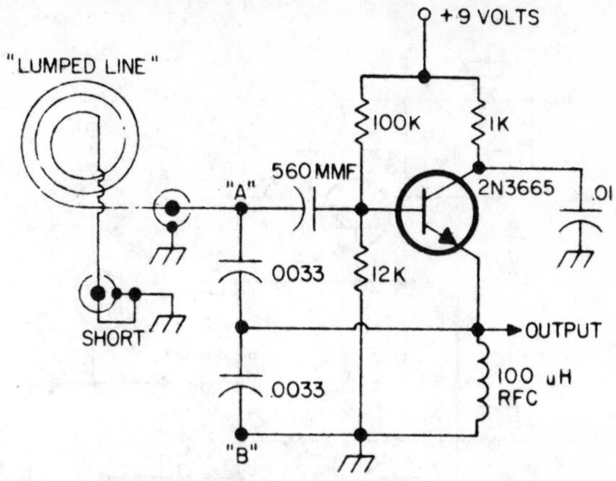


Basic "lumped-line" oscillator. Tank circuit has appearance of conventional coil but functions as full-wave transmission line with no exposed high-impedance or "hot" points.

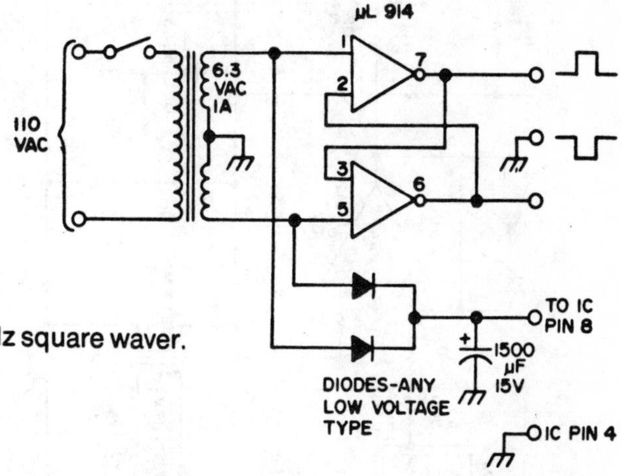


60 Hz square waver.

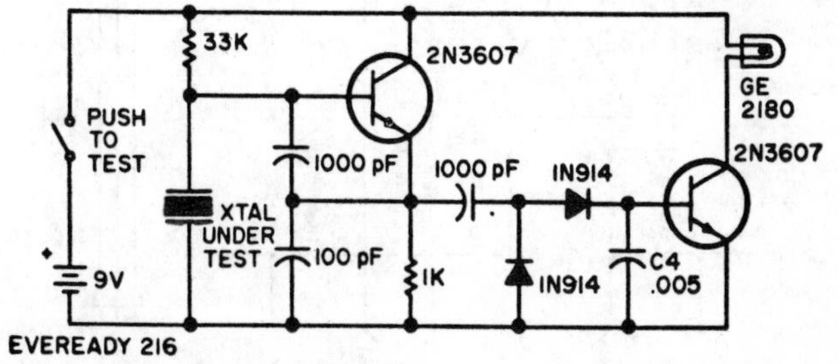


Crystal checker. When the button is pushed, a good crystal should cause the bulb to light.

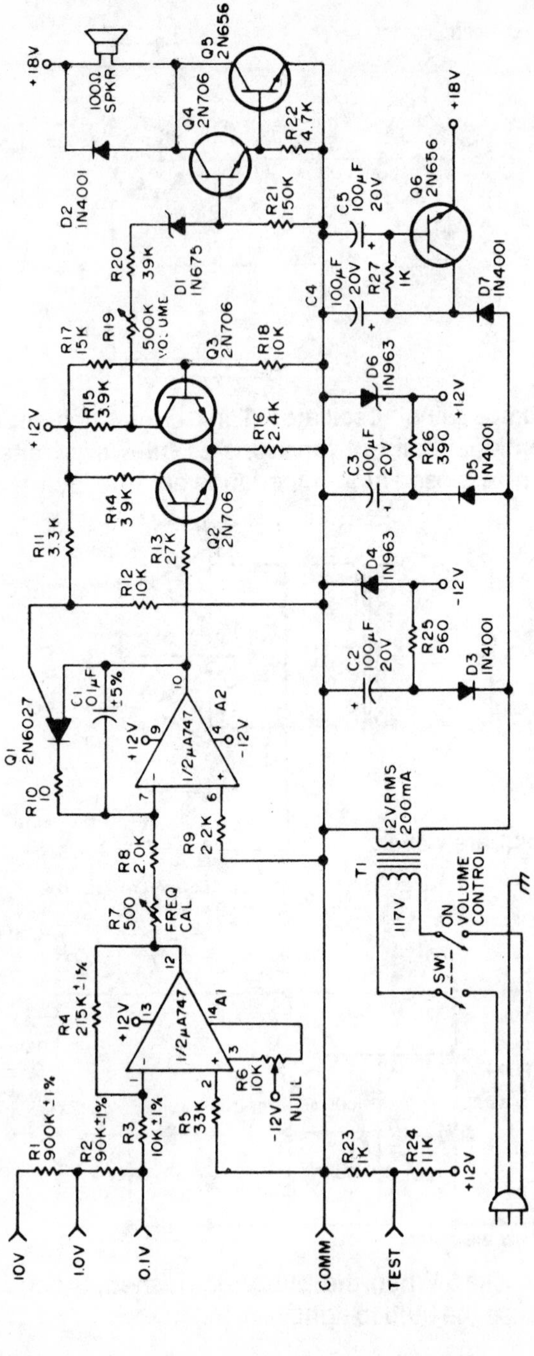
10V
1.0V
0.1V
COMM
TEST
R1 900K ±1%
R2 90K ±1%
R3 10K ±1%
R4 215K ±1%
R5 33K
R6 10K
NULL
-12V
+12V
1/2 μA747
A1
R7 500
FREQ CAL
R8 2.0K
R9 2.2K
R10 10
Q1 2N6027
C1 .01μF ±5%
A2
R11 3.3K
R12 10K
R13 27K
R14 3.9K
Q2 2N706
R15 3.9K
R16 2.4K
R17 15K
R18 10K
Q3 2N706
R19 500K VOLUME
R20 39K
D1 1N675
R21 150K
Q4 2N706
R22 4.7K
Q5 2N656
D2 1N4001
100Ω SPKR
+18V
R23 1K
R24 11K
T1
117V
12VRMS 200mA
SW1 ON
VOLUME CONTROL
C2 100μF 20V
R25 560
D3 1N4001
D4 1N963
C3 100μF 20V
R26 390
D5 1N4001
D6 1N963
C4 100μF 20V
R27 1K
C5 100μF 20V
Q6 2N656
D7 1N4001

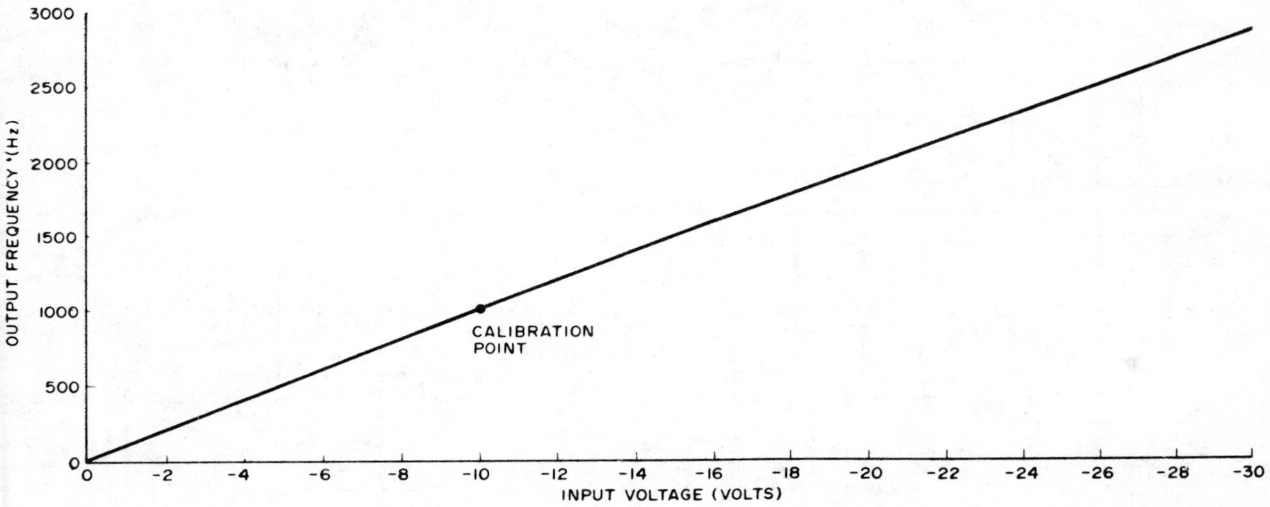


Audible voltmeter produces tone that rises with applied voltage. Chart shows linear relationship of tone to voltage.

WWVB frequency comparator receiver uses a heterodyne process to display error between WWVB at 60 kHz and local rf producing device. ICs are Motorola.

A

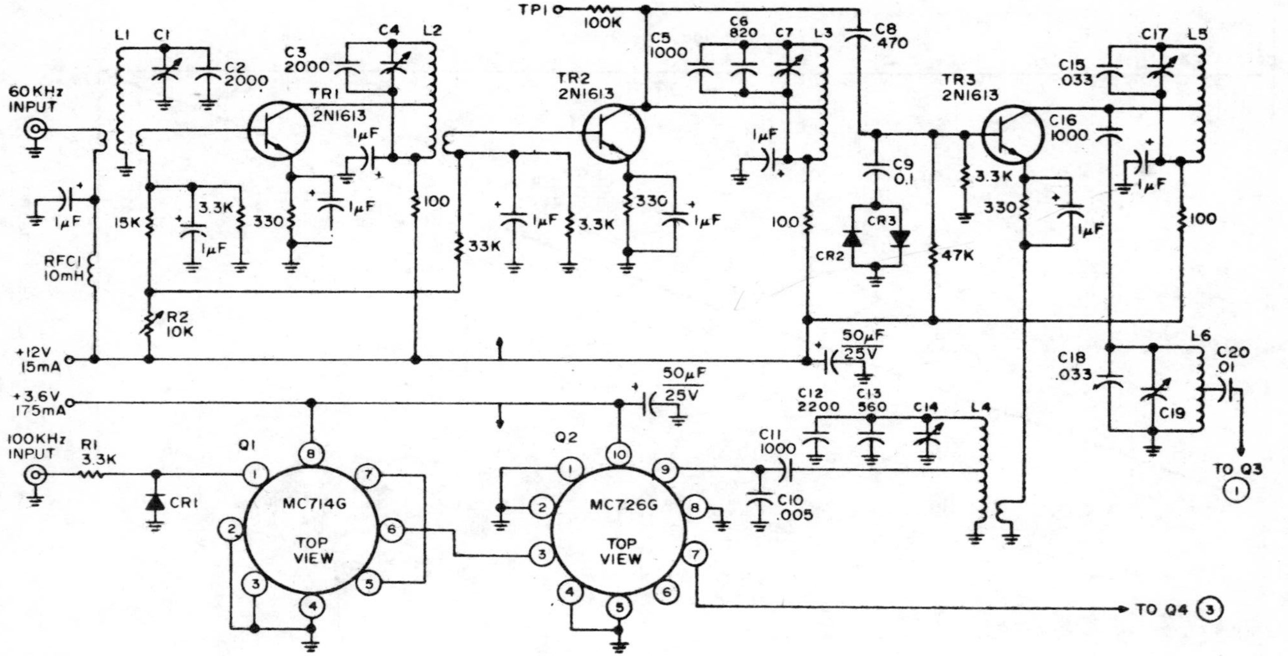

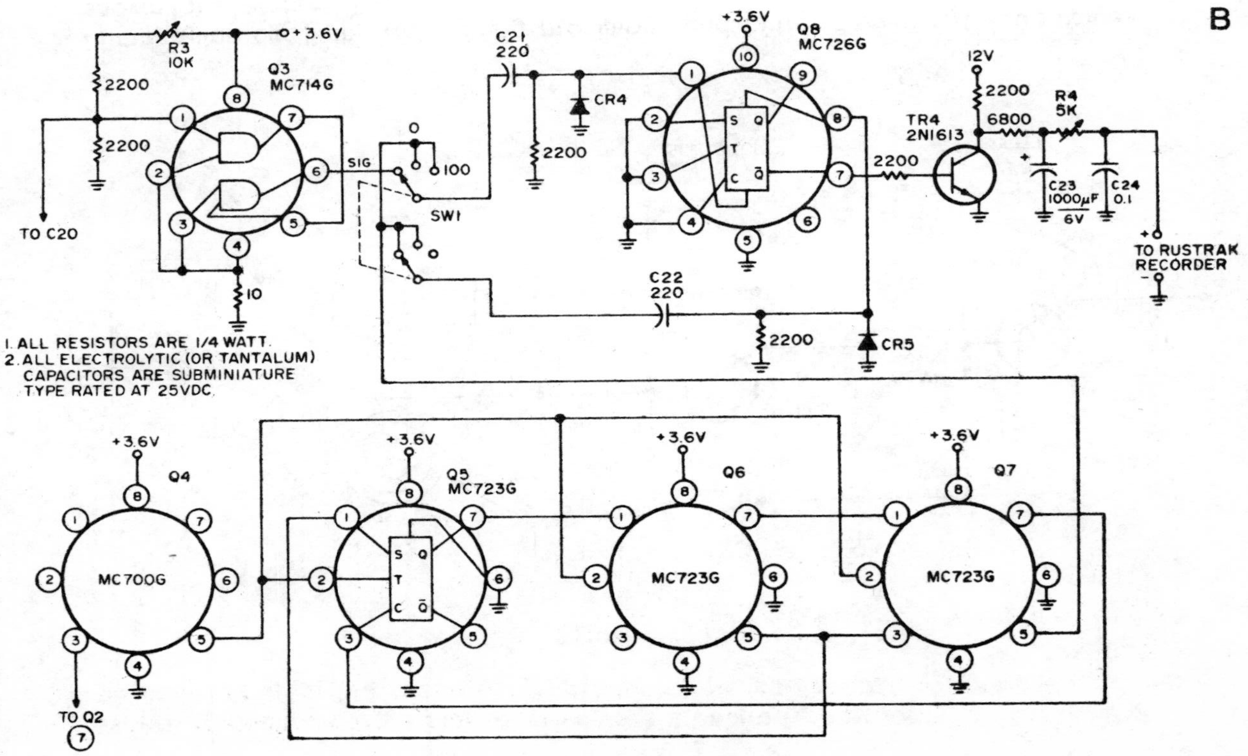
B
+3.6V
R3
10K
2200
2200
Q3
MC714G
TO C20
10
SIG
0
100
SW1
C21
220
CR4
2200
C22
220
2200
CR5
Q8
MC726G
S
Q
T
C
Q̄
12V
2200
6800
TR4
2N1613
2200
R4
5K
C23
1000μF
6V
C24
0.1
+
TO RUSTRAK
RECORDER
-
1. ALL RESISTORS ARE 1/4 WATT.
2. ALL ELECTROLYTIC (OR TANTALUM) CAPACITORS ARE SUBMINIATURE TYPE RATED AT 25VDC.
Q4
MC700G
TO Q2
7
Q5
MC723G
Q6
MC723G
Q7
MC723G

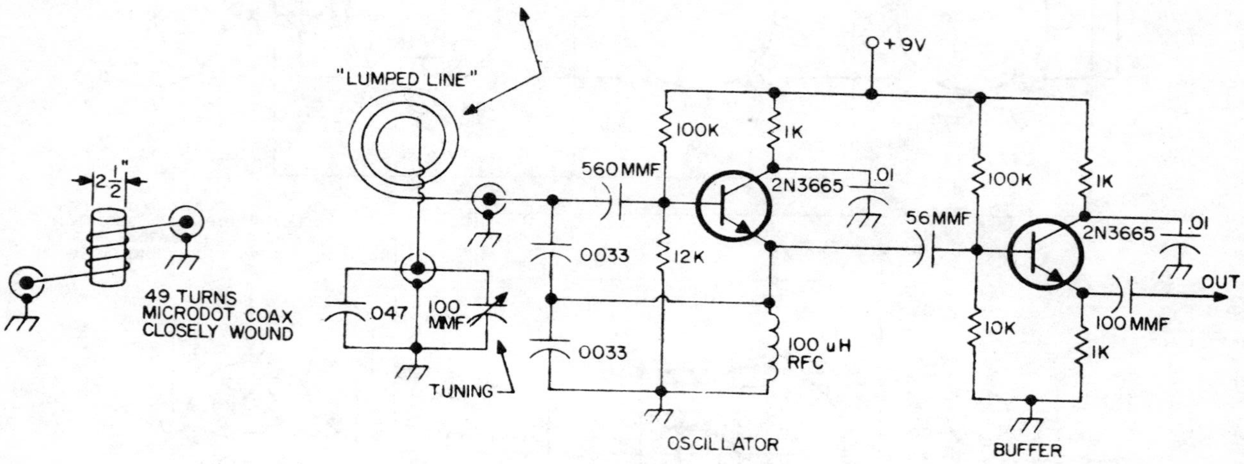


Lumped-line oscillator with tuning provisions and buffer amplifier. Circuit values shown for 21 MHz oscillator.

Postinjection marker circuit with dc gain control of marker size.

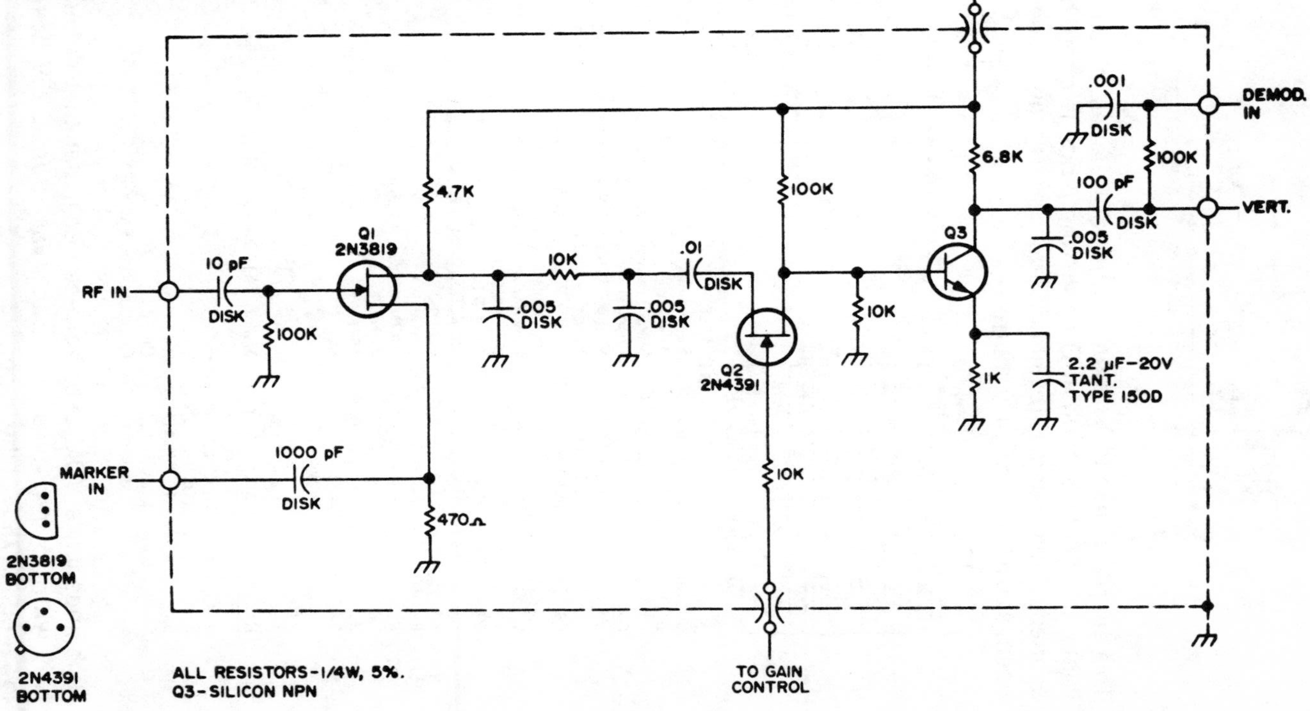

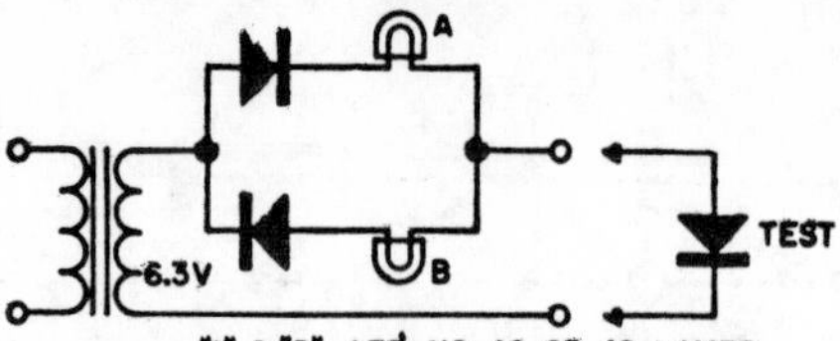


"A" & "B" ARE NO. 48 OR 49 LAMPS

This simple device gives a quick check of diodes. If lamp A lights, the diode is good. If B lights, the diode is good, but connected backwards. If neither lamp lights, the diode is open; and if both light, it is shorted.

| RANGE (VOLTS) | Z (μA) | Rm (ohms) | Rf (ohms) |
|---|---|---|---|
| ± 0.25 | ±0.75 | 333K | 667K |
| 0.75 | • | 1.00M | • |
| 2.50 | • | 3.33M | • |
| 7.50 | • | 10.0M | • |
| 25.0 | • | 33.3M | • |
| 75.0 | 2.25 | • | 222K |
| 250 | 7.50 | • | 66.7K |
| 750 | 22.5 | • | 22.2K |

RANGE RESISTOR CALCULATION

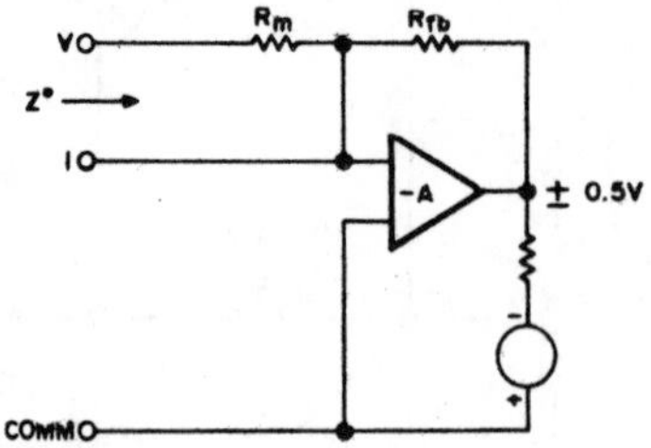


PARTS LIST

| Component | Description | Suggestions |
|---|---|---|
| B1, B2 | 1.5V battery | Any size |
| C1 | 0.1 μF | Mylar or paper any voltage rating |
| CR1 | 1N456, 1N4148 | Low leakage silicon diode |
| M1 | 500-0-500 Microammeter | Simpson Model 1329 43Ω resistance Newark 55F2518 |
| Q1, Q4 | 2N3906, 3N3702 | Low leakage, low current silicon PNP transistor |
| Q2, Q3 | 2N3904, 2N2925 | Low leakage, low current silicon NPN transistor |
| R1 | 333K 1% | 332K R1 thru R10 |
| R2 | 667K 1% | 665K Texas Inst. Type CD 1/2 MR |
| R3 | 2.33M 1% | 2.32M Newark 12F080 (spec. value) |
| R4 | 6.67M 1% | 6.65M |
| R5 | 23.3M 1% | 7.87M + 7.87M + 7.50M |
| R6 | 445K 1% | 442K |
| R7 | 154K 1% | 154K |
| R8 | 44.5K 1% | 44.2K |
| R9 | 22.2K 1% | 22.1K |
| R10 | 957 1% | 953 See text |
| R11 | 22M 10% ½W | |
| R12, R15 | 500K trimpot | Mallory 55L1 Newark 60F2210 |
| R13 | 22K 10% ½W | |
| R14 | 330K 10% ¼W | |
| R16 | 560K 10% ¼W | |
| R17 | 1.5K 10% ¼W | |
| S1 | D.P.S.T. | Optional |
| S2 | Single pole 8 ps. rotary. | Centralab type PA2001 Newark 22F601 |

High-impedance VOM uses zero-center meter that deflects fully with 0.5 mA; full-scale ranges vary from 0.25V at the lower end to 750V at the upper. →

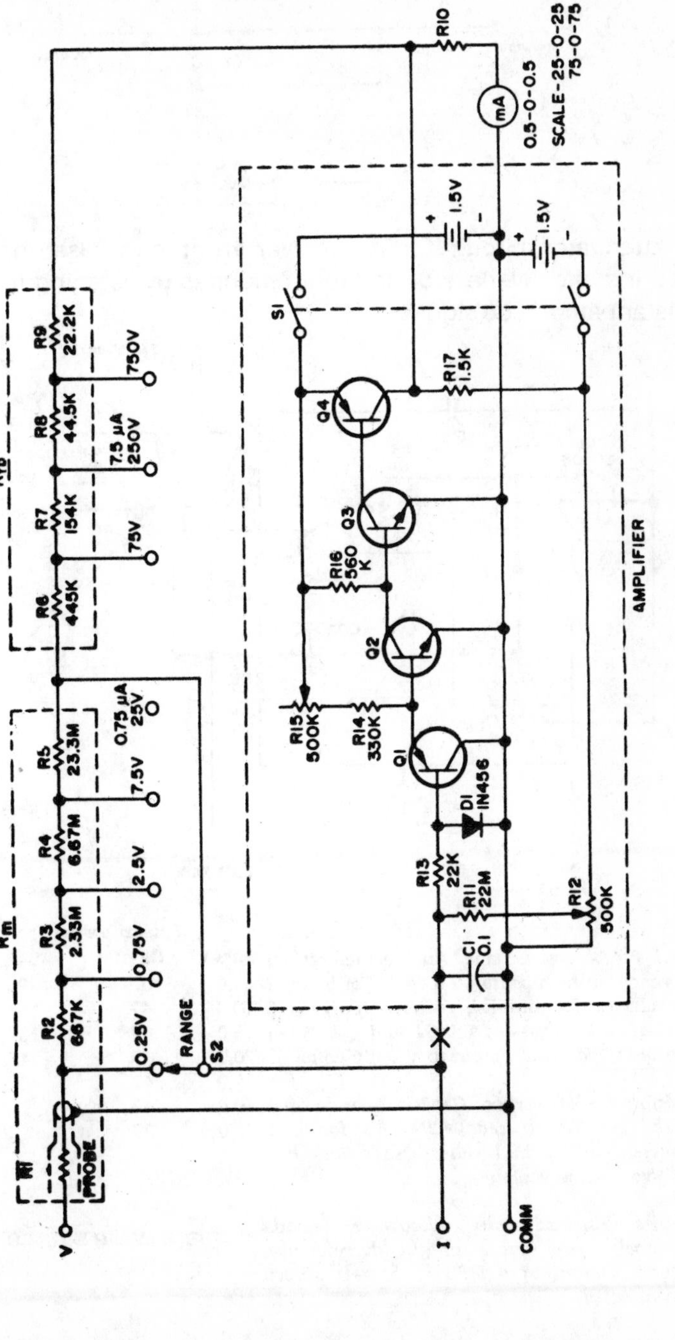
R1b
R6
445K
R7
154K
R8
44.5K
R9
22.2K
75V
7.5 µA
250V
750V
Rm
R2
667K
R3
2.33M
R4
6.67M
R5
23.3M
0.25V
0.75V
2.5V
7.5V
0.75 µA
25V
RANGE
S2
R1
PROBE
V
I
COMM
R10
mA
0.5-0-0.5
SCALE-25-0-25
75-0-75
1.5V
S1
Q1
Q2
Q3
Q4
R15
500K
R14
330K
R16
560
K
R17
1.5K
D1
IN456
R13
22K
R11
22M
R12
500K
C1
0.1
AMPLIFIER

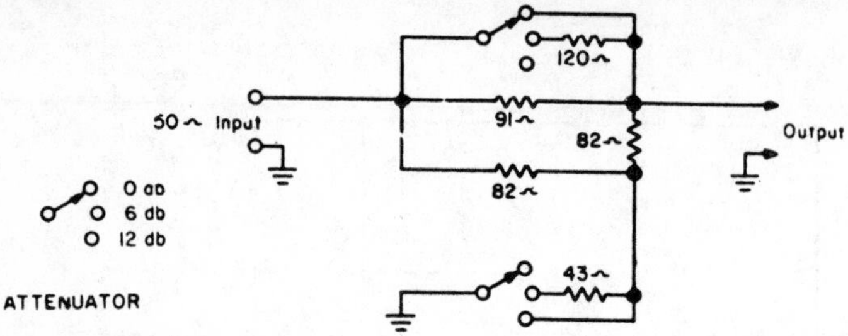


50Ω attenuator circuit for a receiver front end. Use for measuring effectiveness of different antennas by listening to a constant-amplitude signal.

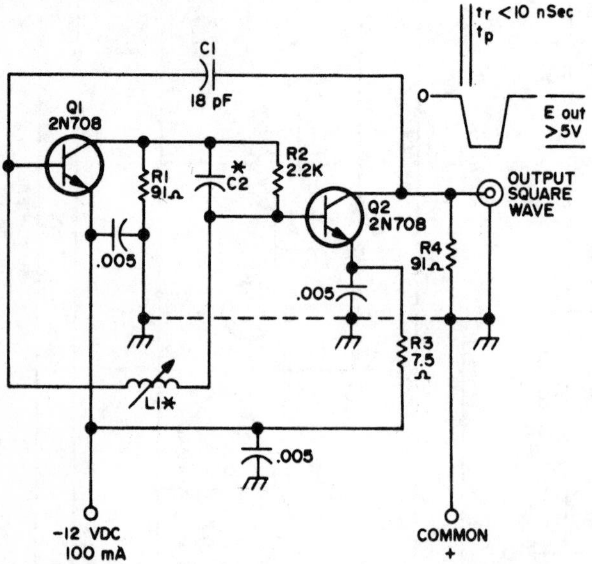


*Trim L1 for best waveshape. An overtone crystal (odd order) can be substituted for C1. For 51 ohm output change R1 and R4, reduce supply to 8 volts. Heat sinks suggested on Q1 and Q2. When cutting and tying, reduce input to about 6 volts.*

*All resistors are ½W carbon. Capacitors in decimal are disc type (short leads). 2N708's for best performance are Fairchild (other brands work, but do not give clean waveshape).*

*Harmonics observed into microwave region. Symmetrical square wave at output; dc reference to ground. Output at 4 MPS ≈ .5 watt. Short circuit protected.*

**Typical Values**

| MPS | C2 pF | L1 |
|---|---|---|
| 4.0 | 100 | 82 μH |
| 10.0 | 47 | 15 |
| 15.0 | 47 | 10 |
| 20.0 | 47 | 4.7 |
| 30.0 | 47 | 2.2 |
| 50.0 | 47 | 2.2 |
| 72.0 | 39 | .22 |

Square-wave source.

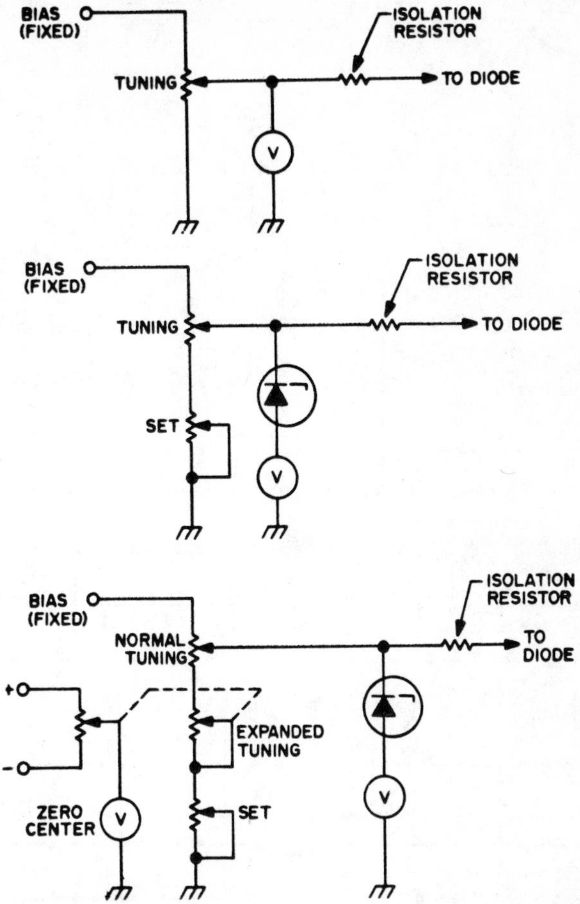


Various ways of using simple voltmeters as frequency readout devices in conjunction with voltage-tuned diodes (varactors).

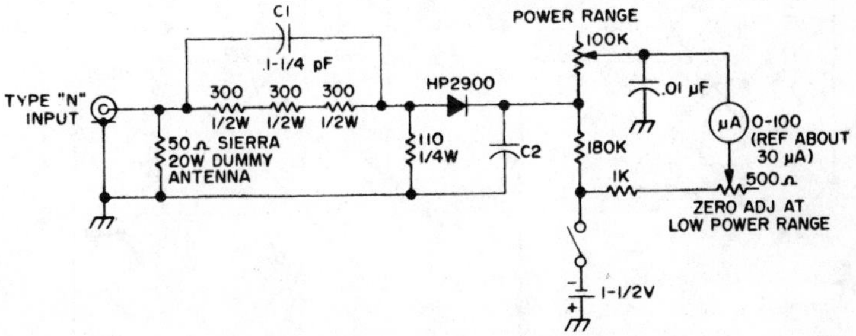


0.025 to 10W RF wattmeter.

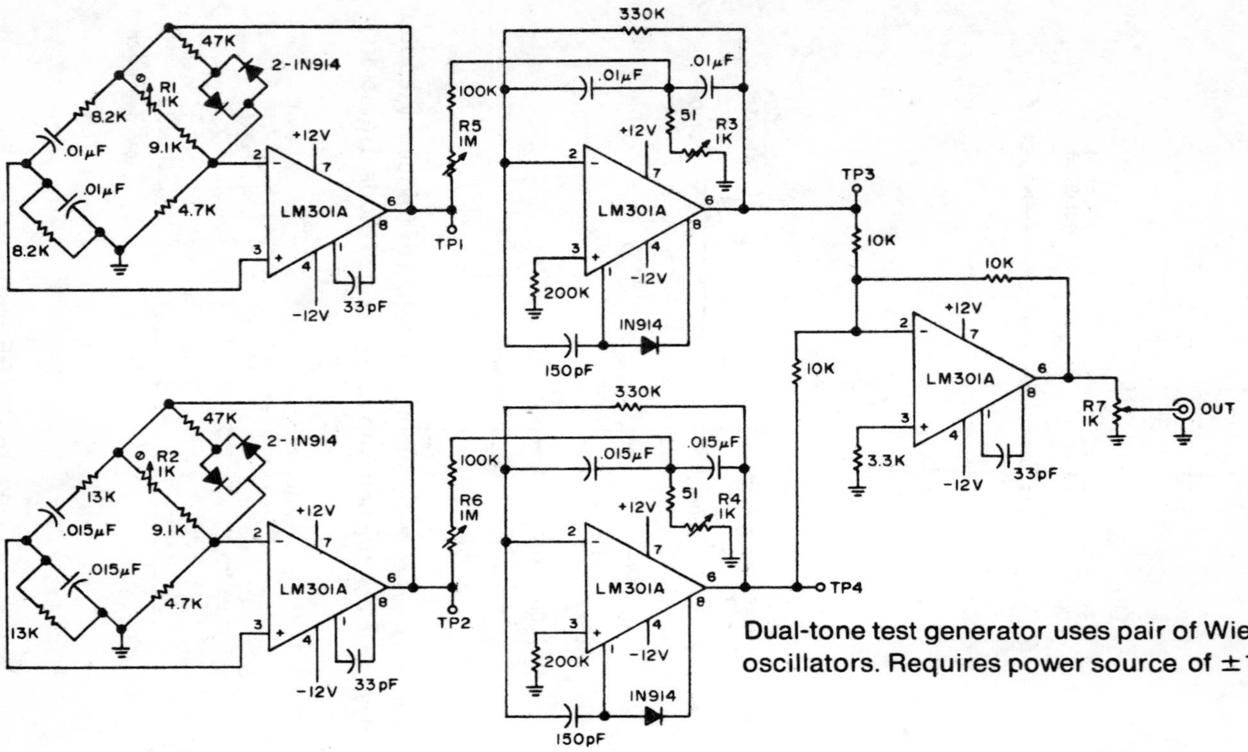


Dual-tone test generator uses pair of Wien bridge oscillators. Requires power source of ±12V.

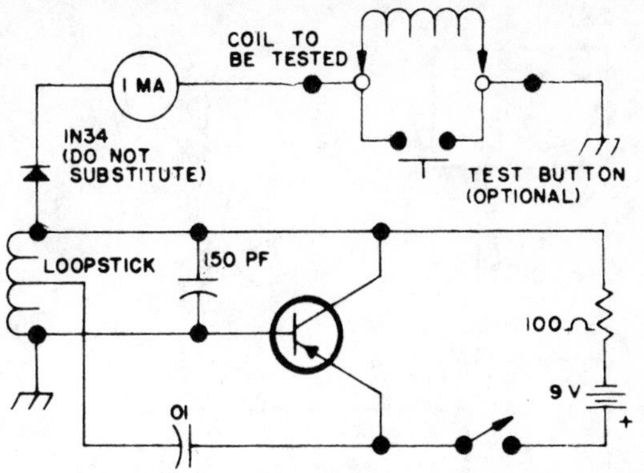


Direct-reading inductance meter. Must be calibrated by user.

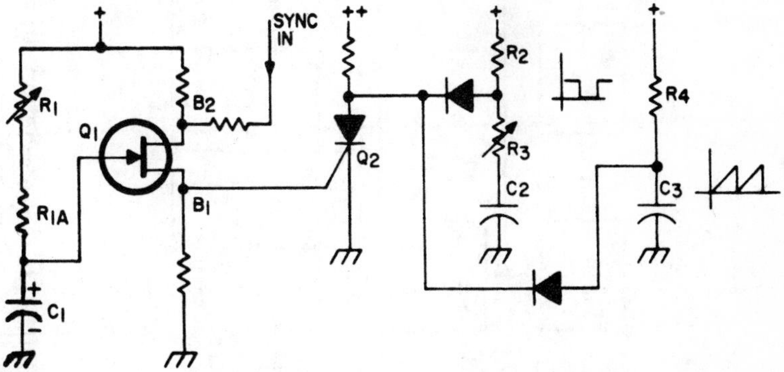


Unijunction-transistor sweep generator offers simplicity and excellent reliability/repeatability.

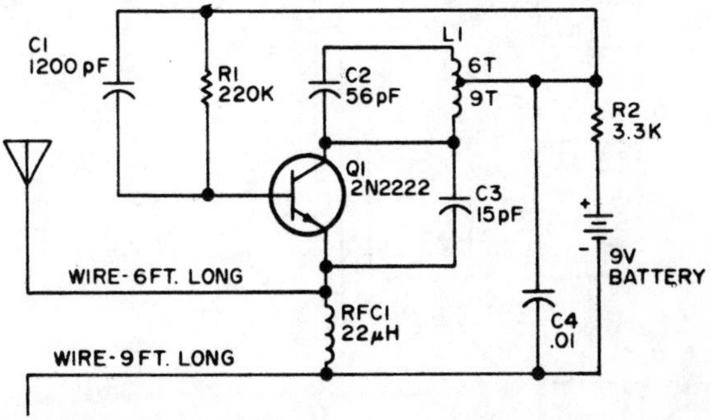


Regenerative detector generates broadband signal on 27 MHz citizen band, for use in tweaking any companion receiver unit.

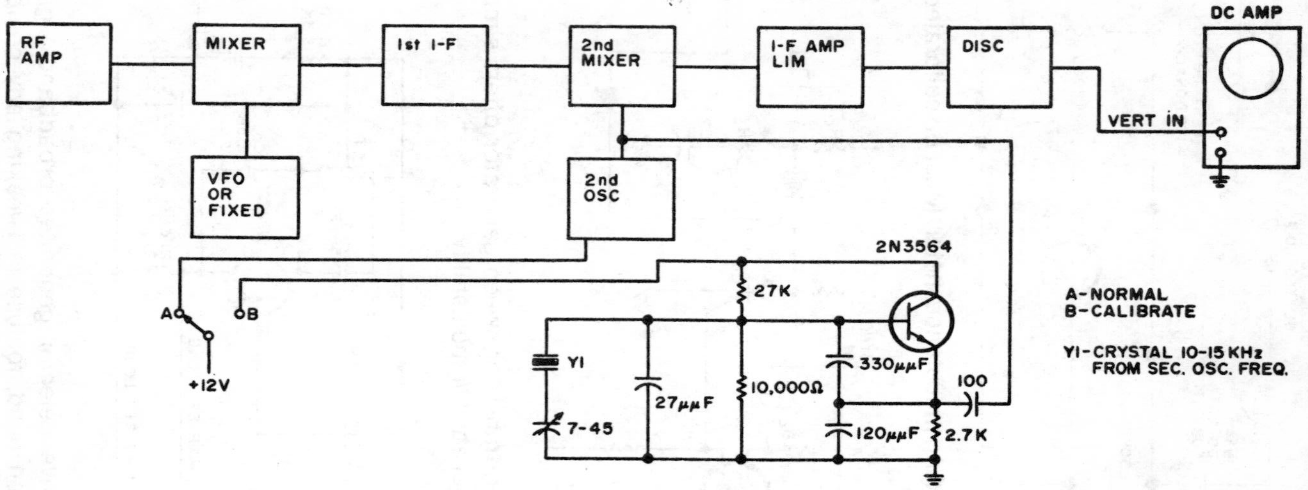


Block diagram of an easily constructed FM deviation meter using an external oscillator in conjunction with an available receiver and oscilloscope.

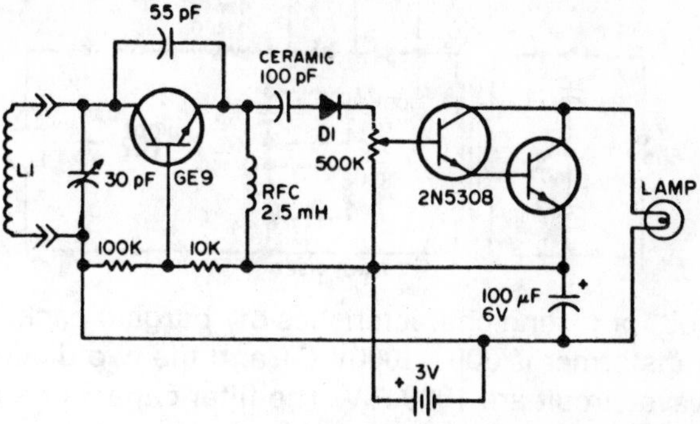


A simple "grid dip" meter that uses a bulb as a resonance indicator. Using the GE-9 transistor will enable the unit to oscillate up to 12 MHz. The indicator lamp should be a No. 48 or 49 bulb. L1 should be wound to cover desired frequency ranges.

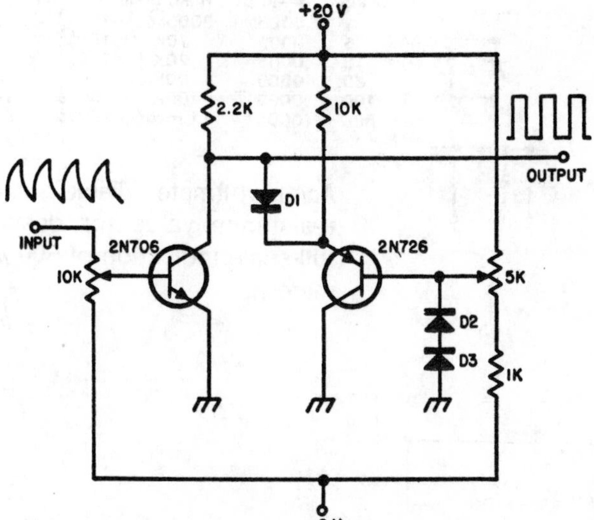


The square-wave output of many inexpensive signal generators deteriorates quite badly at high frequencies, but this circuit will square them off again. The diodes may be any inexpensive computer type such as the 1N914.

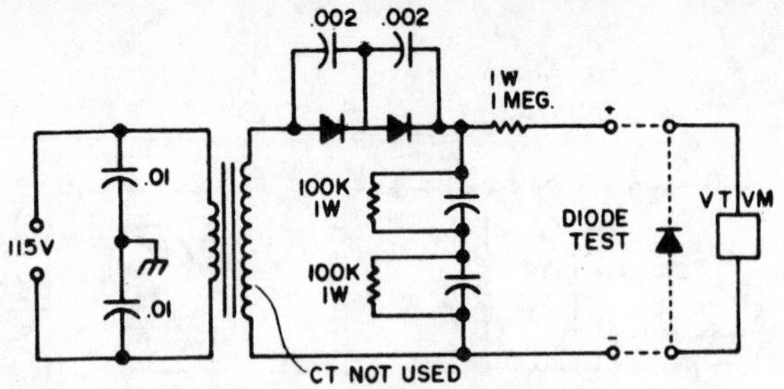


A circuit for testing characteristics of "bargain pack" diodes. The transformer is 600 – 1000V CT and the two diodes in the half-wave circuit are 1000 PIV. The filter capacitor should be approximately 40 $\mu$F apiece with suitable voltage rating. To test an unknown diode, connect as shown and read the voltage on the VTVM Two-thirds of this reading can be interpreted as a safe PIV rating. Reversing the diode reads forward characteristics – a reading of less than 3V indicates a good diode. No voltage indicates a shorted diode.

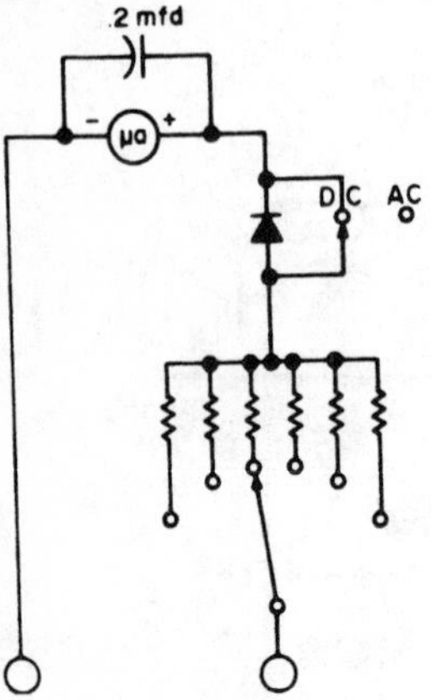


| Desired Voltage Scale | Current Rating of Meter in Amps | Necessary Resistance | Internal Resistance of Meter | Series Resistor |
|---|---|---|---|---|
| 1 | .0005 | 2000Ω | 150Ω | 1850 |
| 5 | .0005 | 10K | 150Ω | 9850 |
| 10 | .0005 | 20K | 150Ω | 19850 |
| 50 | .0005 | 100K | 150Ω | 99850 |
| 100 | .0005 | 200K | 150Ω | 199850 |
| 500 | .0005 | 1 meg. | 150Ω | 999850 |

Ac/dc voltmeter. Table shows resistance value for desired full-scale indication of 500 $\mu$A meter.

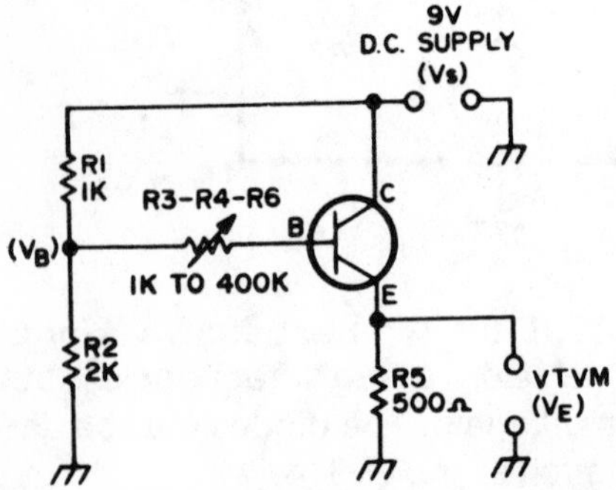


Beta tester.

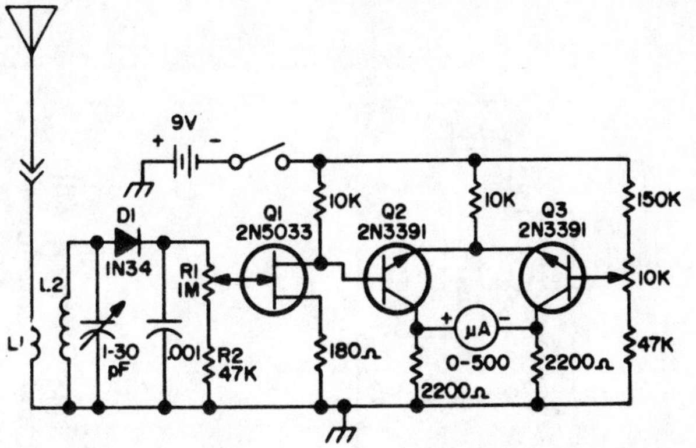


Field-strength meter can be used to tune coils, antennas, and other resonant devices. To cover 13—24 MHz, L1 is to be resonant when used with C1, and should consist of a couple of turns of No. 16 wire wound over L2. L2 is 11 turns No. 16 solid spaced so coil is 1 in. dia by 1 in. long.

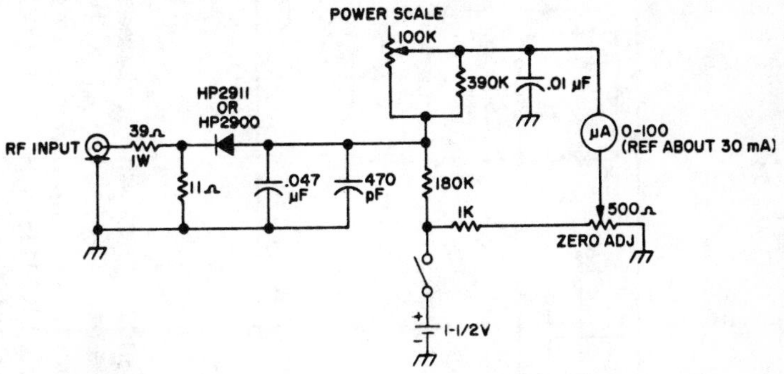


Schematic of milliwattmeter for measuring rf power levels of 1–1000 mW.

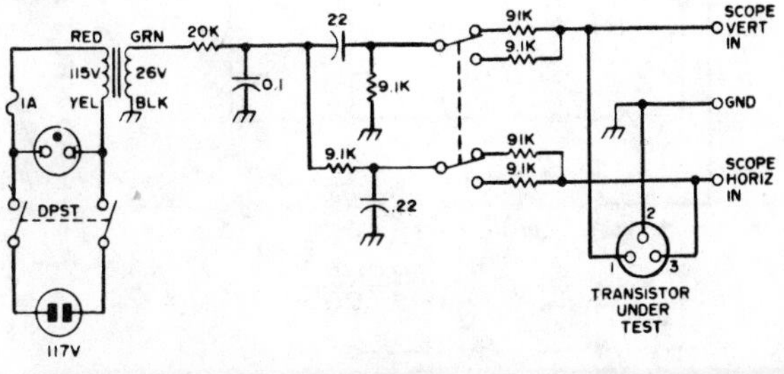


Line quadrature-generator type transistor and SCR tester.

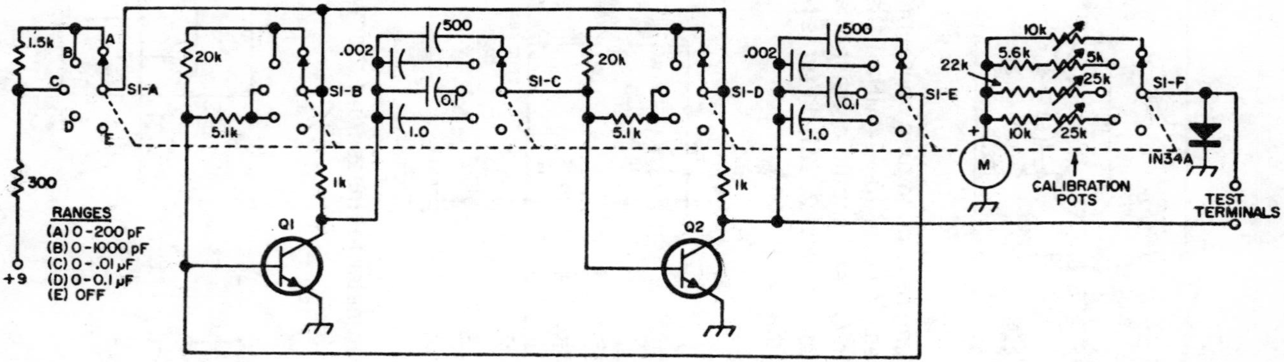


Although his capacitance meter will not measure electrolytic capacitors, it will measure any other type from zero to 0.1 μF with reasonable acfuracy. On the lower end 4 pF can be read accurately and 2 pF easily estimated. Transistors Q1 and Q2 are 2N168, 2N1605, 2N2926, SK3011, or HEP 54; the meter is a 0–50 microampere unit and the range switch a Centralab PA1021.

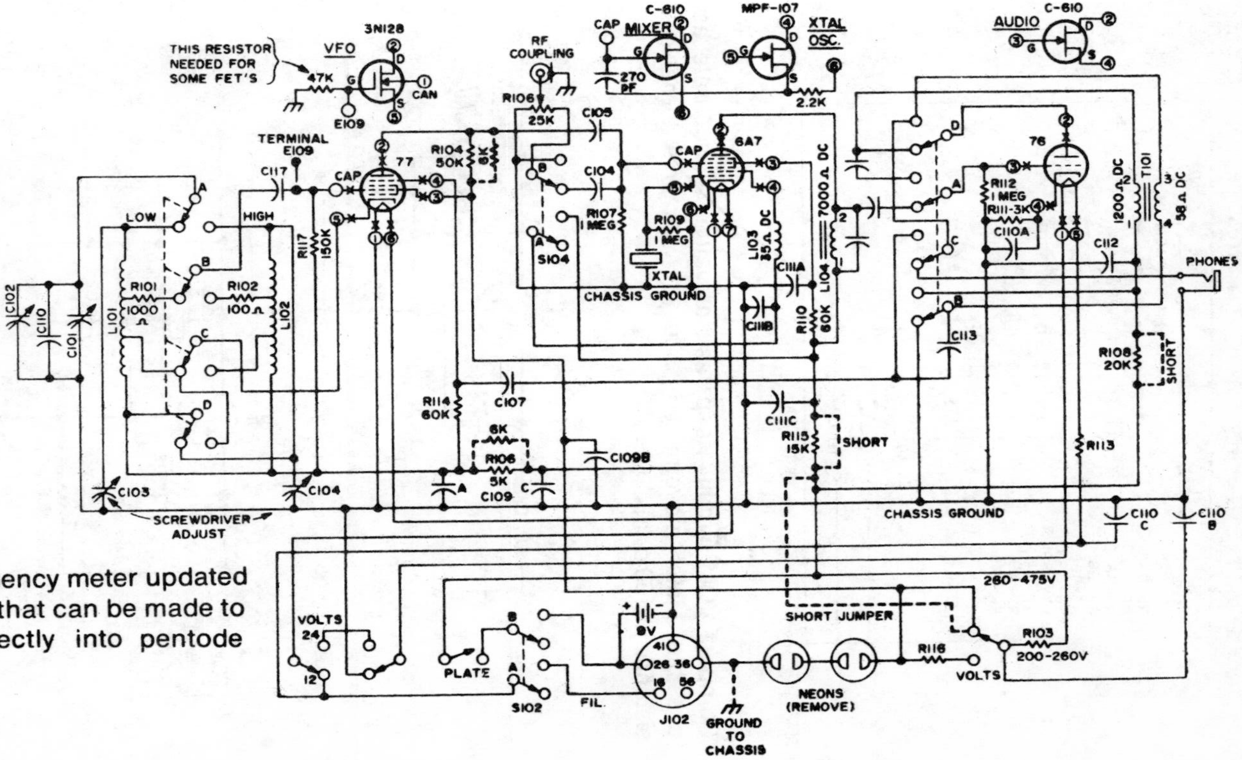


LM frequency meter updated by FETs that can be made to plug directly into pentode sockets.

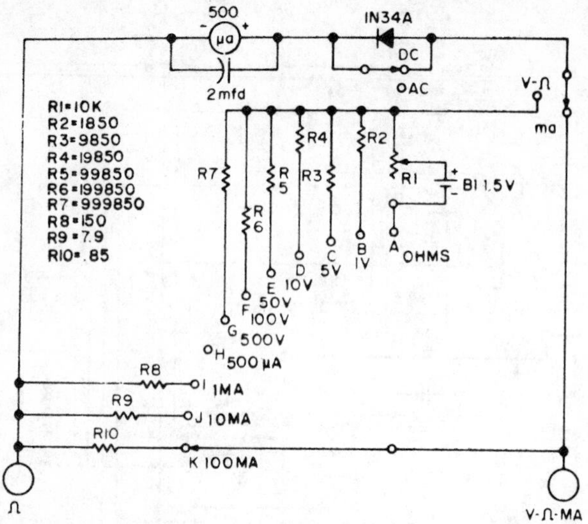


Compact VOM circuit. Keep resistor values as accurate as possible. Meter movement is 500 μA with an internal resistance of 150 ohms.

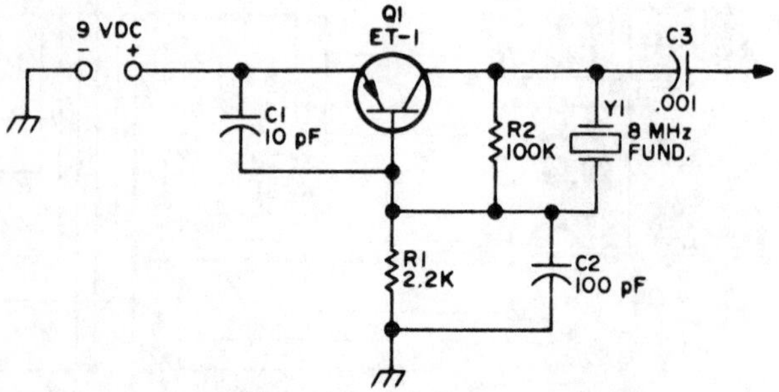


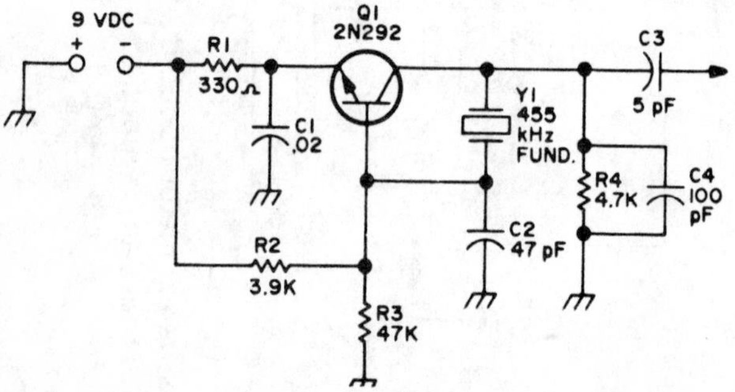


Two oscillator circuits for aligning receivers, anonymously presented originally in FM Bulletin.

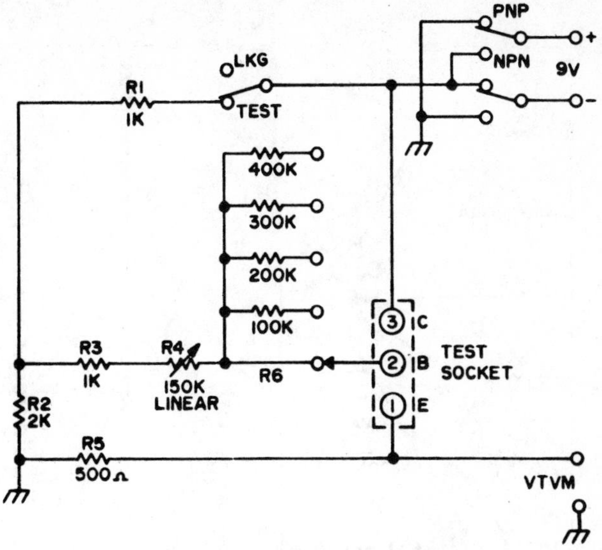


Transistor beta tester.

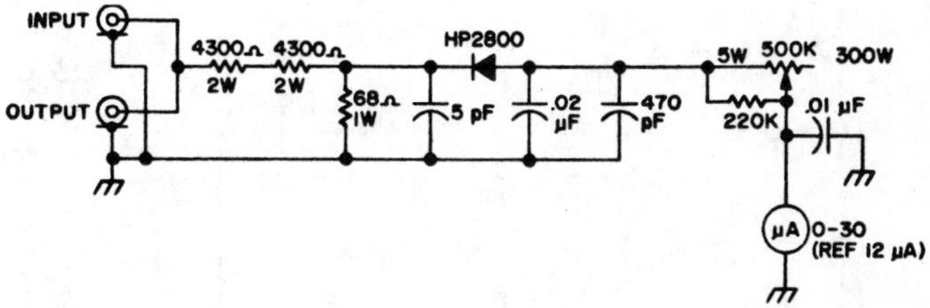


5 to 300W rf wattmeter metering circuit. External 300 or 400W during antenna load.

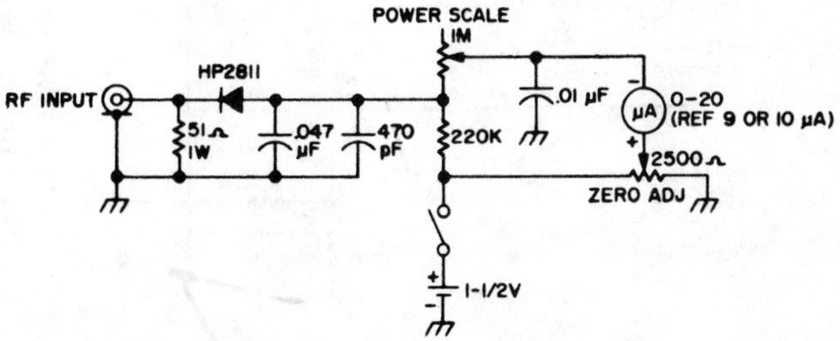


Schematic diagram of 0.05 – 500 mW wattmeter.

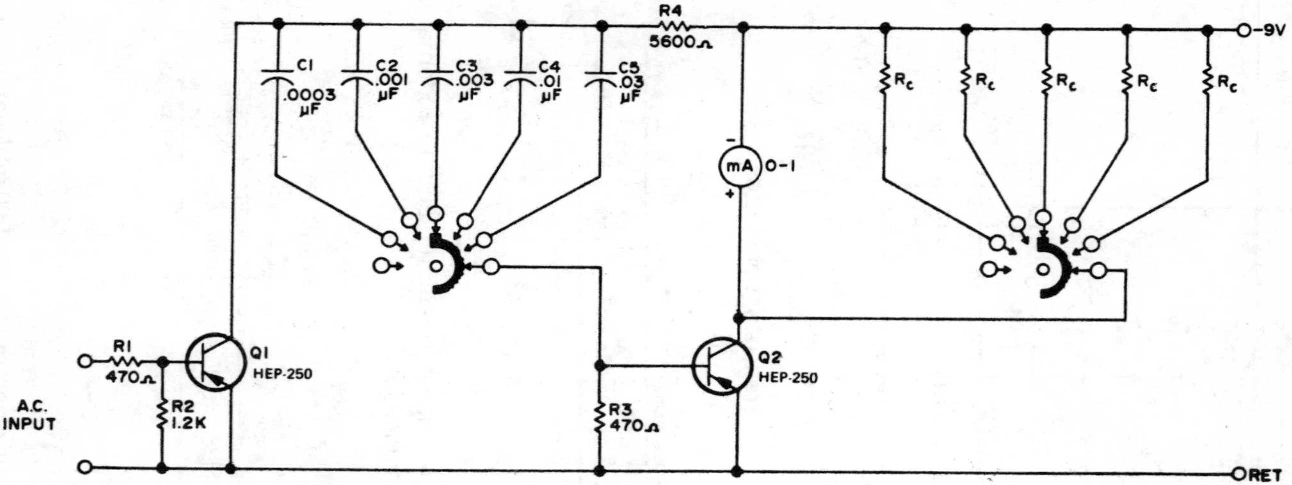


This schematic gives the basics for a simple, inexpensive audio frequency meter. For the cheapie special version, omit all switches and components associated with them. Connect a capacitor of proper value in place of SIA. Ranges are: OFF, 30 kHz, 10 kHz, 3 kHz, 1 kHz, and 300 Hz. Meter is 0 – 1 mA.

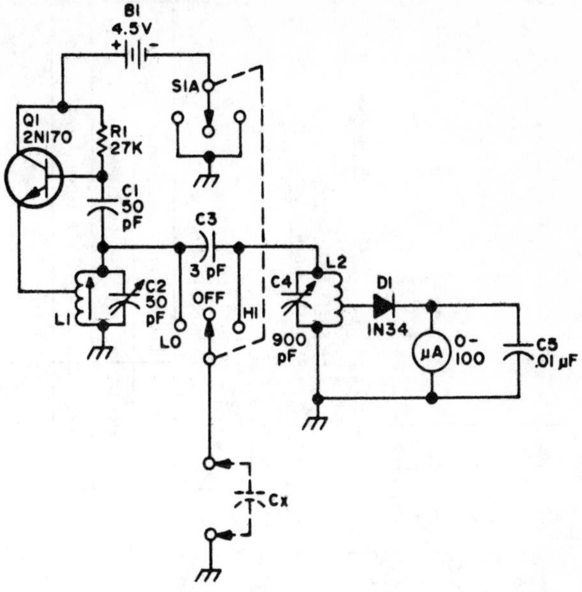


Capacity meter. L1-C2 and L2-C4 should tune to the same frequency, around 1450 kHz. Calibrate by setting C4 near max and mark that "zero." Peak meter with C2, which is zero adjustment. Calibrate with known capacities and mark C4 dial.

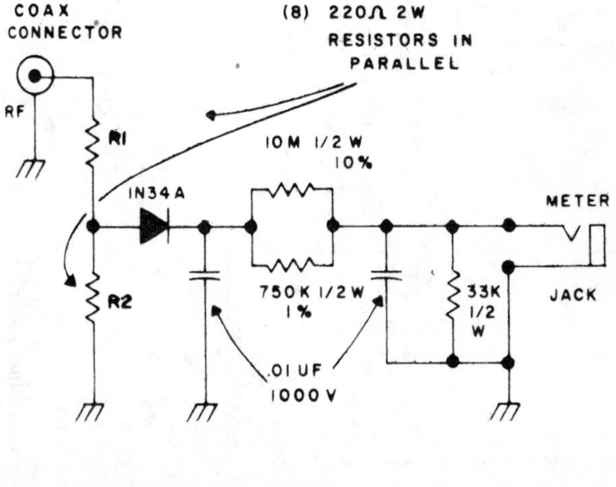


Dummy load and rf wattmeter for 50 ohms. R1 and R2 are each made from 8 paralleled 2-watt resistors of precisely 220Ω each.

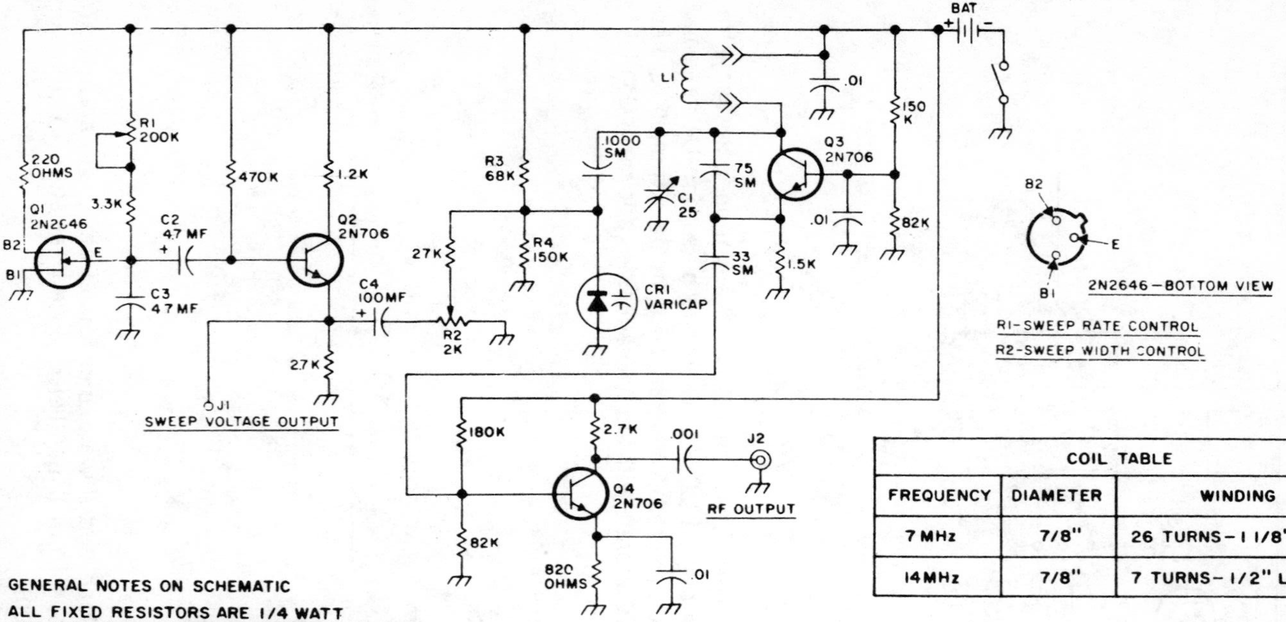


| COIL TABLE | | |
|---|---|---|
| FREQUENCY | DIAMETER | WINDING |
| 7 MHz | 7/8" | 26 TURNS–1 1/8" LONG |
| 14 MHz | 7/8" | 7 TURNS–1/2" LONG |

GENERAL NOTES ON SCHEMATIC

ALL FIXED RESISTORS ARE 1/4 WATT
.001 AND .01 CAPACITORS ARE DISC CERAMIC
CAPACITORS MARKED "SM" ARE SILVER MICA R1
AND R2 STANDARD CARBON ELEMENT POTENTIOMETERS
C2, C3, AND C4 ARE ELECTROLYTICS–25WVDC OR MORE

Sweep oscillator circuit.

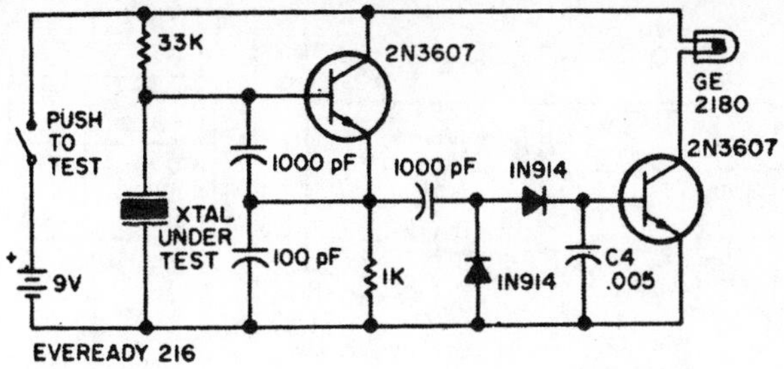


Crystal tester schematic.

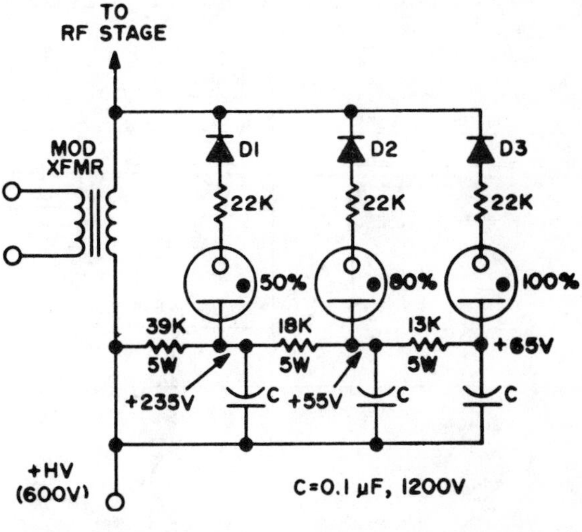


Simple devices can be built to indicate when defined modulation percentages are achieved. This arrangement uses neon bulbs and switching diodes to fire the bulbs when the negative-peak modulation reaches 50%, 80%, and 100%, respectively. Bottom ends of neon bulbs return to voltages which are uniformly 65V more negative than the voltage marking the corresponding modulation percentage. These voltages, and the resistance values in the divider which establishes them, depend upon the voltage supplied to the transmitter (here assumed to be 600V). The 65V offset is the firing voltage of the neon bulb. When the upper end goes 65V more negative than the lower end, the corresponding bulb can fire. In use, the object is to keep the 50% bulb on all the time, the 80% bulb on as much of the time as possible, but never permit the 100% bulb to flash.

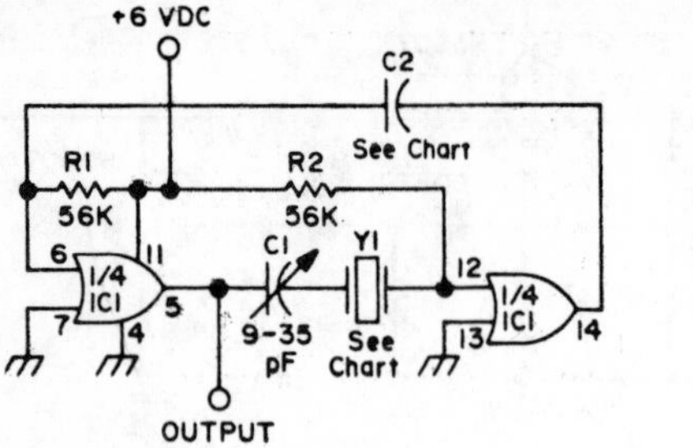


| CRYSTAL FREQ | C2 pF |
|---|---|
| 100 KHz | 1000 |
| 500 KHz | 430 |
| 1 MHz | 39 |

IC1-HEP 570

Crystal oscillator calibrator, high in square-wave output (you will be able to hear it up into the VHF bands), which may be made into two calibrators since the circuit uses only half of the Motorola IC.

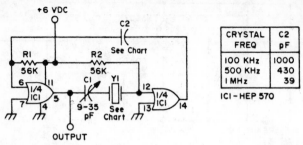

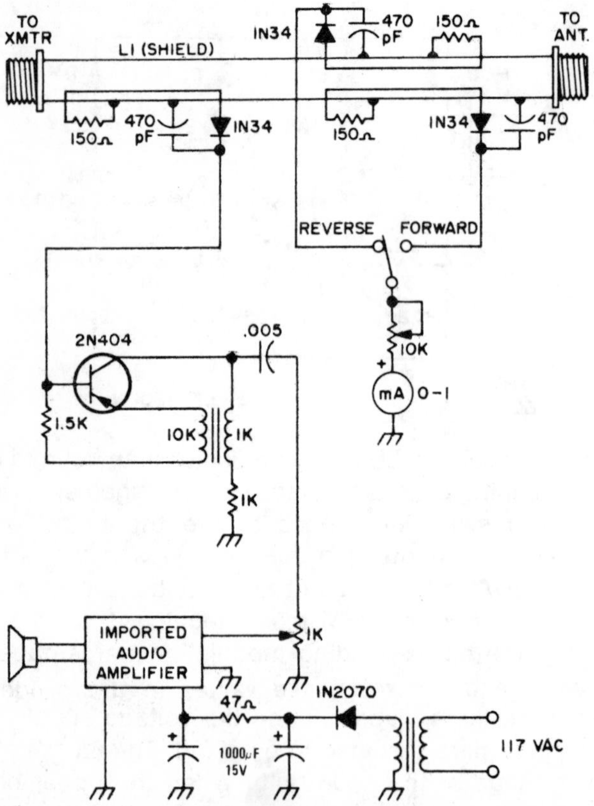


Reflected power meter and CW monitor. L1 is 12 in. RG-11/U coax with plastic coat removed. 5-inch pieces of No. 20 hookup wire inserted under shield for pickup loops.

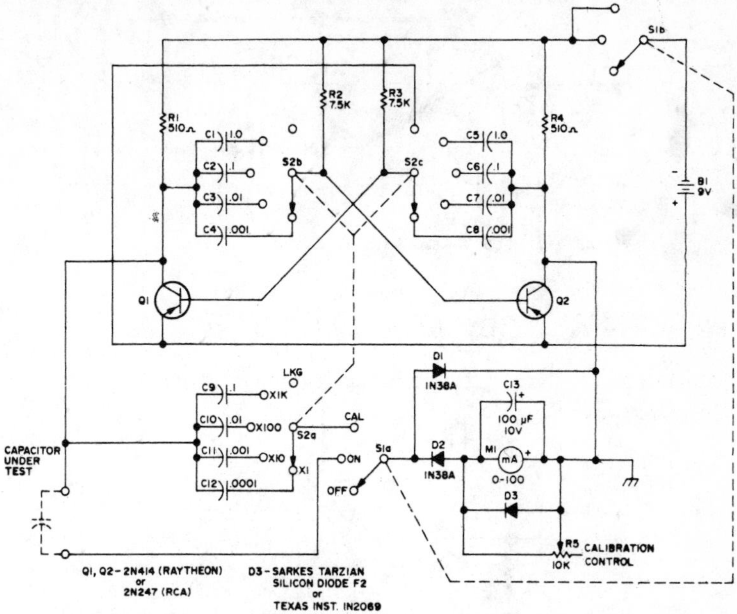


Direct-reading capacitance meter.

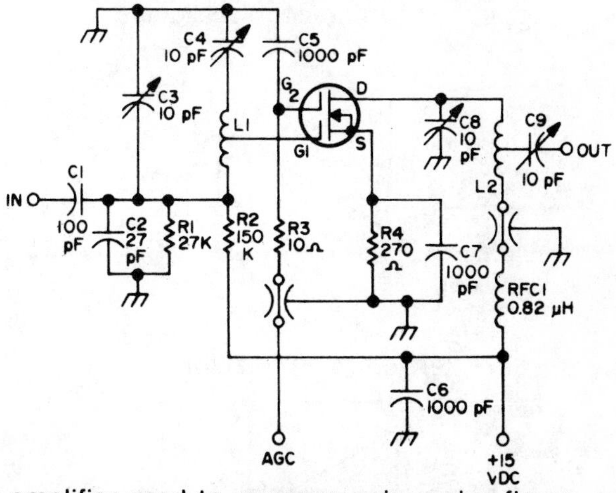


VHF amplifier used to measure gain, noise figure, and agc performance of the Fairchild FT0601 MOSFETS. For agc operation, dual-gate MOSFETS have a built-in advantage owing to their separate gates, especially in the VHF region. Amplifiers such as this one eliminate cross-modulation distortion, decrease receiver noise, and avoid shifting of the receiver's center frequency. Schematic from Fairchild Semiconductor Application Note APP-189.

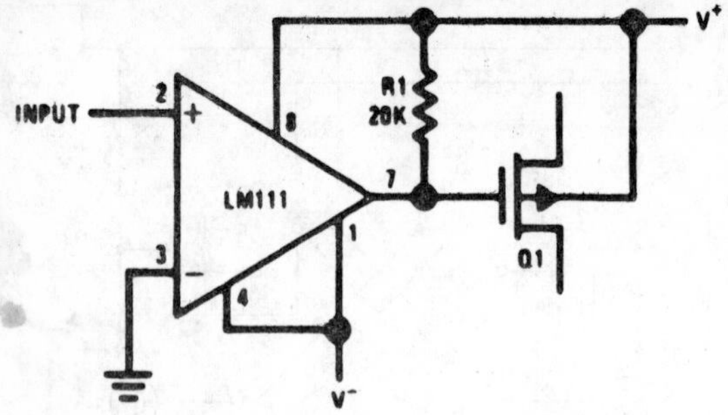


Zero crossing detector driving MOS switch.

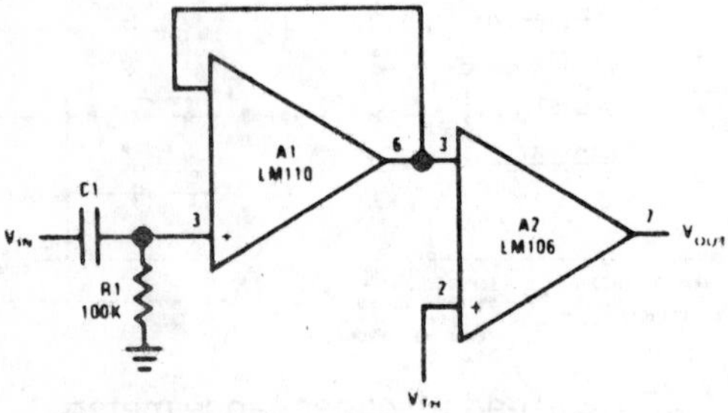


Comparator for ac-coupled signals.

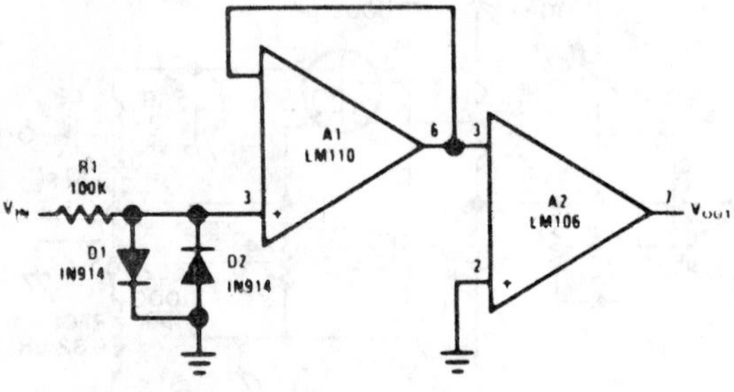


Zero crossing detector.

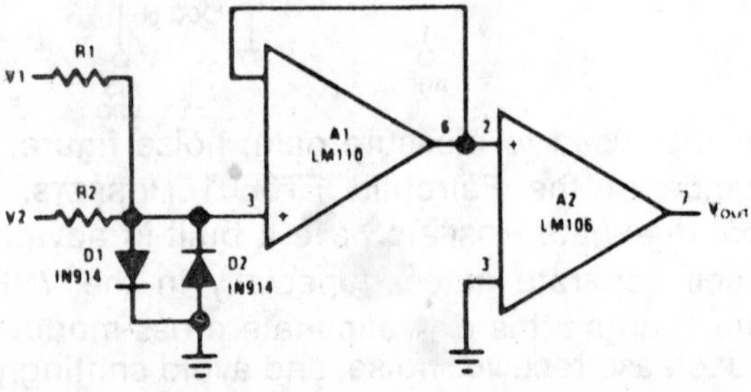


Comparator for signals of opposite polarity.

VHF prescaler for extending counter range to 300 MHz.

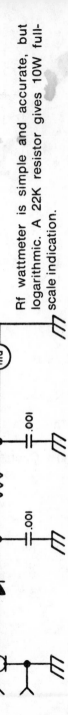

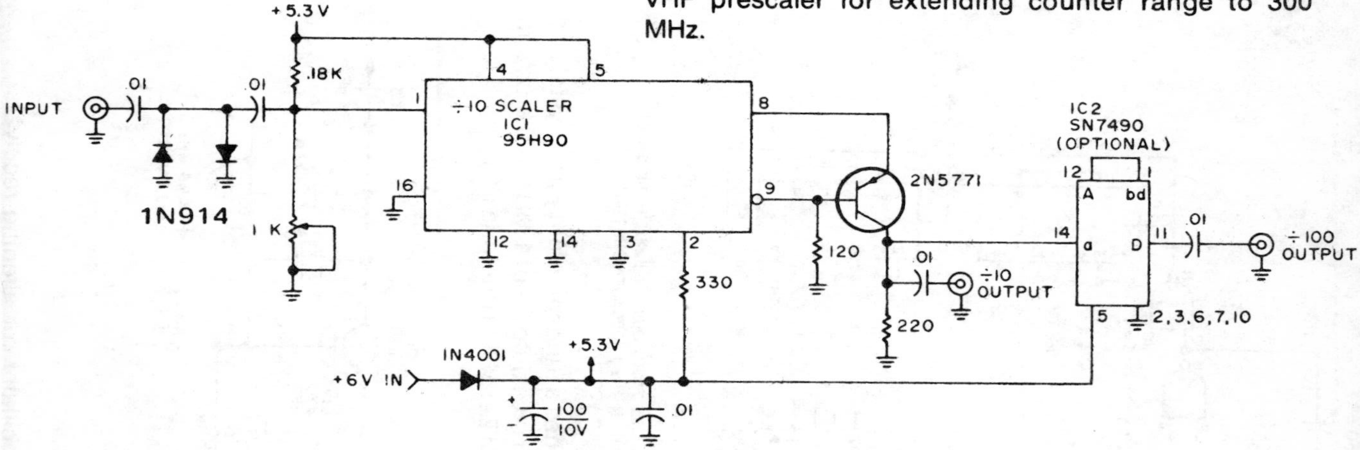


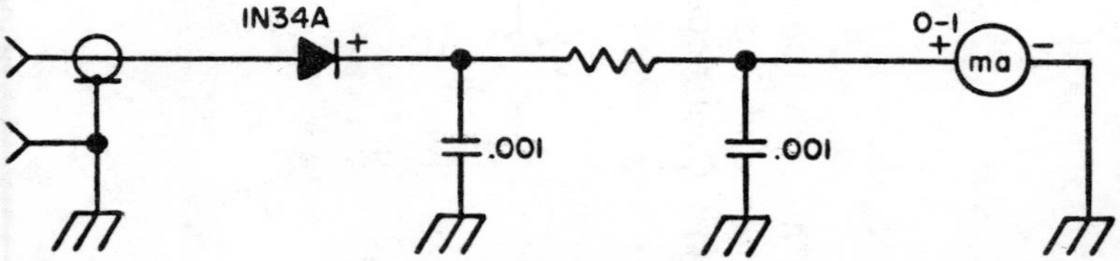


Rf wattmeter is simple and accurate, but logarithmic. A 22K resistor gives 10W full-scale indication.

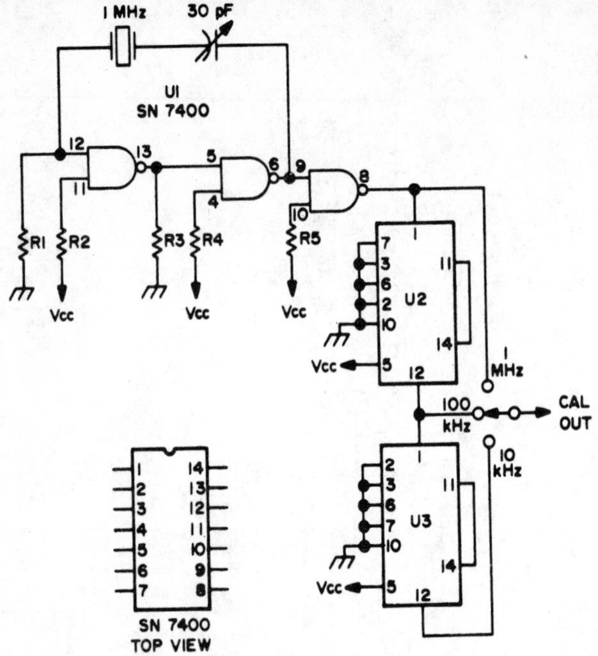


| | | |
|---|---|---|
| R1, R3 | 1.8K | |
| R2, R4, R5 | 1K | |
| U1 | SN 7400 | QUAD 2-INPUT NAND GATE |
| U2, U3 | SN 7490 | DECADE COUNTERS |
| Vcc | 5 VDC | |

This crystal calibrator has a fundamental 1 MHz crystal and its frequency is divided to give outputs of 1 MHz, 100 kHz, and 10 kHz. Circuit courtesy of Zero Beat (Victoria BC).

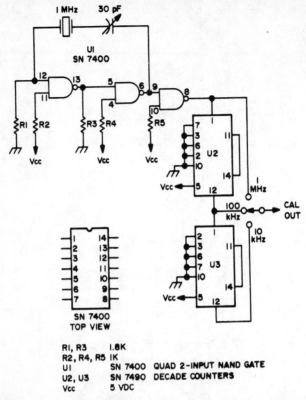

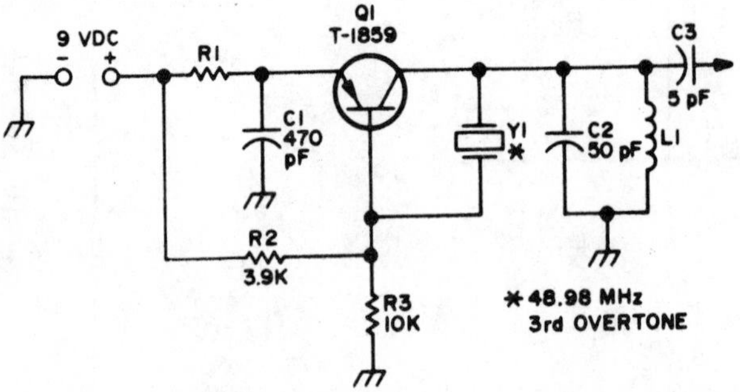


Crystal oscillator for aligning receivers.

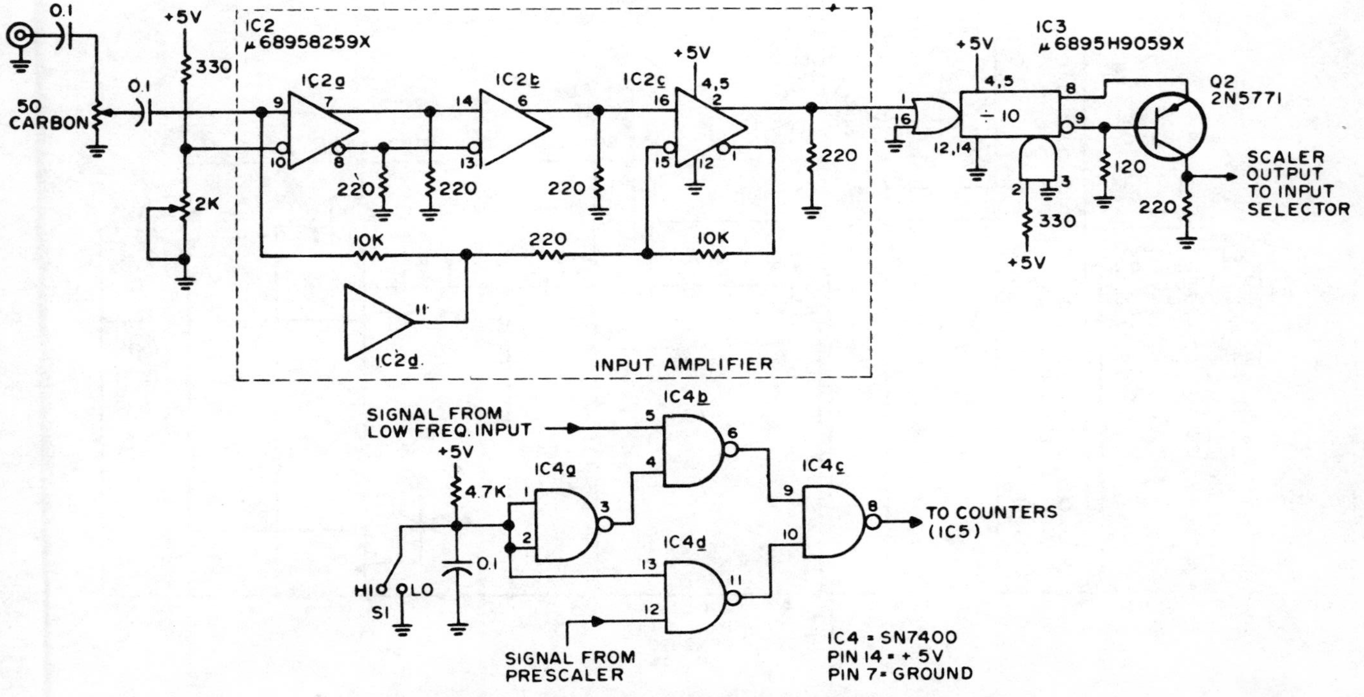


VHF prescaler for frequencies up to 200 MHz.

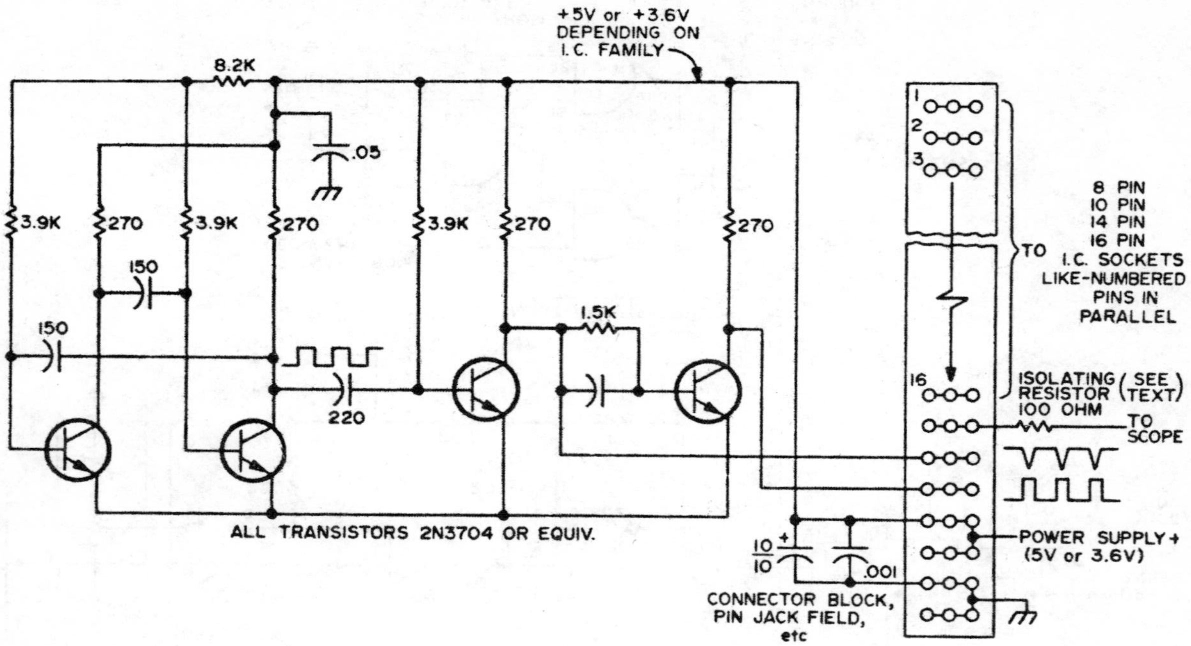


General purpose digital IC tester.

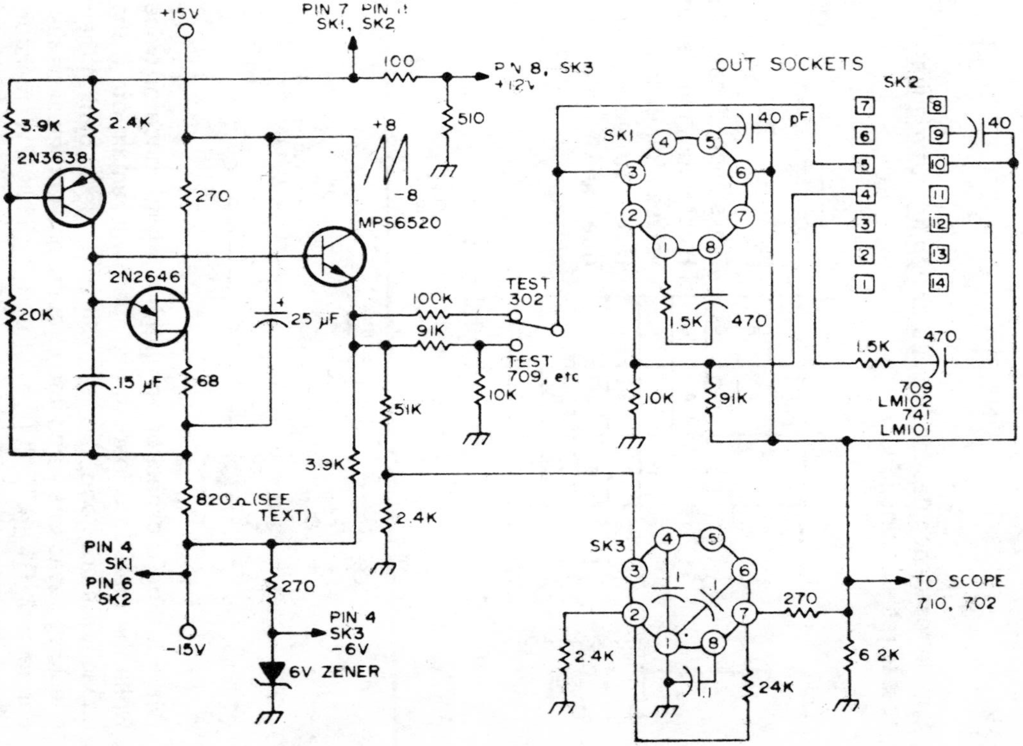


Linear IC tester.

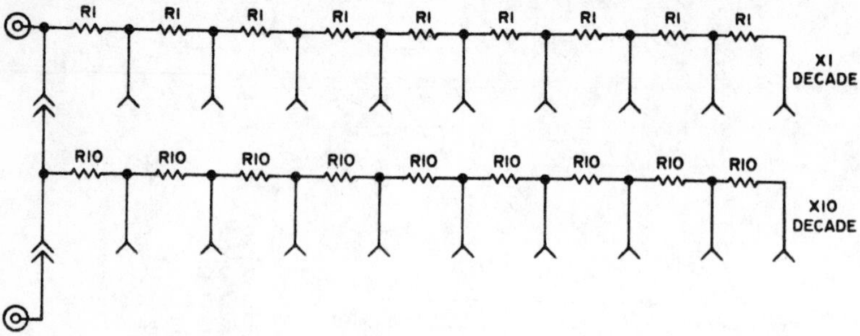


Simple resistance decade uses no switching but requires access to resistor junctions.

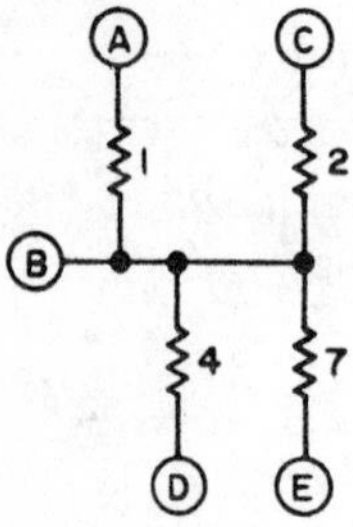


| R | FROM | TO |
|---|---|---|
| I | A | B |
| 2 | B | C |
| 3 | A | C |
| 4 | B | D |
| 5 | A | D |
| 6 | C | D |
| 7 | B | E |
| 8 | A | E |
| 9 | C | E |

Resistance decade in H configuration. Select values after examining table to see a sequence of value increases.

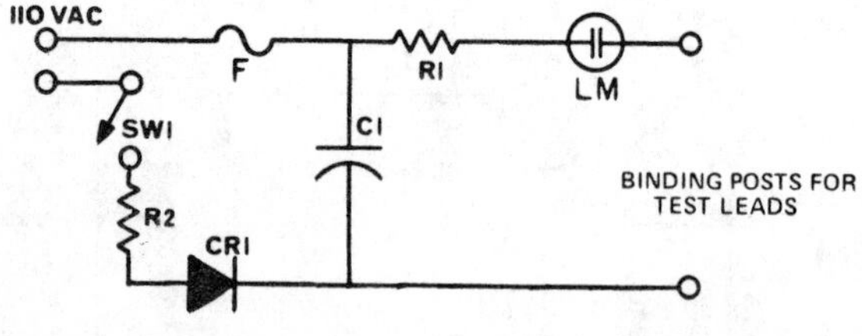


When testing a capacitor, the light will blink once or twice while the plates are loading and then will remain off. If the capacitor is not good, the light will continue to blink. When checking capacitors wired into circuits, one end must be disconnected. R1—47K 2W resistor, R2—330K 2W resistor, LM-1—neon lamp NE 51, C1—8 or 10 $\mu$F capacitor 450V dc, CR1—1M3611 diode (not critical), SWL—ST toggle switch, F—¼A fuse.

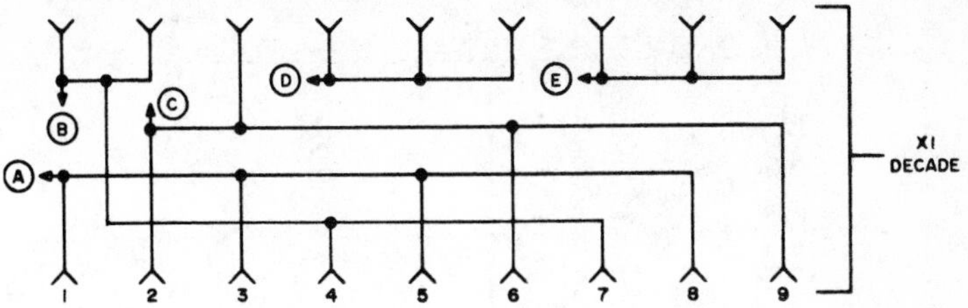


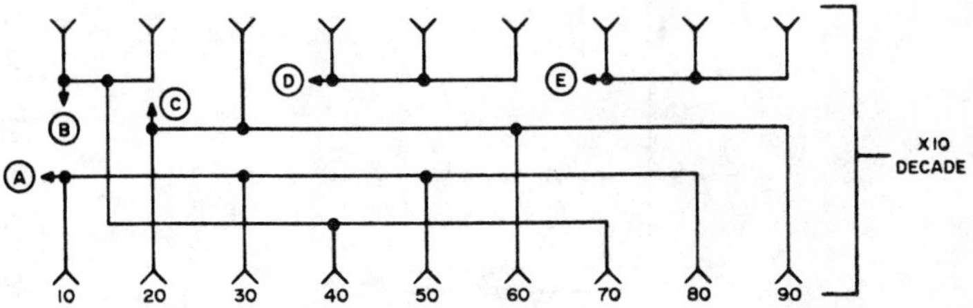


Resistance decade using rotary switch for resistance selection. Circled letters show switch connections.

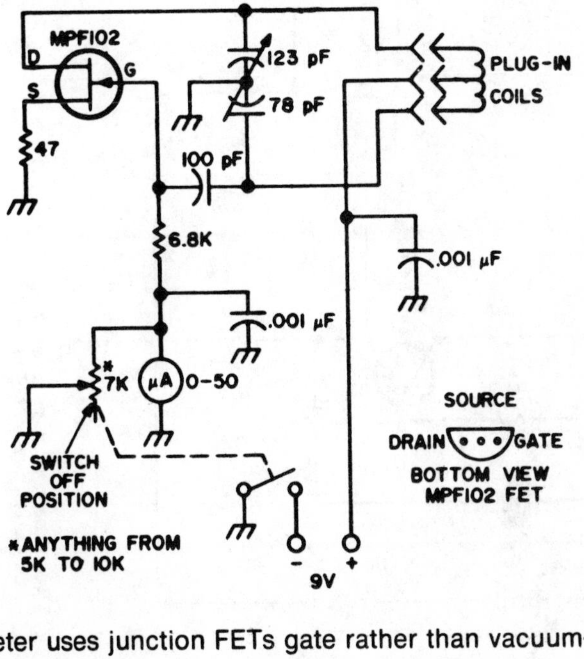


Dip meter uses junction FETs gate rather than vacuum-tube grid for dip function.

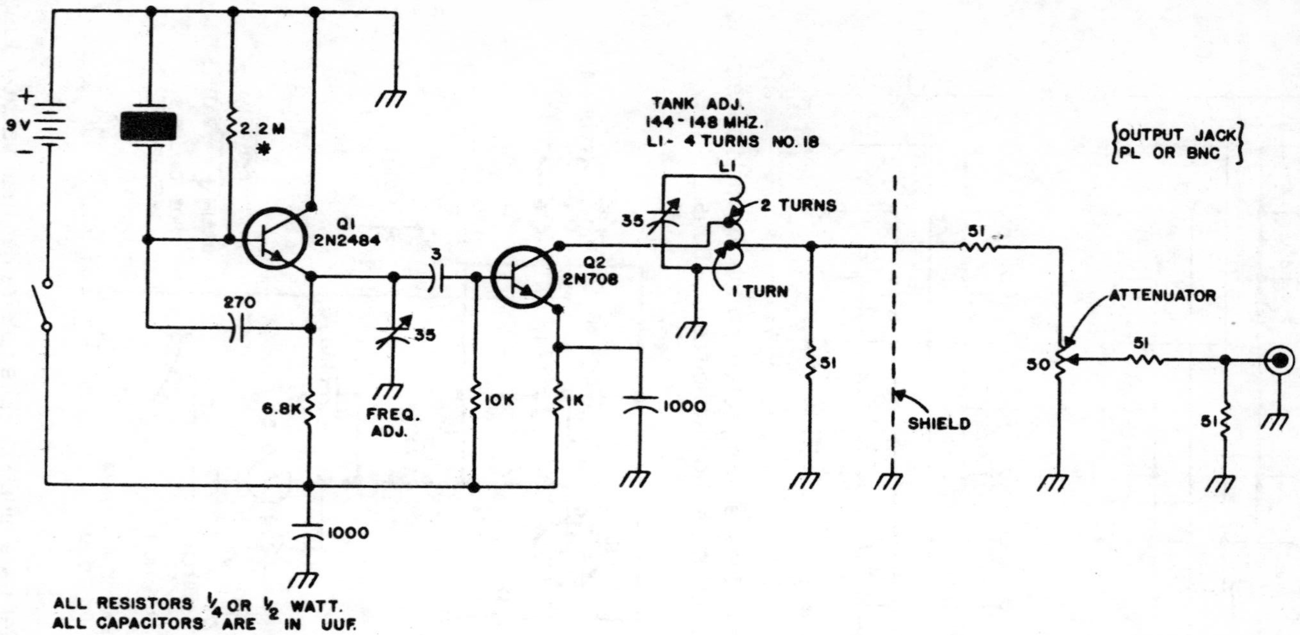


Signal generator for aligning FM receivers in 150 MHz region. Resistor marked with * may have to be changed to lower value if recommended transistor is not used.

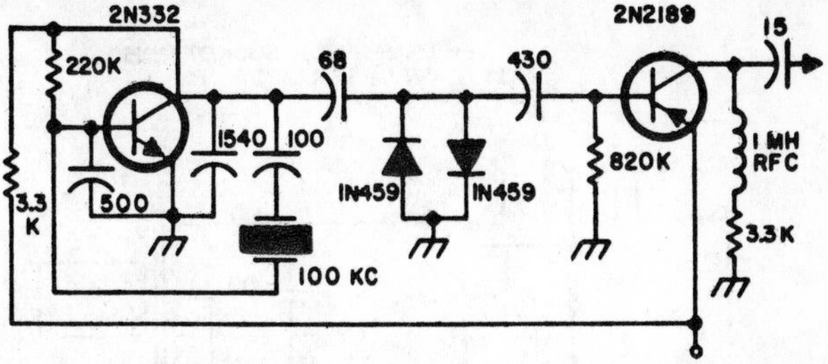


Schematic of 100 kHz calibrator. Parts values are not critical and the transistors do not have to be the ones listed. It is suggested that a late version of the GE transistor manual be consulted for a slightly revised oscillator circuit.

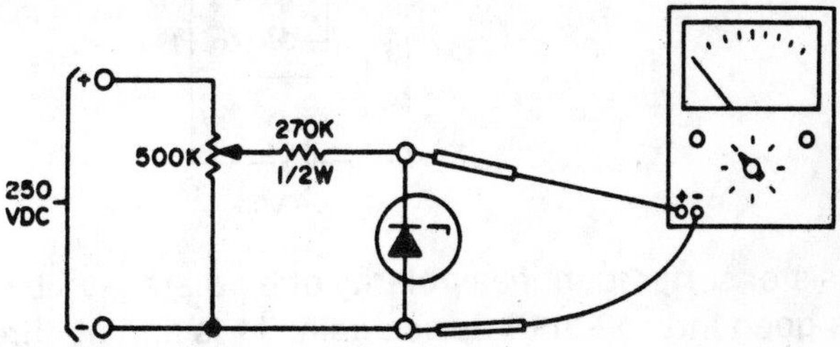


Universal zener diode tester uses voltage divider and VOM to read developed dc voltage.

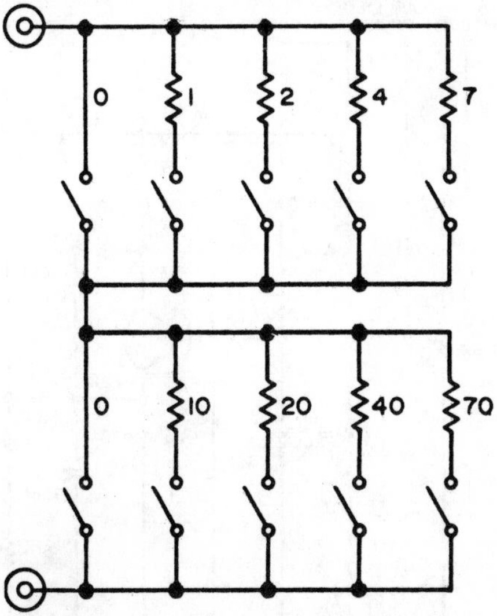


Resistance decade uses two pin jacks and a bank of toggle switches. Note that 2 switches must be closed for each reading.

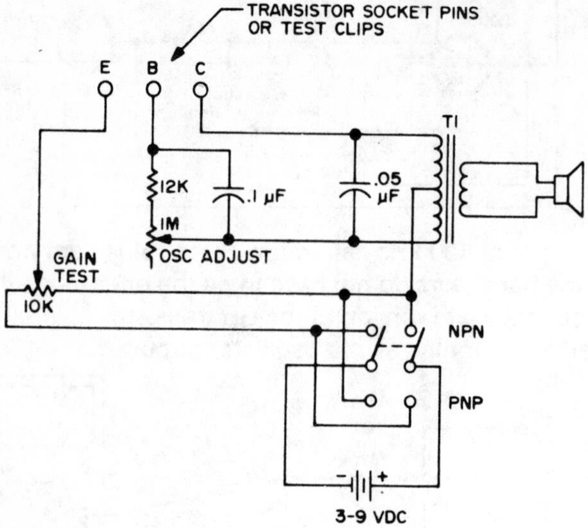


Transistor sorter identifies polarity of bipolars; audible signal gives good indication of device gain. T1 is a miniature 400Ω (..8pCT) to 4 or 8Ω transistor output transformer.

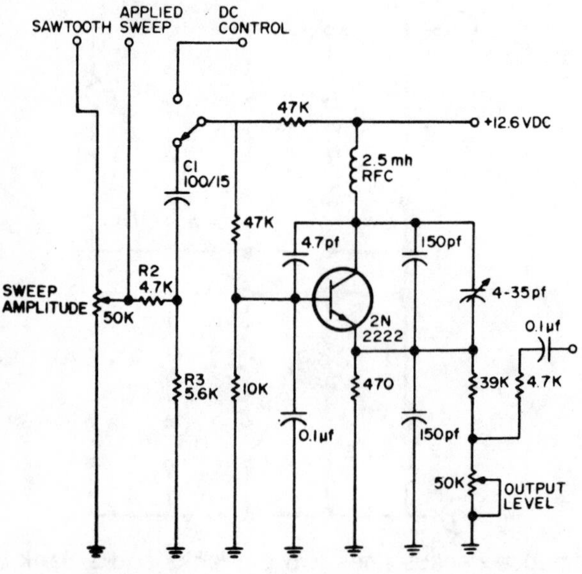


Sweep generator for 430 – 470 kHz.

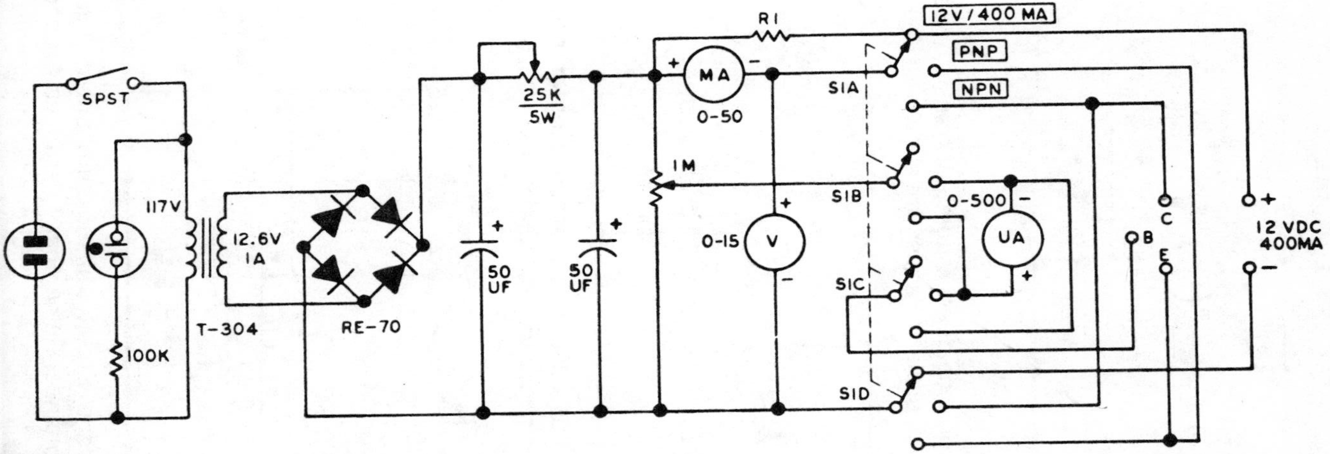


NOTES:
1. UNLESS OTHERWISE INDICATED RESISTANCE IS IN OHMS.
2. S1 = 4-POLE, 3-POSITION SWITCH.

Transistor parameter tracer.

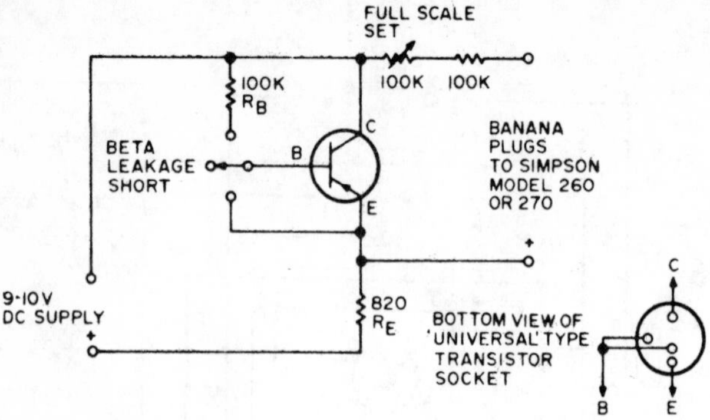


This circuit will provide direct reading of transistor beta right on your VOM ohm scale. Polarities are shown for PNP.

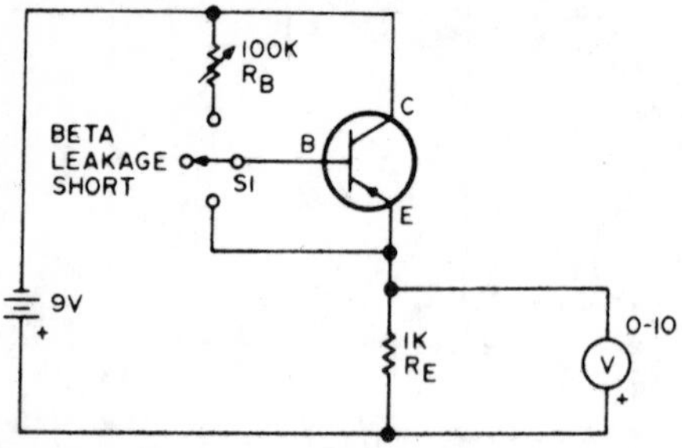


Basic circuit of a popular and simple adapter for testing common "milliwatt" transistors.

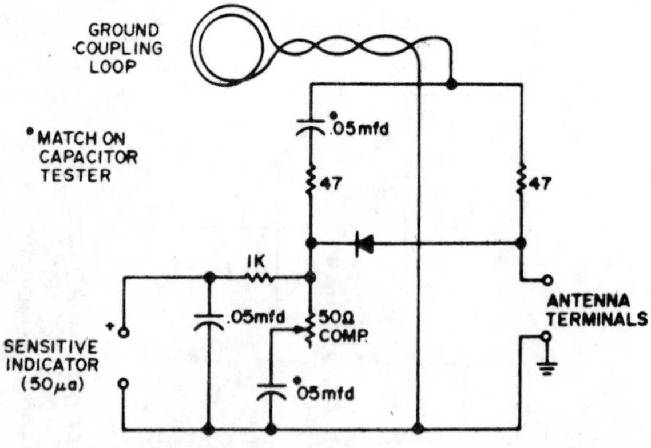


This simple bridge circuit will measure antenna input resistance at resonance. Remember tolerances on capacitors are commonly very loose. A capacitor checker can do an adequate job of matching.

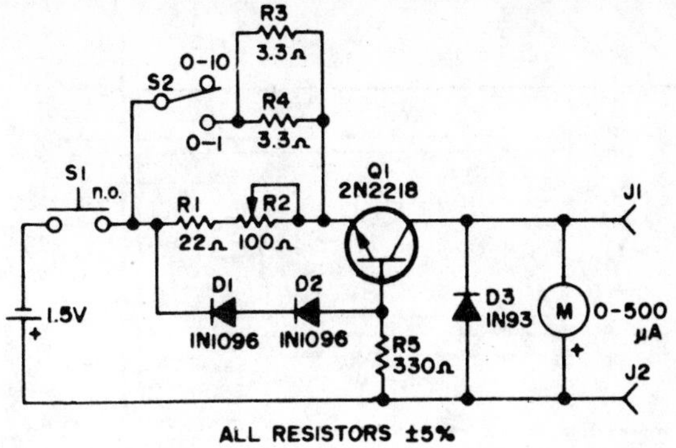


Simple transistor circuit measures resistances from 0 to 1Ω and from 0 to 10Ω in 2 ranges. Your VOM can be substituted for meter.

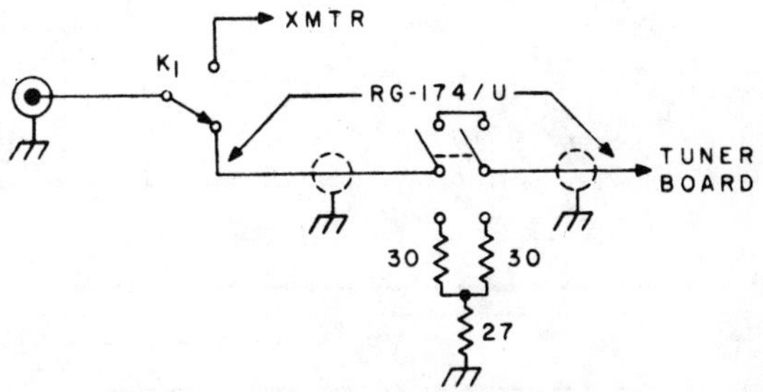


Schematic of 12 dB attenuator added to receiver. Small RG 174/U coax was also added between relay K1 and the tuner board as shown.

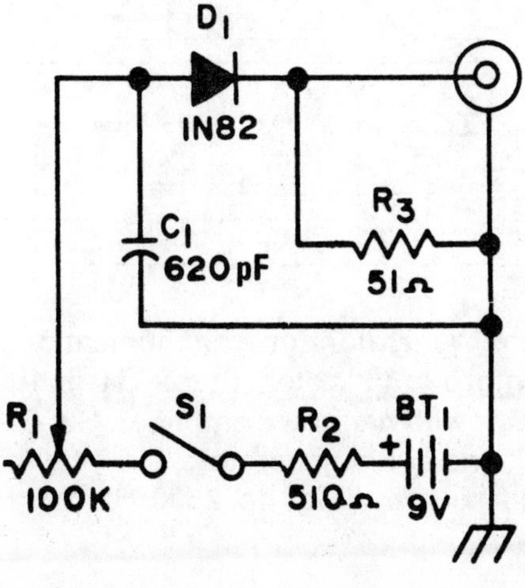


Simple noise generator uses inexpensive components.

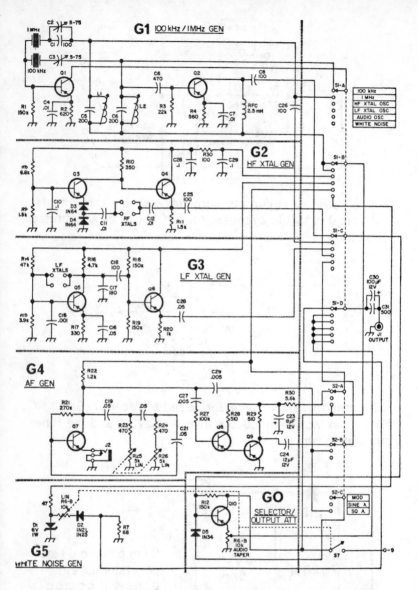

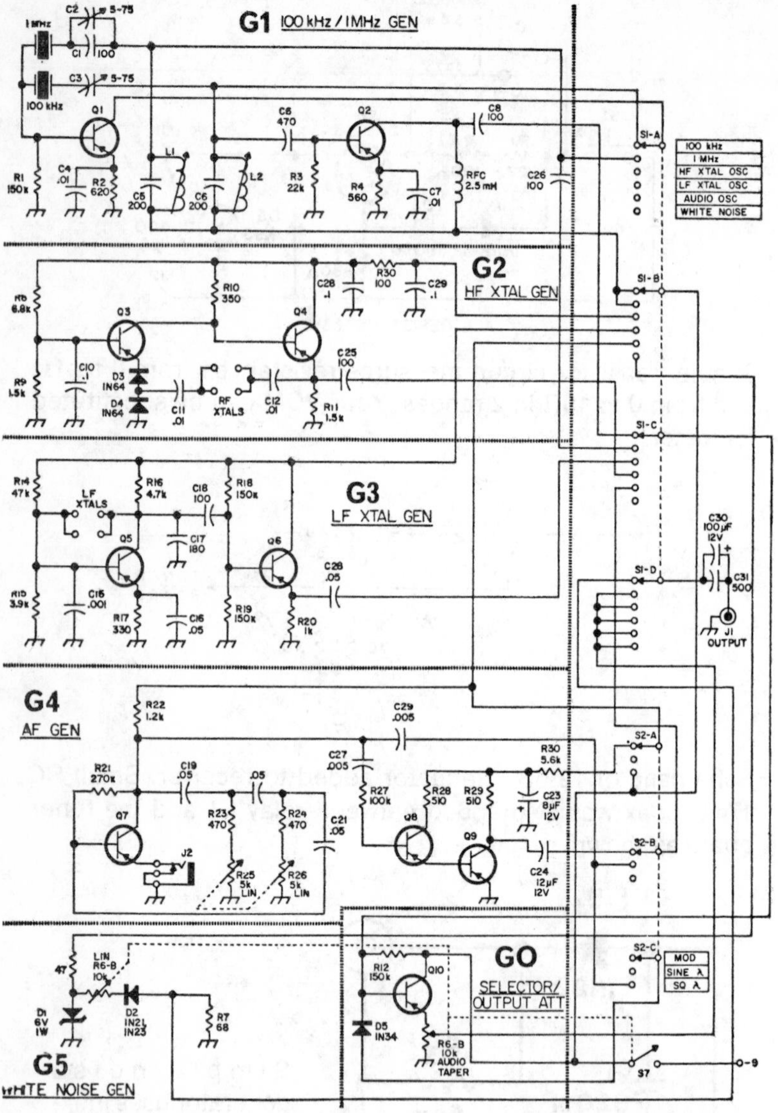


"Gentrac" is multipurpose generator for communications, plus af/rf signal tracer with output meter. Stage G1 is 100 kHz/1 MHz marker generator with harmonic amplifier. G2, G3, G4—respectively—HF, LF, and audio generators, the latter of which produces sine and square waves to 2500 Hz. G5 is white noise generator.

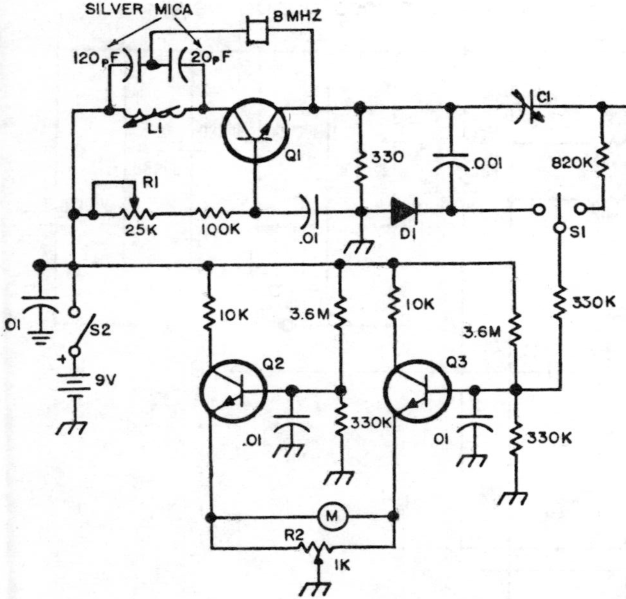


| TURNS | TAP | Q | Cμμf | Lμh |
|---|---|---|---|---|
| 7 | A-B | 135 | 256 | 1.57 |
| 13 | A-C | 210 | 106 | 3.8 |
| 18 | A-D | 235 | 69 | 7 |
| 23 | A-E | 260 | 50 | 8 |
| 6 | B-C | 120 | 310 | 1.28 |
| 11 | B-D | 180 | 134 | 3 |
| 16 | B-E | 215 | 69 | 7 |
| 10 | C-E | 170 | 151 | 2.62 |
| 11 | F-E | 195 | 120 | 3.35 |
| 16 | F-D | 230 | 77 | 5.2 |
| 21 | F-C | 280 | 55 | 7.4 |
| 27 | F-B | 275 | 39 | 10 |
| 34 | A-F | | | |

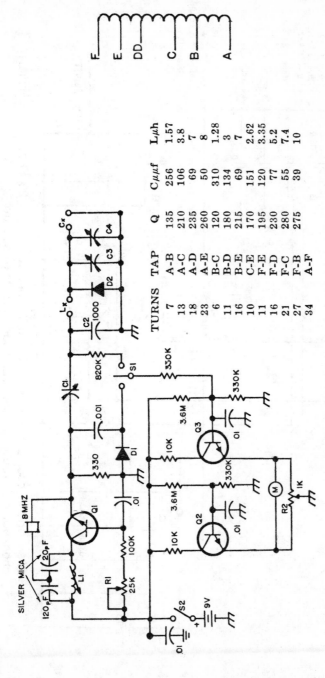

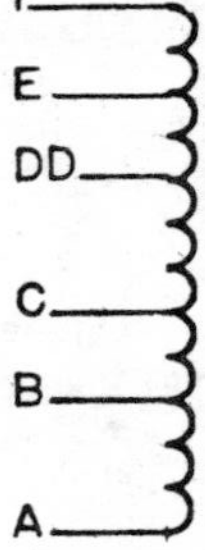


Q meter. A small voltage $E_1$ is introduced in series with a tuned circuit, the unknown coil, and C3. Circuit is tuned to resonance and voltage $E_2$ measured on voltmeter. Q is $E_2/E_1$.

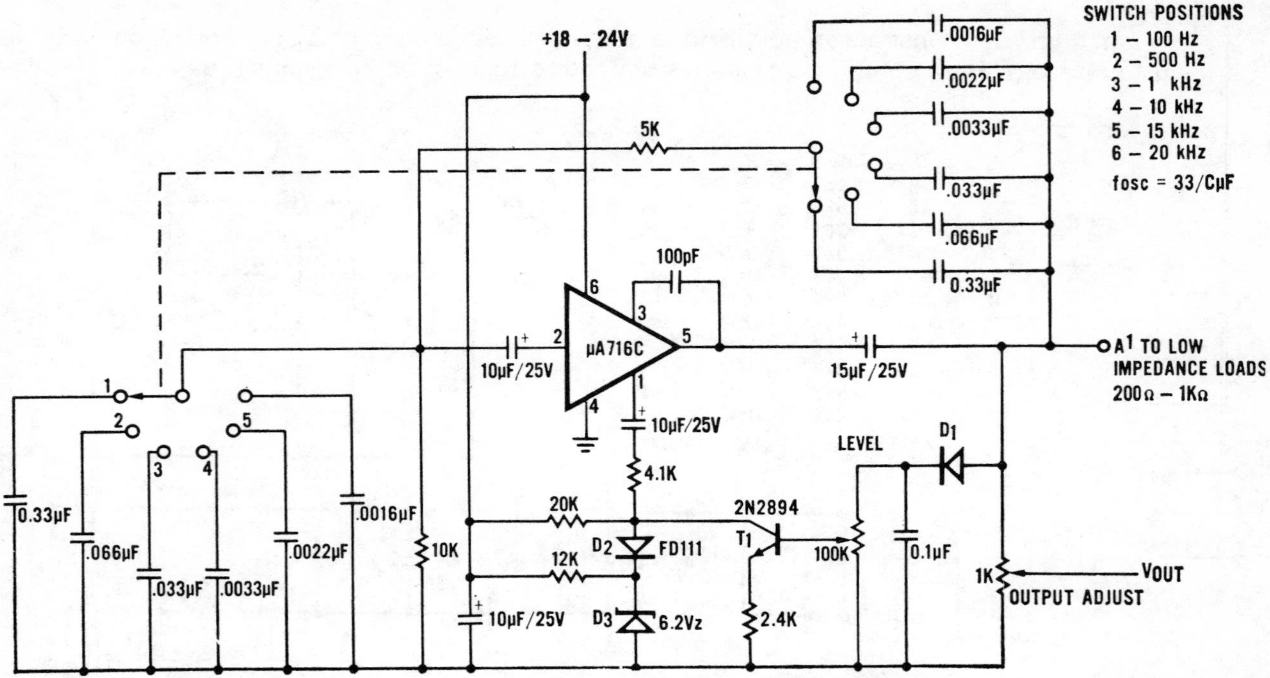


Wien bridge oscillator delivers 100 mW from 100 Hz to 20 kHz, for transmitter checkout, hi-fi troubleshooting, and code-practice oscillator.

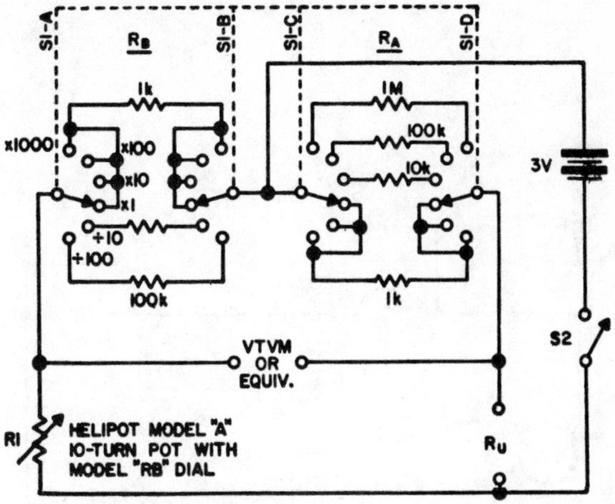


Circuit of the simple resistance bridge. S1 is a Centralab PA2011 rotary switch, S2 is a SPST slide switch, and R1 is a precision, 10-turn 100Ω potentiometer such as the Helipot model A. The unknown resistance is placed across terminals "RU."

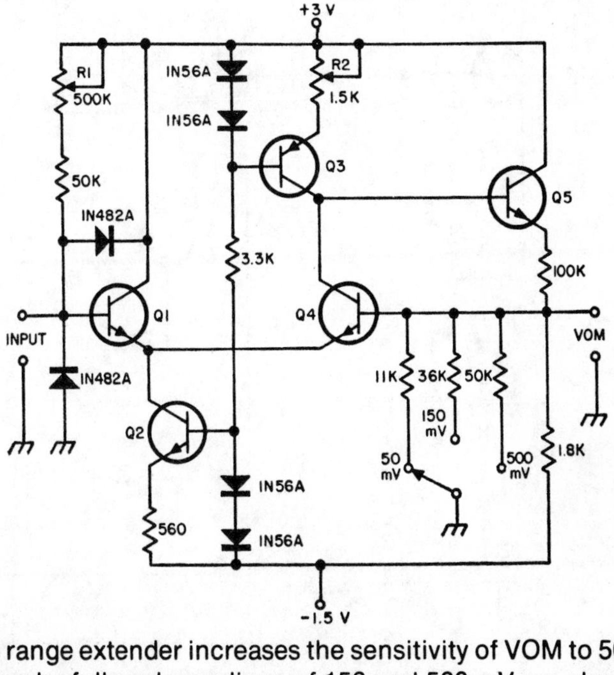


This range extender increases the sensitivity of VOM to 50 mV full scale; full-scale readings of 150 and 500 mV are also provided by the range switch.

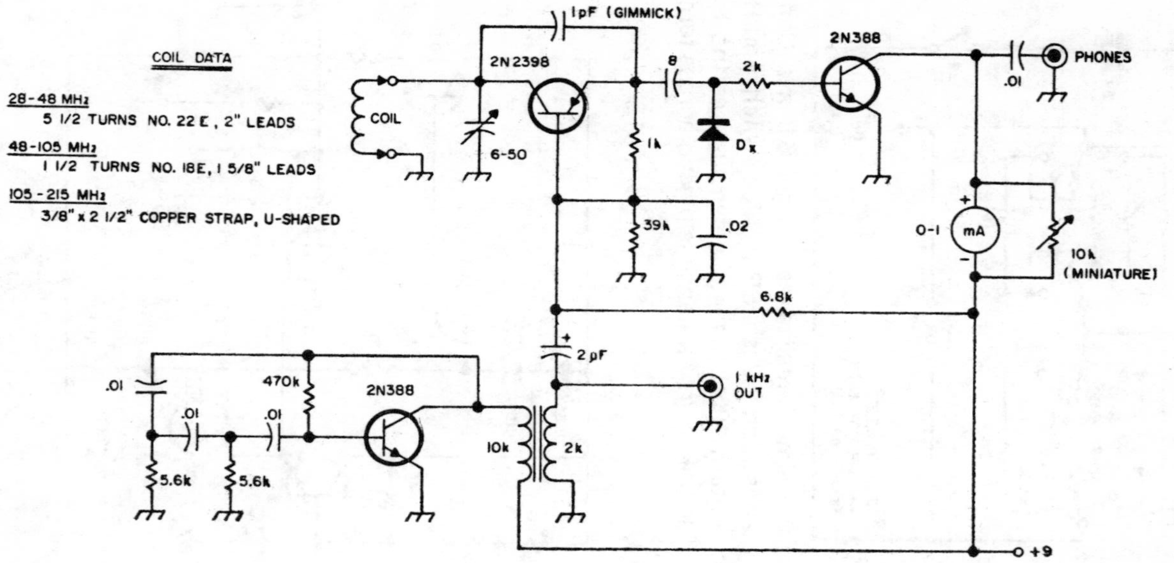


Diode $D_x$ may be almost anything that you have available. The 1 pF gimmick capacitor consists of 1½ in. of twisted wire.

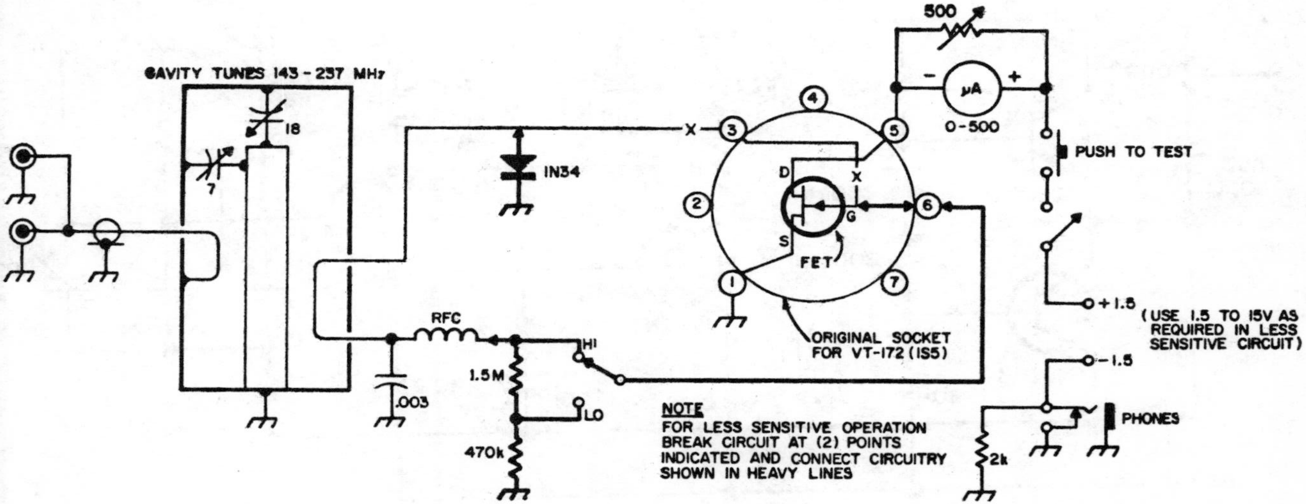


Conversion of the BC-906 frequency meter to use field-effect transistor. This frequency meter will detect rf signals down to a 100 microvolts or so and will tune through both the 2 and 1¼ meter bands.

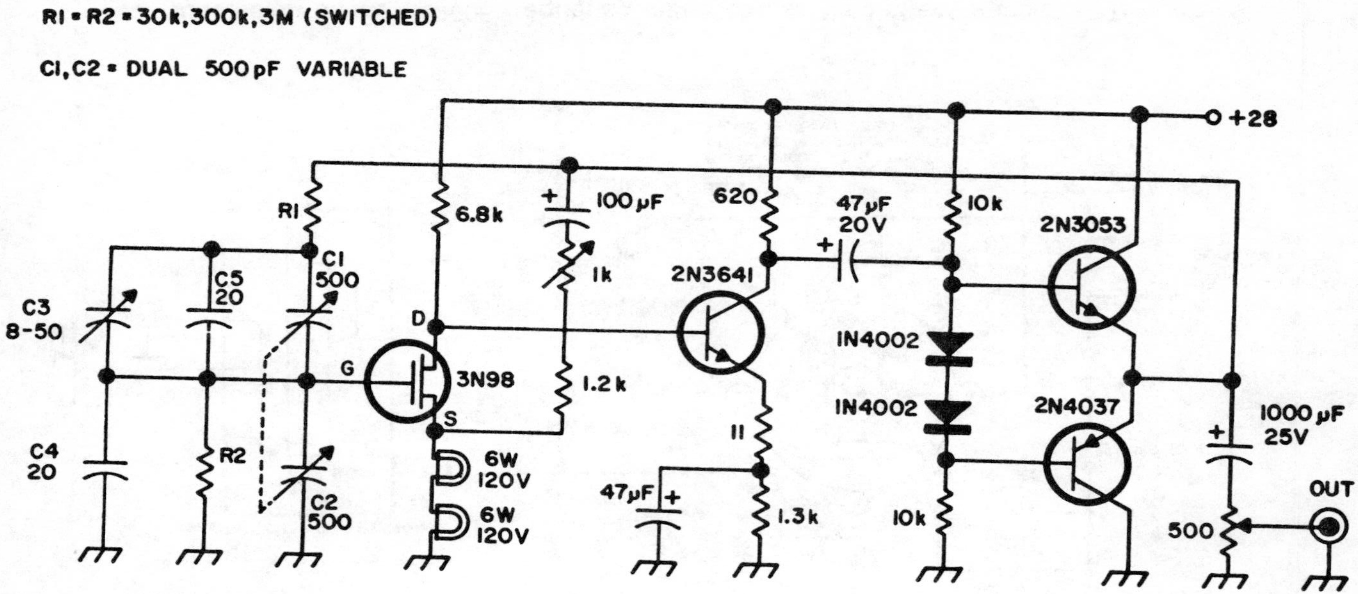


First version of a moderately successful Wien bridge oscillator. For low frequencies, R1 and R2 have to be so large that the circuit is very susceptible to noise and hum.

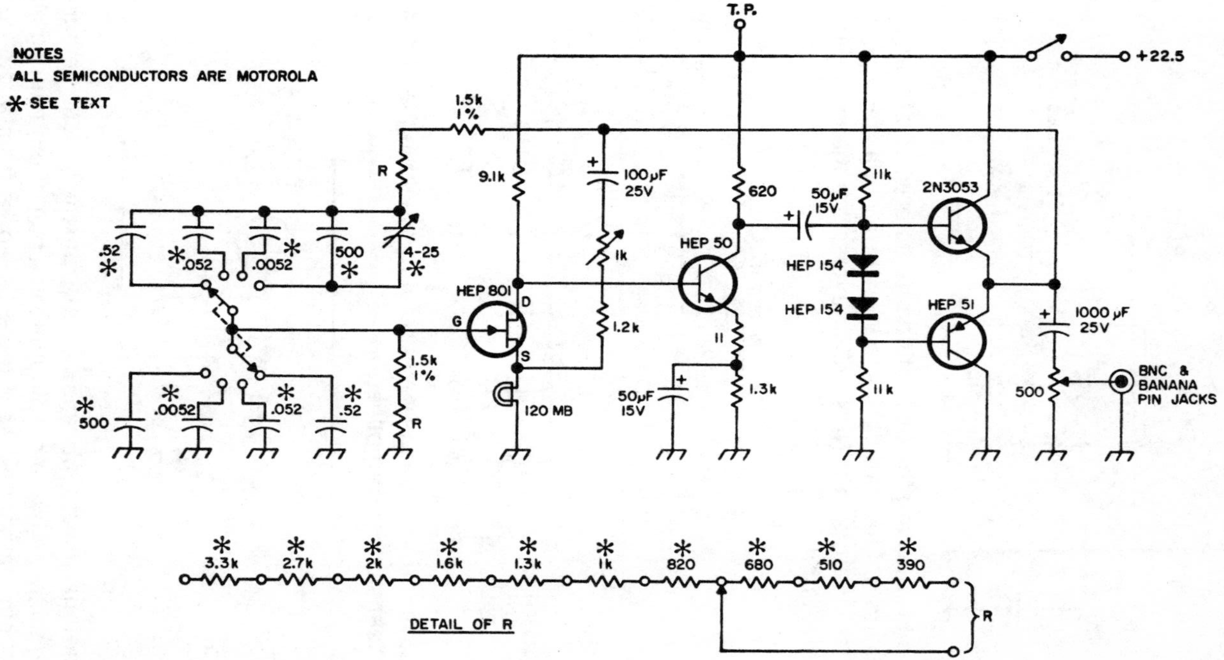


The most satisfactory version of the Wien bridge oscillator.

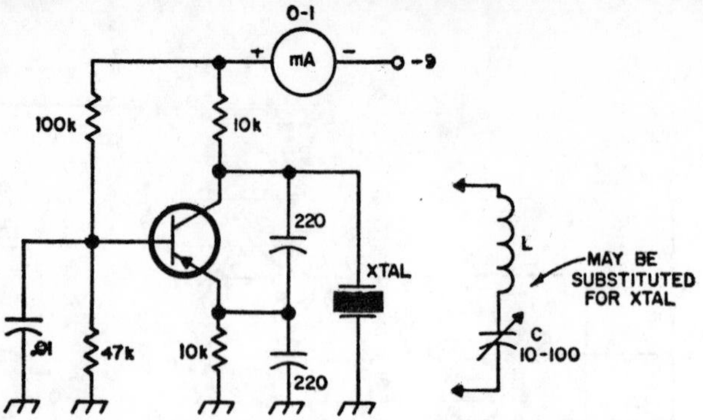


Circuit for testing the oscillating characteristics of unmarked transistors. This circuit is shown with a negative supply for PNP transistors—for NPN units, reverse the meter and use a positive supply.

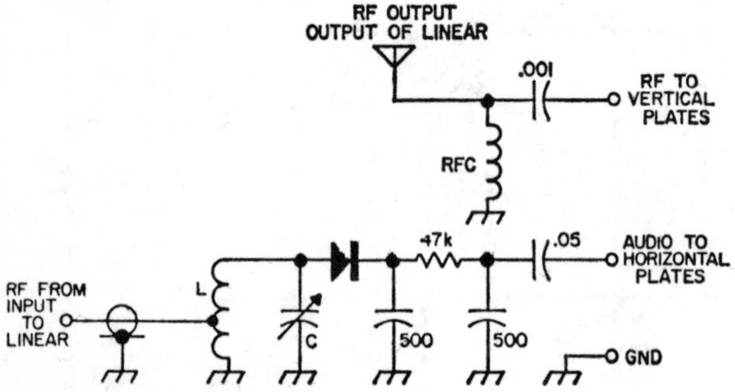


This trapezoid circuit is very useful for checking the linearity of linear sideband amplifiers.

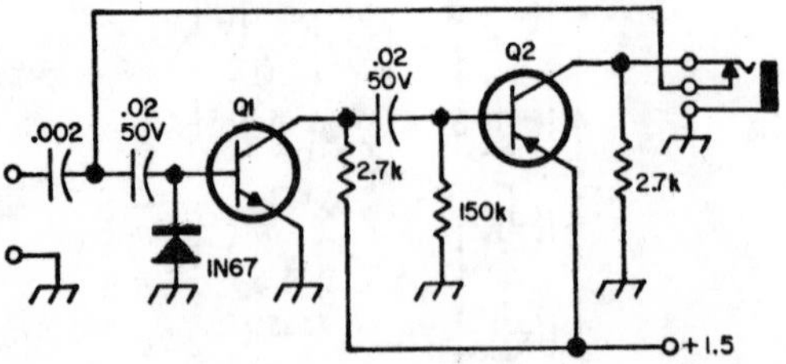


This signal injector/tracer switches from the injection mode to a signal tracer by simply plugging in a pair of high-impedance magnetic earphones. As a tracer is works from audio up to 432 MHz. Transistor Q1 is a 2N170, 2N388A, 2N1605, SK3011 or GE-7; Q2 is a 2N188A, 2N404, 2N2953, SK3004 or HEP 253.

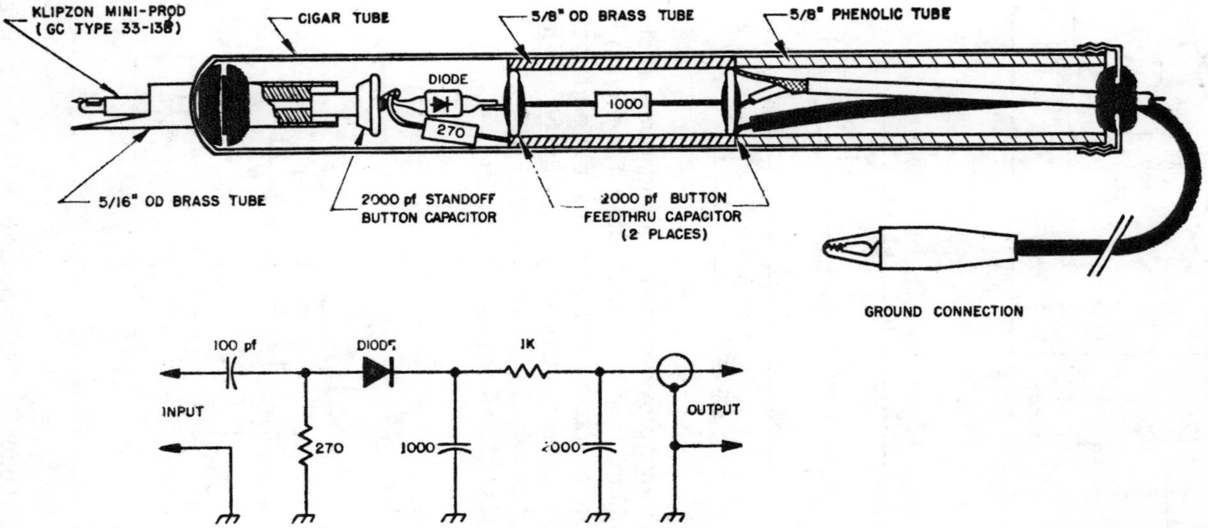


Sensitive rf probe schematic and cutaway view. Cigar tube allows coaxial, shielded construction for flat response to 500 MHz. Capacitors should be button types to minimize inductance if VHF/UHF operation is desired.

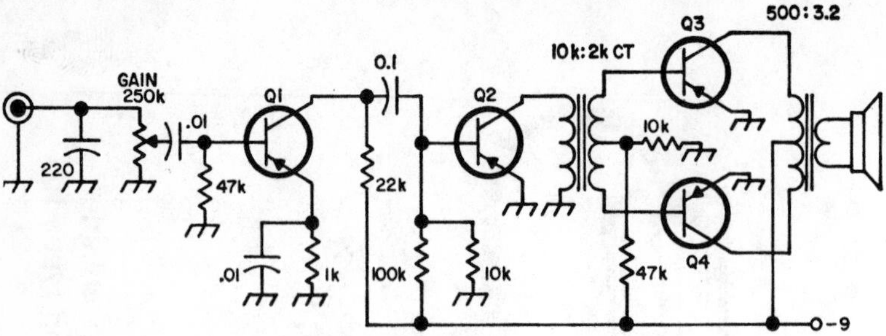


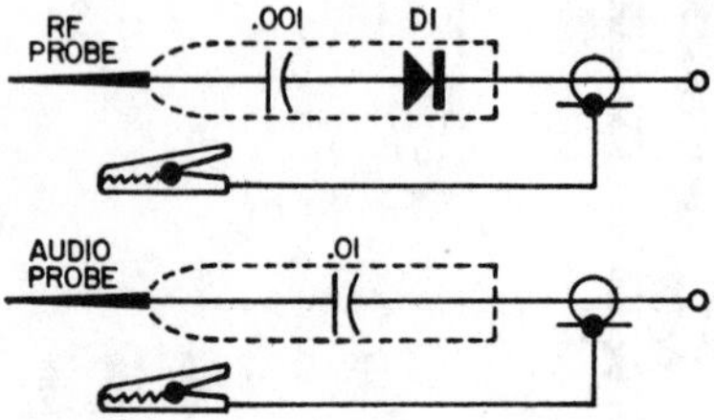


This signal tracer provides more than adequate audio output with only 100 microvolts of modulated rf at the input. It may also be used for tracing audio circuits, but don't depend on its fidelity. All the transistors are germanium types such as the 2N404, 2N1450, 2N2953, SK3004, GE-2 or HEP 253; the diode in the prove is a 1N34A or 1N67A or similar.

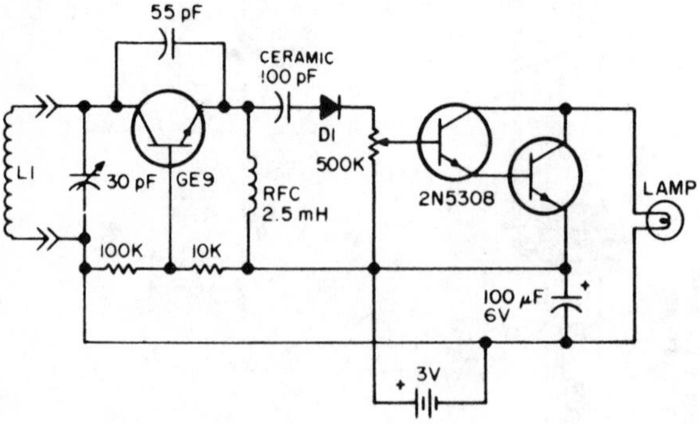


In this "grid dipper," there is no grid, and a lamp dims to indicate resonance. Circuit is an oscillator and Darlington amplifier pair.

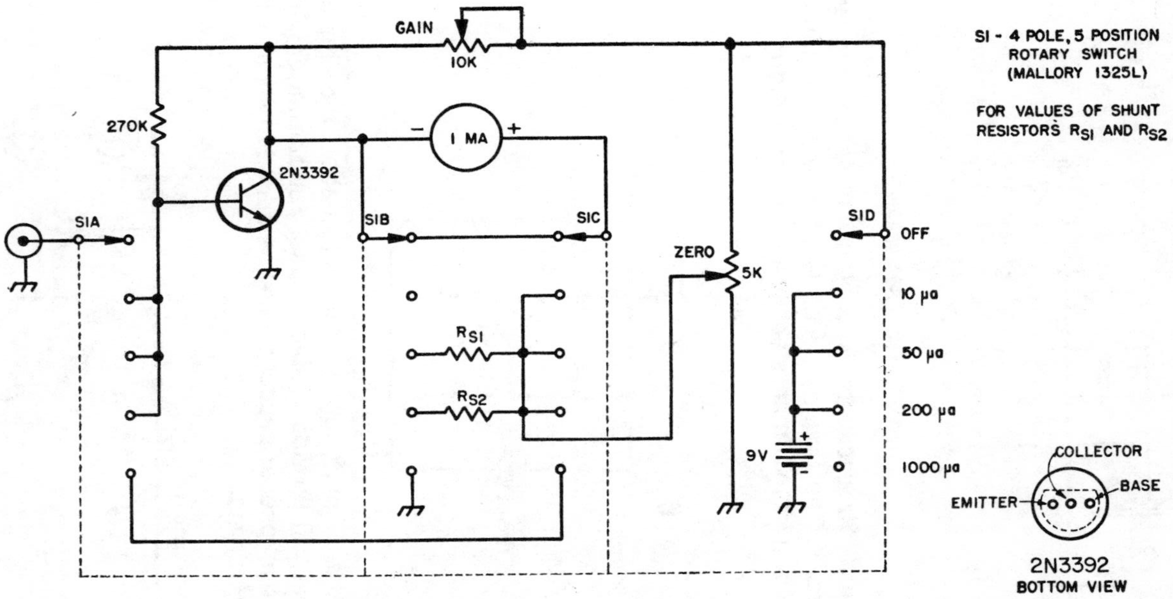


Transistorized microammeter. This instrument will provide full-scale readings down to 10 $\mu$A. Although a 2N3392 was used here, any high-gain silicon transistor that maintains high current gain at low collector-current levels is suitable.

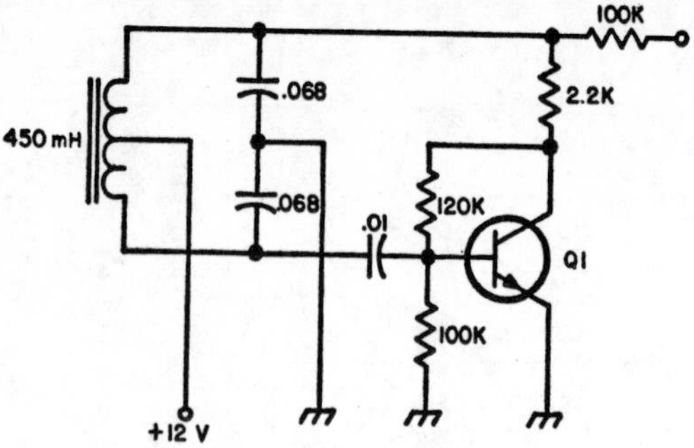


Q1 - 2N2924, 2N3393, 2N3566

Simple 1000 Hz oscillator is very useful for testing and measurement. The Colpitts circuit is used with component values chosen for maximum stability and good waveform.

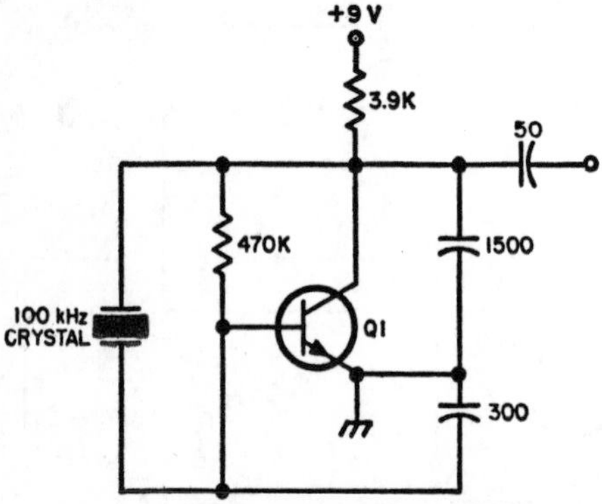


Q1 - 2N708, 2N2711, 2N2923, 2N3691

The 100 kHz calibrator shown here is just about the simplest circuit that will provide usable results. For zeroing in with WWV a small padder capacitor may be added in series with the 100 kHz crystal.

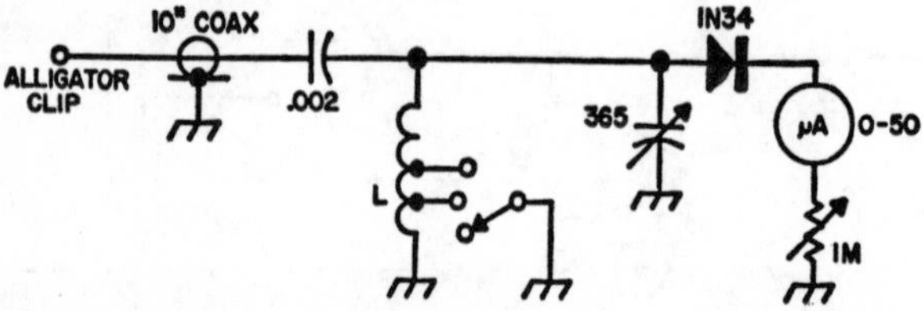


Simple 3–30 MHz wavemeter/detector useful for checking low-power oscillators and amplifiers. L is 30 turns number 20, 16 turns per inch, tapped at 14 and 5 turns from cold end.

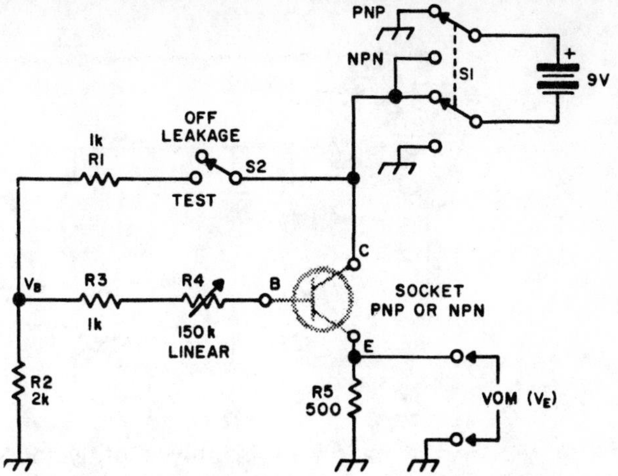


Transistor tester adapter for use with a VOM.

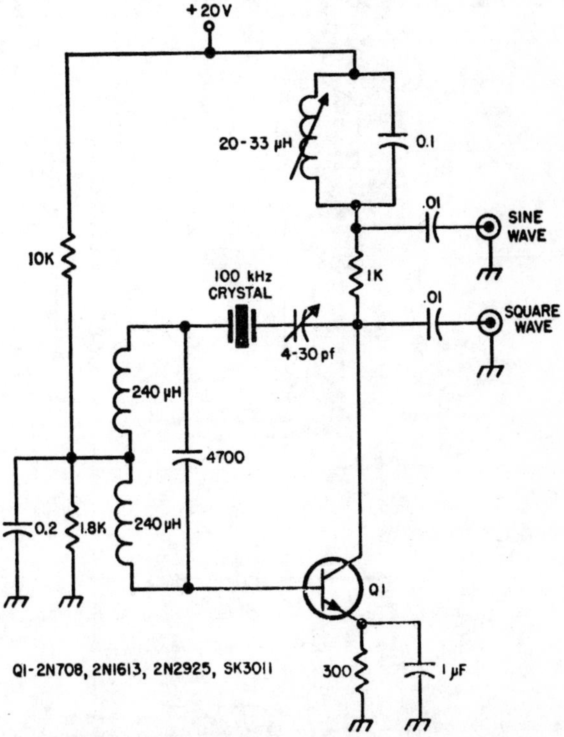


This 100 kHz crystal calibrator uses a crystal in the parallel mode and provides either a sinusoidal or square wave output. The calibrator may be zeroed to WWV with the 4–30 pF trimmer.

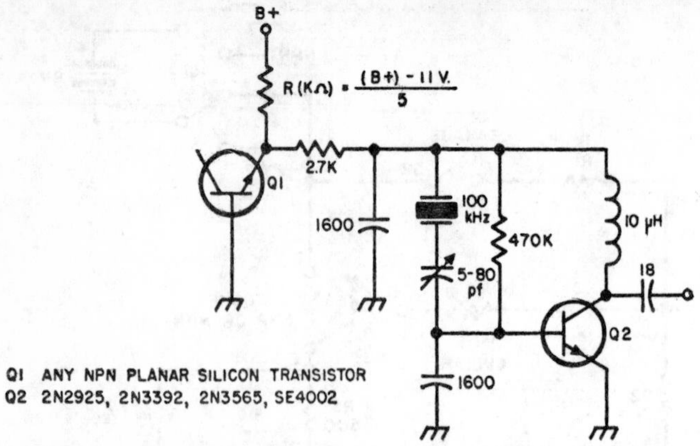


This 100 kHz crystal calibrator is only slightly more complicated, but has a built-in voltage regulator (Q1) and provides usable harmonics up to 150 MHz.

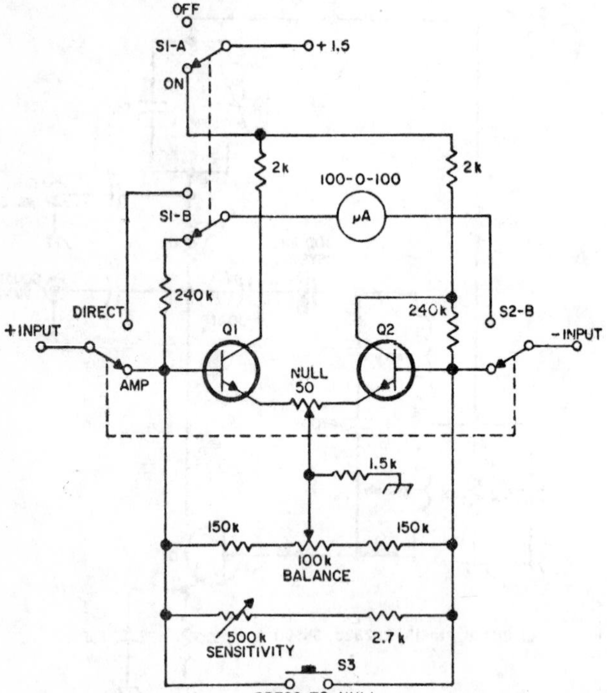


This very sensitive microammeter amplifier may be adjusted from 2 $\mu$A to 100 $\mu$A each side of zero; the input impedance varies from 60K at 2 $\mu$A to 2.5K at 100 $\mu$A. Transistors Q1 and Q2 are 2N930, GE-10, or HEP 50.

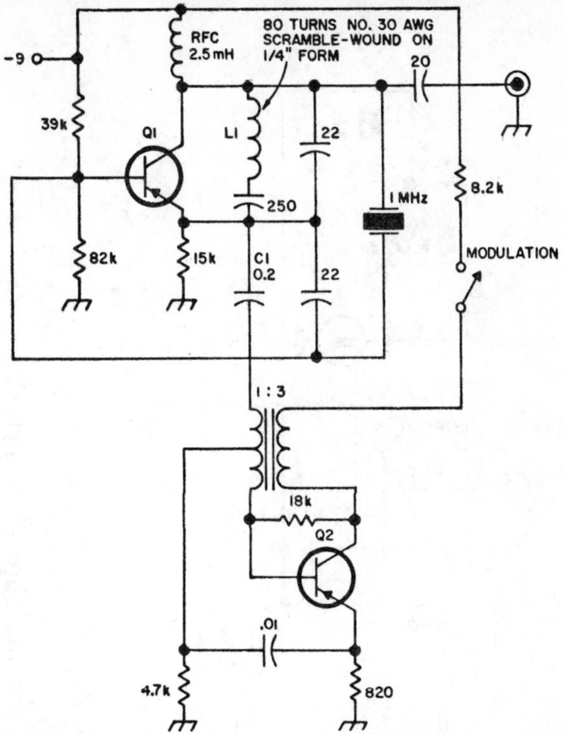


This crystal-controlled oscillator provides very distinctive markers up to 30 MHz. The modulation frequency is approximately 100 Hz, but by changing the value of C1 it may be changed slightly. Q1 is a 2N384, 2N1742, 2N2362, 2N2084, TIN10, GE-9, or HEP 2; Q2 is a 2N2613, 2N2953. 2N1303, SK3004, GE-2, or HEP 254.

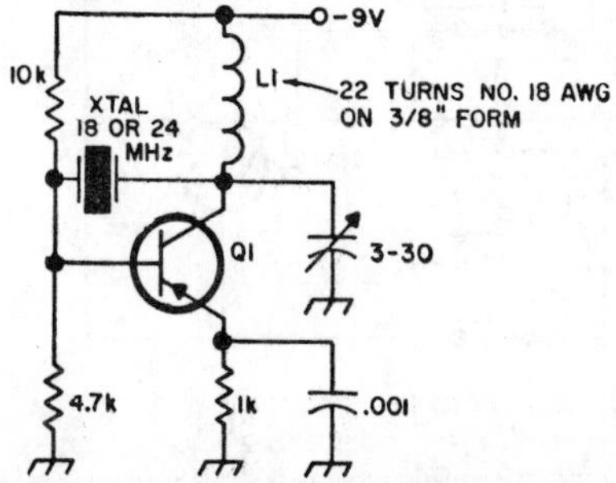


This 2-meter edge marker provides useful harmonics up to several hundred megahertz with 18 ot 24 cyrstals. If modulation is desired, the audio oscillator may be coupled into the base. Q1 is a 2N1745, 2N2362, or HEP 2.

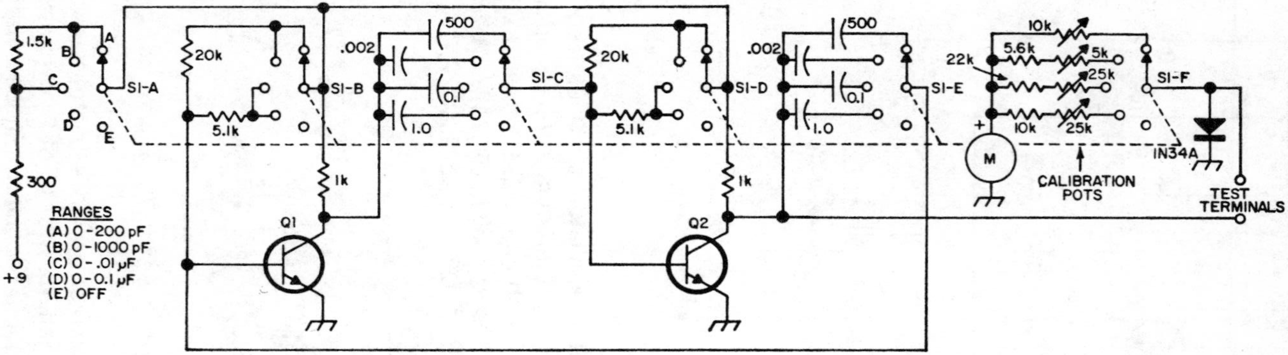


Although this capacitance meter wll not measure electrolytic capacitors, it will measure any other type from zero to 0.1 $\mu$F with reasonable accuracy. On the lower end, 4 pF can be read accurately and 2 pF easily estimated. Transistors Q1 and Q2 are 2N168, 2N1605, 2N2926, SK3011, or HEP 54; the meter is a 0—50 microampere unit and the range switch a Centralab PA1021.

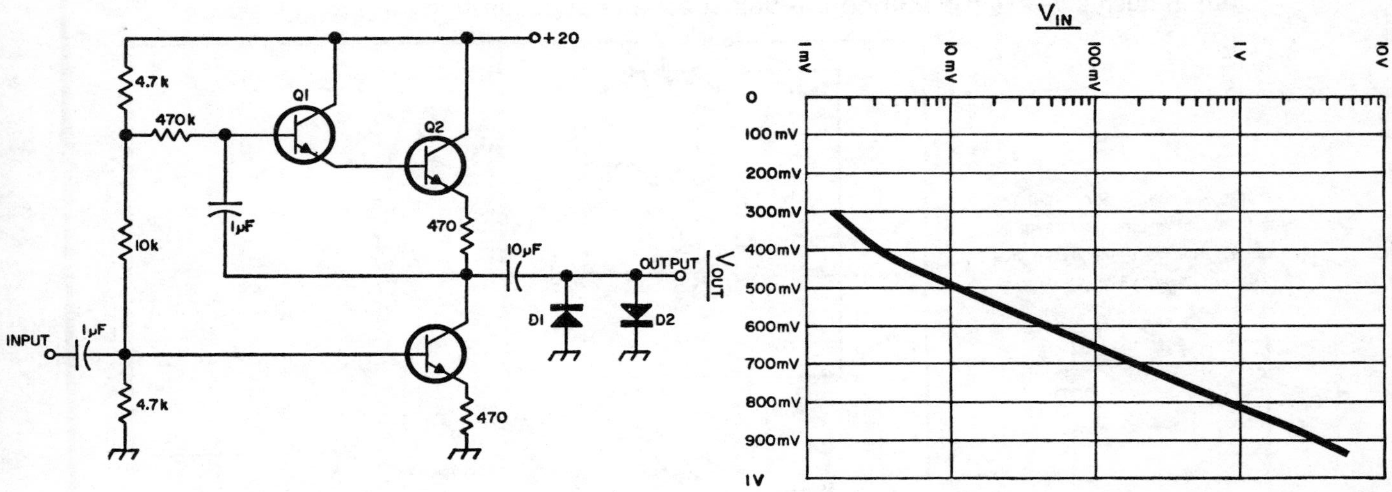


This logarithmic amplifier makes use of the fact that when two back-to-back sioswa are driven by a current generator, they exhibit a logarithmic output of the input signal. With the circuit constants shown, this amplifier follows a nearly perfect logarithmic curve over a 60 dB range; selected diodes will increase this to 80 dB. Q1, Q2, and Q3 are 2N2924, SK3019, GE-10 or HEP 54; D1 and D2 are 1N914.

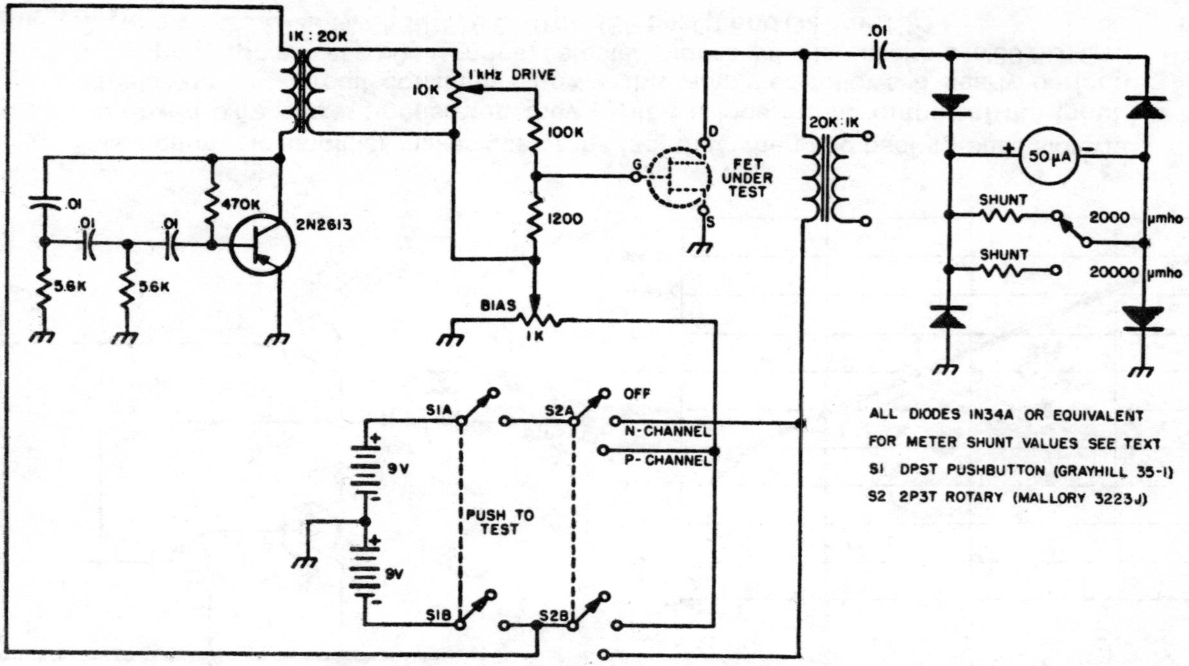


Schematic of the FET transconductance tester. Although a 2N2613 was used in the 1000 Hz oscillator in the original model of this tester, almost any high-gain transistor may be used.

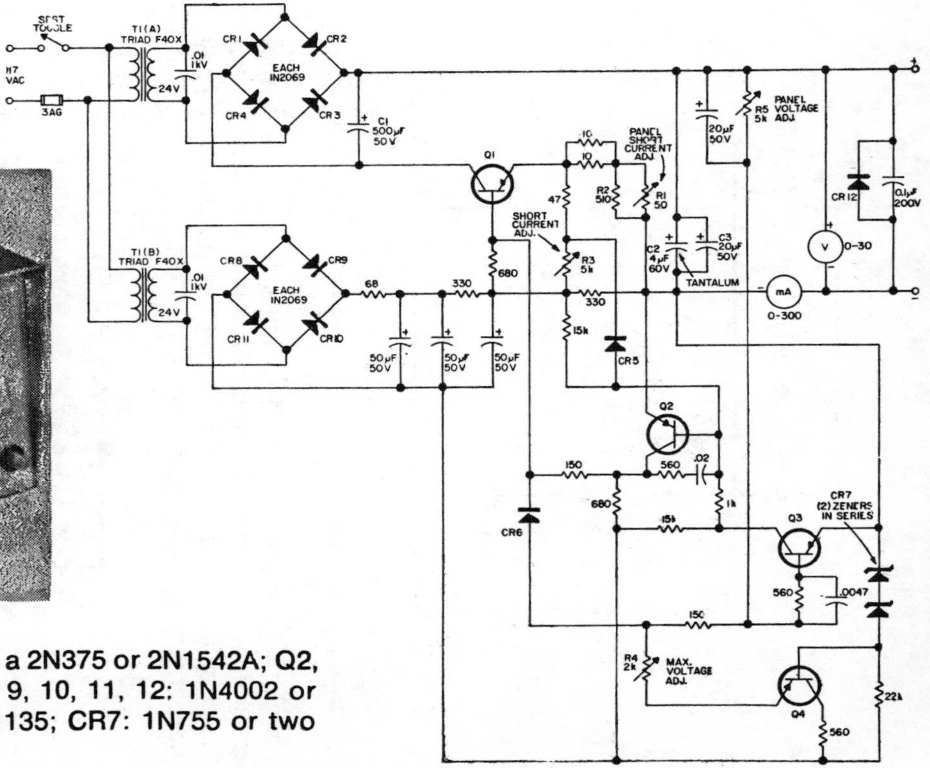


Laboratory-type power supply. Q1 is a 2N375 or 2N1542A; Q2, Q3, and Q4: 2N508; CR1, 2, 3, 4, 8, 9, 10, 11, 12: 1N4002 or 1N2069; CR5 and CR6: HB5 or FD1135; CR7: 1N755 or two 1N468s in series.

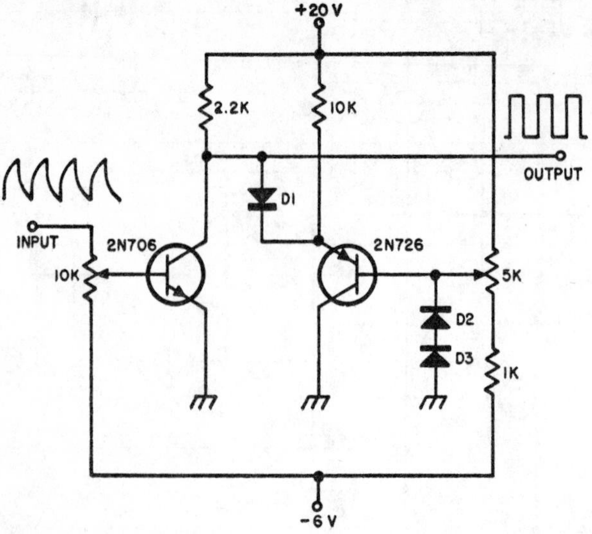


ALL DIODES GENERAL PURPOSE SILICON

The square wave output of many inexpensive signal generators deteriorates quite badly at high frequencies, but this circuit will square them off again. The diodes may be any inexpensive computer type such as the 1N914.

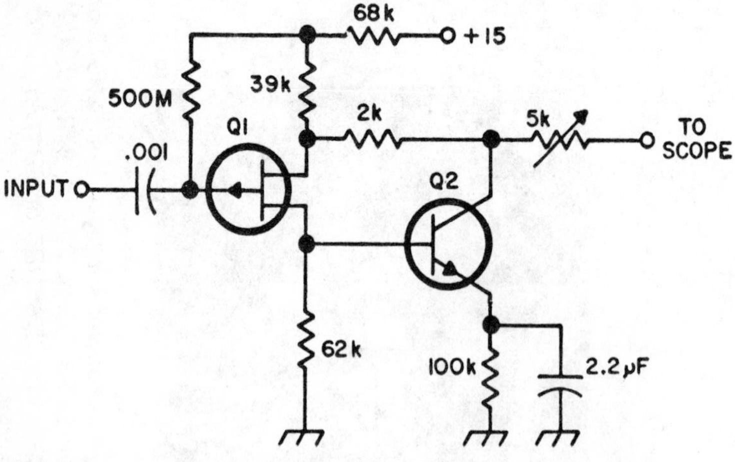


This high-impedance probe provides about 1200 megohms input impedance with unity gain. Upper frequency equalization is provided by the 5K pot. Q1 is a U112, 2N2607, 2N4360; Q2 is a 2N706, 2N708, 2N3394, or HEP 50.

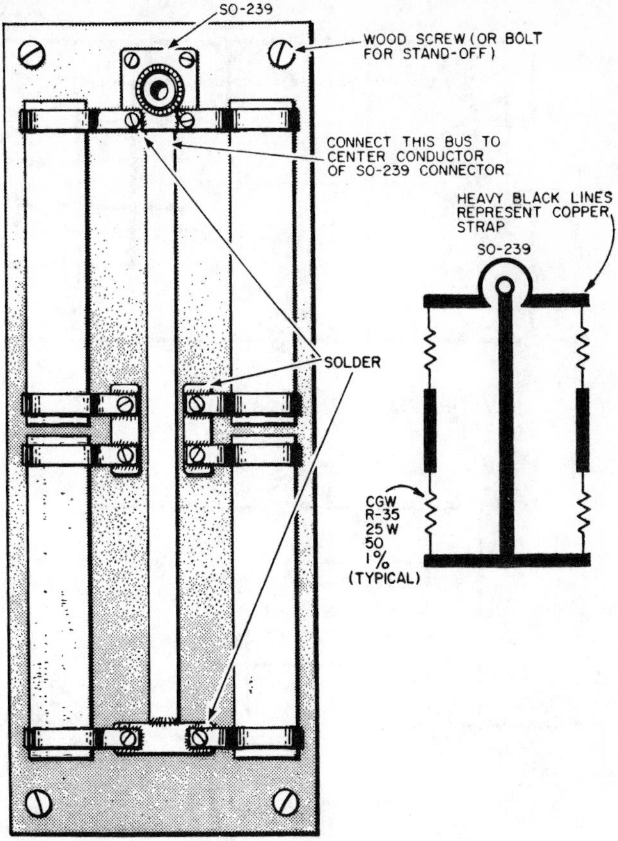


Diagram shows how to connect four noninductive 25W 50-ohm resistors to make a 100W dummy load. Resistors are available surplus; one source is Meshna Electronics, Lynn, Mass. Board shown here is 4 × 10 inches; it may be Bakelite or fiberglass.

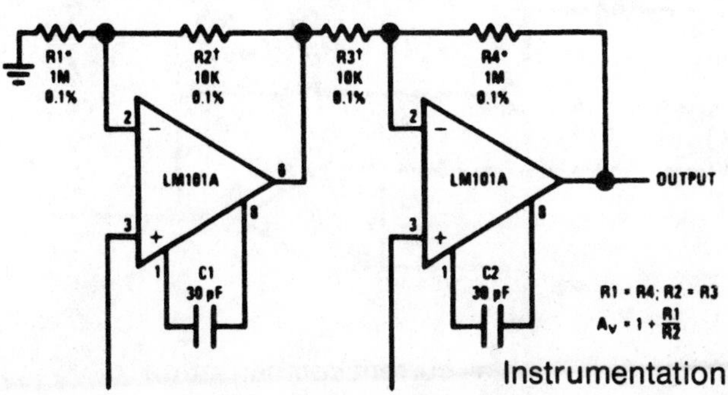


Instrumentation amplifier.

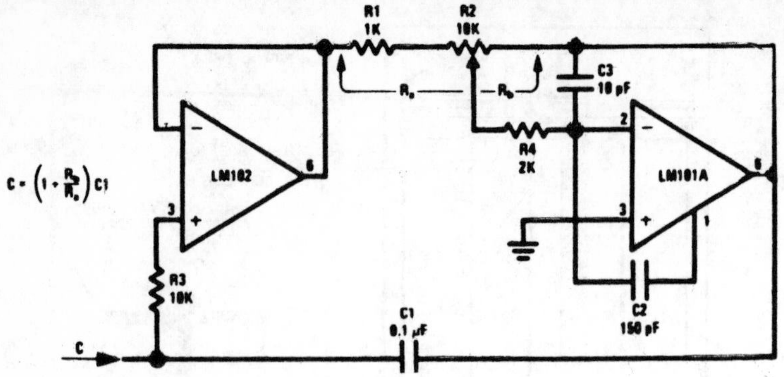


Variable capacitance multiplier.

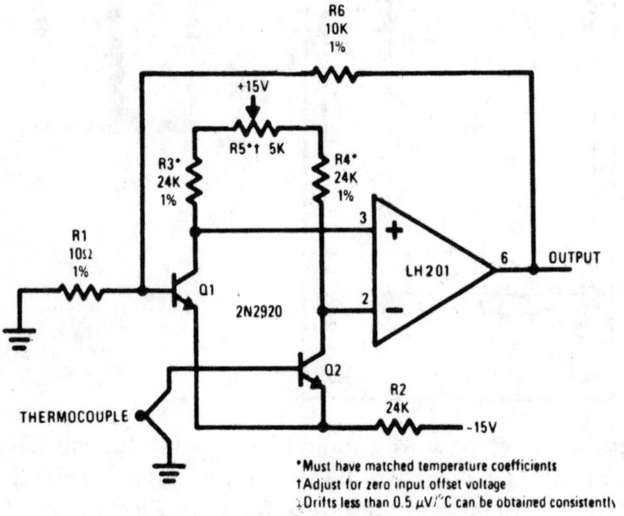


Low-drift thermocouple amplifier.

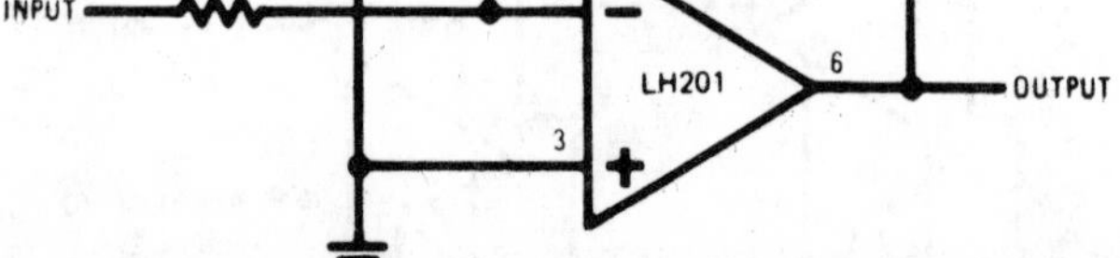

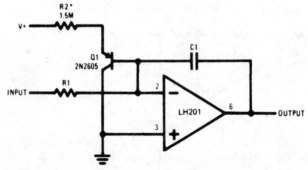

Integrator with bias current compensation

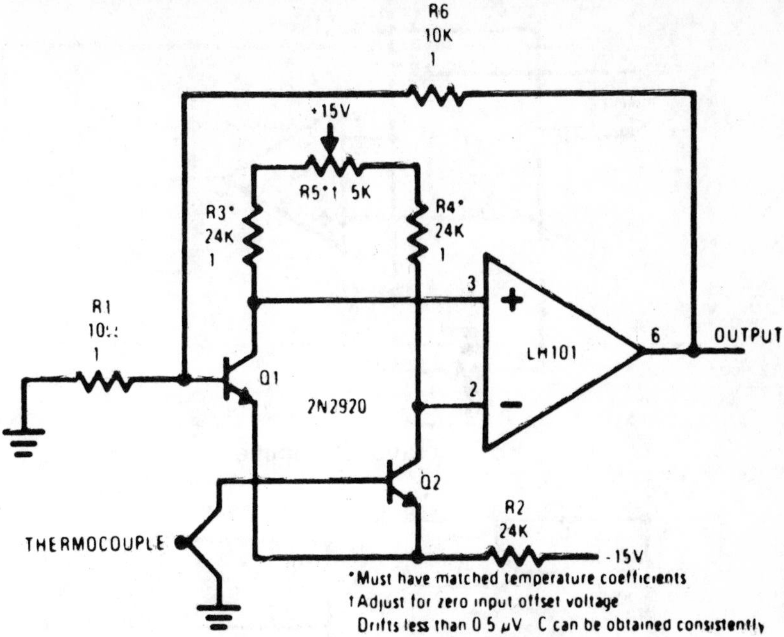


Low-drift thermocouple amplifier.

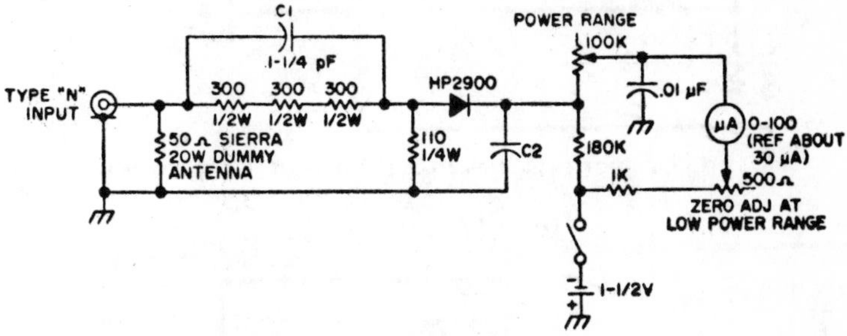


.025 to 10 watt rf wattmeter.

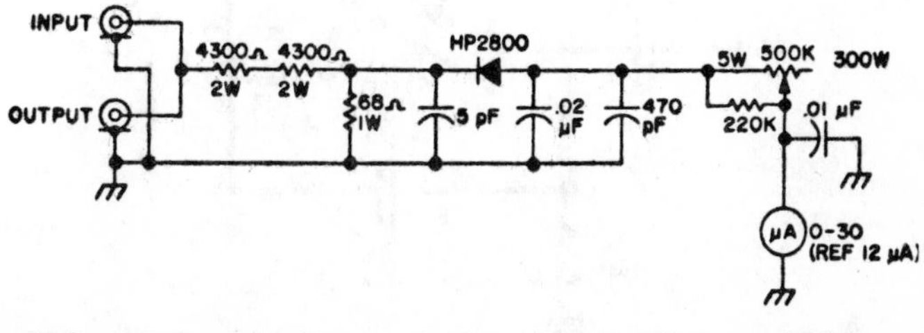


5 to 300 watt rf wattmeter metering circuit. (External 300 or 400 watt dummy antenna load.)

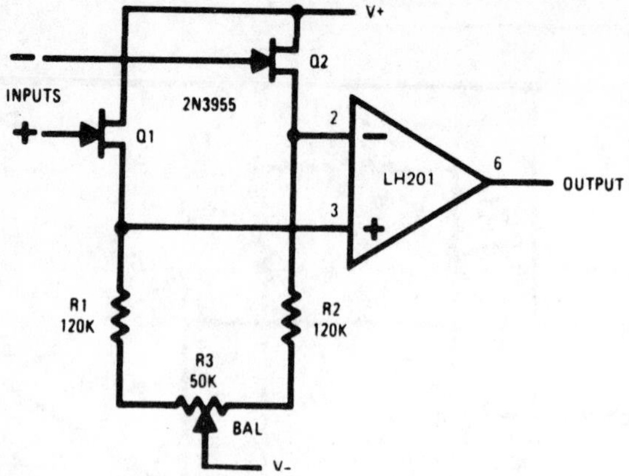


FET operational amplifier.

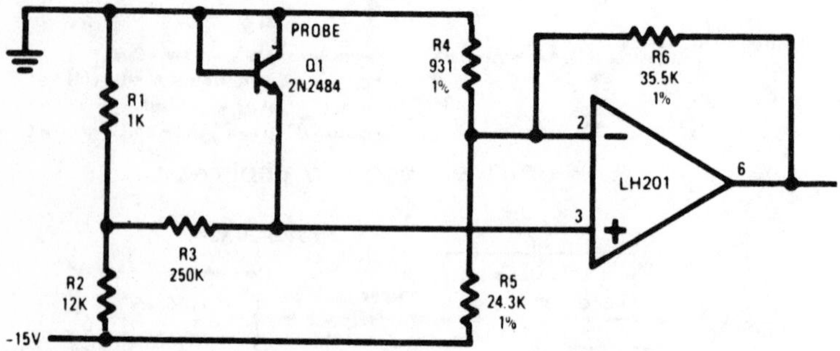


**Pin connections shown are for metal-can package.

Temperature probe.

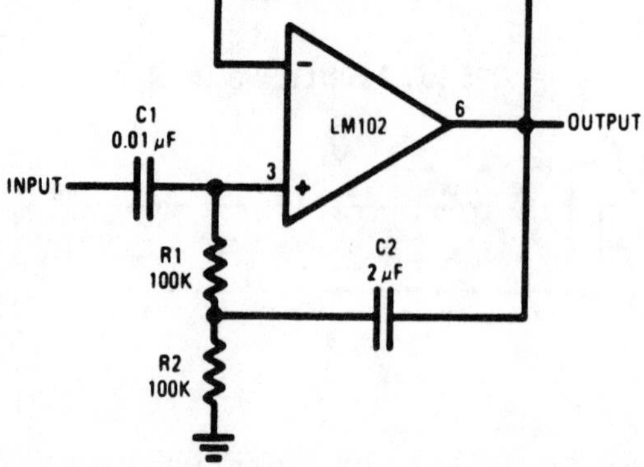


High input impedance ac amplifier.

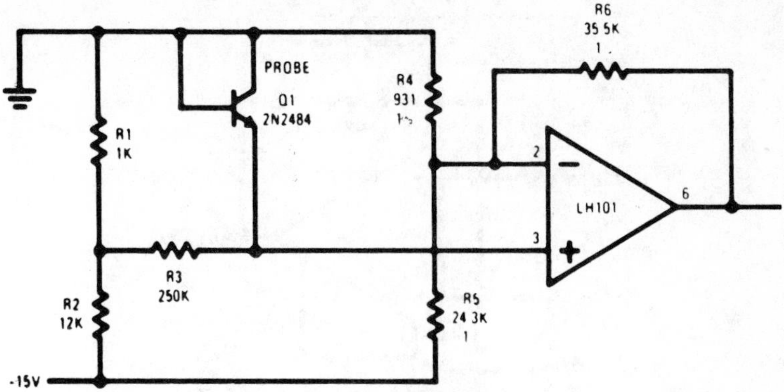


** Pin connections shown are for metal-can package.

Temperature probe.

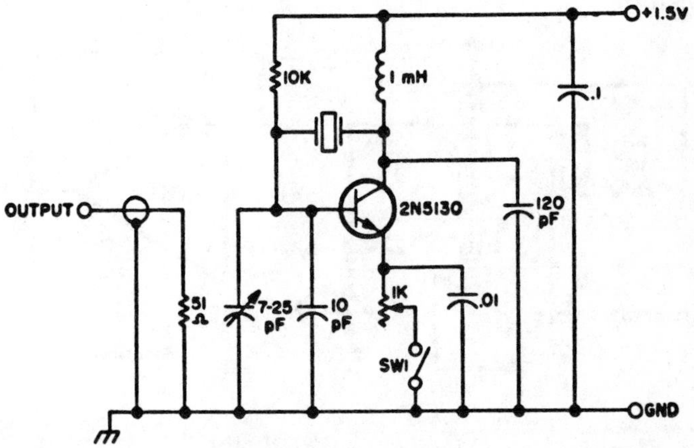


Schematic diagram of the VE3GFW signal source, designed by club members of the Toronto FM Communications Association.

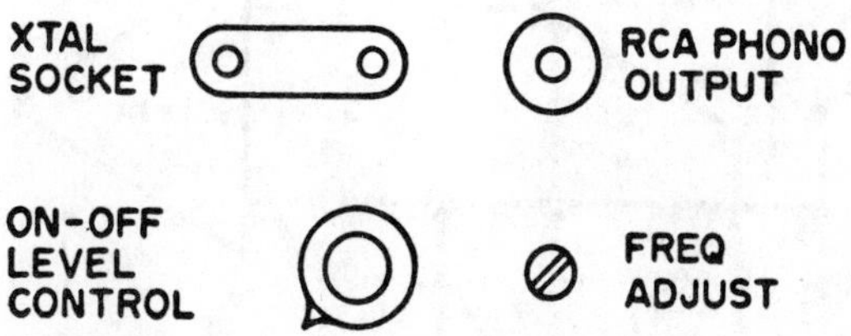


Minibox layout of the VE3GFW signal source panel.

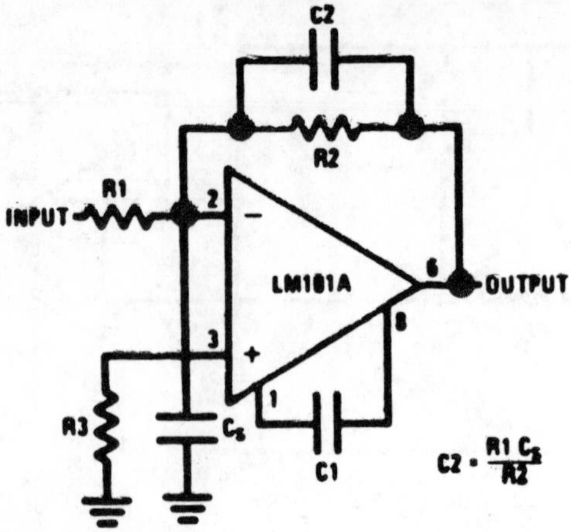


Compensating for stray input capacitances or large feedback resistors.

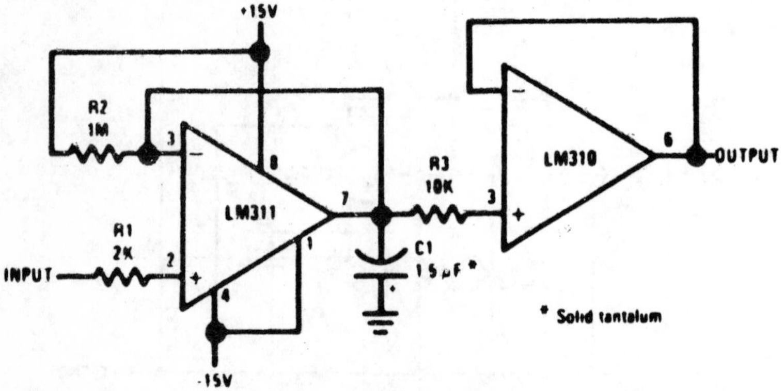


Negative peak detector.

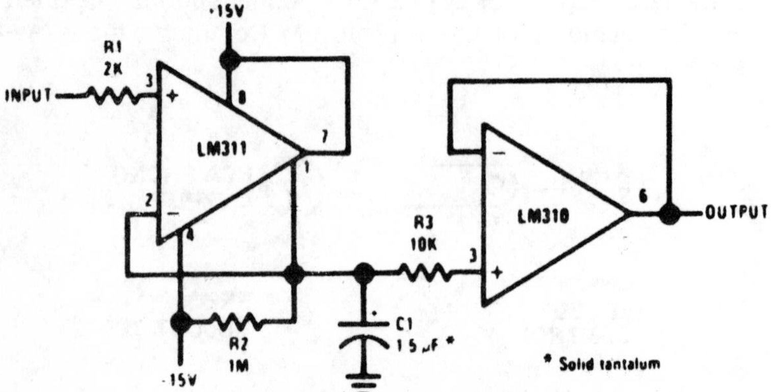


Positive peak detector.

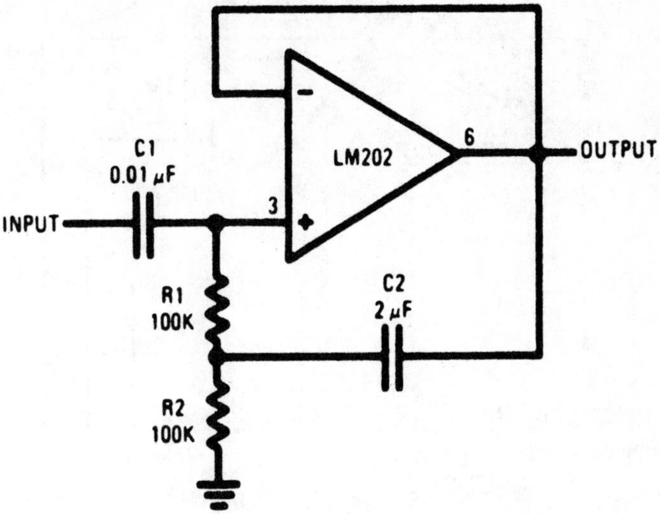


High input impedance ac amplifier.

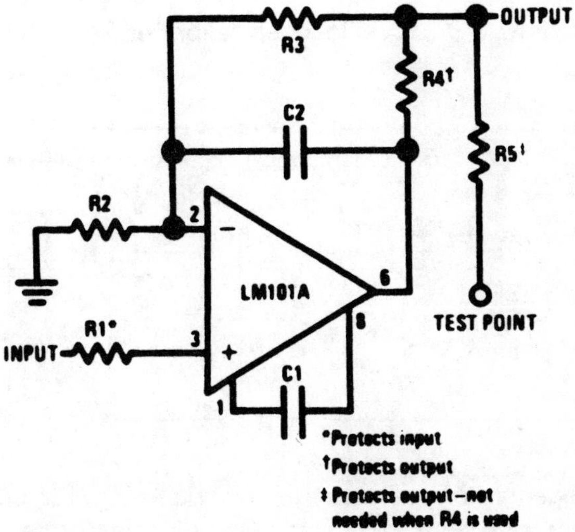


Protecting against gross fault conditions.

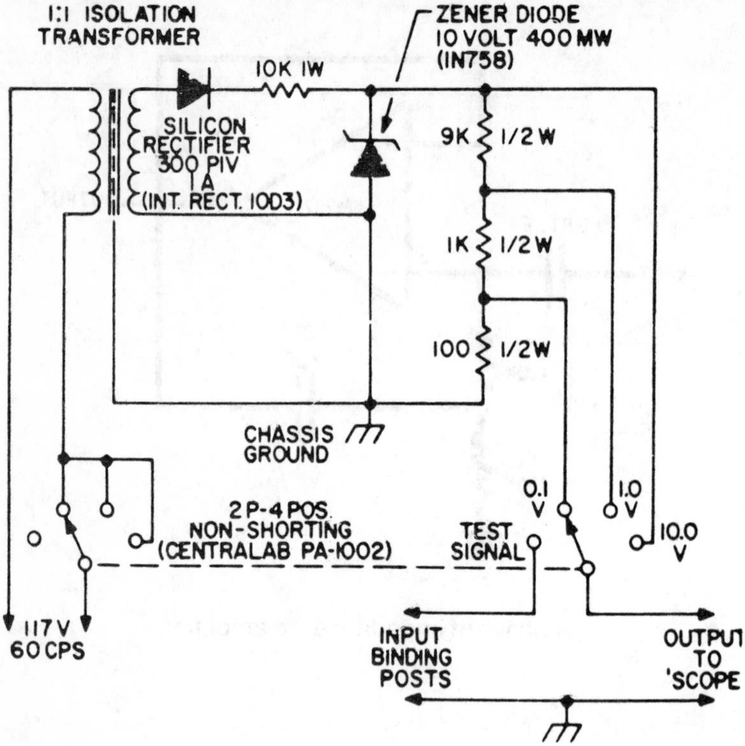


Oscilloscope calibrator.

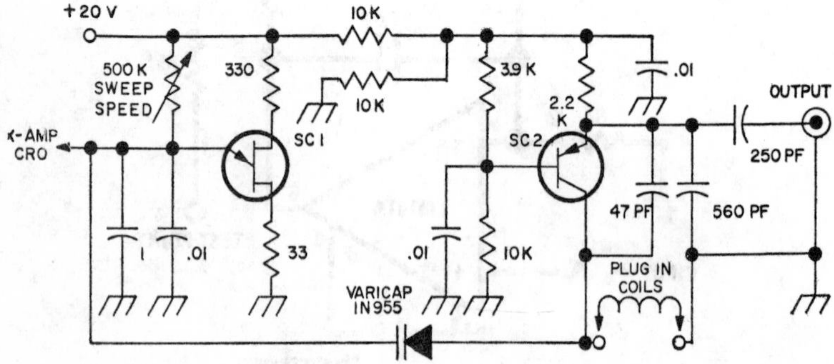


Sweep generator operates from 60 kHz to 60 MHz, depending on coil used. For 455 kHz, use i-f transformer with its capacitor removed. Connect series resistance in output to attenuate signal.

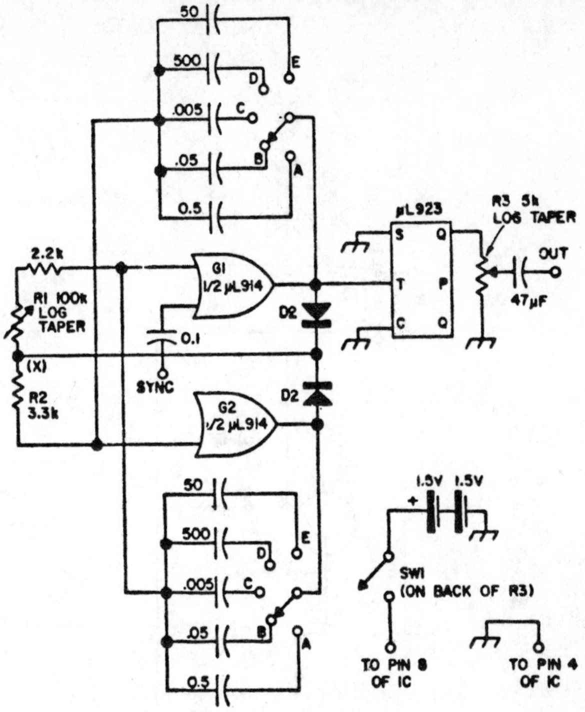


Complete square-wave generator. Bandswitching capacitors are 10%, or better, tolerance. Resistors are ¼W.

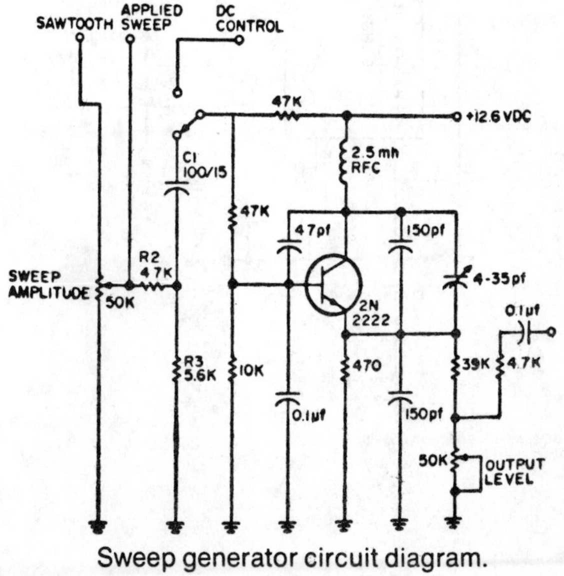


Sweep generator circuit diagram.

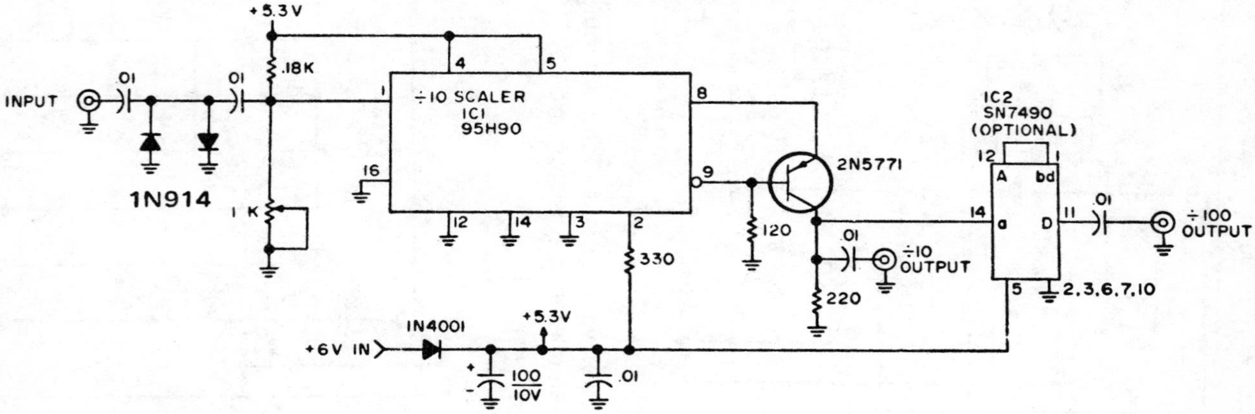


The VHF scaler schematic diagram. All 0.01 capacitors are discs. All resistors are ¼W, 10%.

# Section M

# Control & Tone Circuits, CORs, Repeater Circuits

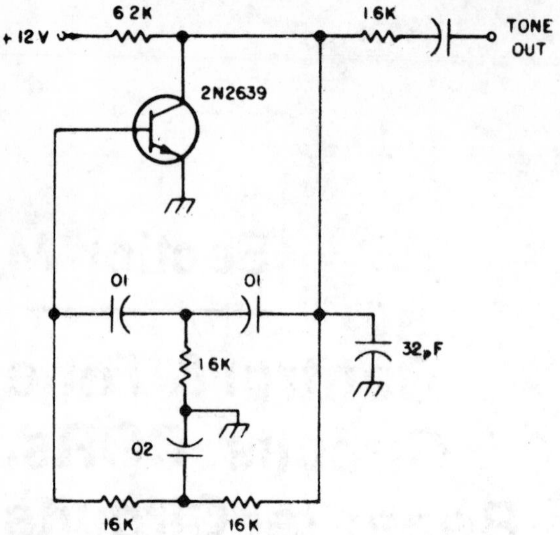


Single-transistor tone oscillator for tone-burst or whistle-on repeater-access use produces 1750 Hz at sufficient amplitude for most transmitters.

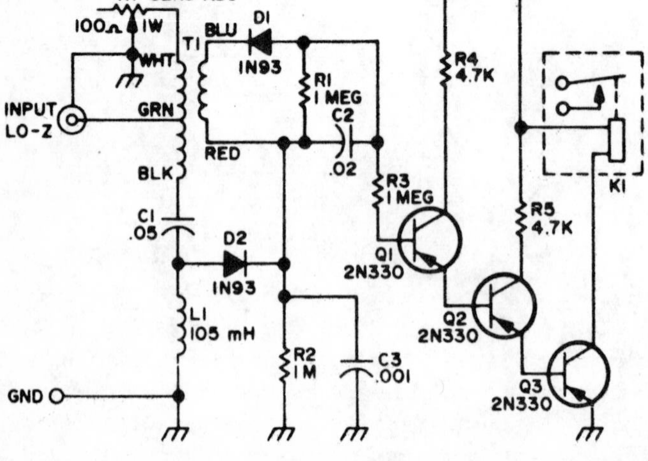


This single-tone decoder has a high degree of selectivity, stability, and reliability.

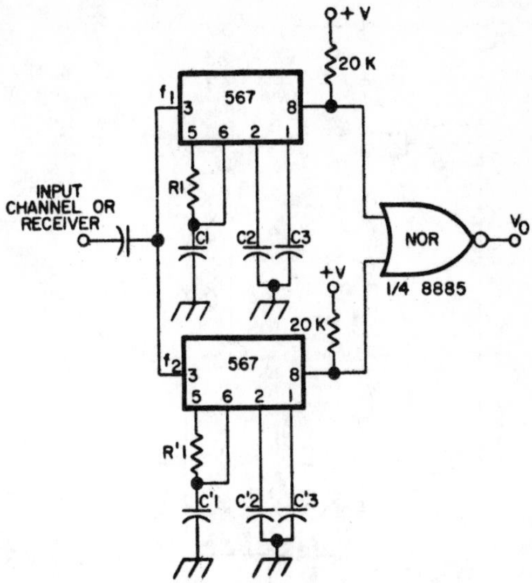


Resistor and capacitor values chosen for desired frequencies and bandwidth. If $C_3$ is made large so as to delay turn-on of the top 567, decoding of sequential ($f_1$, $f_2$) tones is possible.

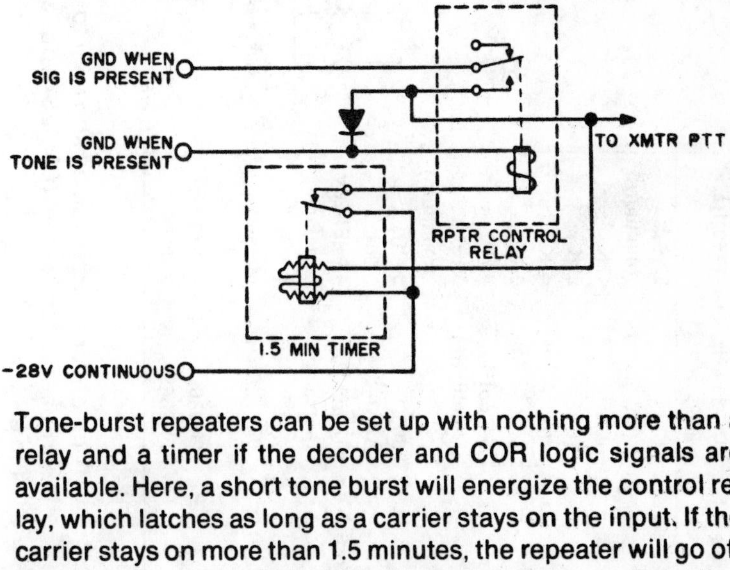


Tone-burst repeaters can be set up with nothing more than a relay and a timer if the decoder and COR logic signals are available. Here, a short tone burst will energize the control relay, which latches as long as a carrier stays on the input. If the carrier stays on more than 1.5 minutes, the repeater will go off the air and a new tone will be required. Each transmission must be accompanied by the proper tone burst.

| Component | 87—112 Hz range | 1750 Hz |
|---|---|---|
| R1 | 10 pot (multiturn) | 50K pot (multiturn) |
| R2, 3, 5 | 10K | 47K |
| R4 | 10K | 22K |
| R5 | 10K | 47K |
| R6 | 5.1K | 5.1K |
| R7 | 5K pot (multiturn) | 4.7K |
| | | 5K pot (multiturn) |
| R8 | 300Ω | 300Ω |
| R9 | 0.1μF | 0.001 μF |
| C1, 2, 3 | 47 μF, 20V tantalum | 47 μF, 20V tantalum |
| C4 | | 1 μF |
| C5 | 0.01 μF | tantalum |
| Q1 | 2N930 | 2N930 |
| Q2 | 2N2369 | 2N2369 |

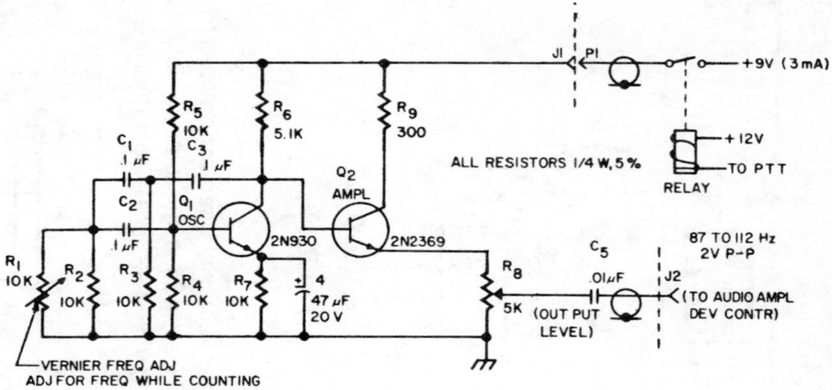


Tone encoder. Part values shown in diagram are for low-frequency continuous-tone carrier squelch system (CTCSS) applications; for single-tone refer to accompanying table.

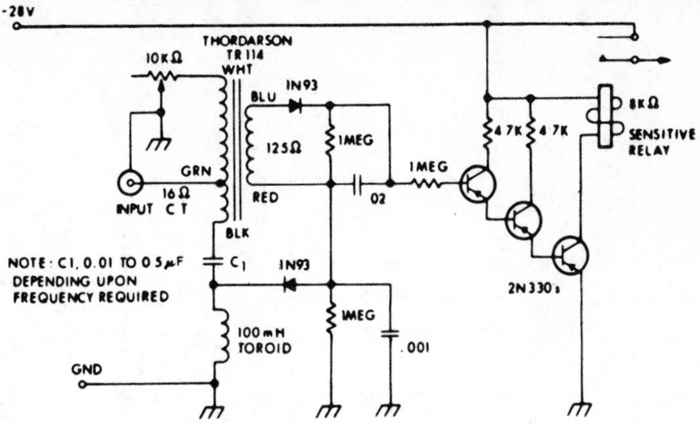


Single-tone decoder.

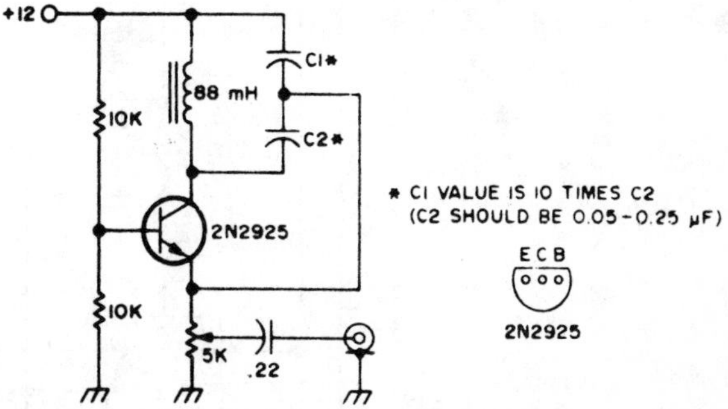


Simple tone encoder for radio remote control applications. C2 is 10 times the value of C1. For 1000 Hz, C1 = 0.22 μF, and C2 is 2 μF. Both capacitors should be nothing less than mica for stability.

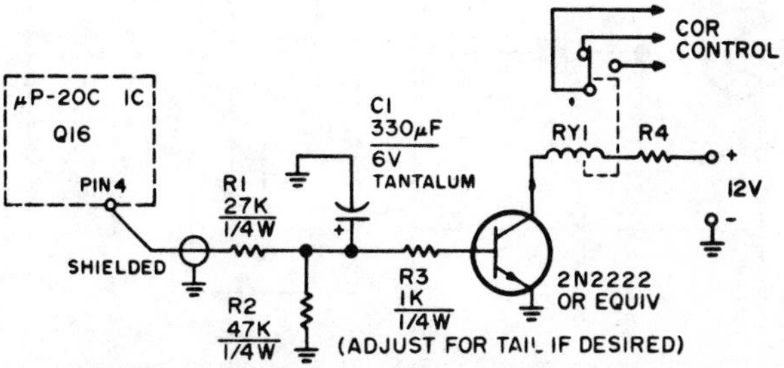


Schematic of a COR circuit for the TR22. R4 must be adjusted to keep the collector current of the transistor less than 750 mA. The value is dependent on the relay resistance. The R2/C1 combination controls tail time.

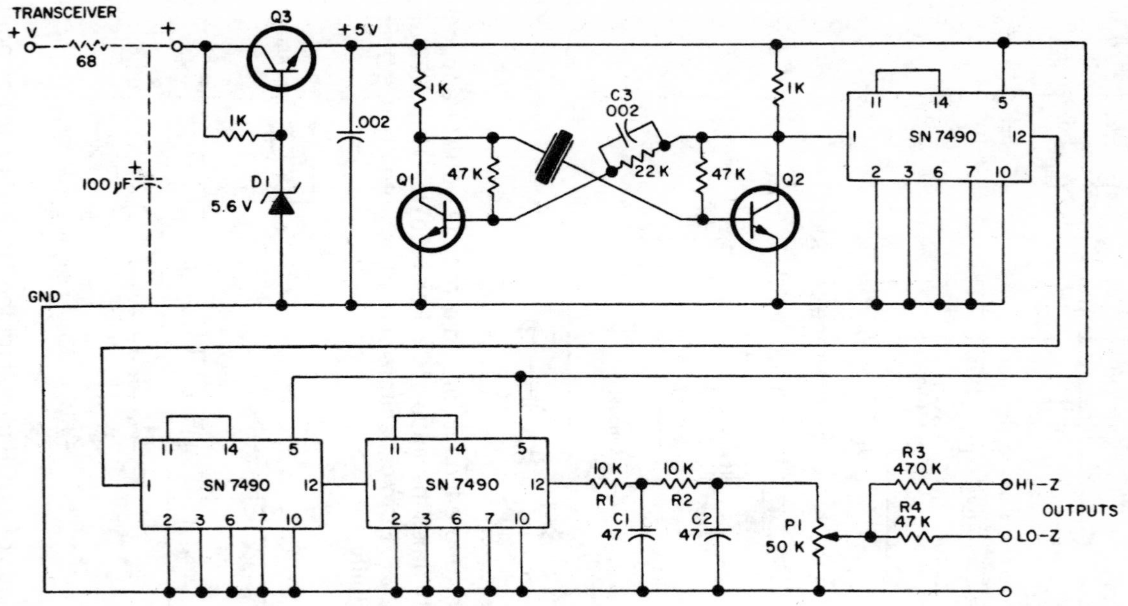


Schematic of subaudible tone generator. Q1, 2—MPS6513; Q3—2N1613.

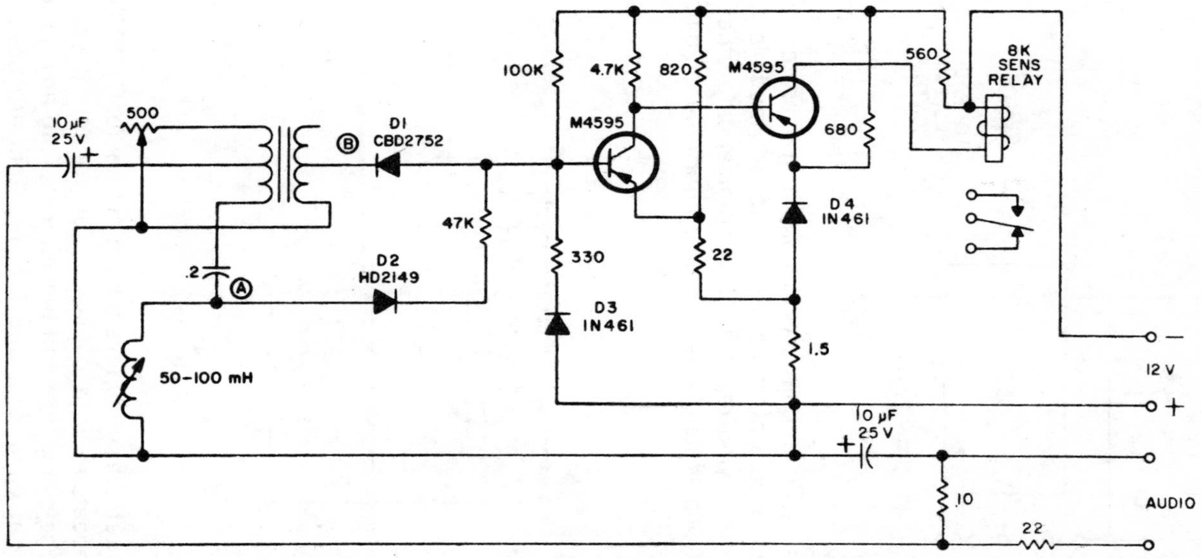


This audio decoder responds to signals in the frequency range of 2200–2900 Hz. Although Motorola part numbers are shown, equivalent values may be substituted.

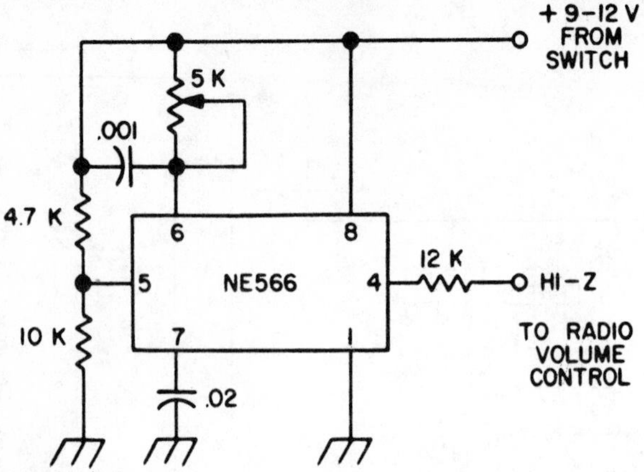


Simple tone oscillator connects to wiper arm of any portable radio's volume control to convert radio into audible tone generator—for keying repeaters and commanding other remotely controllable functions by holding portable radio up to microphone.

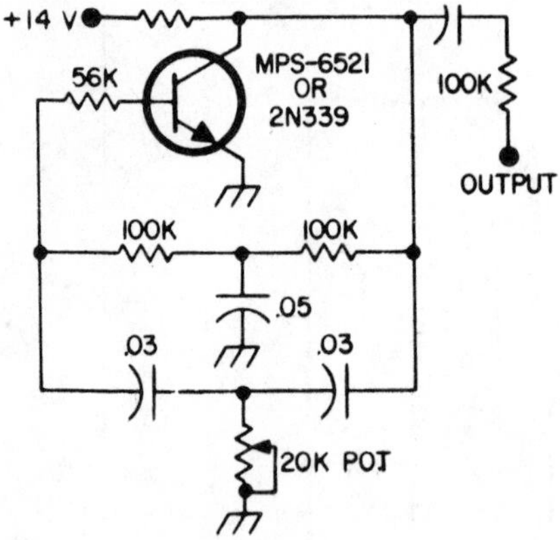


A simple and stable subaudible tone generator for CTCSS use with FM repeaters. With miniature components it can be made postage-stamp size and tucked away into any rig. For stability, Mylar capacitors and film resistors are best, but carbon resistors can be substituted successfully. It was originally designed for 100 Hz output and the 20K potentiometer is used for adjusting this to optimum.

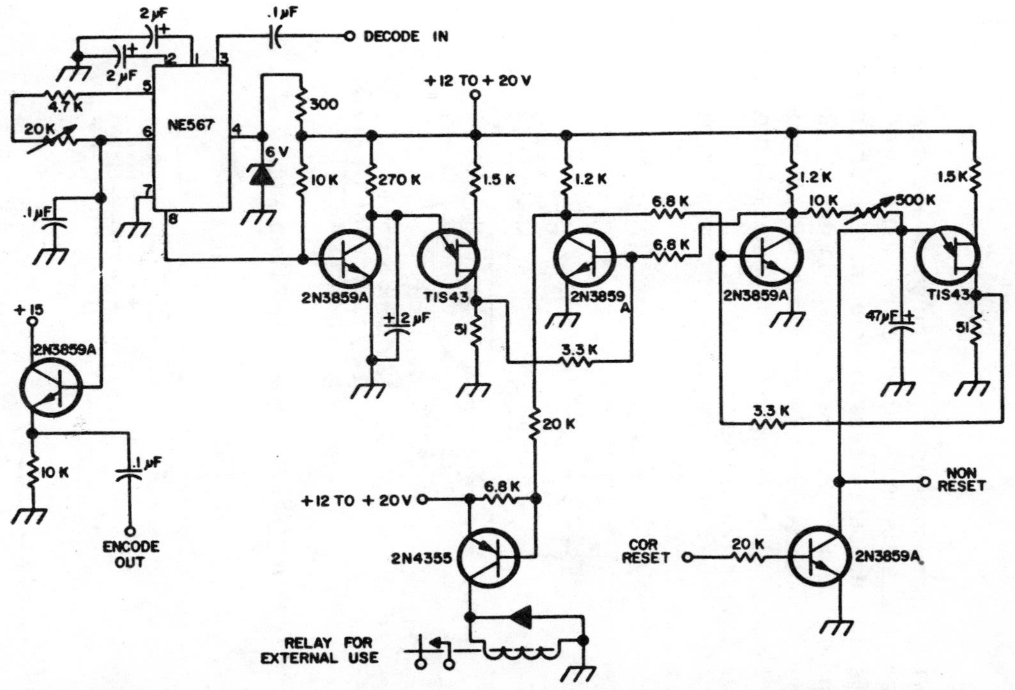


Tone decoder for repeaters does not false-trip. Transistors are 2N3859A types unless otherwise noted.

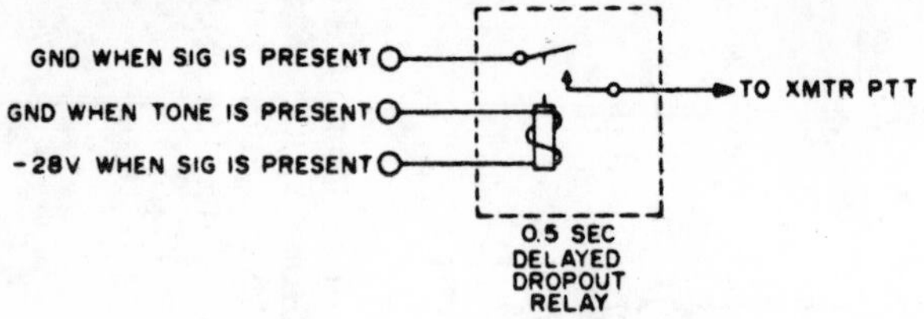


There are many methods for keying continuous-tone-carrier-squelch (PL) systems, but those employing delayed dropout relays are the most successful. Here, the tone and the signal must be present to hold the repeater on the air. Momentary tone variations because of weak signal will not cause "cycling" because of the delay, but the repeater will drop out instantly if the carrier itself drops out.

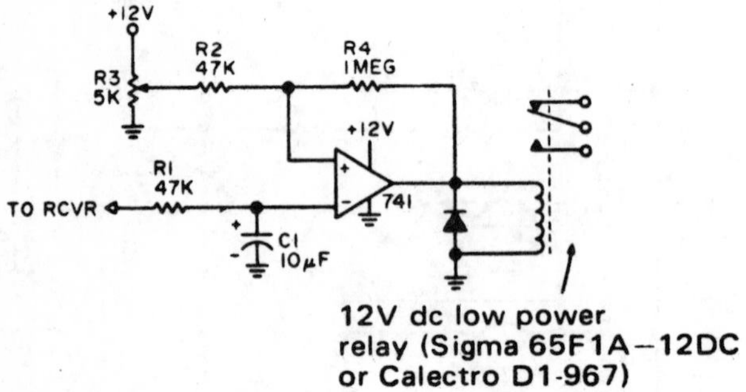


A COR circuit using the 741 op-amp.

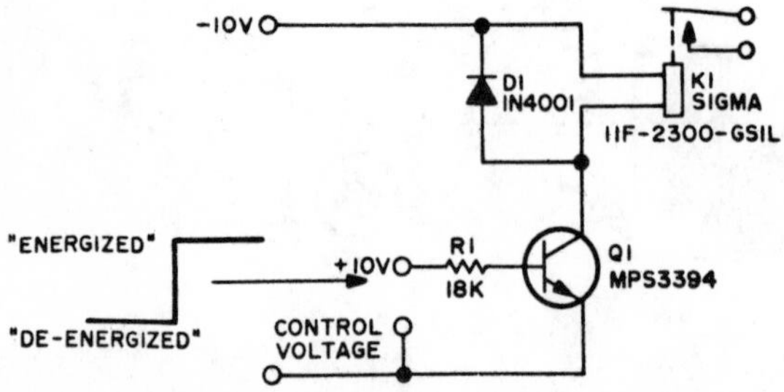


Electronic control of a dc relay. Circuit courtesy Motorola Semiconductor Power Circuits Handbook.

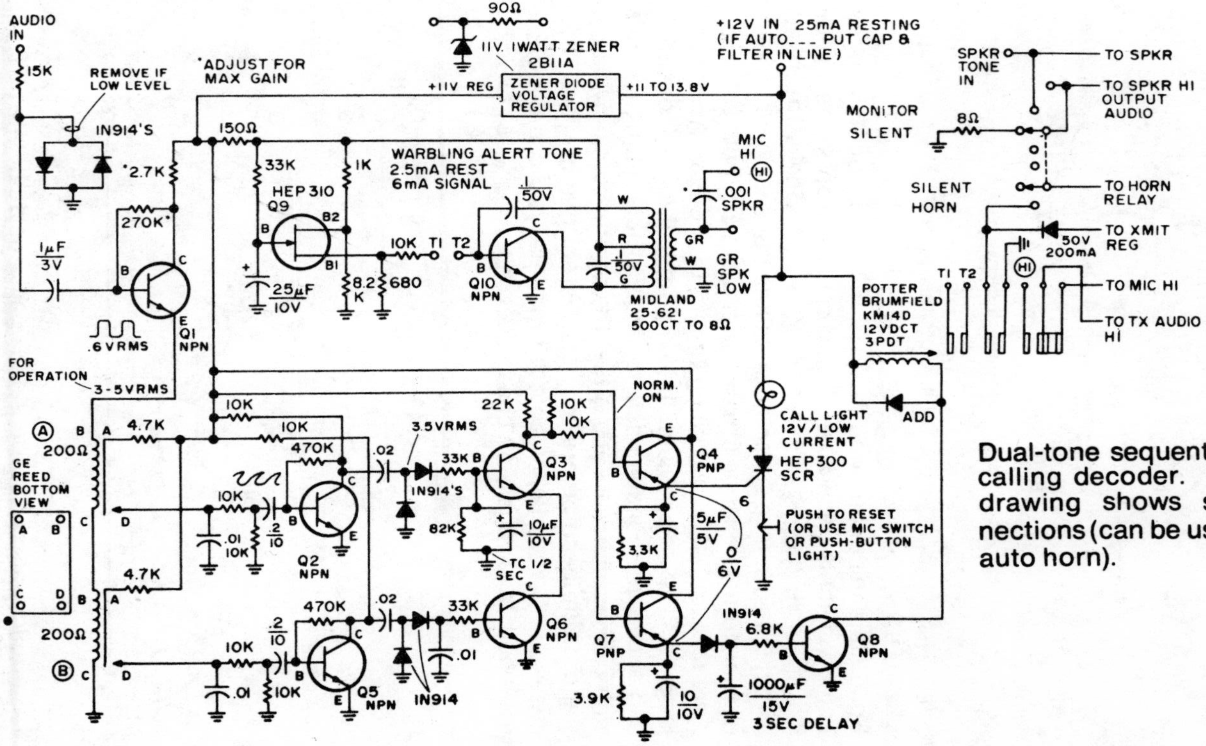


Dual-tone sequential selective calling decoder. Upper-right drawing shows switch connections (can be used to sound auto horn).

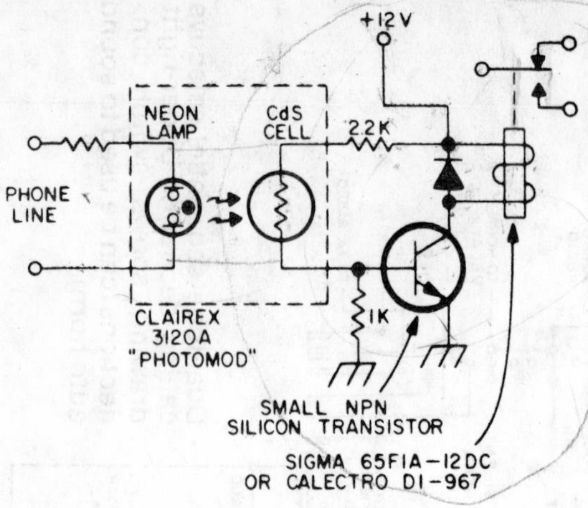


Solid-state telephone ring relay. Ring signal from telephone company lights neon lamp, which causes drop in resistance in CdS cell, turning on transistor, which closes the relay. Circuit has two nice advantages—since there is no direct connection from telephone line to power supply circuit, you don't have to worry about inducing hum into the line. Also, neon lamp acts like an open circuit when not lit (below about 65V), and above that voltage has a resistance quite high (in series with 220K)—all this means is that the phone company has to stand on its head before they can detect this on your line. Instead of Clairex lamp/photocell module, you can use NE-2 neon bulb taped against a cheap CdS cell.

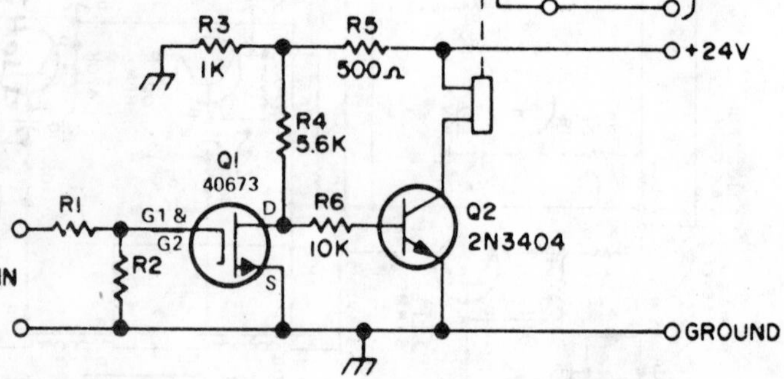


Solid-state carrier-operated relay uses dual-gate FET.

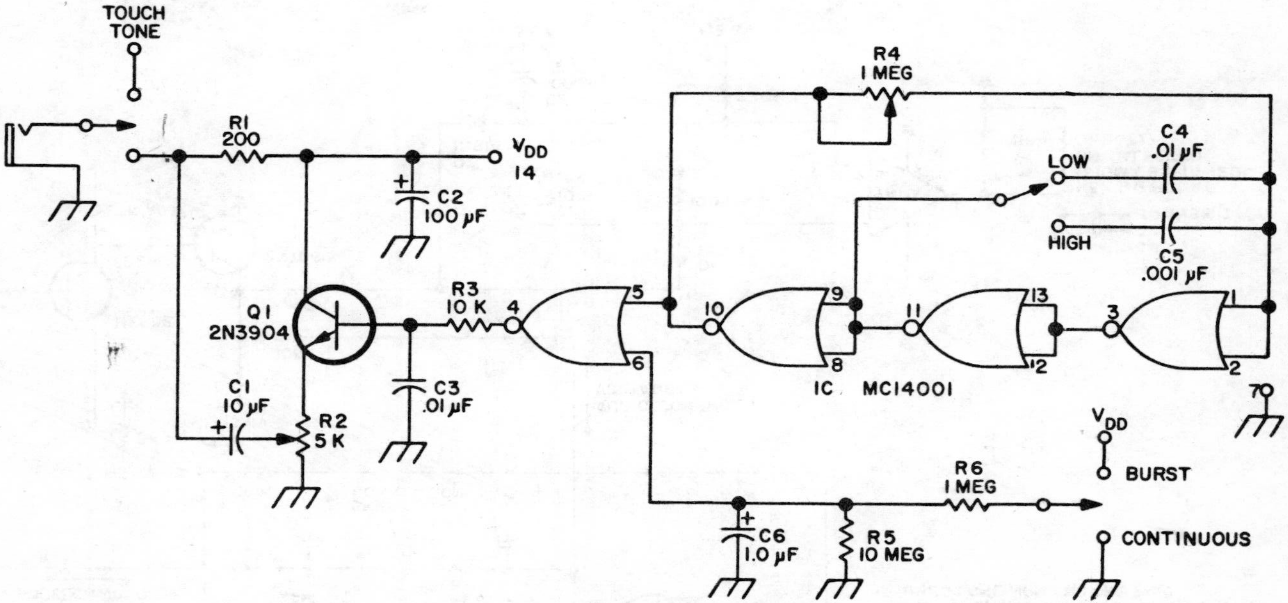


Tone generator features CTCSS (low-frequency) or burst (high-frequency) outputs. All gates are part of Motorola MC14001 integrated circuit.

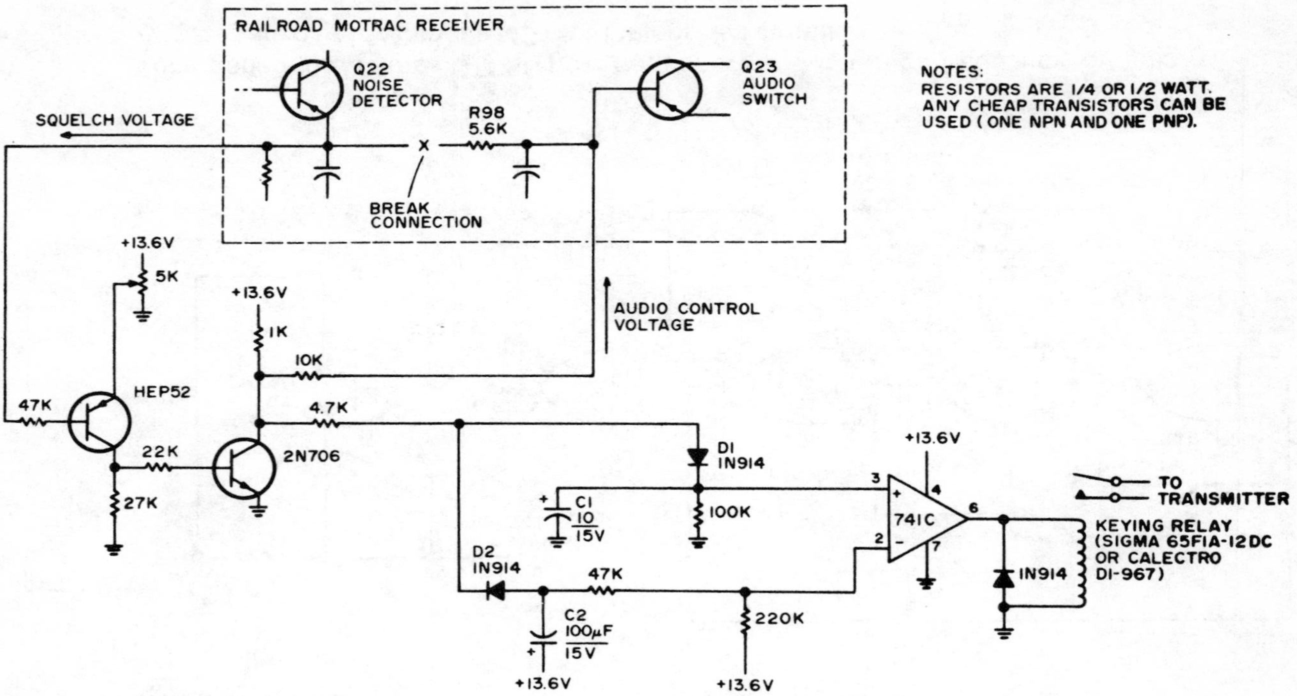


Carrier-operated relay and 3-minute limit timer for repeater control applications.

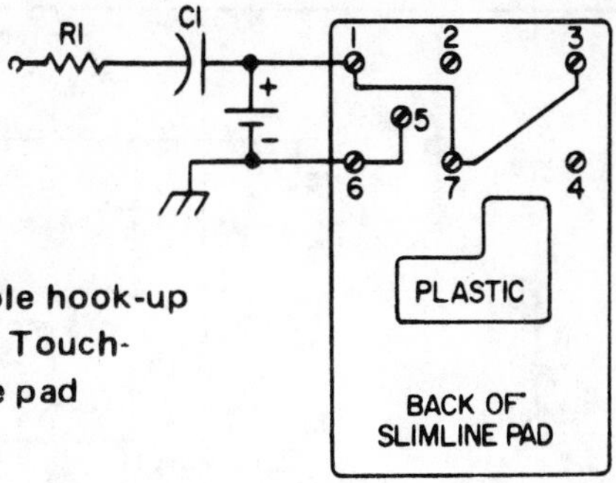


Simple hook-up for a Touch-Tone pad

1½V hearing aid battery mounted inside handset. C1 is approximately a 0.5 – 3 μF, with 1½V.

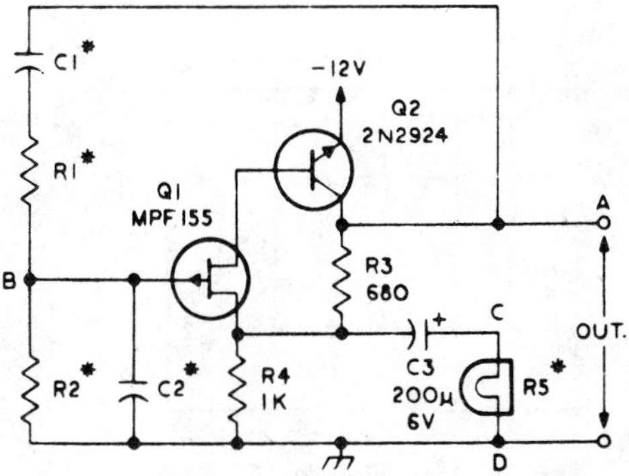


Fixed-frequency oscillator. Components marked with asterisk: C1 & C2 are 0.001 μF, and R1 & R2 are 15.8K for audio frequency of 10 kHz.

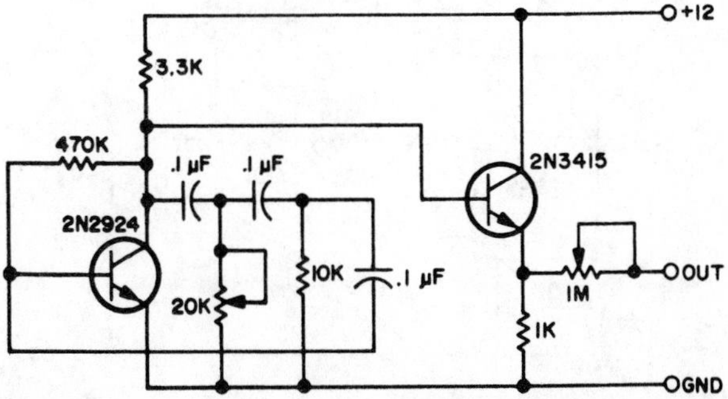


Continuous-tone encoder produces about 100 Hz for superimposition on rf carrier, for CTCSS applications.

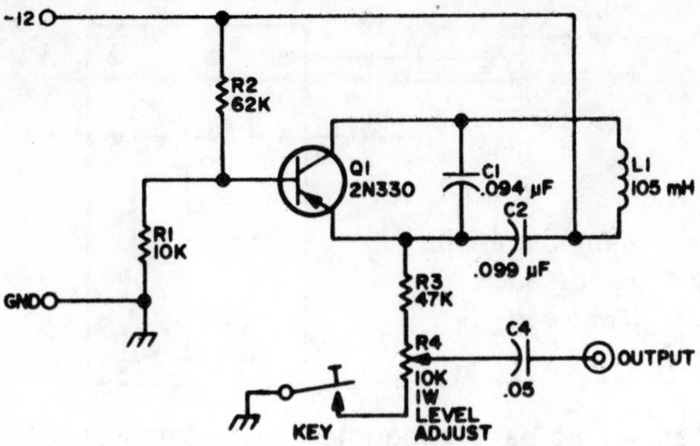


K6ASK single tone encoder.

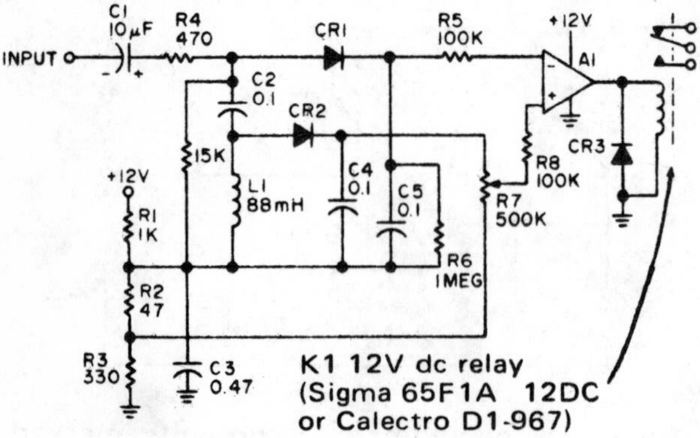


1800 Hz decoder.

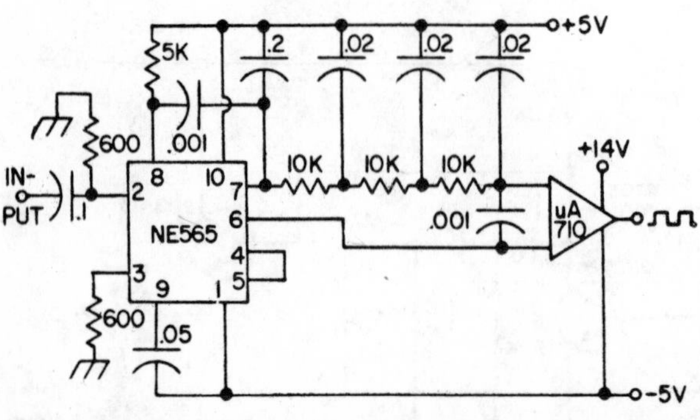


Decoder for frequency-shift keying system uses phase-locked loop and uA710 operational amplifier. Frequency of operation: 1070 & 1270 Hz.

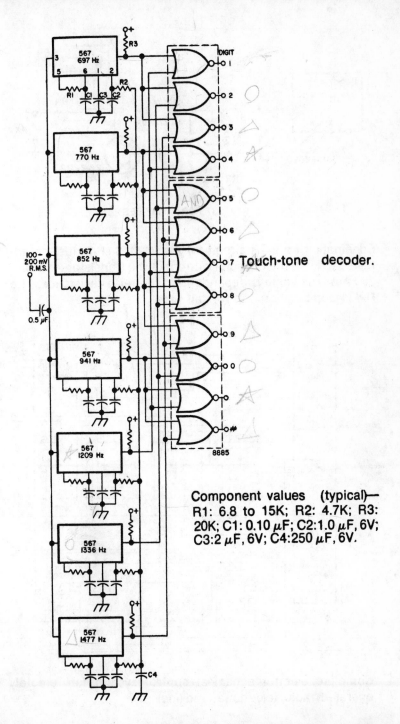

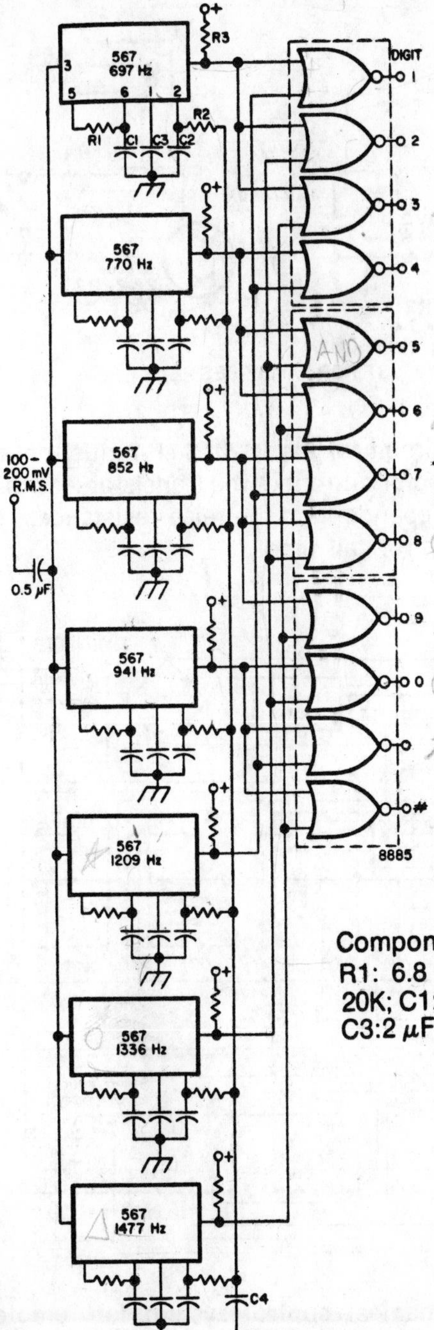


Touch-tone decoder.

Component values (typical)—
R1: 6.8 to 15K; R2: 4.7K; R3: 20K; C1: 0.10 $\mu$F; C2:1.0 $\mu$F, 6V; C3:2 $\mu$F, 6V; C4:250 $\mu$F, 6V.

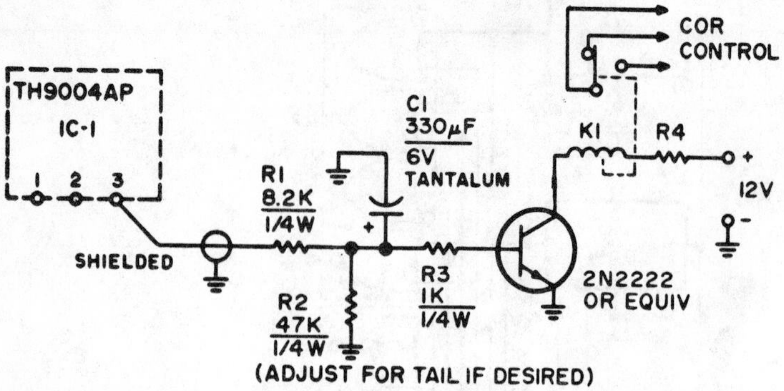


Schematic of a COR circuit for the IC20/21. R4 must be adjusted to keep the collector current of the transistor less than 750 mA. The value is dependent on the relay resistance. The R2/C1 combination control tail time.

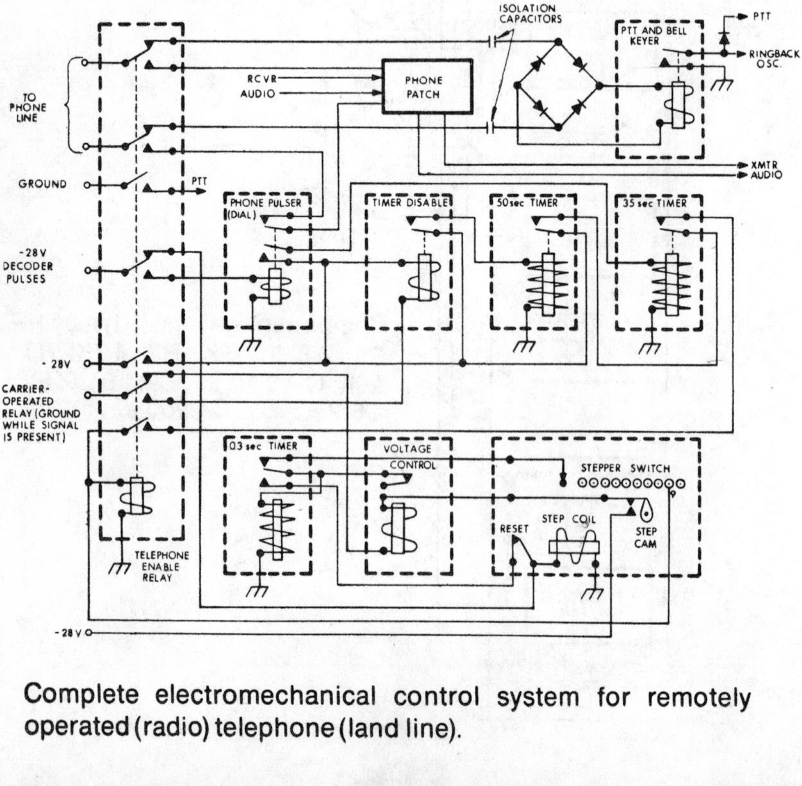


Complete electromechanical control system for remotely operated (radio) telephone (land line).

| Tone (Hz) | R3 (k-Ohms) | Tone (Hz) | R3 (k-Ohms) |
|---|---|---|---|
| 1,800 | 288 | 2,200 | 231 |
| 1,850 | 279 | 2,250 | 226 |
| 1,900 | 271 | 2,300 | 220 |
| 1,950 | 264 | 2,350 | 215 |
| 2,000 | 257 | 2,400 | 210 |
| 2,050 | 250 | 2,450 | 206 |
| 2,100 | 243 | 2,500 | 201 |
| 2,150 | 237 | 2,805 | 178.5 |

A

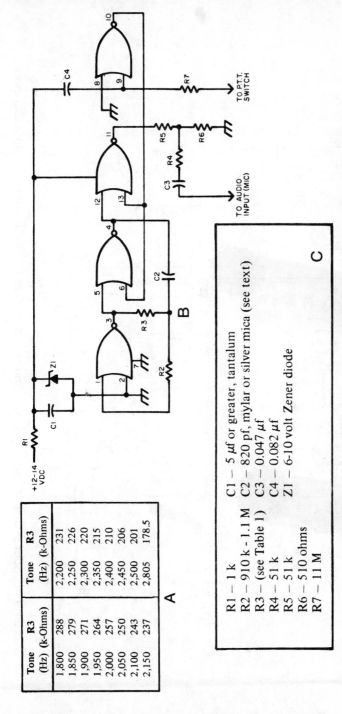

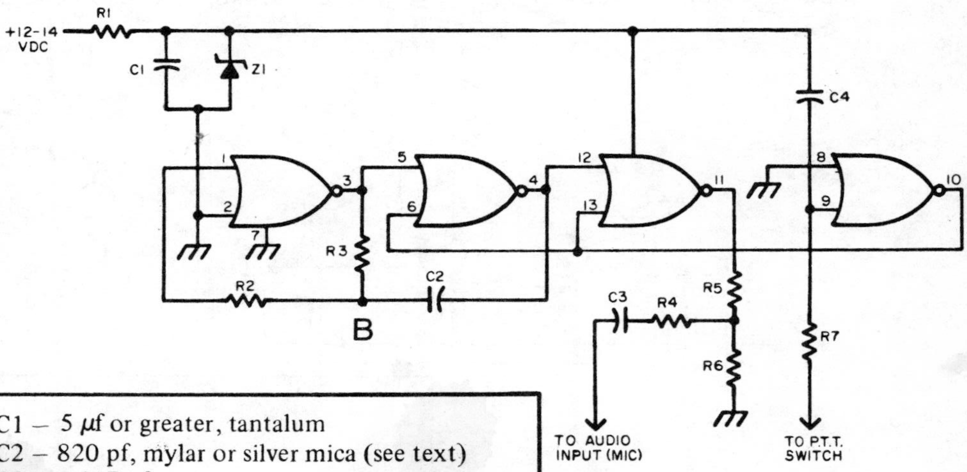


B

| | |
|---|---|
| R1 – 1 k | C1 – 5 μf or greater, tantalum |
| R2 – 910 k - 1.1 M | C2 – 820 pf, mylar or silver mica (see text) |
| R3 – (see Table 1) | C3 – 0.047 μf |
| R4 – 51 k | C4 – 0.082 μf |
| R5 – 51 k | Z1 – 6-10 volt Zener diode |
| R6 – 510 ohms | |
| R7 – 11 M | |

C

Tone generator uses MOS circuitry; gates are all part of RCA CD4001/D or CD4001/E. Unit designed for tone-burst keying when connected to transmitter. Chart shows R3 values when C2 = 820 pF.

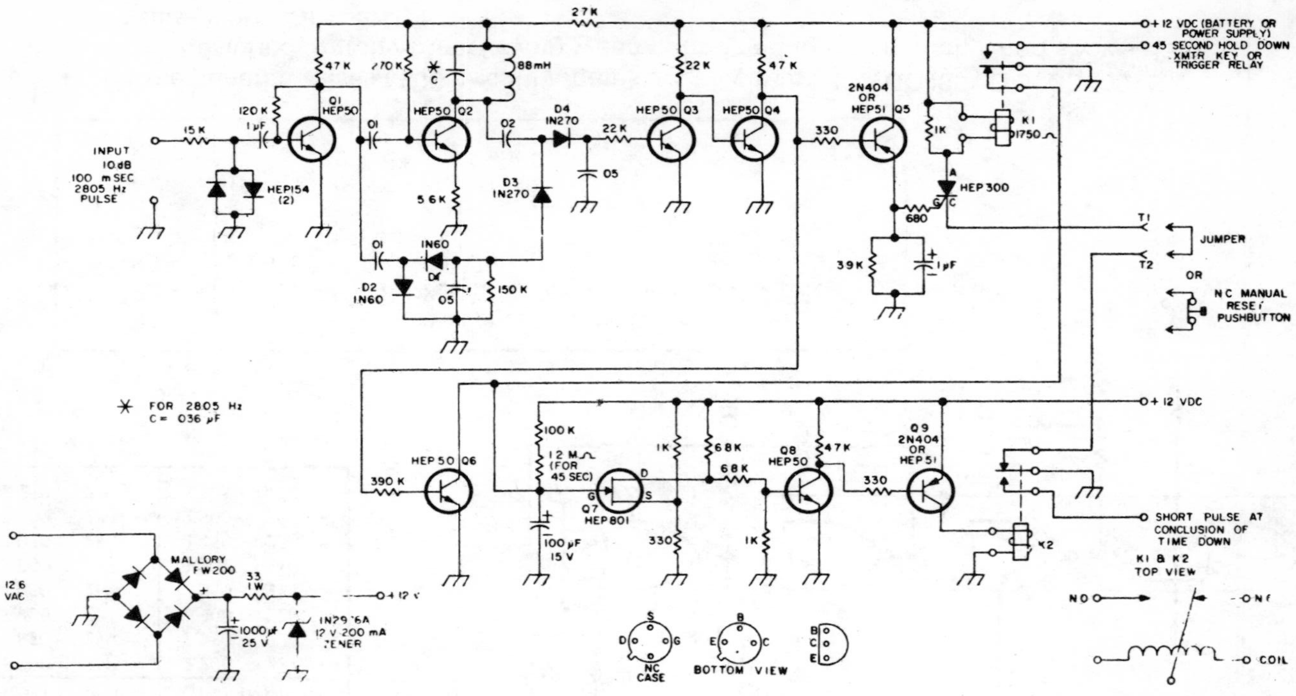


Paging decoder provides relay contact closure when 2805 Hz audio signal is applied at input.

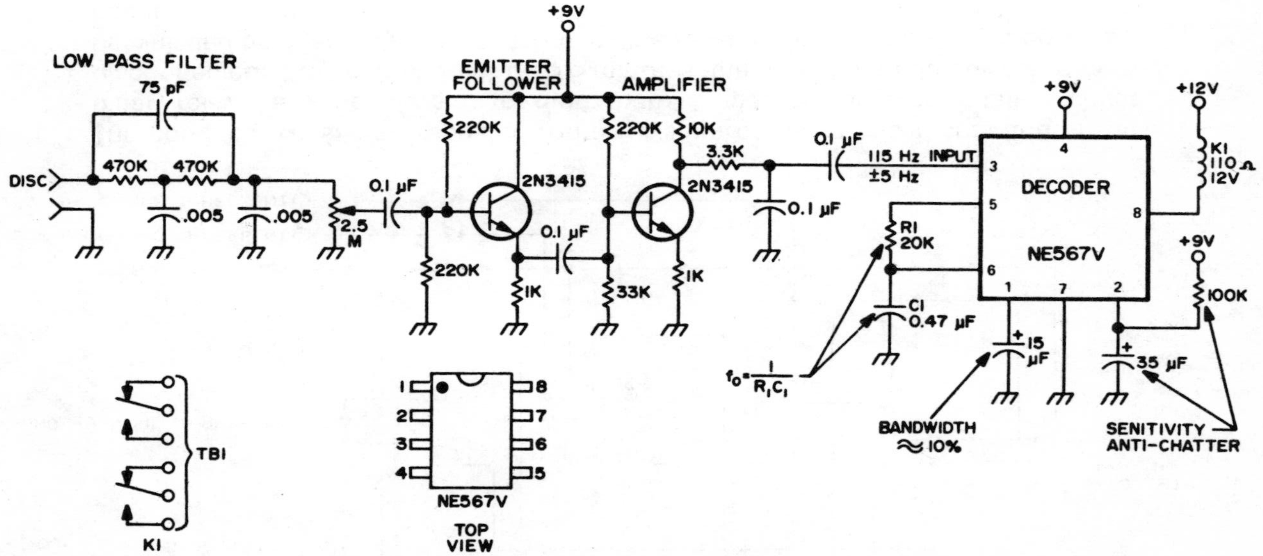


Tone decoder for CTCSS (continuous-tone carrier squelch system) applications operates from frequencies in 100 Hz range.

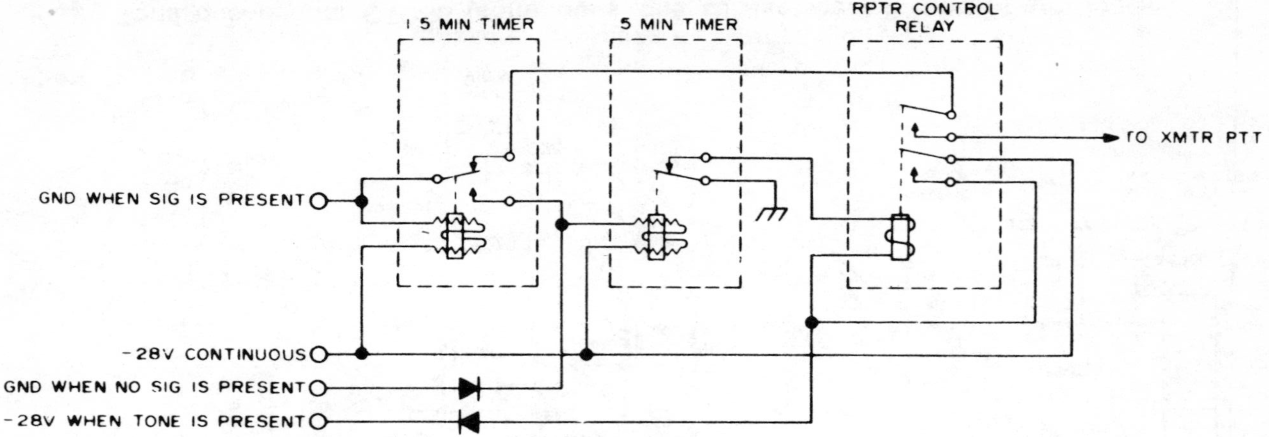


The complete whistle-on repeater control system contains two timers and an ordinary relay. If a carrier stays on for more than 1.5 minutes, the push-to-talk circuit is disconnected. if the carrier stays on 6.5 minutes, the repeater shuts down and must be whistled on again. Also, if nobody uses the repeater for 5 minutes, shutdown will occur.

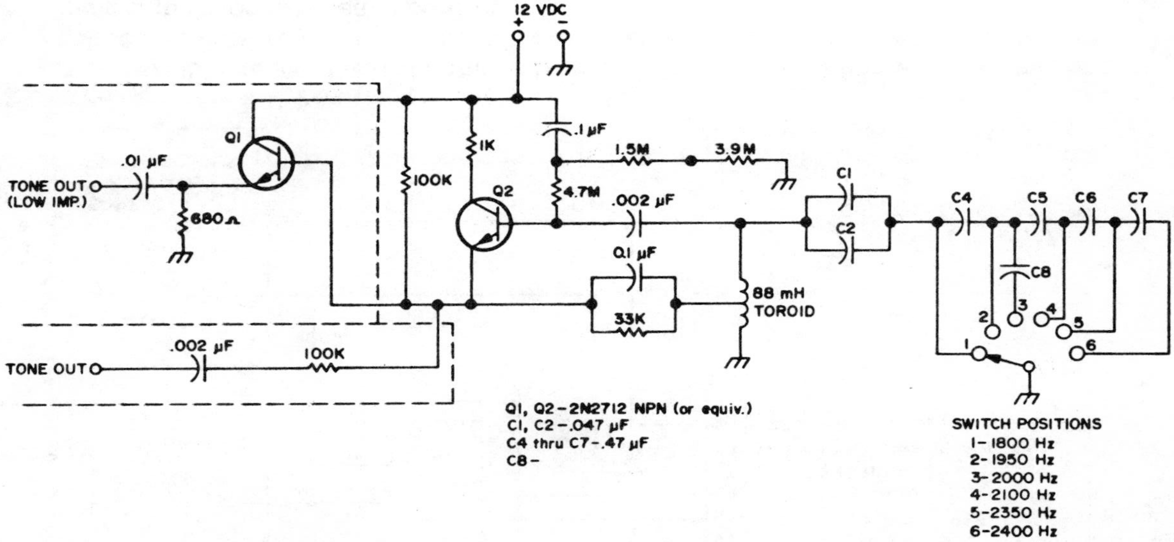


Tone-burst entry is becoming increasingly common as a requirement—repeaters in many parts of the country now require anything from 1800 Hz to 2400 Hz for entry. The circuit above is for a tone-burst generator to meet these requirements. Circuits courtesy of the Central Ohio Radio CLub FM News, September 1971.

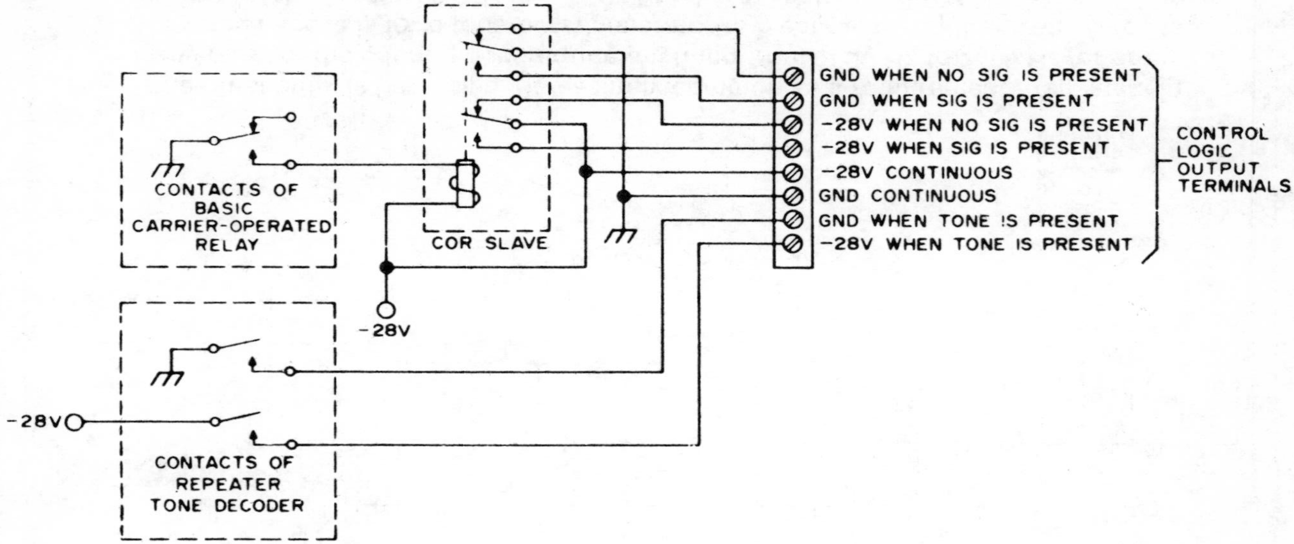


A heavy-duty relay slaved to the carrier-operated relay, along with ground and voltage outputs from the tone decoder, can be used to provide a variety of very useful logic signals for all repeater control functions.

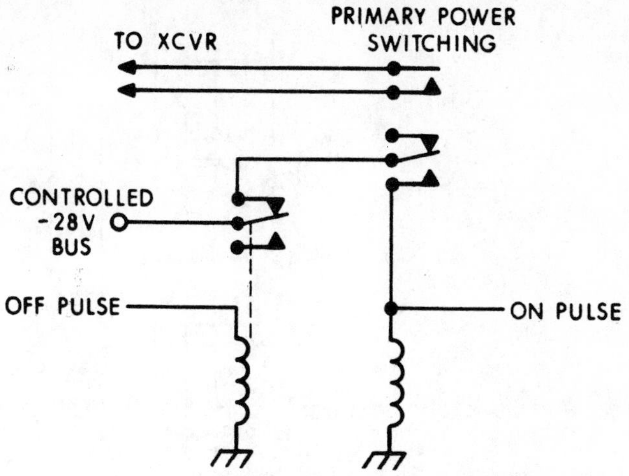


Electromechanical latching relay pair. Relay type is not critical. **On** pulse pulls in primary power switching contacts, which lock because coil voltage is delivered through its own contacts. A short **off** pulse is all that's required to break the circuit.

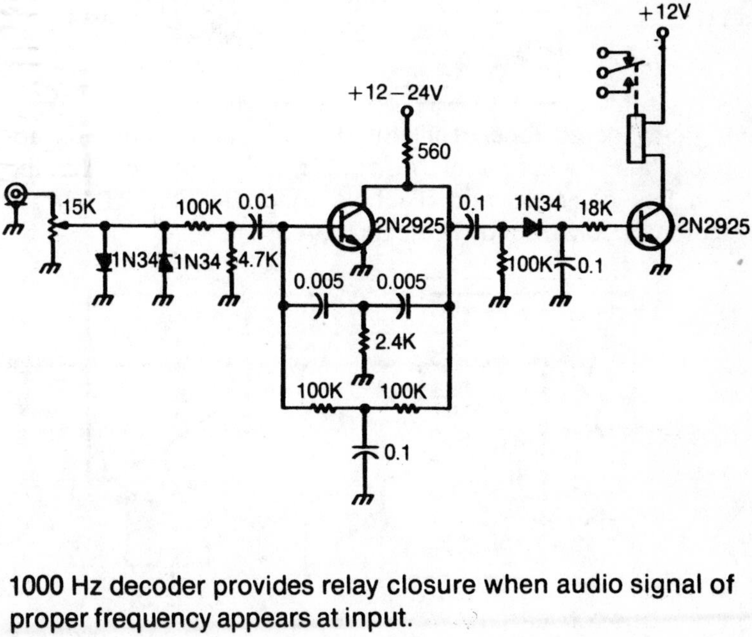


1000 Hz decoder provides relay closure when audio signal of proper frequency appears at input.

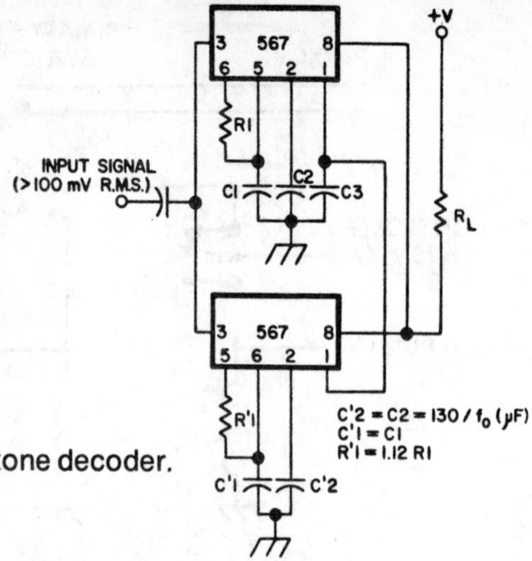


24% bandwidth tone decoder.

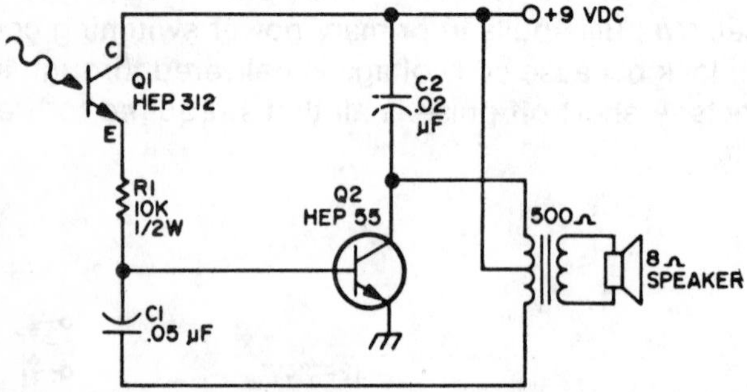


Light-triggered tone oscillator that can find numerous applications as a burglar alarm, or even to let you know that the sun has come up and it's time to go to work. Courtesy Motorola Construction Projects HMA 37.

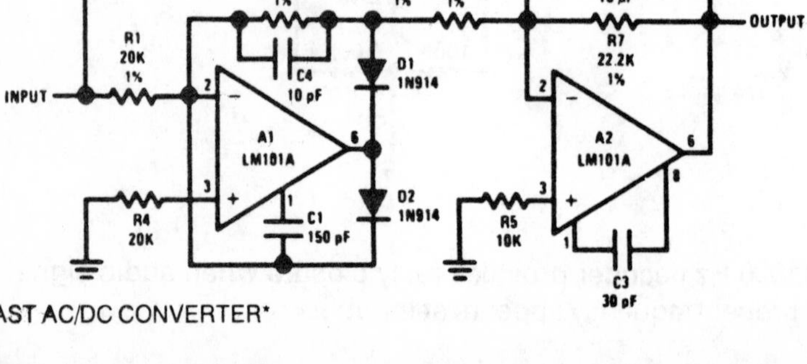


FAST AC/DC CONVERTER*

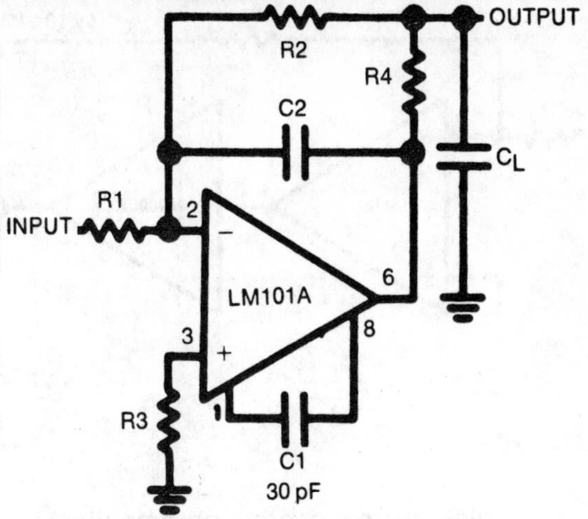


Isolating large capacitive loads.

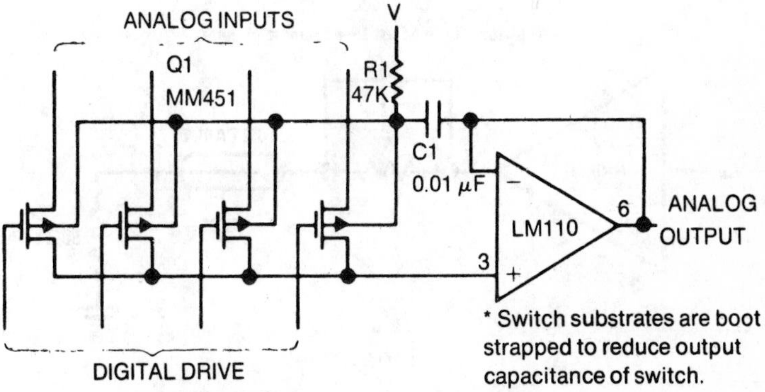


* Switch substrates are boot strapped to reduce output capacitance of switch.

Buffer for analog switch.

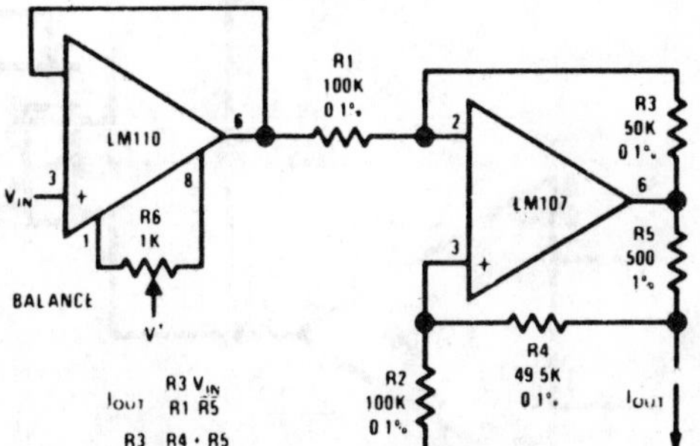


Bilateral current source.

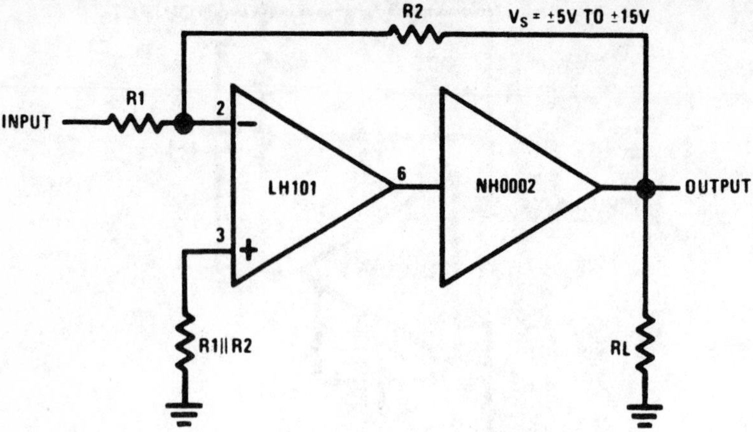


High-current operational amplifier.

Select capacitor to adjust time response of pulse.

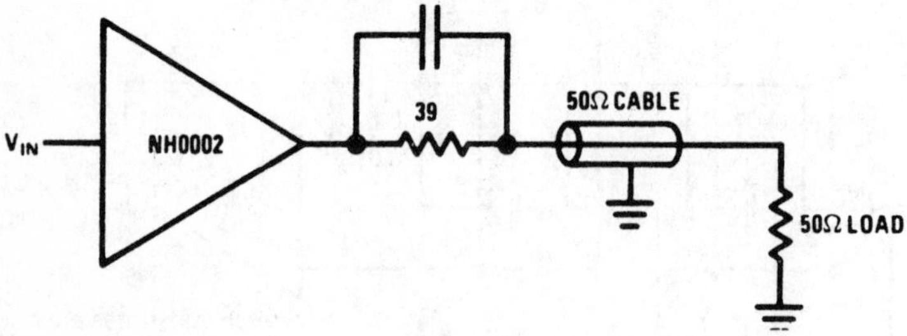


Line driver.

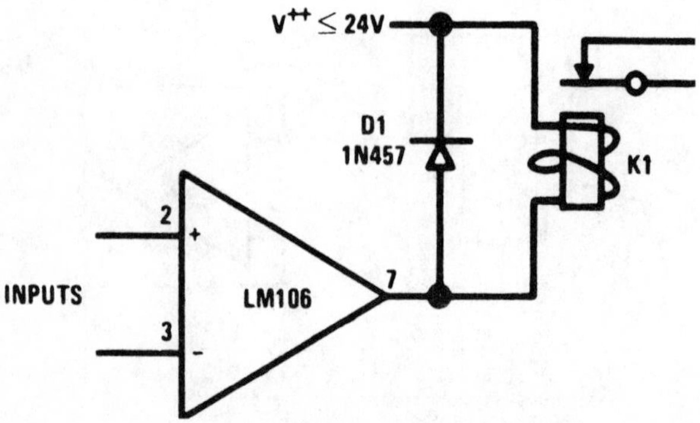


Relay driver.

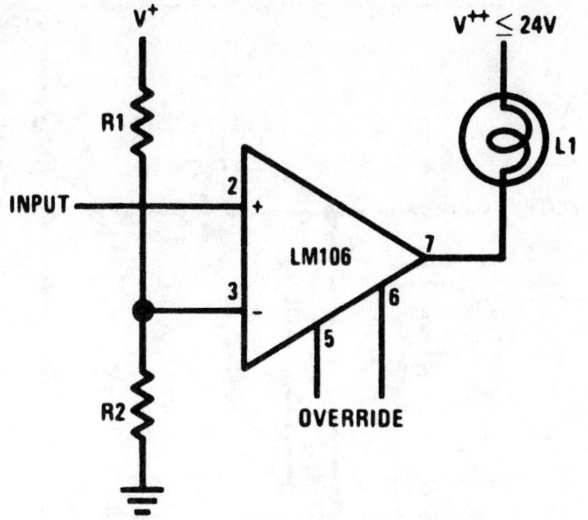


Level detector and lamp driver.

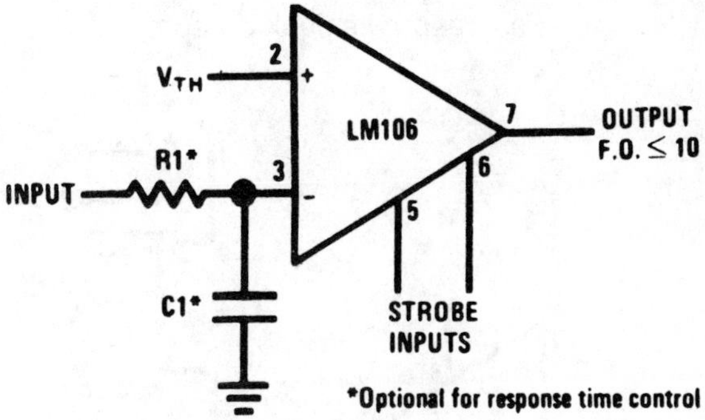


Adjustable threshold line receiver.

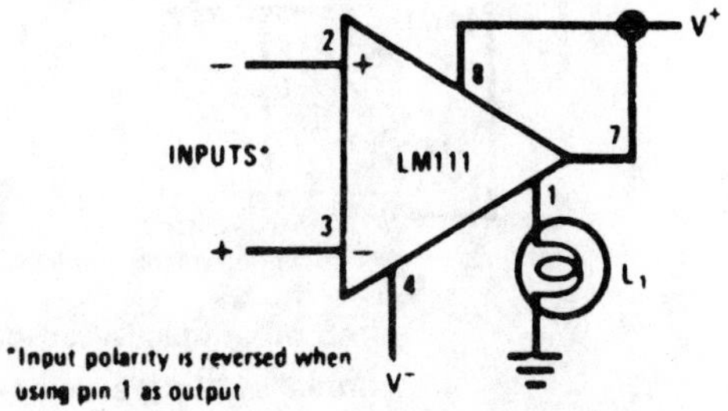


Driving ground-referred load.

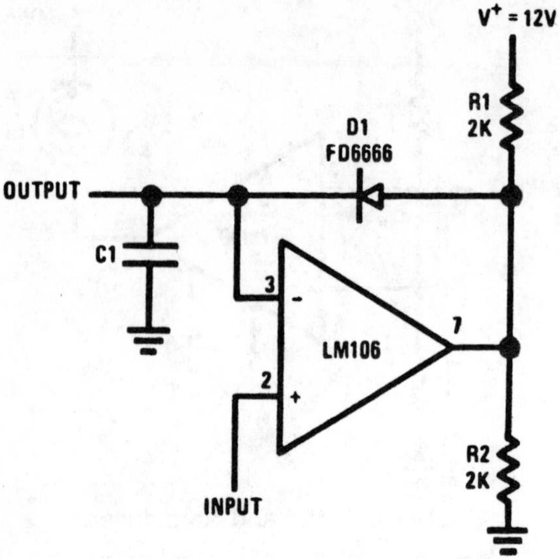


Fast response peak detector.

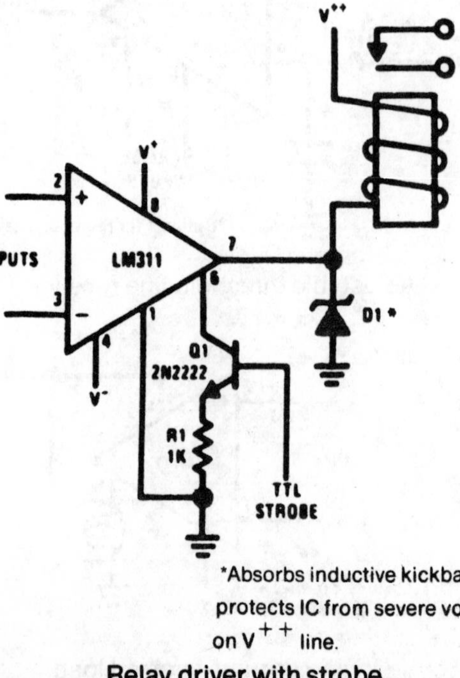


*Absorbs inductive kickback of relay and protects IC from severe voltage transients on $V^{++}$ line.

Relay driver with strobe.

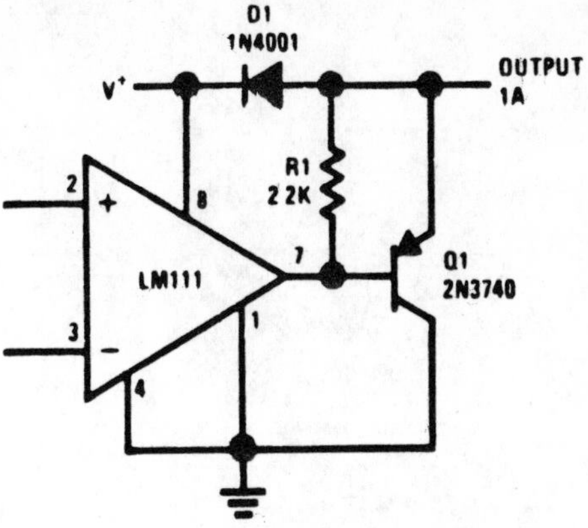


Comparator and solenoid driver.

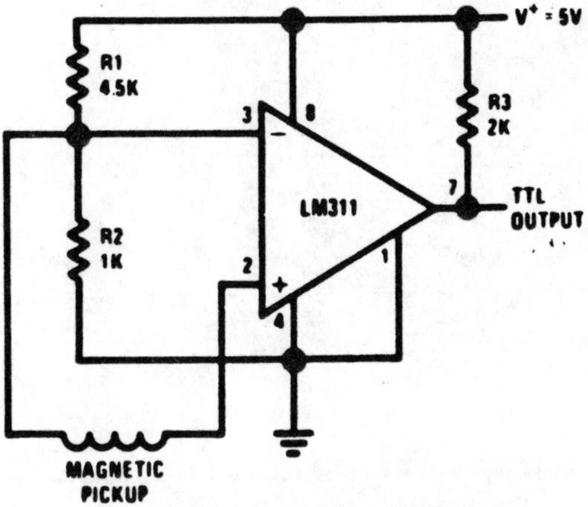


Detector for magnetic transducer.

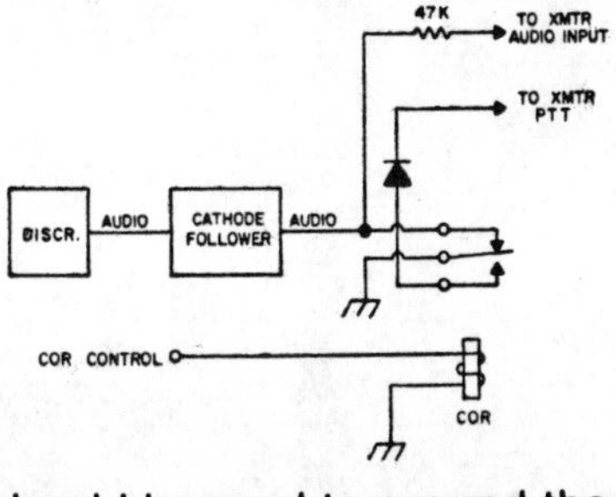


The basic COR should be used to ground the receiver audio when no signals are on the input. The resistor in the audio line provides sufficient isolation so that a local mike can be used even though the audio line is grounded.

# Section N

# Timers

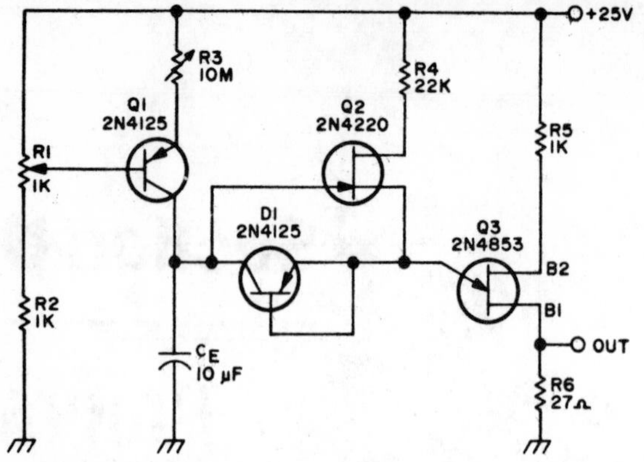


Long-duration FET timer will give a delay up to 10 hours. Circuit courtesy Motorola Semiconductor Power Circuits Handbook.

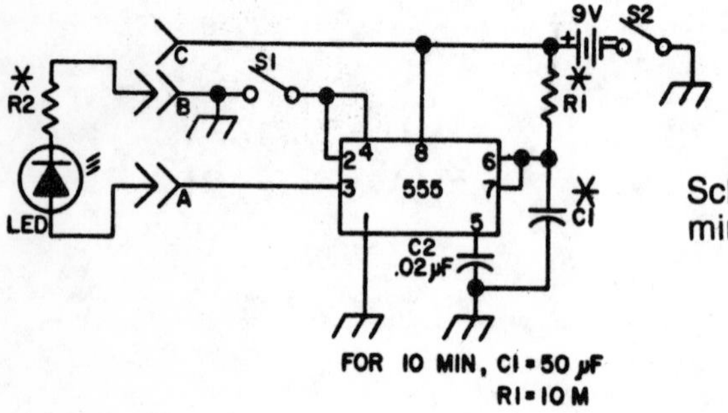


Schematic of 10-minute timer.

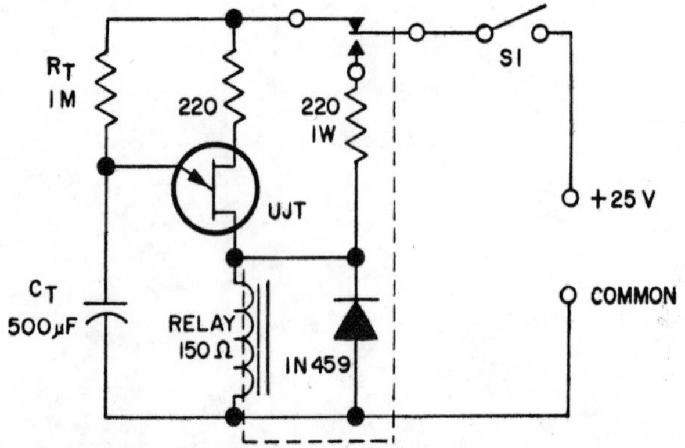


Unijunction timer circuit; relay closes after 15 or 20 sec when S1 is closed, then opens when S1 circuit is broken. Relay should be small low-power type.

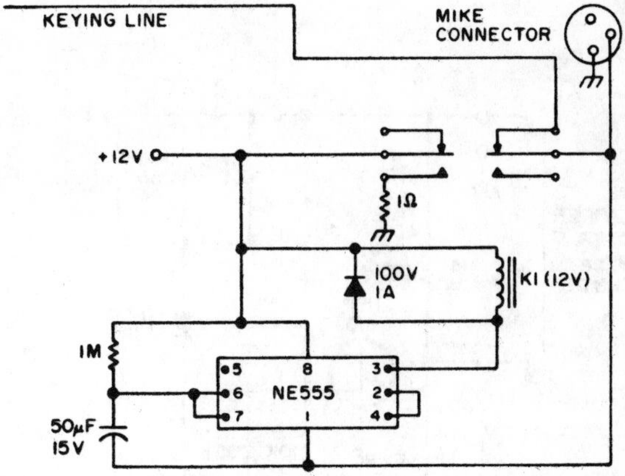


This circuit is designed to time out your rig before it times out the repeater. It uses a simple 1-minute 59-second timer. It is shown wired for a TR22 but can easily be modified to work with any rig.

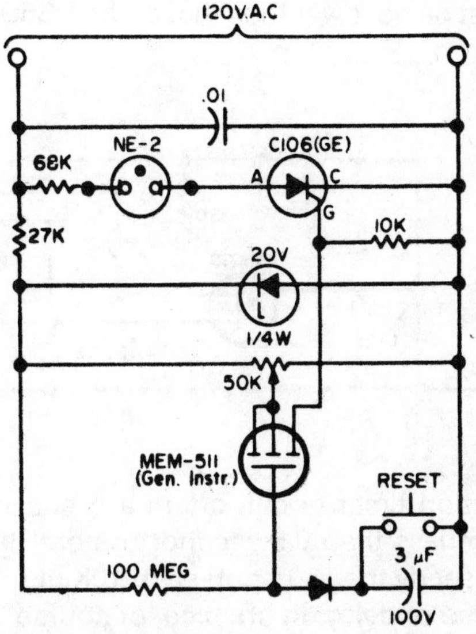


Schematic diagram of the MOSFET 10-minute timer.

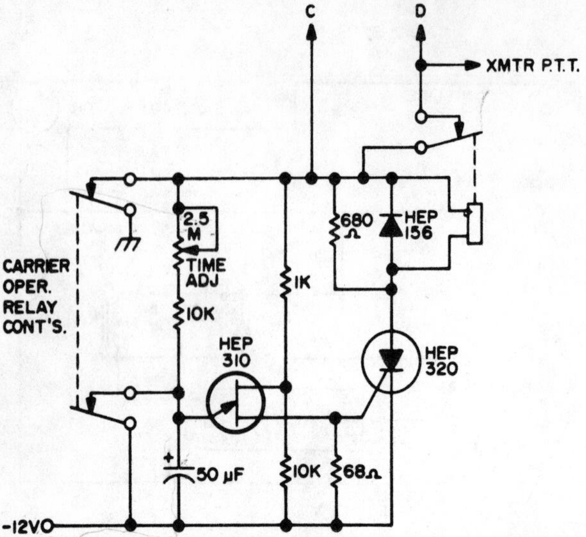


This timer circuit is ideal for repeater applications because its period can be adjusted to any duration from 2 seconds to 4½ minutes. The top set of COR contacts starts the timed sequence and the lower set, on closing, resets the circuit. The relay is a Potter-Brumfield RS5D with a 6V coil having a resistance of 335Ω. Other relays can be used but the coil resistance must be no lower than that of the RS5D.

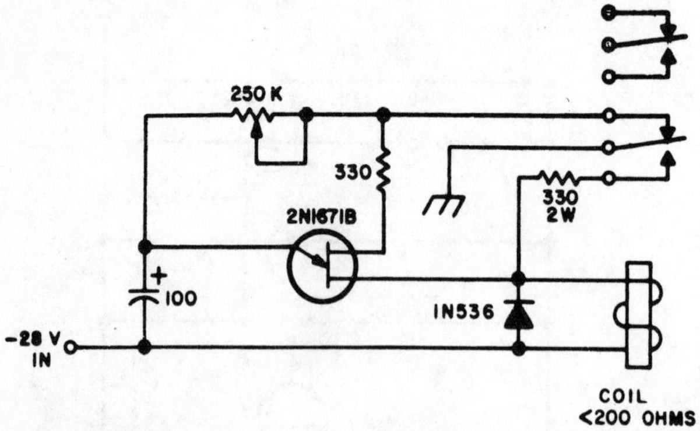


This unijunction timer circuit offers a broad range of delay periods, from less than 0.5 second to more than 3 minutes. The delay is set by the 250K pot; each 10K of resistance gives a second or so of delay. In practice, of course, the pot would be replaced by the proper fixed resistor, because there would be no need for a variable time delay.

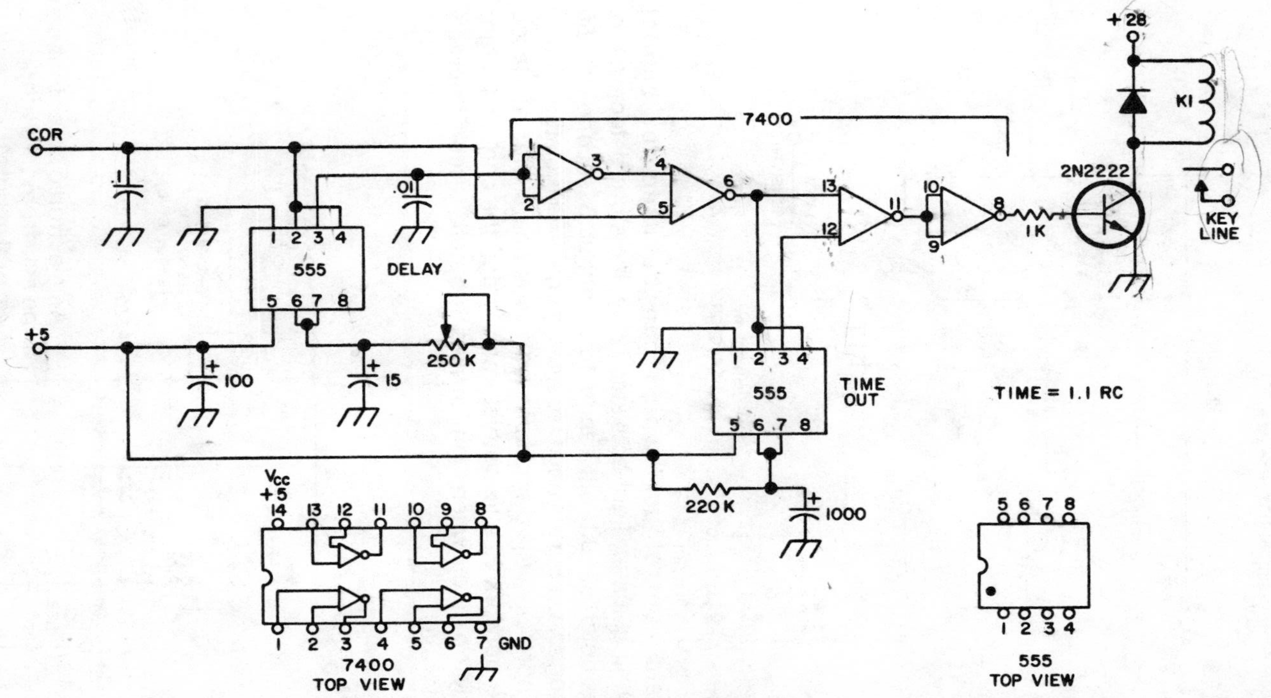


Adjustable time delay offers extra reliability; features delayed dropout as well as delayed pull-in.

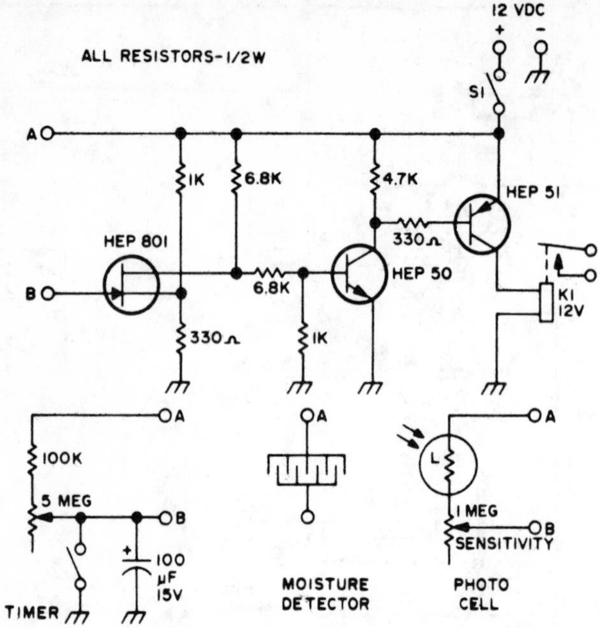


Basic control unit which can be used for almost anything...timer, light-activated relay, moisture detector, etc. The timer circuit will range from about 5 seconds to 50 seconds. Change R and C to alter range. The moisture detector normally uses a small printed circuit board with interlaced wire network to allow moisture to bridge the contacts. Light relay operates when light strikes the cell. Circuit courtesy Motorola.

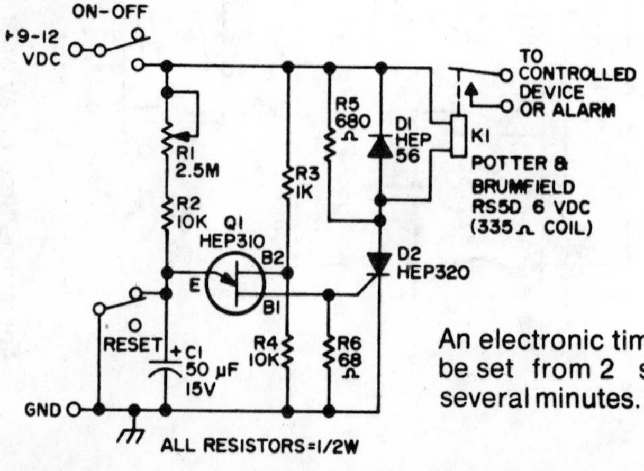


An electronic timer that can be set from 2 seconds to several minutes.

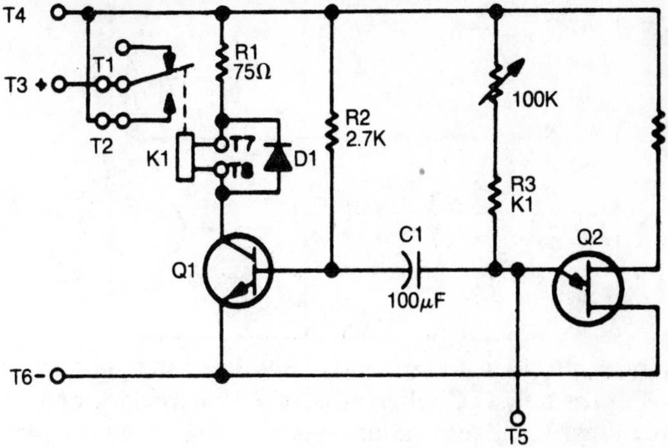


D1 – 1/2A, 100 PRV
K1 – 12V, 150Ω RELAY
Q1 – 2N3416 (PREFERRED) or 2N3393
Q2 – 2N2646

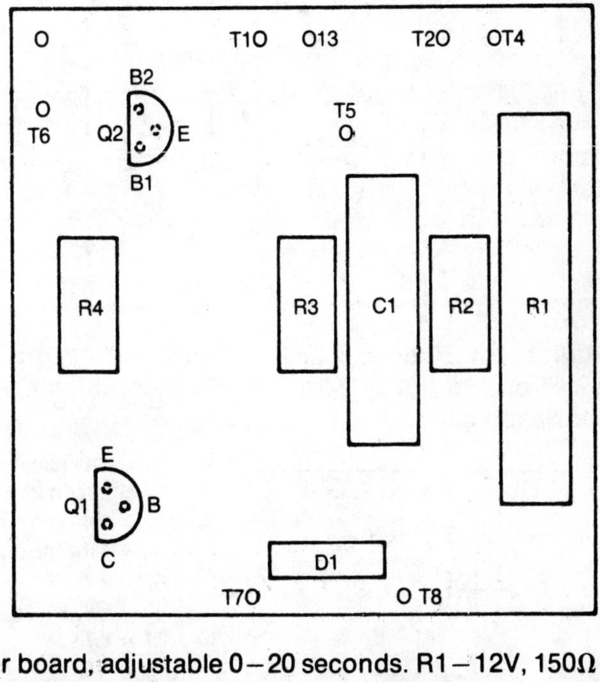


Timer board, adjustable 0 – 20 seconds. R1 – 12V, 150Ω relay; D1 – ½ amp, 100V; Q1 – 2N3416 (preferred) or 2N3393; Q2 – 2N2646.

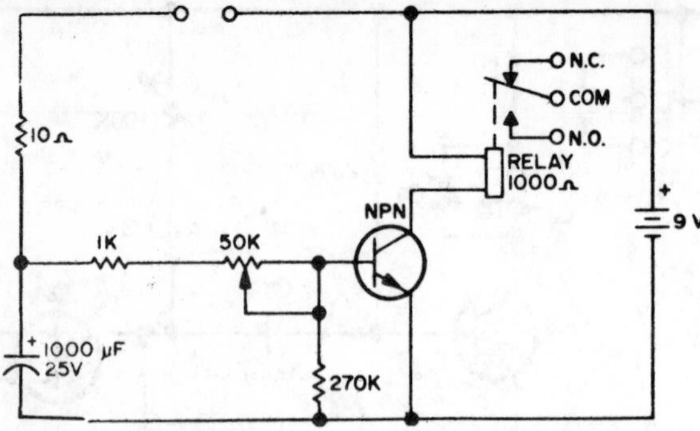


Timing switch, 10 to 100 seconds. No off switch needed since circuit draws only 1 μA when no used. Great for darkroom, 10-minute timer for identification, repeater shutdown for windy talkers, etc. Q1 is a Calectro K4-506 transistor. Circuit courtesy Calectro Handbook.

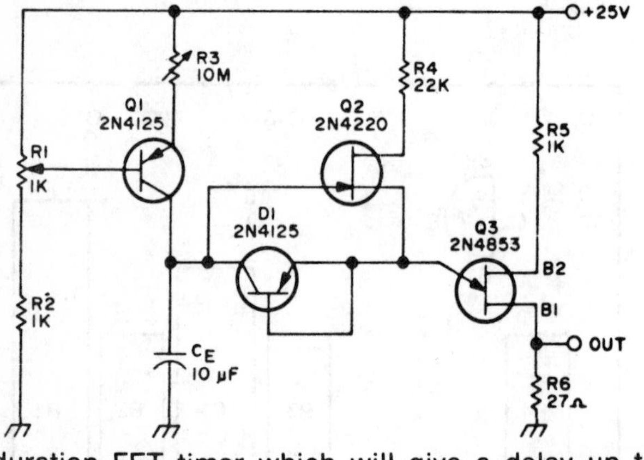


Long-duration FET timer which will give a delay up to 10 hours. Circuit courtesy Motorola Semiconductor Power Circuits Handbook.

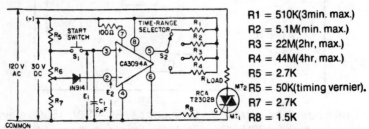

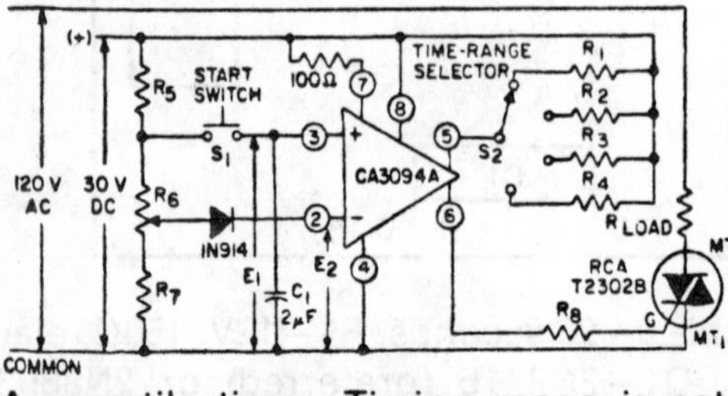


R1 = 510K(3min. max.)
R2 = 5.1M(min. max.)
R3 = 22M(2hr, max.)
R4 = 44M(4hr, max.)
R5 = 2.7K
R5 = 50K(timing vernier).
R7 = 2.7K
R8 = 1.5K

A versatile timer. Timing range is set by switching in R1 through R4; duratior within that time period is set by R6.

# Section O

# Transmitters, Transceivers, Exciters, VFOs

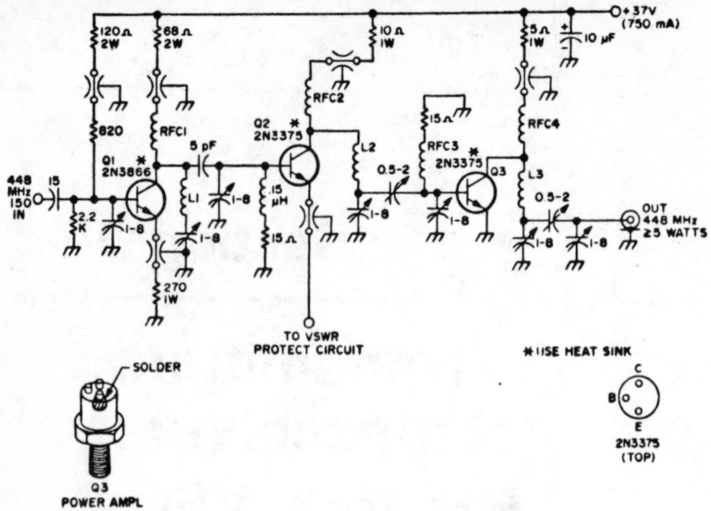


ALL FEED-THRU CAPACITORS = 1000 pF
RFC1, RFC2 = 12t 28 AWG ON 100K, 1/4W RESISTOR
RFC4 = 9t 28 AWG ON 100K, 1/4W RESISTOR
RFC3 = 12t 26 AWG ON 100K, 1/2W RESISTOR
L1 = 3t 18 AWG 3/16 ID, 3/8 LG, 1/8 in. LEADS
L2 = 2-1/2t 18 AWG 3/16 ID, 3/8 LG, 1/8 in LEADS
L3 = 2t 18 AWG 3/16 ID, 3/8 LG, 1/8 in. LEADS

NOTE: EMITTER LEAD OF Q3 IS SOLDER LUG SOLDERED TO TRANSISTOR CASE

RF power amplifier stages. With a 150 mW input, a signal of at least 5W is produced.

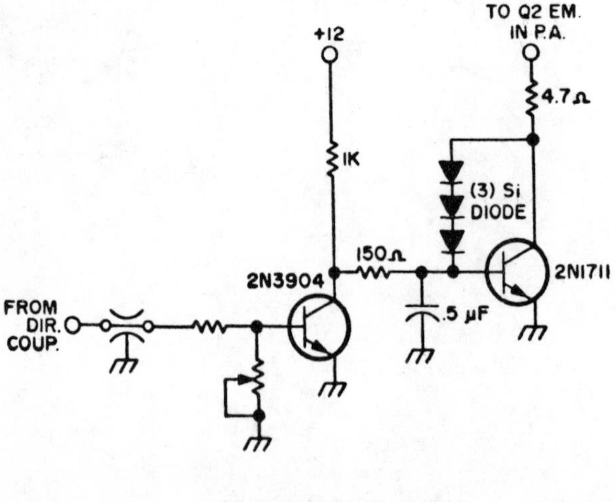


Vswr protect circuit.

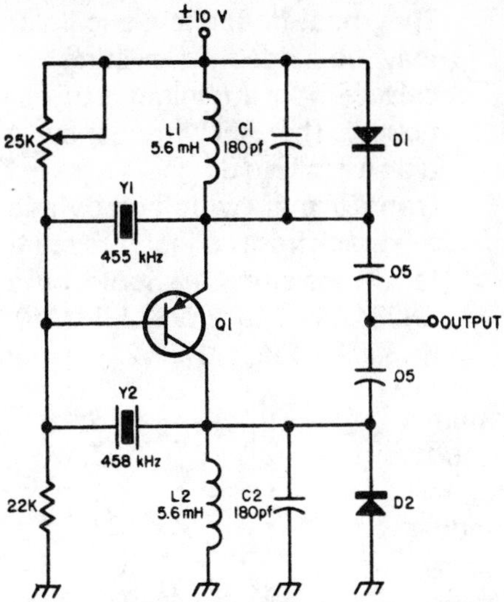


Q1- 2N384, 2N525, SK3004, TI·XM03
D1, D2- GENERAL PURPOSE SILICON DIODES

This two frequency crystal oscillator changes frequency by simply reversing the supply voltage. When the supply voltage is changed, the transistor inverts itself; usually transistors may not be used in the inverted mode, but in an oscillator a gain of only 1 or 2 is needed and this circuit provides a novel and simple way of obtaining two frequencies from a single stage with a minimum of switching.

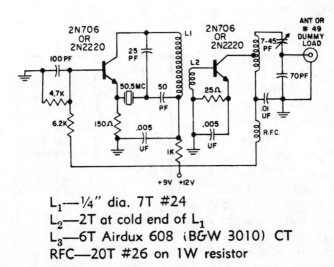

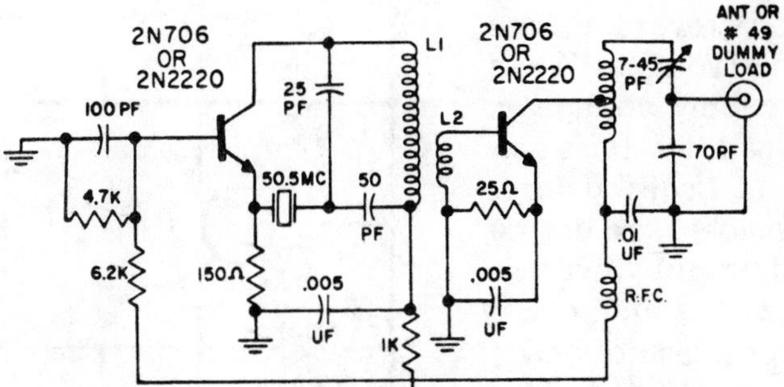


$L_1$—¼" dia. 7T #24
$L_2$—2T at cold end of $L_1$
$L_3$—6T Airdux 608 (B&W 3010) CT
RFC—20T #26 on 1W resistor

100 mW 6 meter transmitter. Modulate AM with a TA300 IC or FM with a couple of diodes.

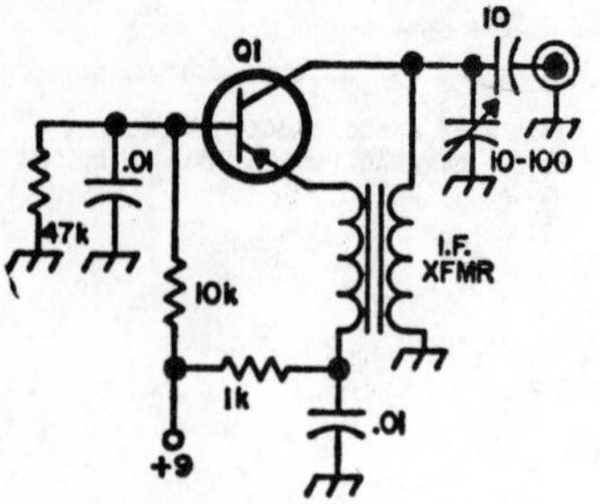


This beat-frequency oscillator may be added to existing receivers with a minimum of difficulty. The BFO frequency is determined by the i-f transformer which provides feedback from collector to emitter. Transistor Q1 should be a 2N384, 2N1749, 2N2362, T1M10, SK3008, GE-9, or HEP 2.

This is the old familiar vacuum tube Pierce oscillator circuit with a field-effect transistor in place of the thermionic triode. Circuit constants shown here are for the 1 MHz region, but the tuned circuit may be adjusted to any frequency desired. Q1 is a 2N4360.

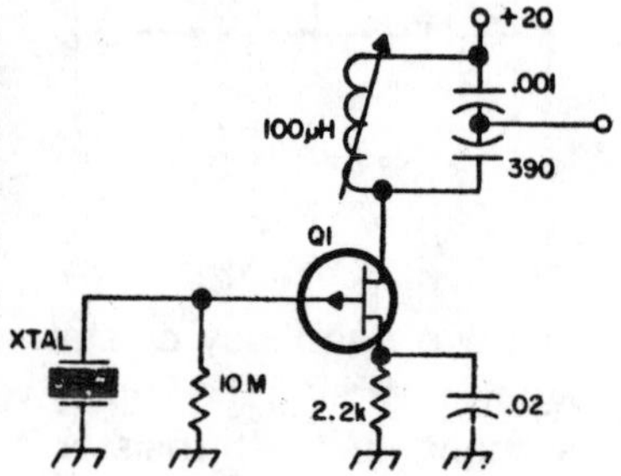


This two-frequency crystal oscillator changes frequency by simply reversing the supply voltage. When the supply voltage is changed, the transistor inverts itself; usually transistors may not be used in the inverted mode, but in an oscillator a gain of only 1 or 2 is needed. This circuit provides a novel and simple way of obtaining two frequencies from a single stage with a minimum of switching. Almost any PNP rf transistor will work as Q1. D1 & 2 are general-purpose silicon diodes.

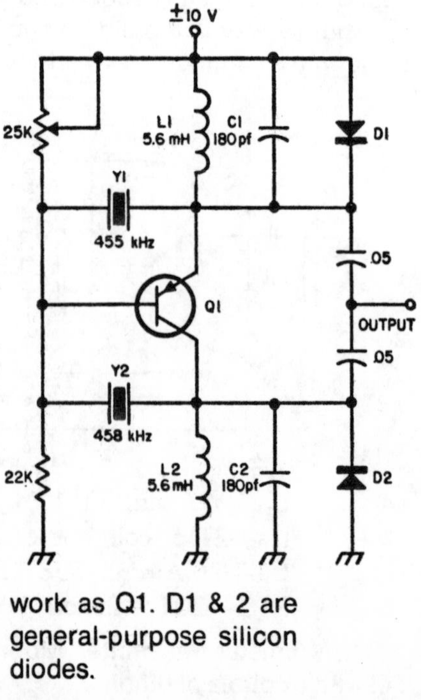

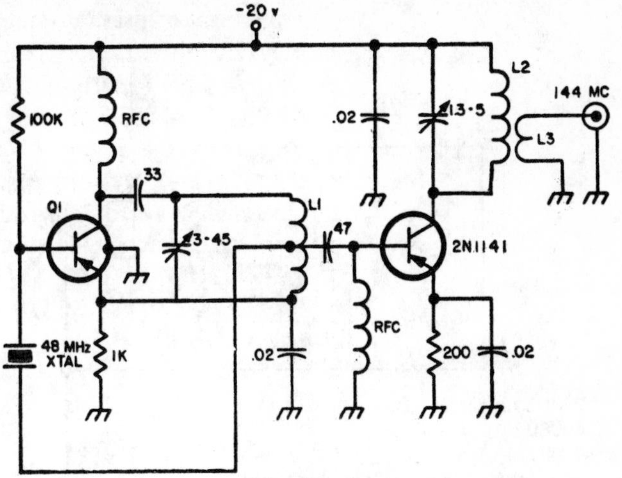


| | |
|---|---|
| L1 | 8 TURNS B&W 3003 (16 TURNS PER INCH, 1/2" DIAM.) TAPPED AT 4 TURNS FROM COLD END. |
| L2 | 8 TURNS NO. 16, 5/16" DIAM, 1" LONG |
| L3 | 3 TURNS NO. 16 BIFILAR WOUND ON COLD END OF L2. |
| Q1 | 2N384, SK3008, TIXM03 |
| RFC | 1.8 μH (OHMITE Z-144) |

This simple 2-meter transmitter may be used as a driver for a larger 144 MHz transmitter or a signal source for testing receivers, converters, and antennas.

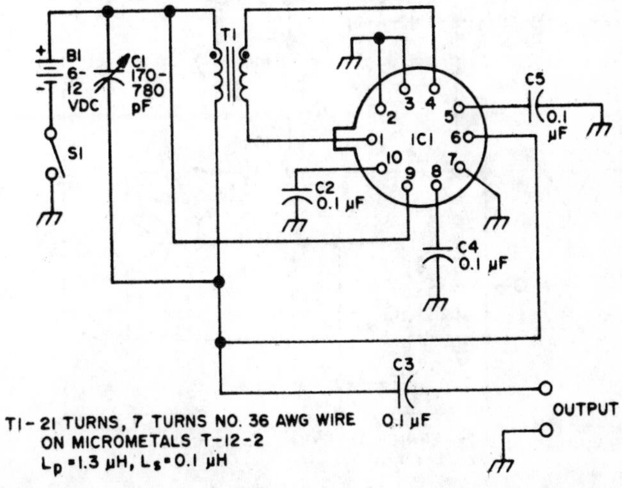


5–10 MHz VFO or 40-meter QRP transmitter; an idea for those backwoods hikes. Courtesy Motorola HMA36, IC Projects for Amateurs.

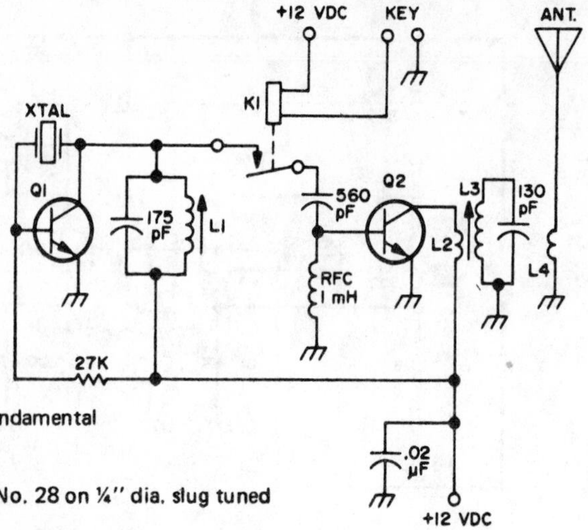


| | |
|---|---|
| XTAL | 7 MHz fundamental |
| Q1 | 40080 |
| Q2 | 40081 |
| L1 | 20 turns No. 28 on ¼" dia. slug tuned form |
| L2, L4 | 5 turns No. 24 on L3 |
| L3 | 28 turns No. 28 on ¼" dia. slug tuned form |
| K1 | sensitive spst relay for 12V (see text) |

Simple low-power transmitter (1 watt) is ideal for CW operation (no modulator).

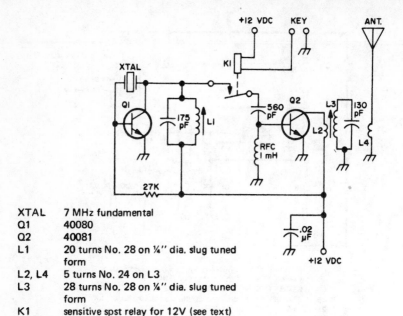

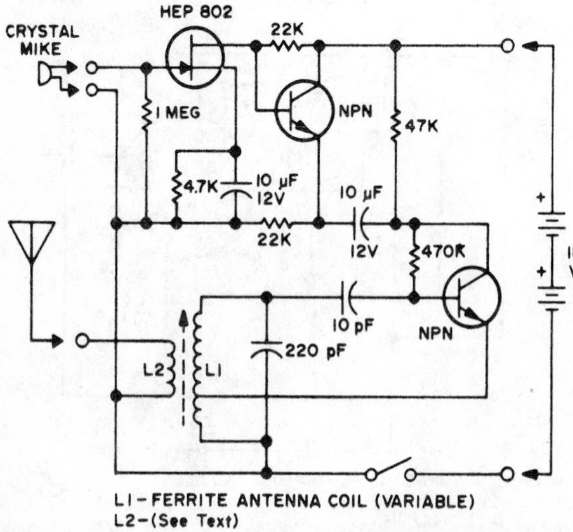


L1–FERRITE ANTENNA COIL (VARIABLE)
L2–(See Text)

AM wireless transmitter. Useful for baby-sitting, sick watch, intercom, etc. L1 is a variable antenna coil, (Calectro D1 841). L2 is four turns of hookup wire wound on top of L1. Q2–3 are NPN silicon transistors (Calectro K4-507).

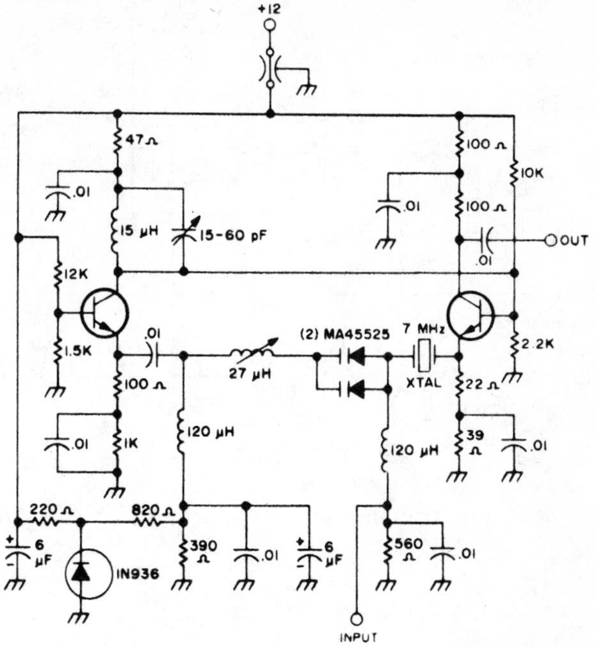


Variable crystal-controlled oscillator circuit.

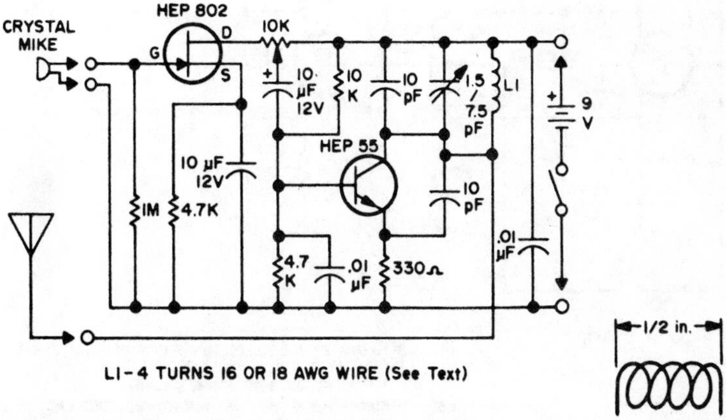


FM wireless transmitter (88–108 MHz). Circuit courtesy Calectro Handbook. C7 is a short length of twisted hookup wire about ½ inch long. L1 is four turns length about ½ inch and ¼ inch dia.

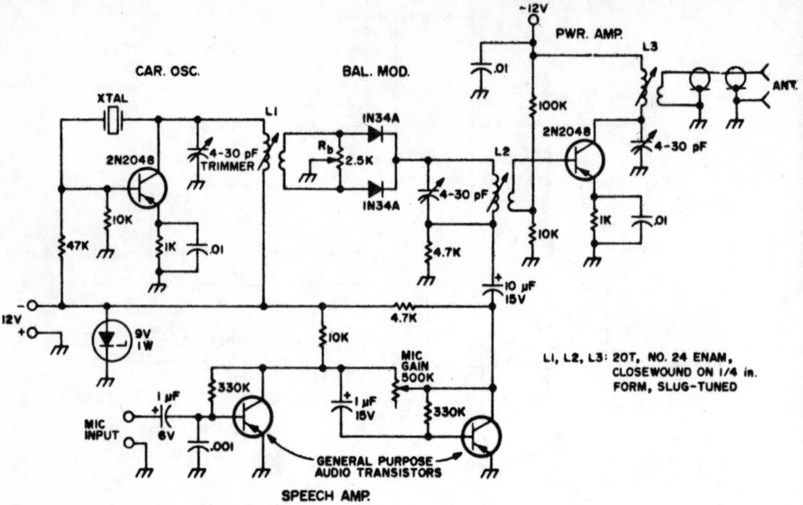


Double-sideband transmitter for operation in 27 – 28 MHz range.

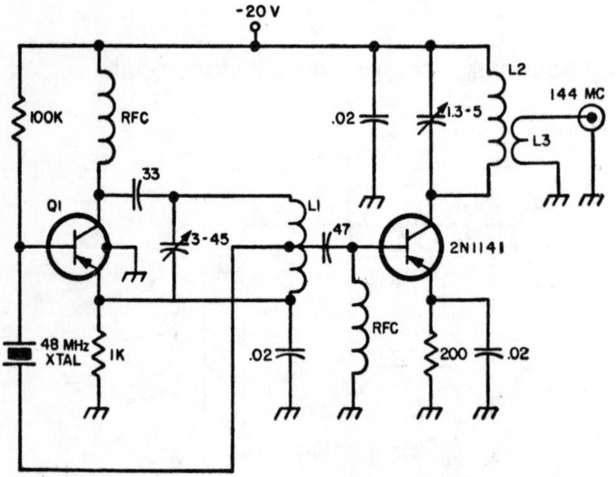


| | |
|---|---|
| L1 | 8 TURNS B&W 3003 (16 TURNS PER INCH, 1/2" DIAM.) TAPPED AT 4 TURNS FROM COLD END. |
| L2 | 8 TURNS NO. 16, 5/16" DIAM, 1" LONG |
| L3 | 3 TURNS NO. 16 BIFILAR WOUND ON COLD END OF L2. |
| Q1 | 2N384, SK3008, TIXM03 |
| RFC | 1.8 μH (OHMITE Z-144) |

This simple 2-meter transmitter may be used as a driver for a larger 144 MHz transmitter or a signal source for testing receivers, converters, and antennas.

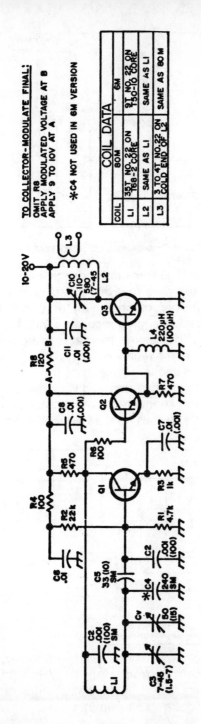

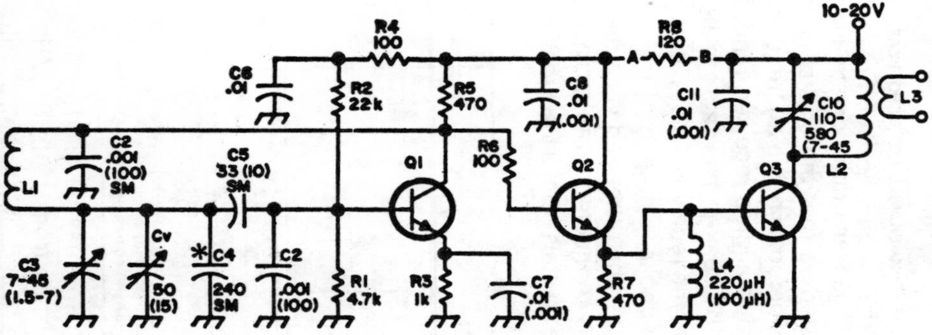


TO COLLECTOR-MODULATE FINAL:
OMIT R8
APPLY MODULATED VOLTAGE AT B
APPLY 9 TO 10V AT A

*C4 NOT USED IN 6M VERSION

| COIL DATA | | |
|---|---|---|
| COIL | 80M | 6M |
| L1 | 35T NO. 22 ON T68-2 CORE | 9T NO. 22 ON T50-10 CORE |
| L2 | SAME AS L1 | SAME AS L1 |
| L3 | 3 TO 4T NO.22 ON COLD END OF L2 | SAME AS 80 M |

Here is a stable VFO that can be assembled in a short time on a piece of Vectorbord. Coil data is supplied for an 80m and 6m version but other bands may be covered with a bit of experimentation. Transistors are all MPS706 but higher output is possible by replacing Q3 with a 2N2270 or 2N3053.

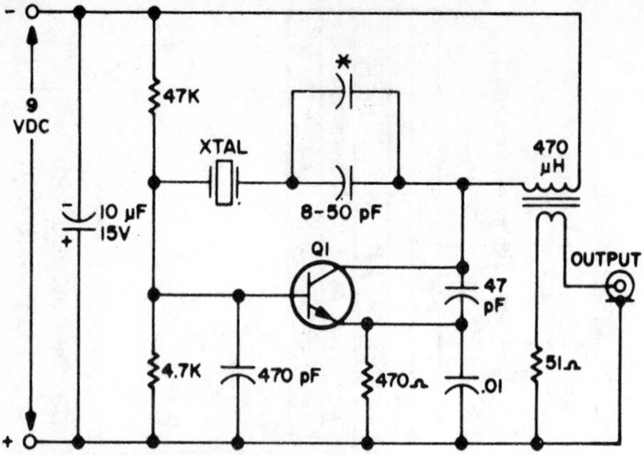


Q1 - VHF PNP (GE-9, ES19, etc)
* - MAY BE NEEDED IF TRIMMER WILL NOT ZERO-BEAT. ADD IN 50 pF STEPS.

2-meter crystal oscillator. The transmitter crystal is used, the netting capacitor is adjusted so that the crystal is zero-beat with someone who is considered as being on frequency. Then the oscillator may be used to align a receiver. Since the output is so high (measured 18V peak-to-peak across the choke), it must be loose-coupled to the receiver. This is done by winding two turns around the choke. One end of the winding is connected to the connector; the other end is connected to a 15Ω resistor which is then connected ground. This circuit is very active and will handle crystals from 2 MHz to 30 MHz.

Q1 - 2N708, 2N3564, 2N3854, 40237

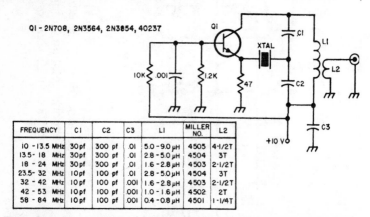

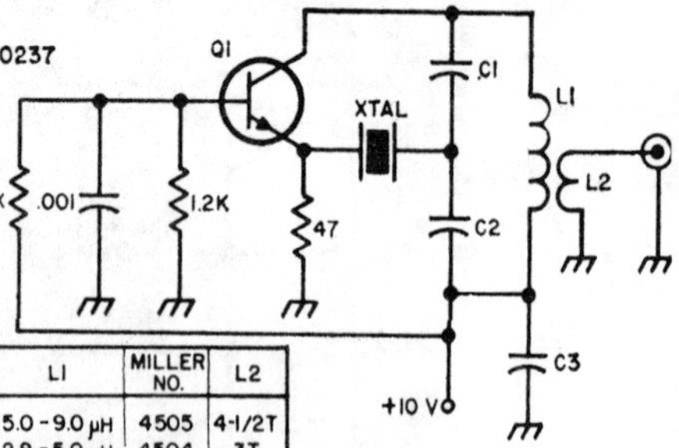


| FREQUENCY | C1 | C2 | C3 | L1 | MILLER NO. | L2 |
|---|---|---|---|---|---|---|
| 10 - 13.5 MHz | 30pf | 300 pf | .01 | 5.0 - 9.0 µH | 4505 | 4-1/2T |
| 13.5 - 18 MHz | 30pf | 300 pf | .01 | 2.8 - 5.0 µH | 4504 | 3T |
| 18 - 24 MHz | 30pf | 300 pf | .01 | 1.6 - 2.8 µH | 4503 | 2-1/2T |
| 23.5 - 32 MHz | 10pf | 100 pf | .01 | 2.8 - 5.0 µH | 4504 | 3T |
| 32 - 42 MHz | 10pf | 100 pf | .001 | 1.6 - 2.8 µH | 4503 | 2-1/2T |
| 42 - 53 MHz | 10pf | 100 pf | .001 | 1.0 - 1.6 µH | 4502 | 2T |
| 58 - 84 MHz | 10pf | 100 pf | .001 | 0.4 - 0.8 µH | 4501 | 1-1/4T |

This Colpitts crystal oscillator may be used with either fundamental or overtone crystals from 10 to 84 MHz with the tuned-circuit components listed. It oscillates quite readily when adjusted and provides a stable output.

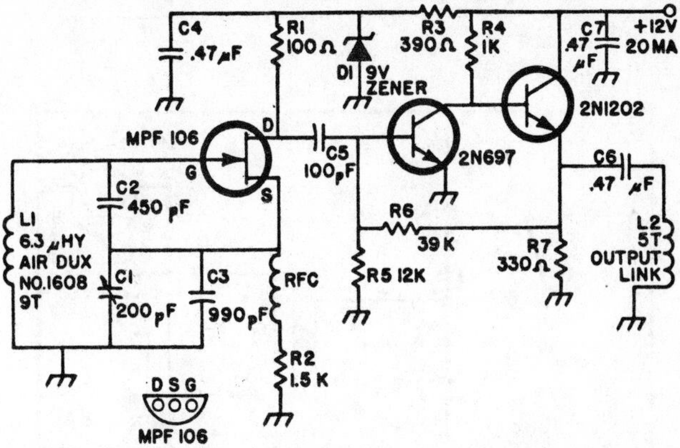


Field-effect-transistor variable frequency oscillator has zero temperature coefficient. Operating frequency in 3.5 MHz range.

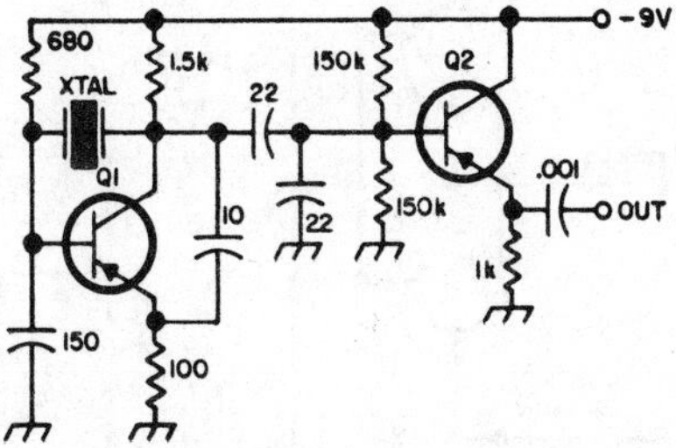


This untuned crystal oscillator will oscillate with any crystal from 300 kHz to 10 MHz. Frequency stability is very good because the emitter-follower buffer amplifier effectively isolates the oscillator from the load. Q1 and Q2 are GE-9, SK30006, or HEP 2.

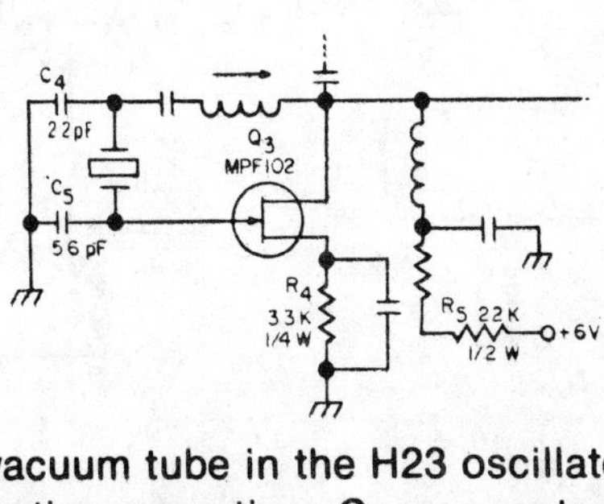


FET replaces vacuum tube in the H23 oscillator to complete the transistorization operation. Components not labeled in the sketch are those components that are already part of the existing oscillator circuit.

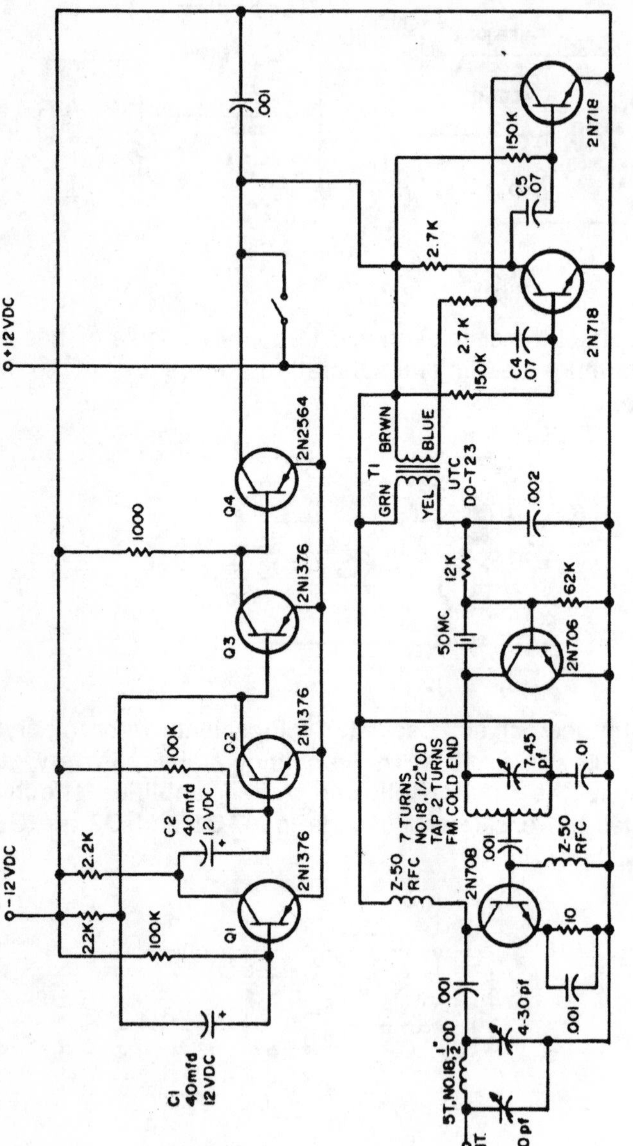


Schematic for beeper. All resistors are ½ watt.

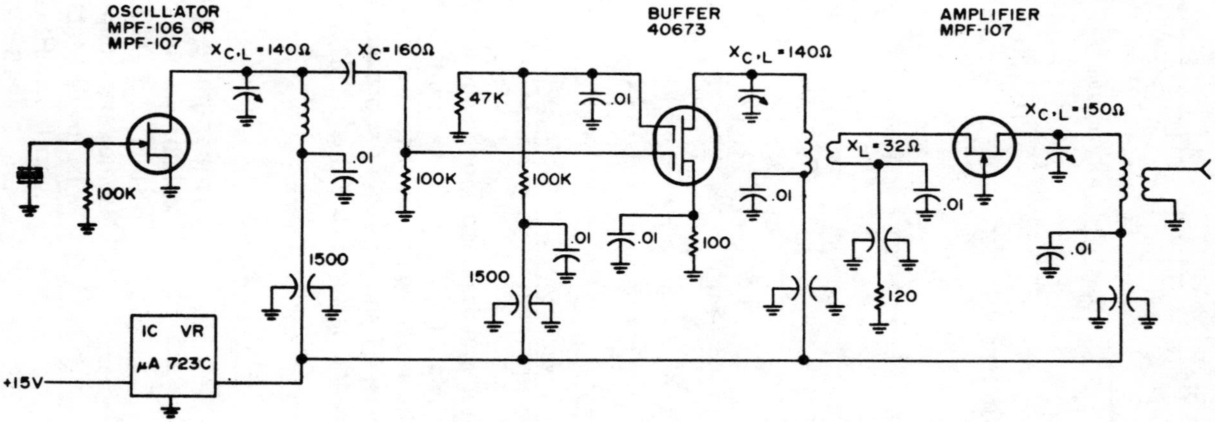


Ultrahigh-stability crystal oscillator circuit that is useful for microwave transmitter frequency control. Use crystals from 1.6 to 160 MHz, fundamental or overtone. For best results get the values of the coils and capacitors for any specific frequency from your friendly enighborhood reactance chart.

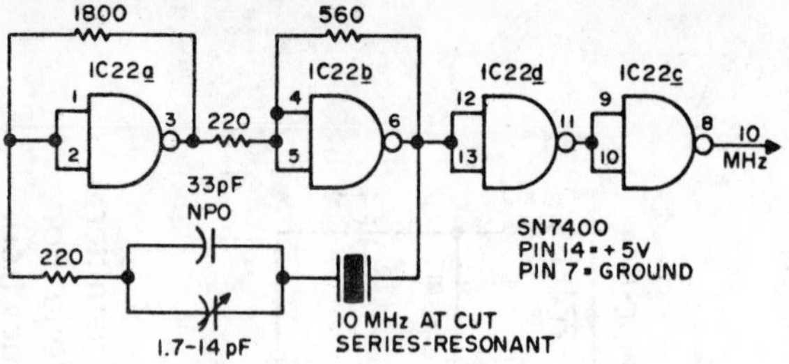


10 MHz crystal oscillator.

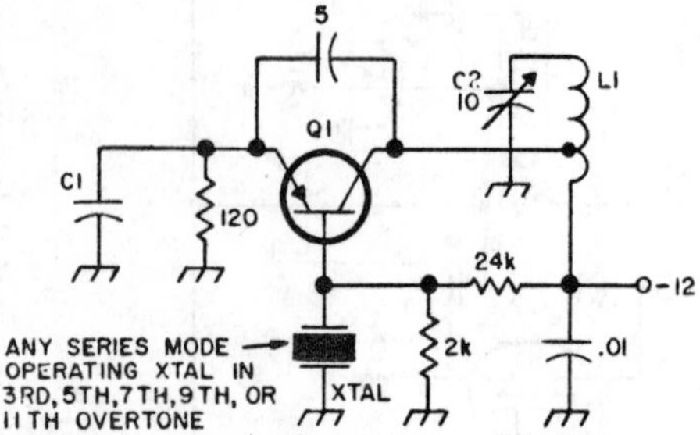


L1 = TUNED TO RESONATE WITH C2 - COLLECTOR TAP 1/3 UP FROM COLD END

C1 = CAPACITIVE REACTANCE OF APPROX. 90 Ω AT OPERATING FREQUENCY

| FREQ (MHz) | 50 | 144 | 220 | 432 |
|---|---|---|---|---|
| C1 (pF) | 36 | 12 | 8.2 | 3.3 |

This crystal oscillator was designed specifically for overtone crystals and will oscillate up to the 11th overtone in the VHF range. Suitable values for C1 are shown for the VHF bands; for other frequencies, C1 should exhibit approximately 90Ω capacitive reactance for best results. Q1 is a TM10, T1400 or HEP 3.

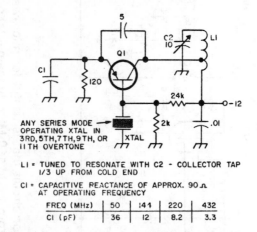

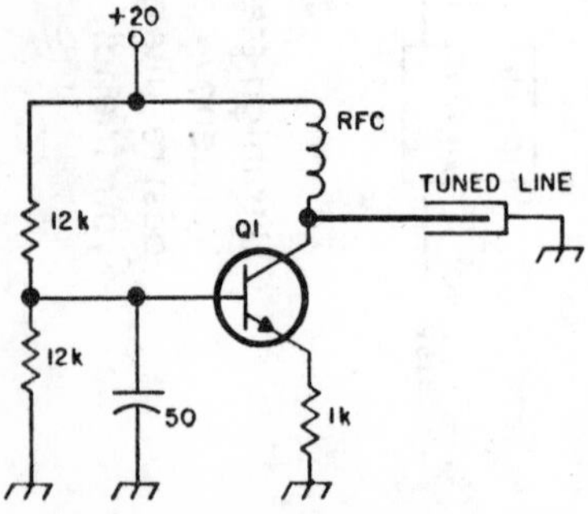


This simple UHF oscillator will provide about 2 mW up to 1000 MHz; some selected transistors will provide usable power up to 1500 MHz or so. Q1 is a 2N918, 2N3478, 2N3564, or HEP 65.

| | |
|---|---|
| L1 | 13 TURNS NO. 26 ON 1/4" SLUG TUNED FORM. TAPPED AT 6-1/2 TURNS. |
| L2 | 10 TURNS NO. 26 ON 1/4" SLUG TUNED FORM. TAPPED AT 5 TURNS. |
| L3 | 9 TURNS NO. 24 ON 3/8" SLUG TUNED FORM. |
| L4 | SAME AS L3. |
| Q1 | 2N708, 2N3009, 2N3646, 2N3856 |
| Q2 | 2N3641, 2N2217 |
| Q3,Q4,Q5 | 2N3642, 2N2218 |

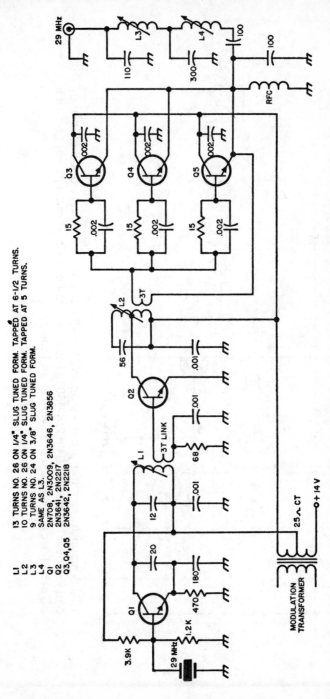

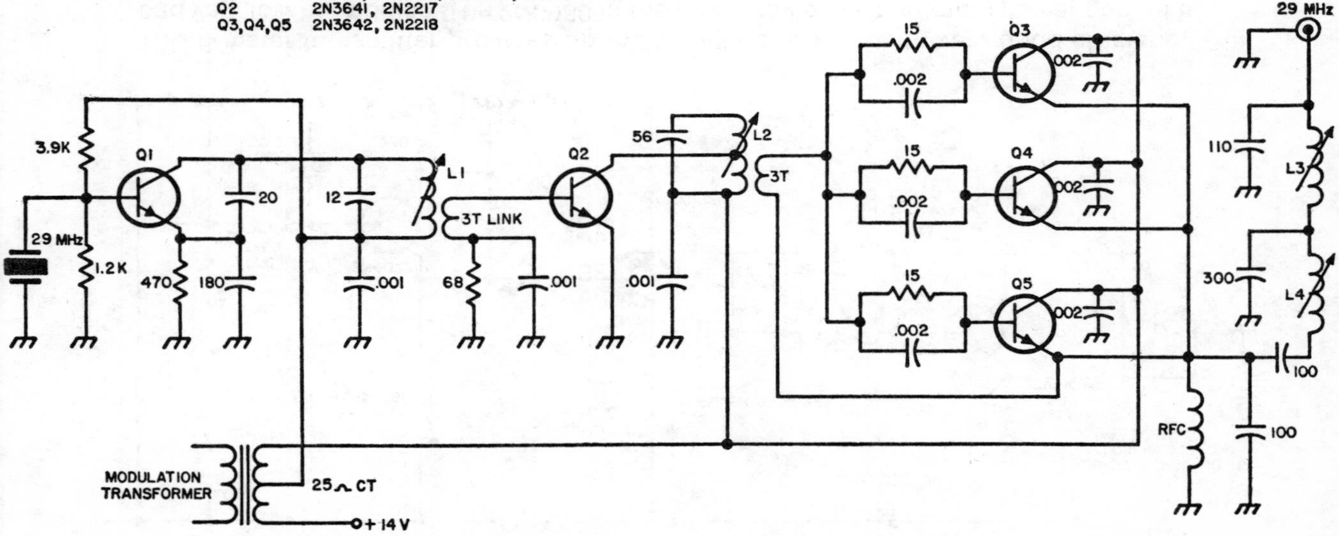


This three watt ten meter transmitter maintains high efficiency and low cost by paralleling three inexpensive silicon transistors in the final stage.

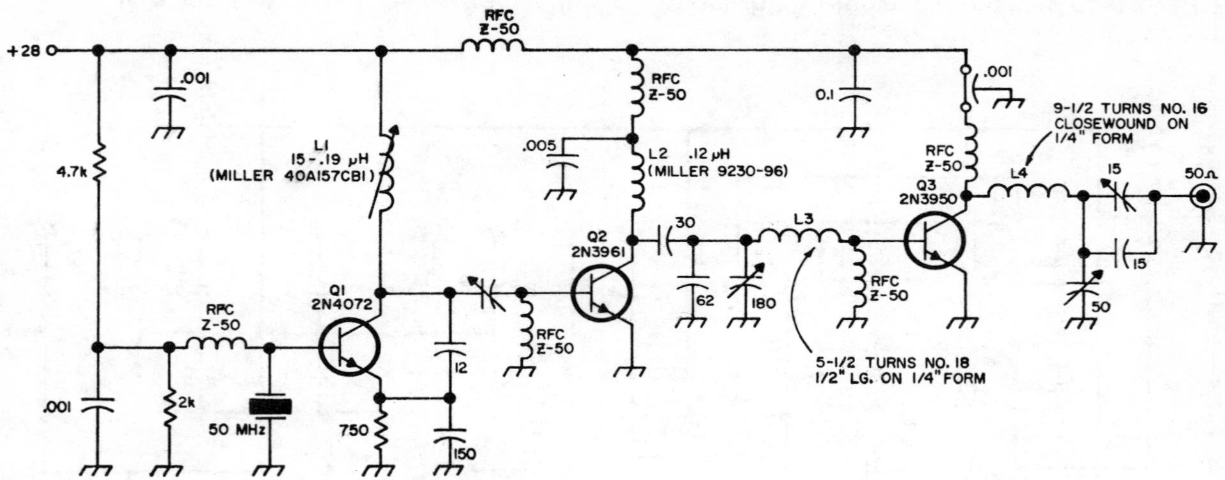


This 6-meter transmitter provides up to 50 watts of power with very good efficiency and very low harmonics. The 2N3950 in the final provides a minimum power gain of 8 dB at 50 MHz and is rated at 50 watts continuous service.

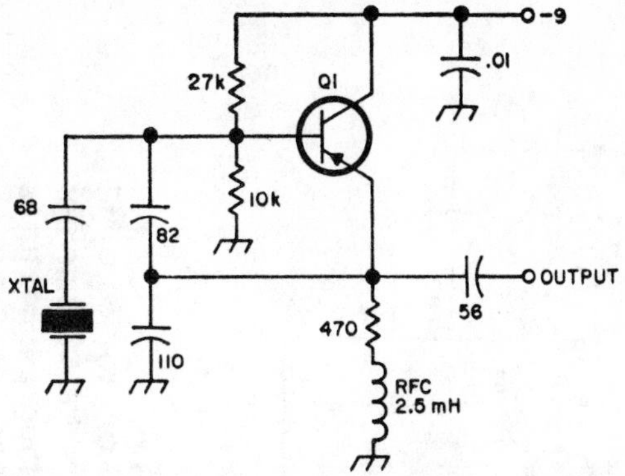


This crystal oscillator will oscillate with any crystal between 3 and 20 MHz with no tuning whatsoever; overtone crystals will oscillate on their fundamental in this circuit. Q1 is a 2N1177, 2N1180, 2N1742, GE-9, SK3006, or HEP 2.

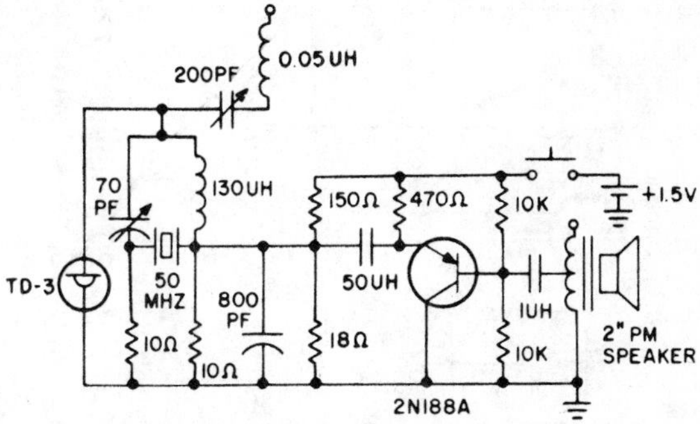


50 MHz transmitter with speaker as AM microphone.

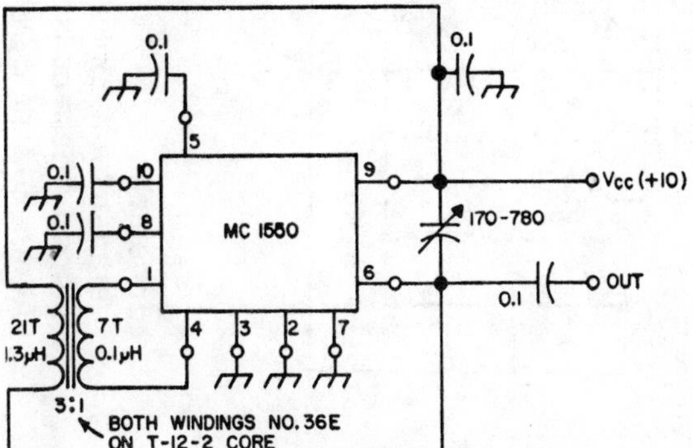


VFO covering 5 to 10 MHz.

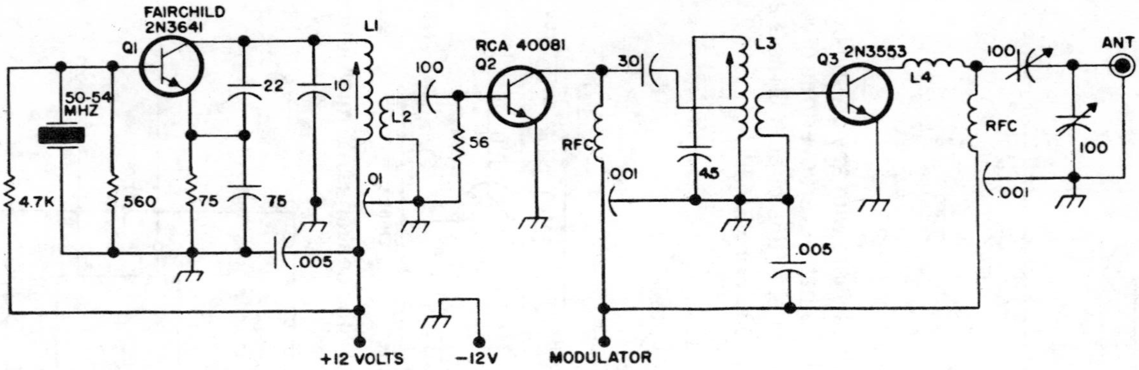


Schematic of 2½W 6m transmitter. L1 and L2 = 6 turns No. 20 on iron-core ⅜″ ceramic form. Links are 2 turns No. 20 insulated at bottom of L1 and L2. L3 is center tapped. These coils are surplus, used, as-is. Both windings are ½″ long. L4 = 6 turns Airdux 516 or B&W 3007. O1 = Fairchild 2N364; Q2 = RCA 40081; Q3 = Bendix B3466 or RCA 2N3553 or 40341 (all heatsinked).

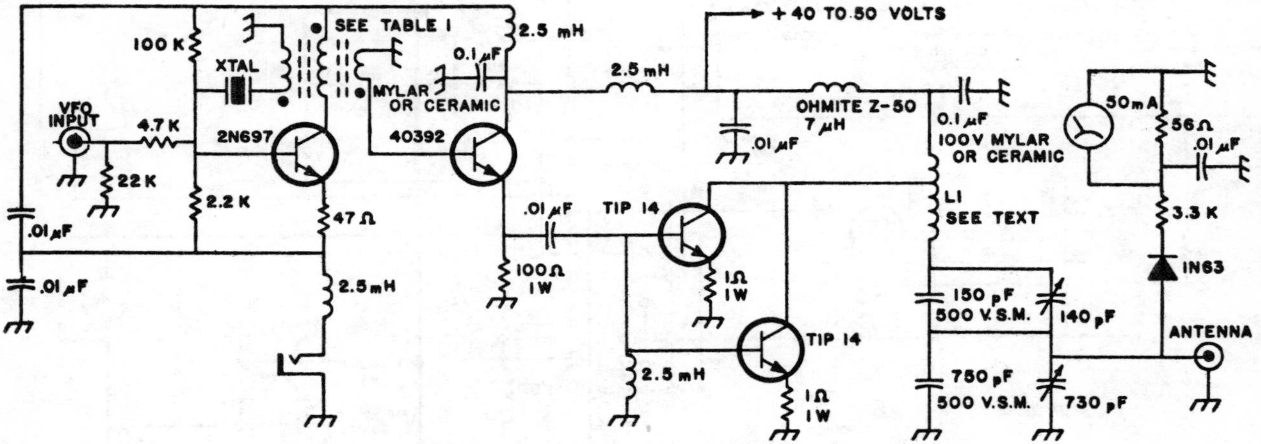


Transistor transmitter puts out 30 watts at 3.5 MHz, and uses Texas Instruments' TIP 14s.

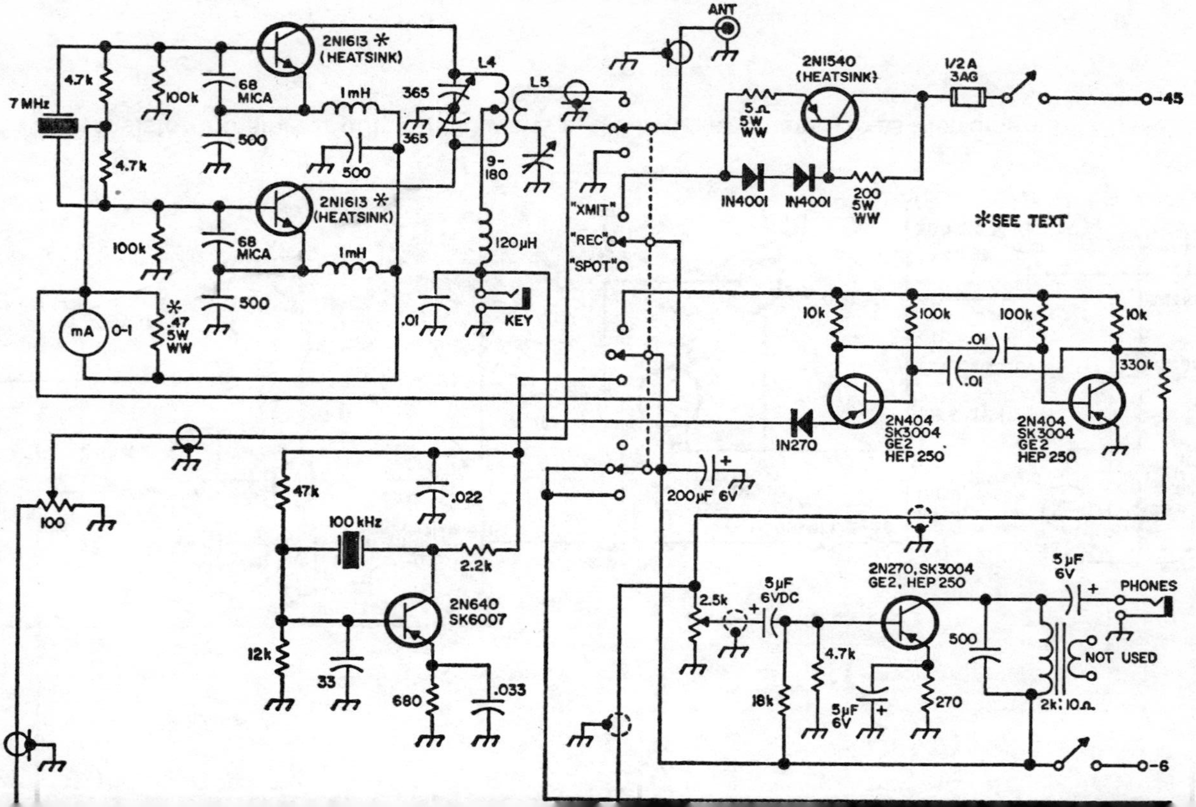
7 MHz
4.7k
4.7k
100k
100k
68 MICA
68 MICA
500
500
500
2N1613 *
(HEATSINK)
2N1613 *
(HEATSINK)
1mH
1mH
365
365
L4
L5
9-180
120µH
.01
KEY
mA
0-1
*
.47
5W
WW
ANT
"XMIT"
"REC"
"SPOT"
2N1540
(HEATSINK)
5Ω
5W
WW
1N4001
1N4001
200
5W
WW
1/2A
3AG
-45
*SEE TEXT
10k
100k
100k
10k
.01
.01
330k
1N270
2N404
SK3004
GE2
HEP 250
2N404
SK3004
GE2
HEP 250
200µF 6V
100
47k
.022
100 kHz
2.2k
2N640
SK6007
12k
33
680
.033
2.5k
5µF
6VDC
18k
4.7k
5µF
6V
2N270, SK3004
GE2, HEP 250
270
500
5µF
6V
PHONES
NOT USED
2k:10Ω
-6

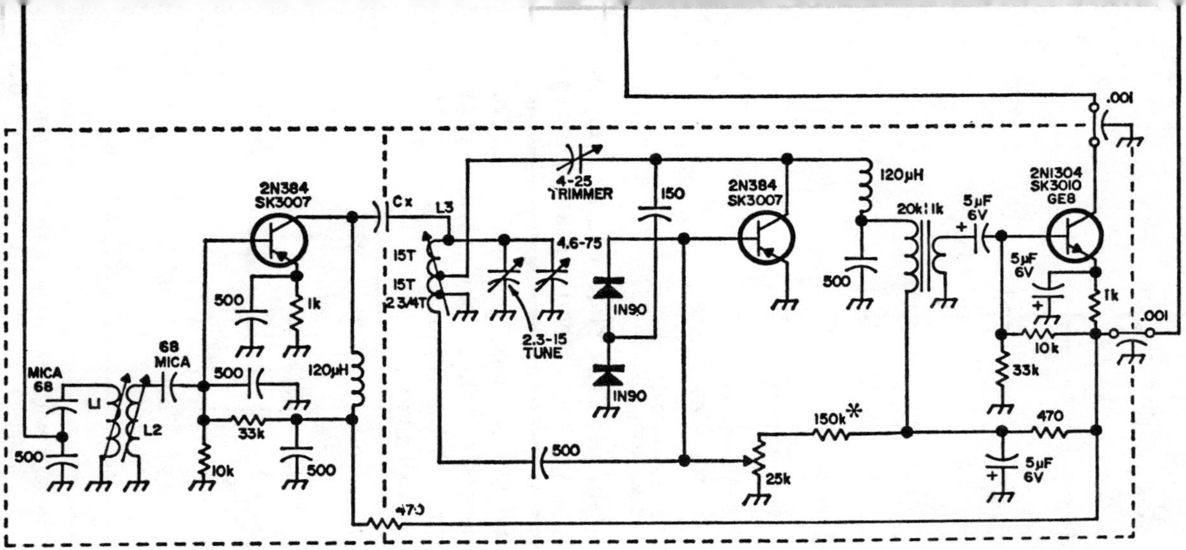


| | |
|---|---|
| $L1$, $L2$ | *30 turns #32 enameled wire close wound on 1/4" slug tuned form. J. W. Miller 46A000CPC or CTC 2206-2-3* |
| $L3$ | *32¾ turns #32 enameled wire close wound on 1/4" slug tuned form. (same form as above)* |
| $L4$ | *11 turns 1" diameter 16 TPI. AirDux or B & W.* |
| $L5$ | *4 turns hookup wire wound around the center of L4.* |

Continuous-wave transceiver for 7 MHz produces 100 mW of rf in transmit; delivers clean notes to earphone in receive mode. Coil data at lower right.

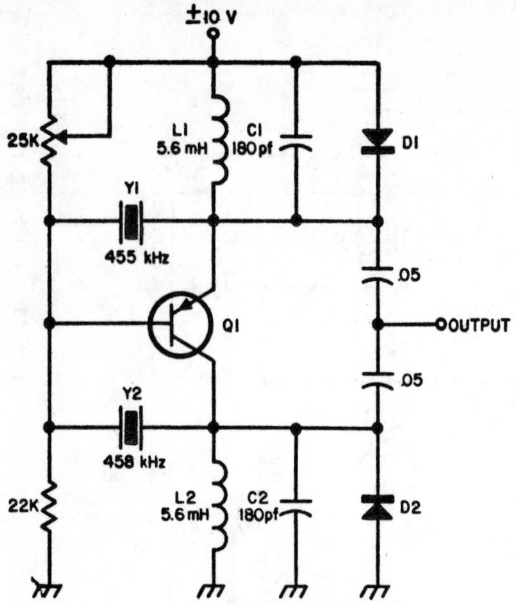


Q1- 2N384, 2N525, SK3004, TI-XM03
D1, D2- GENERAL PURPOSE SILICON DIODES

This two frequency crystal oscillator changes frequency by simply reversing the supply voltage. When the supply voltage is changed, the transistor inverts itself; usually transistors may not be used in the inverted mode, but in an oscillator a gain of only 1 or 2 is needed and this circuit provides a novel and simple way of obtaining two frequencies from a single stage with a minimum of switching.

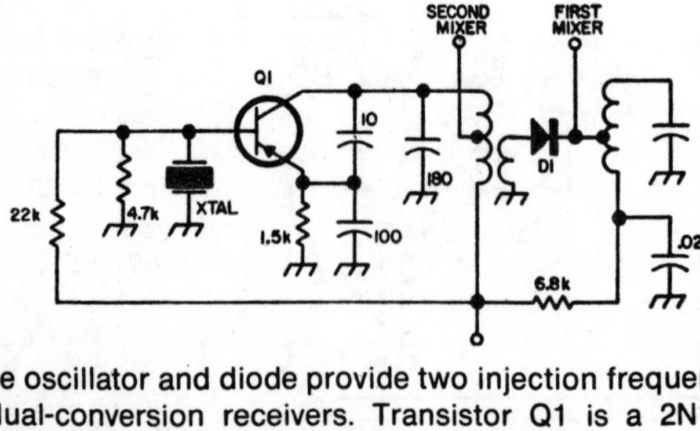


Single oscillator and diode provide two injection frequencies for dual-conversion receivers. Transistor Q1 is a 2N1745, 2N2188, T1M10, GE-9, or HEP 2; the diode should be a 1N82A or similar.

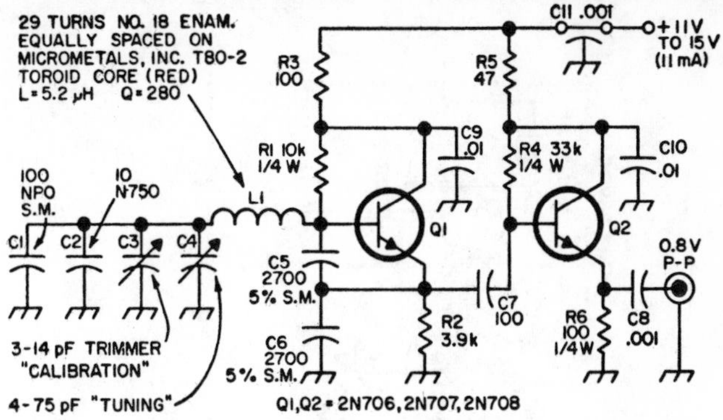


Transistor VFO for 5-6 MHz for SSB mixing service.

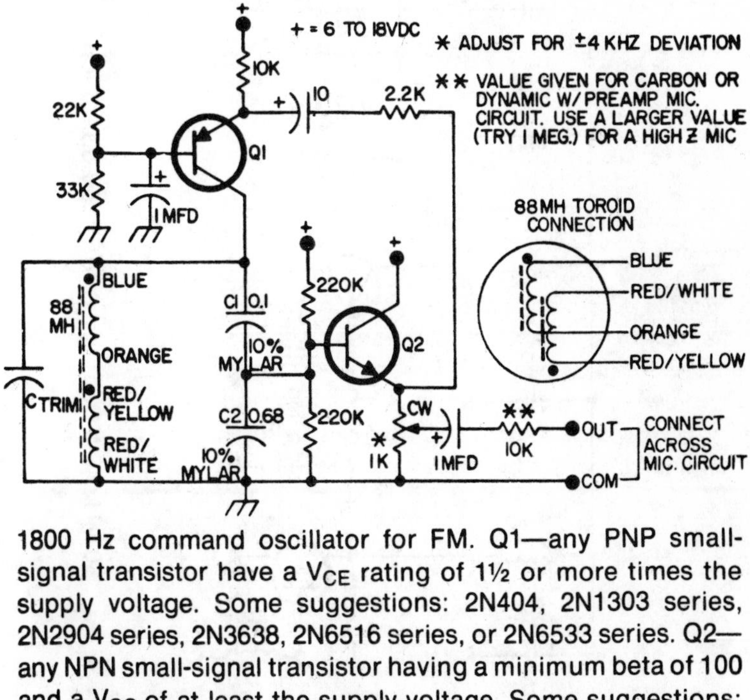


1800 Hz command oscillator for FM. Q1—any PNP small-signal transistor have a $V_{CE}$ rating of 1½ or more times the supply voltage. Some suggestions: 2N404, 2N1303 series, 2N2904 series, 2N3638, 2N6516 series, or 2N6533 series. Q2—any NPN small-signal transistor having a minimum beta of 100 and a $V_{CC}$ of at least the supply voltage. Some suggestions: 2N1308, 2N2712, 14, 16, 2N2916, 18, 20, 2N3565, 2N3569, 2N6513, 14, 15, 20, or 21. $C_{TRIM}$—0.0062 μF was used in the first unit. If Mylar capacitors and the toroid are ± values, $f_1$ = 1817 Hz. If C1 is 10% low, $f_1$ = 1897 Hz. If C1 is 10% high, $f_1$ = 1741 Hz. To find a value for $C_{TRIM}$, measure the frequency ($f_1$) without $C_{TRIM}$. If it is higher than 1800 Hz ($f_o$) calculate $C_{TRIM}$. $C_{TRIM} = 0.1 [f_1/f_o{}^2 - 1] = F$.

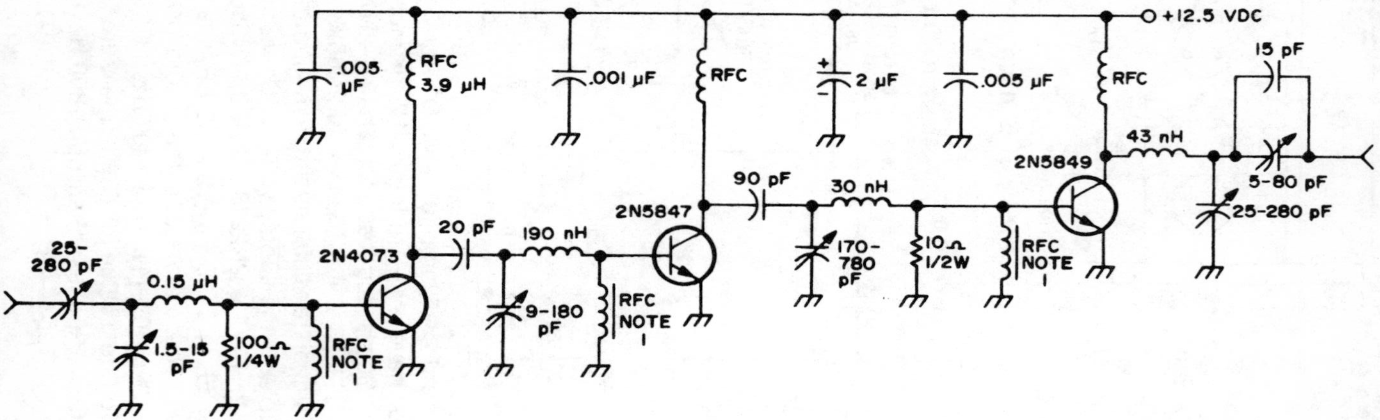


50 MHz 40W transmitter for 12.5V operation.

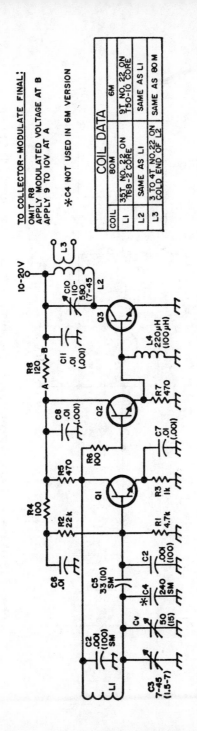

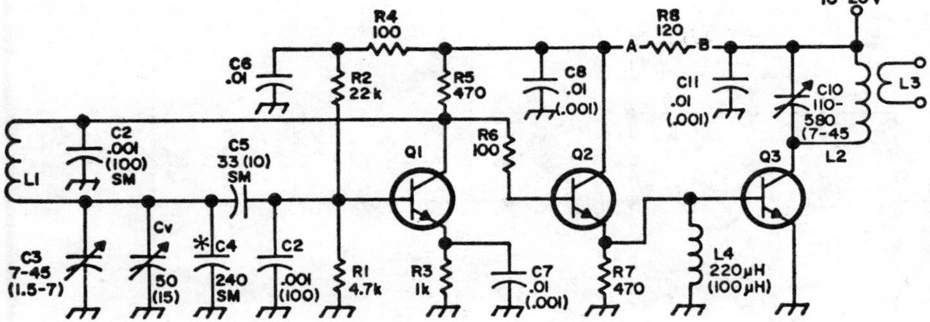


TO COLLECTOR-MODULATE FINAL:
OMIT R8
APPLY MODULATED VOLTAGE AT B
APPLY 9 TO 10V AT A

*C4 NOT USED IN 6M VERSION

| COIL DATA | | |
|---|---|---|
| COIL | 80M | 6M |
| L1 | 35T NO. 22 ON T68-2 CORE | 9T NO. 22 ON T50-10 CORE |
| L2 | SAME AS L1 | SAME AS L1 |
| L3 | 3 TO 4T NO.22 ON COLD END OF L2 | SAME AS 80 M |

Superstable variable-frequency oscillator uses three Motorola MPS706. Unit also serves as a low-power transmitter. Oscillator is little-known Vackar circuit, said to outperform the Clapp.

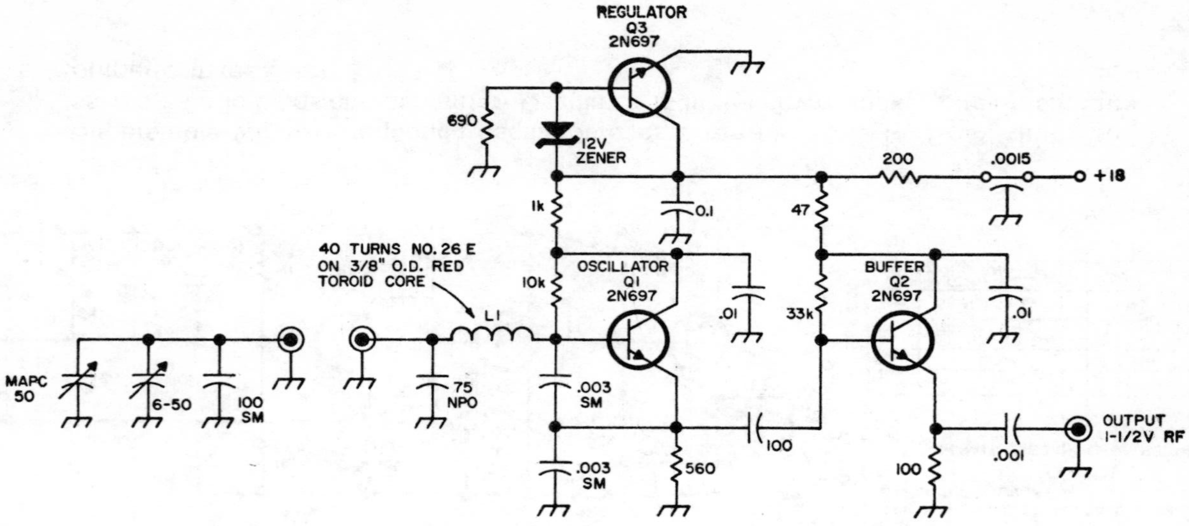


Schematic of the VFO and power supply. Regulation is provided for both the oscillator and buffer. If more rf output is required, a tuned circuit (toroid, of course) can be installed in the emitter of Q2 in lieu of the 1000Ω resistor.

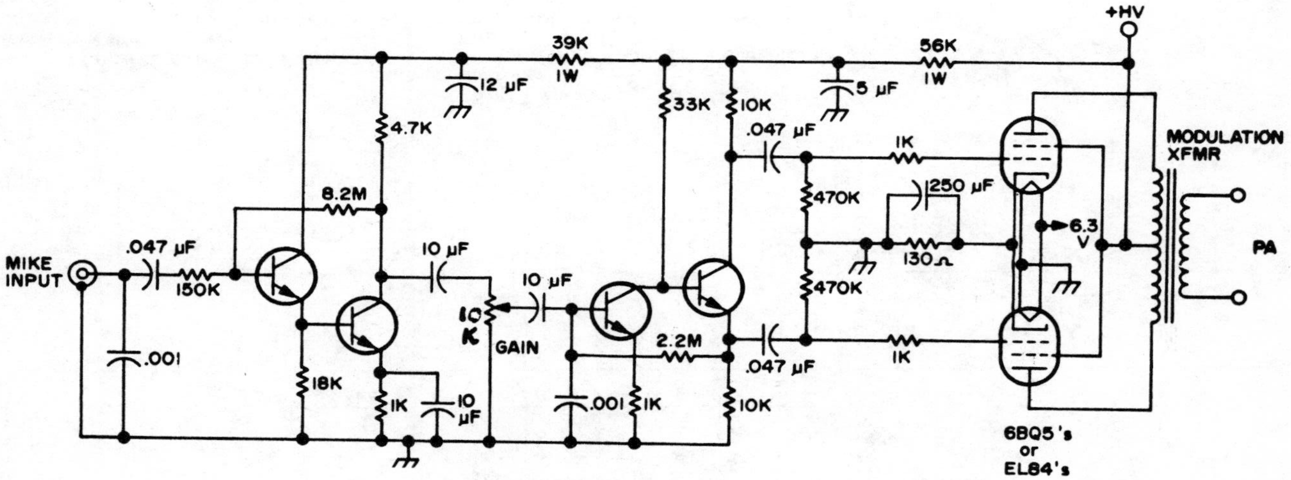


Minimod schematic.

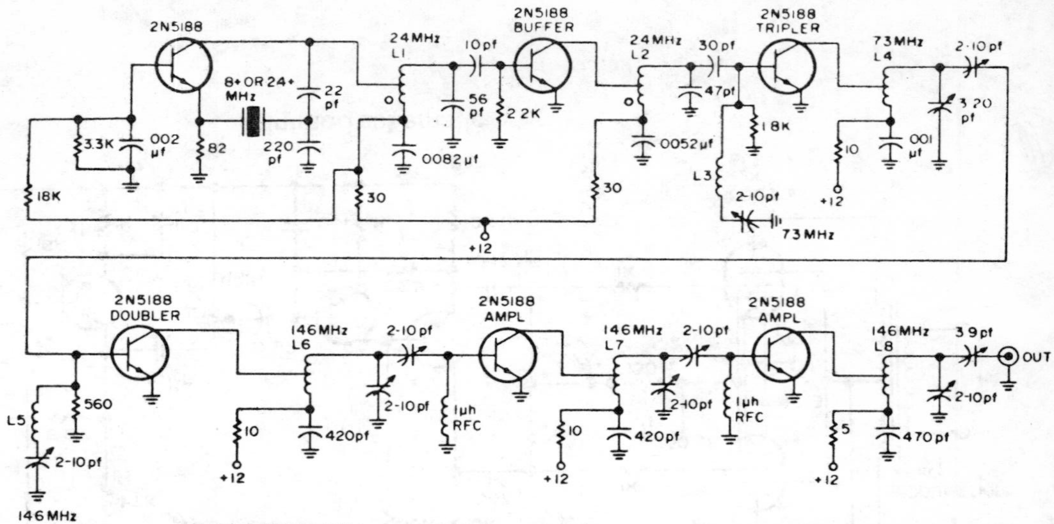


L1-L2 10 turns 22 enamel 5/16'' long, ¼'' diameter. Fr. slug coil form.
L3 20 T 24C, ½'' long by ¼'' diameter poly rod form.
L4 10 T #18, 3/4'' long ¼'' diameter tap 3 turns up.

L5 10 T #24C, ¼'' long by ¼'' diameter poly rod form.
L6 6 T #18, 3/4'' long, ¼'' diameter tap 2½ turns up.
L7 6T #18, 1'' long, ¼'' diameter tap 2 turns up.
L8 5T #14, 5/8'' long, 5/16'' diameter tap 1 turn up.

Unique 150 MHz exciter circuit uses 2N5188's exclusively: Output is about 1 watt unmodulated.

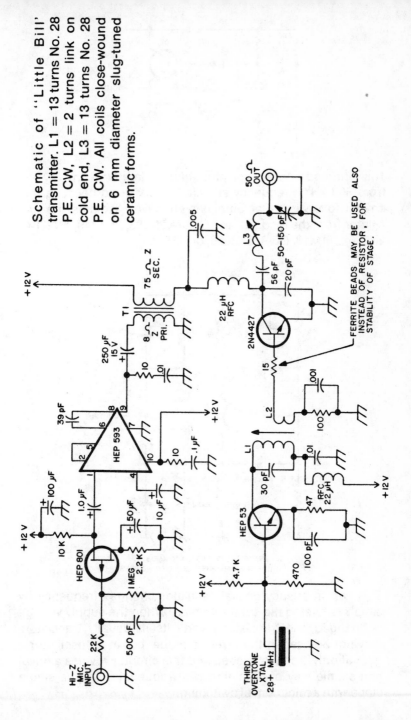

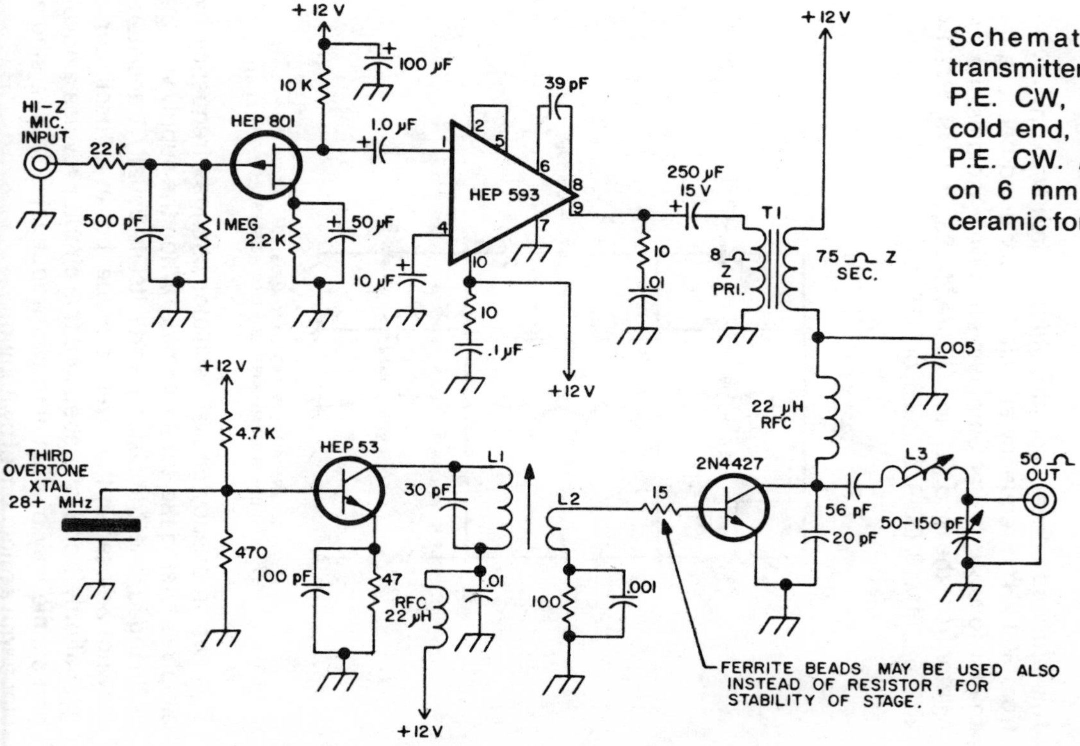


Schematic of "Little Bill' transmitter. L1 = 13 turns No. 28 P.E. CW, L2 = 2 turns link on cold end, L3 = 13 turns No. 28 P.E. CW. All coils close-wound on 6 mm diameter slug-tuned ceramic forms.

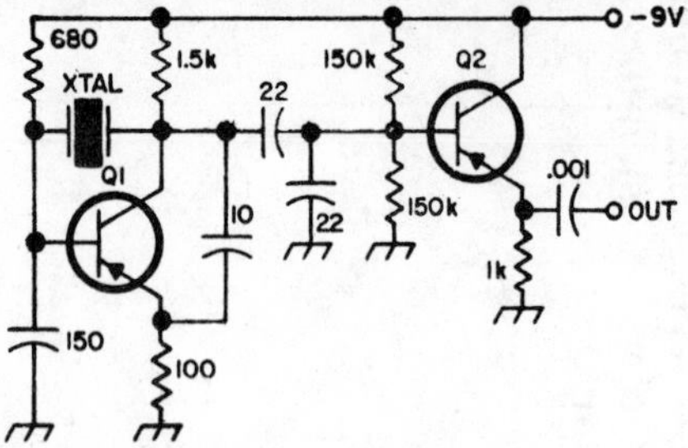


This untuned crystal oscillator will oscillate with any crystal from 300 kHz. Frequency stability is very good because the emitter-follower buffer amplifier effectively isolates the oscillator from the load. Q1 and Q2 are 2N993, 2N1749, 2N2084, 2N2362, T1M10, GE-9, SK3006, or HEP 2.

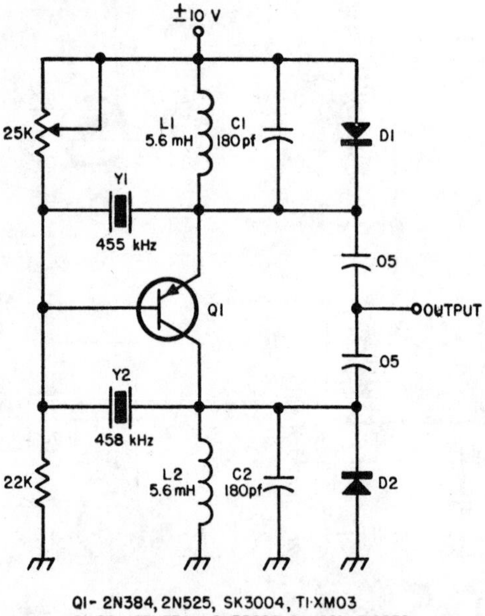


Q1- 2N384, 2N525, SK3004, TI·XM03
D1, D2- GENERAL PURPOSE SILICON DIODES

This two frequency crystal oscillator changes frequency by simply reversing the supply voltage. When the supply voltage is changed, the transistor inverts itself; usually transistors may not be used in the inverted mode, but in an oscillator a gain of only 1 or 2 is needed and this circuit provides a novel and simple way of obtaining two frequencies from a single stage with a minimum of switching.

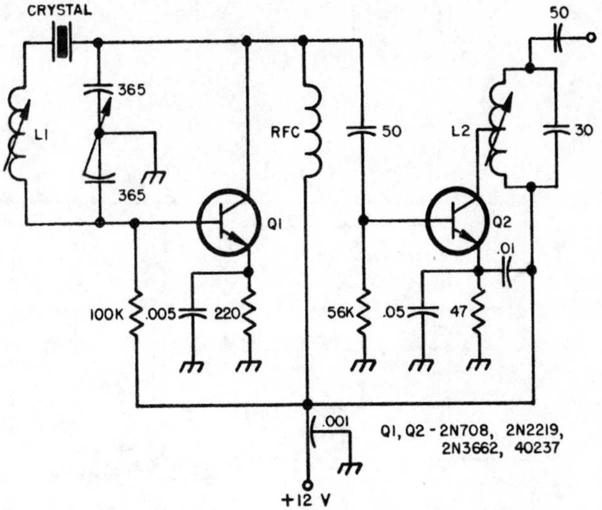


This variable crystal oscillator (VXO) may be used to vary the frequency of an 8 MHz crystal 4 or 5 kHz when the 365 pF dual variable is tuned through its range. When multiplied to 2 meters of 432 MHz, this provides a very stable variable frequency. For 8 MHz crystals, L1 is a 20–25 $\mu$H slug-tuned coil; L2 is chosen to resonate at 8 MHz with the 30 pF capacitor.

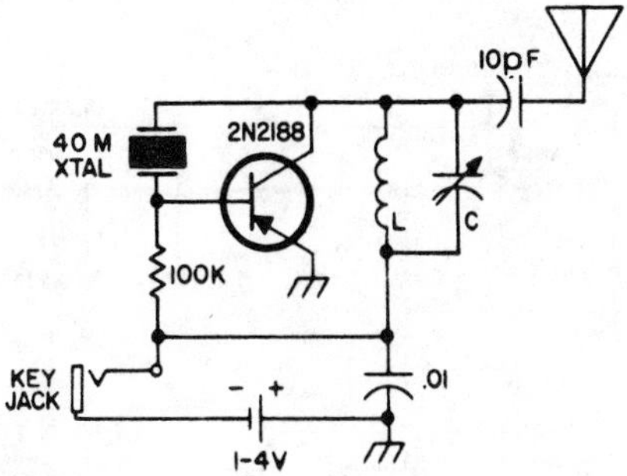


Microminiature CW transmitter may be world's smallest! L and C are selected for resonance at crystal frequency.

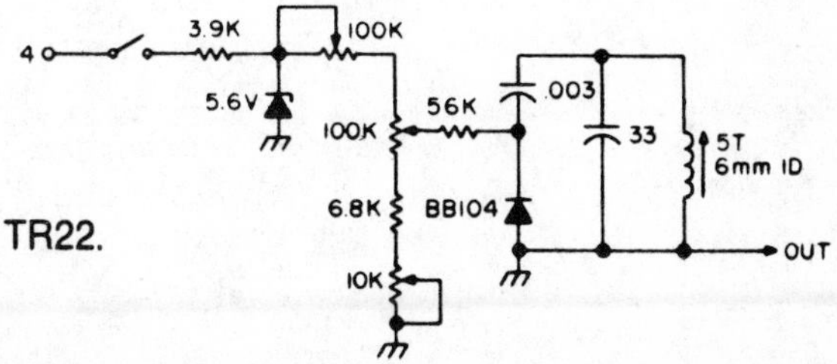


12 MHz VFO for TR22.

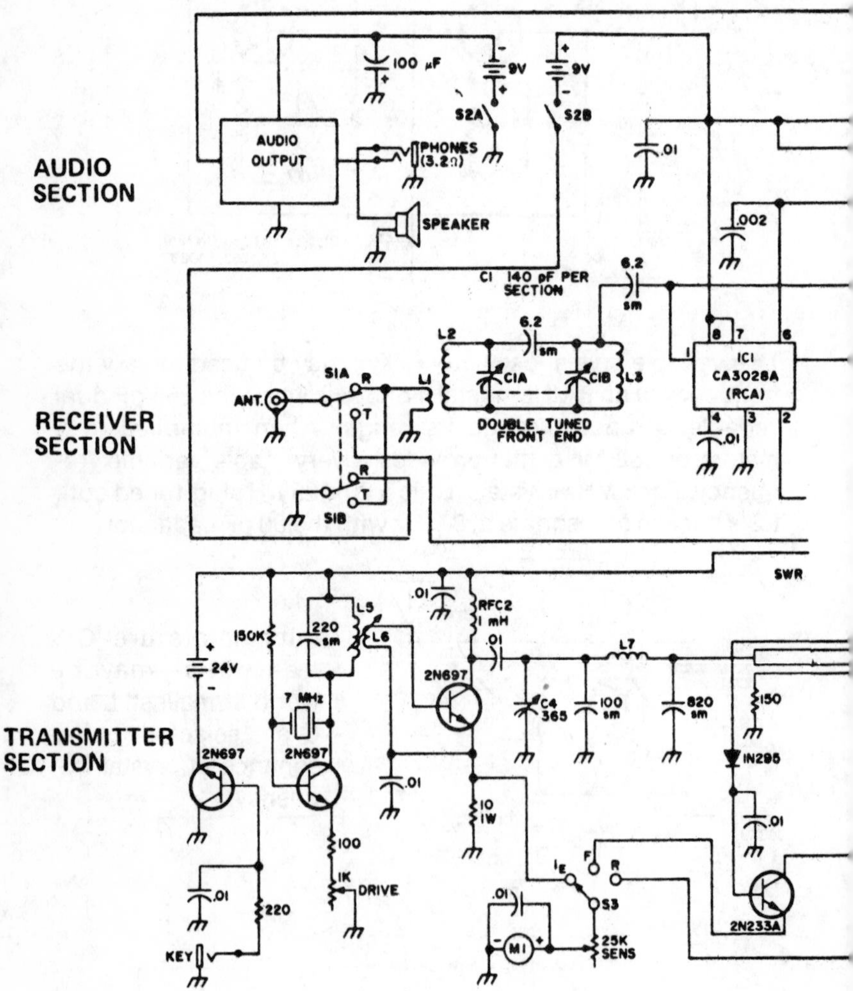

AUDIO SECTION
RECEIVER SECTION
TRANSMITTER SECTION
AUDIO OUTPUT
100 µF
9V
S2A
S2B
PHONES (3.2Ω)
SPEAKER
.01
.002
C1 140 pF PER SECTION
6.2 sm
L2
C1A
C1B
L3
L1
DOUBLE TUNED FRONT END
ANT.
S1A
S1B
R
T
IC1 CA3028A (RCA)
SWR
150K
220 sm
L5
L6
RFC2 1 mH
24V
7 MHz
2N697
C4 365
100 sm
820 sm
L7
150
IN295
10 1W
100
1K DRIVE
220
KEY
M1
25K SENS
S3
F
2N233A

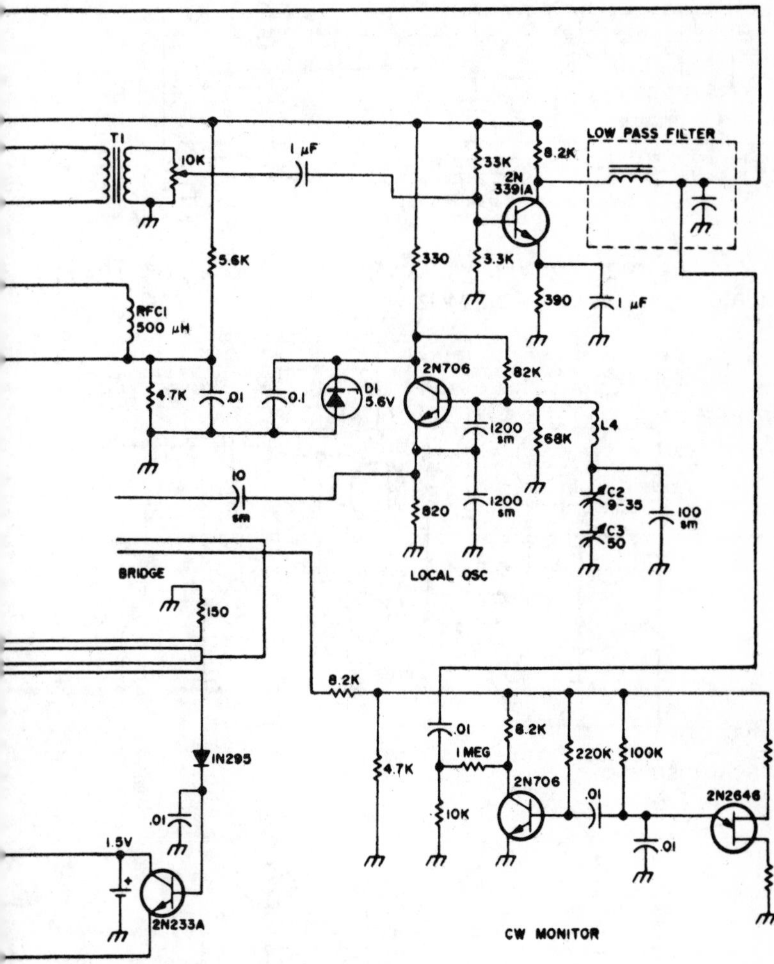


Schematic diagram of solid-state 7 MHz transceiver.

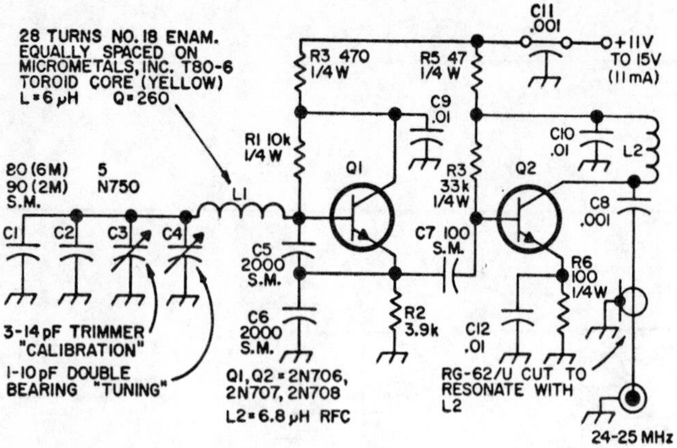


6 or 2 meter transistor VFO with output on 24 – 25 MHz. The oscillator operates on 8 – 8.3 MHz.

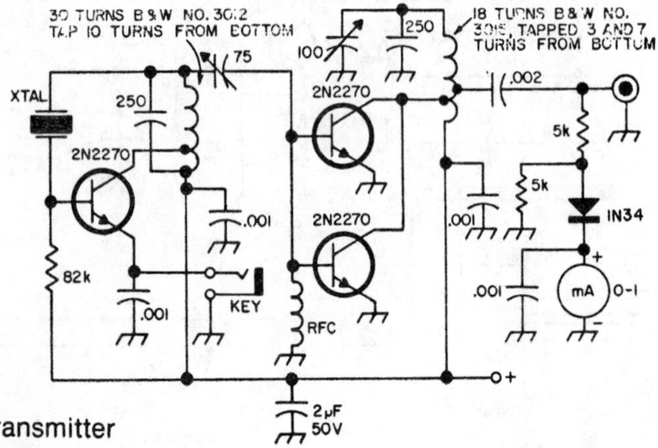


Novice transmitter for 80 meters.

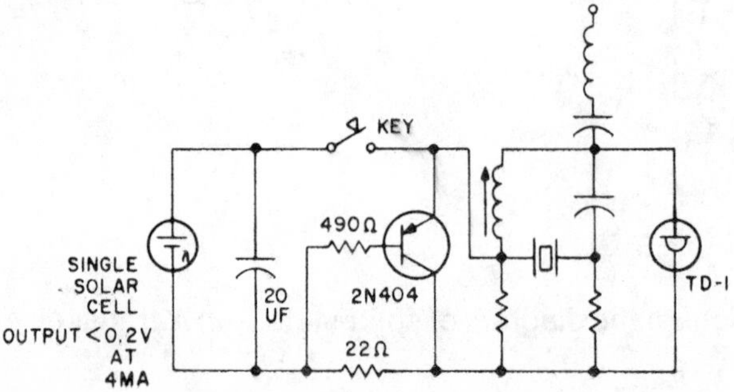


Solar-powered CW transmitter using tunnel diode.

# Section P

# Battery Chargers

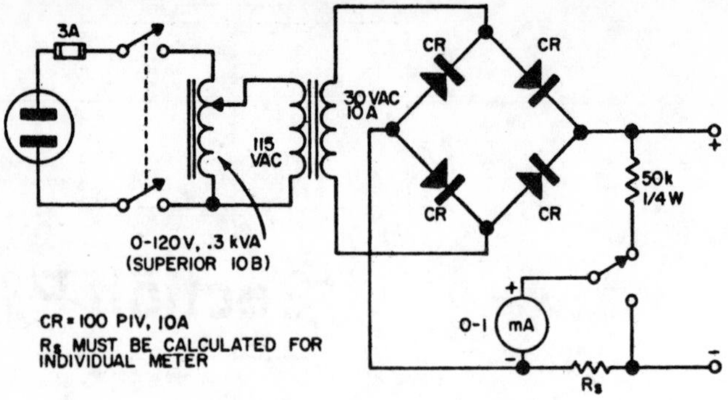


Battery charger schematic.

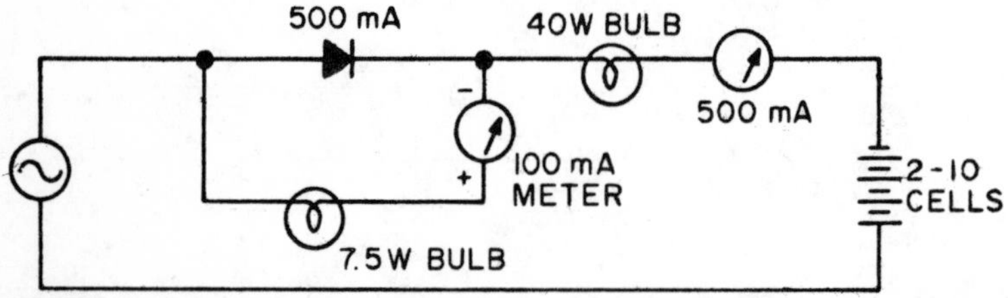


Reverse-current charger for dry cells allows many cells to be charged at one time. Technique allows many times the normal ampere-hour capacity of dry cells. The current through the line meter should be about 10% of that through the 500 mA meter.

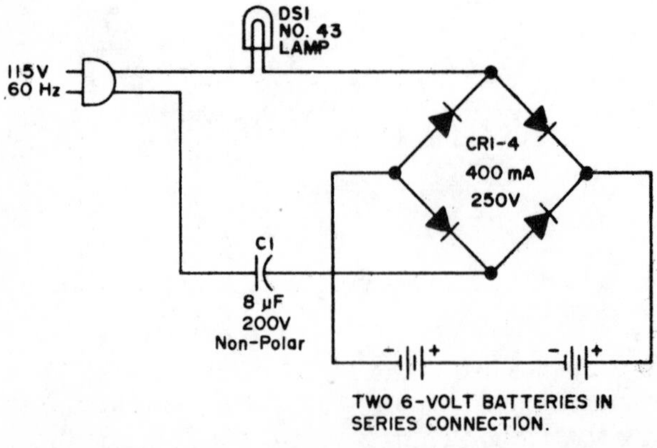


Simple ni-cad battery charger circuit from FM Magazine.

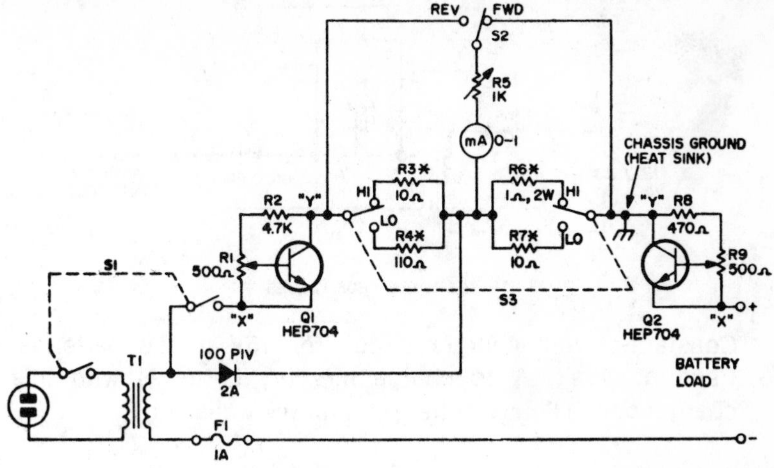


| Switch Setting | Current Ranges Fws. | Rev. |
|---|---|---|
| HI | *1.0A | 100 mA |
| LO | 100 mA | 10 mA |

*Maximum fwd. current available is about 0.7A with battery output terminals shorted.

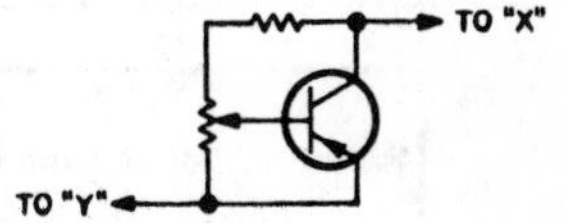


Reverse-current charger is said to offer effective means of rejuvenating dry cells.

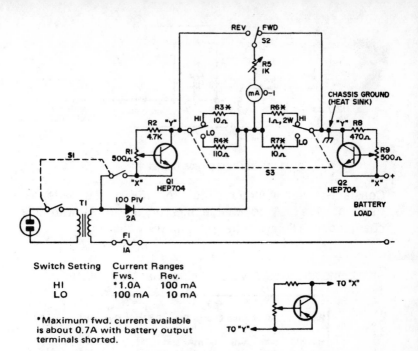

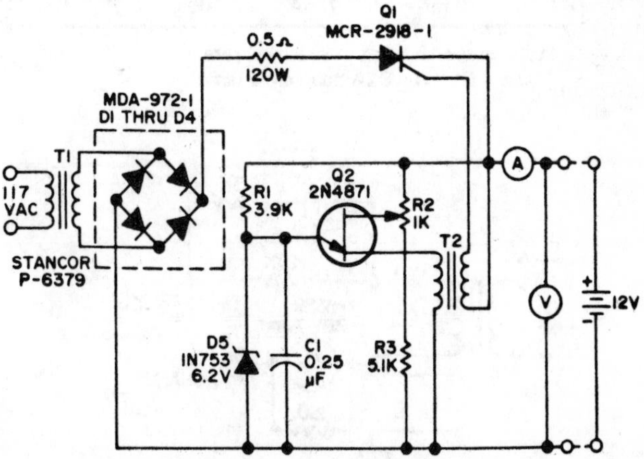


Foolproof Battery Charger. T2 is a Sprague 11Z12. Protects battery against overcharging or reverse charging, also protects itself against shorting or hurting another supply. Will supply 16A. Circuit courtesy Motorola Semiconductor Power Circuits Handbook.

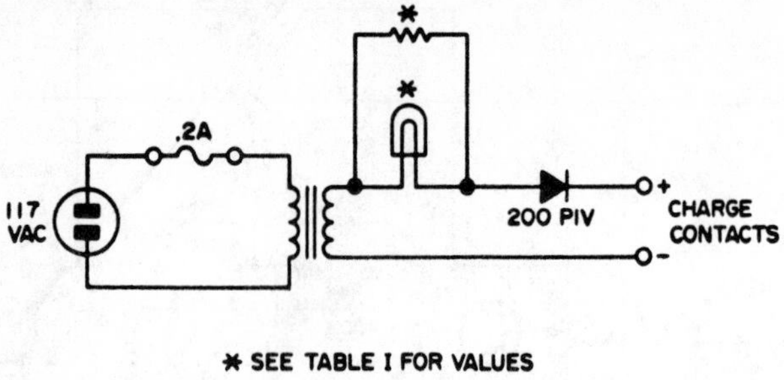


Constant-current charger circuit for 15V ni-cad batteries. Caution: Don't try to charge mercury batteries with this charger or you'll have battery all over the walls.

Lamp and resistor values for popular Motorola units.

| Radio | Batt Rating | 16-hr Charge Rate | Resistor | Lamp Current |
|---|---|---|---|---|
| HT-200 | 500 mA-hr | 50 mA | 120Ω | 70 mA |
| HT-220 | 235 mA-hr | 25 mA* | 270Ω | 40 mA |
| HT-220 OMNI | 450 mA-hr | 45 mA | 180Ω | 70 mA |
| HT-100 | 70 mA-hr | 7 mA | 100K | ** |

*Charge for 15 hours at specified rate.
**Use a GE type B2A neon glow lamp.

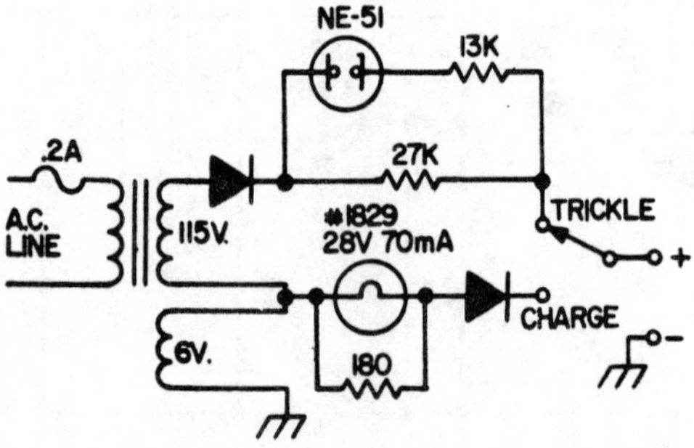


Ni-cad charger suitable for the HT220 or other hand held rigs. The dual indicator lights automatically let you know if you are on trickle or full charge. This is a schematic of the Motorola NLN6804A charger.

# Section Q

# Diode Circuits

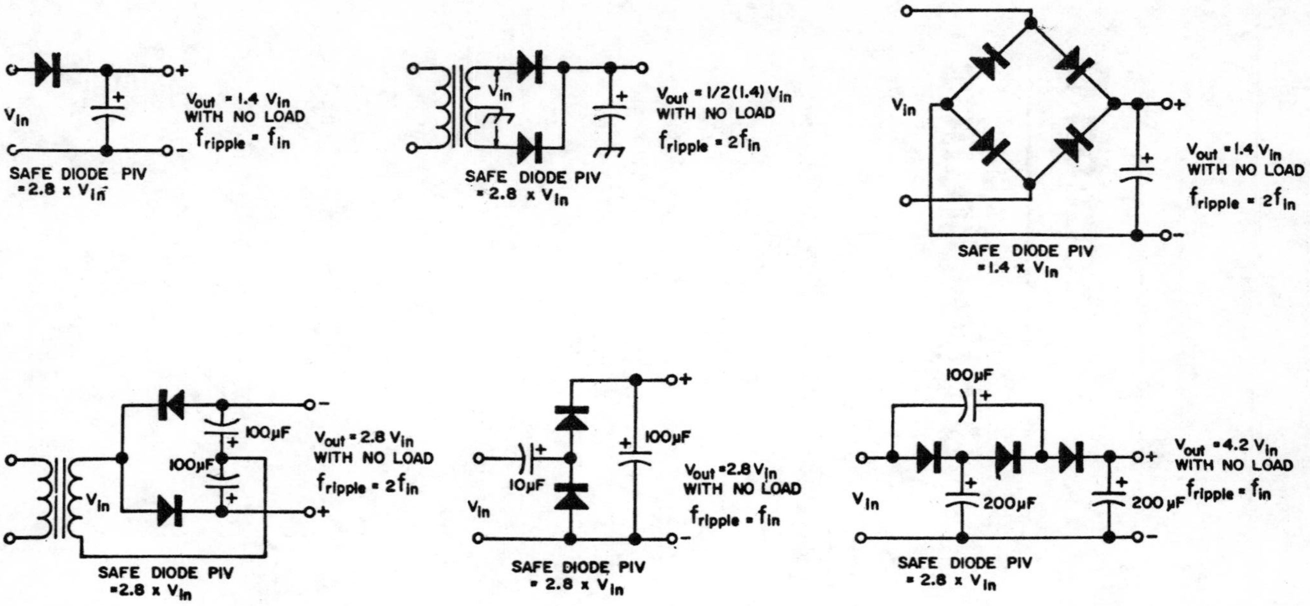


The most popular rectifier circuits with their output voltages and minimum safe diode PIV rating.

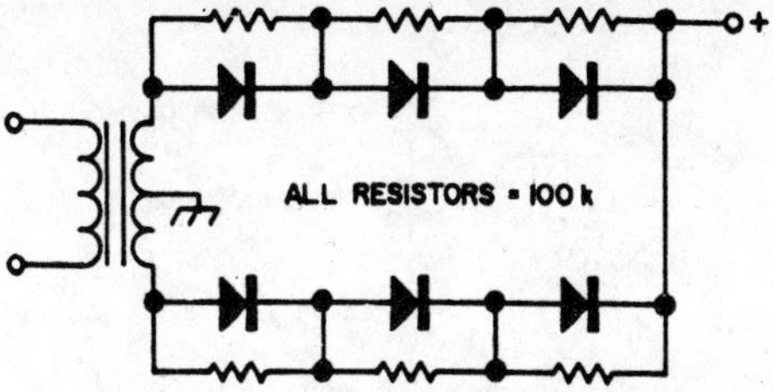


Resistors are used across series diodes to equalize reverse voltage drop.

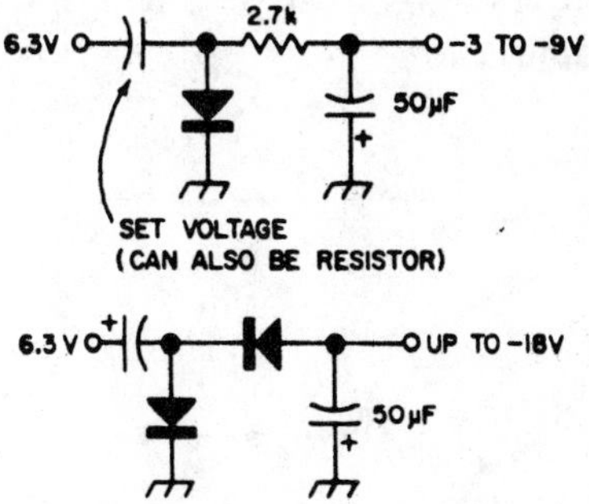


Simple shunt rectifiers can provide low bias voltages.

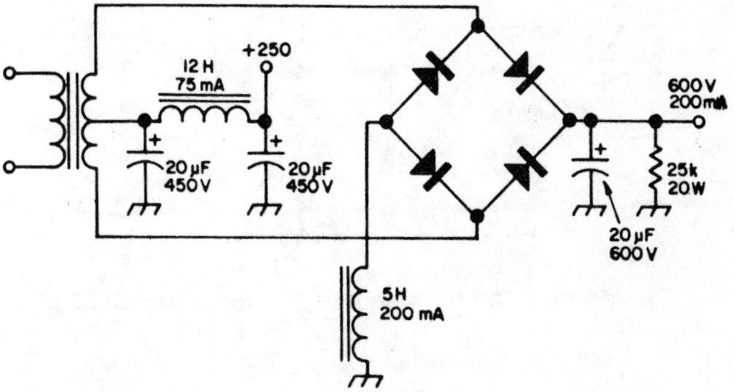


This circuit gives two outputs, 600V and 250V.

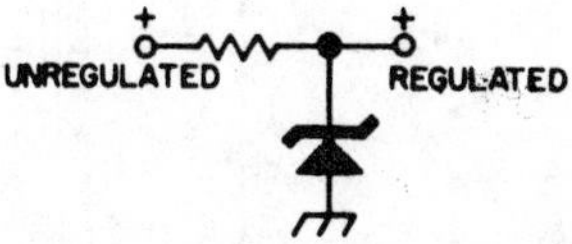


A basic zener regulator. The values depend on input and output voltage, current, etc.

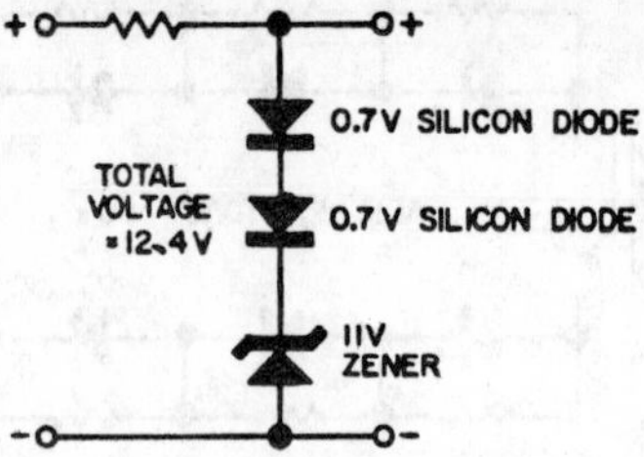


Forward-biased silicon diodes can be used as low-voltage zeners. Their temperature drift is opposite that of regulators with breakdown voltages over 6 V, which is convenient for temperature stabilization.

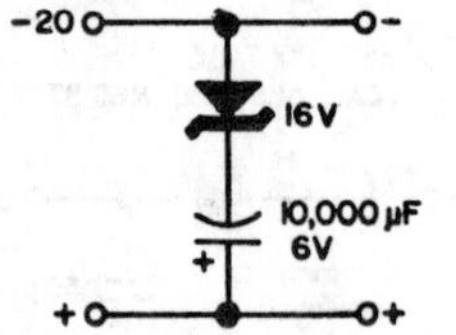


A zener can be used as a ripple filter and to "increase" the voltage rating of a capacitor.

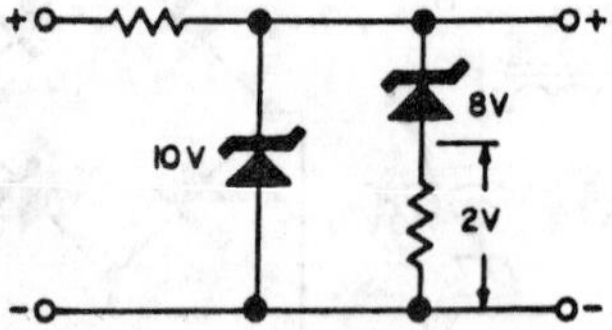


Two zeners can furnish a regulated low voltage.

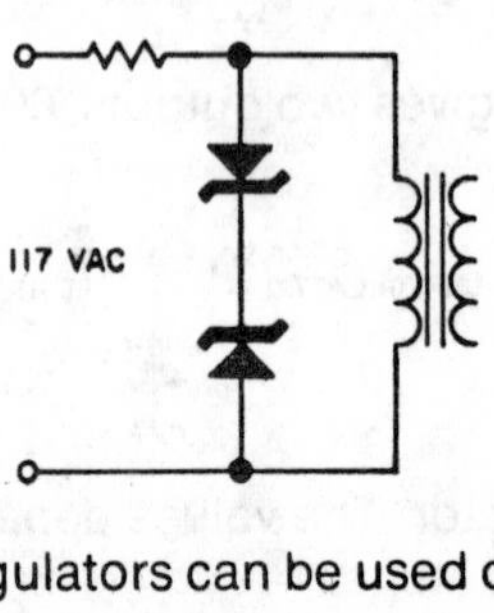


Zener regulators can be used on ac, too.

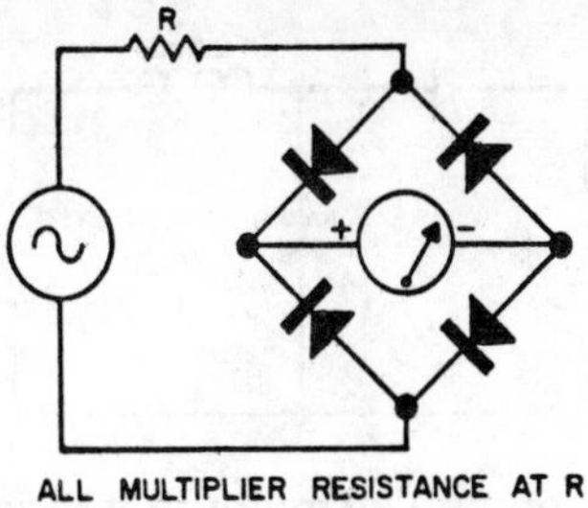


A bridge average-reading ac meter.

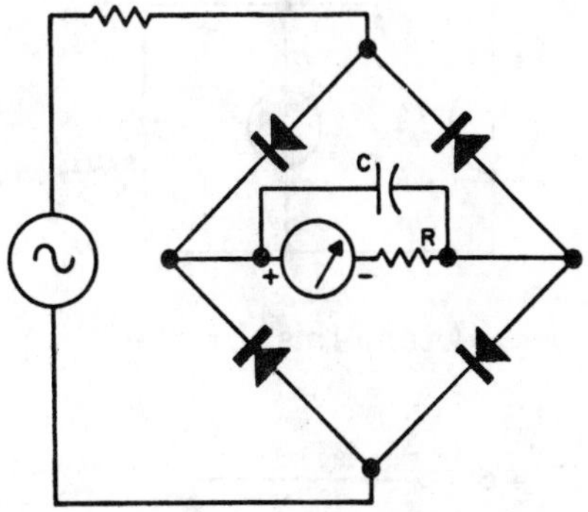


A bridge peak-reading ac meter.

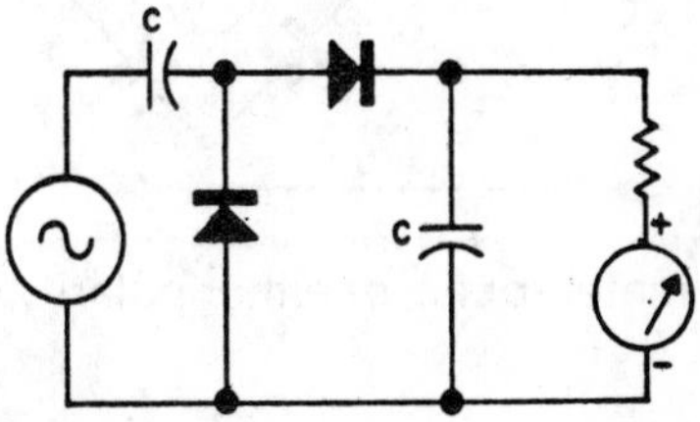


A peak-to-peak ac meter is simply a voltage "doubler."

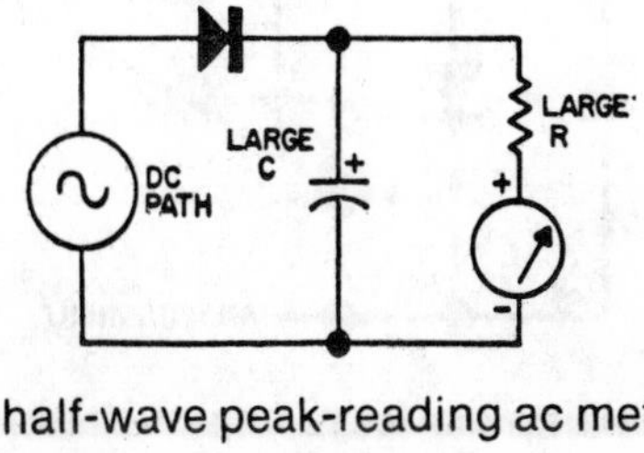


A half-wave peak-reading ac meter.

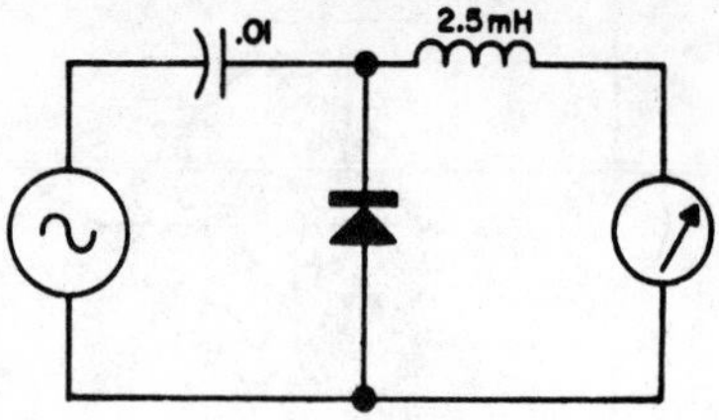


A half-wave peak-reading ac meter that requires no dc path.

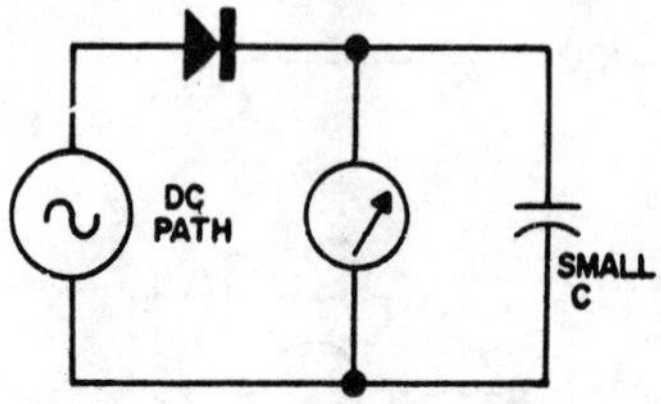


A semi-rms ac meter.

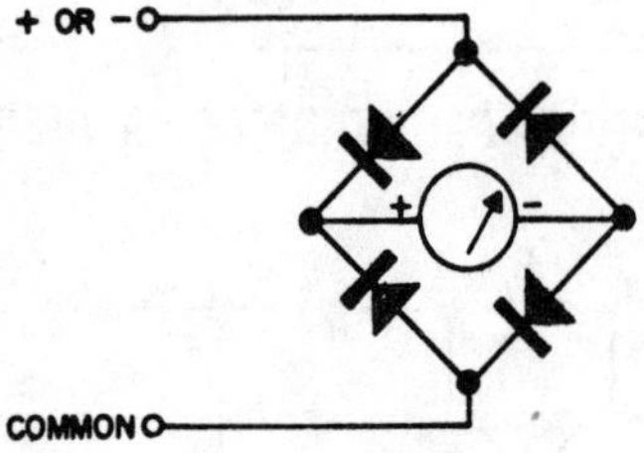


A meter for ac, or either polarity dc.

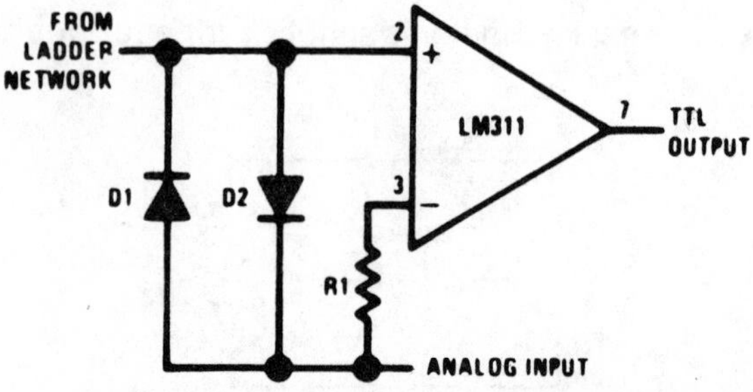


Using clamp diodes to improve response.

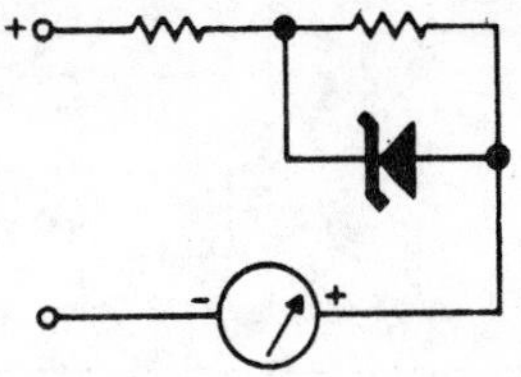

This circuit partially suppresses the low end of a range.

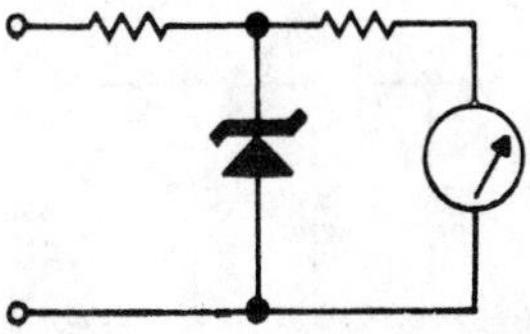

This is a meter-protective circuit. The zener should be tapped on the resistor chain at a point that provides conduction when the meter pointer is pinned.

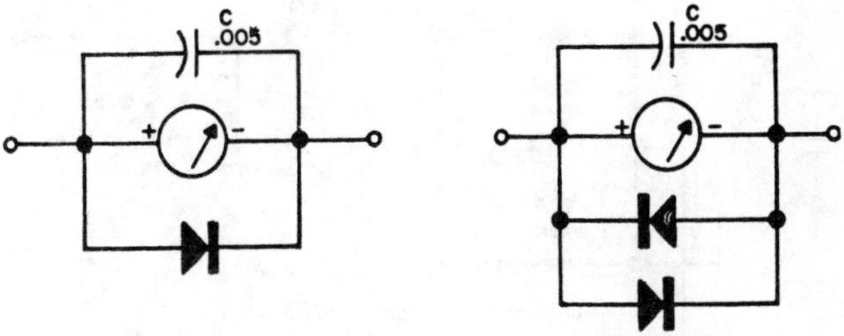


Conventional silicon diodes can protect a meter movement, too. The 0.005 $\mu$F capacitor bypasses rectified rf.

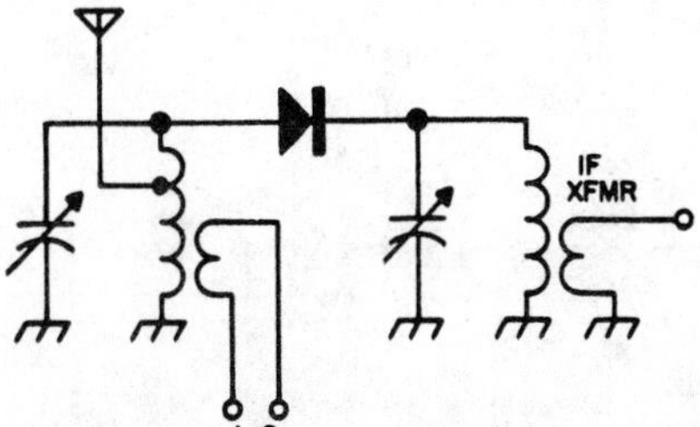


A basic diode mixer as used at UHF and microwave.

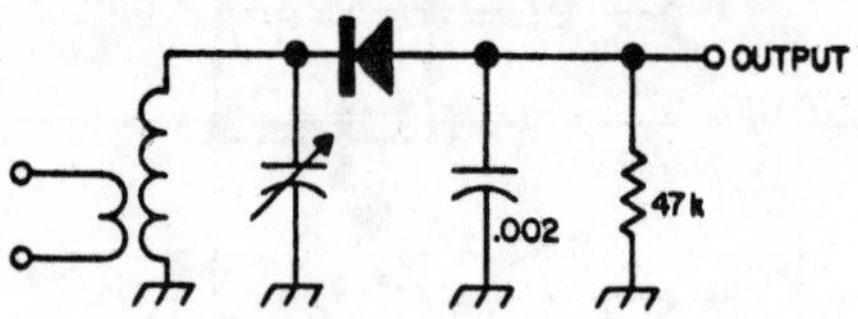


A half-wave detector. This can be used as a crystal set, too.

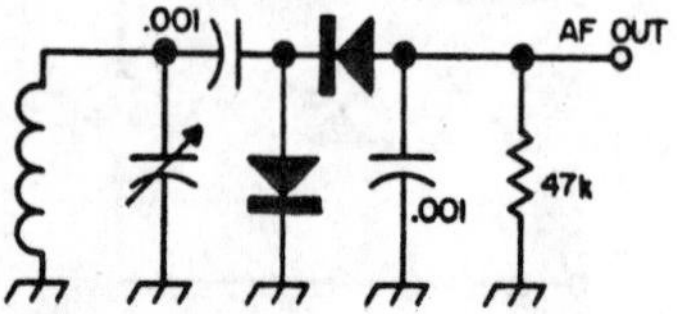


This detector provides better results.

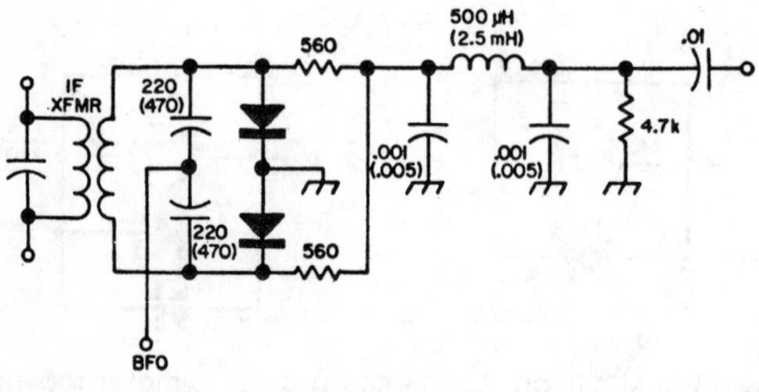


A product detector for 9 MHz SSB. The values in parentheses are for 455 kHz.

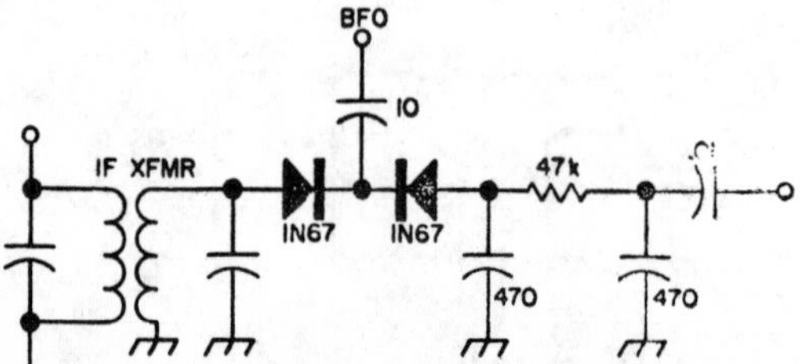


A popular product detector for SSB.

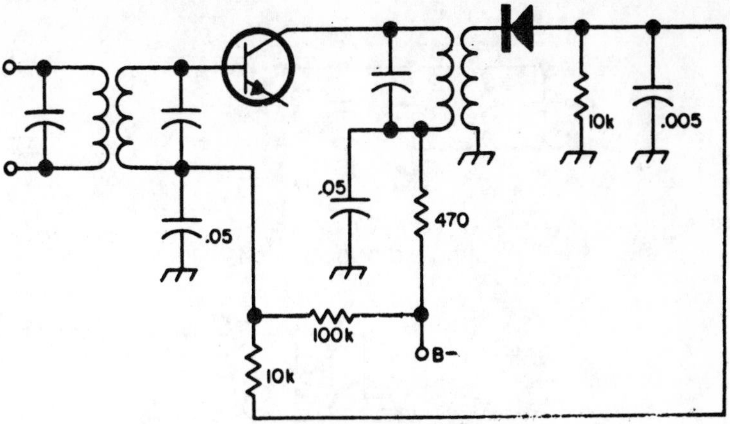


Forward transistor agc.

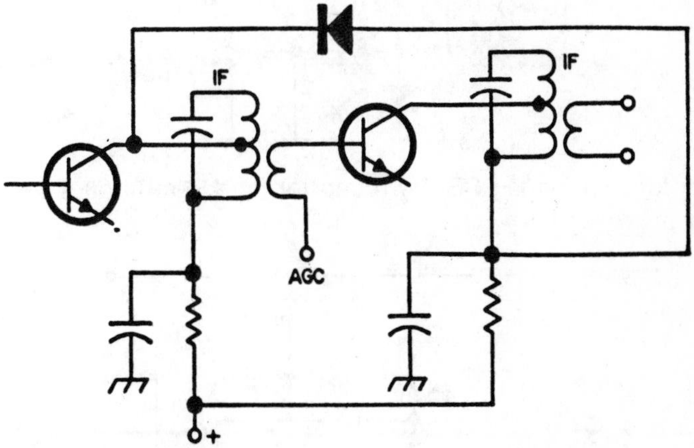


An auxiliary agc diode improves agc action.

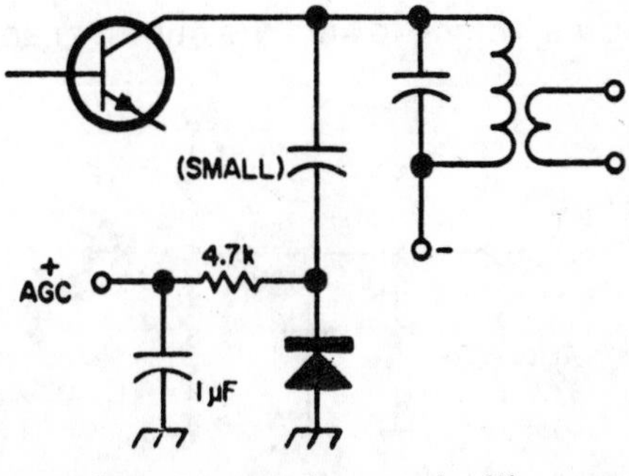


An auxiliary agc detector can be used with a product detector for SSB/CW.

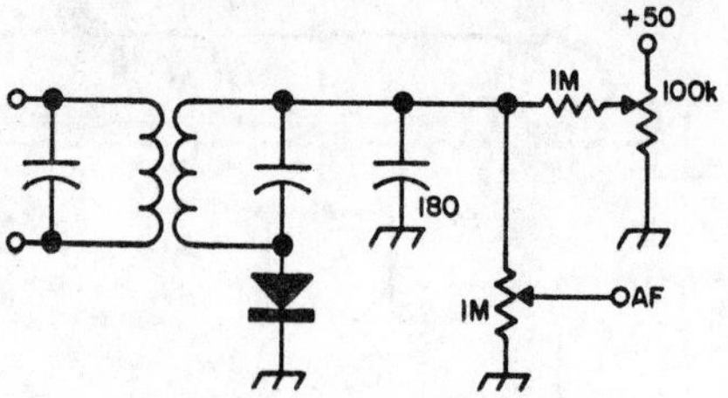


Simple diode squelch.

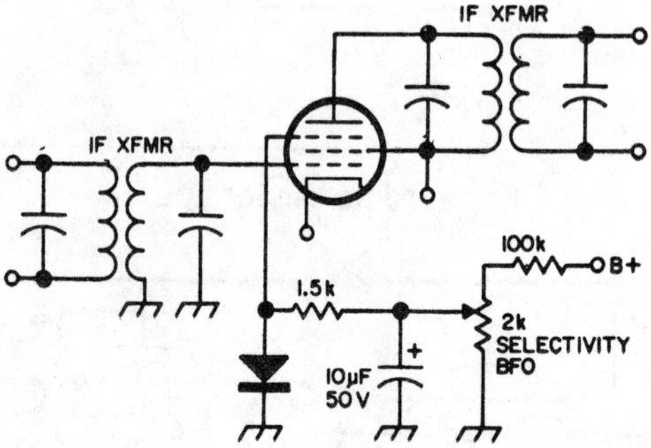


Adapter to provide SSB/CW reception and Q-multiplication in a receiver.

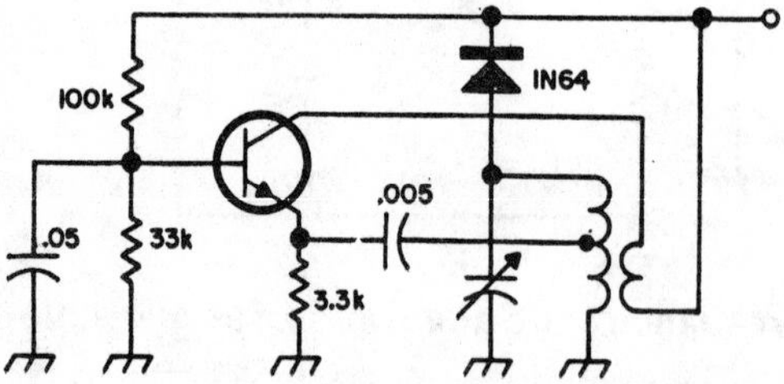


This circuit uses a diode to limit the output of an oscillator.

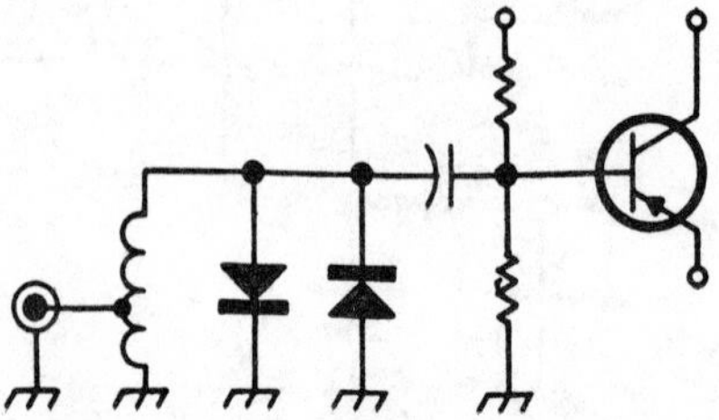

Diodes can be used to protect a transistor rf amplifier from burnout.

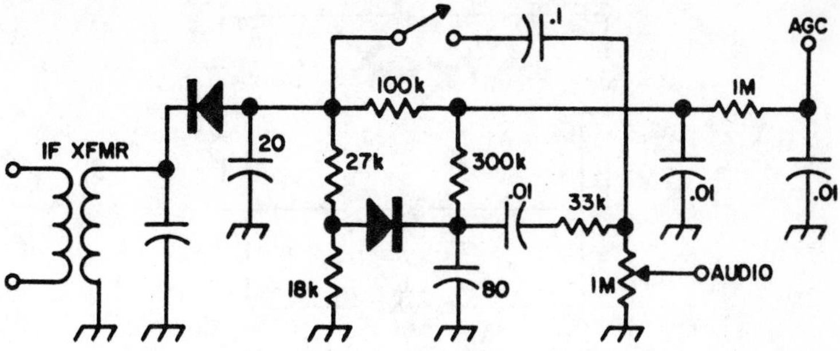


One of the best noise limiters is the rate-of-change limiter designed for TV audio in England.

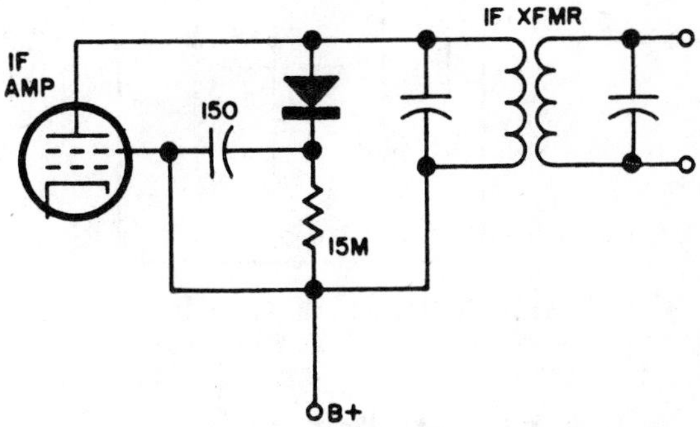


This simple noise limiter is installed in an i-f stage for SSB and CW.

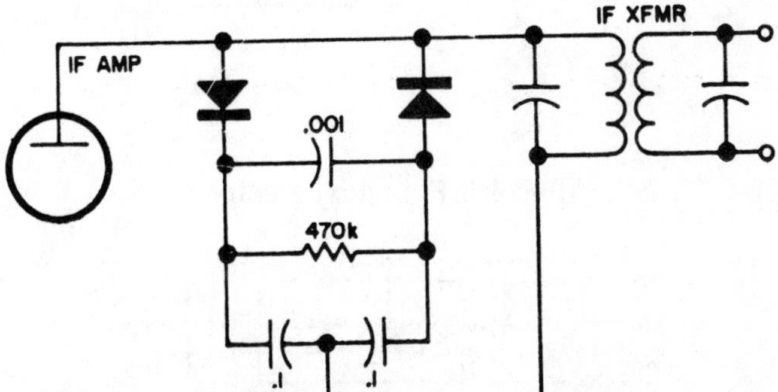


This is an improved version of the SSB i-f noise limiter.

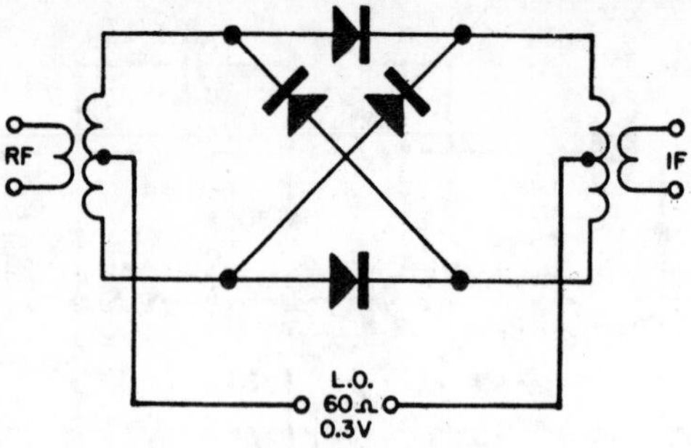


A diode ring balanced modulator.

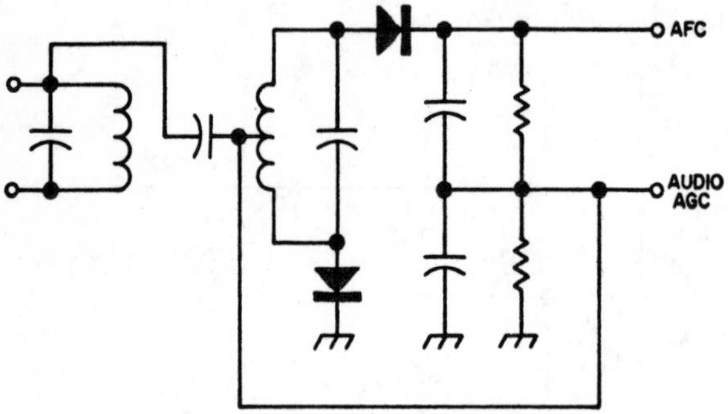


Foster-Seeley FM discriminator.

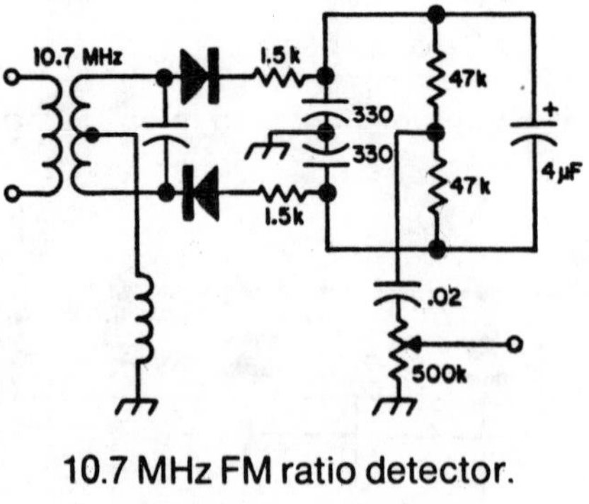


10.7 MHz FM ratio detector.

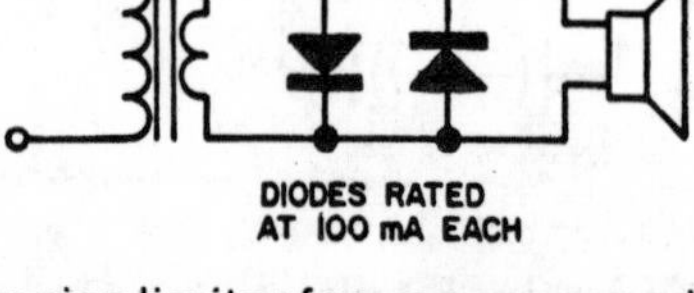


Shunt diode noise limiter for use across a loudspeaker.

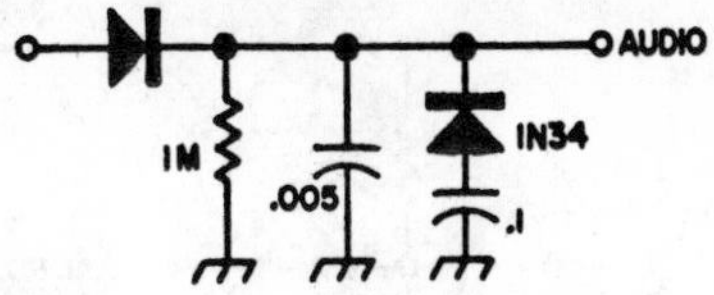


Shunt diode noise limiter that can be easily added to the input of an audio amplifier.

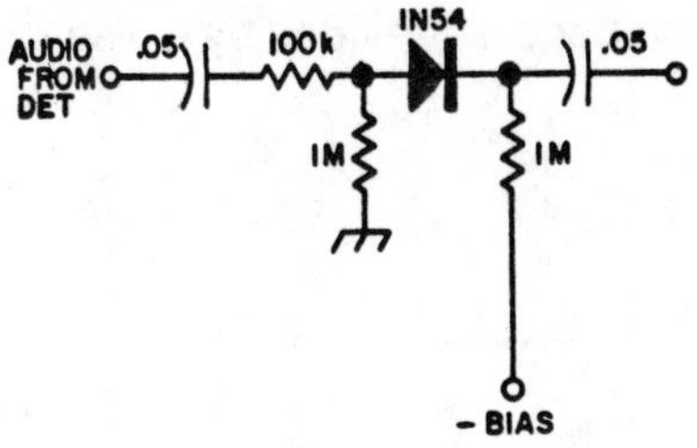


Half-wave series noise limiter with adjustable clipping level.

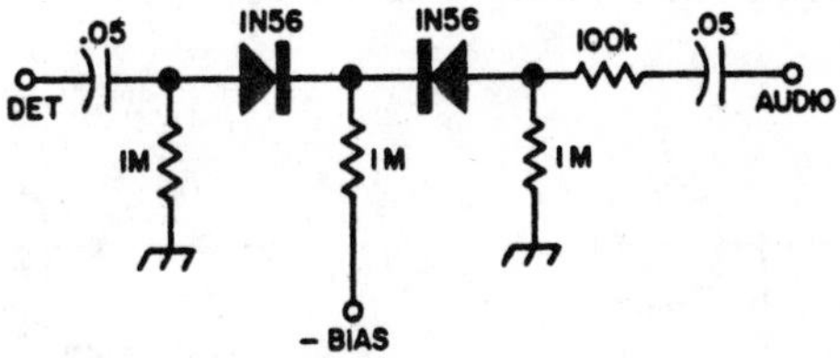


Full-wave series noise limiter.

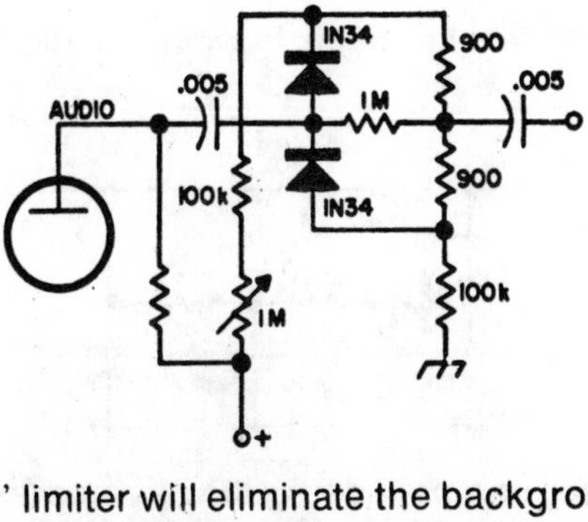


This "trough" limiter will eliminate the background noise that is ignored by conventional limiters.

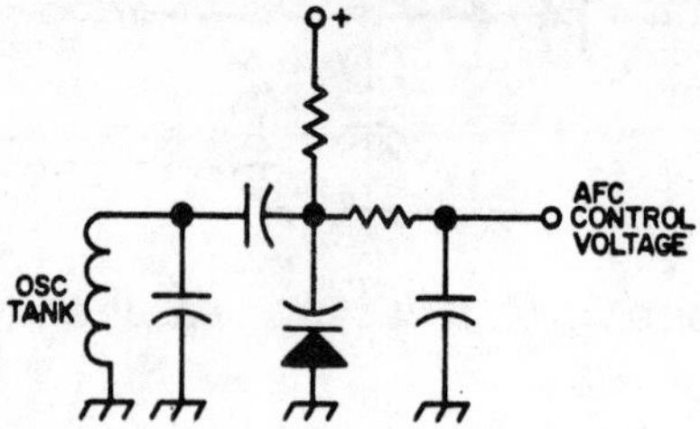


A varactor is often used to provide automatic frequency control. The control voltage is provided by a discriminator.

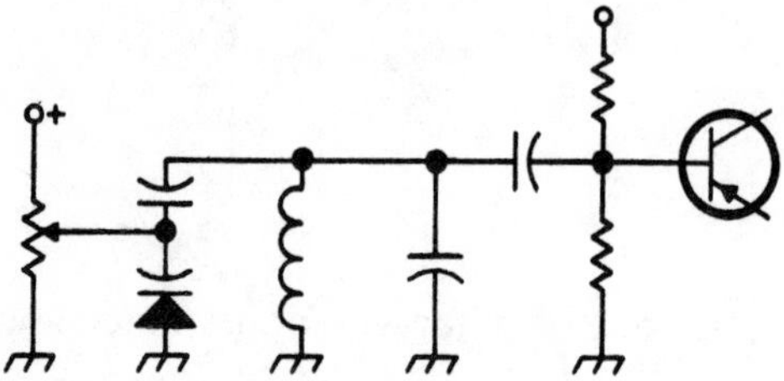


An rf stage or oscillator can be tuned with a varactor.

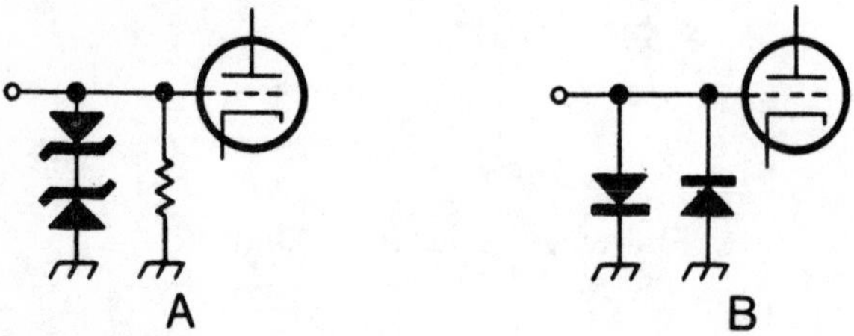


Simple clippers can be made from zener diodes or silicon diodes.

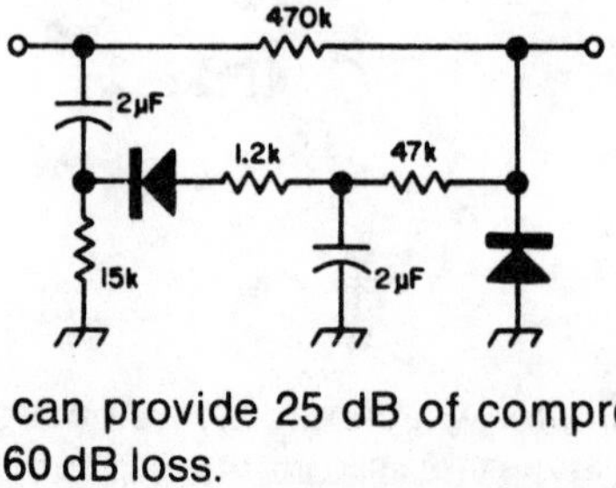


The compressor can provide 25 dB of compression, but the expense of up to 60 dB loss.

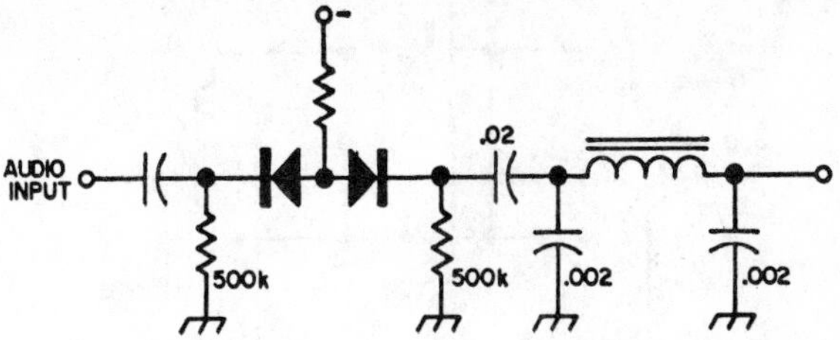


A good clipper for AM or FM use includes adjustable clipping level and a harmonic filter.

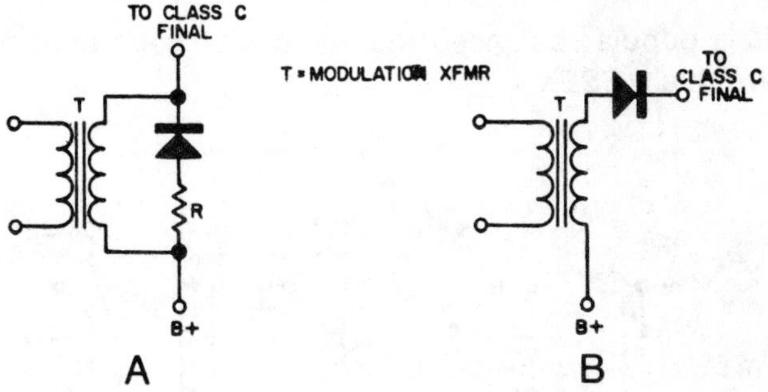


The need for high-level negative-peak clipping is often debated, but its value is championed by many.

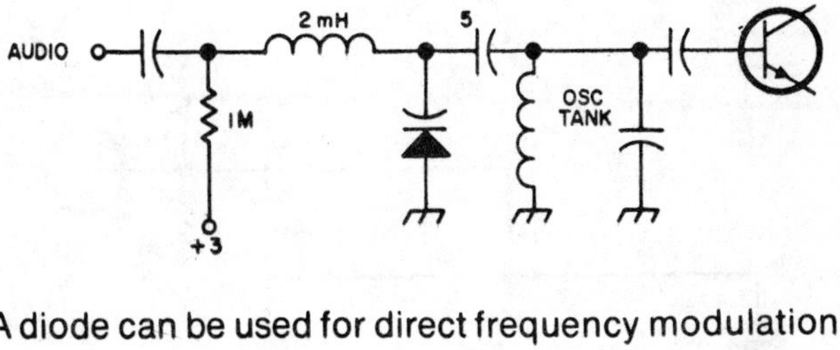


A diode can be used for direct frequency modulation.

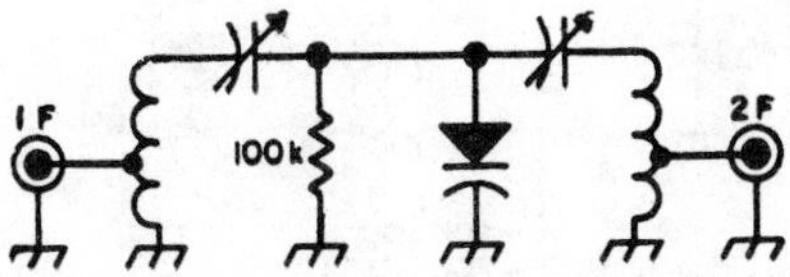


This is a basic varactor doubler.

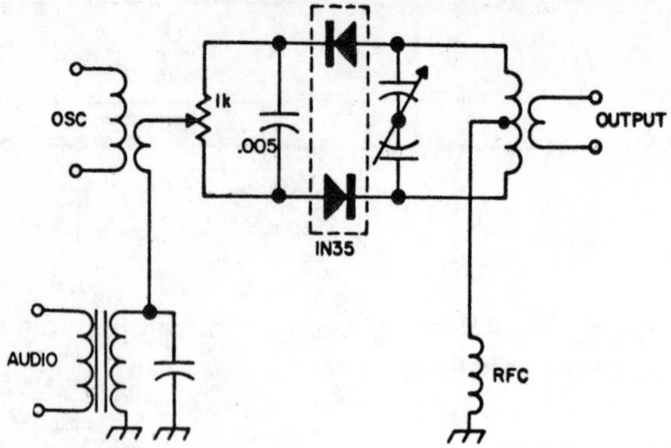


This is a popular balanced modulator for generating DSB (and eventually SSB).

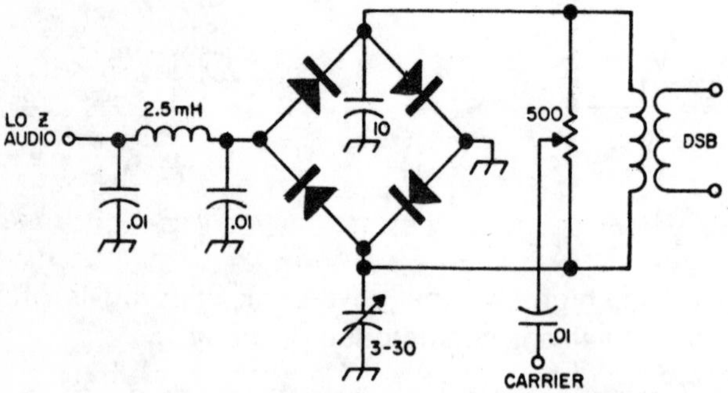


Bridge balanced modulator for SSB.

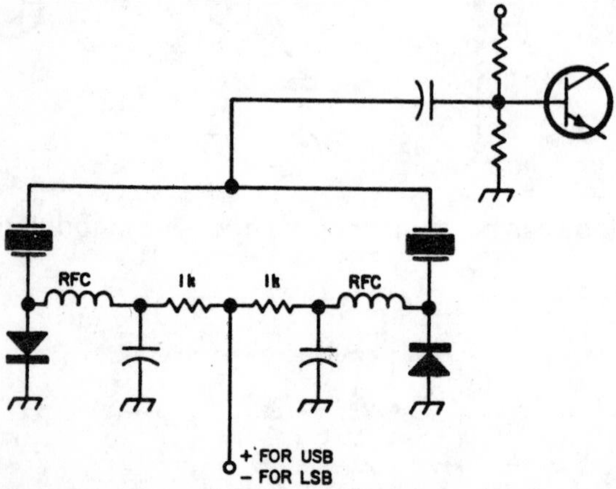


A pair of diode switches can be used to select upper- or lower-sideband-generating crystals.

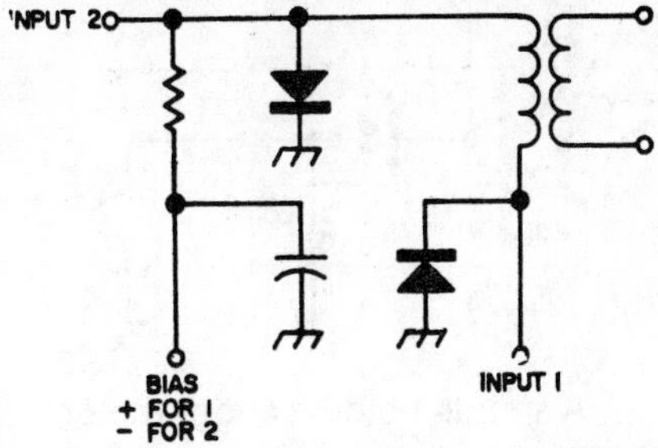


These diode switches can be used in a transceiver or other type of equipment to select either of two inputs.

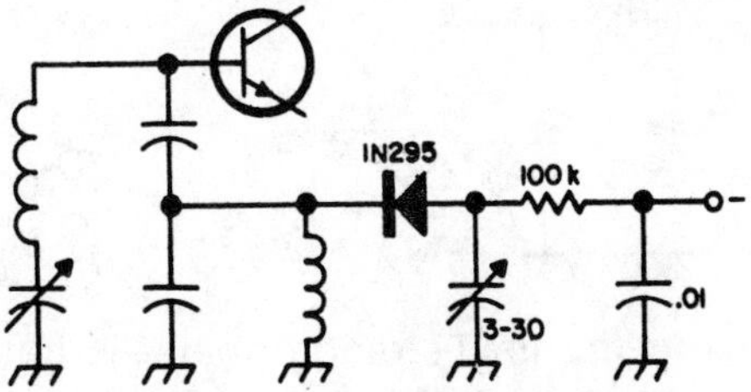


A diode switch is used to connect a small capacitor to a VFO to shift its frequency slightly for radioteletype.

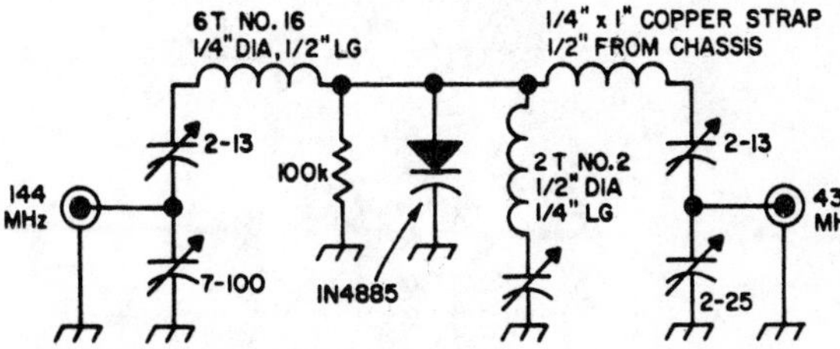


A practical high-pe vractor tripler.

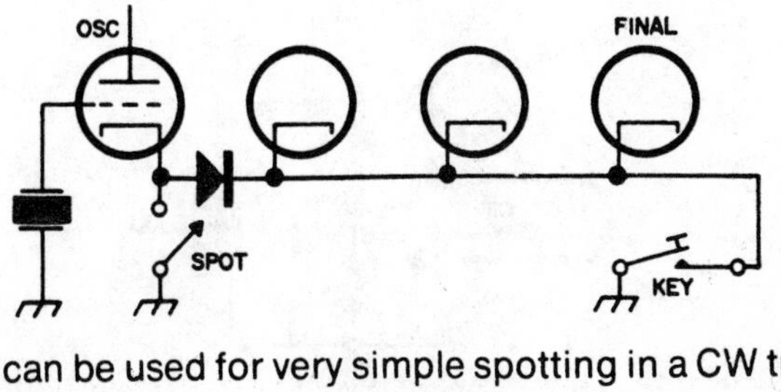


A diode can be used for very simple spotting in a CW transmitter.

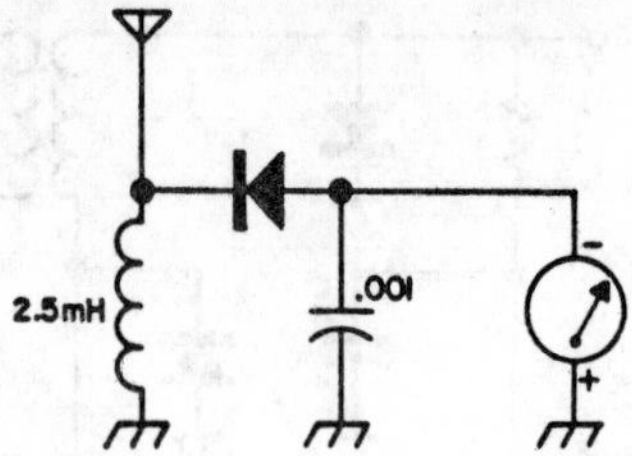


A simple field-strength meter.

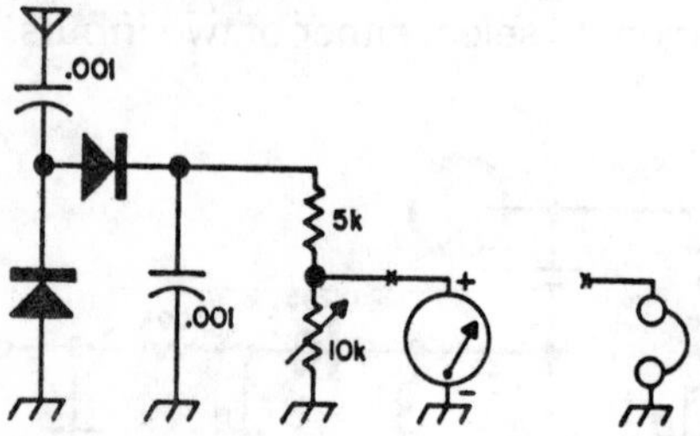


This voltage-doubling field-strength meter is not frequency-selective.

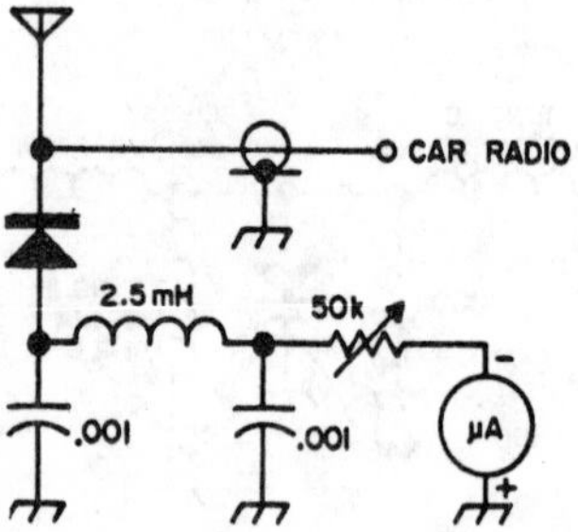


A special type of field-strength meter for use in a car.

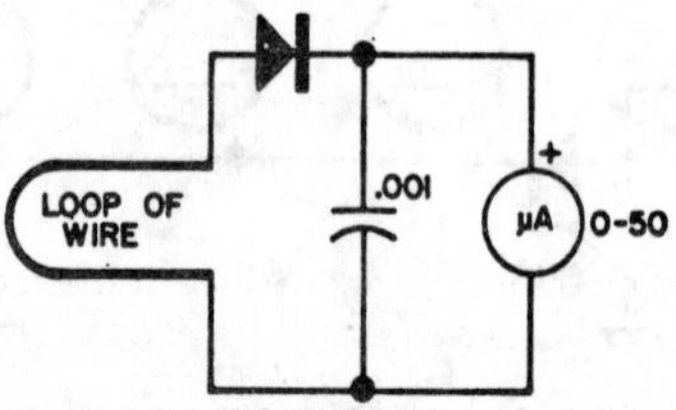


The rf sniffer is a wide-range sensitive rf detector.

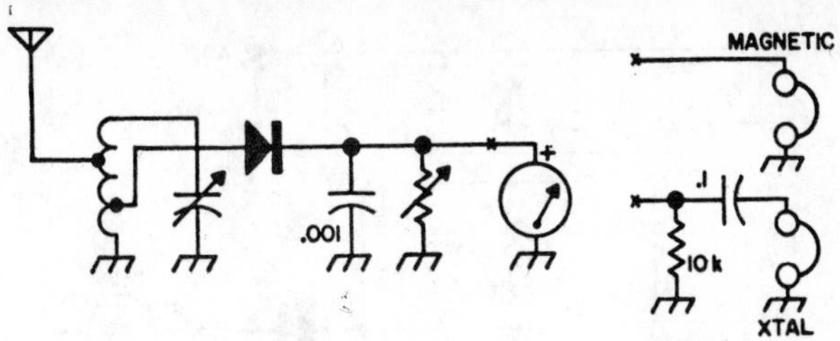


A wavemeter is simply a field-strength meter tunable to frequency. It is especially useful for checking transmitter harmonics.

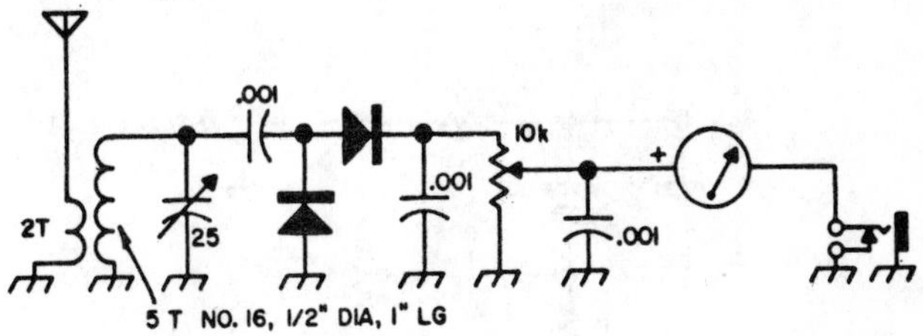


This tunable VHF wavemeter/FSM/monitor covers 6 and 2 meters.

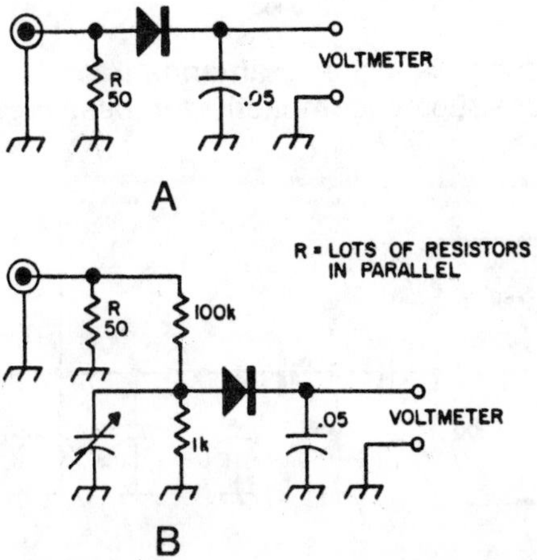


A dummy load should be used for all possible transmitter testing. An rf voltmeter connected to the dummy load makes it a wattmeter. A single diode is limited in voltage rating, so a voltage divider must be used for high power

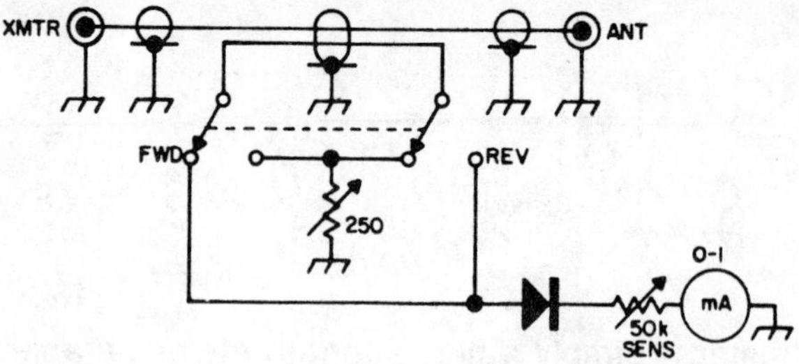


An SWR bridge is valuable for adjusting an antenna. The critical part of the bridge is a piece of coax cable with an extra wire inserted between the cable dielectric and the shield.

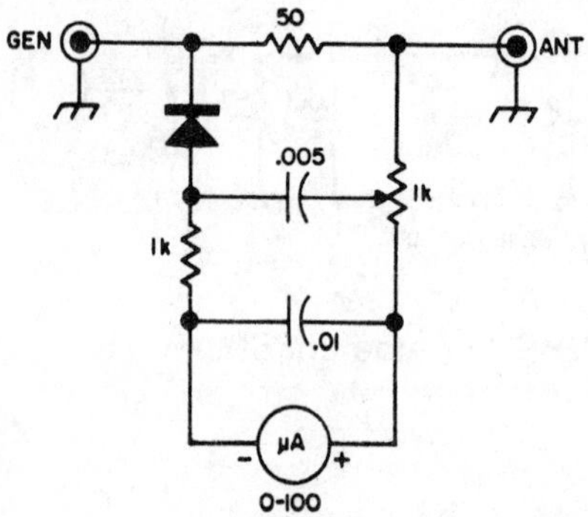


This antennascope is a simple antenna impedance bridge. It should be constructed compactly for best high frequency use.

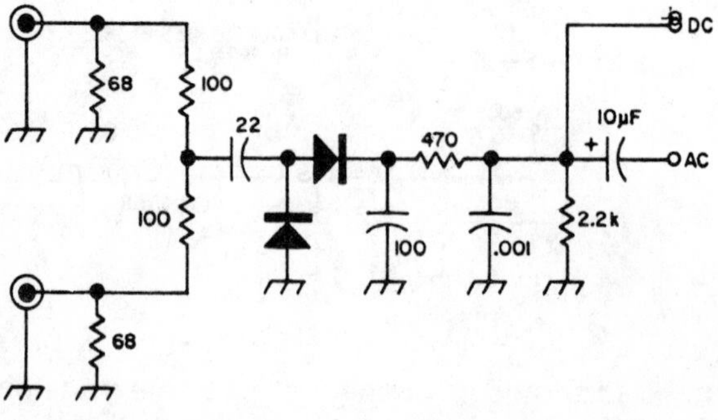


The James Dandy mixer is a general-purpose untuned mixer useful as an impromptu frequency meter, receiver, detector, etc.

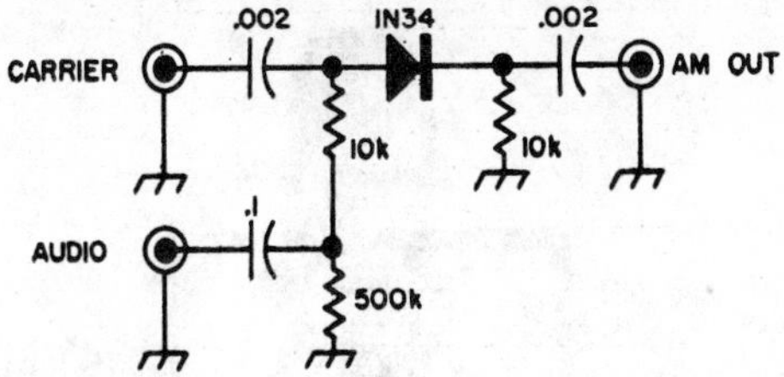


This amplitude modulator can be used to modulate the output of any low-level CW source.

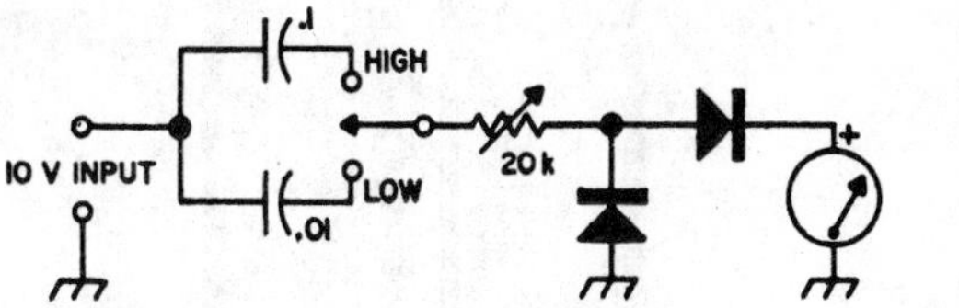


This audio frequency meter must be calibrated before use. It requires an input of 10V.

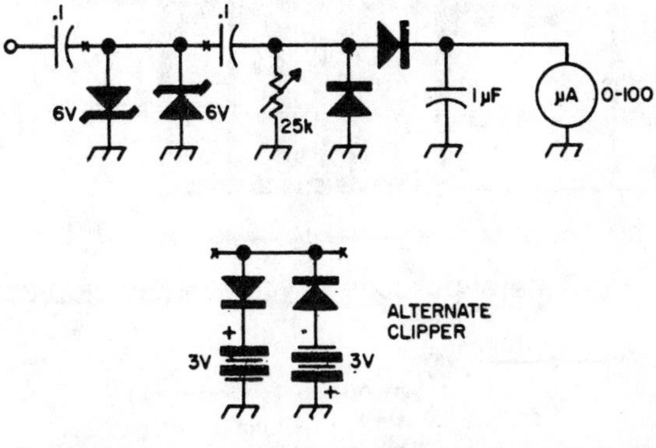


This audio frequency meter/tachometer is self-limiting and linear reading. Either two zeners or two conventional diodes and batteries can be used to set the proper input voltage.

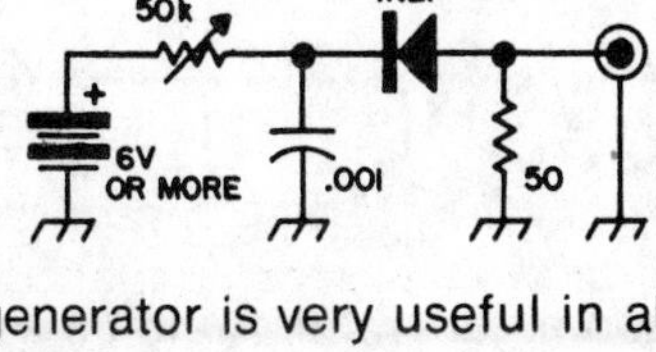


A diode noise generator is very useful in aligning a receiver for lowest noise fiqure.

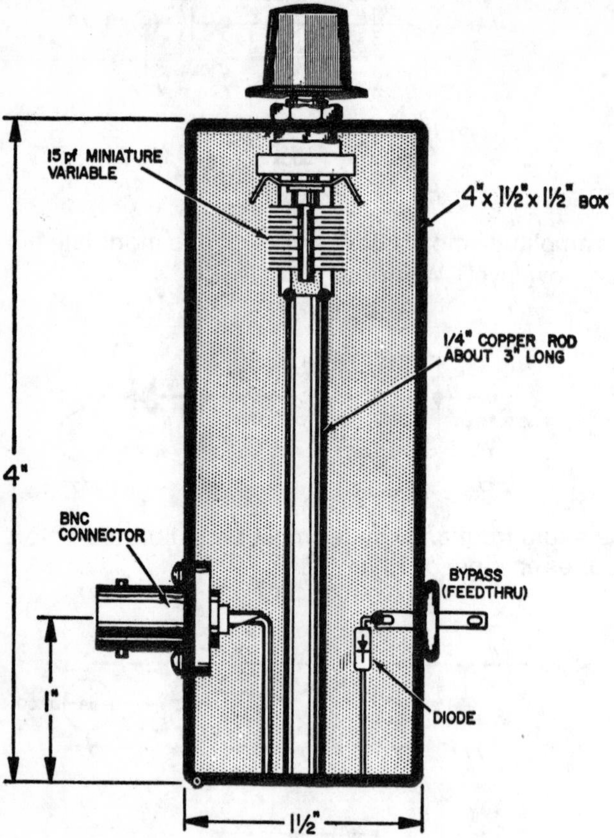


General-purpose wavemeter and monitor.

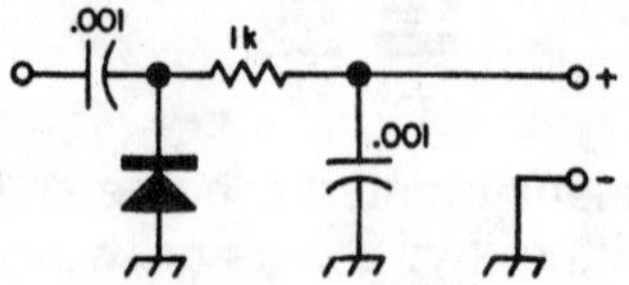


A general-purpose rf detector probe for use with an oscilloscope or voltmeter.

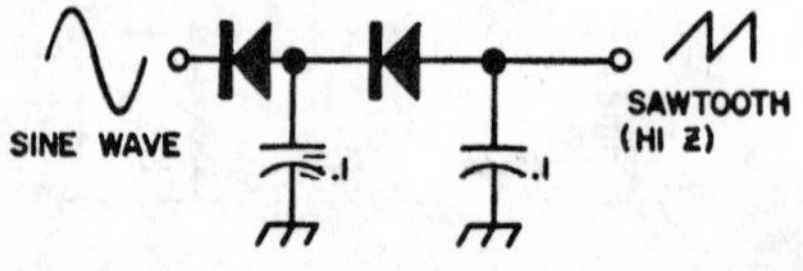


This simple sawtooth generator could be added to a monitor oscilloscope.

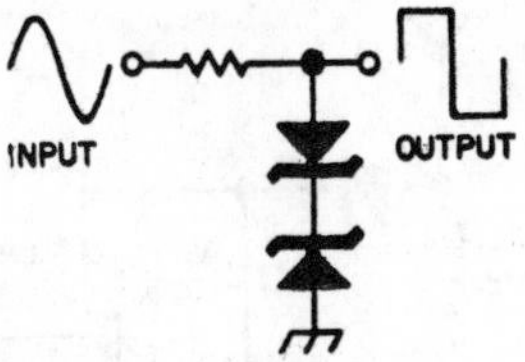


Two zeners can be used to produce a highly clipped sine wave very similar to a square wave.

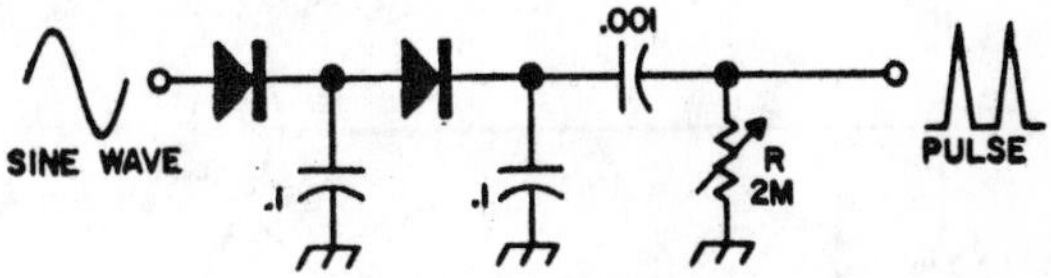


ADJUST R FOR BEST WAVESHAPE

A pulse generator is needed to adjust noise limiters for best results.

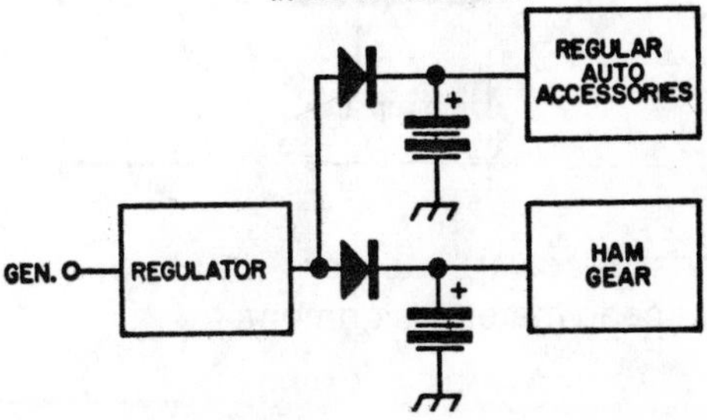


Here's how to use two batteries in your car, one for communications gear and one for the rest of the car needs. The diodes act as oneway switches, keeping the batteries charged, yet preventing any power from flowing from one to the other.

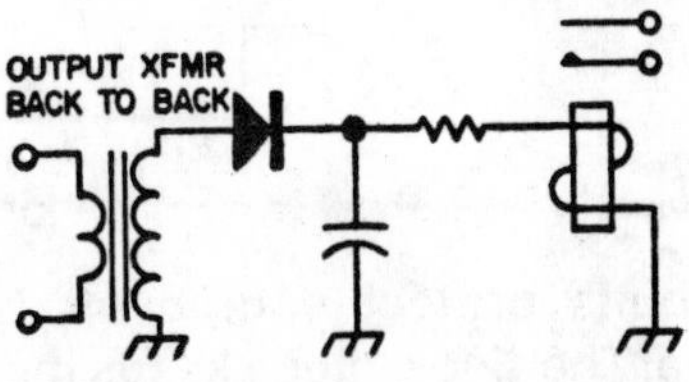


A transmitter can be keyed by a tape recorder for automatic code practice with this circuit.

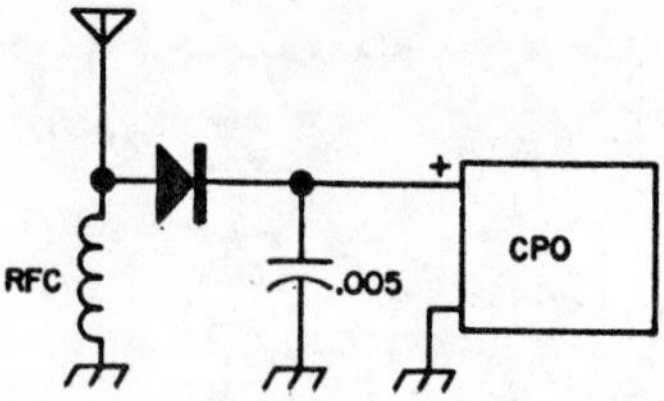


A field-strength meter can key a code oscillator to form a CW monitor.

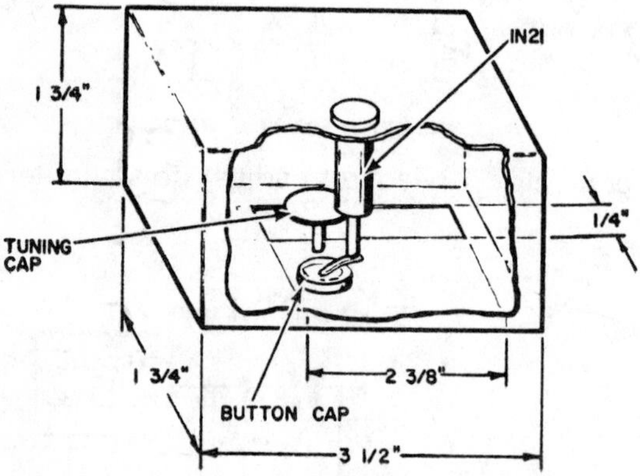


This is a radar receiver; it covers a ham band as well as a police radar speed meter assignment.

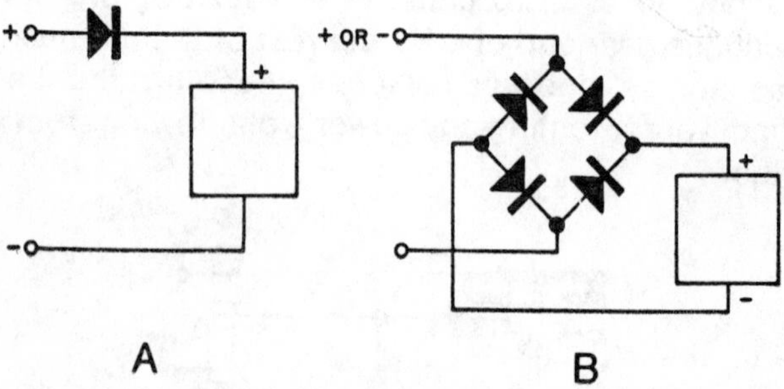


These two circuits protect equipment from incorrectly polarized voltage. The single diode keeps the equipment from working when the polarity is wrong, while the bridge automatically selects the proper polarity.

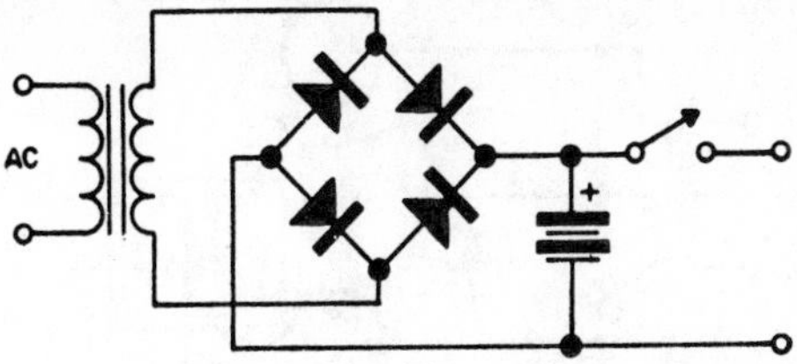


A battery can be floated across a power supply, keeping it charged and providing automatic switching from ac to battery power.

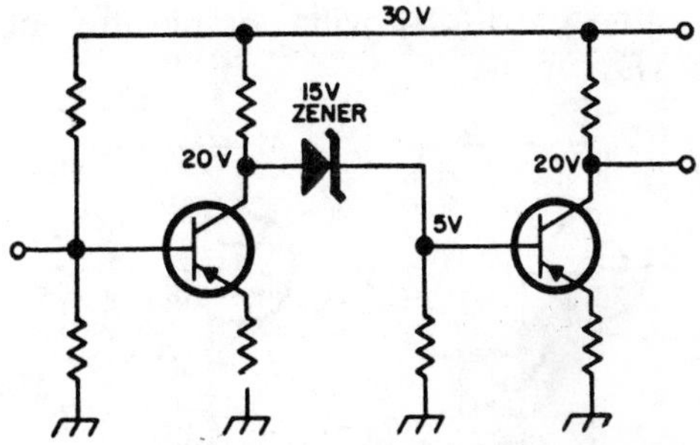


Zeners can be used in dc-coupled amplifiers to replace coupling capacitors.

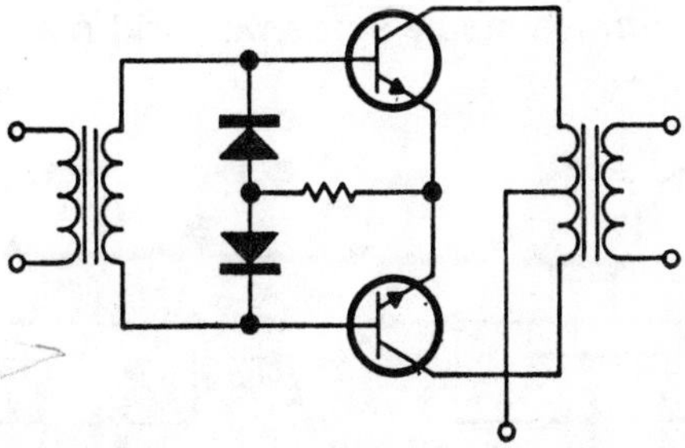

Diodes can provide an artificial centertap for push-pull amplifiers.

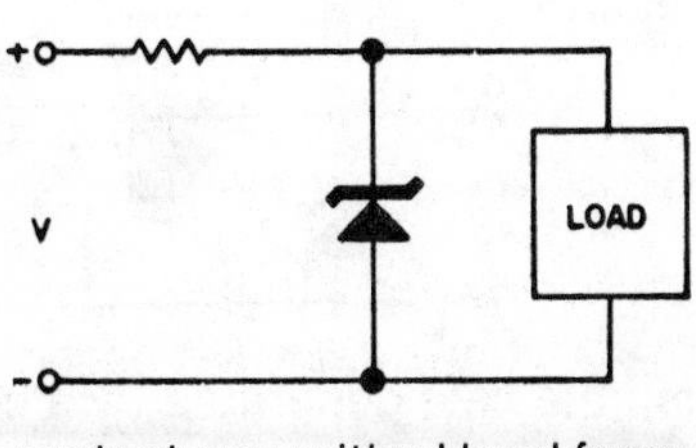


A zener can protect any critical load from overvoltage.

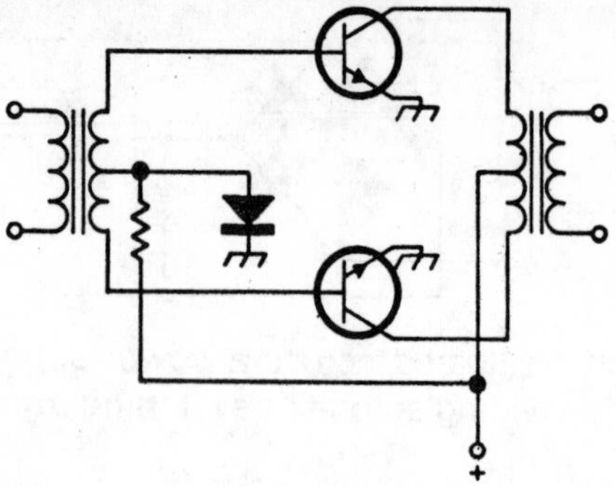


A diode is often used to provide temperature-compensated bias for class B amplifiers.

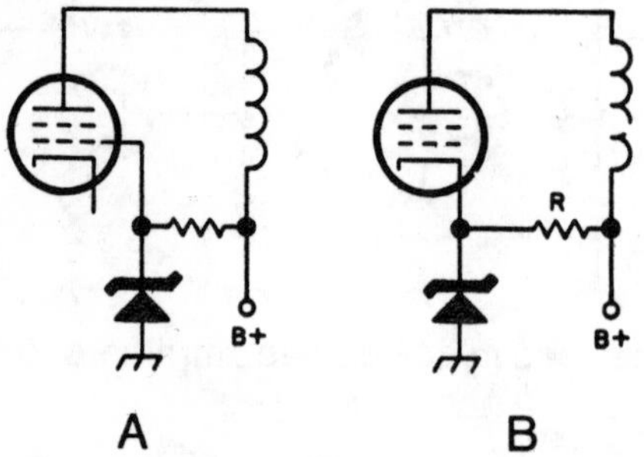


A zener can furnish stable screen or grid bias for a vacuum tube.

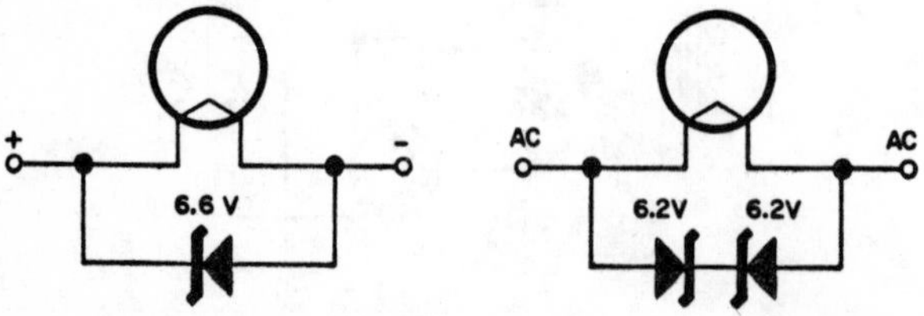


Zeners can protect a delicate filament from overvoltage.

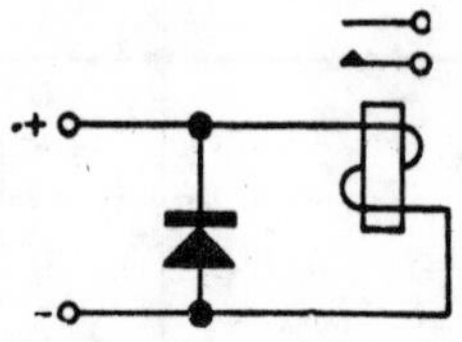


A diode can damp the field generated by a coil when current through it is disconnected.

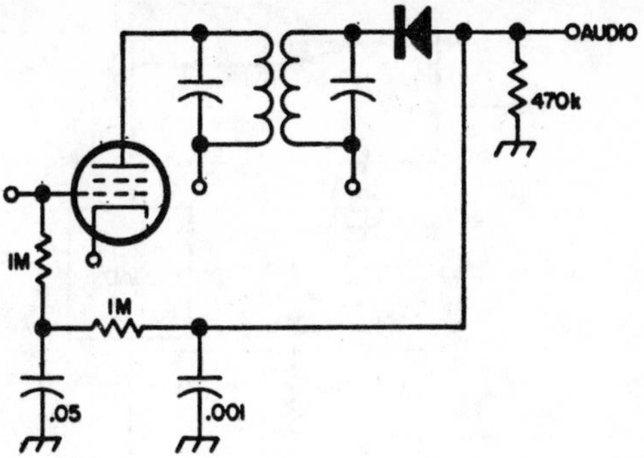


The conventional agc system used in tube-type receivers.

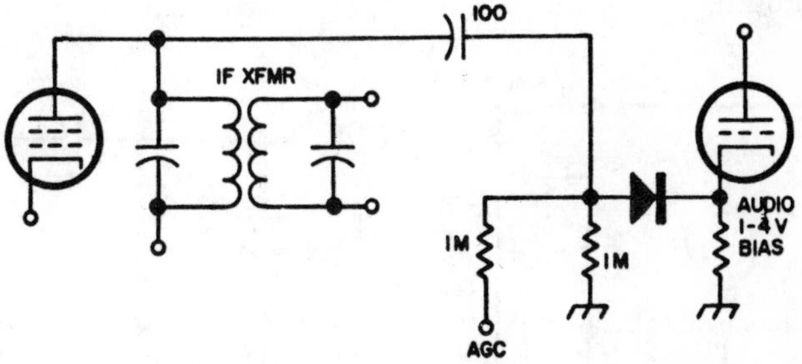


Delayed agc acts only on strong signals.

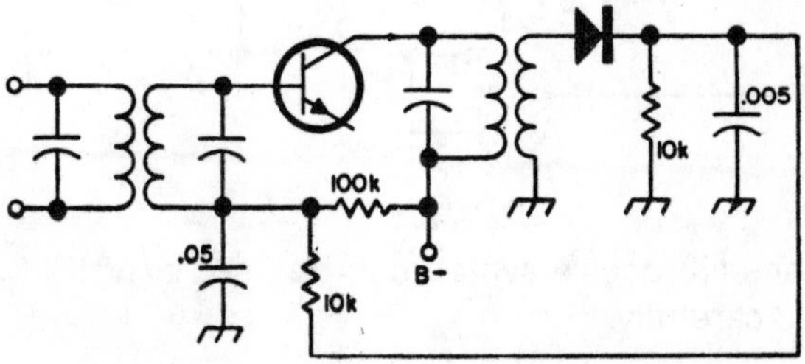


Reverse agc for a transistor/receiver.

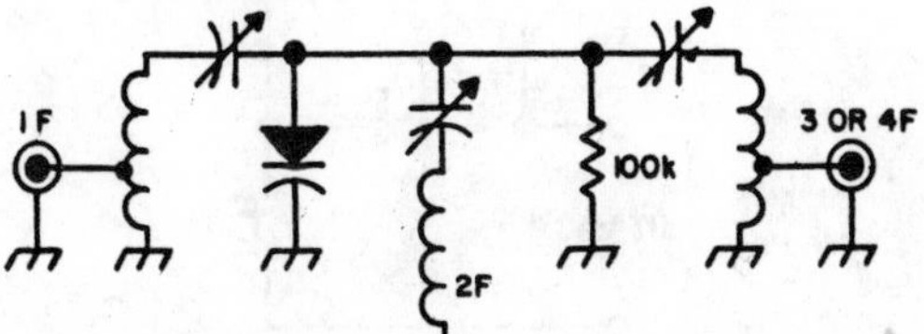


"A" is a varactor tripler or doubler.

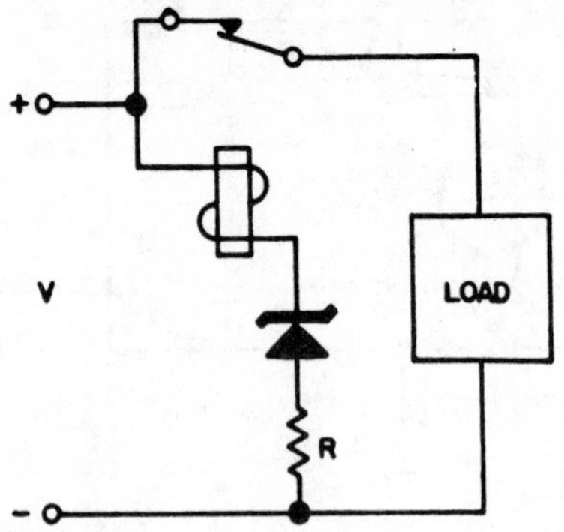


This circuit will disconnect a load when voltage drops below a minimum.

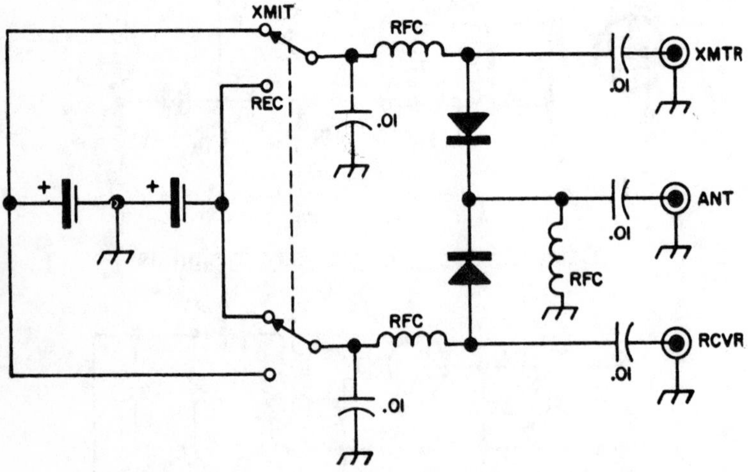


This transmit-receive switch can be used at VHF if it is constructed carefully.

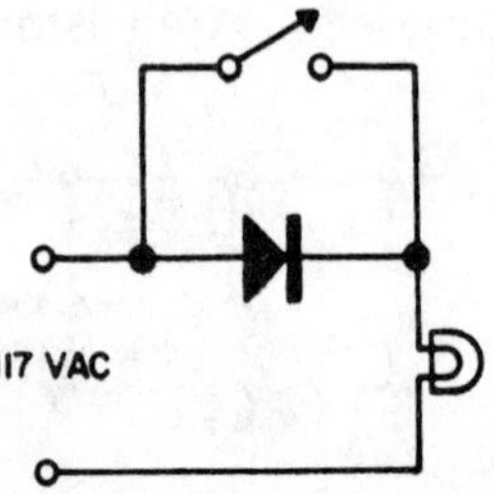


This is a lamp dimmer providing two brillance positions: half on and full on.

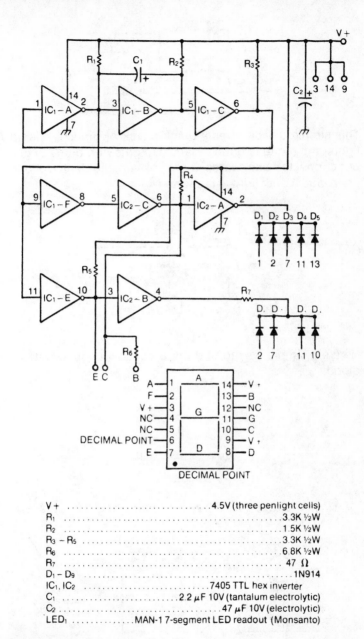

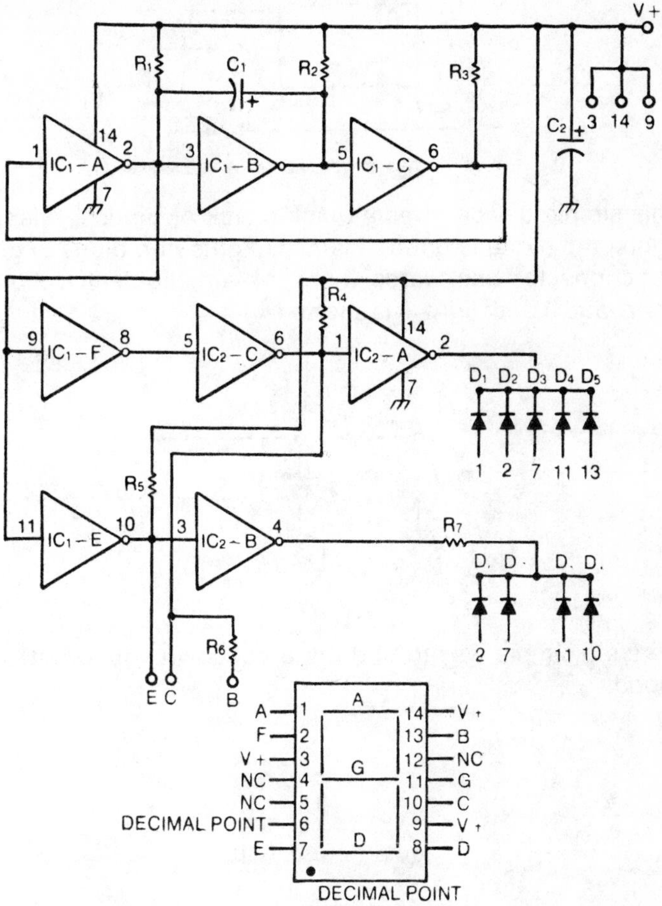


| | |
|---|---|
| V+ | 4.5V (three penlight cells) |
| $R_1$ | 3.3K ½W |
| $R_2$ | 1.5K ½W |
| $R_3 - R_5$ | 3.3K ½W |
| $R_6$ | 6.8K ½W |
| $R_7$ | 47 Ω |
| $D_1 - D_9$ | 1N914 |
| $IC_1$, $IC_2$ | 7405 TTL hex inverter |
| $C_1$ | 2.2 μF 10V (tantalum electrolytic) |
| $C_2$ | 47 μF 10V (electrolytic) |
| $LED_1$ | MAN-1 7-segment LED readout (Monsanto) |

Transistor/diode tester circuit. The readout will actually display letters that stand for the condition of the device being tested, or its type.

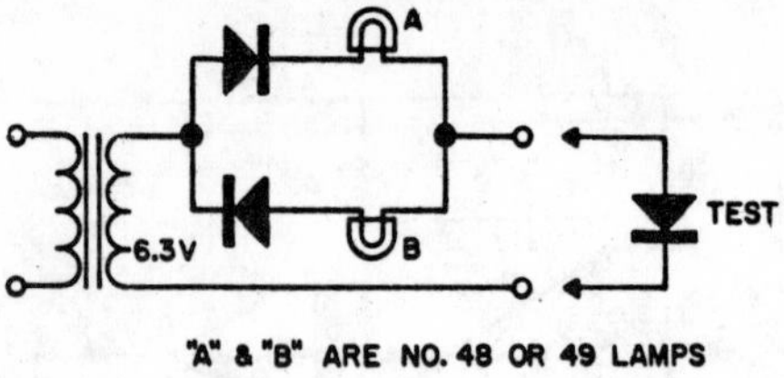


This simple device gives a quick check of diodes. If lamp A lights, the diode is good. If lamp B lights, the diode is good, but connected backwards. If neither lamp lights, the diode is open, and if both light, it is shorted.

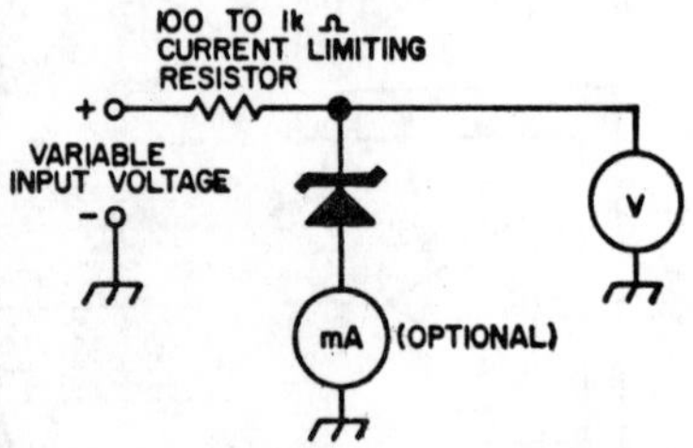


Here's a simple way to find the breakdown voltage of zener diode.

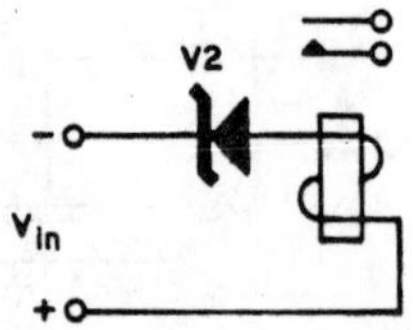


An input voltage over the zener voltage energizes the relay.

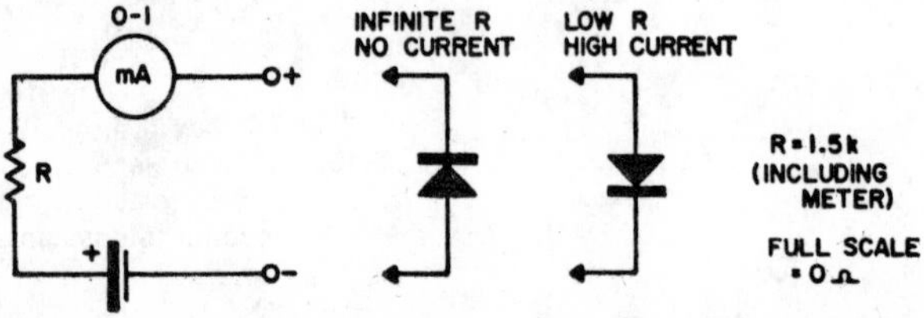


This simple ohmmeter demonstrates how a diode can be checked with an ohmmeter.

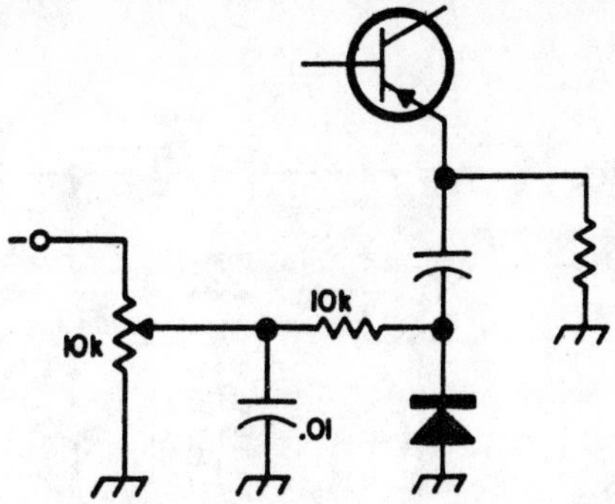


A diode can control the bypassing of an emitter bypass capacitor to change an amplifier's gain.

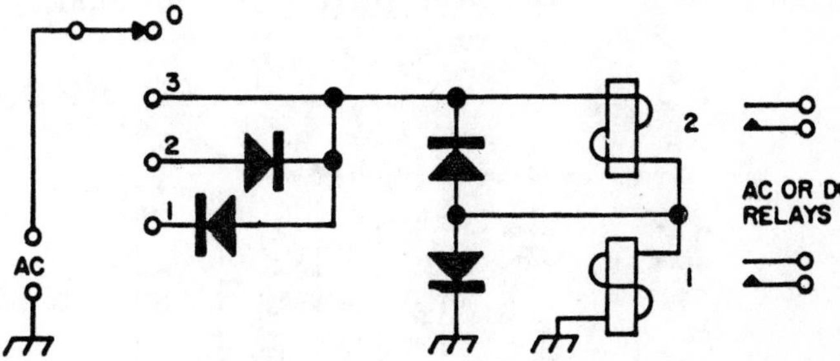


In position 0 neither relay is energized. In position 3 both are energized. In 2, relay 2 is on and in 1, relay 1 is on.

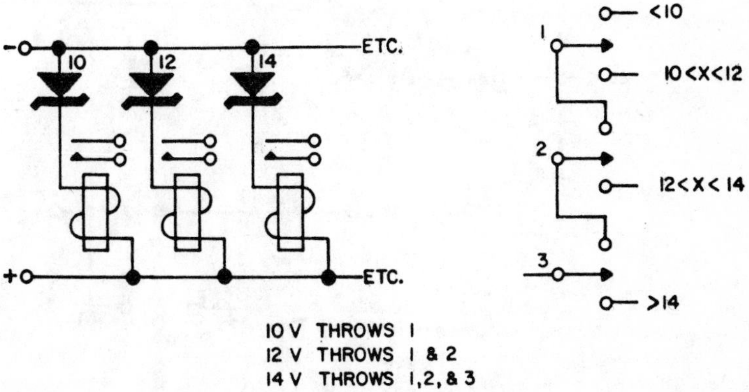


In this scheme, a varying input voltage selects relay contacts in turn.

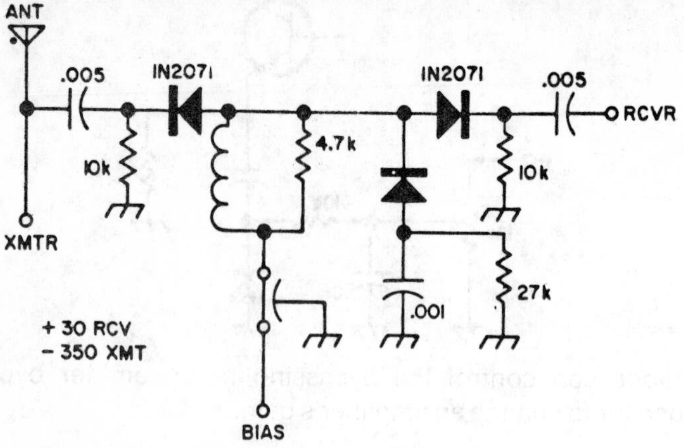


This is a high-frequency antenna switch using diodes.

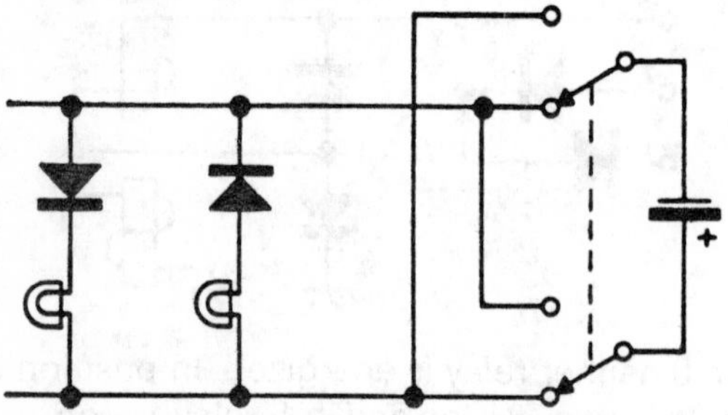


Diodes can be used for mysterious switching of two lamps with one pair of wires.

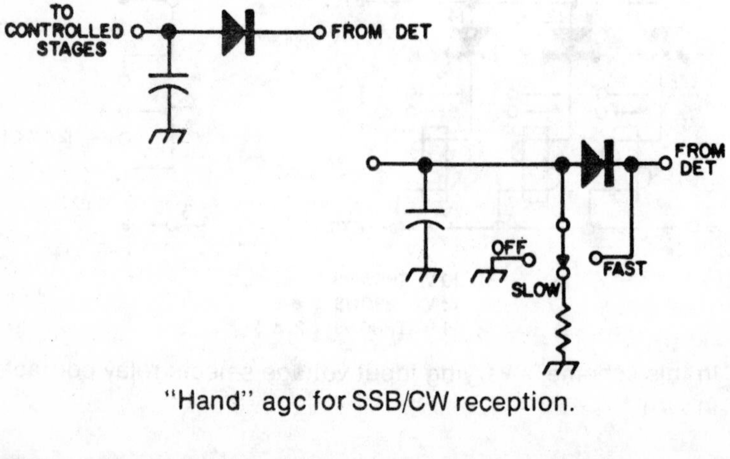


"Hand" agc for SSB/CW reception.

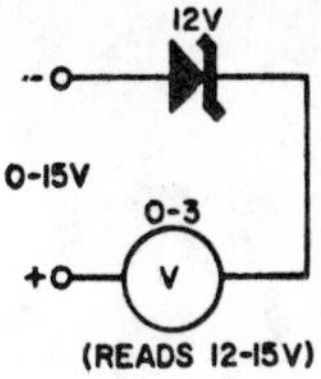


A zener and a low-voltage meter can be used to suppress the low end of a range.

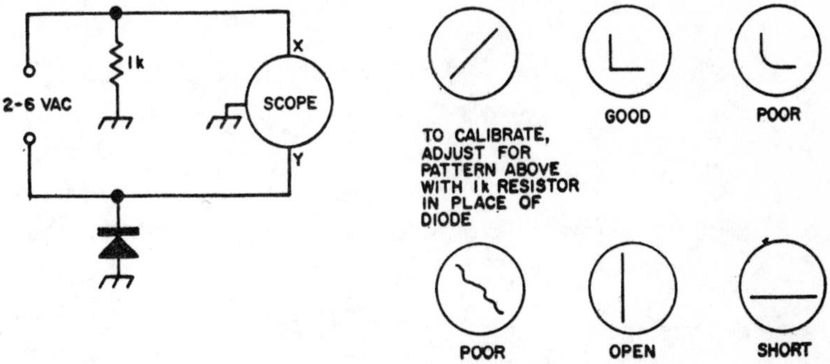


One of the easiest types of diode checks for a person with a scope is this, but it tells nothing about a diode's high voltage performance.

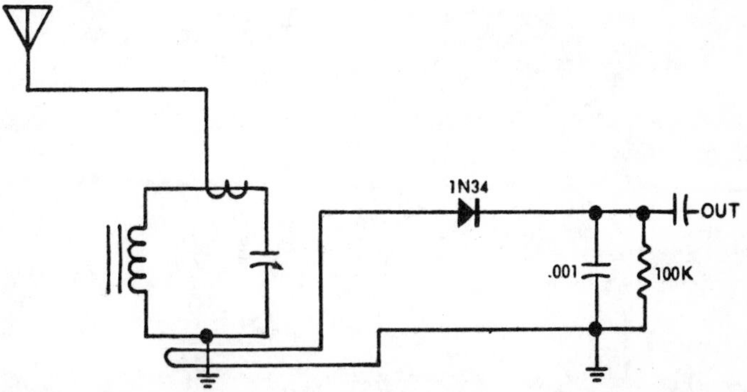


A handful of parts makes a crystal radio set, which can be connected to the battery-saving amplifier. A ferrite loopstick rod and tuning capacitor, both of which may be salvaged from an old transistor radio or purchased inexpensively, tune in the signal. A two-turn link around the loopstick rod couples energy to the diode detector. If you have only one strong station in your vicinity, no tuning arrangement is necessary. Just connect the antenna directly to the 1N34 diode, and listen.

# Section R

# Miscellaneous

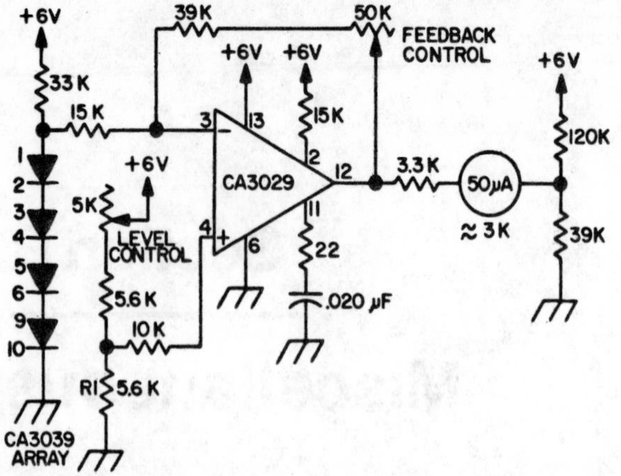


An electronic thermometer using the CA3039 diode array as the sense element. A CA3029 op amp detects the change in voltage fed into it by the diodes against the level control's reference voltage and displays the amplified difference on the meter. The level control should be set so room temperature reads near midscale, while the feedback control adjusts the upper and lower extremes of 120°F and 0°F.

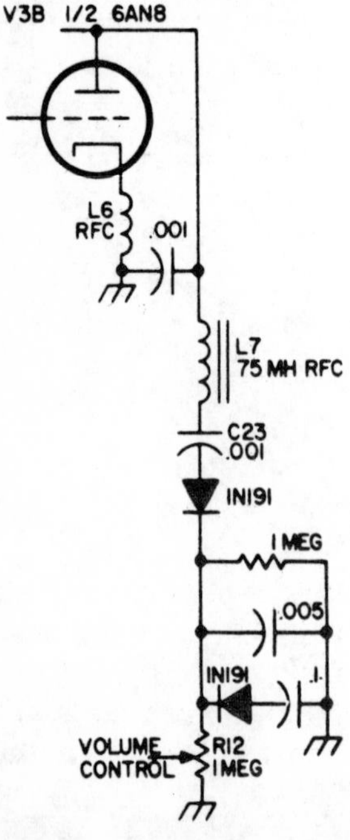


A simple noise limiter that can be used with superregen detectors similar to those in the Heathkit Twoer and Sixer.

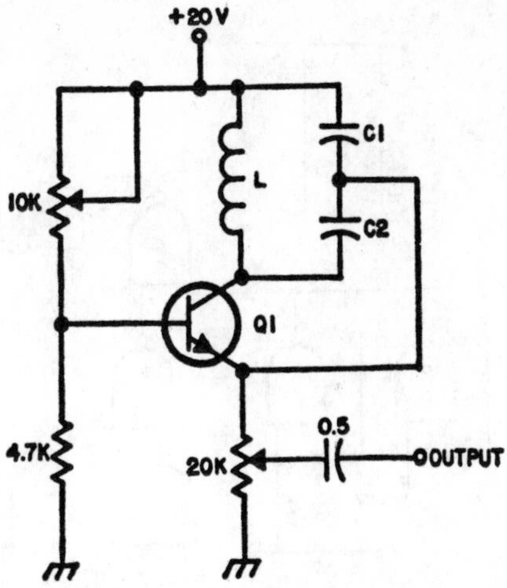


Q1 - 2N2925, 2N3392

| FREQUENCY | C1 | C2 | L |
|---|---|---|---|
| 50 kHz | 3500 pf | 1500 pf | 10 mH |
| 80 kHz | 2200 pf | 910 pf | 6.2 mH |
| 100 kHz | 1800 pf | 750 pf | 4.7 mH |
| 200 kHz | 910 pf | 390 pf | 2.2 mH |
| 455 kHz | 390 pf | 160 pf | 1 mH |
| 1000 kHz | 180 pf | 75 pf | 0.47 mH |

This simple circuit provides an extremely stable BFO. The frequency of oscillation may be tailored to your needs by simply choosing the proper tank components listed in the table.

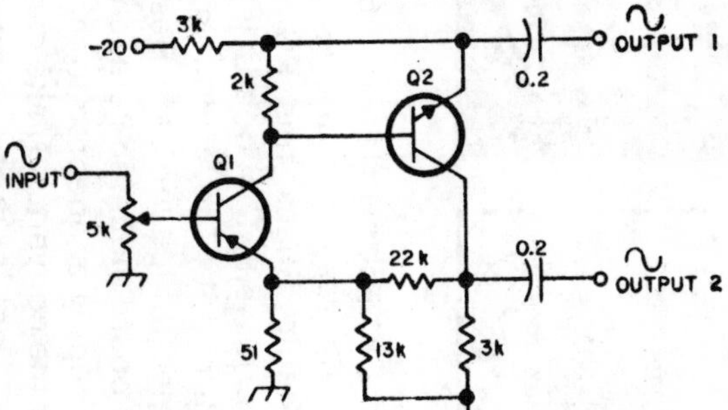


This phase splitting circuit provides two out-of-phase signals for driving a push-pull amplifier without an expensive transformer. The gain of the stage as shown is 150, but this may be adjusted by changing the value of the 22K feedback resistor. Q1 and Q2 are a complementary pair such as the 2N652 and 2N388 or 2N2430 and 2N2706.

TOP VIEW

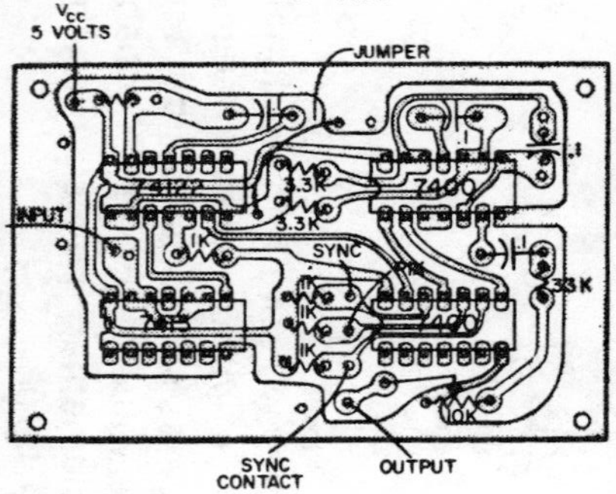


FOIL VIEW

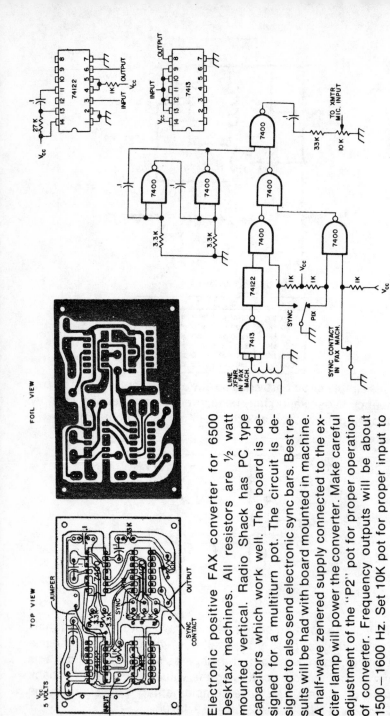

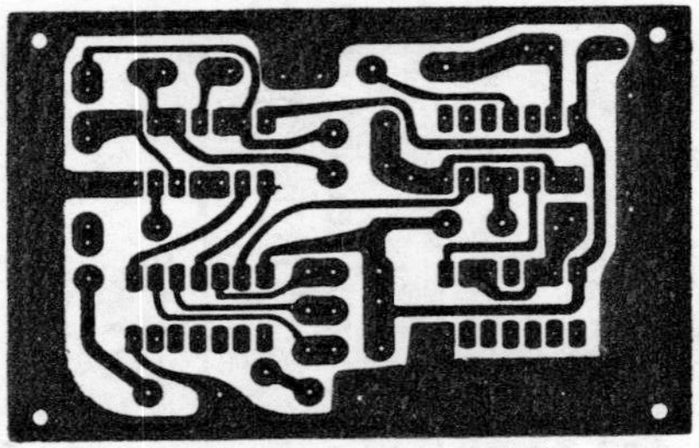

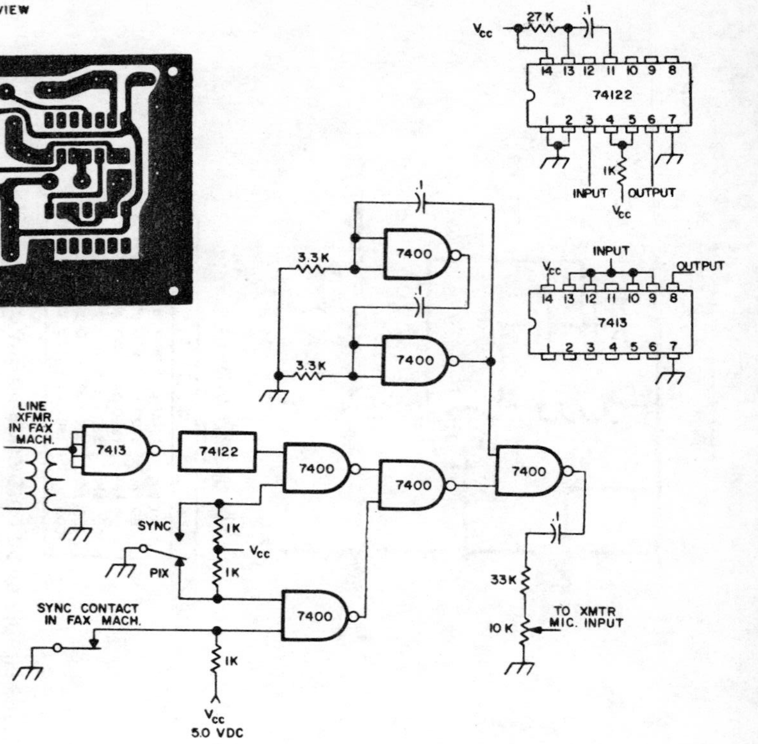


Electronic positive FAX converter for 6500 Deskfax machines. All resistors are ½ watt mounted vertical. Radio Shack has PC type capacitors which work well. The board is designed for a multiturn pot. The circuit is designed to also send electronic sync bars. Best results will be had with board mounted in machine. A half-wave zenered supply connected to the exciter lamp will power the converter. Make careful adjustment of the "P2" pot for proper operation of converter. Frequency outputs will be about 1500 – 1600 Hz. Set 10K pot for proper input to transmitter.

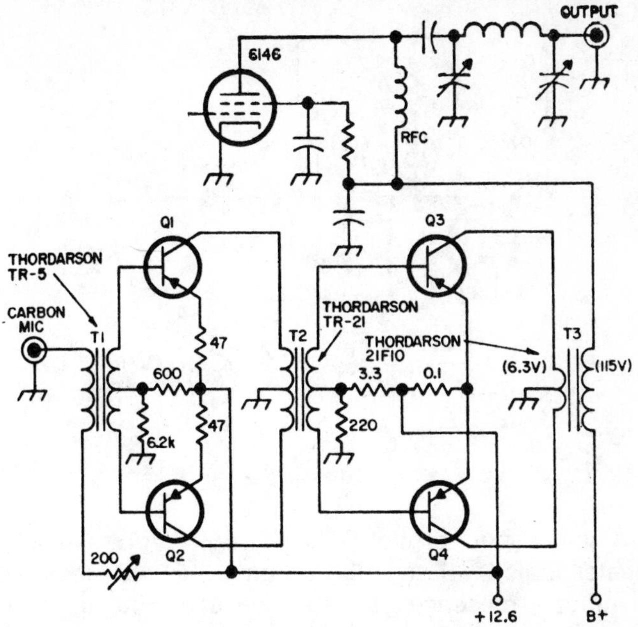


25-watt modulator uses readily available commercial transformers. Transistors Q1 and Q2 are 2N1172, 2N301, 2N1560, SK3009, GE-9, or HEP 232; Q3 and Q4 are 2N174, 2N278, SK3012, GE-4, or Hep 233.

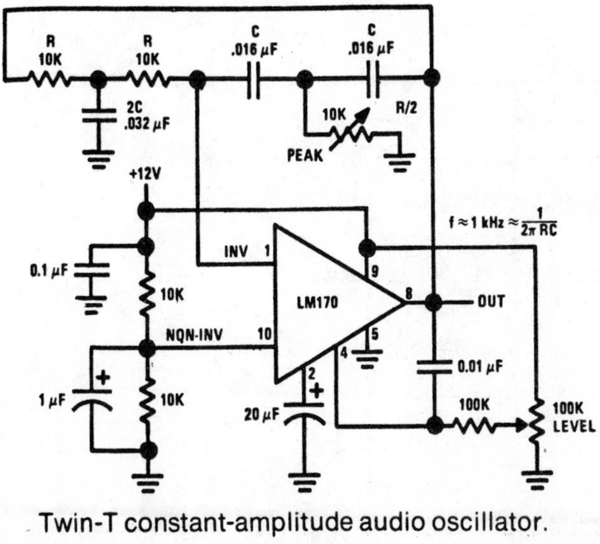


Twin-T constant-amplitude audio oscillator.

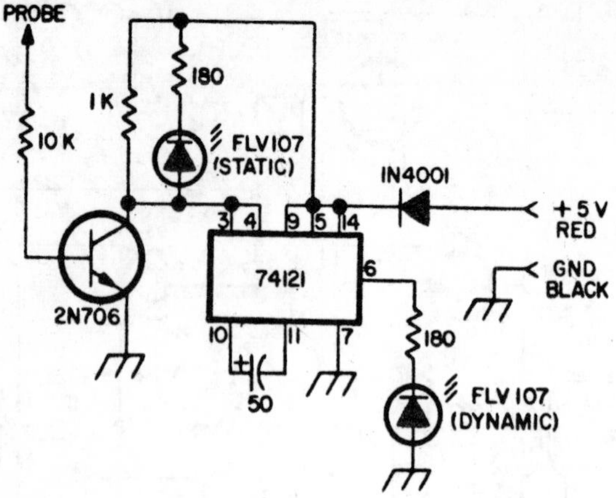


Pulse sniffer logic probe. RTL/TTL logic probe (static) LED indicates logic level at probe tip on = "1." "Dynamic" LED will indicate presence of "1" pulse at probe tip, pulse is stretched to about 50 μsec to be easily visible visible. A 100 nsec 4V pulse will trigger pulse stretcher. 1N4001 idiot-proofs against reversed power leads. Uses +5V from circuit under test. Transistor and LEDs are not critical; basically any high speed switch and indicator LED will work.

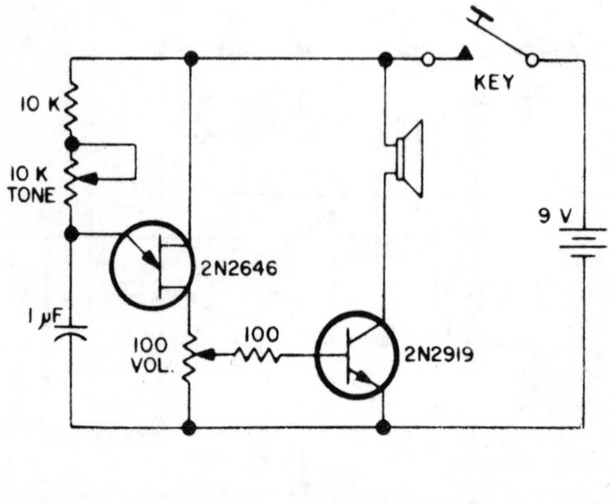


Code-practice oscillator with variable tone and volume.

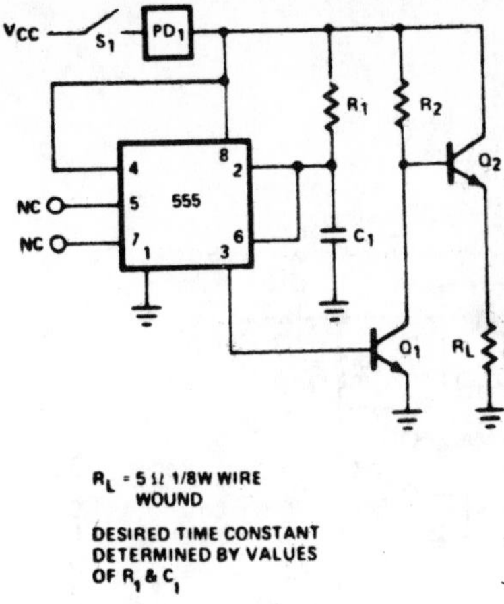


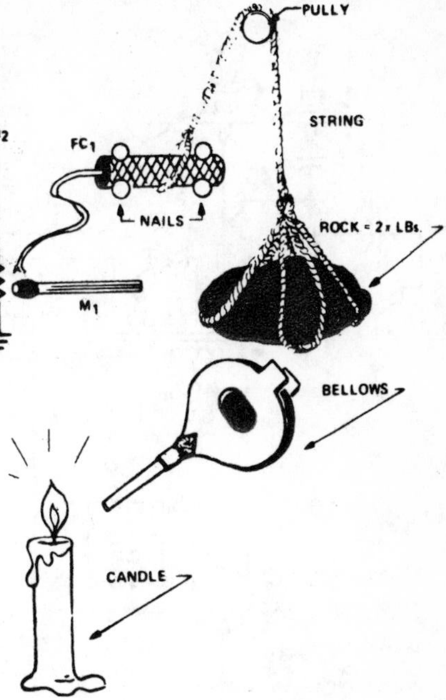


How about a nice timer? Take your pick from these two approaches.

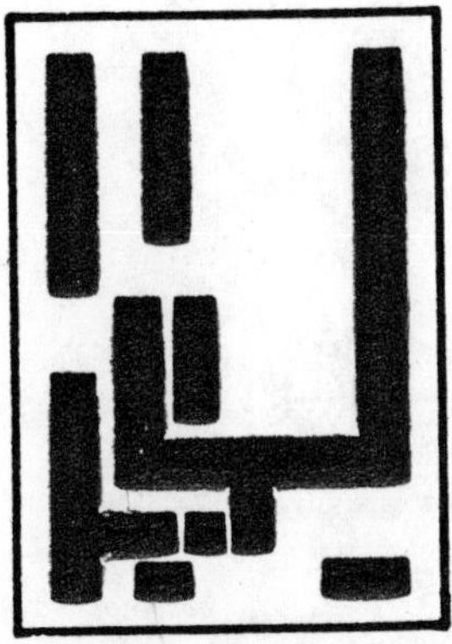

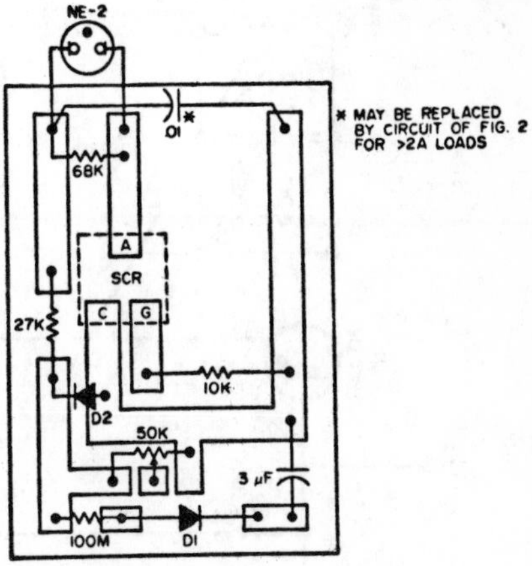


PC layout and schematic of 10-minute timer that switches noninductive loads of up to 2 amperes.

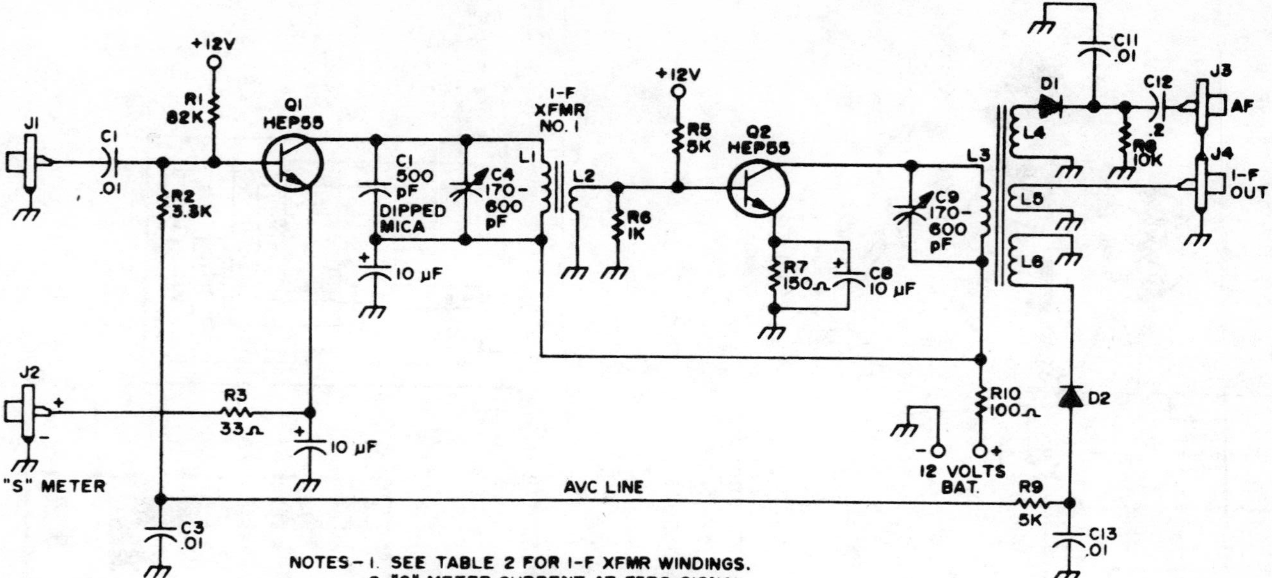


NOTES—1. SEE TABLE 2 FOR I-F XFMR WINDINGS.
2. "S" METER CURRENT AT ZERO SIGNAL IS 4 TO 5 mA.
3. THE POSITIVE VOLTAGE AFTER R10 IS ABOUT 11 VOLTS.

Complete circuit 1.65 MHz i-f. Q1 and Q2 are Motorola HEP 55.

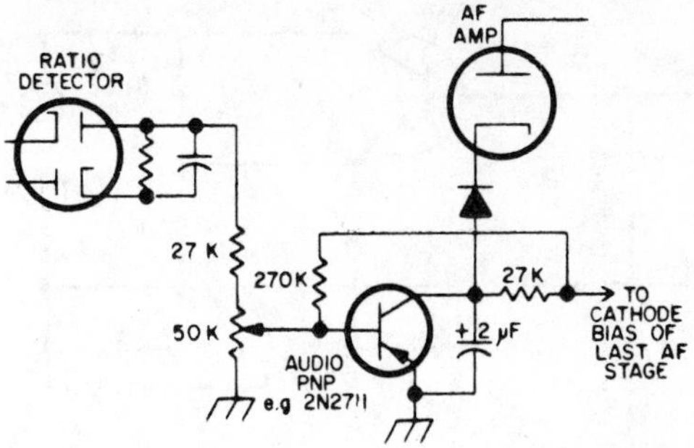


Simple squelch circuit. This circuit is for use with a tube-type FM receiver. The transistor acts as a switch to turn on the first audio stage. Parts not listed are in the receiver. Labeled parts are added.

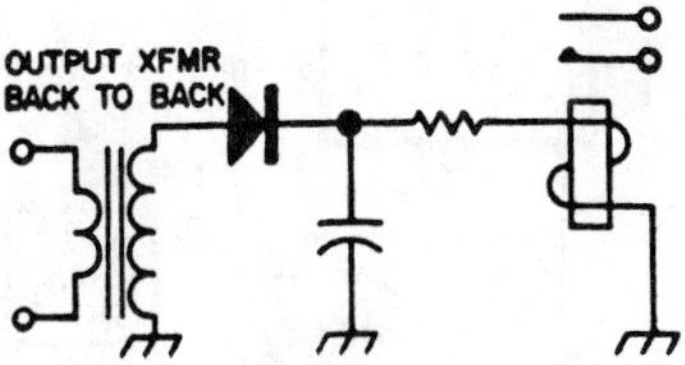


A transmitter can be keyed by a tape recorder for automatic code practice with this circuit.

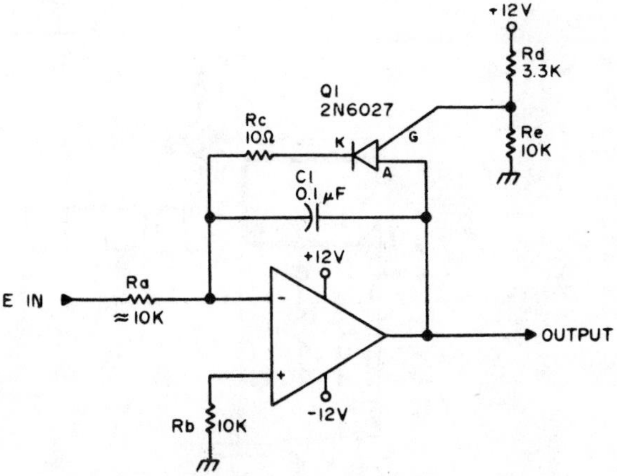


Voltage-controlled oscillator circuit.

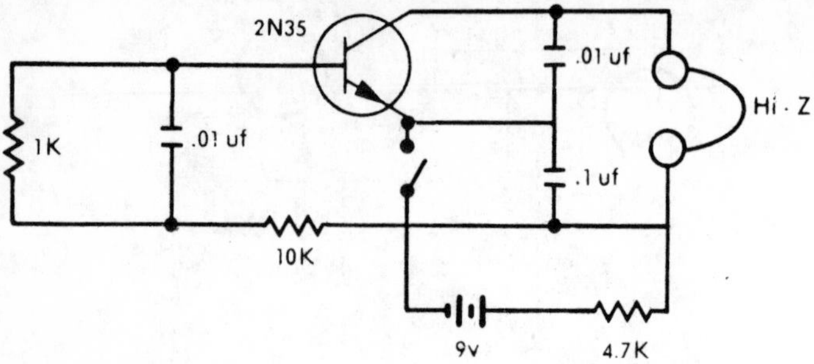


This simple oscillator circuit will not produce a very loud signal, but it will be adequate for code practice if the switch is replaced with a telegraph key.

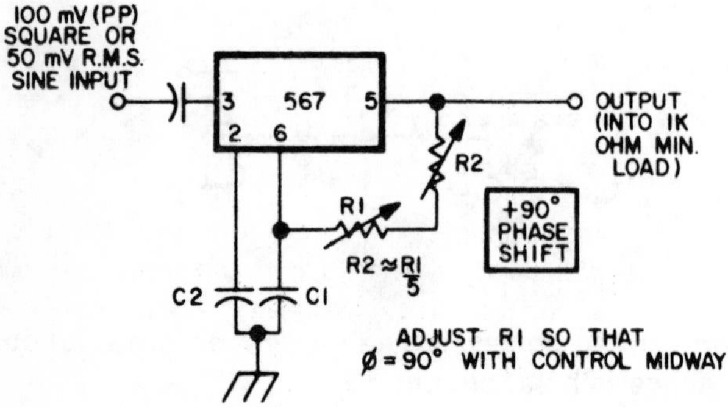


0° to 180° phase shifter.

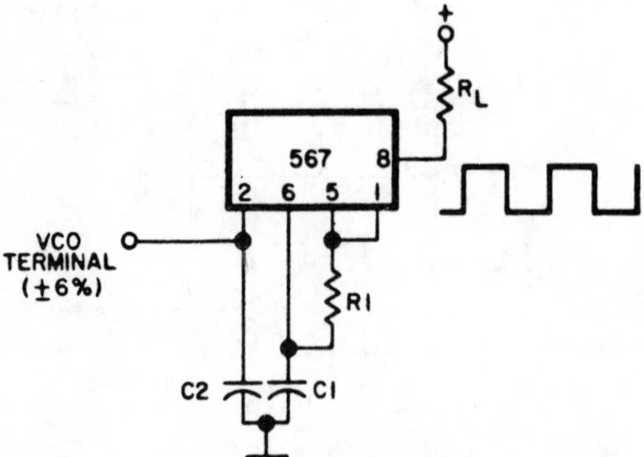


Precision oscillator to switch 100 mA loads.

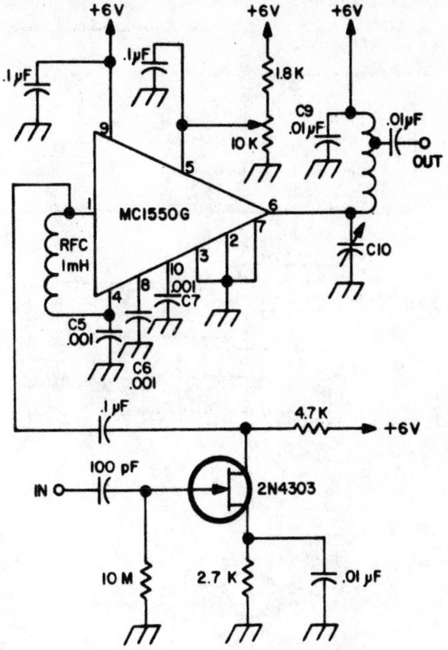


If you cannot install a big receiving antenna, try this substitute. It adapts the very weak signal from a piece of wire, a lamp post, or a mattress to the common 50Ω receiver input terminal.

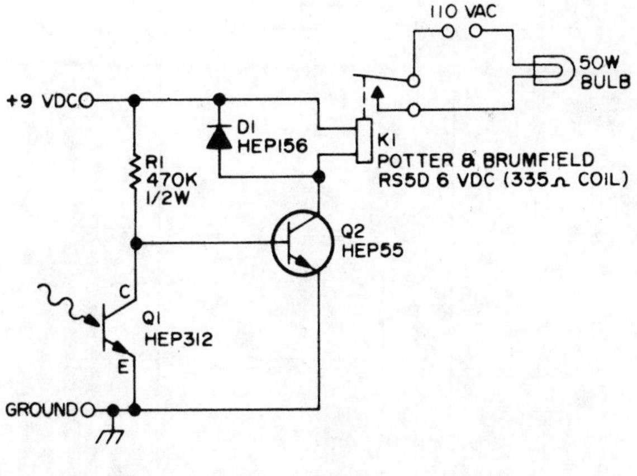


Automatic night light that turns on at dusk and off at sunrise. Courtesy Motorola.

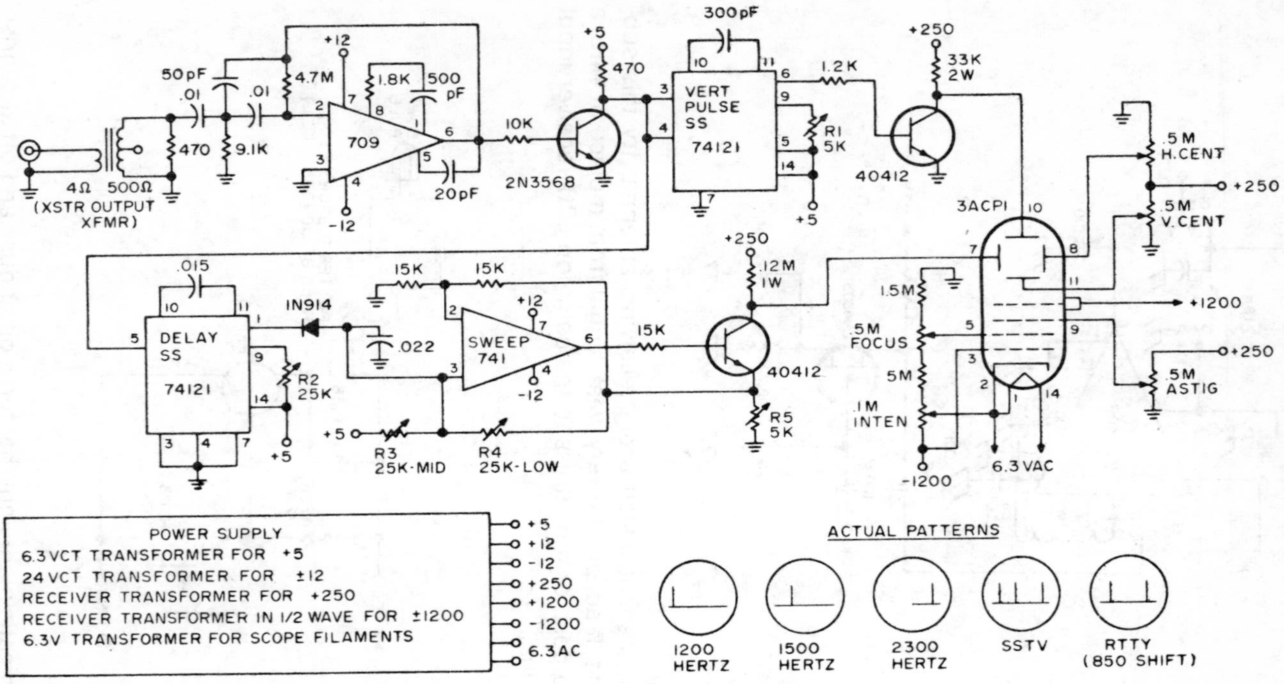


Schematic diagram of the slow-scan TV analyzer.

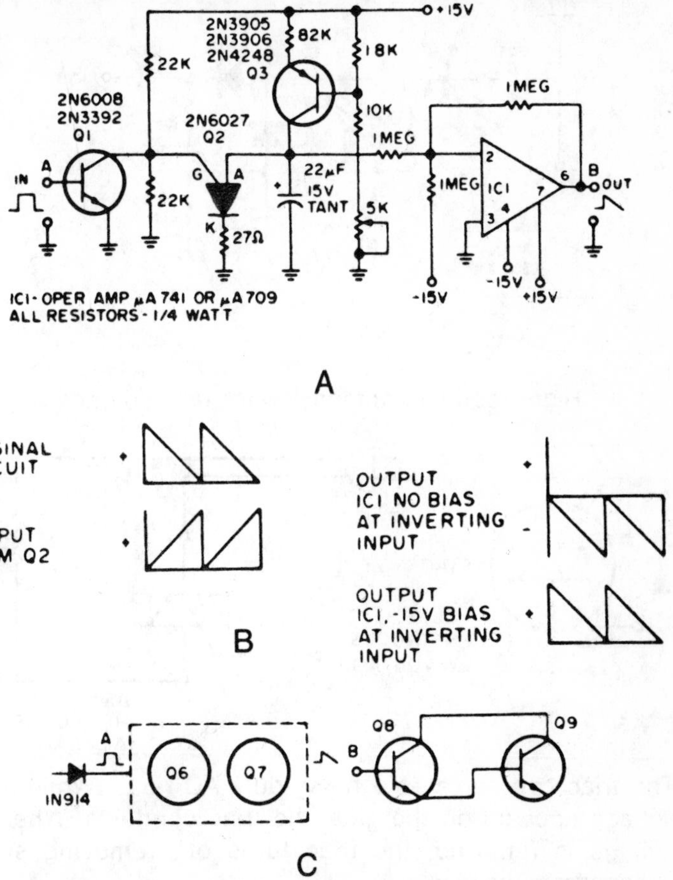


Vertical retrigger improves vertical sweep of slow-scan TV. Output waveforms show comparison.

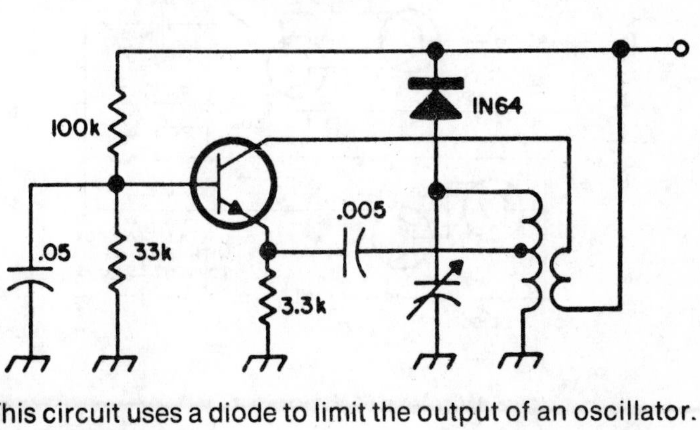


This circuit uses a diode to limit the output of an oscillator.

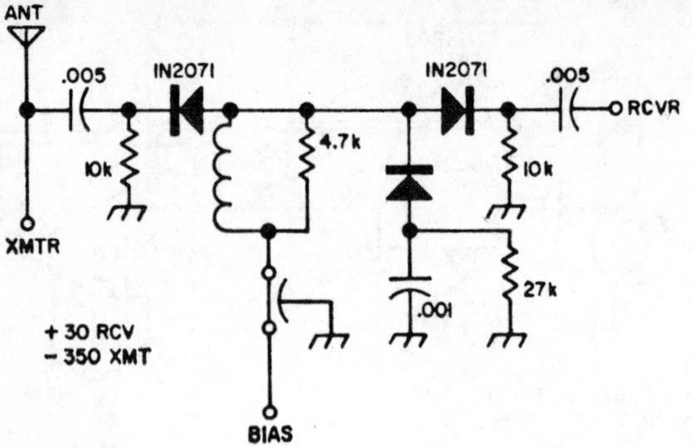


High-frequency antenna switch using diodes.

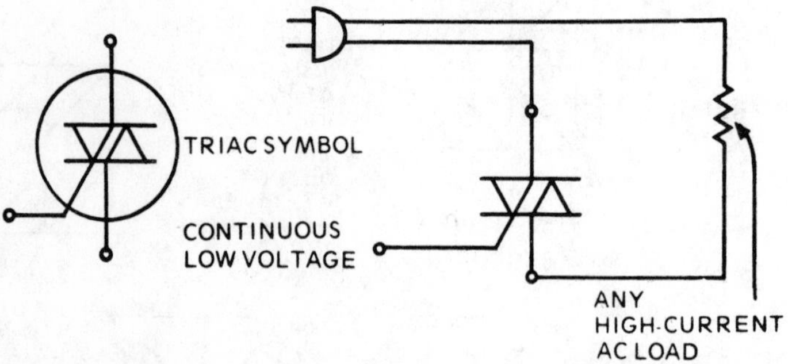


The triac can serve as a heavy-duty AC relay. While a low voltage appears on the gate, the triac conducts. When the voltage is removed, the triac turns off, removing supply voltage from the load.

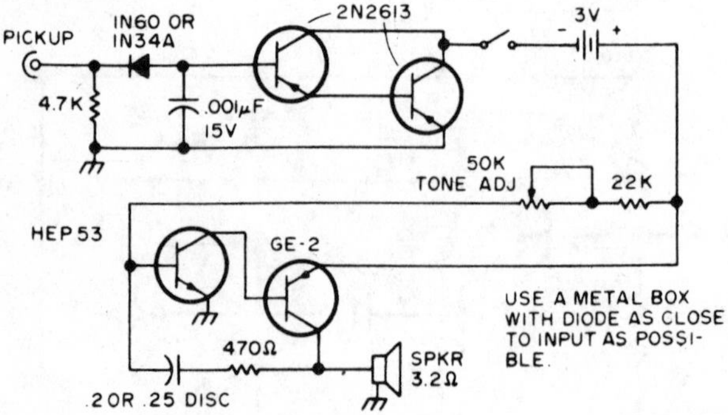


Wireless CW monitor.

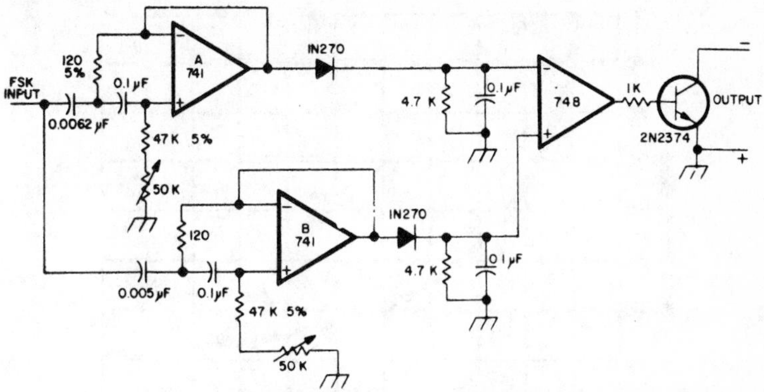


Mark space demodulator. FSK demodulator contains 2 active filters, saving space and improving performance over designs that use conventional LC-tuned circuits. Filter A passes the space frequency of 2025 Hz, while filter B passes the mark frequency of 225 Hz. The top op amp operates open loop, summing the filter outputs. For a mark input, the output transistor saturates so that circuit loop closes. This circuit works fine on a breadboard.

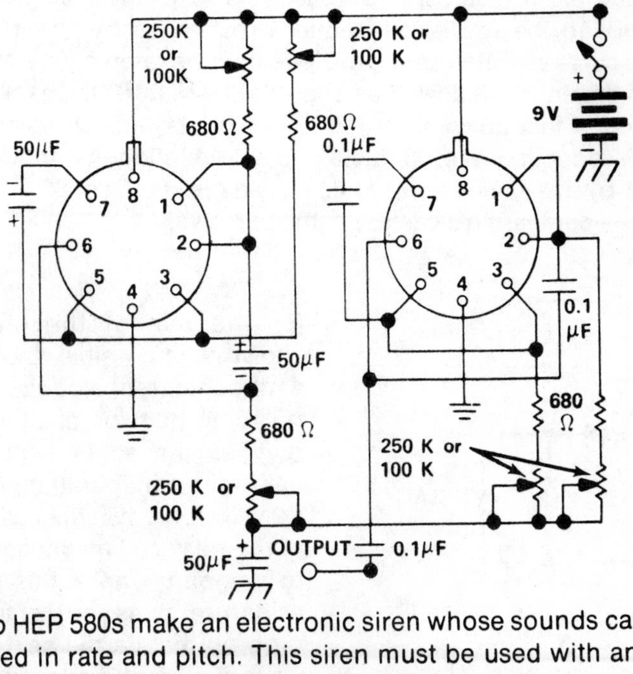


Two HEP 580s make an electronic siren whose sounds can be varied in rate and pitch. This siren must be used with an external amplifier. (Circiut courtesy Motorola.)

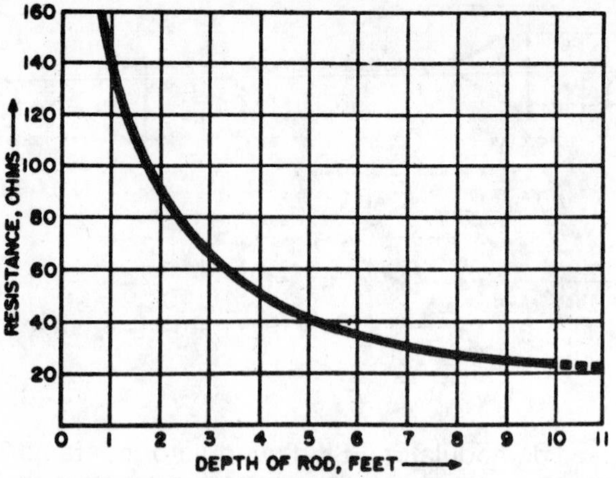


EARTH RESISTANCE DECREASES WITH DEPTH OF ELECTRODE IN EARTH.

When you find that your earth-electrode resistance is not low enough, there are several ways you can improve it: 1) Deepen the earth electrode into the earth; 2) use multiple rods; 3) treat the soil. As you might suspect, driving a longer rod deeper into the earth materially decreases its resistance. In general, doubling the rod length reduces resistance by about 40%. The curve at the left shows this effect. For example, note that a rod driven 2 feet down has a resistance of 88Ω; the same rod driven 4 feet down has a resistance of about 50Ω. Using the 40% reduction rule, $88 \times 0.4 = 35\Omega$ reduction. A 4-foot-deep rod, by this calculation, would have a resistance of $88 - 35$ or 53Ω—comparing closely with the curve values.

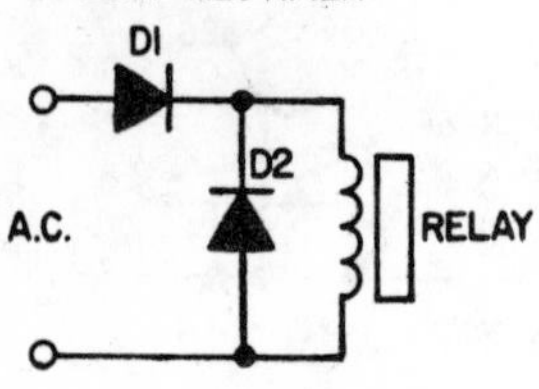


By the use of the circuit shown, it is possible to power a direct-current source, with no relay hum or chatter. D1 rectifies the ac in a normal half-wave configuration, while D2 will slow down the collapse of the relay coil magnetic field to a point where the relay armature does not start to drop out before the next half-wave of current is applied.

Integrated circuit channel scanner. This unit is capable of scanning a series of channels in a receiver by switching crystals in and out of the first oscillator. It works like this. A UJT is used as a clock to produce a series of pulses. This particular UJT is fairly expensive ($2 – $4) but it operates well on 5V. The pulses are of the wrong polarity and quite noisy. To correct both situations, they are fed into one gate of a quad two-input NAND gate, a 7400.The output of this gate is connected through a switch and then to the counter. Note: Bypass 5V supply frequently. 01 – 0.1 μF. 0.1 – .1 μF.

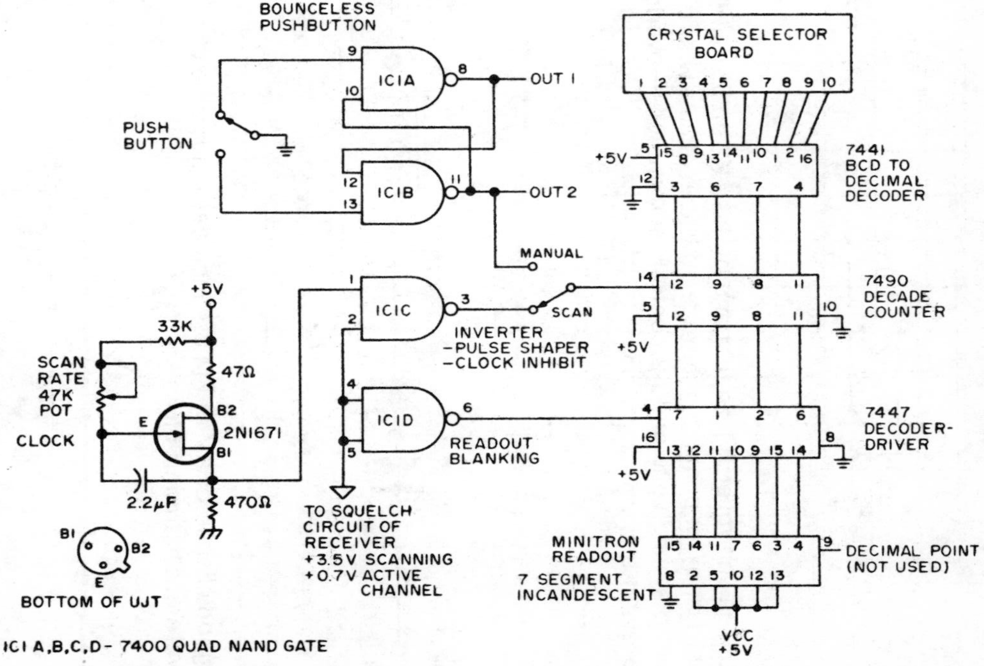

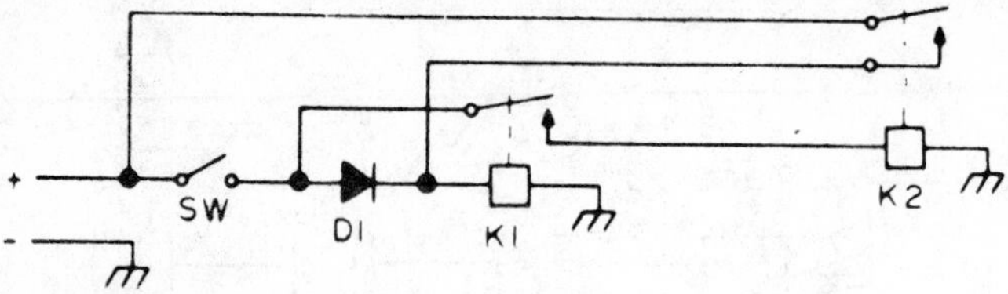


Voltage controller. K1 on first and off last. K2 on last and off first. Handy for pentode or tetrode power amplifiers where K-1 controls plate volts and K2 controls screen volts—overload dekeying, etc.

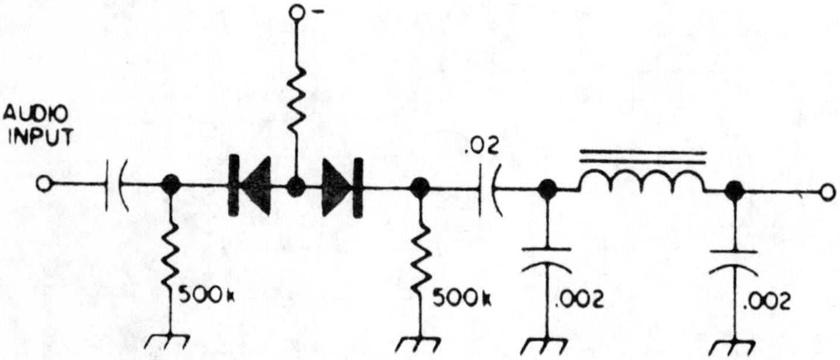


A good clipper for AM or FM use includes adjustable clipping level and a harmonic filter.

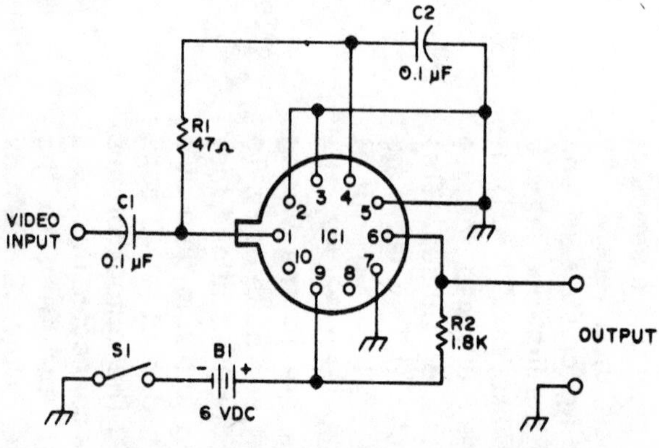


Video amplifier using a Motorola HEP 590 IC.

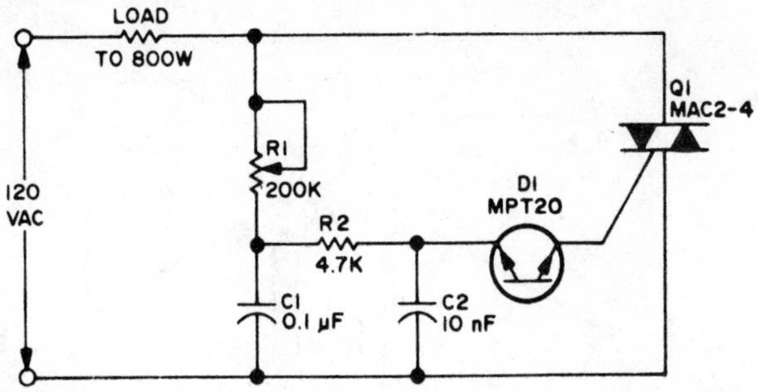


Light dimmer handles up to 800 watts of incandescent lamps. 10 nF = 0.01 μF, by the way. Circuit courtesy of Motorola Semiconductor Power Circuits Handbook.

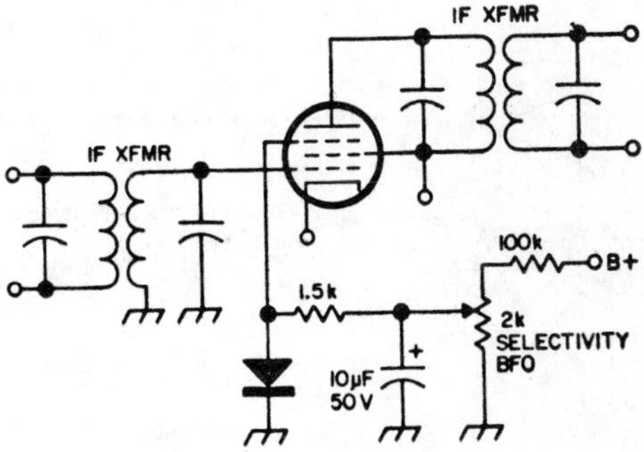


Adapter to provide SSB/CW reception and Q-multiplication in a receiver.

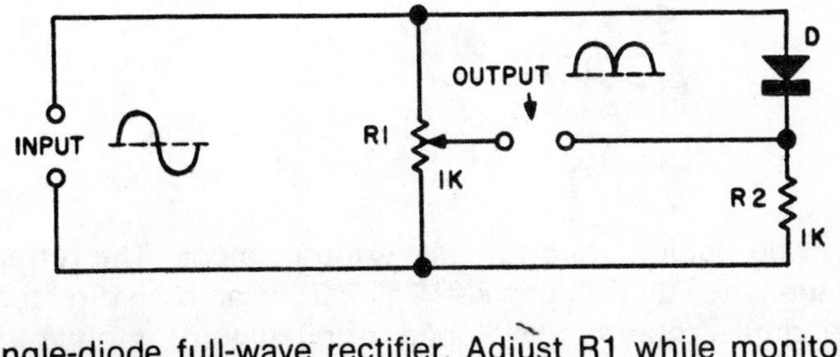


Single-diode full-wave rectifier. Adjust R1 while monitoring the output waveforms on an oscilloscope. When both humps are equal, the setting is correct.

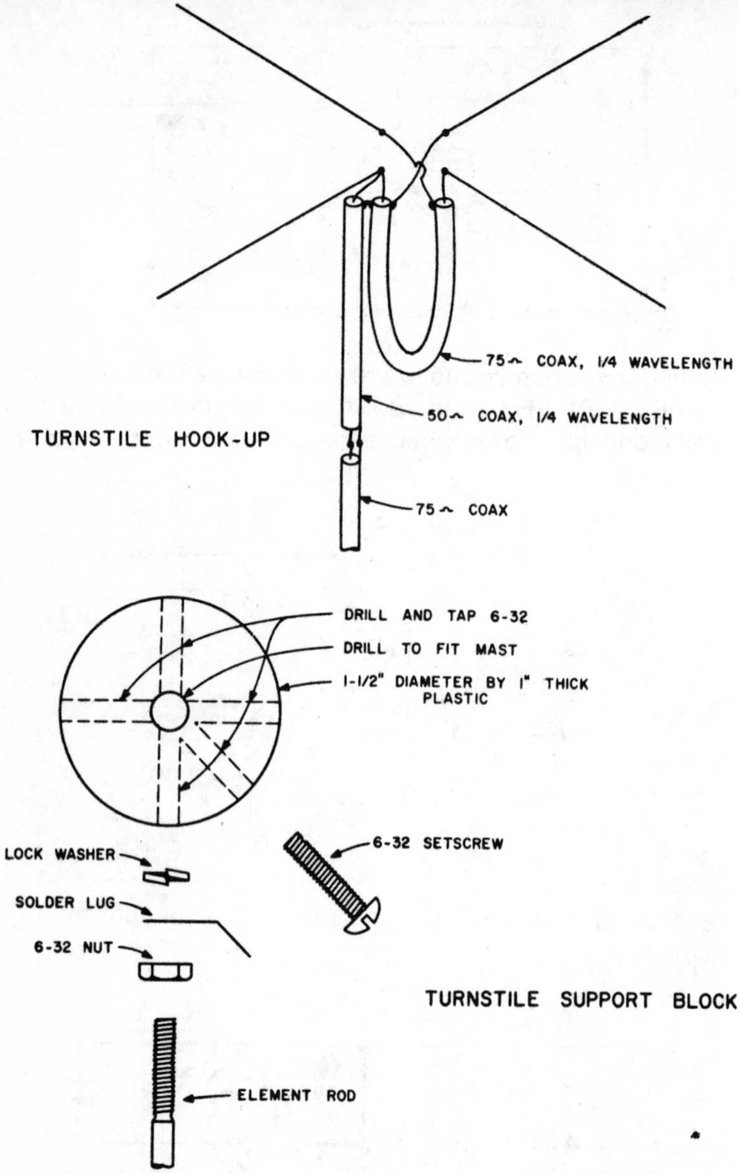


Turnstile hook-up and turnstile support block. The turnstile antenna, two dipoles phased 90° apart, is an omnidirectional horizontal antenna developed originally for mobile use. Dimensions for operation at 145 MHz are: elements-19½″ ⅛″ brass rod; 50Ω coax Q-section RG-59/U, 13½″ long.

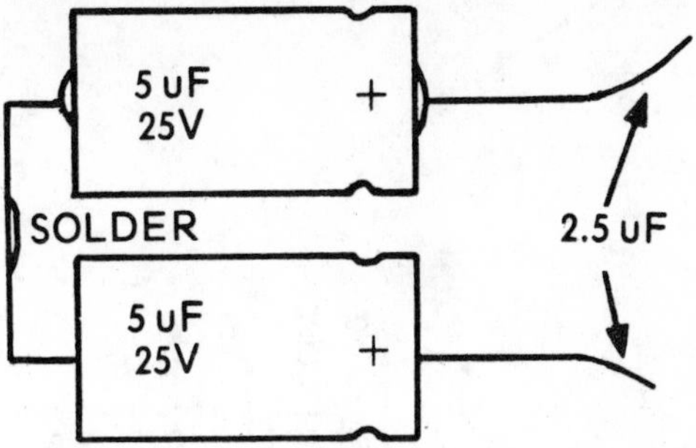


A nonpolarized capacitor can be constructed by connecting two electrolytics in series, back to back. The values must be identical, and each must be twice the value you need for the crossover.

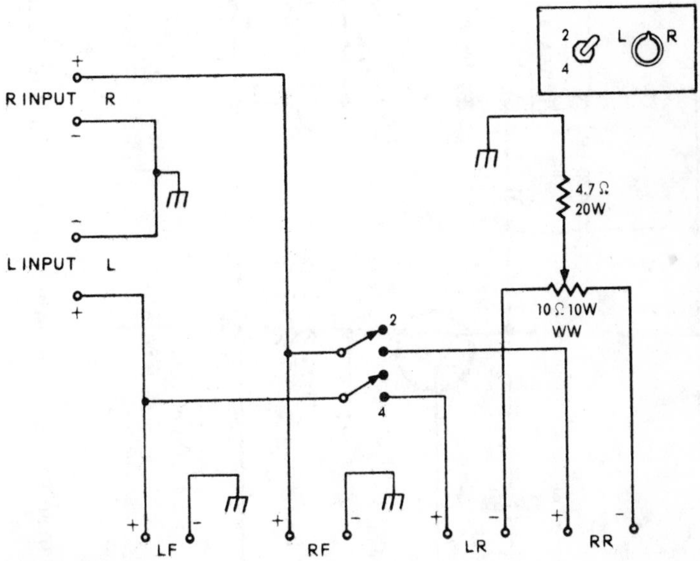


This passive four-channel decoding circuit can be connected to the existing speaker terminals of a stereo amplifier. The double-pole, single-throw switch allows the two rear speakers to be removed from the circuit without affecting the signal normally supplied to the two front speakers. The insert shows how the panel can be laid out.

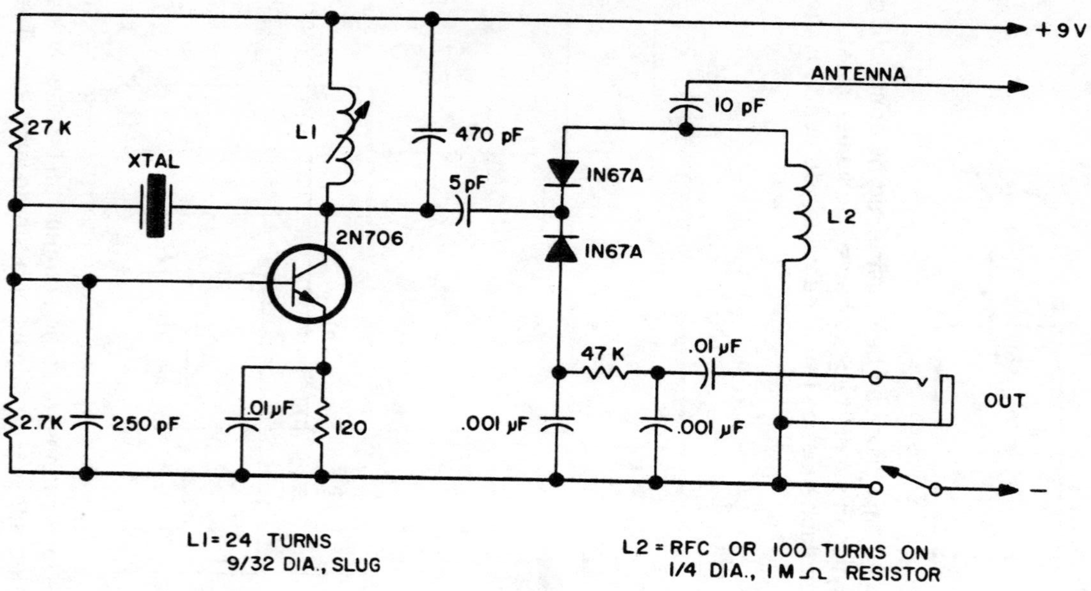


SSB monitor uses ambient rf from transmitter to produce af signal at OUT jack.

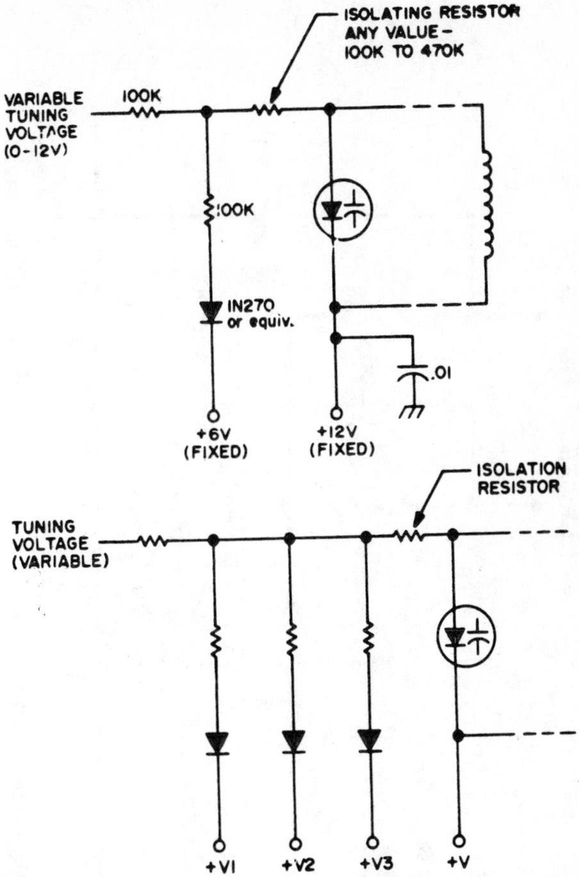


Varactor correction circuit; shaping networks shown here serve to linearize nonlinear tuning range of varactors in rf applications.

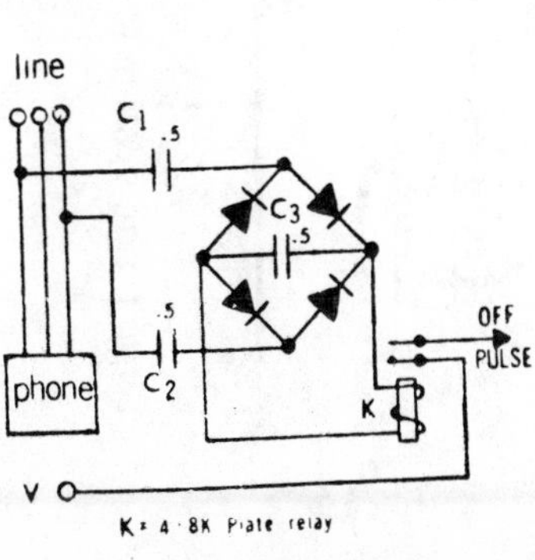


Use this circuit on your unlisted phone at the repeater site. Phone ring pulls in relay, which shuts off (or turns on) repeater.

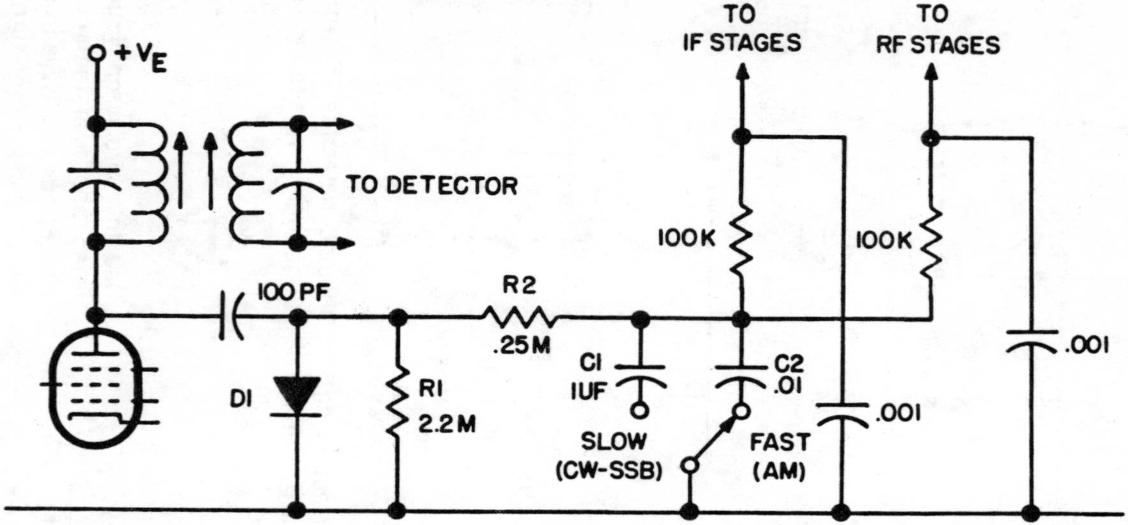


Simple agc system for incorporation into existing tube-type receiver; note switch for attack.

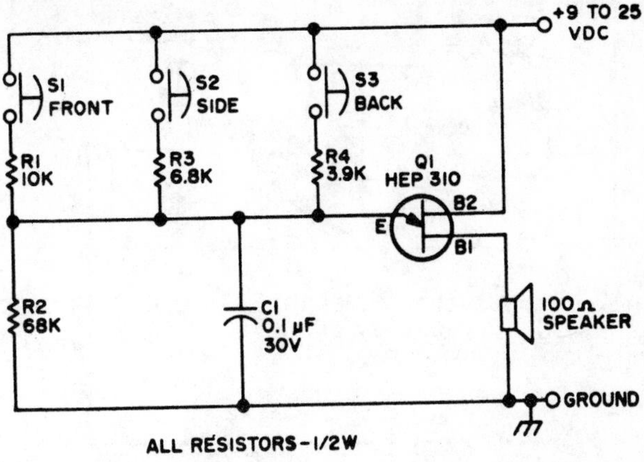


Electronic door buzzer. This circuit features a different tone for each door. The tones may be varied by experimenting with different resistors at each switch. Courtesy Motorola Construction Projects HMA37.

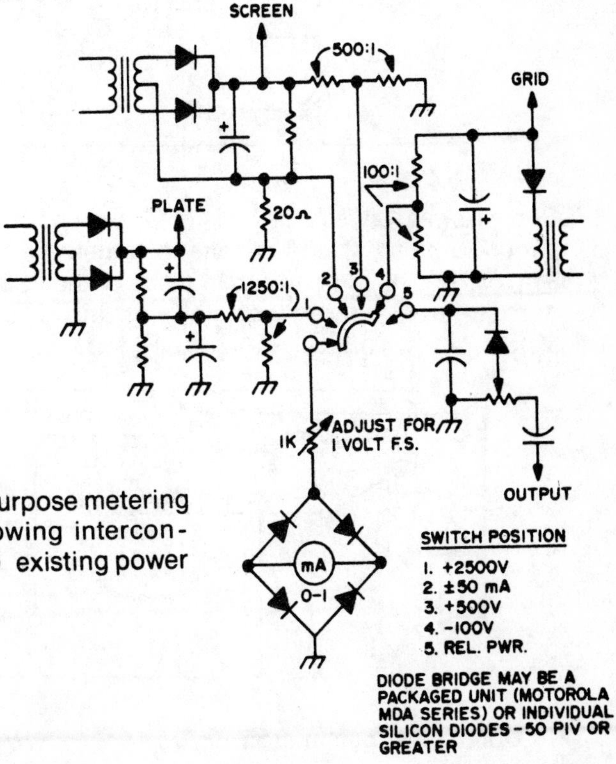


General-purpose metering circuit showing interconnection to existing power supply.

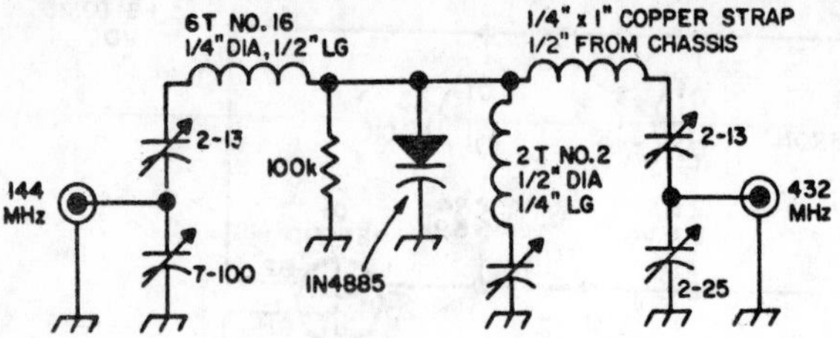


144 to 432 MHz varactor tripler that will give 17W output at 432 when driven by a 25W 2m signal.

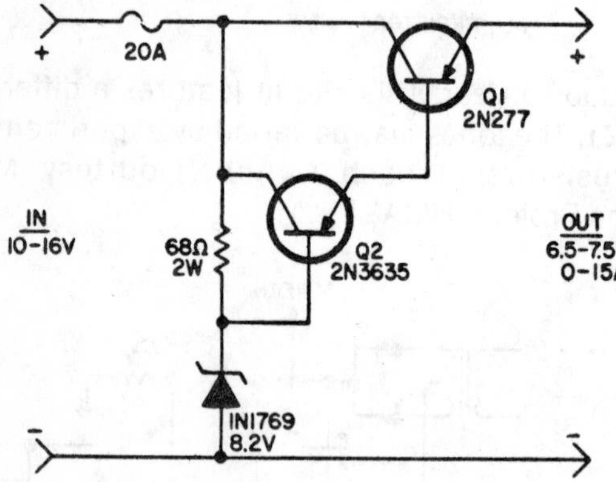


6V supply from a 12V source. This is a handy device to use in case your surplus FM rig happens to require 6V. A 2N2147, 2N4314, or 2N3616 can be used in place of the 2N3635 at Q2. Be sure to use a large heatsink for the 2N277 power transistor.

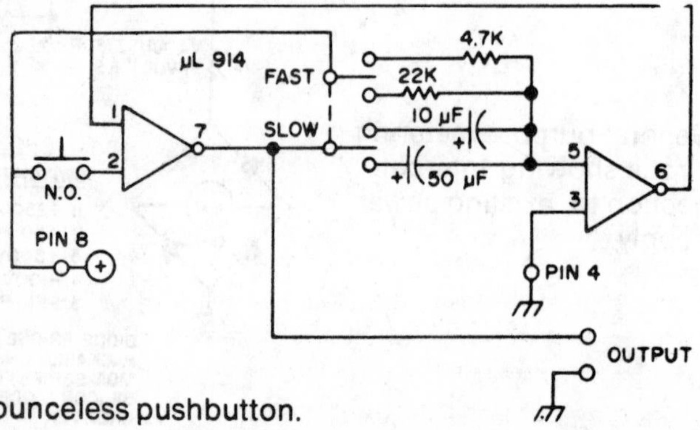


Bounceless pushbutton.

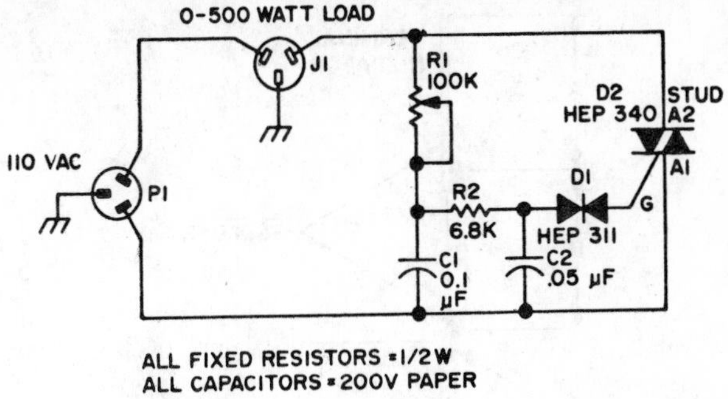


Light dimmer/motor speed control. This circuit is able to control the voltage on loads up to 500 watts. Courtesy Motorola Projects HMA 37.

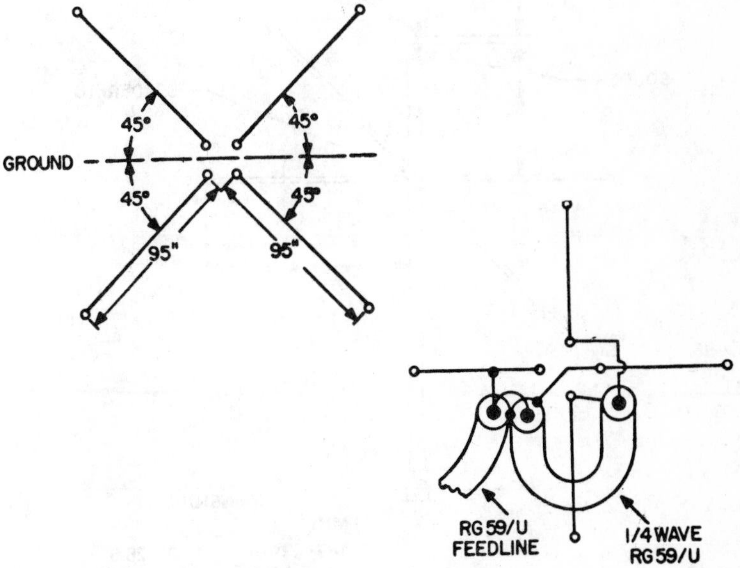


10-meter omnidirectional antenna for OSCAR satellite reception. Mount each dipole antenna at right angles to the other and slant the halves down toward the ground at a 45° angle. Feed one dipole with RG-59 coax, the center conductor goes to one-half of the dipole and the braid goes to the other half. Connect a ¼λ length of RG-59U to the first dipole connections and connect the other end to the 2nd dipole; i.e., braid to one element and the center conductor to the other.

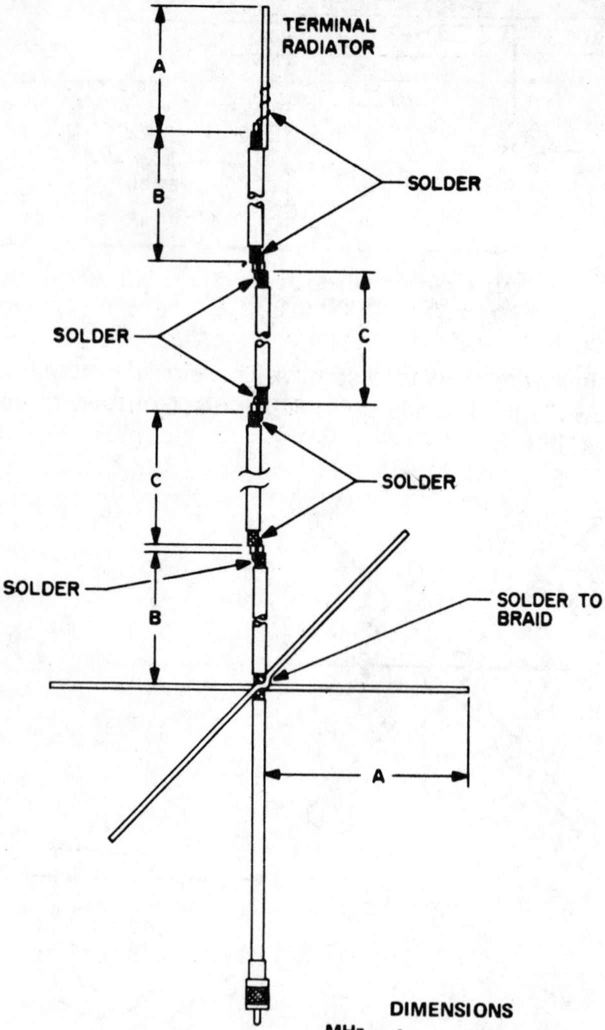


DIMENSIONS

| MHz | A | B | C |
|---|---|---|---|
| 147 | 19" | 13.3" | 26.6" |
| 220 | 12.66' | 8.88" | 17.75" |
| 440 | 6.33" | 4.44" | 8,875" |

Omnidirectional gain antenna for FM home or repeater use. The sections are constructed of measured lengths of RG-8 coax and connected at the junction points by soldering the center conductor of one to the braid of the other and vice versa. The terminal radiator and radials can be lengths of stiff brass rod. A total of 9 "C" sections will give about 5.8 dB gain over a dipole.

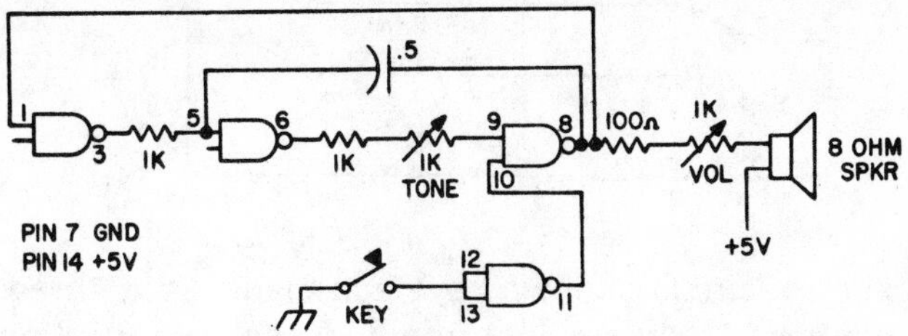


An inexpensive code oscillator circuit which requires one SN7400 quad NAND gate and has tone and volume controls. If the value of the 0.5 μF capacitor is increased, the frequency range of the oscillator will be lowered. The tone output is not the purest dc note, since a square wave is output directly to the speaker. However, it is quite satisfactory for code copy. The SN7400 requires a 5V dc regulated supply. A suitable supply can be built using a bridge rectifier circuit and a LM309K voltage regulator.

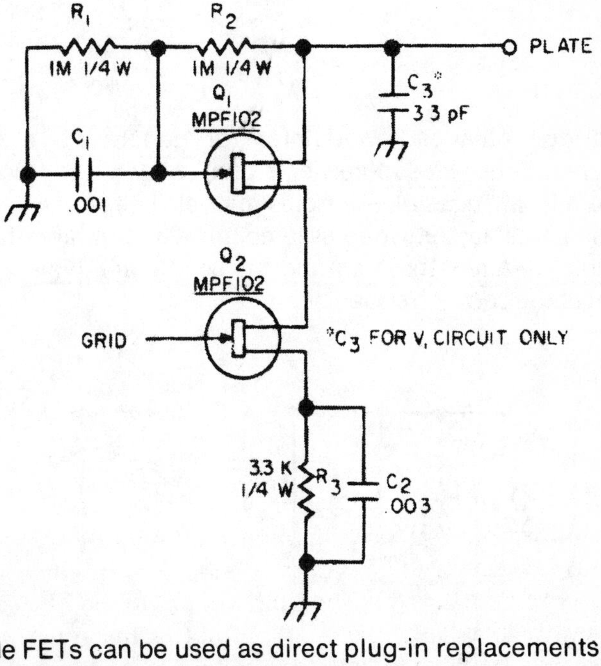


Cascode FETs can be used as direct plug-in replacements for pentode tubes in Motorola hybrid receivers (communications-type walkie-talkies). Will be suitable for replacing various tube-type i-f amplifiers.

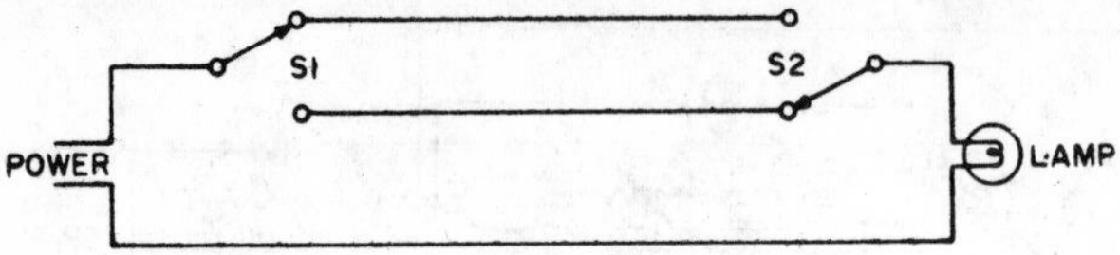


This simple circuit, known to electricians as a three-way switch, for controlling a light from two separate locations, is also a rather exotic digital logic circuit. No gate used in digital logic is any more complex than this circuit, though it may have more control points.

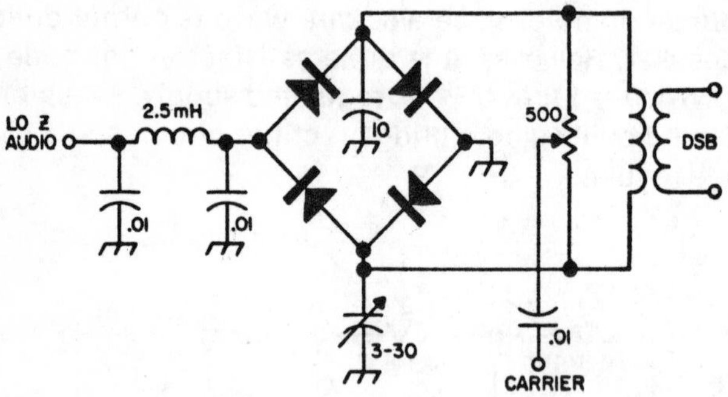


A bridge balanced modulator for generating a double-sideband suppressed-carrier signal. Match the diodes as carefully as possible—surplus diodes will work—and use symmetrical layout when assembling. The primary of the coil should be a few turns around the secondary, which is resonant at the carrier frequency.

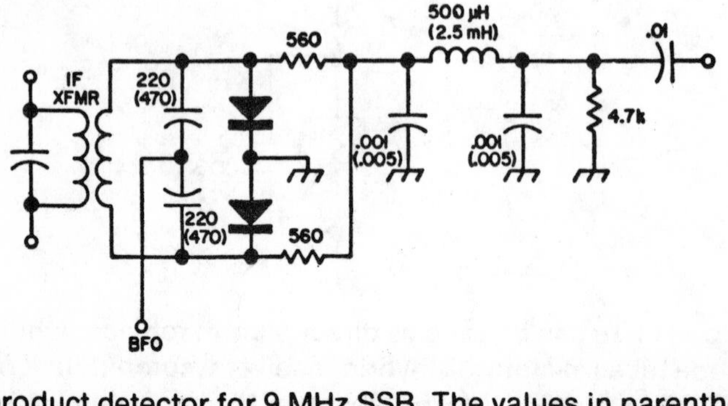


A product detector for 9 MHz SSB. The values in parentheses are for 455 kHz.

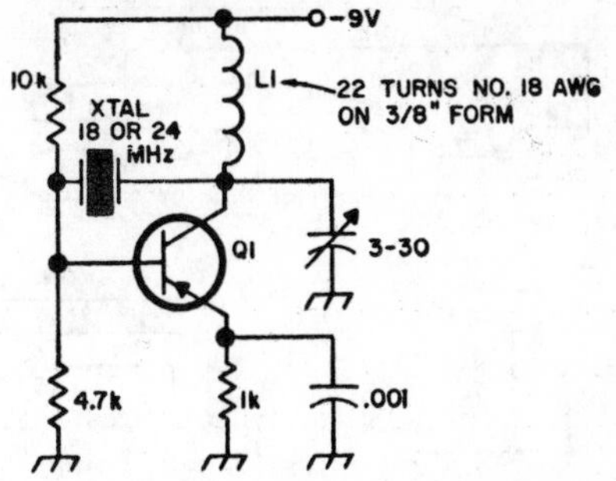


This 2-meter edge marker provides useful harmonics up to several hundred megahertz with 18 or 24 MHz crystals. Though originally designed for a PNP transistor, and NPN type such as the 2N706 or HEP 55 may be substituted if you reverse the supply voltage connections.

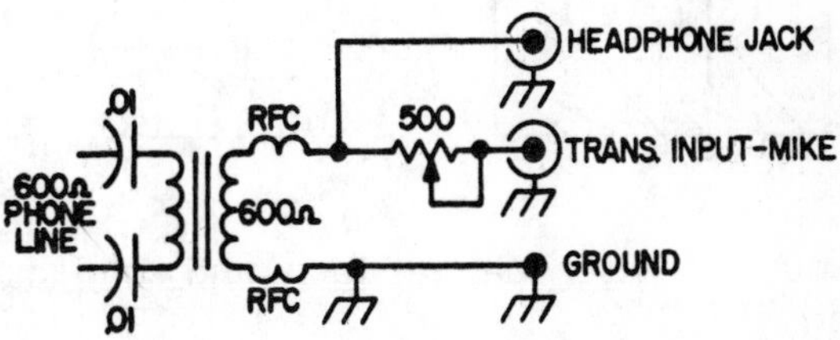


Simple phone patch for experimentation purposes. Adjust the mike gain on the transmitter as you would normally and set the 500Ω potentiometer for proper modulation by the telephone audio.

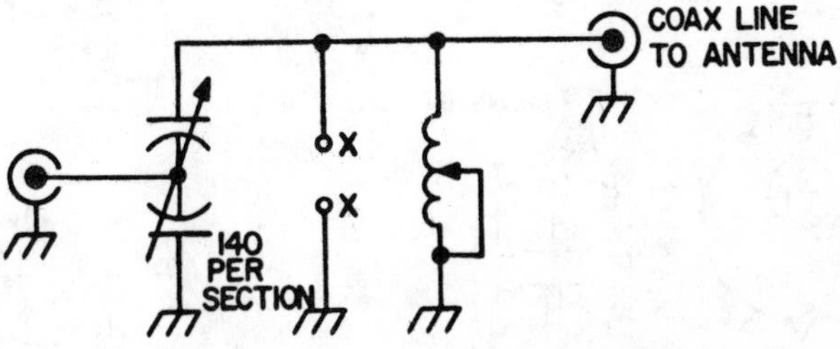


Simple transmission-line tuner to couple coax line operating at a high SWR into a transmitter requiring 50–70Ω load at a low SWR.

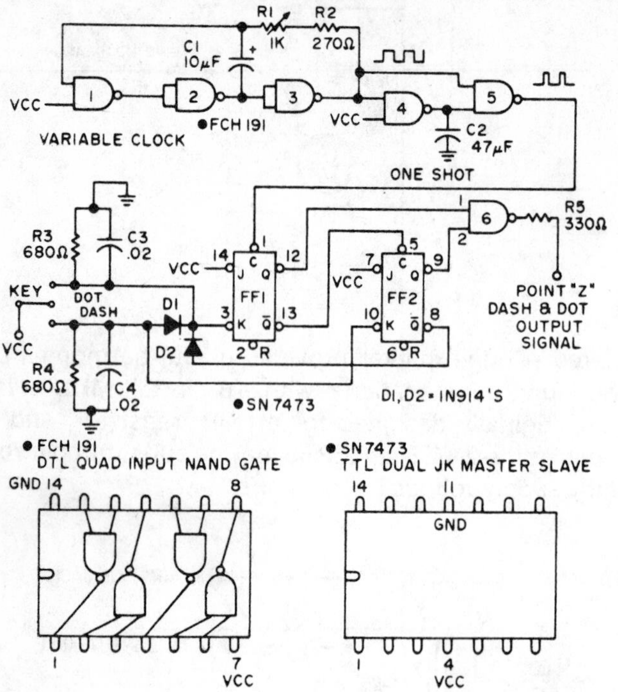


Easy-to-build IC keyer features relay and solid-state output switching.

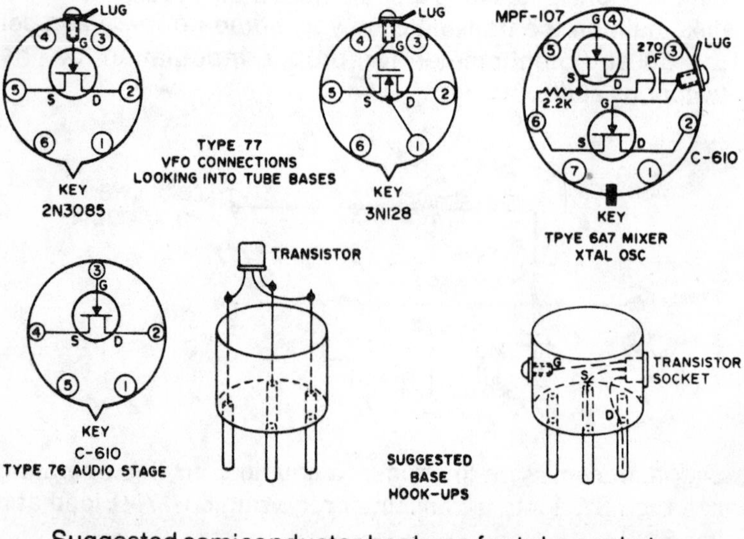


Suggested semiconductor hookups for tube sockets.

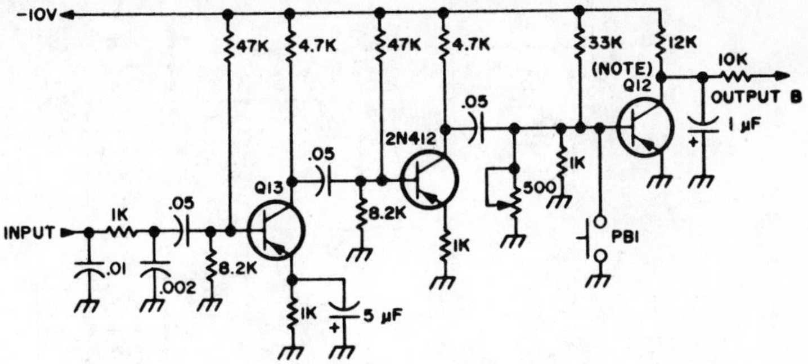


Schematic diagram of noise-actuated squelch unit.

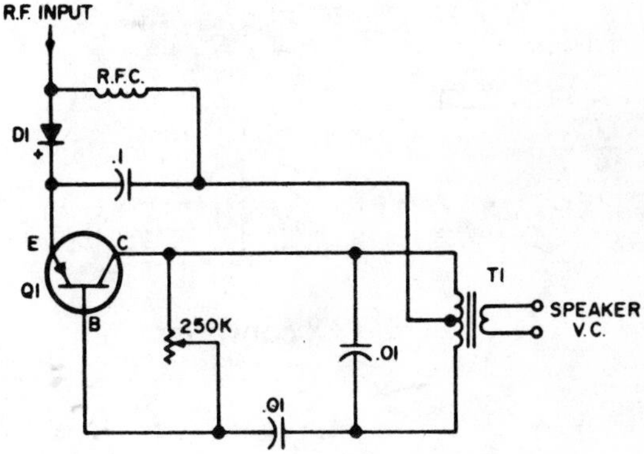


Code monitor said to work with any transmitter. Any general-purpose PNP transistor will work. Connect RF input line to chassis of transmitter to get excitation signal.

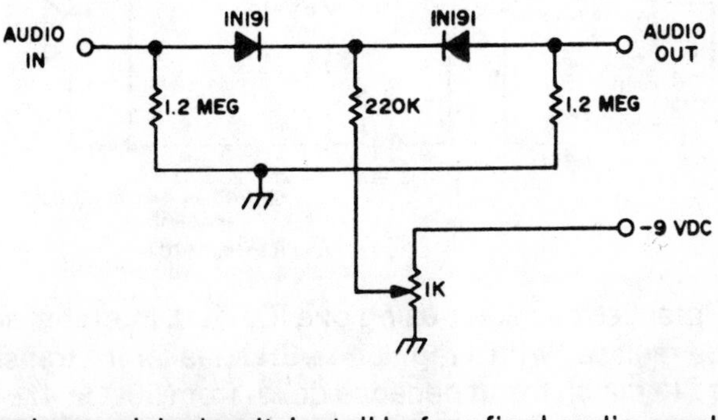


Simple squelch circuit. Install before final audio amplifier.

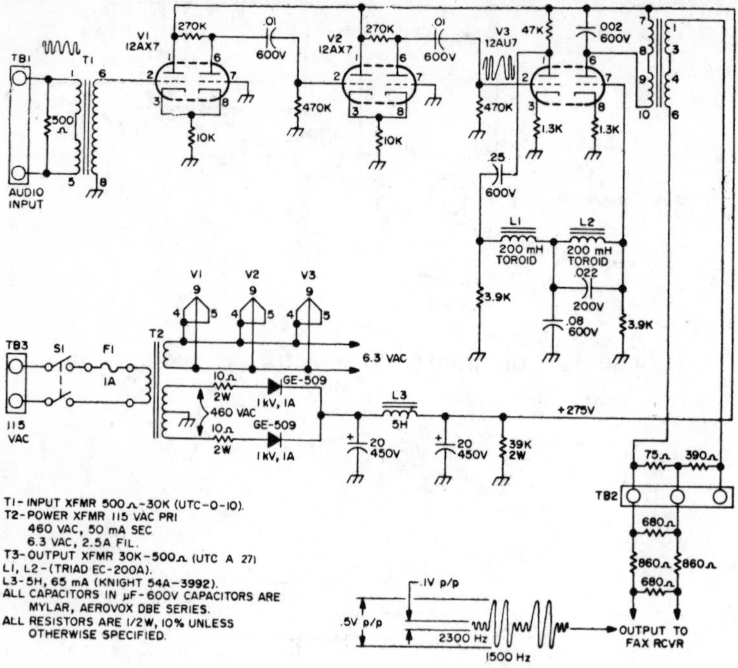


FAX converter.

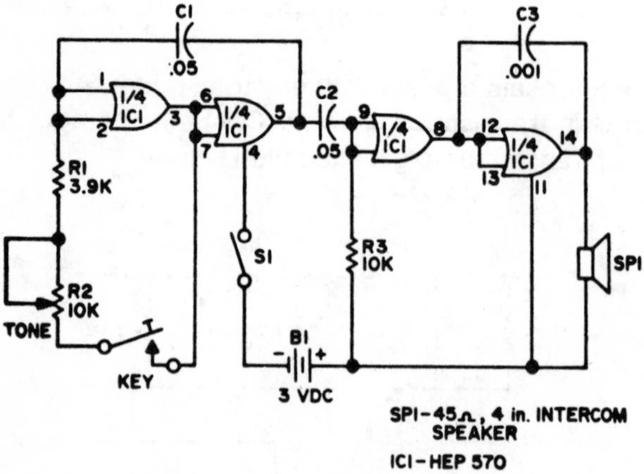


Code-practice oscillator using one IC, loud. Intercom speaker may be replaced with a regular 4 – 8Ω speaker if a transformer is used to match the impedance down from 50Ω to 4Ω. Circuit is from Motorola HMA-36, Radio Amateur's IC Projects.

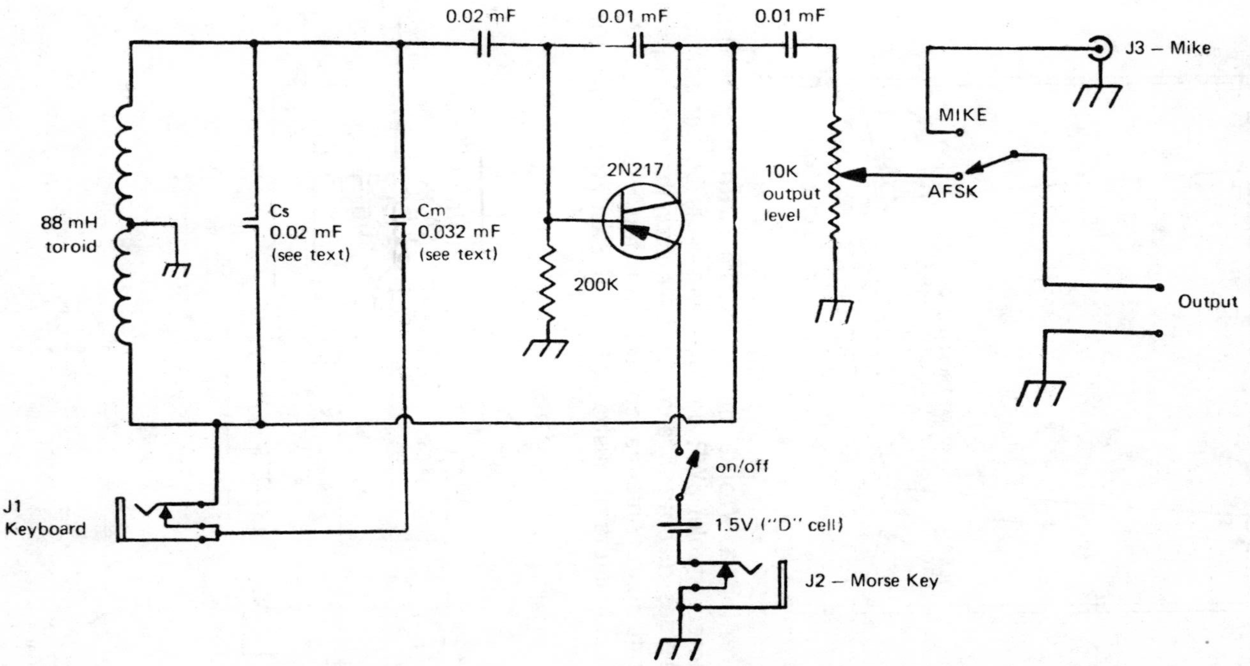


Schematic of AFSK tone generator.

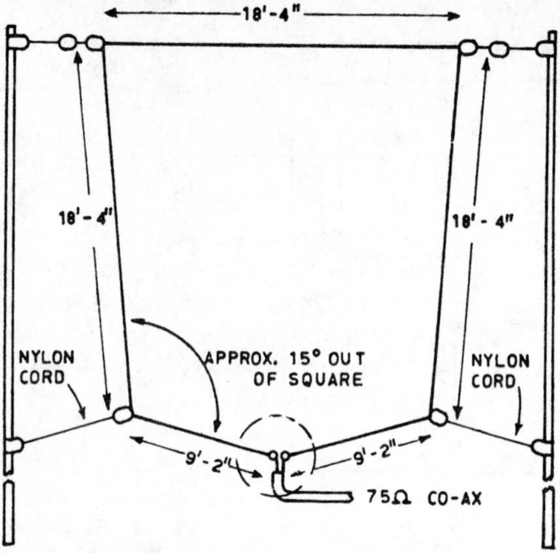


Wire antenna. Works best on 20m, with higher (but not unreasonable) SWR on 40 & 80m. Note bottom section drags sides in about 15, a critical dimension that may require experimenting. Antenna courtesy Amateur Radio July 71. Wireless Institute of Australia, Box 36, East Melbourne, Vic. 3002.

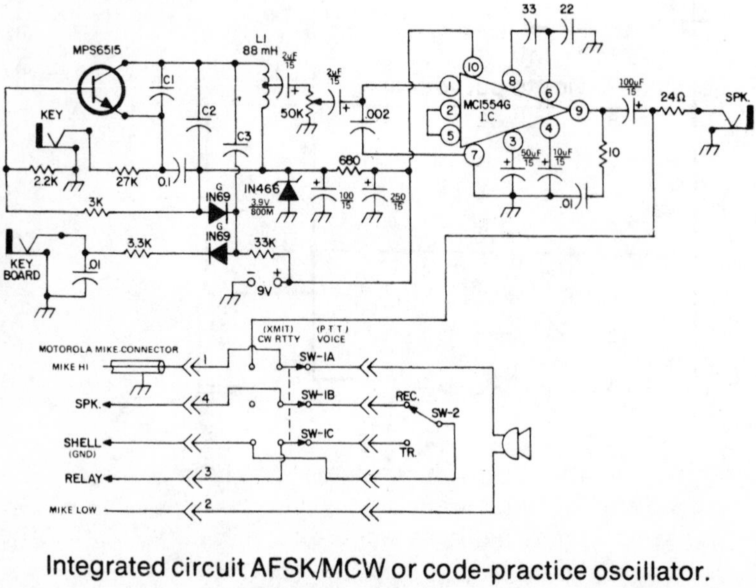


Integrated circuit AFSK/MCW or code-practice oscillator.

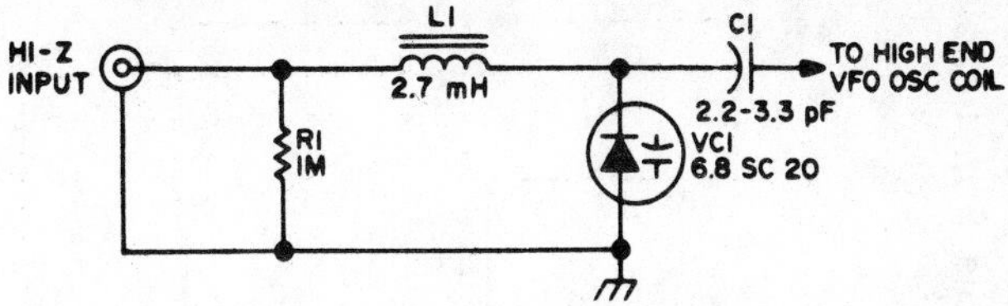


This varactor modulator will put your VFO-operated AM transmitter on FM in ten minutes. Assemble it on a standard three-terminal phenolic tiestrip and tuck it into a corner of your chassis. Both this modification and the one below for crystal-controlled rigs should be driven with a high-output, high-impedance crystal or ceramic mike.

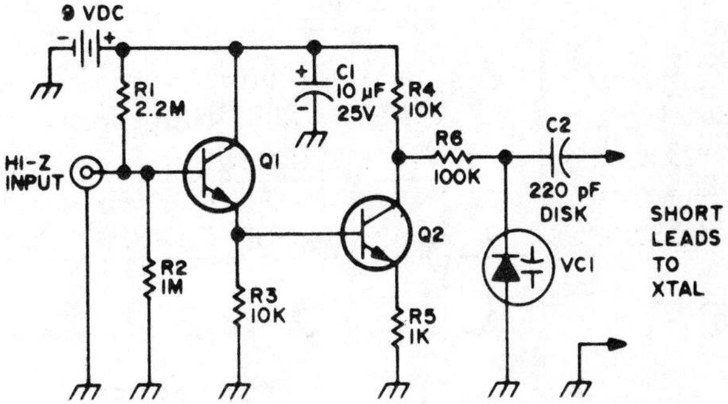


Use this device to put your crystal-controlled AM rig on FM.

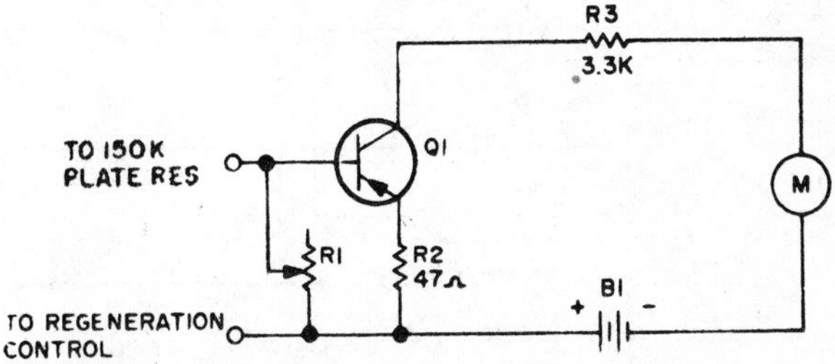


S-meter for the "Sixer." It consists of an inexpensive 0–1 mA meter and a single-transistor meter amplifier. Half-scale deflection is obtained on a signal strong enough to quiet the background hiss.

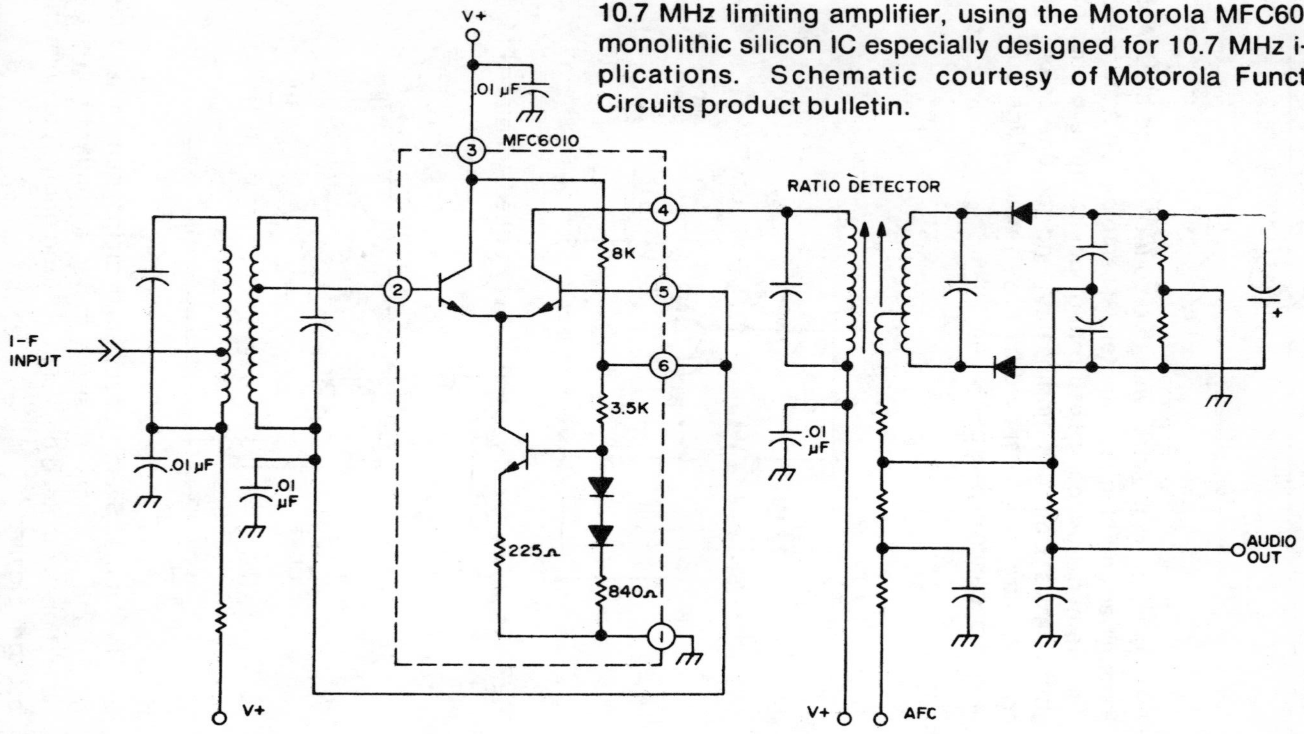


10.7 MHz limiting amplifier, using the Motorola MFC6010, a monolithic silicon IC especially designed for 10.7 MHz i-f applications. Schematic courtesy of Motorola Functional Circuits product bulletin.

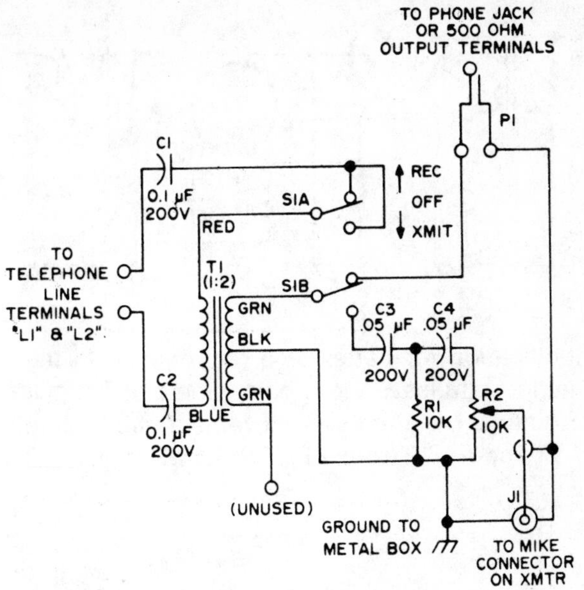


A simple and efficient phone patch.

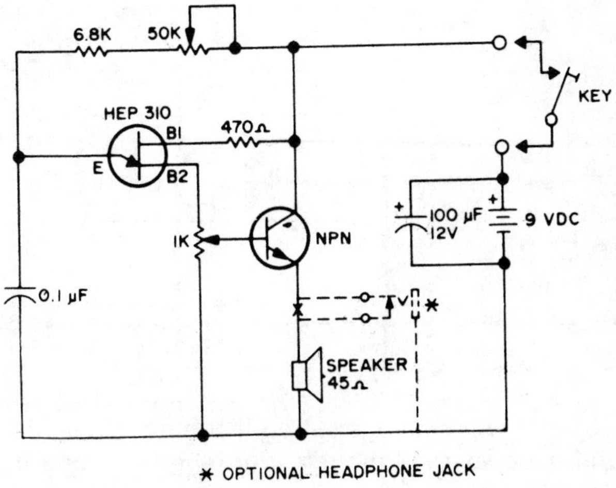


Code-practice oscillator or perhaps a noisy alarm for the home, boat, or car. NPN transistor is Calectro K4-506.

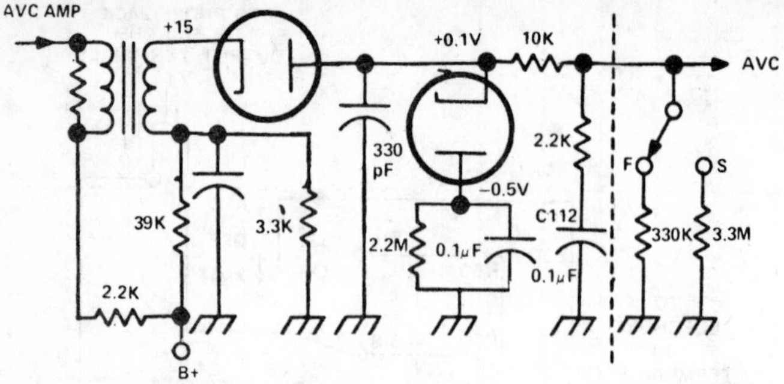


AVC modification for better SBB reception with the Collins 75A4 that lets you leave the rf gain wide open. It gives much longer agc time constant and prevents audio "pumping" on speech. Change C112 from 0.1 μF to 1.5 μF Mylar, 100V.

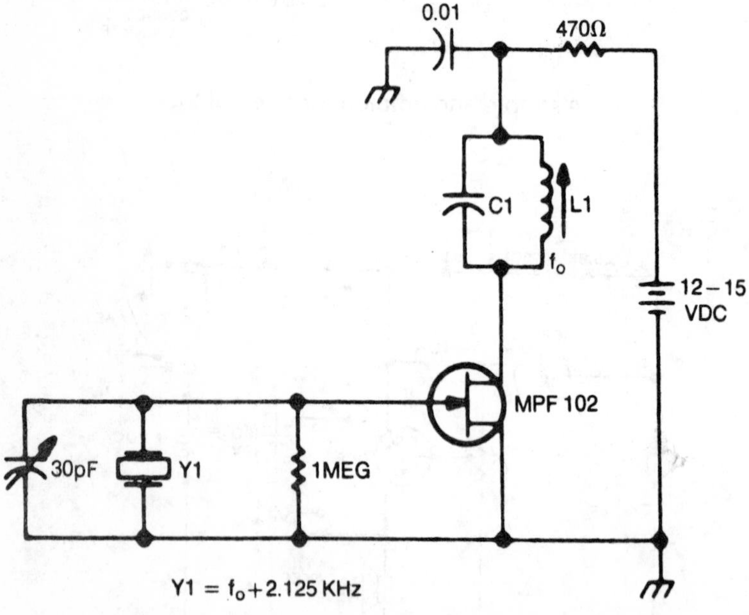


This oscillator will beat against an incoming signal and stay right on frequency even though your receiver may drift a bit. Great for RTTY autostart, net operation, and fixed frequency biz. Circuit courtesy W9ZTK, W9YPS and the RTTY Journal, Box 837, Royal Oak MI 48068. Crystal frequency = net frequency + 2.125 kHz for TT autostart.

Two-stage video amplifier with agc; no tuned circuits required. The curve shows the video amplifier response with agc.

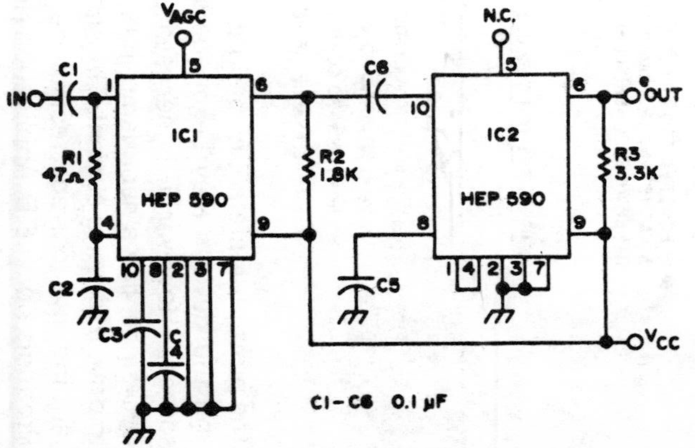


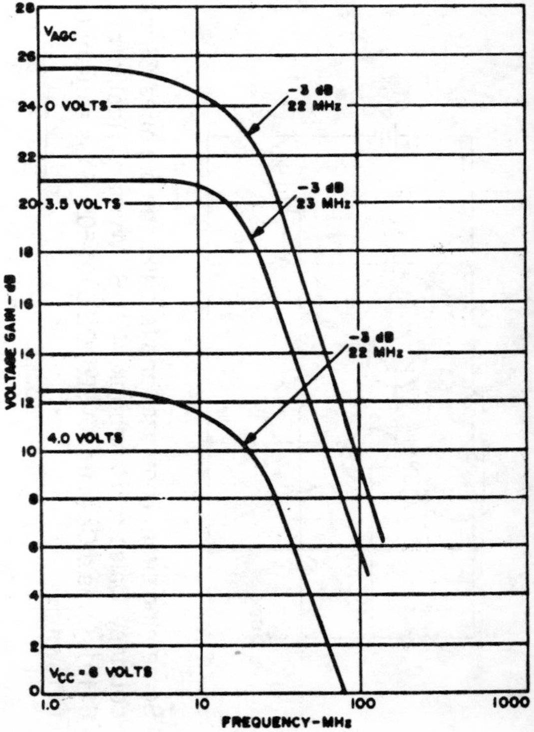

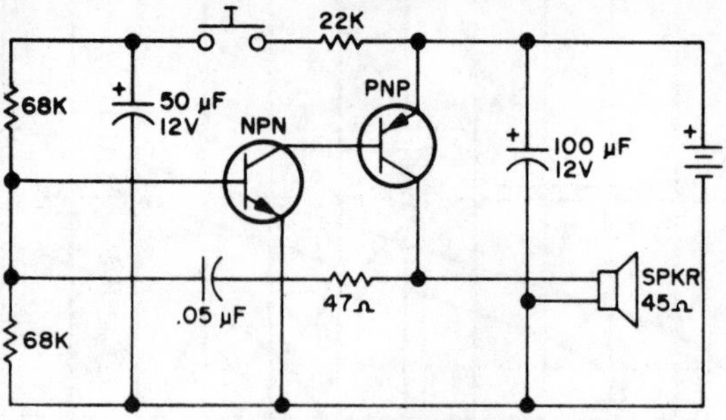


Screaming siren. Tone rises and falls like the big ones. Circuit courtesy Calectro Handbook, 400 S. Wyman, Rockford IL 61101. Q1 is a Calectro K4-506, Q2 a K4-505, S1 is a pushbutton switch.

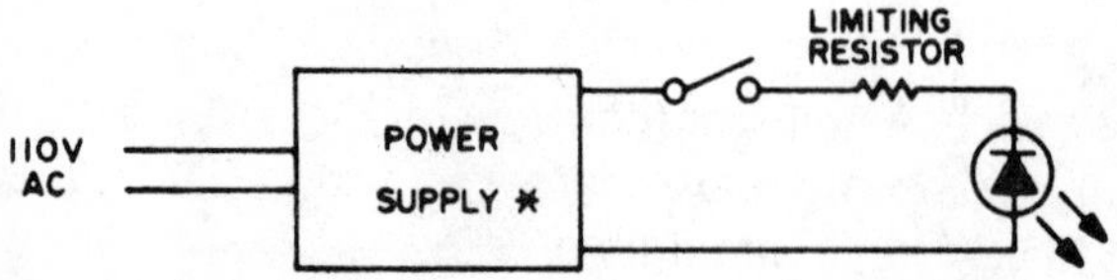


* IF POWERED FROM THE AC LINE, THE OUTPUT MUST BE 6 VAC (STANCOR TRANSFORMER NO. P8385 OR EQUIV). OTHERWISE, POWER SUPPLY IS 6 VDC AT 100 mA.

LIMITING RESITOR CALCULATIONS

$$\frac{V_S - V_F}{I_F} = R\ (\text{min.})$$

$V_S$ = SOURCE VOLTAGE

$V_F$ = LED FORWARD VOLTAGE

$I_F$ = FORWARD CURRENT

R (min.) = VALUE OF RESISTOR (in ohms)

The power requirements for light-emitting diodes are very low and the devices may be operated from a variety of power sources. However, it is necessary to limit the current to an LED since, like a neon lamp, it can be damaged if permitted to draw excessive current. The simple formula will help to determine the correct resistance value. Courtesy Sprague Products Co., L.E.D. Application Notes.

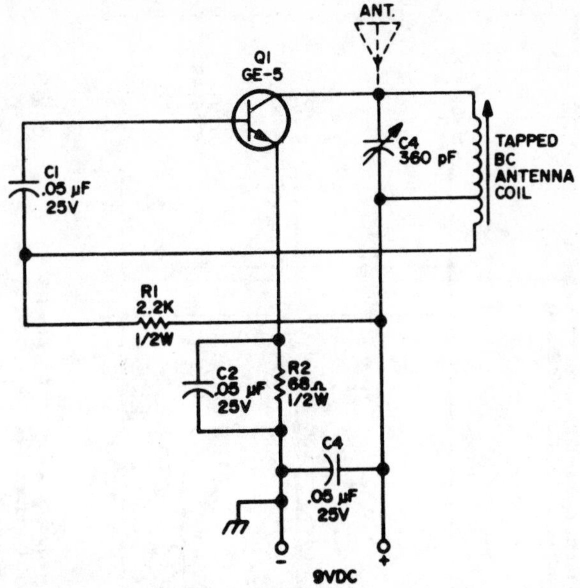


Beat frequency oscillator (courtesy of Gerard Piette, Ontario, Canada). Place unit near receiver and tune C2 until SSB and CW signals become intelligible.

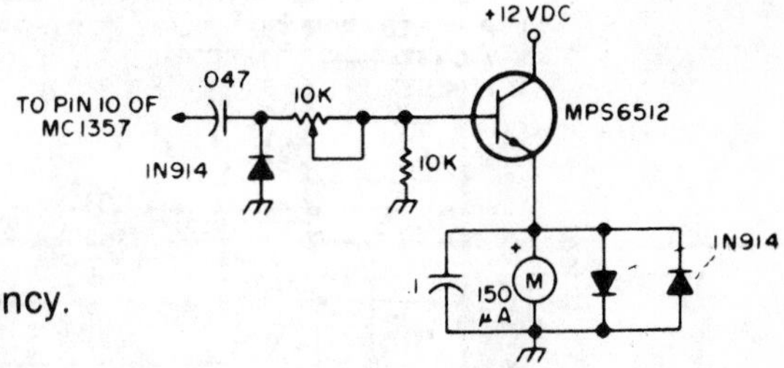


S-meter for Regency.

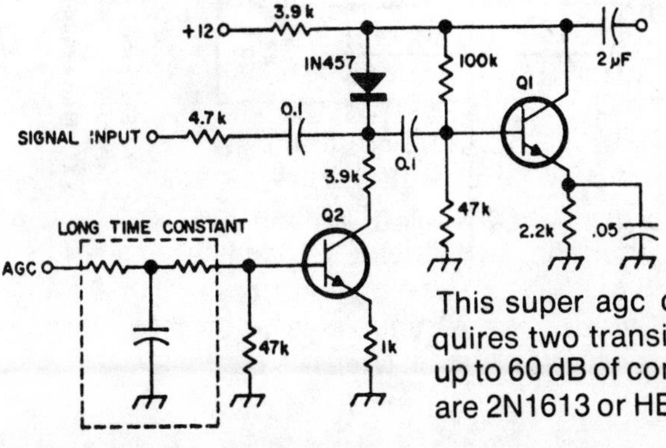


This super agc circuit only requires two transistors to obtain up to 60 dB of control. Q1 and Q2 are 2N1613 or HEP 254s.

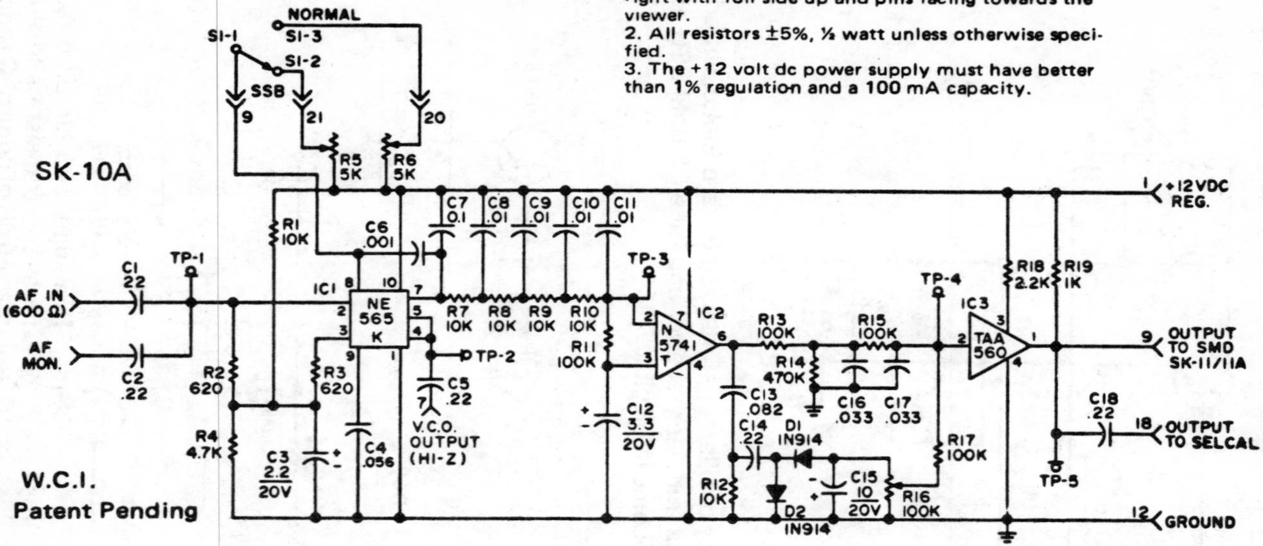


Notes: 1. All P.C. pin connector numbering is from left to right with foil side up and pins facing towards the viewer.
2. All resistors ±5%, ½ watt unless otherwise specified.
3. The +12 volt dc power supply must have better than 1% regulation and a 100 mA capacity.

(A) SSB mode input frequency range: 800–2200 Hz; VCO frequency set to 1500 Hz by R5.
(B) Normal mode input frequency range: 1800–3200 Hz; VCO frequency set to 2500 Hz by R6.
(C) Noise squelch adjustment R16 set for zero output at TP-5 with random noise input.

Phase-locked-loop detector.

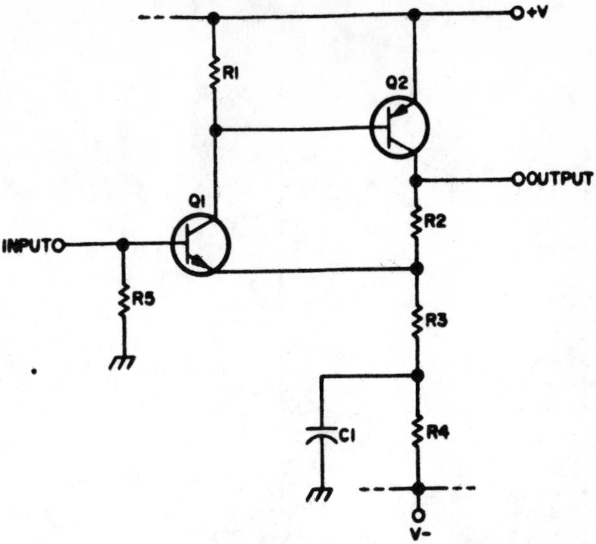


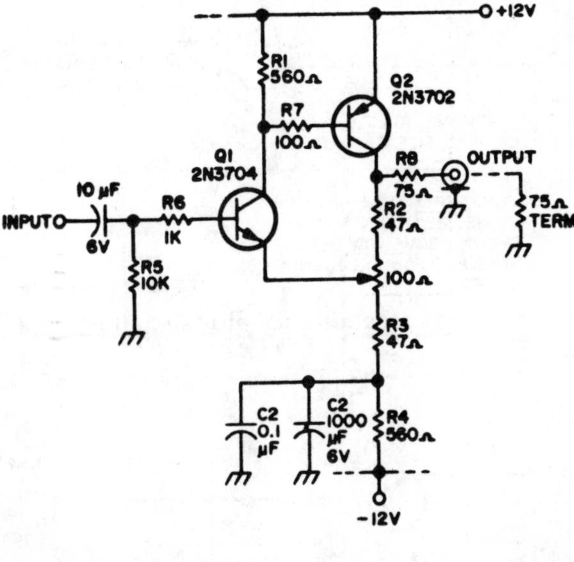


Two-transistor video amplifier, medium input impedance (5K) and low output impedance. Almost any small signal HF transistors will work in the circuit, but ideally they should have a high cut-off frequency, reasonably high current gain and low output capacitance. In deciding on circuit values, the value of R3 should lie between about 22Ω and 100Ω and the value of R2 chosen to suit the gain required. Courtesy, Journal of BATC.

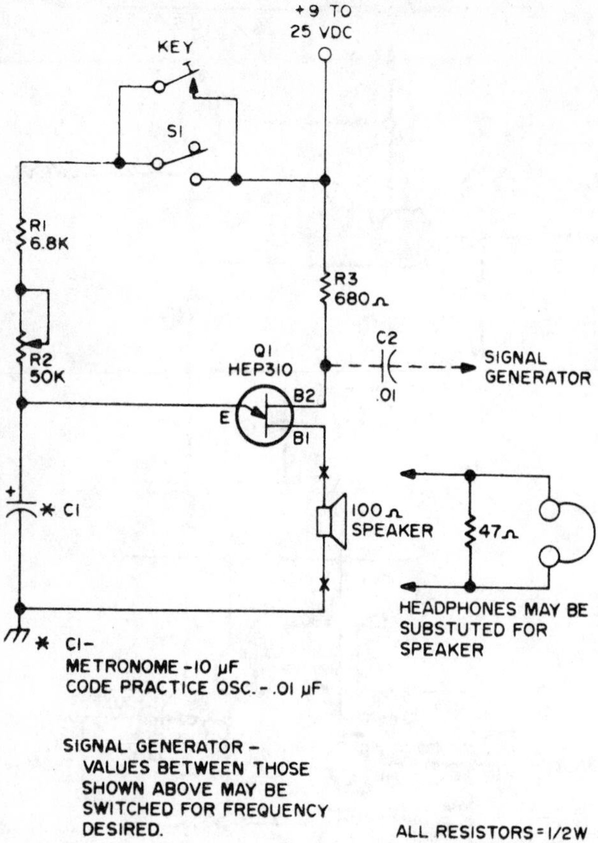


Metronome or code practice oscillator and audio generator.

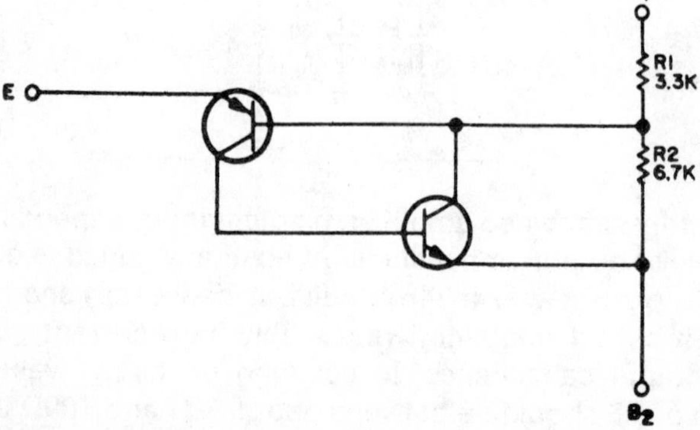


2-transistor, 2-resistor equivalent of a programmable unijunction transistor.

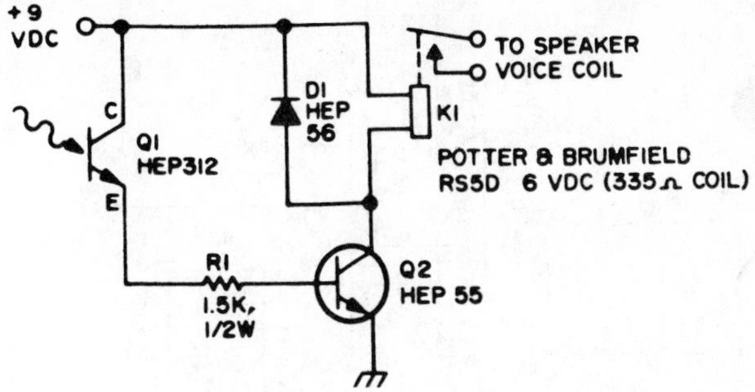


TV commercial killer with no interconnecting wires. Use a flashlight to key relay. Circuit features low battery drain, auto reset. Courtesy Motorola.

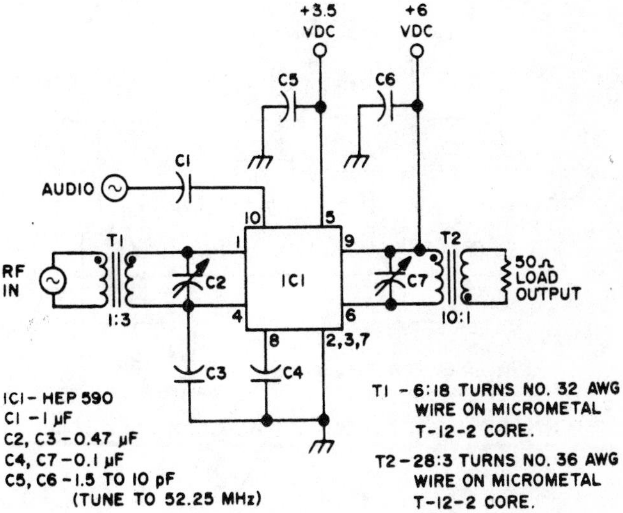


6-meter amplitude modulator using a Motorola HEP 590.

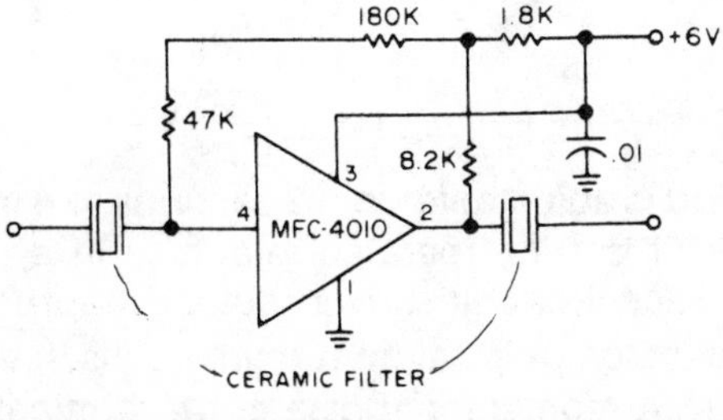


I-f amplifier for 455 kHz uses ceramic crystal filters.

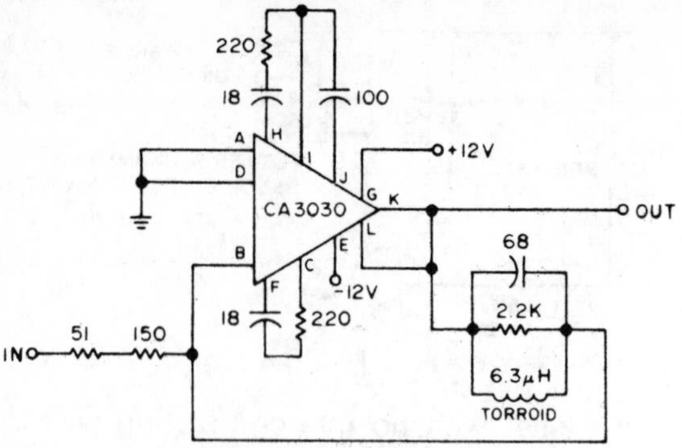


10 MHz bandpass amplifier.

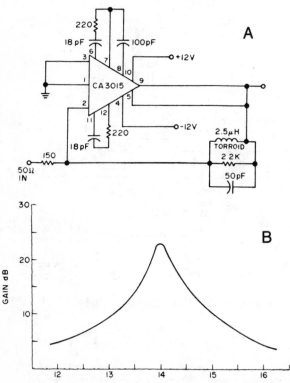

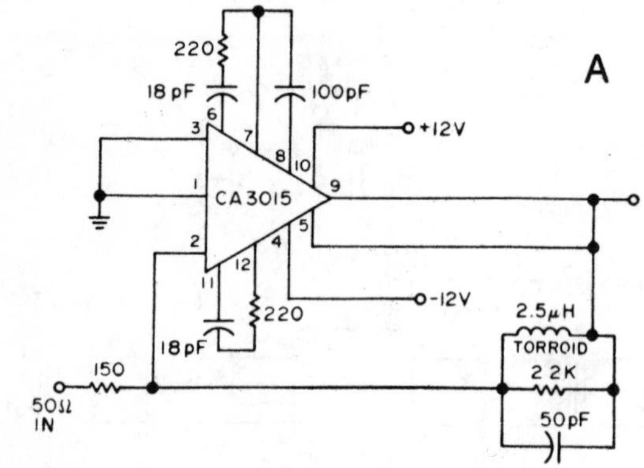


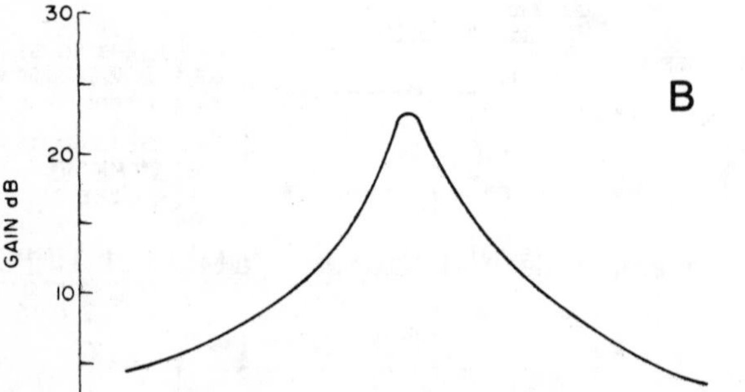


Timer's period is adjustable from 2 seconds to 4 minutes. Unit is designed for carrier-operated-relay use in a repeater application. Adjacent circuit shows "defeat" approach, where a momentary voltage pulse at **on** terminal pulls in the relay and defeats timer. A momentary pulse at **off** terminal returns repeater to timed control.

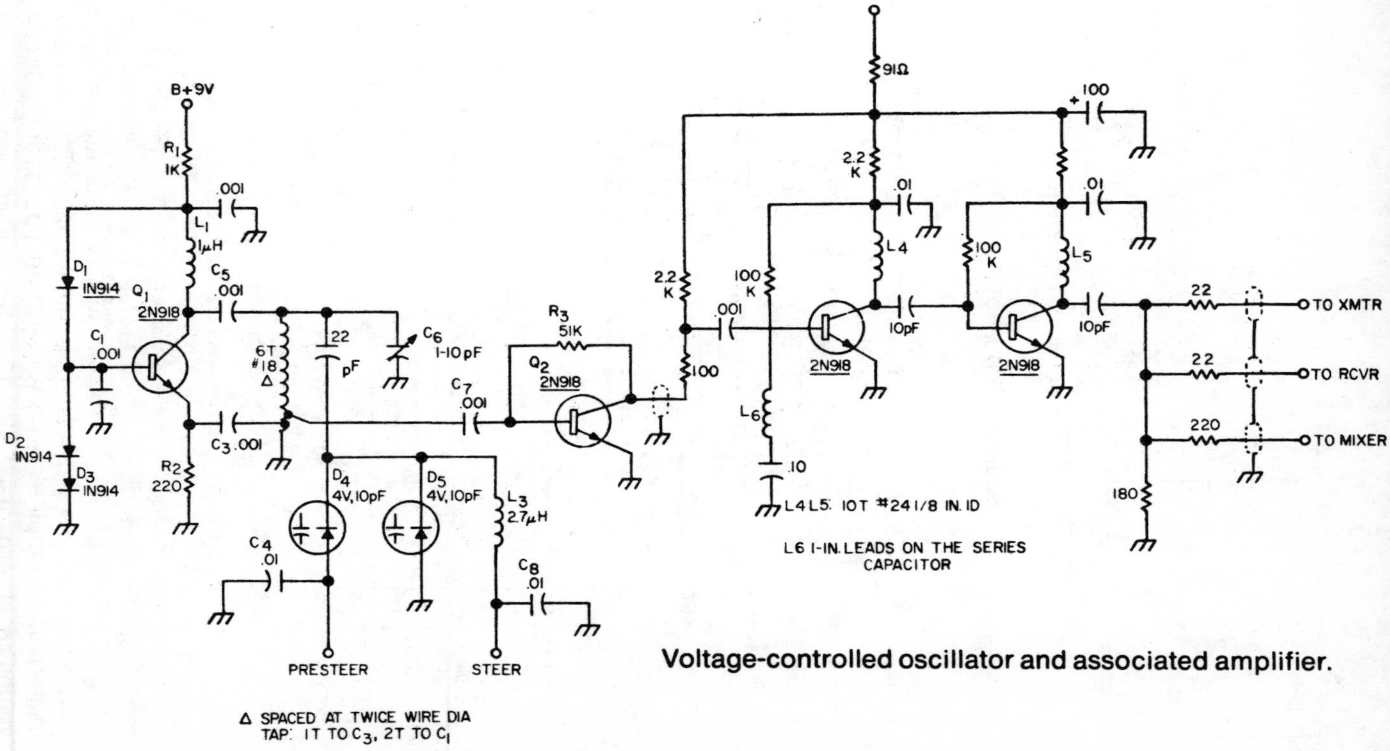


Voltage-controlled oscillator and associated amplifier.

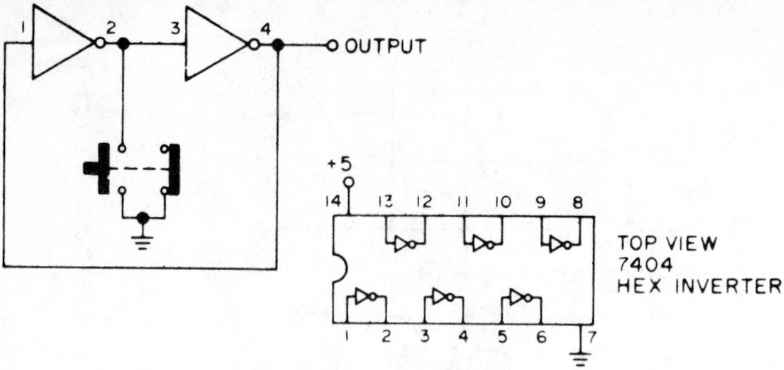


Bounceless pushbutton.

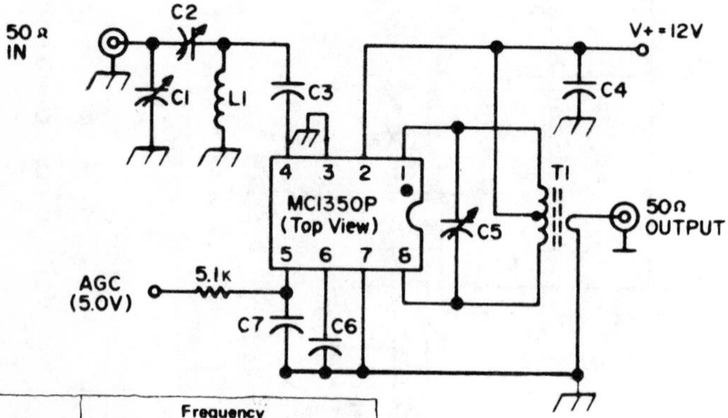


| Component | Frequency 455 kHz | 10.7 MHz | 45 MHz |
|---|---|---|---|
| C1 | | 80–450 pF | 9.0–35 pF |
| C2 | | 5.0–80* pF | 2.0–8.0 pF |
| C3 | 0.05 μF | 0.001 μF | 0.001 μF |
| C4 | 0.05 μF | 0.05 μF | 0.001 μF |
| C5 | 0.001 μF | 36 pF | 1.0–5.0 pF |
| C6 | 0.05 μF | 0.05 μF | 0.001 μF |
| C7 | 0.05 μF | 0.05 μF | 0.001 μF |
| L1 | | 4.6 μH* | 0.8 μH |
| T1 | Note 1 | Note 2 | Note 3 |

*Circuit positions of L1 and C2 are interchanged.

Note 1. Primary: 120 μH (center-tapped)
$Q_u$ - 140 at 455 kHz
Primary: Secondary turns ratio ≈ 13

Note 2. Primary: 6.0 μH
Primary winding - 24 turns #36 AWG (close-wound on 1/4" dia. form)
Core - Arnold Type TH or equiv
Secondary winding - 1-1/2 turns #36 AWG, 1/4" dia. (wound over center-tap)

Note 3. Primary winding = 18 turns #22 AWG (center-tapped)
Secondary winding = 1 turn #22 AWG (over-wound at center of primary)

High-gain i-f amplifier with wide-range agc action. Circuit values are shown for 455 kHz, 10.7 MHz, and 45 MHz.

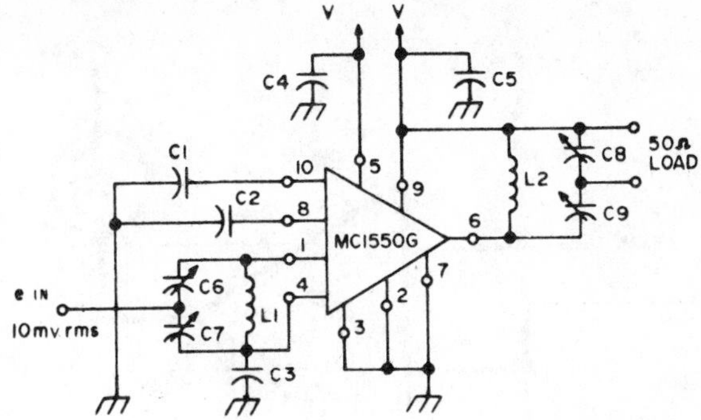


VHF amplifier with 25 dB gain and good agc characteristics. The HEP 590 can also be used in this circuit with similar results.

A

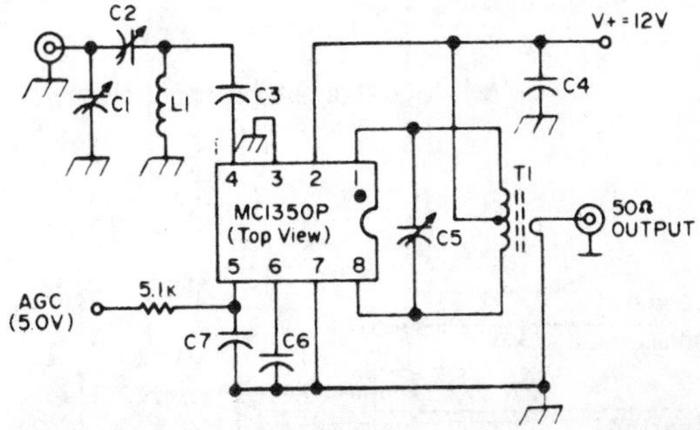


B

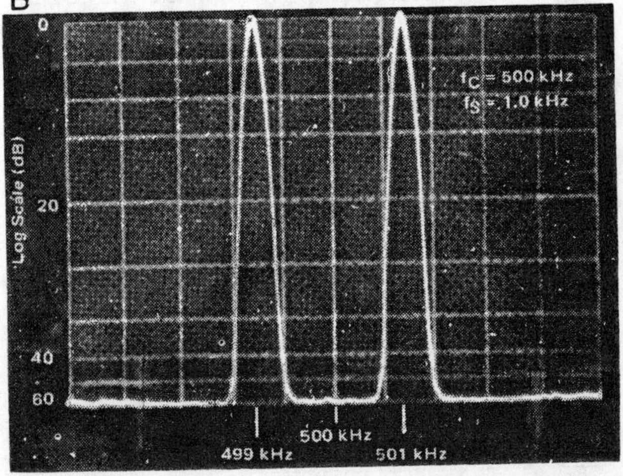


(A) Balanced modulator using MC1596G and (B) double-sideband output spectrum. Carrier rejection is greater than 60 dB without using any coils.

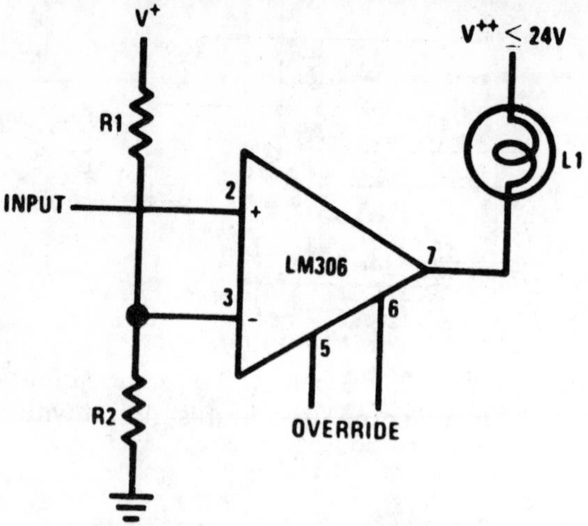


Level detector and lamp driver.

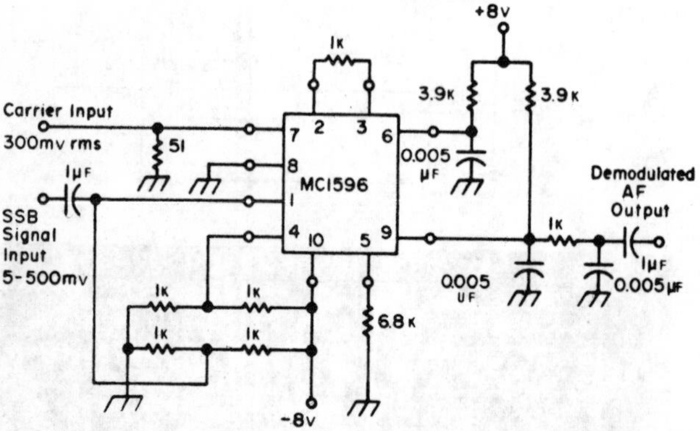


Product detector using MC1596G balanced modulator/ demodulator.

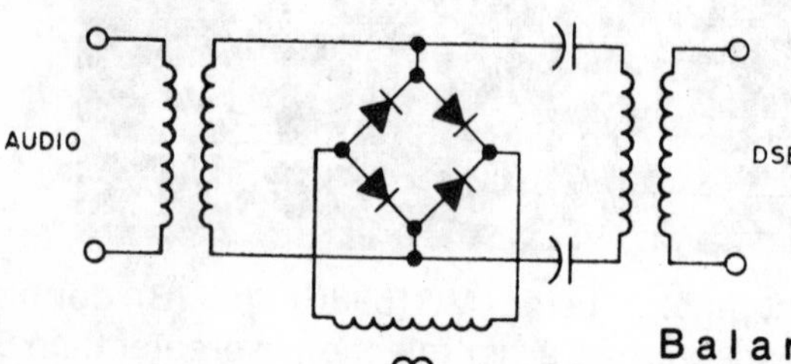


Balanced bridge modulator.

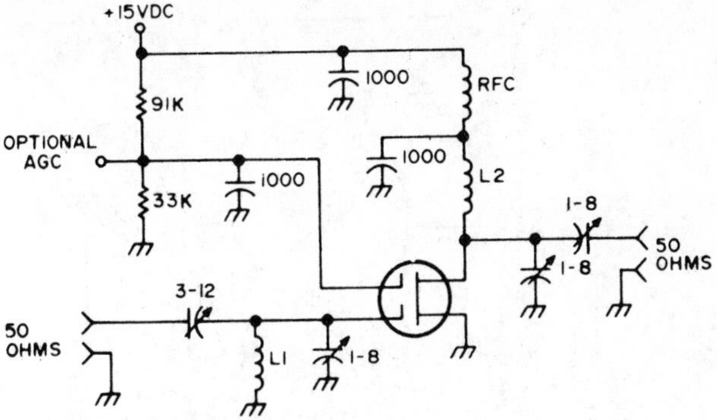


Typical VHF amplifiers using MFE3007 dual-gate MOSFET. Taken from Motorola Application Note AN-478.

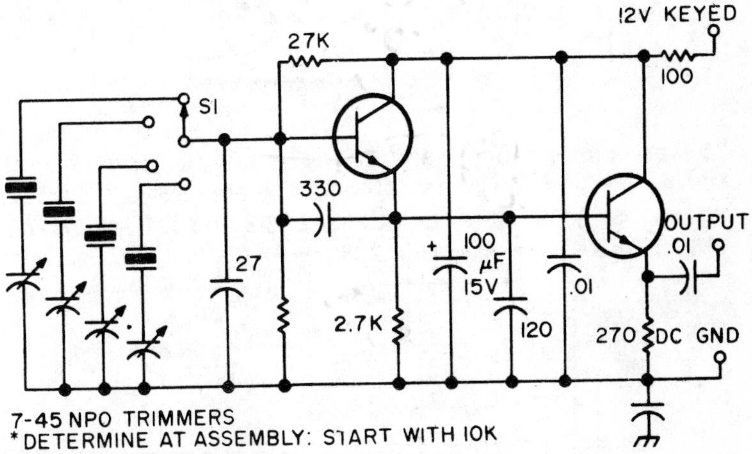


Multifrequency oscillator may be situated remotely from multiplier stages. Unit fits in control head when radio is trunk-mounted.

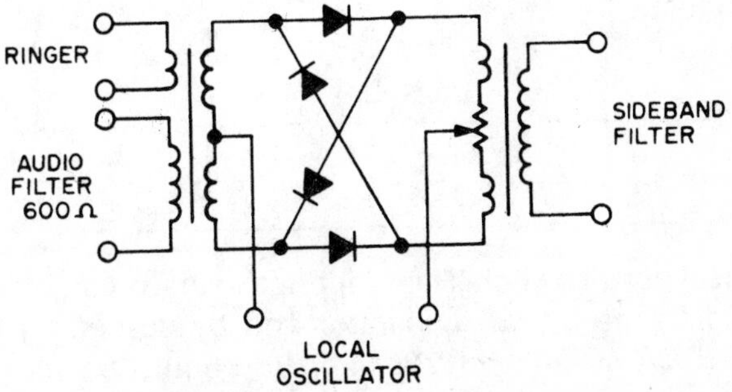


Telephone company ring modulator.

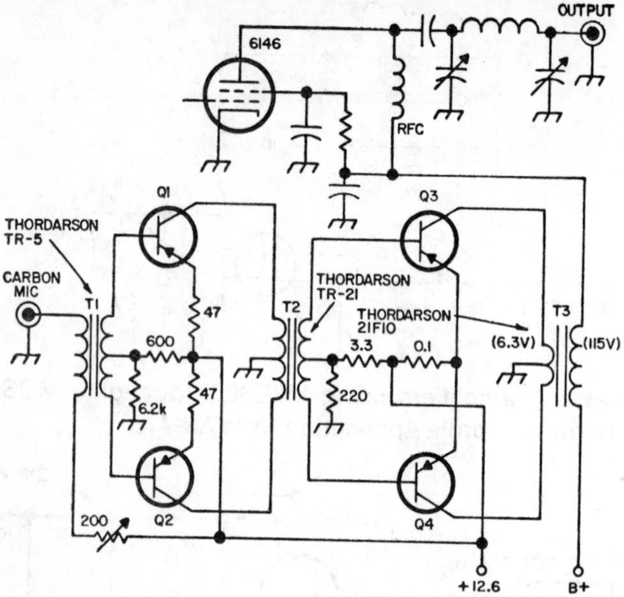


25-watt modulator uses readily available commercial transformers. Transistors Q1 and Q2 are 2N1172, 2N301, 2N1560, SK3009, GE-9, or HEP 232; Q3 and Q4 are 2N174, 2N278, SK3012, GE-4, or HEP 233.

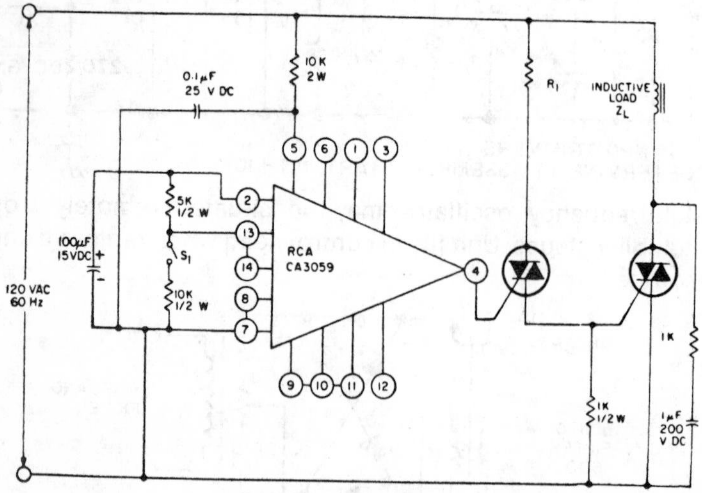


Practical power switch. Heavy inductive loads can be handled by adding the 0.1 $\mu$F capacitor and by cascading the two triacs (RCA 40526 and 2N5444). R1 should be the highest suitable value less than 10K.

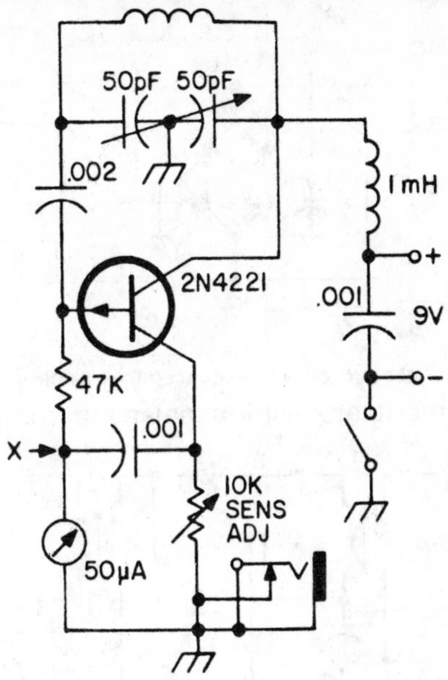


Meter amplifier allows cheap milliammeter to be used where circuit calls for microammeter.

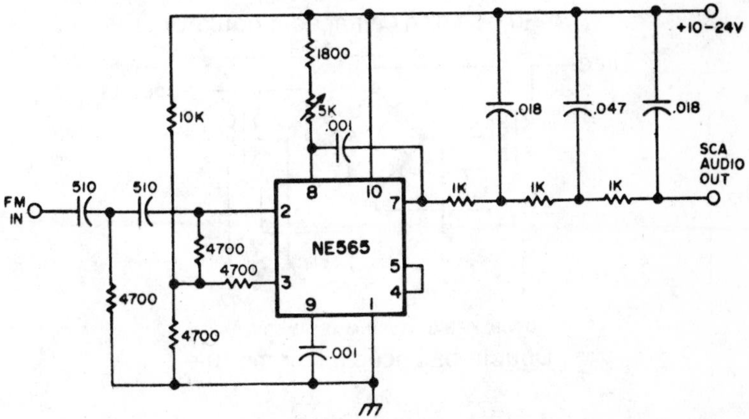


Background music adapter for FM receivers uses type 565 IC PLL. 510 pF capacitors and 4.7K resistors at input form high-pass filter to keep audio from FM set from overloading PLL; ladder filter at output removes everything above about 10 kHz to keep from overloading audio amplifiers follo ing. Pot allows frequency to be adjusted to 67 kHz to pick off SCA subcarrier.

RING MODULATOR W/O BALANCED TRANSFORMERS

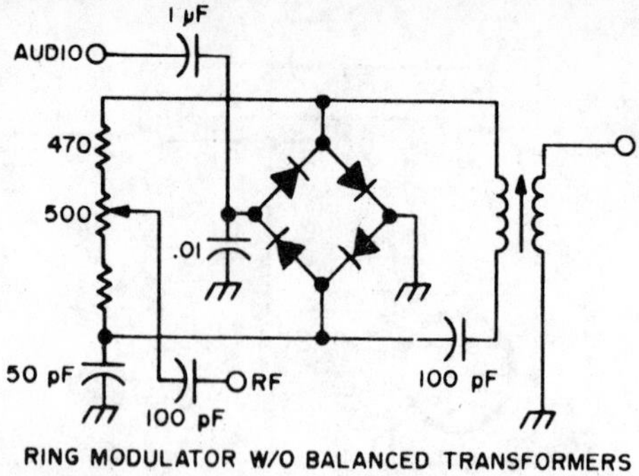

Ring modulator without balanced tranformers.

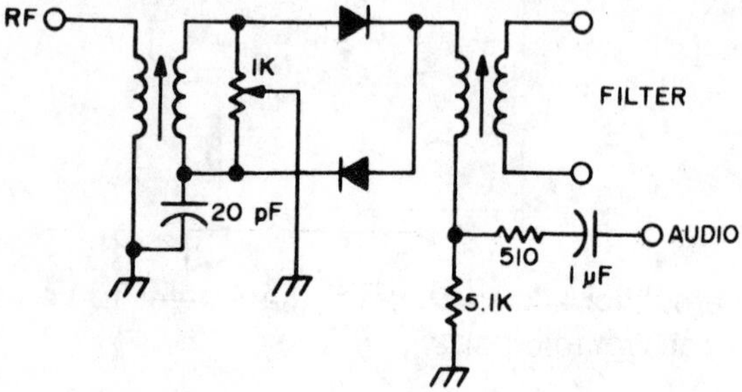

2-diode ring balanced modulator.

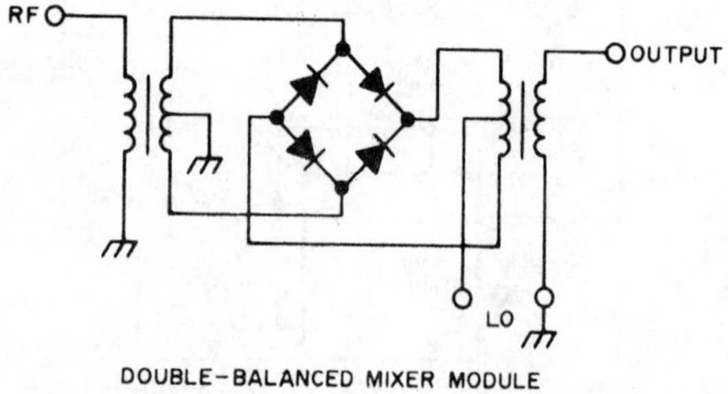

DOUBLE—BALANCED MIXER MODULE

Double-balanced mixer module.

DOUBLE BALANCED MIXER
AS PRODUCT DETECTOR

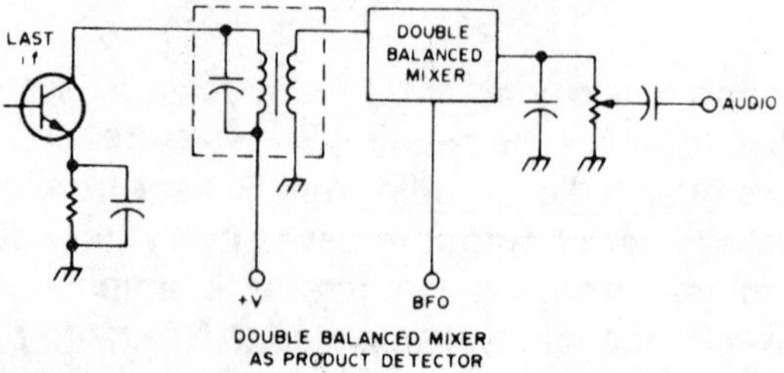

Double-balanced mixer as product detector.

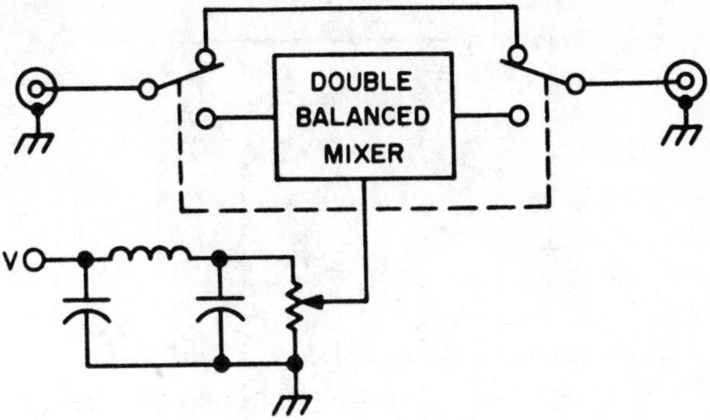


Voltage-variable attenuator.

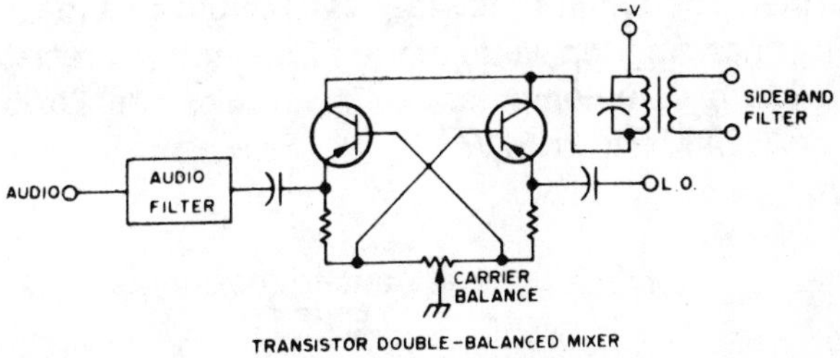


Transistor double-balanced mixer.

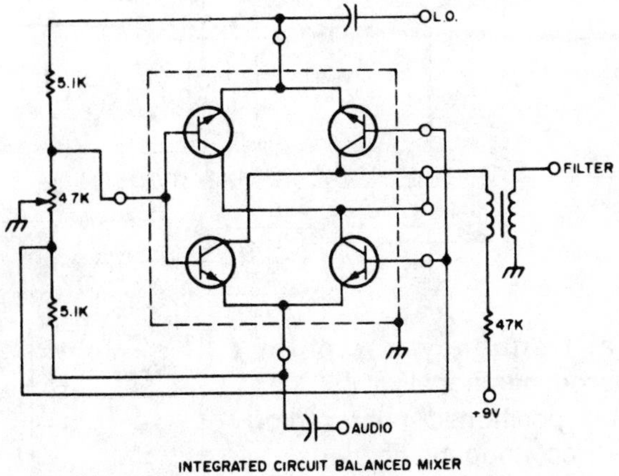


Integrated-circuit balanced mixer.

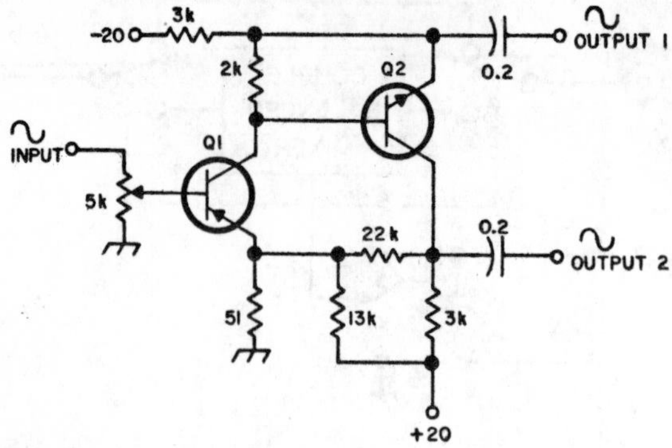


This phase-splitting circuit provides two out-of-phase signals for driving a push-pull amplifier without an expensive transformer. The gain of the stage is 150, but this may be adjusted by changing the value of the 22K feedback resistor. Q1 and Q2 are a complementary pair such as the 2N652 and 2N388 or 2N2430 and 2N2706.

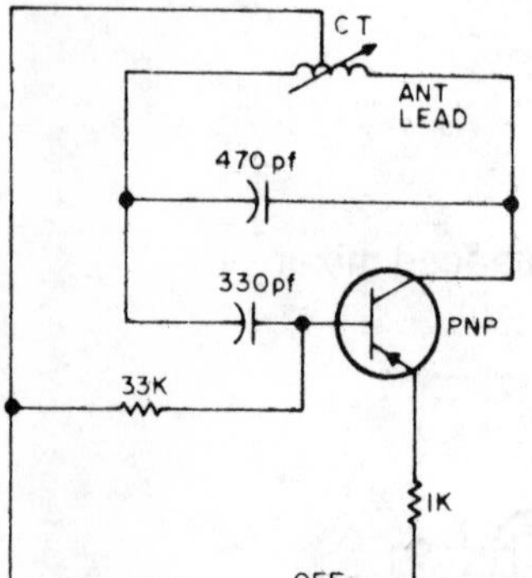


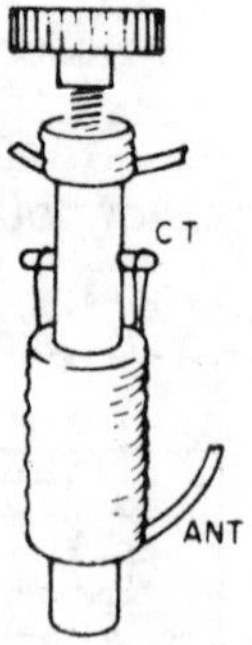


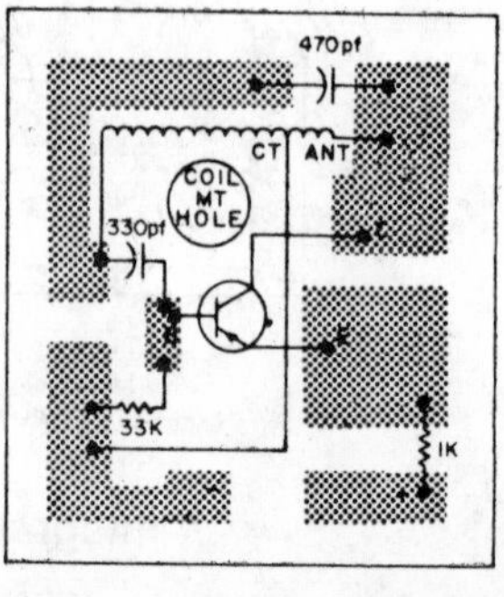


Beat-frequency oscillator heterodynes signals at 455 kHz when positioned near radio. No direct coupling needed.

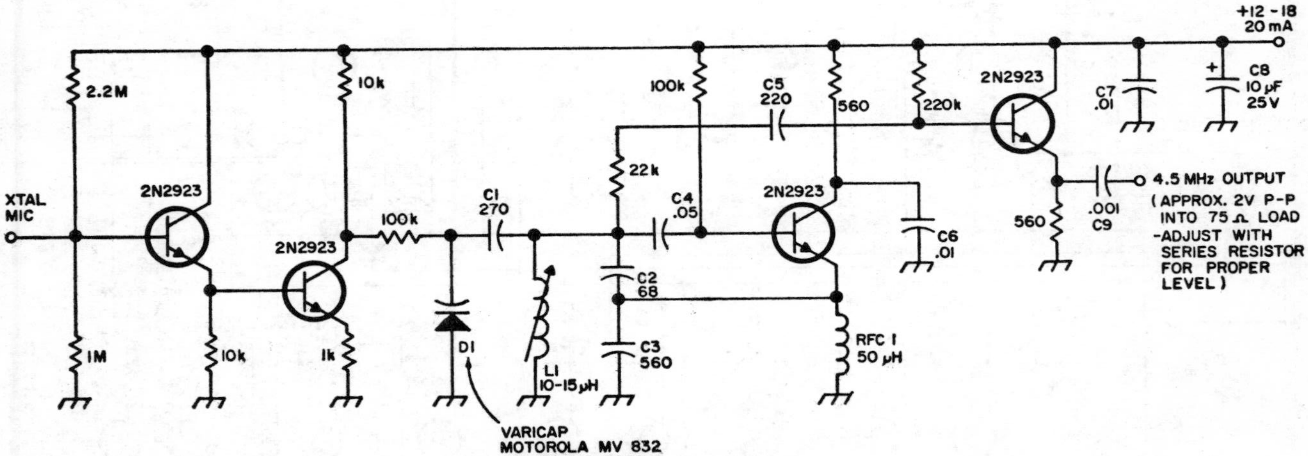


A simple transistorized FM subcarrier generator for transmitting audio on a video transmitter.

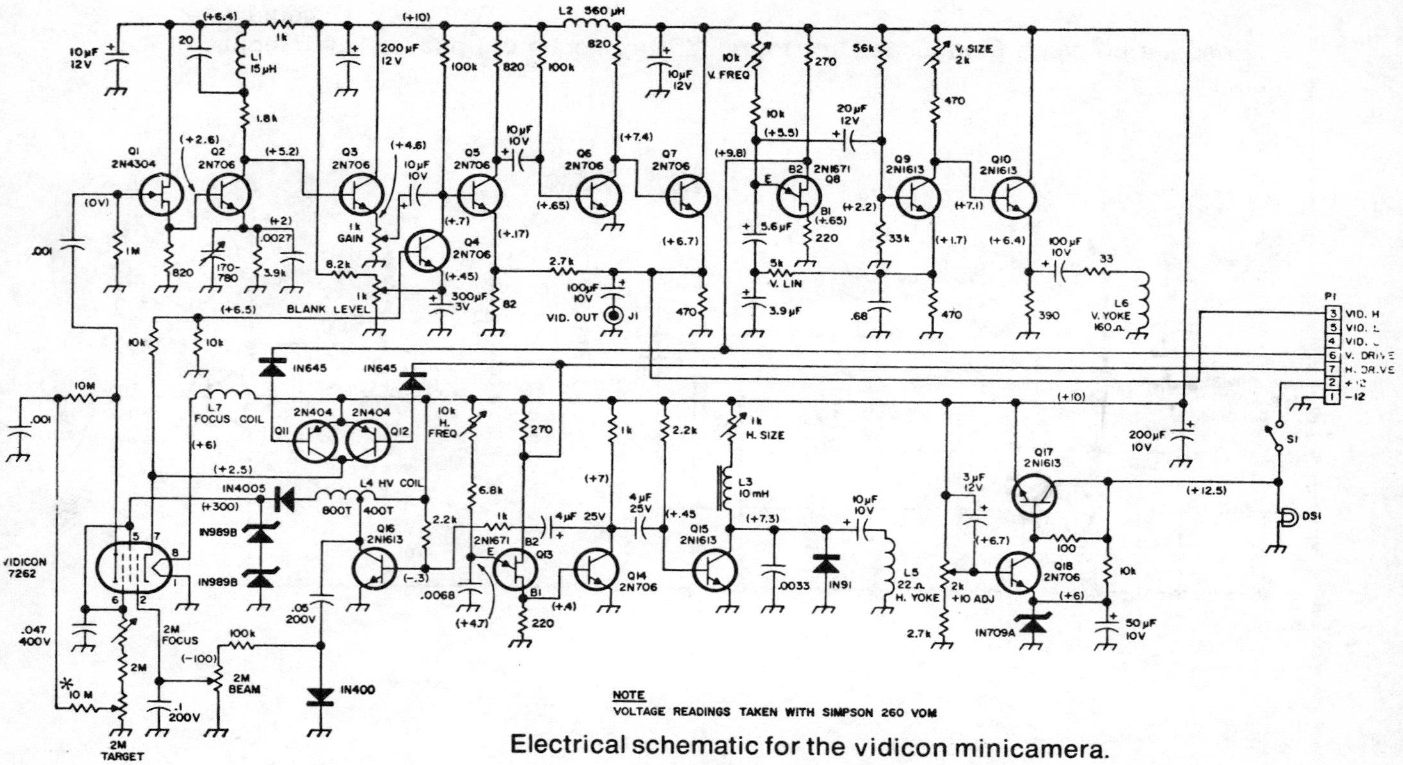


Electrical schematic for the vidicon minicamera.

# Appendix A

# IC Substitution Guide

## HEP 570

**BASE DIAGRAM:**

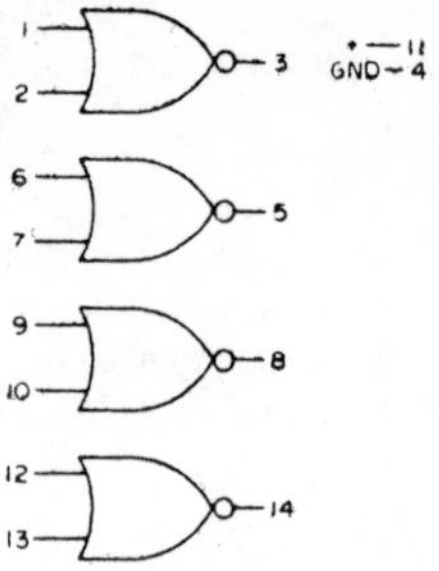


Maximum signal voltage: ±4.0V
Maximum supply voltage: +12V
Operating temperature range: +15 to +55°C
Output current: 2.65 mA per gate element
Package: 14-pin dual inline (socket HEP 453)

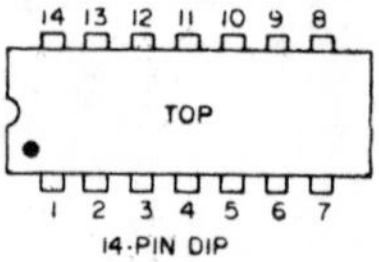


**REPLACES:**

| | |
|---|---|
| Motorola | MC717P (requires more power) |
| | MC724P |
| | MC817P (temperature range is smaller) |
| | MC824P (temperature range is smaller) |
| Fairchild | uL914 (570 has four gates, 914 only 2, and package is different) |
| | U5B991429X (570 has 4 gates, U5B991329X has only 2) |

## HEP 571

**BASE DIAGRAM:**

Maximum signal voltage: ±4.0V
Maximum supply voltage: +12V
Operating temperature range: +15 to +55e +55°CC
Package: 14-pin dual inline (socket HEP 453)

**REPLACES:**

| | |
|---|---|
| Motorola | MC799P |
| | MC899P (temperature range is smaller) |
| Fairchild | uL900 (571 has 2 units, 900 only one) |
| | U5D990029X (two units instead of one) |
| Other | PL990029 (two units instead of one) |

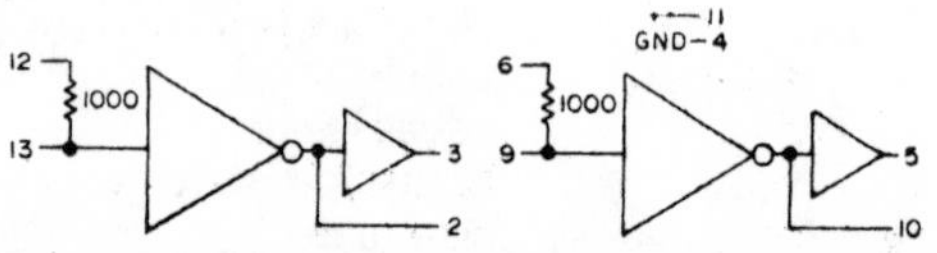

## HEP 572

BASE DIAGRAM:

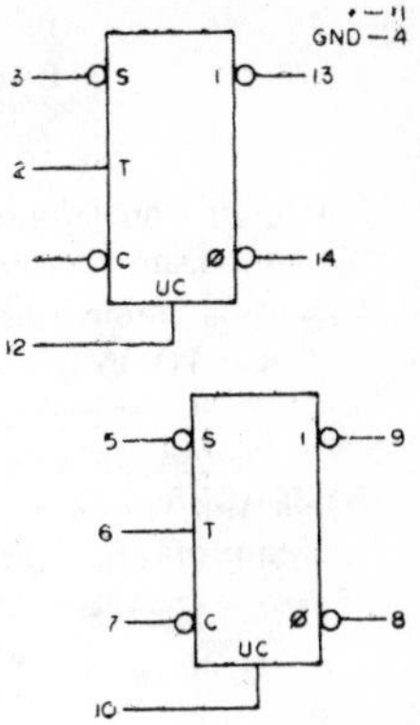


Maximum signal voltage: ± 4.0V
Maximum supply voltage: + 12V
Operating temperature range: +15 to + 55°C
Package: 14-pin dual inline (socket HEP 453)

REPLACES:

| | |
|---|---|
| Motorola | MC776P (requires more power) |
| | MC790P |
| | MC876P (smaller temperature range) |
| | MC890P (smaller temperature range) |
| Fairchild | uL923 (two units instead of one) |
| | U5B992329X (two units instead of one) |
| Other | PL992329 (two units instead of one) |

## HEP 580

BASE DIAGRAM:

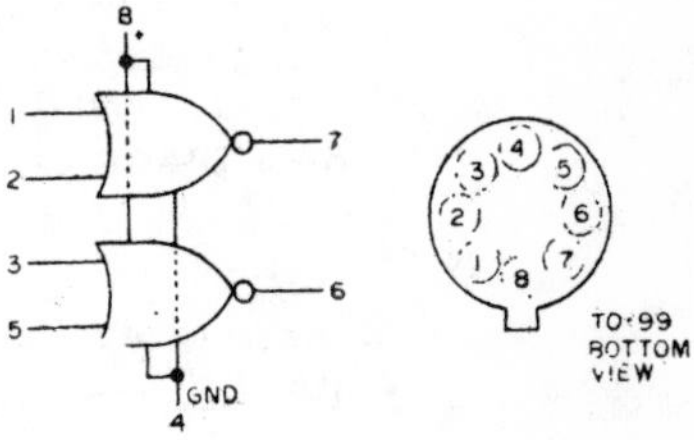


Maximum signal voltage: ±4.0V
Maximum supply voltage: +12V
Operating temperature range: +15 to +55°
Package: TO-99 (8-lead transistor-sized, socket HEP 454)

REPLACES:

| | |
|---|---|
| Motorola | MC710G |
| | MC810G (smaller temperature range) |
| | MC910G (smaller temperature range) |
| | MC710F (different package) |
| | MC810F (different package, smaller temp. range) |
| | MC910F (different package, smaller temp. range) |
| TI | SN17810L (smaller temperature range) |
| | SN17910L (smaller temperature range) |
| Fairchild | U5B991021X |
| | U5B991029X |
| Other | PL991021 (smaller temperature range) |
| | PL991029 |

## HEP 581

BASE DIAGRAM:

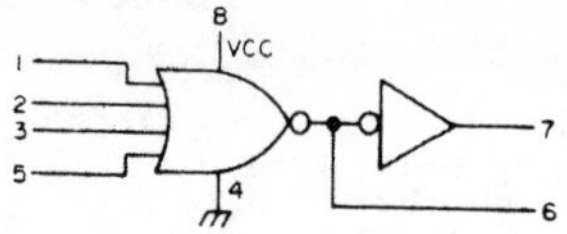


Maximum signal voltage: ±4.0V
Maximum supply voltage: +12V
Operating temperature range: +15 to +55 °C
Package: TO-99 (socket HEP 454)

REPLACES:

| | |
|---|---|
| Motorola | MC711G |
| | MC711F (different package) |
| | MC811G |
| | MC811F (different package) |
| | MC911G (smaller temperature range) |
| | MC911F (different package, smaller temp. range) |
| TI | SN17811L (smaller temperature range) |
| Fairchild | uL911 (lower power ratings) |
| | U3F991129X (different package) |
| | U5B991129X |
| | U5F991121X (different package) |
| Other | PL991129 |

## HEP 582

BASE DIAGRAM:

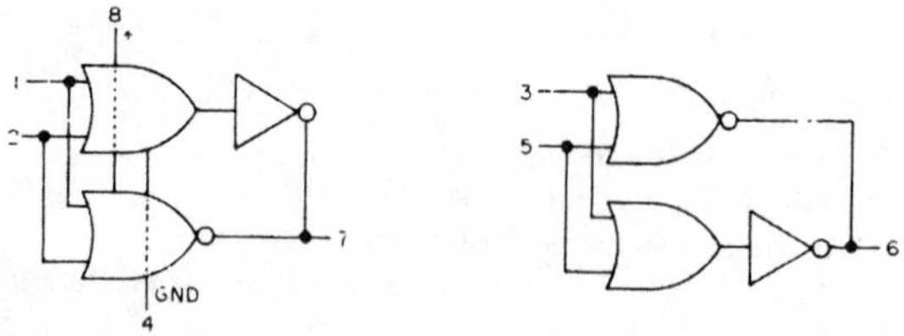


Maximum signal voltage: ±4.0V
Maximum supply voltage: +12V
Operating temperature range: +15 to +55 °C
Package: TO-99 (socket HEP 454)

REPLACES:

| | |
|---|---|
| Motorola | MC781G |
| | MC881G (smaller temperature range) |
| | MC981G (smaller temperature range) |
| Fairchilo | U5D990029X (lower power ratings) |

## HEP 583

BASE DIAGRAM:

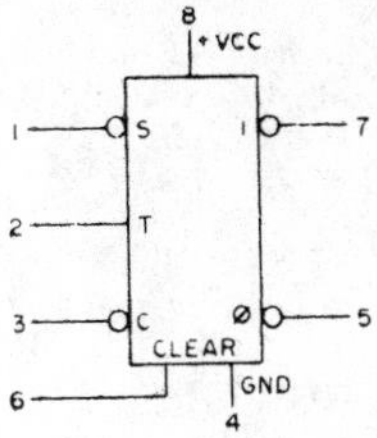


Maximum signal voltage: ±4.0V
Maximum supply voltage: +12V
Operating temperature range: +15 to +55°C
Package: TO-99 (socket HEP 454)

REPLACES:

| | |
|---|---|
| Motorola | MC782G |
| | MC882G (smaller temperature range) |
| | MC982G (smaller temperature range) |
| Fairchild | U5B992329X (lower power ratings) |
| | uL923 (lower power ratings) |
| Other | PL992329 (lower power ratings) |

## HEP 584

BASE DIAGRAM:

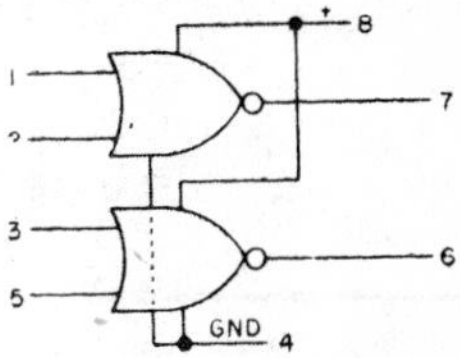


Maximum signal voltage: ±4.0V
Maximum supply voltage: +12V
Operating temperature range: +15 to+55°C
Package: TO-99 (socket HEP 454)

REPLACES:

| | |
|---|---|
| Motorola | MC714G. |
| | MC714F (different package) |
| | MC813G (smaller temperature range) |
| | MC814F (different package, smaller temp. range) |
| | MC914F (different package, smaller temp. range) |
| Fairchild | uL914 |
| | U3F991421X (different package) |
| | U3F991422X (different package) |
| | U5B991421X |
| | UB5991422X |
| | U5B991319X |
| Other | PL991429 |

## HEP 590

### BASE DIAGRAM:

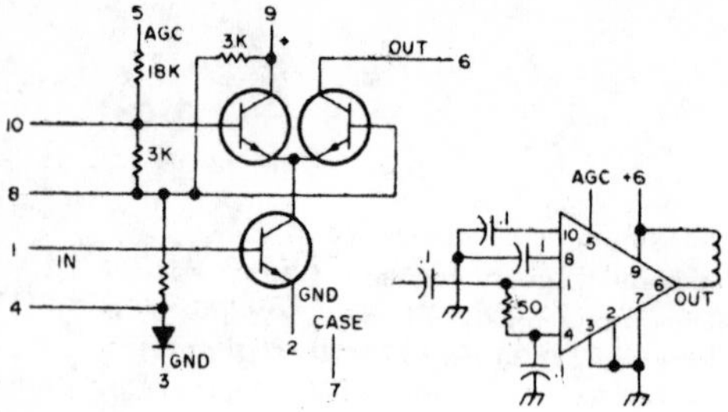


Maximum signal voltage: 5V rms
Maximum supply voltage: +20V
Operating temperature range: −55 to +125 C
AGC supply voltage (max.): 20V
Supply current: 2.5 mA dc
Package: 10-lead TO-5 (socket HEP 451)

REPLACES:

| | |
|---|---|
| RCA | CA3002 |
| | CA3003 |
| | CA3004 |
| Motorola | MC1550G |
| GE | PA713 |
| | PA7601 |
| Fairchild | U5D770339X |

## HEP 591

### BASE DIAGRAM:

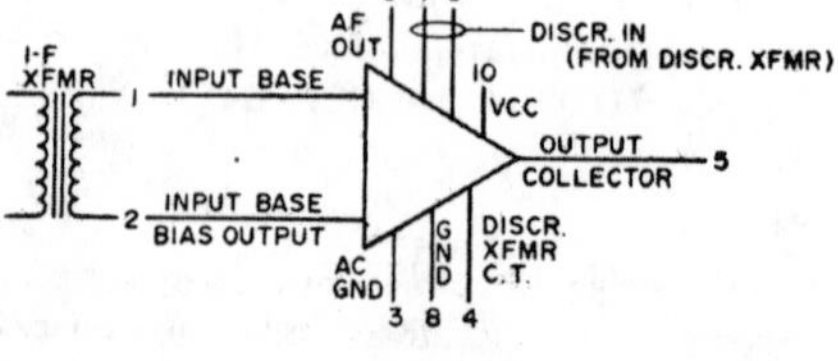


**Maximum signal voltage: ±3V p-p**
**Maximum supply voltage: +10V**
**Supply current: 27 mA max., 12 mA min.**
**Operating temperature range: −55 to +125°C**
**Typical voltage gain: 60 dB min.**
**Package: 10-lead TO-5 (socket HEP 451)**

**REPLACES:**

| | |
|---|---|
| RCA | CA3013 |
| | CA3014 |
| Motorola | MC1314G |

## HEP 592

### BASE DIAGRAM:

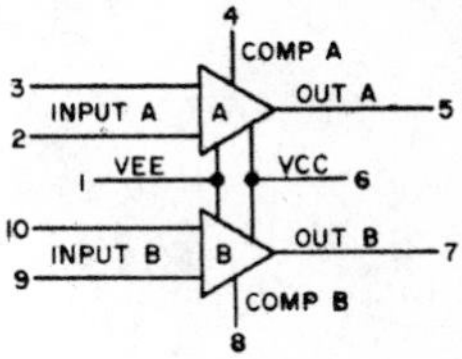


Maximum signal voltage: ±2V p-p
Maximum supply voltage: +16V pin 6 to pin 1
Voltage gain (each channel): 10,000 typical
Output voltage swing: 4.5V p-p min., 5.5V p-p typical
Operating temperature range: 0 to +75°C
Package: 10-lead TO-5 (socket HEP 451)

REPLACES:

| | |
|---|---|
| Motorola | MC1302G |
| | MC1302P (different package) |
| | MC1303P (different package, differnt power level) |

May replace type 709 opamps with modifications of compensation network values; each 592 is equivalent of two 709s.

## HEP 593

### BASE DIAGRAM:

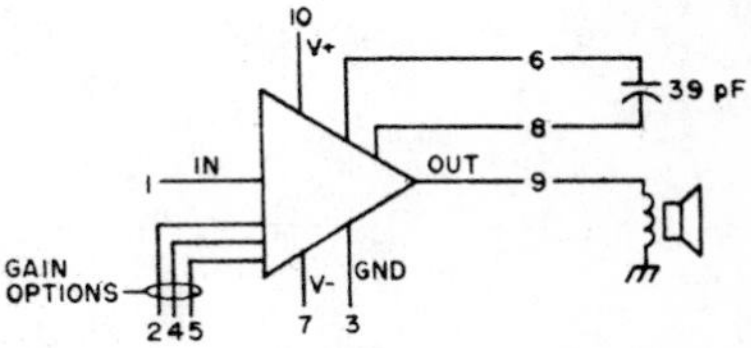


Maximum signal voltage: not rated; has 3—way gain option
Maximum supply voltage: 18V
Supply current: 15 mA max. (no signal), 0.5A (peak signal)

Audio output power: 1.8W
Operating temperature range: −55 to +125°C
Package: 10-lead TO-5 (socket HEP 451)
REPLACES:

| | |
|---|---|
| Motorola | MC1554G |

HEP 553

BASE DIAGRAM:

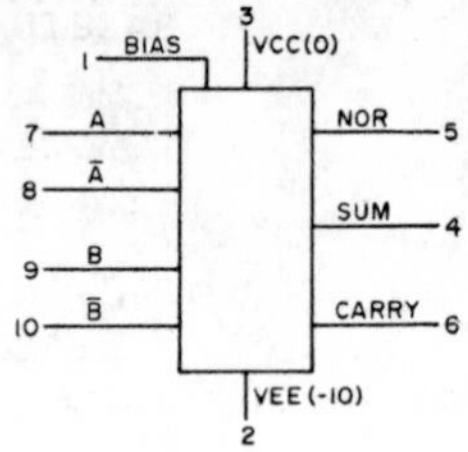


Maximum signal voltage: −10V
Maximum supply voltage: −10V
Operating temperature range: +15 to +55°C
Package: 10-lead TO-5 (socket HEP 451)

REPLACES:

| | |
|---|---|
| Motorola | MC303G |
| | MC303F (different package) |
| | MC353G |
| | MC353F (different package) |
| Other | SW303F (different package) |
| | SW303T |
| | SW353F (different package) |
| | SW353T |

HEP 554

BASE DIAGRAM:

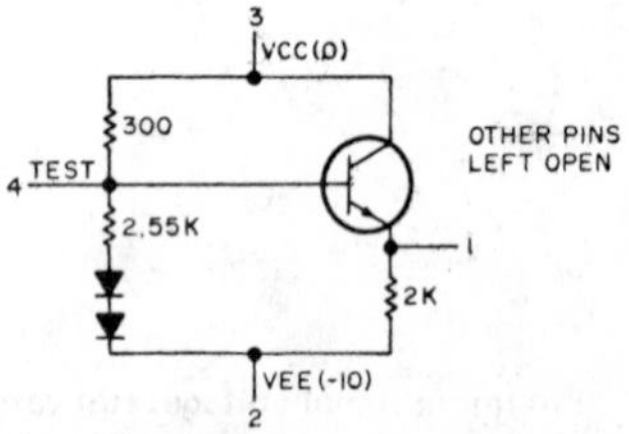


Maximum supply voltage: −10V
Operating temperature range: †15 to +55°C
Package: 10-lead TO-5 (socket HEP 451)

REPLACES:

| | |
|---|---|
| Motorola | MC304G |
| | MC304F (different package) |
| | MC354G |
| | MC354F (different package) |

Other SW304T
SW304F (different package)
SW354T
SW354F (different package)

## HEP 556

BASE DIAGRAM:

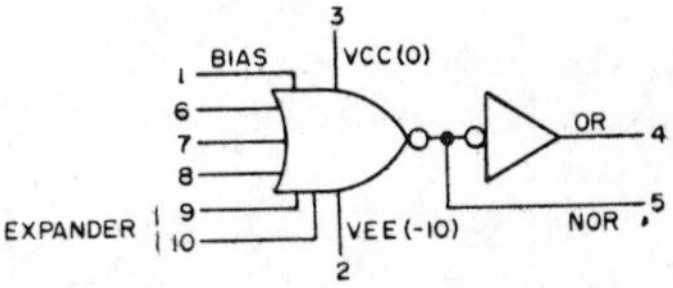


Maximum signal voltage: −10V
Maximum supply voltage: −10V
Operating temperature range: +25 to +55°C
Package: 10-lead TO-5 (socket HEP 451)

REPLACES:

| | |
|---|---|
| Motorola | MC306G |
| | MC306F (different package) |
| | MC356G |
| | MC356F (different package) |
| Other | SW306T |
| | SW306F (different package) |
| | SW356T |
| | SW356F (different package) |

## HEP 558

BASE DIAGRAM:

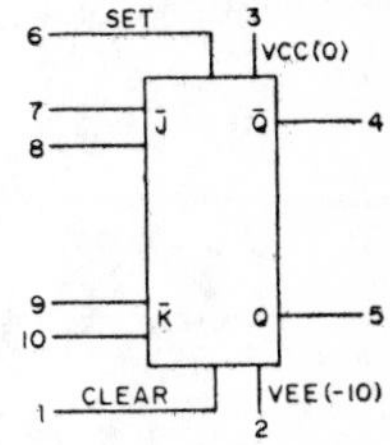


Maximum signal voltage: −10V
Maximum supply voltage: −10V
Supply current: 21mA
Operating temperature range: +15 to +55°C
Package: 10-lead TO-5 (socket HEP 451)

REPLACES:

| | |
|---|---|
| Motorola | MC308G |
| | MC308F (different package) |
| | MC358G |
| | MC358 (different package) |
| Other | SW308T |
| | SW308F (different package) |
| | SW358T |
| | SW358F (different package) |

# Appendix B

# Electronic Symbols

# ELECTRONIC SYMBOLS

## SHIELDING

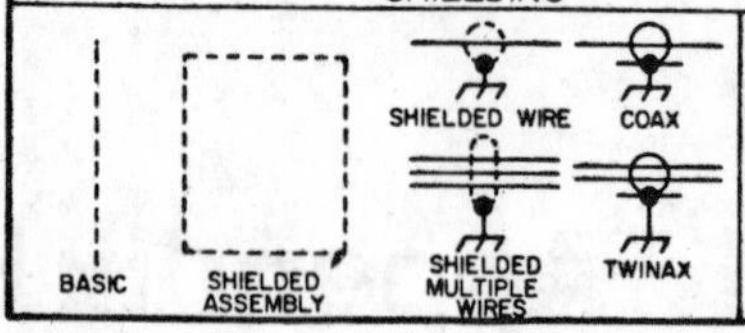


SHIELDS ARE ASSUMED TO BE GROUNDED UNLESS OTHERWISE NOTED

INDICATE IF SHIELD ON CONDUCTOR IS TO BE GROUNDED ONE END ONLY, & SPECIFY WHICH END

## SWITCHES

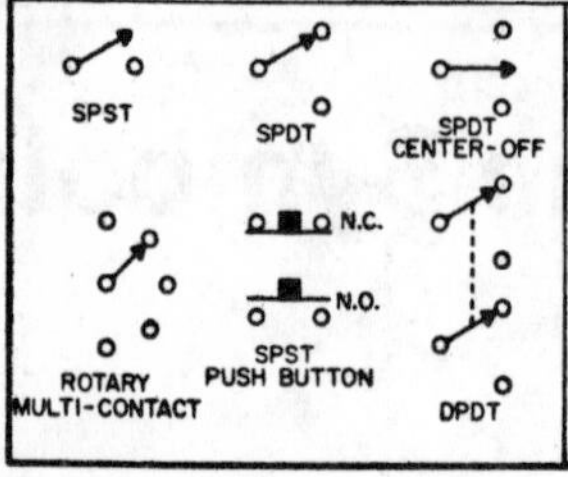


ONLY BASIC CONFIGURATIONS ARE SHOWN. COMBINE OR EXPAND AS NECESSARY TO SUIT

SELF-HOLDING SWITCH OPERATION IS ASSUMED UNLESS SPRING-RETURN IS SPECIFIED

NON-SHORTING TYPE IS ASSUMED UNLESS SHORTING TYPE IS SPECIFIED

SWITCH SECTIONS WHICH ARE SIMULTANEOUSLY OPERATED OR GANGED SHOULD BE JOINED WITH A DOTTED LINE

## TRANSFORMERS

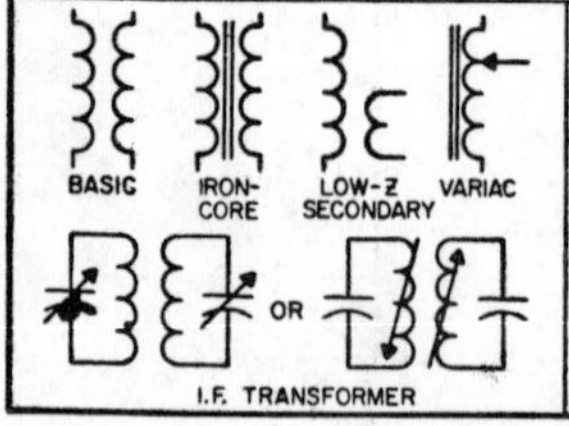


SHOW (4) LOOPS EACH WINDING EXCEPT (6) LOOPS ON HIGH-VOLTAGE SECONDARIES AND (2) ON FILAMENT WINDINGS AND NON-RESONANT LOW-Z RF COUPLING LINKS

TRANSFORMER DESCRIPTION SHOULD INCLUDE MANUFACTURER'S PART NUMBER AND RATINGS, OR CONSTRUCTION DATA

## TRANSISTORS

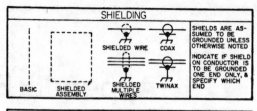

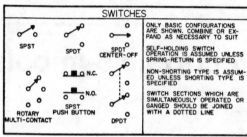

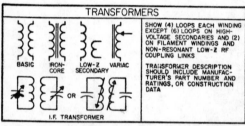

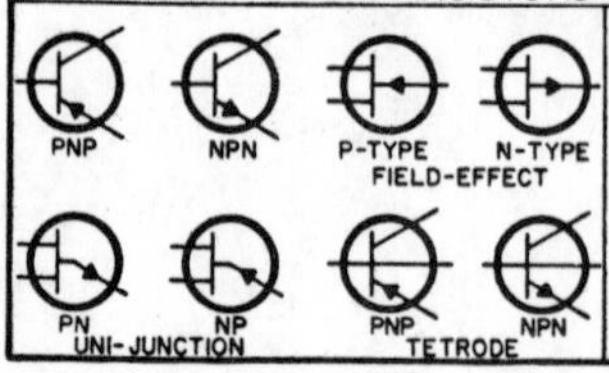


SPECIFY MANUFACTURER'S PART NUMBER OR RATINGS

REFER TO TRANSISTOR MANUALS FOR TYPES NOT SHOWN HERE.

# ELECTRONIC ABBREVIATIONS

(AS USED ON DRAWINGS AND SCHEMATICS)

| NOMENCLATURE | ABBREVIATION(S) |
|---|---|
| ALTERNATING CURRENT | AC |
| AMPERE | A |
| AMPLIFIER | AMP |
| AMPLITUDE MODULATION | AM |
| ANTENNA | ANT |
| AUDIO FREQUENCY | AF |
| AUTOMATIC FREQUENCY CONTROL | AFC |
| AUTOMATIC GAIN CONTROL | AGC |
| AUTOMATIC VOLUME CONTROL | AVC |
| BATTERY | B |
| BEAT FREQUENCY OSCILLATOR | BFO |
| BROADCAST | BC |

| NOMENCLATURE | ABBREVIATION(S) |
|---|---|
| CAPACITANCE, CAPACITOR | C |
| CONTINUOUS WAVE | CW |
| CRYSTAL | X, XTAL |
| CURRENT | I |
| DECIBEL | dB |
| DIODE, SEMICONDUCTOR (ALL TYPES) | D |
| DIRECT CURRENT | DC |
| DOUBLE COTTON COVERED | D.C.C. |
| DOUBLE POLE DOUBLE THROW | DPDT |
| DOUBLE POLE SINGLE THROW | DPST |
| DOUBLE SILK COVERED | D.S.C. |
| ELECTRON TUBE (ALL TYPES) | V |
| ENAMEL COVERED | ENAM |
| FILAMENT | FIL |
| FREQUENCY | FREQ, f |
| FREQUENCY MODULATION | FM |
| FUSE | F |
| GROUND | GND |
| HENRY | H |
| HERTZ (CYCLES PER SECOND) | Hz |
| IMPEDANCE | Z |
| INDUCTANCE, INDUCTOR | L |
| INSIDE DIAMETER | I.D. |
| INTERMEDIATE FREQUENCY | I.F. |
| JACK | J |
| KILOHERTZ (KILOCYCLES PER SECOND) | kHz |
| KILOHM | k, kΩ |
| KILOVOLT | kV |
| KILOWATT | kW |
| LAMP | I |
| LOUDSPEAKER | SPKR |
| MEGAHERTZ (MEGACYCLES PER SECOND) | MHz |
| MEGOHM | M, MΩ |
| METER | M |
| MICROAMPERE | μA |
| MICROFARAD | μF |
| MICROHENRY | μH |
| MICROPHONE | MIC |
| MICROVOLT | μV |
| MICROWATT | μW |
| MILLIAMPERE | mA |
| MILLIHENRY | mH |
| MILLIVOLT | mV |
| MILLIWATT | mW |
| NEGATIVE (POLARITY) | –, NEG |
| NORMALLY CLOSED | NC |
| NORMALLY OPEN | NO |
| OHM | Ω |
| OSCILLATOR | OSC |
| OUTSIDE DIAMETER | O.D. |
| PICOFARAD | pF |
| PLUG | P |
| POSITIVE (POLARITY) | +, POS |
| POWER AMPLIFIER | PA |
| PRIMARY | PRI |
| PUSHBUTTON | PB |
| RADIO FREQUENCY | RF |
| RADIO FREQUENCY CHOKE | RFC |
| RECEIVE | REC |
| RECEIVER | RCVR |
| RELAY | K |
| RESISTANCE, RESISTOR (ALL TYPES) | R |
| ROOT MEAN SQUARE | RMS |
| SECONDARY | SEC |
| SHORTWAVE | SW |
| SINGLE COTTON COVERED | S.C.C. |
| SINGLE POLE DOUBLE THROW | SPDT |

| NOMENCLATURE | ABBREVIATION(S) |
|---|---|
| SINGLE POLE SINGLE THROW | SPST |
| SINGLE SILK COVERED | S.S.C. |
| SWITCH | S |
| TIME | t |
| TRANSFORMER | XFMR, T |
| TRANSISTOR (ALL TYPES) | Q |
| TRANSMIT | XMIT |
| TRANSMITTER | XMTR |
| ULTRA HIGH FREQUENCY | UHF |
| VACUUM TUBE VOLTMETER | VTVM |
| VERY HIGH FREQUENCY | VHF |
| VOLT OHM METER | VOM |
| VOLT, VOLTS | V |
| VOLTAGE | E |
| WATT | W |
| WAVELENGTH | λ |

## ELECTRON TUBES

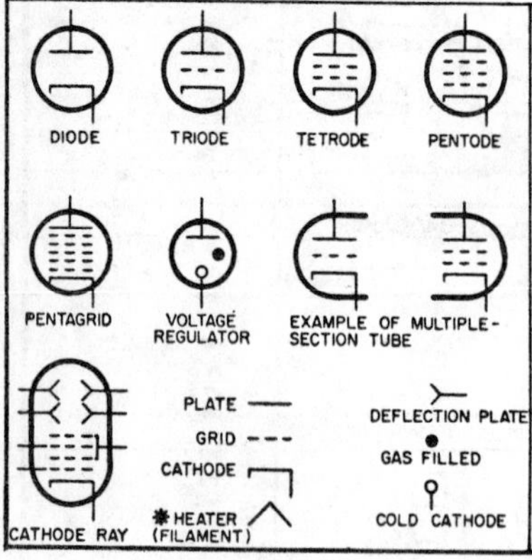


ALWAYS LABEL ELEMENTS WITH TUBE PIN NUMBERS

REFER TO TUBE MANUAL FOR DATA ON INDIVIDUAL TUBE TYPES

✱ FILAMENTS OR HEATERS (WITH THE EXCEPTION OF DIRECTLY-HEATED CATHODES) SHOULD BE SHOWN EXTERNAL TO TUBE CIRCLE, AND PREFERABLY IN THE POWER SUPPLY

## FUSE

INDICATE CURRENT, VOLTAGE RATINGS, AND SLO-BLO, ETC., AS APPROPRIATE

## GROUND CONNECTIONS

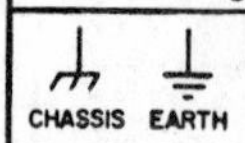


CHASSIS GROUND SYMBOL IS NORMALLY THE ONLY TYPE USED IN SCHEMATICS

EACH GROUNDED CIRCUIT COMPONENT WILL BE SHOWN CONNECTED TO AN *INDIVIDUAL* CHASSIS GROUND, UNLESS A COMMON GROUND BUS IS ESSENTIAL TO PROPER CIRCUIT OPERATION

## HEADSET

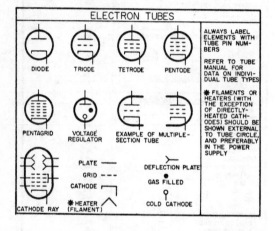

NORMALLY USED IN BLOCK DIAGRAMS, BUT MAY BE USED IN ANY SCHEMATIC WHERE CONNECTED DIRECTLY INTO CIRCUIT WITHOUT PHONE PLUG

INDICATE IMPEDANCE IF VALUE IS CRITICAL

# ELECTRONIC SYMBOLS

## ANTENNA

NORMALLY USED IN BLOCK DIAGRAMS, BUT MAY BE USED IN ANY SCHEMATIC WHERE ANTENNA IS CONNECTED DIRECTLY TO CIRCUIT WITHOUT BENEFIT OF RF CONNECTOR

## BATTERY

DO NOT FORGET TO INDICATE VOLTAGE AND POLARITY

## CAPACITORS

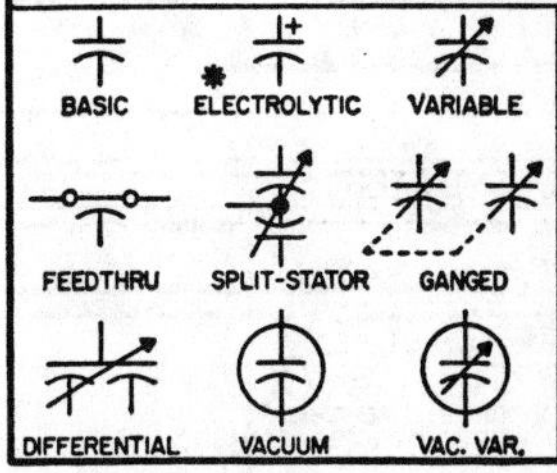


NOTE THAT CURVED PORTION OF SYMBOL ALWAYS DESIGNATES OUTSIDE FOIL OF FIXED CAPACITORS (EXCEPT ELECTROLYTICS, WHERE IT INDICATES THE NEGATIVE TERMINAL)

THE CURVED PORTION IN THE CASE OF A VARIABLE WILL INDICATE THE MOVABLE PART

✱ INDICATE POLARITY, AND VALUE IN μF

WHEN OTHER THAN ELECTROLYTIC, VALUES ARE ASSUMED TO BE pF WHEN 1 OR GREATER, AND μF WHEN LESS THAN 1

## CONDUCTORS

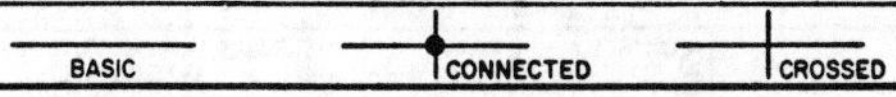


## CONNECTORS

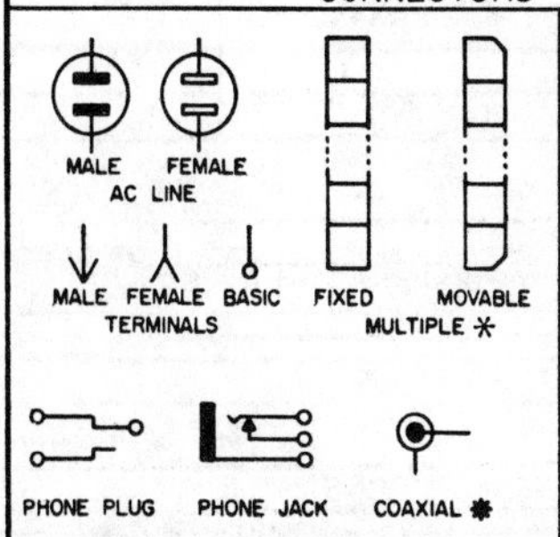


SHOULD NONE OF THE SYMBOLS DESCRIBED HERE SEEM TO MATCH YOUR SITUATION, DESCRIBE THE CONNECTOR AND/OR LIST THE MANUFACTURER'S PART NUMBER

✱ FOR ANY COAXIAL-TYPE CONNECTOR, SUCH AS RF, MICROPHONE, PHONO, ETC.

✳ NUMBER THE BLOCKS TO CORRESPOND TO TERMINAL MARKINGS, WHEN APPROPRIATE

## CRYSTAL

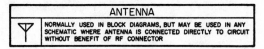

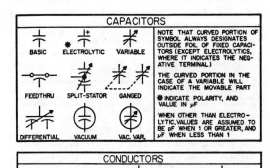

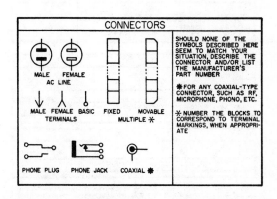

ALWAYS INDICATE CRYSTAL FREQUENCY (IN kHz, MHz, ETC.)

# ELECTRONIC SYMBOLS

## INDUCTORS

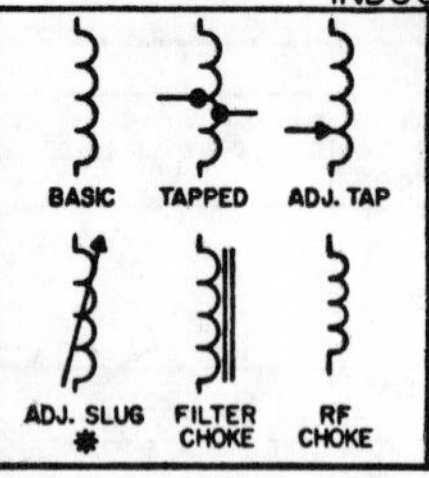


INCLUDE ALL NECESSARY DATA INCLUDING ANY OF FOLLOWING INFORMATION WHICH IS APPLICABLE:
- WIRE SIZE & TYPE
- COIL OR FORM O.D. OR I.D.
- NUMBER OF TURNS AND/OR LENGTH
- MANUFACTURERS PART NUMBER
- TAP POSITION ABOVE COLD END

✱ FERRITE CORE WILL BE ASSUMED UNLESS BRASS IS SPECIFIED. INDICATE TYPE OF FERRITE, IF CRITICAL

## KEYS

✱ BE SURE TO DESIGNATE "DIT" & "DAH" CONTACTS

## LAMPS

INDICATE MANUFACTURER'S PART NUMBER AND/OR VOLTAGE & CURRENT RATING

## LOUDSPEAKER

INDICATE VOICE COIL IMPEDANCE & POWER RATING, ETC., WHEN CRITICAL

## METERS

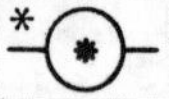

✱ INDICATE TYPE OF METER HERE (μA, mA, V, ETC.)
✳ INDICATE SCALE RANGE HERE (0-1, 0-50, ETC.)
DON'T FORGET TO INDICATE PROPER POLARITY

## MICROPHONE

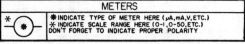

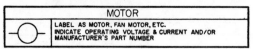

NORMALLY USED IN BLOCK DIAGRAMS BUT MAY BE USED IN SCHEMATIC WHEN WIRED DIRECTLY INTO CIRCUIT WITHOUT CONNECTOR

INDICATE TYPE (CARBON, XTAL, ETC.)

## MOTOR

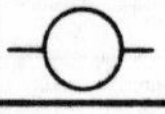

LABEL AS MOTOR, FAN MOTOR, ETC.

INDICATE OPERATING VOLTAGE & CURRENT AND/OR MANUFACTURER'S PART NUMBER

## RELAYS

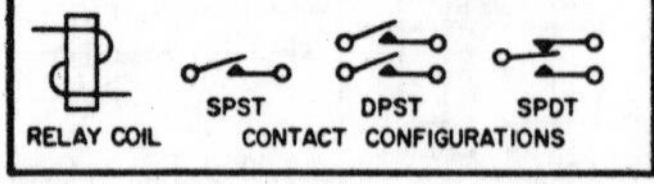


SPECIFY COIL VOLTAGE, RESISTANCE, ETC., AND/OR MANUFACTURER'S PART NUMBER

CONTACT CONFIGURATIONS SHOWN ARE BASIC AND MAY BE EXPANDED

## RESISTORS

INDICATE VALUE, IN OHMS (Ω), KILOHMS (k), OR MEGOHMS (M), AND/OR MANUFACTURER'S PART NUMBER.
1/2 W 10% IS ASSUMED UNLESS OTHERWISE NOTED

## SEMICONDUCTOR DIODES

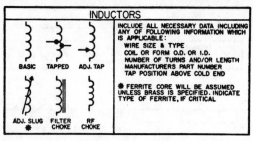

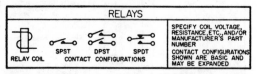

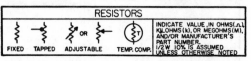

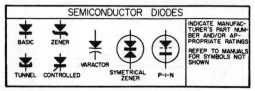

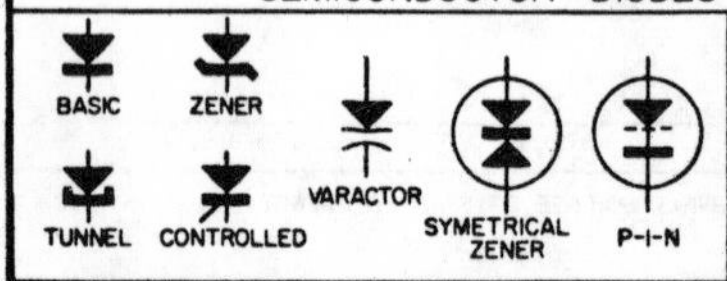


INDICATE MANUFACTURER'S PART NUMBER AND/OR APPROPRIATE RATINGS

REFER TO MANUALS FOR SYMBOLS NOT SHOWN

# Index

## A